RISK ANALYSIS BASED ON DATA AND CRISIS RESPONSE BEYOND KNOWLEDGE

PROCEEDINGS OF THE 7TH INTERNATIONAL CONFERENCE ON RISK ANALYSIS AND CRISIS RESPONSE (RACR 2019), ATHENS, GREECE, 15-19 OCTOBER 2019

Risk Analysis Based on Data and Crisis Response Beyond Knowledge

Editors

Chongfu Huang

Academy of Disaster Reduction and Emergency Management, Beijing Normal University, Beijing, China

Zoe Nivolianitou

Institute of Nuclear & Radiological Sciences and Technology, Energy & Safety, National Center of Scientific Research "Demokritos", Athens, Greece

CRC Press
Taylor & Francis Group
Boca Raton London New York

CRC Press is an imprint of the
Taylor & Francis Group, an **informa** business

A BALKEMA BOOK

Published by:
CRC Press/Balkema
Schipholweg 107C, 2316 XC Leiden, The Netherlands

First issued in paperback 2023

© 2020 by Taylor & Francis Group, LLC
CRC Press/Balkema is an imprint of the Taylor & Francis Group, an informa business

No claim to original U.S. Government works

ISBN 13: 978-1-03-257106-5 (pbk)
ISBN 13: 978-0-367-25146-8 (hbk)
ISBN 13: 978-0-429-28634-6 (ebk)

DOI: https://doi.org/10.1201/9780429286346

Typeset by Integra Software Services Pvt. Ltd., Pondicherry, India

Visit the Taylor & Francis Web site at
http://www.taylorandfrancis.com

and the CRC Press Web site at
http://www.crcpress.com

Table of contents

ix

Preface

In the Internet age, more and more data can support a large number of models for risk analysis. The simple, transparent, and reliable risk models have been favored by researchers. The assessment of integrated risks in complex systems is towards practical use. Risk analysis based on data is winning subjective judgment. Meanwhile, the world becomes increasingly turbulent. The black-swan events occur more frequently, and a crisis of under-coordination, such as the debt crisis of 2008, might suddenly erupt. Crisis response beyond knowledge is increasingly testing people's intelligence.

As the changing world is always uncertain, various risks like natural disasters, terrorism, and so on exist everywhere and endless crises always happen. Risk and development are coexisting. The world needs more intelligent technologies to analyze risks and govern crises in with high-efficiency and at low-cost. RACR (International Conference on Risk Analysis and Crisis Response), launched by the Risk Analysis Council of China Association for Disaster Prevention in 2007 and taken over by SRA-China (Society for Risk Analysis—China) since 2011, provides a unique international forum to discuss these issues from a scientific and technical point of view and also in terms of management, services and usages. We, the organizers, are very much looking forward to a vivid exchange of ideas and visions in this context, with a broad thematic coverage.

RACR is a series of biennial international conferences on risk analysis, crisis response, and disaster prevention for specialists and stakeholders. RACR-2019, on October 15-19, 2019 at the National Centre for Scientific Research "Demokritos"(NCSRD), situated in Greece at the Athens greater area, is the seventh edition in this series, following the successful RACR-2007 in Shanghai (China), RACR-2009 in Beijing (China), RACR-2011 in Laredo (USA), RACR-2013 in Istanbul (Turkey), RACR-2015 in Tangier (Morocco) and RACR-2017 in Ostrava-Prague (Czech).

To raise risk awareness within all countries and emerging economies, RACR-2019, with the theme: Risk Analysis Based on Data and Crisis Response Beyond Knowledge, is held in Athens, the cradle of western civilization, which is the blue and white city amidst the Mediterranean basin with a significant port nearby, Piraeus, also of Chinese interest. RACR-2019 is hosted by the Laboratory of Systems Reliability and Industrial Safety, of the National Centre for Scientific Research "DEMOKRITOS", the largest multidisciplinary research centre in Greece, since the early '60s.

This volume is composed of 96 high-quality papers submitted to RACR-2019, through a rigorous peer-review procedure, reflecting the status of the art researches in the world. The proceedings covers almost all risk fields such as applying risk science based on data, stakeholder engagement to manage risks, risk analysis related to black-swan events, responding to a crisis of undercoordination, reliability and safety in industrial systems, modern trends in crisis management, internet of intelligences and risk radar, nanotechnology safety, safety in transport domain, progress in occupational health and safety, natural hazards inducing technological accidents (natechs), human factors in the industrial environment, life-cycle analysis of units, legal aspects in major accidents prevention, disaster risks in line with "belt and road", risk analysis in project investment and finance, terrorist attack and crisis response and many more.

Finally, special thanks should be forwarded to the conference host! We would like to thank all contributors of the papers submitted to the conference, all reviewers involved in the paper review, and all those who gave kind concerns and support to RACR-2019. Particularly, thanks to Dr. Stelios Karozis, Prof. Gordon Huang, Dr. Nikolas Ventikos, Mr. Rundong Wang and Miss Wen Tian who invested a lot of time and energy to help us edit this proceedings. We are sure that your selfless participation may not only ensure the quality of the conference papers, but also give a strong impetus to the promotion of RACR and SRA-China.

Sincerely
Chongfu Huang
Zoe Nivolianitou
July 1, 2019

Organization

ORGANIZED BY: Society for Risk Analysis - China

HOSTED BY: National Centre for Scientific Research "Demokritos" (NCSRD), Athens Greece

CO-ORGANIZED BY: Beijing Cazl Technology Service Co., Ltd.

General Chair: Chongfu Huang, Beijing Normal University, China

Co-Chairs: Dr. Zoe Nivolianitou, National Centre for Scientific Research "Demokritos"(NCSRD), Institute of Nuclear & Radiological Sciences and Technology, Energy & Safety, Athens Greece

TECHNICAL COMMITTEE

Chair: Prof. Nikolas Ventikos, National Technical University of Athens, School of Naval Engineering, Greece

Co-Chair: Prof. Gordon Huang, University of Regina, Canada

Prof. Sen Qiao, Institute of Geophysics, China Earthquake Administration, China

Members

Dr. Olga Aneziris, NCSR « DEMOKRITOS », Greece

Dr. Olivier Salvi, INERIS Developpement SAS, France

Prof. Ortwin Renn, University of Stuttgart, Germany

Prof. Kalliopi Sapountzaki, HUA, Greece

Prof. George Wang, East Carolina University, USA

Dr. Yundong Huang, St. Edward's University, USA

Prof. Bernatik Ales, VŠB – Technical University of Ostrava, Czech

Prof. Cengiz Kahraman, Istanbul Technical University, Turkey

Prof. Dong Wang, Nanjing University, China

Dr. Nesrin Benhayoun, Abdelmalek Essaadi University, Morocco

Dr. Vincent Ho, Hong Kong Association of Risk Management and Safety, Hong Kong, China

Prof. Xiaobing Hu, Civil Aviation University of China, China

Prof. Guofang Zhai, Nanjing University, China

Dr. Effie Marcoulaki, NCSR « DEMOKRITOS », Greece

Dr. Konstantinos Kirytopoulos, National Technical University of Athens, Greece

Dr. Athanassios Sfetsos, NCSR « DEMOKRITOS », Greece

Prof. K. Spyrou, National Technical University of Athens, Greece

Prof. J. Dokas, Democritus University of Thrace, Greece

Dr. Sofia Karma, NTUA, Greece

Dr. Homer Papadopoulos, NCSR « DEMOKRITOS », Greece

Dr. Miranda Dandoulaki, EKDDA, Greece

Mr. Ioannis Tafyllis (MSc), GSCP, Greece

Dr. Michail Chalaris, Hellenic Fire Academy, FB, Greece

Prof. Ben Ale, Delft University of Technology, Netherlands

LOCAL ORGANIZING COMMITTEE

Dr. Stelios Karozis, Institute of Nuclear & Radiological Sciences and Technology, Energy & Safety (INRASTES), Athens, Greece

KEYNOTE SPEAKERS

Prof. Tsuchida Shoji, Kansai University, Japan
Prof. Chongfu Huang, Beijing Normal University, China

Prof. Ben Ale, TUDelft, The Netherlands

ACKNOWLEDGMENTS OF GREECE-CHINA RACR2019

Special thanks are due to the Greek Republic as represented by the President of the NCSRD, Dr. G. Nounesis and all other institutions and establishments which have greatly and tangibly contributed to the success of this conference.

Risk Analysis Based on Data and Crisis Response Beyond Knowledge – Huang & Nivolianitou (eds)
© *2020 Taylor & Francis Group, London, ISBN 978-0-367-25146-8*

The validity of geospatial information diffusion

Chongfu Huang
Key Laboratory of Environmental Change and Natural Disaster, Ministry of Education, Beijing Normal University, Beijing, China
State Key Laboratory of Earth Surface Processes and Resources Ecology, Beijing Normal University, Beijing, China
Academy of Disaster Reduction and Emergency Management, Faculty of Geographical Science, Beijing Normal University, Beijing, China

Yundong Huang*
The Bill Munday School of Business, St. Edward's University, Austin, TX, USA

ABSTRACT: In this article, two criteria are suggested to inspect whether a model of geospatial information diffusion is valid to fill in the missing observation in a gap unit. The first criterion serves for inspecting a gap unit in the study area, from which a sample is taken to train the model. The second criterion serves for a gap unit outside the area but nearby. The standard deviation of the trained model plays a key role in the former. The valid radiance determined by the maximum distance between the units of the sample point passed the test and the center of the study area plays an important role in the latter.

Keywords: geospatial information diffusion, gap unit, observed unit, background data, inspection criterion

1 INTRODUCTION

When we study an earth surface phenomenon such as "economic development level" and "earthquake risk", in a study area like a province, city, or county, we might miss observations on some geographic units in the area. For example, one day after the 2008 great Wenchuan earthquake (Zhang et al., 2009), the rescuers did not have disaster information for places where communication was lost, and this lack of disaster information seriously affected the efficiency of the rescue because the phenomenon "spatial distribution of disaster" was unknown.

The most popular method to fill in the missing observations is the interpolation method, which estimates the values of a curve at any position between known points and is widely used in the Geographic Information System (GIS) (Eldrandaly and Abu-Zaid, 2011) and risk assessment (Stavrou and Ventikos, 2014).

Geographically weighted regression (GWR) is a potentially well-suited spatial predictive model (Lieske and Bender, 2011). GWR allows the actual parameters for each location in space to be estimated and mapped as opposed to fitting of a trend surface to the parameters. Essentially, any GWR model is a statistical regression model and the least squares method is commonly used to estimate the coefficients. GWR also offers extensions of generalized linear models, including logistic and Poisson regressions (Fotheringham et al., 2002).

However, interpolation and GWR fail in many cases, because any interpolation is based on the mathematical hypothesis that the corresponding interpolation space is continuous, and the

---

*Corresponding author: hydmail2000@gmail.com;

GWR model depends on whether we clearly know what type of function can express the relationship between response variable and predictors based on data.

As a universal approximation, the geospatial information diffusion (GID) technique based on self-learning discrete regression was suggested by Huang (2019). With the help of background data as the medium, the method can fill in the missing observations more accurately than interpolation and GWR for filling in flood observations.

In theory, GID can universally approximate any continuous function to any degree of accuracy, given that observed units are sufficient and the geographic units are very small. Otherwise, it is important to inspect whether a trained GID is valid to supplement and complete the incomplete geospatial data with the estimated observations. In particular, it is also important to inspect if the GID is valid to use in the units outside the study area whose data of the observed units have been employed to train the GID.

2 INFORMATION DIFFUSION IN GEOGRAPHICAL SPACE

The concept of information diffusion was proposed in function learning from a small sample of data (Huang, 1997). A convenient model of the information diffusion technique is normal diffusion, which can be illustrated in a fuzzy set, as shown (Huang, 2002):

Let $X = \{x_i \mid i = 1,2,\ldots,n\}$ be a given sample, and let $U = \{u\}$ be its universe of discourse. The function in Eq. (1) is known as a normal diffusion function, which diffuses the information carried by observation x to the monitoring point u in the normal approach.

$$\mu(x,u) = \exp[-\frac{(x-u)^2}{2h^2}], \quad x \in X, u \in U \tag{1}$$

The diffusion coefficient h can be calculated using Eq. (2).

$$h = \begin{cases} 0.8146\ (b-a), & n = 5; \\ 0.5690\ (b-a), & n = 6; \\ 0.4560\ (b-a), & n = 7; \\ 0.3860\ (b-a), & n = 8; \\ 0.3362\ (b-a), & n = 9; \\ 0.2986\ (b-a), & n = 10; \\ 2.6851\ (b-a)/(n-1), & n \geq 11. \end{cases} \tag{2}$$

where $b = \max\limits_{1 \leq i \leq n}\{x_i\}, \quad a = \min\limits_{1 \leq i \leq n}\{x_i\}$.

Using $\mu(x,u)$, we change a given sample point x (observation) into a fuzzy set with membership function $\mu_x(u) = \mu(x,u)$ on universe U. The principle of information diffusion guarantees that reasonable diffusion functions exist to improve the non-diffusion estimates if the given samples are incomplete.

The simplest model of GID to fill in the missing observations is self-learning discrete regression (SLDR) (Huang, 2019), which can be illustrated with approximate reasoning:

Let G be a study area where a phenomenon denoted as F is under study. Suppose that G is composed of n geographic units $g_1, g_2,\ldots, g_n$, i.e.,

$$G = \{g_1, g_2, \ldots, g_n\} \tag{3}$$

Furthermore, suppose that the phenomenon F could be recognized using n observations taken from the n geographic units, respectively. An observation, denoted as w, is a number or a vector.

Definition 1: Let G be an area composed of geographic units $g_1, g_2,\ldots, g_n$, and w_i be an observation of g_i. If a phenomenon F on G can be recognized using the set of observations taken from the geographic units,

2

$$W = \{w_1, w_2, \ldots, w_n\} \tag{4}$$

when all w_i, $i = 1, 2, \ldots, n$, are assigned, we say that W is a *complete data set* with respect to F on G, and otherwise, it is *incomplete*.

Definition 2: Let g and o be two geographic units in a study area. If g is observed and assigned, but o is not, to recognize phenomenon F, g is known as an *observed unit* and o is known as *gap unit* to recognize F.

Definition 3: Let o be a gap unit. With data ω and a series of observed units, if a model exists to assign o to a value for the purpose of recognizing phenomenon F, ω are known as *media*.

Definition 4: Let o be a gap unit of area G,

$$G = \{g_1, g_2, \ldots, g_{n-1}, o\}. \tag{5}$$

If a set Z_G of the attribute values describing selected geographic features,

$$Z_G = \{z_{g_1}, z_{g_2}, \cdots, z_{g_{n-1}}, z_o\} \tag{6}$$

are media, then Z_G is known as a *background data set*.

For example, the attribute values of the population and per capita GDP features are background data used to recognize a natural disaster.

Thus far, we can formally give a definition of information diffusion in a geographical space.

Definition 5: Let W be an incomplete data set for recognizing phenomenon F on area G, and let Z_G be a background data set. If a model γ can use Z_G to make W complete, it is said that γ uses Z_G to diffuse the information of W in G for recognition of F.

If there is no loss in generality, suppose that a study area G is composed of n-q observed units $g_1, g_2, \ldots, g_{n-q}$ and q gap units $g_{n-q+1}, \ldots, g_n$, i.e.,

$$G = \{g_1, g_2, \ldots, g_{n-q}, g_{n-q+1}, \ldots, g_n\} \tag{7}$$

Furthermore, suppose that the attribute values of t geographic features are background data. In unit g_i, the attribute value of the j th feature is z_{ij}.

The SLDR model to fill in the gap units with background data consists of two parts: constructing a relationship matrix and approximately infering values in the gap units with background data.

2.1 *Constructing a relationship matrix*

Let $U_1, U_2, \ldots, U_t$ be t monitoring spaces that serve to diffuse background data of t features, respectively, and let U_{t+1} be a monitoring space that diffuses observations obtained from the observed units. Letting $\lambda = t + 1$, we define a λ-dimensional monitoring space,

$$U_1 \times U_2 \cdots \times U_\lambda. \tag{8}$$

where $U_j = \{u_{j1}, u_{j2}, \cdots, u_{jm_j}\}$, $j = 1, 2, \cdots, \lambda$.

Let $\tau = n$-q. A λ-dimensional sample X with size τ is written as,

$$X = \{(x_{i1}, x_{i2}, \cdots, x_{i\lambda-1}, x_{i\lambda}) | i = 1, 2, \cdots, \tau\} \tag{9}$$

where

$$x_{i1} = z_{i1}, \; x_{i2} = z_{i2}, \; \cdots, \; x_{i\lambda-1} = z_{it}, \; x_{i\lambda} = w_i, \; i = 1, 2, \cdots, \tau.$$

For the λ-dimensional sample point,

3

$$x_i = (x_{i1}, x_{i2}, \cdots, x_{i\lambda}) \in X,$$ (10)

and the λ-dimensional monitoring point,

$$u = (u_{1k_1}, u_{2k_2}, \cdots, u_{\lambda k_\lambda}) \in U_1 \times U_2 \cdots \times U_\lambda,$$ (11)

where $k_j \in \{1, 2, \cdots, m_j\}$, $j = 1, 2, \cdots, \lambda$, we use the λ-dimensional normal diffusion formula in Eq. (12) to diffuse the information of x to point u.

$$\mu(x_i, u) = \prod_{j=1}^{\lambda} \exp[-\frac{(x_{ij} - u_{jk_j})^2}{2h_j^2}].$$ (12)

where the diffusion coefficient h_j can be calculated using Eq. (2) with the attribute values of the j th feature and the observations in X.

Let

$$Q_{k_1 k_2 \cdots k_\lambda} = \sum_{i=1}^{\tau} \prod_{j=1}^{\lambda} \exp[-\frac{(x_{ij} - u_{jk_j})^2}{2h_j^2}].$$ (13)

We obtain an information matrix of X on $U_1 \times U_2 \cdots \times U_\lambda$, as shown in Eq. (14).

$$Q = \{Q_{k_1 k_2 \cdots k_{\lambda-1} k_\lambda}\}_{m_1 \times m_2 \times \cdots \times m_{\lambda-1} \times m_\lambda}$$ (14)

$\forall k_\lambda \in \{1, 2, \cdots, m_\lambda\}$, let

$$S_{k_\lambda} = \max_{\substack{1 \leq k_j \leq m_j \\ 1 \leq j \leq m_{\lambda-1}}} \{Q_{k_1 k_2 \cdots k_{\lambda-1} k_\lambda}\},$$ (15)

and

$$r_{k_1 k_2 \cdots k_{\lambda-1} k_\lambda} = \frac{Q_{k_1 k_2 \cdots k_{\lambda-1} k_\lambda}}{S_{k_\lambda}}.$$ (16)

We obtain a relationship matrix between background data and observations in X, written as shown

$$R = \{r_{k_1 k_2 \cdots k_{\lambda-1} k_\lambda}\}_{m_1 \times m_2 \times \cdots \times m_{\lambda-1} \times m_\lambda}$$ (17)

2.2 *Inferring with background data*

Let $z = (z_1, z_2, \cdots, z_t)$ be the background data of a gap unit, and

$$u_{\lambda-1} = (u_{1k_1}, u_{2k_2}, \cdots, u_{\lambda-1 k_{\lambda-1}}) \in U_1 \times U_2 \cdots \times U_{\lambda-1}.$$

The z can be changed into a fuzzy set on the universe of discourse $U_1 \times U_2 \cdots \times U_{\lambda-1}$ using the λ-1-dimensional normal diffusion formula in Eq. (18) and normalizing with the maximum value, as shown in Eq. (19).

$$\mu(z, u_{\lambda-1}) = \prod_{j=1}^{\lambda-1} \exp[-\frac{(z_j - u_{jk_j})^2}{2h_j^2}].$$ (18)

$$\begin{cases} a_{k_1 k_2 \cdots k_{\lambda-1}} = \frac{q_{k_1 k_2 \cdots k_{\lambda-1}}}{s}, \\[2ex] s = \max_{\substack{1 \le k_j \le m_j \\ 1 \le j \le \lambda-1}} \{q_{k_1 k_2 \cdots k_{\lambda-1}}\}, \\[2ex] q_{k_1 k_2 \cdots k_{\lambda-1}} = \prod_{j=1}^{\lambda-1} \exp\left[-\frac{(z_j - u_{jk_j})^2}{2h_j^2} \right]. \end{cases} \tag{19}$$

The fuzzy set is written as $\tilde{A}$ with memberships $a_{k_1 k_2 \cdots k_{\lambda-1}}$, $k_j = 1, 2, \ldots, m_j$, $j = 1, 2, \ldots, \lambda - 1$.

For the fuzzy input $\tilde{A}$, using the approximate reasoning operator represented in Eq. (20), we can obtain a fuzzy output $\tilde{B}$ with membership function $\mu_B(u_{\lambda k_\lambda})$.

$$\mu_B(u_{\lambda k_\lambda}) = \max_{\substack{1 \le k_j \le m_j \\ 1 \le j \le \lambda-1}} \min\{a_{k_1 k_2 \cdots k_{\lambda-1}}, r_{k_1 k_2 \cdots k_{\lambda-1} k_\lambda}\}. \tag{20}$$

Finally, using the center-of-gravity method in Eq. (21), we obtain a crisp value w:

$$w = \frac{\sum\limits_{k_\lambda=1}^{m_\lambda} \mu_B(u_{\lambda k_\lambda}) u_{\lambda k_\lambda}}{\sum\limits_{k_\lambda=1}^{m_\lambda} \mu_B(u_{\lambda k_\lambda})}. \tag{21}$$

Whether the above SLDR model consisting of Eqs. (12) to (21) and trained with the sample in Eq. (9) is valid, we can judge it by the following two criteria.

3   THE FIRST INSPECTION CRITERION

Any trained estimator to infer output with input has an accuracy problem, and the model of the GID is no exception.

In theory, the accuracy of the model determines the validity of the model and whether the results calculated by the model are usable. The higher the accuracy of the model, the more you can use it with confidence.

In practice, the model is an abstract representation of the system under study. It is extremely difficult to inspect whether a model is 100% valid. In addition, whether the model is valid is relative to the research purpose and user needs. In some cases, the model achieves 60% confidence that it meets the requirements; in other cases, the model reaches 99% and may be unsatisfactory.

This section does not discuss the accuracy of the trained model, because this is not only related to the model architecture, but also subject to the size of the sample used to train the model and the scattering degree of the sample. This section only discusses what criterion would be used to inspect if a trained SLDR model is valid, i.e., the difference between estimated observation and real observation of a test sample point is not too big.

The criterion suggested in this section serves for inspecting if a trained SLDR model is valid to fill in the missing observation of a gap in the study area, from which a sample is taken to train the model.

If removing around 10% elements from X in Eq. (9), and the set of elements reserved in the X, this called the learning sample and is denoted as X_L. The collection of elements that are removed, called the test sample, is denoted as X_T.

There is no loss in generality, so suppose that X_L and X_T consist of τ_1 and τ_2 points, respectively, which are written as

$$X_L = \{(x_{i1}, x_{i2}, \cdots, x_{i\lambda-1}, x_{i\lambda}) | i = 1, 2, \cdots, \tau_1\} \tag{22}$$

$$X_T = \{(x_{j1}, x_{j2}, \cdots, x_{j\lambda-1}, x_{j\lambda}) | j = 1, 2, \cdots, \tau_2\} \tag{23}$$

Employing Eqs. (12) to (21) to deal with X_L, we can estimate $x_{i\lambda}$ to be output $\hat{x}_{i\lambda}$ with input $(x_{i1}, x_{i2}, \cdots, x_{i\lambda-1})$, $i = 1, 2, \cdots, \tau_1$.

The standard deviation of the trained SLDR model is defined by

$$\sigma = \sqrt{\frac{\sum_{i=1}^{\tau_1} (\hat{x}_{i\lambda} - x_{i\lambda})^2}{\tau_1}} \tag{24}$$

Employing the trained SLDR model, we also can estimate $x_{j\lambda}$ to be output $\hat{x}_{j\lambda}$ with input $(x_{j1}, x_{j2}, \cdots, x_{j\lambda-1})$, $j = 1, 2, \cdots, \tau_2$, in X_T, and obtain a standard deviation:

$$\sigma' = \sqrt{\frac{\sum_{j=1}^{\tau_2} (\hat{x}_{j\lambda} - x_{j\lambda})^2}{\tau_2}} \tag{25}$$

When $\sigma' \leq \sigma$, the trained model is valid to fill in the missing observations of the gap units in the study area, with respect to this test.

This should be repeated several times to randomly construct X_L and X_T and compare their standard deviations. If the standard deviation from X_T is always no bigger than the standard deviation from X_L, the SLDR model trained by X in Eq. (9), i.e. $X_L \cup X_T$, is almost valid.

The above criterion by error comparison is called the *deviation criterion*.

4 THE SECOND INSPECTION CRITERION

That a SLDR model is valid in the study area does not mean it is also valid outside the area. For example, ATC-13 is valid to evaluate loss due to building damage struck by an earthquake in California, USA (ATC, 1985), but is invalid in Yunnan, China.

This section discusses what surrounding area of the study area can be classified as the neighborhood where a trained SLDR model is still valid.

Let G be a study area with τ observed units:

$$G = \{g_i | i = 1, 2, ..., \tau\} \tag{26}$$

The vector of the background data of unit g_i is denoted as x_i, and the observation of the unit to study a phenomenon F is denoted as y_i. The sample consisting of background data and observations of G is denoted as

$$W = \{(x, y_i) | i = 1, 2, ..., \tau\} \tag{27}$$

Let ζ_i, ξ_i be the coordinates of the center of unit g_i. The center of gravity of the τ observed units is

$$\zeta_0 = \frac{\zeta_1 + \zeta_2 + \cdots + \zeta_\tau}{\tau}, \quad \xi_0 = \frac{\xi_1 + \xi_2 + \cdots + \xi_\tau}{\tau} \tag{28}$$

If neglecting the influence of the Earth's curvature, the distance between (ζ_i, ξ_i) and (ζ_0, ξ_0) is

$$d_i = \sqrt{(\zeta_i - \zeta_o)^2 + (\xi_i - \xi_o)^2}, \quad i = 1, 2, \cdots, \tau \tag{29}$$

Let

$$\tau_0 = \begin{cases} \frac{\tau}{2}, & \text{if } \tau \text{ is even} \\ \frac{\tau+1}{2}, & \text{if } \tau \text{ is odd} \end{cases} \tag{30}$$

and G_0 be a sub-area of G, consisting of τ_0 units nearest point (ζ_o, ξ_o), written as

$$G_0 = \{g_i | i = 1, 2, ..., t_0\} \tag{31}$$

The maximum distance between units of G_0 and (ζ_o, ξ_o) is $\psi_0 = \max_{1 \leq i \leq \tau_0} \{d_i\}$.

The sample taken from G_0 is denoted as W_0. The complements of G_0 and W_0 are denoted as G_0^c and W_0^c, respectively: for G in Eq. (26), $G = G_0 \cup G_0^c$; and for W in Eq. (27), $W = W_0 \cup W_0^c$. The size of W_0^c is $\tau_c = \tau - \tau_0$. W_0, W_0^c are written as

$$W_0 = \{(x_i, y_i) | i = 1, 2, \cdots, \tau_0\} \tag{32}$$

$$W_0^c = \{(x_j, y_j) | j = 1, 2, \cdots, \tau_c\} \tag{33}$$

Employing Eqs. (12) to (21) to deal with W_0, we can estimate y_i to be output $\hat{y}_i$ with input x_i in W_0.

The standard deviation of the trained SLDR model is defined by

$$\sigma_0 = \sqrt{\frac{\sum\limits_{i=1}^{\tau_0} (\hat{y}_i - y_i)^2}{\tau_0}} \tag{34}$$

Employing the trained SLDR model, we also can estimate y_j to be output $\hat{y}_j$ with input x_j in W_0^c, with error

$$\varepsilon_j = ||\hat{y}_j - y_j|| \tag{35}$$

Considering the above deviation criterion, we say that the trained model is valid to estimate y_j by $\hat{y}_j$ when $\varepsilon_j \leq \sigma_o$. Let

$$W_{\sigma_0} = \{(x_k, y_k) | (x_k, y_k) \in W_0^c | \, ||\hat{y}_k - y_k|| \leq \sigma_0\} \tag{36}$$

where $\hat{y}_k$ is estimated by the trained SLDR model with x_k in W_0^c. The geographic unit with sample point (x_k, y_k) in W_{σ_o} is denoted as g_k, whose center coordinate is denoted as (ζ_k, ξ_k).

If there is no loss in generality, suppose that there are K elements in W_{σ_o}. The distance d_k between g_k and (ζ_0, ξ_0) can be calculated by using Eq. (29). The maximum distance between units corresponding to W_{σ_o} and (ζ_o, ξ_o) is $\psi_{\sigma_o} = \max_{1 \leq k \leq K} \{d_k\}$. In other words, according to the above first inspection criterion, an SLDR model trained by sample W_0 will be valid to fill in the missing observation of a unit in G_0^c when the distance between the unit and (ζ_0, ξ_0) is not greater than ψ_{σ_o}.

Obviously, $\psi_{\sigma_o} > \psi_0$ when $W_{\sigma_o} \neq \varphi$ (empty set). In this case, let $\psi = \psi_{\sigma_o} - \psi_0$, called the *valid radiance* of G_0.

When we spread the learning area from G_0 to G, the valid radiance of G should not be less than ψ. Therefore, the SLDR model, trained with the sample from G, should be valid in the neighborhood constituted with the units around G so that their distances to the nearest units of G, respectively, are not more than ψ.

The criterion using the radiance ψ to inspect if a trained GID is valid is called the *neighborhood criterion*.

5 CONCLUSION AND DISCUSSION

It is important to inspect if a trained model of the GID is valid to fill in the missing observation of a gap unit. We can employ two inspection criteria for this: deviation criterion and neighborhood criterion.

Both of the two criteria are based on dividing a sample taken from a study area into two parts: one for training a GID model, the other for testing validity.

The deviation criterion is that the model is valid in the study area when the average error of the model, with respect to the test units, is not greater than the standard deviation of the model trained by using about 90% of the sample points.

The neighborhood criterion is that the model is valid for a gap unit outside the study area, to fill in the missing observation when the gap unit is in a neighborhood of the area bounded by a valid radiance. The radiance is determined by the maximum distance between the units of the sample point that passed the test and the center of the study area. The test is done with the deviation criterion, and the model is trained by using the sample points of the units nearest the center of the study area and account for around 50% of the total.

In scientific research and engineering practice, inspecting the validity of a model or method is more difficult than proposing a model or method. This paper suggests two criteria to inspect whether a trained GID model is valid, but it is not given any confident estimation. Research in this topic must be carried out in response to specific issues.

ACKNOWLEDGMENTS

This project was supported by the National Key Research and Development Plan (2017YFC1502902) and the National Natural Science Foundation of China (41671502).

REFERENCES

Applied Technology Council (ATC). 1985. *Earthquake Damage Evaluation Data for California*. ATC-13 Report, Redwood City, USA.

Eldrandaly, K.A. and Abu-Zaid, M.S. 2011. Comparison of six GIS-based spatial interpolation methods for estimating air temperature in western Saudi Arabia. *Journal of Environmental Informatics* 18(1): 38–45.

Fotheringham, A.S., Brunsdon, C. and Charlton, M. 2002. *Geographically Weighted Regression: The Analysis of Spatially Varying Relationships*. Chichester: John Wiley & Sons.

Huang, C.F. 1997. Principle of information diffusion. *Fuzzy Sets and Systems* 91(1): 69–90.

Huang, C.F. 2002. Information diffusion techniques and small sample problem. *International Journal of Information Technology and Decision Making* 1(2): 229–249.

Huang, C.F. 2019. Geospatial information diffusion technology supporting by background data. *Journal of Risk Analysis and Crisis Response* 9(1): 1–9.

Lieske, D.J. and Bender, D.J. 2011. A robust test of spatial predictive models: geographic cross-validation. *Journal of Environmental Informatics* 17(2): 91–101.

Stavrou, I.D. and Ventikos N.P. 2014. Ship to ship transfer of cargo operations: risk assessment applying a fuzzy inference system. *Journal of Risk Analysis and Crisis Response* 4(4): 214–227.

Zhang, Y., Xu, L.S., Chen, Y.T. 2009. Spatio-temporal variation of the source mechanism of the 2008 great Wenchuan earthquake. *Chinese Journal of Geophysics* 52(2): 379–389.

Risk Analysis Based on Data and Crisis Response Beyond Knowledge – Huang & Nivolianitou (eds)
© 2020 Taylor & Francis Group, London, ISBN 978-0-367-25146-8

Black swans and dragons

Ben Ale
Technical University Delft, Delft, The Netherlands

Des Hartford
BC Hydro, Burnaby, BC, Canada

David Slater
School of Engineering, Cardiff University, Cardiff, UK

ABSTRACT: Since Nassim Taleb coined black swan as an event that occurred as a complete surprise for everybody, the metaphor of the black swan has been applied to a much wider variety of events. Black swan events now comprise events that are a surprise for some but not for others, events that have a low likelihood, events that are not believed to be possible but still proved to be possible, events that were dismissed as being too improbable to worry about but happened anyway. For a decision maker the black swan problem is choosing where to put effort to prevent, or mitigate events for which there are warnings, or for which the possibility has been put forward. Does the fact that there are thousands of books written about fire breathing dragons warrant the development of an Anti-Dragon Defense Shield? The black swan may have been a surprise for Willem de Vlamingh in 1697, it was not a surprise for the inhabitants of Australia, for which the appearance of tall white humans was their "black swan event" .In this paper we explore the options available to decision makers when confronted with the various sorts of swans or dragons.

Keywords: black swan, mitigate event, warning, dragon, decision make, Australia

1 INTRODUCTION

A Black Swan event was the term that Nassim Taleb coined for events that came as a complete surprise (Taleb, 2010). Before Willem de Vlamingh encountered a bunch of black swans in Australia in 1697 all known swans were white. That is all swans known in the northern hemisphere of the world. This was about 200 years before Darwin (1859) published his "Origin of Species". The current belief was that the world has been created perfectly and therefore unchanging. Australia just did not exist. For people in the Northern Hemisphere world, the idea of a black swan just had never occurred to them. Encountering a black swan was the pure Black Swan event.

Taleb devotes a large portion of his book and of his further publications to random events, probability distribution curves and discussions about rare events. He attempts to explain the difficulties people have in understanding the difference between the improbable and the impossible; and in estimating probabilities in general. This watered down the pure definition of the black swan event to that of an event of which the probability was underestimated, or of which the possibility was dismissed because the probability was thought to be too low to consider.

The main example on which Taleb's reasoning is founded is that of the financial crisis of 2007. It can be argued that this crisis was not unexpected at all, or should not have been because it was completely predictable. Selling mortgages and other loans to people who cannot pay the loan back creates a hole in the finances somewhere. As long as people continue to believe that the hole will be filled by somebody, that hole can remain. But it is like a game of black peter. Sooner or later somebody, or some bank, will be confronted with having this debt and not being able to refinance it, at which time the dominos fall over. The financial crisis therefore was not a Black Swan event at all.

As a result of Taleb's analyses and descriptions what might be termed a "cygnology" has emerged by which more and more sorts of events have been designated as belonging to the genus swan (Aven, 2015). In the remainder of the paper we will try to unpick this "cygnology" and analyse the importance of the difference between the various genera for making decisions.

2 "CYGNOLOGY"

"Cygnology" plays an important role in the current discussions about risk management. The discussion is being held in words and the definition of words seems to be central in the discourse. Many words are used in ordinary speech and in writings with only loose definition.

2.1 Chance

Probability, chance, odds and likelihood are primary examples. In the "normal" use of language these words are used as synonyms. In mathematics they have precise and distinct definitions.

Unfortunately, these exact definitions present decision makers and those who philosophize about risk some serious problems. Probability is defined as the number of a particular outcome (often called success) divided by the total number of tries, if the number of tries is infinite. In the real world the number of tries cannot be infinite. If the particular outcome is an accident, if one looks carefully enough, no accident is really the same, nor are the circumstances.

What confounds the issue even further is that more often than not the "probability" is really a frequency. It is the number of occurrences in a certain timeframe, usually a year. Even if events do not occur every year, the cumulative number over a longer period can rise to a number above 1, which violates the rule about probability, which is a number between 0 and 1 by definition. When a frequency is non-zero the relevant question is no longer whether the event can happen, but when it will happen, making the probability equal to 1, thus fulfilling Murphy's law.

Finally all definitions by means other than mathematical formulae tend to be imprecise and circular, such as that probability is the likelihood, or chance (http-1, 2019).

For decision makers the philosophy around terminology may not be exciting, but it is important to know whether any number or wordy expression about chance is a probability, a frequency and whether that has any meaning at all. For this paper probability is independent of the time frame. It is the chance (!) that something will happen ever. Frequency is the number of events per year.

2.2 Improbable or impossible

Another set of notions that play an important role in "cygnology" are those of improbability and impossibility. In decision making surprises often stem from confusing the improbable with the impossible.

An improbable event is usually understood to be an event of which the probability is low. Low in this respect is not really defined. Just as a small elephant is a large animal, what constitutes a low probability depends on the judgement of the beholder and is often contextual. A probability of 1 in 6 may be low if it pertains to the chance of heads coming up in the throw of a coin, it is high if it where the chance of an airplane not completing its journey. When the numbers get really small and the costs of making the improbable impossible are high, it is tempting to decide that a 1 in a million per year event "will not happen to us" and decision makers often cherish the illusion.

On the other hand some analogies cause the impossible to be deemed probable. The demonstrated possibility of a nuclear mushroom explosion played a significant role in the discussion about the safety of nuclear power plants, even though in such a plant a mushroom explosion is physically impossible. This led Kaplan & Garrick (1981) to redefine risk as $R = f(s,p,c)$, with p probability, c consequence and s scenario, their message being that there should be a plausible scenario by which the event could materialize in order to make it part of the total risk.

2.3 *Science and fiction*

It is not always easy to decide what is possible and what is impossible. Jules Verne described how to get to the moon by firing a large bullet from an enormous gun. Although the bullet had the size of what later was a lunar orbiter, this method was impossible. Hergé (1953) used a rocket that looked like a German V2 rocket. Still not possible, but only 16 years later people actually landed on the moon. The impossible proved to be possible after all. Decision makers face the problem whether scenario's that, prima facie seem impossible, are merely improbable; and whether the probability may increase over time, or is already underestimated. In terms of swans the analogy may be a fire breathing dragon. There are whole libraries of books written about them and movies made. If they exist their effect could be devastating and sooner or later they will appear. The question whether an Anti-Dragon Defense Shield is worth investigating may with our current understanding of dragonhood sound ridiculous, but there are many other stories of mass destruction by floods, asteroids, weapons and other natural and manmade disasters that have been described in fiction and are much less improbable that dragons. Later in the paper we will return to the problem of decision making.

3 KNOWLEDGE

Another parameter that is important in "cygnology" and distinguishes one type of swan to another is the level of knowledge. The United States Secretary of Defense Donald Rumsfeld made an idea famous that was first created by Luft & Ingham (1955)., in the Johari window.

3.1 *Known and unknown*

In this Johari diagram, the relationship between peoples' conscious knowledge and the hidden knowledge, or between peoples' knowledge and knowledge in general, was divided up into four quadrants: the things people know they know: the known knowns; the things people know they do not know: the known unknowns; the things people do not know that they know, or more generally the things that people do not know that it is known somewhere in the world: the unknown knowns and finally the things that people do not know that it is unknown; the unknown unknowns.

Of these four quadrants the known knowns are the easiest to deal with.

The known unknowns are only slightly more complicated to deal with. One could try and resolve the problem by further investigations to make the unknown known. However when time or money runs out one could decide not to. It is also possible that people just do not want to know. We will discuss later how a decision maker can take advantage of sustained ignorance. Then there are things that are known to be unknown but have to remain unknown. The future temperature rise of the earth is an example. There is no way of doing a controlled experiment to determine the temperature rise depending on the concentration of greenhouse gasses in the atmosphere. There is only one earth so we have to be resigned to having an estimate that is the maximum achievable. A special case is the case of pseudo randomness. If it were possible to measure exactly all the variables and parameters that determine the where a gunshot would hit, they could be made to pass through the same hole. In practice this is too difficult and hence a certain spread needs to be accepted. Finally there are the fundamental unknowns. Although it is known that the probability of a 6 coming up in the casting of a die

11

is 1/6, there is no way of knowing what the result of the next throw will be. Neither is it possible to know whether Schrödinger's cat is alive or dead. This aleatory uncertainty cannot be reduced.

The unknown knowns comprise the things we used to know but have forgotten and more importantly the things we do not know but others do and most importantly the things others have tried to bring to our attention but we dismissed them. In this realm resides the infamous memo that surfaces after the accident, which contains the warning that the accident is imminent. Also the miscommunications between managers and technicians constitute unknown knowns, such as the behavior of the booster rockets of the Challenger under low ambient temperatures. Many of the unknown knowns can be resolved by asking around or research. If William de Vlamingh would have asked an Australian, the black swan would not have been such a surprise, but he did not know Australians existed.

The unknown unknowns are the most problematic and the most discussed. These are the real black swans. There is no way to know such an unknown unknown. Although one can argue that we should be prepared for a surprise, there is not really a method to do this. The essence of a surprise is that it is a surprise. There are however many instances where surprises are claimed where they should not have been a surprise at all. In many cases the possibility of the event was dismissed. One reason can be the low probability in the mistaken belief that improbable and impossible are synonymous. Beyond design-basis accidents are an example. Another is the difficulty in assessing the probability of large consequence low probability events, making estimates uncertain and the mistaken belief that uncertain is synonymous with wrong. Yet another is the hesitation to admit to potentially catastrophic consequences where denial is much less expensive than facing the problem. In the Netherlands it took a major flood in 1953 to make the authorities take the warnings of experts that the dikes were not strong enough seriously. A special class of surprise is accidents that are the result of not taking known safety precautions. Prescribed safety precautions are there to prevent an event. Not taking them makes an event unavoidable. This claim of surprise is akin to claiming that throwing heads with a coin came as a surprise (unless one assumed the coin was tampered with and had tails only).

3.2 *Uncertainty*

The black swan metaphor led to the realization that risks may be underestimated because the risk assessor is not aware of knowledge that is available elsewhere. This led authors to replace the triplet of Kaplan and Garrick by $R = f(K, p, c)$, where K is knowledge (Aven, 2012).

It should be noted that pure aleatory uncertainty can be integrated in the estimate of probability when the probability distribution is known. Long before Taleb it was an established fact that in real life, many probability distributions have a much larger portion of the probability mass away from the center, than would be predicted by the familiar Gauss distribution. The difference in probability of a 4-sigma event between a normal distribution and a Cauchy distribution is two orders of magnitude. Distributions with so-called fat tails can be observed for instance in the distribution of the location of crashed aircraft relative to the intended flight path. Many of these distributions are power law distributions. If it is assumed a priori that the probability distribution of the outcomes of an activity is normal, the occurrence of an event may be more of a "surprise" when the real distribution follows a power law. Because of its relatively fast decline, the normal distribution makes it more inviting to declare the improbable impossible.

Unfortunately the advice of Taleb not to make systems too large to fail, cannot be followed for many technological systems such as the flood protection of a nation, or hydropower dams chemical production sites and the transport of hazardous materials. A failure of these systems has in common that the failure rate is small but the consequences catastrophic. Many of these systems have been in existence for a long time and there is not much opportunity to reduce the size. Because of the low probability of accidents the data base for assessing the probability of failure remains small making the estimates uncertain resulting in the confusion of improbability and impossibility, which is only broken from time to time by an event.

12

3.3 *Ignorance*

The advantages of ignorance were already described in the beginning of the previous century (Keedy, 1908). Ignorance is bliss seems especially to hold for managers of larger companies, who increasingly call accidents and incidents, black swan events. Ignorance also supports the claim that causes of accidents, incidents and events are identified in hindsight, looking to accident causation as a linear process, that the world is really too complicated to model accident causation in a fault tree and therefore nobody is to blame, except may be the operator who made the final mistake.

The sub-prime-mortgage crisis is described as being the result of an unknown unknown: Although the risk of home owners defaulting on the mortgage was known, it was completely underestimated by the unknown unknown that they could all default simultaneously (http-2-2019). However, what such a description fails to recognize is that a basic safety barrier in the mortgage market was broken deliberately: one should not sell mortgages to people of which it is certain that they cannot pay their loan back and thus defaulting is unavoidable. Because it was deliberate it should have been known and there is actually no reason to assume that it was not (http-3, 2019). Denial and the black swan metaphor however were extremely helpful for bank managers to avoid jail sentences.

Safety barriers and safety systems that are built into technical systems are there for a reason. It is therefore asking for trouble when they are removed especially when nobody remembers what they were for and never seem to be challenged. One could claim that it is hindsight to conclude that a safety valve should not have been welded shut, that bolts should have been tightened using a torque wrench, that a new catalyst should have been tested by an adiabatic heating test (OVV, 2015) and that bellows should be supported (http-4, 2019), but this was all known in advance when the system and the procedures were designed and for a reason. That the current CEO does not know that, neither is it a reason to deviate from the design, nor should it be an excuse. Nevertheless, black swans are on the move as they appear to be a convenient escape for those who do not want to spend money on safety before the event or to be blamed after.

4 DECISION MAKING

How to make decision in uncertainty has been the subject of countless books and papers for decades (Smithson, 1988; Hirshleifer, 1992; Etner et al 2009). As it has now become impossible to read everything that ever has been written on the subject itself, presents a situation of unknown knowns. In this blissful ignorance there are a number of observation that can be made.

Although the known knowns may look unproblematic, there are still some issues to consider. First of all there is the problem that the general understanding may in fact be erroneous. It is the general understanding that the general understanding that the earth was stationary was contradicted at first by a single person Galileo with the words. "Eppur, si muove", but according to Boller & George (1990) he never said that. On the other hand the myth that sunlight could start a fire was proven to be false on 1920s Lambert & Butler English Cigarette cards, while it is in fact true. Which raises the Fire Breathing Dragon question: given the abundance of books and movies about them, should an anti-dragon-defense-shield be developed. That such a question is not trivial is proven by the question whether the –later proven to be false–allegation of the existence of weapons of mass destruction warranted the start of a war, which prompted Rumsfeld's discussion on knowledge. On the other hand climate change has been given the dragon treatment for decades.

In a situation of known unknowns it remains a good strategy to try and reduce the extent of ignorance or uncertainty by further research. However this may be time consuming and also may be expensive. Most decisions have a constraint in time. A decision not made in time usually also constitutes a decision, which is to let events take their course uninfluenced. If there is a constraint in money or other means one has to decide on this uncertainty, in both cases one needs to be prepared to accept that one could get into a situation where after an – usually adverse – event somebody will say that you could have known. A good secondary strategy therefor is to try and identify those facts, or those pieces of information that would change the decision and try to resolve those issues with priority.

The situation of unknown knowns is more problematic. Fortunately, seeking whether there is actually information out there that one is not aware of can be supported by the internet, although there are caveats there too. Not everything that has been written is valid and a significant part of the information gathered before 1990 is not on the internet. For accidents and disasters history books can be eye openers (Nash, 1976). It remains an undeniable fact that the Australians knew about black swans and that in the absence of the knowledge that Australia existed there was no way for William the Vlamingh to know. However in the majority of accidents one could have known if one had followed procedures, read the instructions, listened to an operator who tried to convey a message (http-5, 2019) or search the literature, whether on internet or in a library. After the event the knowns usually surface in the post-accident inquests in the form of memo's, emails and notes. Therefor it is a good strategy to have an organized way of dealing with unorganized incoming information so that weak signals are not missed (Guillaume, 2011).

The unknown unknowns are perceived to be the most frightening. But there is nothing that can be done about them because it is unknown that they are unknown. It never occurred to William the Vlamingh nor to anybody in the northern hemisphere at the time that swans could be anything other than white. Until July 1940, when the Tacoma bridge collapsed, it never occurred to engineers that the effect of the wind could make a bridge flutter and collapse. These are the real Black Swans. Strangely enough the bridge problem proved to be an unknown known to the architects who designed the Erasmus Bridge in Rotterdam, which had to be closed shortly after the opening in 1996, to take remedial action against the same effect (http-6, 2019).

5 DISCUSSION AND CONCLUSION

The evolution of the genus of black swan has led to a number of undesirable side-effects. The first is that an event that is the result of taking a protective mechanism, which under circumstances will be the last line of defense, out of commission is no longer seen as a direct cause that should have been known in advance, but as an explanation in hindsight and falling in the black swan family. The second is that ignorance can be used as an excuse for failed management supervision. The unknown known (the organization is so big I cannot know what all the operators are doing) and the known unknown (I cannot read all the incoming warnings) can be used to make the operator culpable and give management a "get out of jail free" card. These two mechanisms combine in the denial of the potential for catastrophic events, which takes away the need to organize contingency and disaster plans. Being sent to the realm of myths and dragons has been the fate of many disaster scenarios. Unfortunately fantasy may become reality. A serious accident in a nuclear power plant was the subject of a movie released on March 16, 1979 and only 12 days later the Three Mile Island event occurred. Therefor decision makers should be aware whether their choices are based on facts, reason, belief or disbelief and that new signals may arise that should change their course of action. Finally we should be aware that nothing will prepare us for the "Big White Men" event, because its possibility never will occur to us until after the fact.

REFERENCES

Aven, T. (2012) The risk concept – historical and recent development trends. *Reliability Engineering and System Safety* 99, 33–44.
Aven, T, (2015) Implications of black swans to the foundations and practice of risk assessment and management, *Reliability Engineering and System Safety* 134, 83–89.

Boller, Paul, F., John George, (1990) *They Never Said It*, p. 3.0.

Guillaume, Eve M.E. (2011), *Identifying and Responding to Weak Signals to Improve Learning from Experiences in High-Risk Industry*. Boxpress BV, Oisterwijk, Netherlands, ISBN: 978-90-8891-264-1.

Darwin, Charles (1859) On the origin of Species, reprinted by Dover ISBN 10: 0486450066.

Etner, J., M. Jeleva, J. –M. Tallon. (2009) *Decision theory under uncertainty*. Documents de travailduCentred'EconomiedelaSorbonne2009.64-ISSN:1955-611X.2009.

Herge (1953) Destination Moon, French and European Publications Inc, ISBN 9781405208154.

Hirshleifer, Jack, John G Riley (1992) *The analytics of uncertainty and information* Jack Cambridge University Press ISBN 0 521 23959 7.

http-1, https://learn.problemgambling.ca/probability-odds-random-chance(last vsited 09-03-2019).

http-2, https://www.quora.com/What-are-some-of-the-most-important-black-swan-events-in-history (last vsited 09-03-2019).

http-3, http://archive.economonitor.com/nouriel/2008/02/05/the-rising-risk-of-a-systemic-financial-melt down-the- twelve-steps-to-financial-disaster/ (last vsited 09-03-2019).

http-4 https://en.wikipedia.org/wiki/Flixborough_disaster (last visited 10/03/2019).

http-5, https://www.nola.com/news/gulf-oil-spill/2010/08/bp_manager_boss_both_ignored_w.html (last visited 09-03-2019).

http-6://ta.twi.tudelft.nl/users/vuik/information/erasmus.html (last visited 09-03-2019).

Luft, J.; Ingham, H. (1955) *The Johari window, a graphic model of interpersonal awareness*. Proceedings of the western training laboratory in group development. University of California, Los Angeles.

Kaplan, S and B.J. Garrick, (1981) On the quantitative definition of risk, Risk Analysis, vol 1, no 1, pp 11–27.

Keedy, Edwin R. (1908) Ignorance and Mistake in the Criminal Law, *Harvard Law Review* Vol. 22, No. 2, pp. 75–96.

Nash Jay R. (1976) *Darkest Hours*, Nelson-Hall, Chicago ISBN 0 88229-140-8.

OVV (2015) *Explosies PSOP2 Shell Moerdijk*, Onderzoeksraad voor Veiligheid, Den Haag, Netherlands.

Smithson, M. (1988) *Ignorance and Uncertainty, emerging paradigms*, Springer, New York, ISBN 978 0 387-96945-9.

Taleb, N.N., (2010) *The Black Swan*. Penguin, ISBN 978-0-1410-3459-1.

Risk Analysis Based on Data and Crisis Response Beyond Knowledge – Huang & Nivolianitou (eds)
© 2020 Taylor & Francis Group, London, ISBN 978-0-367-25146-8

Risk analysis and preliminary countermeasures of earthquake disasters in Yunnan region

Sen Qiao*, Aiwen Liu, Xueliang Chen, Zongchao Li, Changlong Li & Tiefei Li
Institute of Geophysics, China Earthquake Administration, Beijing, China

ABSTRACT: Yunnan region is in the eastern side of the collision zone between the Indian plate and the Eurasian plate. Earthquake disasters that occurred in this area are very serious. Due to the special geographical location of Yunnan, it was included in the two major economic corridors in the strategy called "one belt and one road": China-Indochina Peninsula corridors and Bangladesh-China-India-Burma corridors. This paper systematically analyses the seismicity and tectonic background in Yunnan region: earthquake hazard analysis and earthquake disaster risk analysis were carried out and the preliminary countermeasures for emergency response were proposed.

Keywords: Yunnan region, earthquake disaster, risk analysis, emergency countermeasures

1 INTRODUCTION

Yunnan region is located in the eastern side of the collision zone between the Indian plate and the Eurasian plate, and in the southern part of the famous north–south seismic belt in the mainland of China. Modern crustal movement and tectonic activity are very intense, with high frequency and intensity seismic activity. Since the 1900s, there have been three earthquakes with a magnitude greater than five every year, two earthquakes with a magnitude greater than six every 3 years, and one earthquake with a magnitude greater than seven every 10 years. Yunnan is one of the provinces with the largest number of earthquakes, the strongest earthquake disasters, and the highest risk of earthquake disasters in China.

In the Global Risk Report released at the World Economic Forum in 2019, major natural disasters were ranked in the top three for probability and the top five for influence among the 30 risk factors in five major fields (economy, environment, geoenvironment, society, and science and technology), which will affect the world economy in the next 10 years (Global Risks report, 2019). The special geographical location of Yunnan region includes the China-Indochina Peninsula economic corridor and the Bangladesh-China-India-Myanmar economic corridor in the strategic layout of the "one belt and one road" strategy. Therefore, it has great scientific and social significance for studying the seismic hazard and risk analysis of earthquake disasters. This study systematically analyses the distribution and tectonic background of major earthquake disasters in Yunnan region.

We carried out seismic hazard analysis, exposure analysis, and risk analysis of earthquake disasters in Yunnan region. In addition, suggestions and measures for emergency response to earthquake disasters were proposed.

---

*Corresponding author: qiaosen@cea-igp.ac.cn

2 SEISMIC ACTIVITY CHARACTERISTICS AND TECTONIC BACKGROUND IN YUNNAN REGION

The earthquakes in Yunnan are basically tectonic earthquakes. The pushing of the Indian plate in a north-east direction to the mainland of China is the main cause of the frequent occurrence of strong earthquakes. This is mainly reflected in two aspects: the strong direct compression of the Indian plate from west to east to Yunnan region, and the penetration of the Sichuan-Yunnan rhombic block derived from the positive push of the Indian plate toward the Tibet block, from north to south. Therefore, the earthquakes in Yunnan region have the dual characteristics of both marginal earthquakes and intraplate earthquakes. Seismic activities in Yunnan region are mainly distributed in the eight major seismic zones (Minzi Qun, 1989) (Figure 1): Xiaojiang, Tonghai-Shiping, Zhongdian-Dali, Tengchong-Longling, Lancang-Gengma, Daguan-Mabian, Simao-Puer, and Nanhua-Chuxiong. Xiaojiang fault zone and Puduhe fault zone are the main fault zones affecting seismic activity in Kunming area.

There have also been many destructive earthquakes in the history of Yunnan, such as the Songming 8.0 earthquake in 1833, the Tonghai 7.7 earthquake in 1970 (which caused more than 15,000 deaths), the Daguan 7.2 earthquake in 1974, the Longling 7.3 and 7.4 earthquakes in 1976, the Lancang-Gengma 7.6 and 7.2 earthquakes in 1988, and the Lijiang 7.0 earthquake in 1996. These earthquakes caused great disasters and the frequent occurrence of large earthquakes in Yunnan indicates that this region has a high seismic risk.

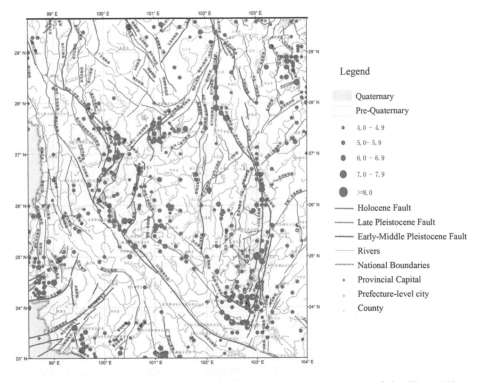

Figure 1.   The active structural and epicenter of Yunnan region, showing part of the Chuan-Dian area. Source: www.ncepe.cn/syccn/index.php?r=site/articleRead2&bid=25030100&aid=182.

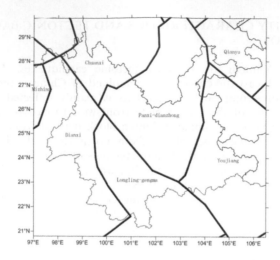

Figure 2. Seismotectonic zones in Yunnan region.

3 SEISMIC HAZARD ASSESSMENT FOR YUNNAN REGION

3.1 *Seismic belts in Yunnan region*

The fifth generation of China Seismic Hazard Map (GB 18306-2015) uses new methods for seismic belts zonation. It uses three-level seismogenic sources division, and 29 seismic belts were divided across the country, in which the southern part of the Xianshuihe – Eastern Yunnan Seismic Belt, the western part of the Youjiang River Seismic Belt, and the northern part of the Southwestern Yunnan Seismic Belt are located in Yunnan Province. Seismotectonic zones are divided in seismic belts, and there are seven seismotectonic zones in Yunnan and surrounding areas, as shown in Figure 2: Chuanxi, Panxi-dianzhong, Dianxi, Longling-gengma, Youjiang, Qianyu and Mizhina.

3.2 *Seismogenic sources in Yunnan region*

Zhou et al. (2013) deeply analyzed earthquake activity characteristics in China and the surrounding areas, and established the principle of three-level delineation of seismogenic sources. First, zoning seismic belts as units of seismicity parameters statistics, where earthquake magnitude-frequency relation accords to G-R relation (Gutenberg and Richter, 1944). Second, in seismic belts, zoning seismotectonic provinces (sources of background, SBG) with different background seismicity characteristics. Third, in SBGs, zoning potential seismic sources (sources of seismic tectonism, SST) around active tectonic faults. The three-level delineation of seismogenic sources is a main feature of seismogenic sources zoning in China. There are 29 seismic belts, 77 SBGs and about 1200 SSTs zoned nationwide. Seismogenic sources in Yunnan are shown in Figure 3. In this seismogenic sources model, seismogenic sources are the surface projection of the macro-epicenter of potential earthquakes.

3.3 *Seismicity of seismogenic sources in Yunnan region*

The difference of tectonic activity environment between east and west China was considered in the determination of the seismicity of the seismogenic sources. The methods and basis of the zoning of SBGs and SSTs in the east and west focused on different points. In western areas, we focused on the boundaries of different levels of active blocks controlling major earthquake activity (Zhang et al, 2003). Different types of seismogenic sources models are created, and tectonic analogy principles are fully used under the seismogenic sources framework to zone new potential sources and avoid underestimating earthquake capacity in high-magnitude SSTs. In eastern areas, we focus on the recognition of strong earthquake tectonic belts and enrich the basis of zoning strong earthquake SSTs (Xie and Pan, 2011).

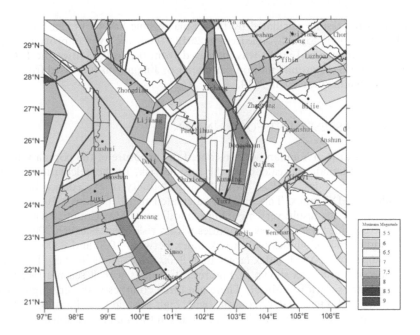

Figure 3.   Seismogenic sources in Yunnan region.

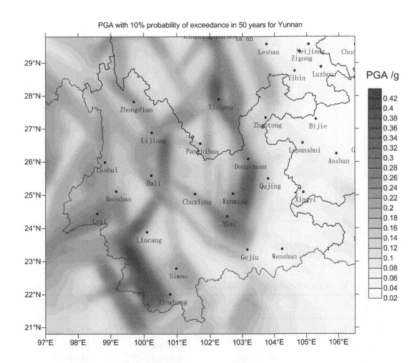

Figure 4.   Seismic hazard map for Yunnan region.

3.4   *Seismic hazard of Yunnan region*

A seismic hazard map of PGA with 10% probability of exceedance in 50 years for Yunnan region is shown in Figure 4. This figure shows that western and central Yunnan has a high seismic hazard, while the seismic hazard in eastern Yunnan is low.

19

Figure 4 shows the results calculated by the China seismic hazard model built in 2011. In the updated hazard map formally released in 2016, there are two improvements: (1) the ground motion parameter of Ludian was increased from 0.1 to 0.15 after the Ludian M6.5 earthquake in 2014; (2) the ground motion parameter of Jinggu was increased from 0.1 to 0.15 after the Jinggu M6.6 Earthquake in 2014.

4 SEISMIC RISK ASSESSMENT FOR YUNNAN

Based on seismic hazards in Yunnan, we considered the exposure of the population and GDP in every county of Yunnan and calculated the seismic risk. In the calculation, we used the Potential Seismic Risk Index as PGA (with 10% probability of exceedance in 50 years) × population index × GDP index. This index is calculated in 10^4 persons • 10^8 Yuan • g. We leveled four classes for the result: very high, high, middle, and low. Counties with a risk index less than 200 are considered "low" risk. Counties with a risk index between 200 and 1000 are considered "middle" risk. Counties with a risk index between 1000 and 8000 are considered "high" risk. Counties with a risk index larger than 8000 are considered "very high" risk. The seismic risk map of Yunnan is shown in Figure 5. Counties with a seismic risk level "very high" and "high" are shown in Table 1. There are three counties with very high seismic risk (Kunming, Qujing and Yuxi) and 17 counties with high seismic risk. Because of the limited data, we did not consider the vulnerability of buildings in the Yunnan regions.

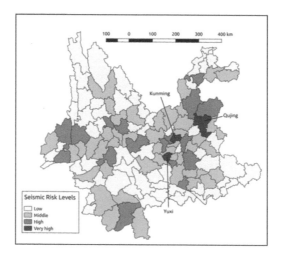

Figure 5. Seismic risk map of Yunnan.

Table 1. Counties with "very high" and "high" seismic risks.

County	Risk level	County	Risk level
Kunming	Very high	Mile	High
Qujing	Very high	Xuanwei	High
Yuxi	Very high	Anning	High
Xishan	High	Xiangyun	High
Dali	High	Gejiu	High
Chuxiong	High	Luxi	High
Huize	High	Jinghong	High
Baoshan	High	Tonghai	High
Zhaotong	High	Yiliang	High
Tengchong	High	Yunxian	High

5 PRELIMINARY COUNTERMEASURES FOR REDUCING THE RISK OF EARTHQUAKE DISASTERS AND EMERGENCY RESPONSES

We should strictly implement the national standards of the People's Republic of China. Earthquake-resistant fortification of various projects is carried out according to the zoning map of China's ground motion parameters (GB18306-2015). For new construction projects, we should standardize site selection, design, construction, supervision and acceptance. For old buildings, seismic strengthening should be undertaken.

Emergency response plans should be carefully formulated. The people's governments of the state, provinces, prefectures (cities) and counties should formulate corresponding earthquake emergency response plans, according to the classification and starting conditions of earthquake emergency response. Emergency rescue work should be carried out according to the four levels that are particularly significant: earthquake disaster, major earthquake disasters, greater earthquake disasters, and general earthquake disasters. All large enterprises should also formulate corresponding earthquake emergency response plans and carry out self-rescue and mutual rescue work according to the level of earthquake emergency response.

We also should offer good service for "one belt and one road" construction. Due to the special geographical location of Yunnan region, there two major economic corridors: the China-Indochina Peninsula corridors and the Bangladesh-China-India-Burma economic corridors. Seismic fortification of major projects and major open port infrastructure projects should be done well, which can minimize the impact of earthquake disasters on the construction of "one belt and one road".

Earthquake emergency drills and popular science publicity and education for earthquake prevention and disaster reduction should also be carefully organized. The monitoring and research of earthquake events should also be escalated.

6 CONCLUSION AND DISCUSSION

Through the analysis and comparisons above, we can draw the following conclusions. 1) Western and Central Yunnan have higher seismic hazards than the eastern Yunnan region. 2) Kunming, Yuxi and Qujing have high seismic risks in Yunnan. 3) We should strictly implement the national standards of the People's Republic of China (GB18306-2015). 4) The government should formulate an earthquake emergency plan to guide emergency rescue after the earthquake. 5) Enterprises should formulate an earthquake emergency plan to carry out self-help and mutual rescue after the earthquake.

ACKNOWLEDGMENTS

This research work was supported by the National Key Research and Development Program (2017YEC1500405), the National Natural Science Foundation of China (51678537, 51278470), the National Key Research and Development Program (2017YFC1500205), and the Special Fund of the Institute of Geophysics, China Earthquake Administration (DQJB19B06).

REFERENCES

General Administration of Quality Supervision, Inspection and Quarantine of the People's Republic of China, China National Standardization Management Committee. 2015. *China seismic zoning map*. GB 18306-2015. (in Chinese).

Ziqun M, 1989. *Seismic Area and Belts in YUNNAN and GUIZHOU region. Seismic hazard research in Yunnan and GUIZHOU regions.* Yunnan Science and Technology Press: Kunming, pp. 87–92. (in Chinese).

Global Risks Report. 2019. World Economic Forum 2019. Davos.

Gutenberg B, Richter CF. 1944. Frequency of earthquakes in California. *Bulletin of the Seismological Society of America* 34(4): 185–188.

Xie FR, Pan H. 2011. *Report on determination of seismicity parameters in seismic hazard map of China.* Working group on National Seismic Hazard Mapping. (in Chinese).

Zhang PZ, Deng QD, Zhang GM, et al. 2003. Active tectonic blocks and strong earthquakes in continental China. *Science in China* (Series D), 46 (Supplement): 13–24.

Zhou BG, Chen GX, Gao ZW, et al. 2013. The technical highlights in identifying the potential seismic sources for the update of national seismic zoning map of China. *Technology for Earthquake Disaster Prevention* 8(2): 113–124. (in Chinese).

Risk Analysis Based on Data and Crisis Response Beyond Knowledge – Huang & Nivolianitou (eds)
© *2020 Taylor & Francis Group, London, ISBN 978-0-367-25146-8*

Source-attributed QALYs lost, induced by $PM_{2.5}$-bound PAHs in Beijing, based on a PMF model

Hongbin Cao*, Sihong Chao, Jianwei Liu, Yanjiao Chen & Xuemin Liu
Faculty of Geographical Science, Beijing Normal University, Beijing, China

ABSTRACT: Polycyclic aromatic hydrocarbons (PAHs) are a group of organic compounds mainly from incomplete combustion and pyrogenic decomposition. High-molecular-weight (HMW, 4–6 rings) PAHs usually have carcinogenic toxicity and tend to exist in the particulate phase. In this study, daily $PM_{2.5}$ samples were collected and 16 priority PAHs were detected for one year in the urban district of Beijing, China. Three sources were identified by the positive matrix factorization model (PMF) model: vehicle emission, coal combustion, and petroleum volatilization, natural gas and biomass combustion. With the increase in $PM_{2.5}$ concentration, the mass contribution of coal combustion and vehicle emission increased; while the mass contribution of petroleum volatilization, natural gas and biomass combustion decreased. The source-attributed cancer risk was further evaluated by Ba Pequivalent concentrations (BaPeq) and source profiles as a result of 30 years of exposure for local residents. Furthermore, contribution to the QALYs (quality adjusted life years) lost by those suffering from lung cancer was estimated for each source. Although vehicle emissions contributed only 50.5% of the mass concentration of $PM_{2.5}$-bound PAHs, it induced 66.3% QALYs lost (338.9 QALYs). Therefore, the source-attributed QALYs lost was suggested as a better index for the determination of priority control sources rather than the source-attributed mass contribution, in view of protecting human health.

Keywords: polycyclic aromatic hydrocarbons, $PM_{2.5}$, sources, lung cancer, quality adjusted life years

1 INTRODUCTION

Polycyclic aromatic hydrocarbons (PAHs) are a group of organic compounds mainly resulting from incomplete combustion and pyrogenic decomposition. High-molecular-weight (HMW, 4–6 rings) PAHs usually have carcinogenic toxicity and tend to exist in the particulate phase. Many studies have investigated the $PM_{2.5}$-bound PAHs in Beijing, and the total concentration of 16 priority PAHs was between 61.2 ng/m^3 (Chen et al., 2017) and 199.7 ng/m^3 (Li et al., 2016) in the last five years. The sources were mainly attributed to coal combustion, vehicle exhaust, biomass burning, coking, and petroleum volatilization (Chen et al., 2017; Chao et al., 2019). However, few studies have examined the relationship of $PM_{2.5}$ concentration and PAHs profiles and the difference in source contributions between different $PM_{2.5}$ levels.

Source apportionment should be conducted for source control. However, different sources emit different combinations of congeners (profiles) and different PAHs congeners have different toxicity levels. For example, vehicle emissions often have more high-rings PAHs with higher toxicity, and diesel engines emit more 4–6 ring PAHs than gasoline engines, for example, BbF, BkF, BaP, DahA, IcdP and BghiP (Harrison et al., 1996); Coal combustion emits more 3–4

---

*Corresponding author: caohongbin@bnu.edu.cn

rings PAHs (Larsen et al., 2003). Because different sources have different profiles on 2–3rings with non-cancer risk and 4–6 rings with cancer risk, even if the total emission concentration is the same the toxicity and subsequent health risk may be quite different. Thus, in terms of protecting human health, priority control sources should be determined based on their contribution to human health risk. Our previous study evaluated source-attributed cancer risk (Chao et al., 2019), however, QALYs may be a better index to evaluate the burden of disease.

Quality adjusted life years (QALYs) is a comparative risk index, which measures the life expectancy weighted by quality of life (QOL). According to health state, QOL is set between 1 (perfect healthy state) and 0 (death). QALYs have been used to evaluate the disease burden of several pollutants in the environment, such as lead, sulfate and $PM_{2.5}$ (Chuang et al., 2005; Coyle et al., 2011; Hamilton et al., 2015), but their application in evaluating the health risk contribution of emission sources has not been reported.

This study aimed to examine the correlation between the concentration of $PM_{2.5}$ and $PM_{2.5}$ -bound PAHs, to investigate the distribution characteristics, sources of $PM_{2.5}$-bound PAHs, to evaluate the QALYs lost by lung cancer induced by the sources of $PM_{2.5}$-bound PAHs under different $PM_{2.5}$ levels.

2 MATERIALS AND METHODS

2.1 *Sampling and chemical analysis*

In total, 218 daily $PM_{2.5}$ samples were collected from January 14 to December 312016 using a high-volume aerosol sampler (TH-1000C II, Wuhan Tianhong Co. Wuhan, China) in the urban district of Beijing, China. Quartz-fiber filters (QFFs) were used for daily sampling. $PM_{2.5}$ mass was determined by weighing the QFFs before and after sampling by electronic analytical balance with an accuracy of 0.01mg (AX205, Mettler-Toledo International Trading Co., Ltd.). The 16 PAHs (Table 1) were detected using a gas chromatograph with a mass spectrometer detector (GC-MS) (Bruker 450GC-320MS). A mixture of deuterated surrogate compound, including Anthracene-D10, Chrysene-D12 and Perylene-D12, were added into all samples for the determination of the recovery ratio. The PAH recoveries of the standard spiked matrix range from 70% to 130% (Chao et al., 2019).

2.2 *Source apportionment by the PMF model*

The positive matrix factorization model (PMF 5.0 software) was used for source apportionment. Fifteen PAHs concentrations of one-year samples (n = 218) were put into the PMF model to identify the main PAH source profiles and quantify their contributions. Factors ranging from 3 to 7 were attempted. Finally, a 3-factor run with the lowest Q (robust) was chosen, which also has meaningful physical interpretation for each factor (Chao et al., 2019).

2.3 *Source-attributed cancer risk of PAHs*

Exposure to PAHs and their associated health risks were evaluated for local residents. PAHs have reproductive and developmental toxicity and can lead to lung cancer. Due to the availability of relative data, only lung cancer was discussed in this article. The carcinogenic potency of a PAH mixture is often expressed by its BaP equivalent concentration (*BEC*), which can be calculated using Eq. (1) (Nisbet and LaGoy, 1992):

Table 1. The TEFs value of 16 priority PAHs (Nisbet and LaGoy, 1992).

PAHs	NAP	ACE	ACY	FLO	PHE	ANT	FLA	PYR
TEFs	0.001	0.001	0.001	0.001	0.001	0.01	0.001	0.001
PAHs	BaA	CHR	BbF	BkF	BaP	DahA	IcdP	BghiP
TEFs	0.1	0.01	0.1	0.1	1	5*	0.1	0.01

$$BEC = \sum_{i=1}^{n} C_i \times TEF_i \tag{1}$$

where, C_i is the concentration of PAH congener i, μg/m³ and TEF_i is the toxicity equivalency factor (TEF) of PAH congener i. The TEF values proposed by Nisbet and LaGoy (Nisbet and LaGoy, 1992) were adopted in this study.

The incremental lifetime cancer risk (ILCR) via an inhalation pathway was estimated by Eqs (2) and (3) (US EPA,1989):

$$EC = \frac{BEC \times EF \times ED}{LT} \tag{2}$$

$$ILCR = EC \times UR \tag{3}$$

where, EC is life time exposure BaP equivalent concentration of PAHs, μg/m³, BEC is the BaP equivalent concentration of PAHs, μg/m³, EF is the exposure frequency (350 days/a) (USEPA, 1989), ED is the exposure duration (30 a), LT is the lifetime exposure time (82 years, the average life expectancy of the Beijing population) (http://xxzx.bjchfp.gov.cn/), and UR is the unit risk of BaP (0.0006(μg/m³)$^{-1}$) (IRIS (Integrated Risk Information System), https://www.epa.gov/iris).

When the concentration in Eq. (1) was set as the profiles of each source obtained by PMF, the health risks posed by each emission source were calculated.

2.4 *Source-attributed diseased population and QALYs lost*

The health risk is an increased probability for developing a disease over a person's lifetime due to exposure to chemicals. The incremental number of the diseased population was calculated by multiplying the health risk by the total number of the exposed population. The population QALYs lost was then evaluated from the QALYs lost per person and the incremental number of the diseased population. For an individual, QALYs lost includes the timedue to the loss of life quality during disease duration and the time due to premature death. Diseased population and QALYs lost were calculated following Eqs (4)–(6):

$$P_{dis} = ILCR \times P_{exp} \tag{4}$$

$$\Delta QALY_{individual} = DD \times \left(QoL_{gen} - QoL_{dis}\right) + EYLL \times QoL_{gen} \tag{5}$$

$$\Delta QALY_{population} = P_{dis} \times \Delta QALY_{individual} \tag{6}$$

where, P_{exp} is the exposure population (1.25×10^7 for the population in Beijing), DD is disease duration (1.14 y for lung cancer) (Wu et al., 2009), $EYLL$ is the expected years of life lost from lung cancer (15.127 y) (Liu et al., 2019), QoL_{dis} is the quality of life for people with lung cancer (0.677) (Nafees et al., 2017), and QoL_{gen} is the quality of life for the general population (0.914) (Wang et al., 2010).

In Eqs(4)–(6), when the ILCR was set as the health risk posed by each emission source, the QALYs lost for each emission source was obtained.

3 RESULTS AND DISCUSSIONS

3.1 *PM$_{2.5}$-bound PAHs concentrations for different PM$_{2.5}$ level*

Correlation analysis was conducted between PM$_{2.5}$ and PM$_{2.5}$-bound PAHs. The concentration of PAHs was significantly positively correlated with the levels of total PAHs and individual PAH congeners, except for NAP and FLO ($p<0.05$). In terms of the contamination extent and the hazardous effects on human health, PM$_{2.5}$ was classified into five levels (Table 2) and the

Table 2. PM$_{2.5}$ classification standard (μg/m^3).

PM$_{2.5}$ level	Level 1	Level 2	Level 3	Level 4	Level 5
Range	0–75	75–115	115–150	150–250	250–
Health effect	Excellent +Fine	Mild contamination	Moderate contamination	Serious contamination	Severe contamination

concentration distribution of PAHs for different PM$_{2.5}$ level was analyzed (Figure 1). The mean concentration of PAHs was 25.9, 47.4, 96.0, 173.0, and 115.7 ng/m^3 for PM$_{2.5}$ levels 1–5, respectively. The largest mean concentration of PAHs occurred in level 4. PAHs have similar profiles between different PM$_{2.5}$ levels, with higher concentrations for middle-ring and high-ring congeners. Level 4 has higher concentrations of most PAHs congeners than other levels.

3.2 Mass contribution of three sources for different PM$_{2.5}$ levels

The data of 15 PAHs of 218 samples were introduced into the PMF model. Three factors were identified with the least Q value and meaningful physical interpretations, and were interpreted

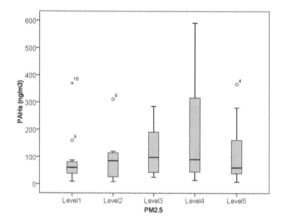

Figure 1. PAHs concentrations for different PM$_{2.5}$ levels.

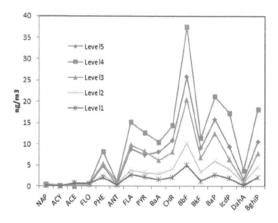

Figure 2. PAHs profiles for different PM$_{2.5}$ levels.

26

Table 3. Mass contribution of three sources to PAHs for different PM$_{2.5}$ levels.

PM$_{2.5}$	n	Coal combustion		Vehicle emission		Volatilization of petrolum, natural gas and biomass combustion		Sum
		Mass (ng/m^3)	Ratio	Mass (ng/m^3)	Ratio	Mass (ng/m^3)	Ratio	Mass (ng/m^3)
Level 1	230	9.1	34.0%	9.7	36.2%	8.0	29.8%	26.8
Level 2	68	12.5	25.2%	28.5	57.5%	8.6	17.3%	49.6
Level 3	28	38.0	38.8%	43.8	44.7%	16.2	16.6%	98.0
Level 4	29	45.5	24.6%	118.8	64.2%	20.7	11.2%	184.9
Level 5	10	48.3	39.6%	67.5	55.4%	6.0	4.9%	121.8
Mean	365	15.9	30.8%	26.1	50.5%	9.7	18.7%	51.7

as vehicle emission, coal combustion, and volatilization of petroleum, natural gas and biomass combustion. Please see our previous study for details (Chao et al., 2019). Applying the source apportionment results from 218 days to 365 days in Beijing, vehicle emissions contributed 50.5% mass concentration of PM$_{2.5}$-bound PAHs, coal combustion contributed 30.8%, and volatilization of petroleum, natural gas and biomass combustion contributed 18.7% (Table 3). With the increase of PM$_{2.5}$ concentration, the mass contribution of coal combustion and vehicle emission increased, while the mass contribution of petroleum volatilization, natural gas and biomass combustion decreased.

3.3 Source-specific lung cancer risk and QALYs for different PM$_{2.5}$ levels

Beijing urban residents were the exposure population. It is assumed that no emission sources control was conducted and the air PAHs was kept at the current level. Total exposure to PM$_{2.5}$-bound PAHs was 4.84 ng/m^{3}BaP equivalent concentration, among which levels 1–5 contributed 1.44, 0.90, 0.68, 1.49 and 0.32 ng/m^3, respectively. Correspondingly, the total lung cancer risk was 2.9×10^{-6}, which is higher than the acceptable level of 1×10^{-6} (US EPA,1989). The cancer risk contributions by levels 1–5 were 8.6×10^{-7}, 5.4×10^{-7}, 4.1×10^{-7}, 9.0×10^{-7} and 1.9×10^{-7}, respectively. The cancer risk of coal combustion, vehicle emission and volatilization of petroleum, natural gas and biomass combustion was 6.0×10^{-7}, 1.9×10^{-6} and 3.8×10^{-7}, respectively. The cancer risk induced by vehicle emissions was the largest, and higher than the acceptable level of 1×10^{-6}. The individual QALY lost by lung cancer was 14.37 years (disease duration, 0.27 years; premature death, 14.1 years). The population QALYs lost by Beijing urban residents due to lung cancer induced by exposure to PM$_{2.5}$-bound PAHs was 511.5 years, among which the days of levels 1–5 contributed 152.2, 95.6, 71.6, 158 and 34 years, respectively (Table 4). Level 1 and level 4 days made much higher contributions: level 1

Table 4. QALYs lost forthree sources at different PM$_{2.5}$ levels.

PM$_{2.5}$	n	Coal combustion		Vehicle emission		Volatilization of petrolum, natural gas and biomasscombustion		Sum
		QALYs lost (yr)	Ratio	QALY lost (yr)	Ratio	QALY lost (yr)	Ratio	QALY lost (yr)
Level 1	230	38.2	25.1%	79.3	52.1%	34.6	22.7%	152.2
Level 2	68	15.5	16.2%	69.1	72.3%	11.0	11.5%	95.6
Level 3	28	19.4	27.1%	43.7	60.9%	8.6	11.9%	71.6
Level 4	29	24.1	15.2%	122.7	77.6%	11.3	7.1%	158.0
Level 5	10	8.8	25.9%	24.1	70.8%	1.1	3.3%	34.0
Total	365	106.0	20.7%	338.9	66.3%	66.6	13.0%	511.5

because more days were included and level 4 because of higher PAHs concentrations. Vehicle exhaust contributed the largest QALYs lost (338.9 QALYs, 66.3%), which is higher than its mass contribution (50.5%).

4 CONCLUSION

Beijing urban residents will endure lung cancer risks exceeding the acceptable level of 1×10^{-6} if no emission control countermeasures are conducted and the $PM_{2.5}$-bound PAHs at kept at the current level for 30 years. The QALYs lost for PAHs-induced lung cancer was 511.5 years. Vehicle exhaust contributed the largest QALYs lost (338.9 QALYs, 66.3%), which is higher than its mass contribution (50.5%). When $PM_{2.5}$ concentration is at serious contamination levels (150–250µg/ m^3), the bound PAHs have the highest concentration and correspondingly the largest lung cancer risk and QALYs lost. The QALYs method is a good index for evaluating sources-attributed health risks. The QALYs lost by non-cancer toxic effects will be studied in a future study.

ACKNOWLEDGMENTS

This project was supported by Beijing Natural Science Foundation (8192026).

REFERENCES

Chao, S.-H., Liu, J.-W., Chen, Y.-J., Cao, H.-B., Zhang, A.-C. 2019. Implications of seasonal control of $PM_{2.5}$-bound PAHs: An integrated approach for source apportionment, source region identification and health risk assessment. *Environmental Pollution*, 247: 685–695.

Chen, Y., Li, X.-H., Zhu, T.-L., Han, Y.-J., Lv, D. 2017. $PM_{2.5}$-bound PAHs in three indoor and one outdoor air in Beijing: Concentration, source and health risk assessment. *Science of the Total Environment* 586: 255–264.

Chuang, H., Chao, K., Wang, J., 2005. Estimation of burden of lead for offspring of female lead workers: a quality-adjusted life year (QALY) assessment. *Journal of Toxicology & Environmental Health Part A*, 68(17–18): 1485–1496.

Coyle, D., Stieb, D., Burnett, R., et al. Thun M. 2011. Impact of particulate air pollution on quality-adjusted life expectancy in Canada. *Journal of Toxicology & Environmental Health Part A*, 66(19): 1847–1864.

Hamilton, I., Milner, J., Chalabi, Z., et al., 2015. Health effects of home energy efficiency interventions in England: a modelling study. *BMJ Open*, 5:e007298. doi: 10.1136/bmjopen-2014-007298.

Harrison, R.M., Smith, D.J.T., Luhana, L. 1996. Source Apportionment of Atmospheric Polycyclic Aromatic Hydrocarbons Collected from an Urban Location in Birmingham, U.K. *Environmental Science & Technology*, 30: 825–832.

Larsen, R.K., Baker, J.E. 2003. Source Apportionment of Polycyclic Aromatic Hydrocarbons in the Urban Atmosphere: A Comparison of Three Methods. *Environmental Science & Technology*. 37: 1873–1881.

Li, Y., Liu, X.R., Liu, M., Li, X.F., Meng, F., Wang, J., Yan, W.J., Lin, X.B., Zhu, J.M., Qin, Y.K., 2016. Investigation into atmospheric $PM_{2.5}$-borne PAHs in Eastern cities of China: concentration, source diagnosis and health risk assessment. *Environmental Science: Processes & Impacts*, 18: 529–537.

Liu, J.-W. 2019. Sources-attributed burden of disease of $PM_{2.5}$-bound metals based on source apportionment and quality adjusted life year. PhD thesis. Beijing Normal University, 2019. [in Chinese].

Nafees, B., Lloyd, A.J, Dewilde S, 2017. Health state utilities in non–small cell lung cancer: An international study. *Asia-Pacific Journal of Clinical Oncology*, 13(5): e195–e203.

US EPA, 1989. Risk Assessment Guidance for Superfund: Volume 1. *Human Health Evaluation Manual* (Part A); Interim Final, EPA/540/1-89/002. U.S. Environmental ProtectionAgency, Office of Emergency and Remedial Response, Development, Washington, DC.

Wang, Y. 2010. The study of health-related quality of life and its effect on health service utilization in Chinese population. PhD thesis. Peking Union Medical College, Chinese Academy of Medical Sciences. [in Chinese].

Wu, X.Y., Zhang C.M., Ge, X.P., 2009. Investigation and analysis on the survival duration and influencing factors of primary lung cancer in the urban area of Beijing. *Beijing Medical Journal*, 31(1): 20–23.

Risk Analysis Based on Data and Crisis Response Beyond Knowledge – Huang & Nivolianitou (eds)
© *2020 Taylor & Francis Group, London, ISBN 978-0-367-25146-8*

Risk governance challenges in the Greek search and rescue region

Athanasios Liaropoulos & Kalliopi Sapountzaki
Department of Geography, Harokopio University, Athens, Greece

Zoe Nivolianitou
Institute of Nuclear & Radiological Sciences and Technology, Energy & Safety, National Center of Scientific Research "Demokritos", Athens, Greece

ABSTRACT: The recent activity about research on submersed hydrocarbon reservoirs in the Greek territory and the oncoming construction of several offshore platforms in the area demand the establishing of a new and powerful risk governance protocol for search and rescue operations in this area. Greece, owing to its great number of islands, has obligations over an extended Search and Rescue Region (SRR). This authority is daily exercised on different occasions ranging from emergency medical help to rescue operations in endangered ships sailing in both the Aegean and Ionian Sea, together with the saving of hundreds of immigrants trying to trespass Greek waters. Regarding the oil & gas industry, Greece is having three offshore platforms in the northern part of the Aegean Sea, situated on deep waters away from the coast and prospect for more oil drills. This paper aims at proposing some new methodological tools in order to evaluate the adequacy of the emergency preparedness arrangements of the Greek offshore sector. In doing so, the paper presents a summary of the operational procedures and the conditions that exist and prevail in the Search and Rescue (SAR) at sea for endangered offshore platforms (oil rigs) in Greece, focusing on the use of a well-established methodology such as the HAZOR in the Risk Governance scheme.

Keywords: risk governance, emergency preparedness, operational procedure, offshore platform, Greece

1 INTRODUCTION

The offshore industry for oil and gas extraction is almost 50 years old since its modest start in the early '60s. During the first 20 years of the offshore industry, there were several major accidents on off-shore installations and during helicopter transportation of personnel to/from shore. Later on, the multi fatality major accident in the North Sea in July 1988 on Piper Alpha gave significant push to the development of the Search and Rescue operations, known as SAR, in the offshore oil installations. This discipline has grown in the Northern part of Europe (Norway and the UK) and has led the EU to publish the Directive 2013/30/EU on June 28th, (known as OSD) on the safety of offshore oil and gas with the aim to reduce as far as possible the occurrence of major accidents related to offshore oil and gas operations and to limit their consequences. This Directive has been also amended by the Greek state with Law 4409/2016. Greece, owing to its great number of islands, has obligations over an extended Search and Rescue Region (SRR) together with Evacuation Escape and Rescue (EER) operations, overlapping in many cases with those of its neighboring countries.

The offshore oil industry in order to assess the risk in the Evacuation Escape and Rescue (EER) operations is using various methods stemming from the hazardous process industry Risk Analysis; HAZOP methodology is one of them, well-known since the early 90's (Kletz, 1997). This paper examines the potential adaptation of the HAZOP methodology in the above mentioned types of SAR, EER operations and its capability to assess all types of risks including systemic risks arising from geopolitical issues. In order to apply the HAZOP methodology in

SAR operations taking place in offshore oil installations, it is necessary to analyze the relation between EER and SAR.

2 SEARCH AND RESCUE OPERATIONS IN GREECE

Greece has obligations over an extended Search and Rescue Region (SRR) and this authority is daily exercised on different occasions ranging from emergency medical help to rescue operations in endangered ships sailing in both the Aegean and Ionian Sea together with the saving of hundreds of thousands of immigrants trying to trespass Greek waters. Regarding the oil & gas industry, Greece is having three offshore platforms in the northern part of the Aegean Sea, situated on deep waters away from the coast. To that end Greece has founded the Joint Rescue Coordination Center (JRCC) in Piraeus in order to comply with the obligations arising from international conventions in providing SAR services for air and maritime accidents; this is done in accordance to the rights given to the country by the ICAO Regional Air Navigation Agreement (1952) making the Greek Search and Rescue Region (SRR) to coincides with the Athens FIR (Flight Information Region). Another International Organisation, IMO (International Maritime Organisation) through its SOLAS and International Convention on Maritime Search and Rescue in 1979 gives the possibility Greece to coordinate maritime SAR incidents in the same area.

Greece has signed agreements for SAR services in cases of accidents at sea with Italy (2000), Malta (2008) and Cyprus (2014); the situation remains still unclear with Turkey; the latter has declared responsibility for SAR operations in part that overlaps the Athens FIR almost up to the middle of the Aegean. This area includes Greek islands, Greek territorial waters and Greek airspace, constituting for Greece a violation of its sovereignty and a source of friction between the two countries, while producing problems in the efficiency of Greek SAR operations (Hellenic Ministry of Foreign Affairs, 2019).

3 EVACUATION AND ESCAPE VS SEARCH AND RESCUE IN OFFSHORE OIL INSTALLATIONS

Part of the Health and Safety (H&S) operations in the offshore oil installations is the Evacuation Escape and Rescue (EER) process. The overall H & S discipline, and in particular the EER, has been analyzed with various methodologies and approached by different researchers. Evacuation is the abandonment of an installation and the surrounding area, during an emergency, in a systematic manner. Escape is the abandonment of an installation and the surrounding area in an emergency when the evacuation system has failed. Rescue (recovery and rescue) is the process of recovering individuals after evacuation or escape from the facility, rescue people near their establishment and transport them in a safe place (OGP, 2010).

All three stages of evacuation, escape and rescue are normally approached in the legal framework all together as a single process consisting of distinct stages; these can be triggered by an initial catastrophic event that will affect not only the start of the EER process, but it will also influence all of its subsequent stages.

According to Health and Safety Executive (HSE), Approved Code of Practice (ACOP) (Au, et al., 1995) the "external emergency response plan" of an offshore oil installation is the SAR Framework in the UK. As described above, SAR is an attempt to rescue the human life that is at stake and, even if it seems to be identical with the Recovery-Rescue stage of EER, however, it is not limited to this stage, as it is possible that SAR operations include the stages of Evacuation and Escape. Additionally, the type of the initial catastrophic event affects the way SAR is executed.

4 APPLICATION OF HAZOP ON SAR FOR OFFSHORE OIL INSTALLATIONS

An experimental application of the HAZOP methodology has been attempted in order to assess the various hazards that occur in SAR operations in the offshore oil installations. HAZOP is a methodology of hazard intensification in the process industry introduced by Professor Kletz (1997); it is applied by a team of experts that create a full description of the process and

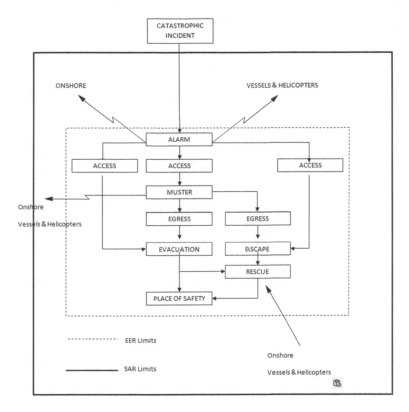

Figure 1. The EER and the SAR process (source: writers elaboration, based on HSE, OTH 95 446).

examine what deviations from the process are present and what their consequences are. This is done by applying keywords at each stage of the process and by using the brainstorming method. The main elements to be investigated are: Intention, Deviation, Causes, and Consequences. There are specific words that define the processes, or the human activities involved in the stages of the process to be studied that are called "property words". Keywords are applied to the property words making the choice of keywords and property words quite important. HAZOP methodology has been used by the offshore oil industry for decades, while the HSE Offshore Safety Division's (HSE, 1995) made the application of HAZOP in the EER easier and, moreover, adapted it to the SAR requirements. For a complete assessment it is necessary to apply the method for each catastrophic event that may threaten each individual installation, which, however, is very time-consuming and produces repeated results. For this reason, all initial catastrophic events were grouped into one single application of the method, while the individual characteristics of the installations were generalized in order to be applied for each type of installation. Special events and situations, such as the underwater operations by divers, are not covered and should be assessed separately.

Both terms EER and SAR refer to the management of the same event, the first one is an internal installation process that is carried out either by means of the installation or by resources not present on the instillation, in cooperation or not, with the "external" sources. It is the way the installation approaches this process internally; however, SAR is an "external" procedure, towards the installation. The term "external" resources can describe private entities (leased by the company) or governmental agencies. Both terms are interlinked because they describe the same operation from a different point of view, namely, the stages of EER are a) Alarm, b) Access, c) Muster, d) Egress, e) Evacuation/Escape, f) Rescue and g) Place of safety. Figure 1 illustrates the various stages of EER and their relation to SAR depicting the various stages of the whole process and the four deferent ways that the incident can evolve. Additionally, Figure 2 presents the Hierarchical task analysis tree for the basic SAR model in an offshore installation.

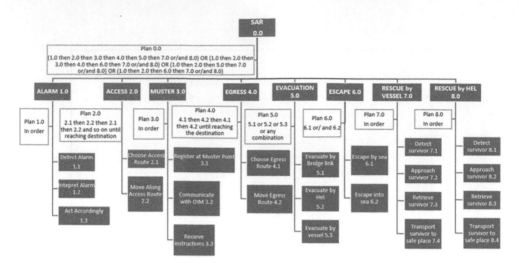

Figure 2.   The Hierarchical task analysis tree for the basic SAR model in an offshore installation.

5   RESULTS AND DISCUSSION

The objective of the study was to investigate the use of risk assessment methods and in particular the HAZOP technique on SAR operations. The experimental application demonstrated that it is feasible to apply the technique on SAR operations in offshore oil installations. Guidewords and property words have been ad hoc suggested and a significant number of hazards, connected

Table 1 Property Words

SAR STAGE	Property Words
ALARM	Alarm system: system information
	Response
	Communication: system information
ACCESS	Escape route
	Decision
	Movement
MUSTER	Muster point
	Communication: system information
	Registration
	Survival equipment: equipment use of equipment
EGRESS	Escape route
	Decision
	Movement
EVACUATION BY BRIDGE	Escape route
	Decision
	Movement
EVACUATION BY HELICOPTER	Availability
	Approach
	Landing
	Take off
	Helideck
	Boarding
	Communication: system information
	Equipment: use of equipment

EVACUATION/ ESCAPE BY RESCUE BOAT/ LIFE RAFT	Boat: Availability
	    Unmoving equipment (e.g. hulls, seats, etc.)
	    Mobile equipment (engine, ventilation system etc.)

	Launch system: system procedure
	Crew
	Communication: system information
	Navigation
	Drop zone
	Survival equipment: equipment use of equipment
ESCAPE DIRECT INTO SEA	Escape devises: equipment use of equipment
	Decision
	Movement

	Survival equipment: equipment use of equipment
	Drop zone
	Availability
RESCUE BY HELICOPTER	Response time: Readiness Trip duration
	Number of helicopters
	Search area
	Navigation
	Meteorological conditions
	Communication
	Obstacles / debris
	Autonomy / total time in the area
	Retrieve of survivals
	Preservation of life
	Helicopter equipment: equipment use of equipment
	Survival equipment: equipment use of equipment
RESCUE BY BOAT	Availability
	Response time: Readiness Trip duration
	Search area
	Navigation
	Meteorological conditions
	Communication
	Obstacles / debris
	Autonomy / total time in the area
	Retrieve of survivals
	Preservation of life
	Survival equipment: equipment use of equipment

Table 2 Guidewords

Guidewords
Failed
Damage / Injury
Fails during
Not done
Inadequate / Insufficient
Too late/ Too soon
Congested / Overloaded

Figure 3.   Tables of property and guidewords. source: elaborated by the authors, based on HSE, OTH 95 446.

with SAR, have been identified. More accurate results are likely to be produced, if the methodology is applied to a specific installation or area of SAR operations. The hazards identified were divided into three main categories: a) physical hazards, which refer to equipment (design, damage or failure) and to physical conditions (weather, fire, etc.), b) administrative Risk-Command & Control hazards, referring to inadequate procedures, inadequate communications and failure of security systems, c) behavioral hazards, referring to human mistakes or adverse behavior. This experimental application of the technique failed to assess the systemic risks that arise from an incident in an offshore oil installation and in particular problems arising from complex geopolitical situations (e.g. disputes over the SRR boundaries like in the Greek case).

Figure 3 below presents the Property and Guide words that have been used in the analysis, while Figure 4 presents the hazards that have been identified by the methodology application in the Recue by Helicopter stage of the process and their potential consequences

Stage: Rescue by Helicopter (Hel)

Physical Hazard	Command and Control Hazard	Behavioral Hazard
Low visibility Result: difficulty in locating the Person in Water (PIW)	No available Hel or Hel do not reach the area Result: Unable to rescue PIW by Hel	The PIW did not wear lifejackets/rescue equipment Result: reduced survival time, Injuries to PIW, Difficulty in locating the PIW, Difficulty in retrieving the PIW
Increased ripple, Result: difficulty in locating the PIW, Difficulty in retrieving the PIW, Fatigue of Hel rescuers	Delayed Hel arrival Result: reduced survival time, Injuries to PIW	PIW wounded/unconscious Result: difficulty in locating the PIW, Difficulty in retrieving the PIW, reduced survival time
Low temperatures Result: reduced survival time, Fatigue of Hel rescuers	Insufficient number of Hel Result: reduced survival time, Injuries to PIW	Incorrect procedures of PIW upon arrival of Hel Result: Damage to the boat / raft, Injury of rescuers/PIW, Unable to retrieve PIW
Fire in the sea Result: injury of PIW, Inability of Hel to approach the area	Delay in the detection of PIW Result: reduced survival time, Injuries to PIW	Inappropriate recovery technique Result: Injury of rescuer/PIW
Debris in the sea Result: injury to rescuers/PIW, Inability of rescuers to approach	Increased Hel number Result: Risk of collision, need for traffic co-ordination	Provision of medical care not available or available with delay Result: reduced life expectancy of PIW
Damage to the Hel hoist Result: reduced rescue capabilities, Only search capability	Crew fatigue Result: reduced performance, Injury of rescuers/PIW, Hel accident	The Hel searches in the wrong area Result: Negative search results
Hel/PIW communication system failure Result: Impossibility or reduced rescue capabilities, Unable to co-ordinate with resources	Was given incorrect Research Area Result: Negative search results	Incorrect terminology in Hel communication Result: misinterpretation, non-execution of instructions
SAR operation in the night Result: reduced detection capabilities, increased rescue risk	Reduced autonomy of Hel Result: reduced time the area	Injured / sickness of rescuer Result: Impossibility or reduced rescue capabilities
Damage to a major Hel system Result: probability of falling/ditching of the Hel, Withdrawal of the Hel		Overloading the rescue rope/ rescuer Result: injury of rescuer / PIW, Inability to rescue, Damage of the equipment

Figure 4. The hazards for the recue by helicopter stage.

6 CONCLUSIONS

It has been shown that the application of HAZOP on SAR operations is feasible and produces a significant amount of data. SAR planners should take into account the potentials of this technique, while procedures and manuals should be reevaluated based on those finding. Nevertheless, it should be taken into account that there are significant indications that HAZOP technique cannot assess the systemic risks that arise from an incident in an offshore oil installation.

REFERENCES

Au, S.Y.Z. & Gould, G.W. 1995. *A Methodology for Hazard Identification on EER Assessments*. Health and Safety Executive (HSE), OTH 95 446, HSE Books.

Brachner, M. 2015. A simulation model to evaluate an emergency response system for offshore helicopter ditches. *In L.* Yilmaz, *W.K.V.* Chan, *I.* Moon, *T.M.K.* Roeder, *C.* Macal, *and M.D.* Rossetti, *(eds.)*, *Proceedings of the 2015 Winter Simulation Conference.*

British Standard BS: IEC61882:2002. Hazard and operability studies (HAZOP studies): Application Guide British Standards Institution.

Hellenic Republic, Ministry of Foreign Affairs. 2019. Foreing Policy Issues, Issues of Greek-Turkish Relations, in https://www.mfa.gr/en/issues-of-greek-turkish-relations.

Jemli, J. 2012. *Maritime Search and Rescue in the Russian part of the Gulf of Finland.* MSc Thesis, School of Engineering. Aalto University, Espoo, Finland.

Kletz, T.A. 1997. Hazop-past and future. *Reliability Engineering & System Safety*55(3): 263–266.

OGP. 2010. Evacuation, Escape & Rescue. International Association of Oil & Gas Producers. London, OGP.

Risk Analysis Based on Data and Crisis Response Beyond Knowledge – Huang & Nivolianitou (eds)
© *2020 Taylor & Francis Group, London, ISBN 978-0-367-25146-8*

Mutual information-based approach for vine copula selection for hydrological dependence modeling

Lingling Ni, Dong Wang* & Jianfeng Wu*
Key Laboratory of Surficial Geochemistry, Ministry of Education, Department of Hydrosciences, School of Earth Sciences and Engineering, Nanjing University, Nanjing, China

ABSTRACT: Hydrological risk analysis and management entails multivariate modeling which requires modeling the structure of dependence among different variables. Vine copulas have been increasing applied in multivariate modeling wherein the selection of vine copula structure plays a critical role. This study develops a mutual information (MI)-based sequential approach to select a vine structure which is model-independent. Then, an MI-based approach for hydrological dependence modeling is developed. An application of drought characterization is utilized to show the performance of the proposed approach. Results indicate that the MI-based approach satisfactorily models different kind of dependence structure.

Keywords: Vine copula, mutual information, multivariate modeling

1 INTRODUCTION

Recent years have been witnessing increasing occurrences of hydrological events, such as droughts, flood, windstorms, and hot days, over many areas, perhaps due to climate change (Dai, 2013; Field *et al.*, 2012; Hirabayashi *et al.*, 2013) and their aggravated impacts on natural and human systems are casing an increasing concern (Lehner et al., 2006; Mishra and Singh, 2011). Properties of these hydrological events are incorporated in their probabilistic modeling needed for risk analysis and management.

However, hydrological events, such as drought and flood, are not only characterized by several properties or variables but by the dependence between them and even their temporal and spatial structure. In recent decades, considerable attention has been paid to model dependence structures.

Copulas have been a natural choice, because they are flexible for constructing multivariate distribution by separating the construction into univariate marginal distributions and interdependence structure (Nelsen, 2006). As a result, there has been a flurry of applications of multivariate analysis in hydrology. Although copulas are flexible in dependence modeling, building higher-dimensional copulas is generally a difficult task (Aas *et al.*, 2009; Hao and Singh, 2016). A *d*-dimensional copula is constructed using a set of d(d-1)/2 bivariate copulas sequentially in constructing conditional distributions which constitutes the vine copula (or pair copula construction) (Joe, 2014). From this point of view, vine copulas are hierarchical models as they sequentially apply bivariate copulas as the building blocks for constructing a higher-dimensional copula. The higher flexibility of vine copulas enables to model a wider range of complex multivariate dependence than other traditional copulas. Recently, vine copulas have been used in hydrological studies on frequency analysis (Xiong et al., 2014), flood characterization (Daneshkhah et al., 2016), rainfall simulation (Gyasi-Agyei and Melching, 2012), and streamflow prediction (Liu *et al.*, 2015).

The specification of a vine copula generally entails three primary steps: (1) selection of the structure of vine copulas (i.e., selecting which two variates to be a pair); (2) choice of a bivariate copula

---

*Corresponding authors: Dong Wang (wangdong@nju.edu.cn); Jianfeng Wu (jfwu@nju.edu.cn)

family for each pair selected in (1); and (3) estimation of the corresponding parameters for each given copula. As steps (2) and (3) depend on the vine tree structure, the selection of adequate tree structure is critical to the specification of vine copulas (Czado et al., 2012). Since the number of possible R-vines grows exponentially as d (number of variates) increases, it is not feasible to simply try and accomplish steps (2) and (3) for all possible R-vine copulas and then choose the "best" one. The most used strategy is to proceed sequentially tree-by-tree, starting by defining the first (last) tree for the R-vine, continuing with the second (next to the last) tree, and so on.

Mutual information (MI) is a measure of dependence between two random variables (RVs), and conditional mutual information (CMI) is mutual information conditioned on given variables (Cover and Thomas, 1991). MI and CMI have been used in selecting inputs for artificial neural networks (ANN) for uncertainty analysis, and identifying the optimal rain gauge density.

Hydrological systems are complex and their variables have large uncertainties. This study therefore proposes a sequential selection approach based on MI and CMI for selecting important pairwise dependence in the vine copula. An MI-based approach for vine copula construction for hydrological dependence modeling is developed. To evaluate the performance of the proposed approach, drought characterization is modeled, and the constructed copula is evaluated based on simulation.

2   VINE COPULA AND MUTUAL INFORMATION

Copulas are functions that "couple" multivariate distribution functions to their one-dimensional marginal distribution functions (Nelsen, 2006) which are obtained from Sklar's theorem. Let H be an n-dimensional distribution function with margins F_1, F_2,..., F_n. Then there exists an n-copula C such that for all x in R^n,

$$H(x_1, x_2, ..., x_n) = C(F_1(x_1), F_2(x_2), ..., F_n(x_n)) \tag{1}$$

The multivariate density can be obtained as follows:

$$f(x_1, x_2, ..., x_n) = \left[\prod_{i=1}^{n} f_i(x_i) \right] c(u_1, u_2, ..., u_n) \tag{2}$$

where c is the copula density, and $f_i(x_i)$ is the marginal density. We denote $F_i(x_i)$ as u_i ($i = 1, 2,..., n$).

Let S be a non-empty subset of $\{1, 2,..., n\}$, which will be the conditioning set of variables, and let T be a subset of S^C with cardinality of at least 2, which will be the conditioned set of variables, Sklar's theorem implies that there is a copula $C_{T;S}(\cdot; x_s)$ such that:

$$F_{T|S}(x_T | x_S) = C_{T;S}(F_{j|S}(x_j | x_S) : j \in T; x_S) \tag{3}$$

where $F_{T|S}(x_T | x_S)$ is the conditional distribution, and $C_{T;S}$ is a $|T|$-dimensional copula applied to univariate margins $F_{j|S}(\cdot; x_s)$, not a conditional distribution obtained from $C_{S|T}$. If $T = \{i, j\}$, then, using Sklar's theorem, there are copula densities $c_{i,j;S}(\cdot; x_s)$ such that:

$$f_{i,j|S}(x_i, x_j | x_S) = c_{i,j;S}(F_{i|S}(x_i | x_S), F_{j|S}(x_j | x_S); x_S) \cdot f_{i|S}(x_i | x_S) f_{j|S}(x_j | x_S) \tag{4}$$

where $c_{i,j;S}(\cdot; x_s)$ is the pair-copula density. Then the joint density function can be factorized as:

$$f_{1, 2, ..., n}(x_1, x_2, ..., x_n) = \prod_{k=1}^{n} f_k \cdot \prod c_{i, j;S(i, j)}(F_{i|S(i,j)}, F_{j|S(i,j)}) \tag{5}$$

Shannon (1948) defined entropy as in Eq. (7), which is a measure of uncertainty of a random variable.

$$H(X) = -\sum_{i=1}^{n} p(x_i) \log p(x_i) \tag{6}$$

where X is a discrete random variable, and $p(x_i) = P(X = x_i)$ is the probability mass function.

This notion is extended to define mutual information, which is a measure of dependence of between two random variables.

$$I(X; Y) = \sum \sum p(x, y) \log \frac{p(x, y)}{p(x)p(y)} \tag{7}$$

where $p(x, y)$ is the joint probability mass function of X and Y, and $p(x)$ and $p(y)$ denote the marginal probability mass functions.

Given conditioning set of variables S, the conditional mutual information can be defined by,

$$I(X_i; X_j | \boldsymbol{X}_S) = \sum p(\boldsymbol{x}_S) \sum p(x_i, x_j | \boldsymbol{x}_S) \log \frac{p(x_i, x_j | \boldsymbol{x}_S)}{p(x_i | \boldsymbol{x}_S) p(x_j | \boldsymbol{x}_S)} \tag{8}$$

3 MUTUAL INFORMATION-BASED APPROACH FOR VINE COPULA SELECTION FOR DEPENDENCE MODELING

Since a vine copula consists of a sequence of pair-copulas, it is necessary to determine a specific factorization where a choice of selection is important. We develop a sequential approach based on mutual information.

The selection procedure can be summarized as follows:

(1) Starting from the first tree, calculate the mutual information of all possible variable pairs, then choose the tree that maximizes the sum of MI.
(2) After the first tree is chosen, according to the proximity condition, calculate the conditional mutual information of all possibilities, then choose the tree that maximizes the sum of CMI.
(3) Repeat step (2) until tree n–1 is specified.

Based on this iterative selection strategy, we develop a hydrological dependence model applying vine copula based on this MI-based selection approach, following the similar steps of dependence modeling. The procedure, is described as follows:

(1) Vine copula structure selection: Using the proposed MI-based selection approach, determine the vine structure.
(2) Margins inference: Fit variables with several candidate marginal distributions, and use statistical criteria to select the best fitted one. Then, perform the goodness-of-fit test to verify whether the observations are appropriately modeled by chosen distributions.
(3) Vine copula determination: Following the determined vine structure in step 1, starting from the first tree, select the bivariate copula family by commonly used criteria, then estimate the corresponding parameters, finally, use a goodness-of-fit test to verify the performance of specified pair-copula. Follow for tree 2, construct pair-copulas to couple with conditional univariate margins, repeat what we have done in tree 1, until the n–1 tree is specified.

4 APPILICATION

Drought has been receiving increasing attention for their frequent occurrence and rising aggravated impacts on water demand. Drought characterization is a prerequisite for drought frequency analysis to help drought resistance planning and risk management.

Monthly streamflow data covering a period of 1950-2003 from Lijin station, located in Yellow River basin, China, are used. Four main variables are used to characterize each drought episode: namely, drought severity peak (P), severity (S), duration (D), inter-arrival time (I). The drought event is defined as a period when the average monthly flow is lower than the corresponding truncation level. A total of 70 drought events have been observed.

To construct the 4-dimensional distribution, the proposed mutual information-based approach was utilized to select the vine copula structure. Following the selection algorithm in Section 3, results are given in Table 1. According to the strategy that each tree chosen should be the maximum sum of MI, we fitted the copulas $C_{P,S}$, $C_{S,D}$, $C_{D,I}$ in tree 1 of the vine. This means that the selected vine copula was a D-vine copula, which had path structures.

The first step was to derive the marginal distribution. We fitted a univariate with commonly used distributions using the maximum likelihood estimation. The candidates were gamma, exponential, log-normal, Weibull, generalized extreme value (GEV), and generalized Pareto (GP). The Akaike information criterion (AIC) was used to select the best fitted distribution. Accordingly, the best fitted marginal distributions of peak, severity, duration, and inter-arrival time were exponential, log-normal, GEV, and GEV, respectively. They all passed the Kolmogorov-Smirnov (KS) test.

The following step was to choose the pair-copulas and estimate the corresponding parameters. The commonly used meta-elliptical copula (Gaussian and Student t), Archimedean copula (Clayton, Gumbel, Frank, and Joe) and their corresponding rotated versions with 90, 180, and 270 degrees, were selected as the candidate copulas. The maximum pseudo-likelihood estimator was used to estimate the parameters of pair-copulas, AIC was used to

Table 1.   Mutual information of each pair.

	Peak	Severity	Duration	Interarrival time
Peak	2.30	**1.48**	0.94	0.68
Severity	**1.48**	2.30	**1.14**	0.70
Duration	0.94	**1.14**	2.30	**0.92**
Interarrival time	0.68	0.70	**0.92**	2.30

Table 2.   Candidate copulas and corresponding parameters, AIC, and log-likelihood in each tree.

	Gaussian			Student t			Clayton		
	Par	AIC	L-lik	Par	AIC	L-lik	Par	AIC	L-lik
CP,S	0.97	-185.33	93.66	(0.96,2.15)	-187.57	95.78	10.03	-201.58	101.79
CS,D	0.84	-76.79	39.40	(0.86,2.09)	-85.57	44.79	4.57*	-116.98	59.49
CD,I	0.66	-34.99	18.49	(0.67,4.45)	-35.80	19.90	1.72*	-42.59	22.29
CP,D;S	-0.70	-43.87	22.94	(-0.69,30.00)	-41.39	22.69	-1.12#	-32.04	17.02
CS,I;D	-0.26	-2.5	2.25	(-0.25,30.00)	-0.40	2.20	-0.33※	-3.72	2.86
CP,I;S,D	-0.03	1.96	0.02	(-0.03,17.84)	3.76	0.12	-0.17#	0.72	0.64

	Gumbel			Frank			Joe		
	Par	AIC	L-lik	Par	AIC	L-lik	Par	AIC	L-lik
CP,S	6.87*	-203.06	102.53	21.30	-168.55	85.28	10.78*	-201.50	101.75
CS,D	3.29	-102.87	52.44	9.61	-80.04	41.02	5.31	-117.40	59.70
CD,I	1.95	-43.25	22.63	5.09	-33.80	17.90	2.50	-42.73	22.36
CP,D;S	-1.82#	-38.37	20.18	-5.38	-39.09	20.55	-1.97※	-28.89	15.45
CS,I;D	-1.18#	-3.49	2.74	-1.47	-1.44	1.72	-1.25	-3.43	2.72
CP,I;S,D	-1.06※	1.20	0.40	0.01	2.00	0.00	-1.14※	0.42	0.79

*: rotated 180 degrees; #: rotated 90 degrees; ※: rotated 270 degrees

select the best fitted copula, and Cramer-von Mises (CM) test was used for testing the goodness of fit. Table 2 shows the estimated parameters, log-likelihood, and AIC values. We selected the copula with the smallest AIC to construct the vine copula, and these selected pair-copulas were an acceptable model as the *p*-value were relatively large and the statistics of CM values were small.

To investigate the performance of this model, repeated simulations based on the constructed vine copula were made, and simulated results were compared with the observed ones. Figure 1 compares scatterplots of observed data and simulated data for each pair. The Box-plots of Kendall' tau calculated from the generated data are illustrated in Figure 2. The data length in each simulation varied as 100, 200, and 500 and the number of simulations was fixed at 100. That is, we calculated tau from 100, 200, and 500 samples, respectively, and finally, for each length, we got 100 values of tau. The dotted line denotes the empirical tau derived from the observed data. Figure 2 shows that almost all observed tau fell within boxplots indicating

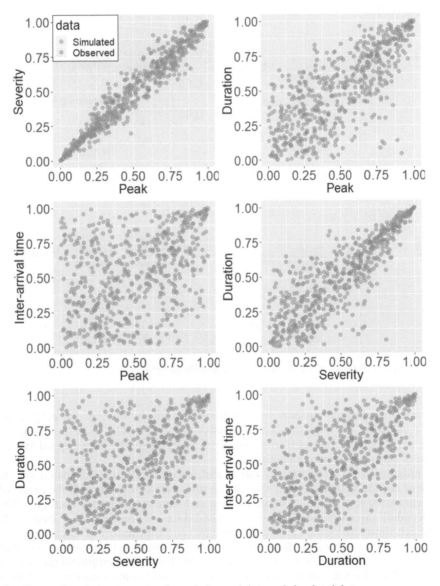

Figure 1. Comparison between scatterplots of observed data and simulated data.

39

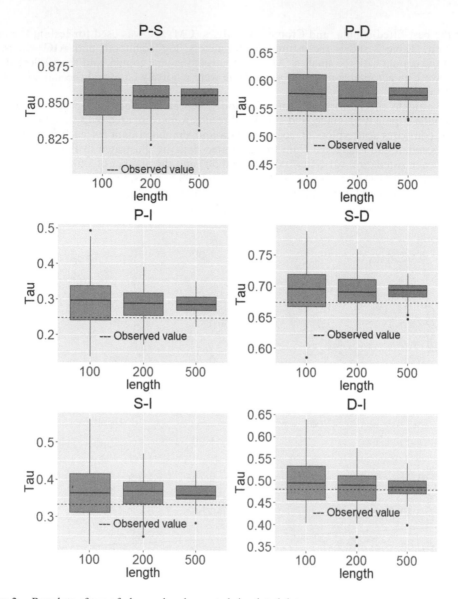

Figure 2. Boxplots of tau of observed and repeated simulated data.

that the inter-dependence patterns were preserved relatively well by the constructed vine copula. In all pairs, the average absolute relative errors (AARE) were less than 3%. That shows the constructed copula performed excellent in capturing the inter-dependence of pairs, even some of them not chosen in the first.

To investigate the performance of this model, repeated simulations based on the constructed vine copula were made, and simulated results were compared with the observed ones. Figure 1 compares scatterplots of observed data and simulated data for each pair. The Boxplots of Kendall' tau calculated from the generated data are illustrated in Figure 2. The data length in each simulation varied as 100, 200, and 500 and the number of simulations was fixed at 100. That is, we calculated tau from 100, 200, and 500 samples, respectively, and finally, for each length, we got 100 values of tau. The dotted line denotes the empirical tau derived from the observed data. Figure 2 shows that almost all observed tau fell within boxplots indicating that the inter-dependence patterns were preserved relatively well by the constructed vine copula. In all pairs, the average absolute relative errors (AARE) were less than 3%. That shows the constructed copula performed excellent in capturing the inter-dependence of pairs, even some of them not chosen in the first.

5 CONCLUSIONS

This study proposes an MI-based sequential approach to select R-vine structure which is model-independent, and then develops an approach for hydrological dependence modeling. A hydrological processes is utilized to evaluate the performance of the proposed approach. By simulation, the results show that the constructed vine copulas based on the proposed approach satisfactorily model different types of dependence structure of hydrological variables.

ACKNOWLEDGMENTS

This study was supported by the National Key Research and Development Program of China (2016YFC0401501,2017YFC1502704), and the National Natural Science Fund of China (41571017, 51679118 and 91647203), and Jiangsu Province"333 Project" (BRA2018060).

REFERENCES

Aas, K., C. Czado, A. Frigessi, and H. Bakken (2009), Pair-copula constructions of multiple dependence, *Insurance Mathematics & Economics*, *44*(2), 182–198.

Czado, C., E.C. Brechmann, and L. Gruber (2012), Selection of vine copulas, edited, pp. 17–37, Springer.

Dai, A. (2013), Increasing drought under global warming in observations and models, *Nature Climate Change*, *3*(1), 52.

Daneshkhah, A., R. Remesan, O. Chatrabgoun, and I. P. Holman (2016), Probabilistic modeling of flood characterizations with parametric and minimum information pair-copula model, *Journal of Hydrology*, *540*, 469–487.

Field, C.B., V. Barros, T.F. Stocker, and Q. Dahe (2012), *Managing the risks of extreme events and disasters to advance climate change adaptation: special report of the intergovernmental panel on climate change*, Cambridge University Press.

Gyasi-Agyei, Y., and C.S. Melching (2012), Modelling the dependence and internal structure of storm events for continuous rainfall simulation, *Journal of Hydrology*, *464*, 249–261.

Hao, Z., and V.P. Singh (2016), Review of dependence modeling in hydrology and water resources, *Progress in Physical Geography*, *40*(4), 549–578.

Hirabayashi, Y., R. Mahendran, S. Koirala, L. Konoshima, D. Yamazaki, S. Watanabe, H. Kim, and S. Kanae (2013), Global flood risk under climate change, *Nature Climate Change*, *3*(9), 816.

Lehner, B., P. Döll, J. Alcamo, T. Henrichs, and F. Kaspar (2006), Estimating the impact of global change on flood and drought risks in Europe: a continental, integrated analysis, *Climatic Change*, *75*(3), 273–299.

Liu, Z., P. Zhou, X. Chen, and Y. Guan (2015), A multivariate conditional model for streamflow prediction and spatial precipitation refinement, *Journal of Geophysical Research-Atmospheres*, *120*(19), 10116–10129.

Mishra, A.K., and V.P. Singh (2011), Drought modeling–A review, *Journal of Hydrology*, *403*(1–2), 157–175.

Nelsen (2006), *An introduction to copulas*, Second Edition ed., Springer, New York.

Shannon, C.E. (1948), A Mathematical Theory of Communication, *Bell System Technical Journal*, *27*(4), 379–423.

Xiong, L., K. Yu, and L. Gottschalk (2014), Estimation of the distribution of annual runoff from climatic variables using copulas, *Water Resources Research*, *50*(9), 7134–7152.

Risk Analysis Based on Data and Crisis Response Beyond Knowledge – Huang & Nivolianitou (eds)
© *2020 Taylor & Francis Group, London, ISBN 978-0-367-25146-8*

Influence factors analysis of tornado-destroyed houses in Funing, Jiangsu, China

Jingyi Gao
School of Architecture and Urban Planning, Nanjing University, Nanjing, China

Peng Qiao
School of Geography and Ocean Sciences, Nanjing University, Nanjing, China
Department of Housing and Urban-Rural Development of Jiangsu Province, Nanjing, China

Guofang Zhai*
School of Architecture and Urban Planning, Nanjing University, Nanjing, China

Dalin Cheng
Institute of Urban Planning and Design, Nanjing University, Nanjing, China

ABSTRACT: Tornado is one of the most serious meteorological disasters and causes large damage within a short time. In this study, we investigated the level of damage in houses caused by the 2016 tornado in Funing, Jiangsu Province, China. Then we analyzed the correlation among the damage levels and other factors such as the strength of tornado, the height and area of the damaged houses, and the building structure by using the multinomial logistic regression model. We found that in the scale of a town, the damage level caused by tornado did not relate strongly to the intensity of the tornado hazard. Our findings demonstrated the correlations among the hazard consequences, the height, and the area of houses: in general, the taller and bigger (area) a house is, the worse damage it will suffer. We hope this study will help the reduction of loss caused by meteorological disaster like tornado.

Keywords: tornado hazard, destroyed houses, influence factors

1 INTRODUCTION

Tornadoes are the most violent atmospheric hazards (Shen & Hwang 2015), and cause a large number of fatalities, great economic loss and significant environmental damages. Tornadoes are the strongest vortices in the atmosphere, and often occur in thunderstorms in summer, especially in the afternoon and evening. While they cover limited affected areas, they are strongly destructive and disastrous. Many statistics show that the loss caused by windstorms is more than for other disasters. Tornadoes have become a focus, as they occur more than 2000 times per year, and there were more than 1200 tornadoes in the United States (Zheng 2016). However, compared with other disasters such as earthquakes, typhoon and flooding, there is little research on tornadoes.

Tornadoes are small probability events and they seldom reach or exceed level F4 (EF4). It is hard to predict a tornado's path but the damage is usually caused in a limited area at an

---

*Corresponding author: guofang_zhai@nju.edu.cn

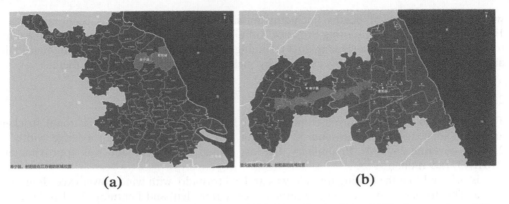

(a)   (b)

Figure 1.   The location of the Funing tornado in 2016. The red area shows (a) Funing county and Don-ghai county, and (b) the tornado disaster area.

extremely high speed. The complex reasons for this—building structure, vegetation condition, and corresponding environment—determine the uncertainty in estimating wind disaster inten-sity, leading to the difficulty in estimating its strength.

Tornadoes are mostly studied in the United States because of the frequent occurrence there. The US has built a tornado database, and the field observation experiments, environmental characteristics research, the climate characteristics analysis, and impact assessments are all world leading. European countries have also studied and built databases to analyze tornadoes (Gao 2018). In China, a developing country, research on tornadoes started much later than in developed countries like the United States and Japan. Since the 1980s, studies on tornadoes in China have tended to focus on the description of hazard itself and since the 1990s, the study of tornadoes has been mainly climate characteristics analysis. However, many scholars are beginning to explore the mechanism of tornadoes, and the monitoring of weather events like tornadoes by using weather radars is increasing gradually. This study is also expanding from the regional scale to the national scale. (Huang 2016)

Compared to globally, tornadoes seldom happen in China, but nationally they still cause huge losses and casualties in some areas. In addition to tornado study in the perspective of meteorology, some Chinese researchers assess the environmental vulnerability and risk evalu-ation before the tornadoes happened. The study after a tornado always focuses on the statis-tics of losses and the environmental damage.

The 2016 tornadoes in Funing county and Donghai county are valuable cases to study (Figure 1), especially regarding the damage caused to houses in Funing. This should help enhance the defense capabilities of houses and avoid future risks by building houses to with-stand tornadoes. By investigating and analyzing the disaster situation of the Funing tornado, this paper introduces the losses caused by this tornado hazard, and investigates the influencing factors for tornado-destroyed houses. This should provide suggestions on how to reduce the loss of tornado hazards in China in the future.

2   MATERIALS AND METHODS

2.1   *Site analysis*

Funing County is in Yancheng city, Jiangsu Province. The county is 1439 km^2 with a population of 111,000. There are 13 towns and four sub-district offices, including Funing Economic Development Zone, High-tech Zone, Jinsha Lake Tourism Resort and the Modern Service Park.

Jiangsu is a province with flat terrain. Hot and humid summer provides favorable condi-tions for severe convective weather and tornadoes. Additionally, according to statistics for China in the past 30 years (Huang 2016), Jiangsu Province has a high frequency of tornadoes

43

with a large number of deaths (including missing persons) and serious economic losses. Jiangsu Province and neighboring Province Anhui are the two provinces with the most serious tornado disasters in China (Huang 2016). Funing is located in the middle of Jianghuai Plain and Jiangsu Province, which may explain why such a severe tornado happened there.

2.2 Tornado in Funing, 2016

The tornado in Funing occurred from 14.00 to 15.00 on June 23, 2016. The zonal stricken area in Yancheng city was around 52 km long and the tornado had an average width of 2.5 km, covering 269 km². Preliminary, the loss caused by this tornado was approximately RMB 4377.39 million.

It is believed that the Funing tornado was an EF4 tornado, with wind power exceeding 17th grade. This tornado led to extreme weather—heavy rain, hail and lightning. It also flattened house, damaged public facilities, leading to casualties and road obstructions. The Funing tornado is the worst for 40 years, killing 99 people and injuring at least 846.

2.3 Data source

The data used in our study comes from the field investigation of tornado-damaged houses organized by Department of Housing and Urban-Rural Development of Jiangsu Province after the disaster. The investigation collected information on 5381 damaged houses from 19 villages in six towns.

We investigated the function of tornado-damaged houses, the owners of the houses, and the the buildings' areas, structures and number of stories. Then we classified the degree of house damage into four levels to classify whether a house was suitable for repair.

2.4 Methods

This study mainly used quantitative analysis based on the data from the investigation. Additionally, we tried to determine the correlation between the tornado strength and destruction level. It is hard to measure the explicit windspeed of a tornado, so we described the strength of the Funing tornado on different villages by referring to research by Zhiyong Meng in 2018. (Meng 2018).

First, we used PivotTab to analyze data on the 5381 houses. Second, we used CrossTab in SPSS to determine whether there are correlations among the classifications of the damaged houses and other factors. In this study, CrossTab showed that the building structure and the function of houses seem strongly unrelated to the classifications, which may need more analysis.

Third, in order to avoid subjectivity by investigators in collecting the data, we took the dependent variable as a categorical variable and used a multinomial logistic regression model in the SPSS software.

We removed the factors that were less correlated to the dependent variable according to the CrossTab results and removed the data that was not statistically significant. This resulted in 2557 complete statistics that were used in the regression models.

3 FINDINGS AND DISCUSSION

3.1 The general characteristics of the survey

The statistics for 5381 tornado-damaged houses were collected from six towns and 19 villages, mainly from Banhu, Chenliang, Shuoji and Xingou. According to the function of the buildings, more than 98% of houses were residential rather than commercial constructions, pigsties, industrial constructions and other public service facilities. In addition, more than 99% of the houses were owned by the occupiers.

Table 1. Case processing summary.

		N	Marginal percentage
Damage classification	Slight damage	1025	40.1%
	Dangerous and needs to be repaired immediately	525	20.5%
	Unable to be repaired at all	1007	39.4%
Number of stories	1	1980	77.4%
	2	577	22.6%
Building structure	Masony-concrete	1728	67.6%
	Masonry	556	21.7%
	other hybrid	273	10.7%
Tornado strength estimation	EF1	789	30.9%
	EF2	99	3.9%
	EF3	1211	47.4%
	EF4	458	17.9%
Valid		2557	100.0%
Missing		0	
Total		2557	

Nearly half of the houses were between 50 m^2 and 100 m^2, and 32.69% were between 100 m^2 and 200 m^2. The structures consisted of 3362 houses with a masonry-concrete or other hybrid structure, and 1408 houses in masonry. More than 4000 houses had one story and 1266 houses had two stories—some were totally destroyed so could not be measured. After the field investigation, 2227 houses were classified as 'slight damage', 884 houses were classified as 'dangerous and need to be repaired immediately', and 1844 houses were classified as 'unable to be repaired at all' (Table 1).

3.2 *The main results of the current study*

After analyzing statistics for 2556 houses, the results of the multinomial logistic regression model and subsequent findings and speculation are given below.

First, two-story houses were more likely to be classified as 'slight damage' or 'dangerous and need to be repaired' than a one-story house, suggesting that two-story houses are more likely to suffer severe damage.

Second, masonry houses are more likely to be classified as 'slight damage' than the masonry-concrete houses and other hybrid structures. However, the 'other hybrid structure' houses and masonry-concrete houses are more likely to be classified as 'dangerous and need to be repaired' than masonry houses, which means that the masonry houses will suffer less damage than hybrid structure houses.

Third, the building's area is positively related to both classifications, which means that the larger a house is, the more severely damaged it will be.

Finally, the level of houses damage is not in linear correlation with the tornado strength at our site. There are two classifications: 'slight damage' and 'dangerous and needs to be repaired'. The results of the regression model show that the houses classified as 'slight damage' were more likely to have been hit by the EF4 than EF3 tornado. However, when classified as 'slight damage' or 'dangerous and needs to be repaired', the house is more likely to have been damaged by the EF1 or EF2 tornado, rather than the EF4 tornado. In other words, the EF3 tornado will cause less damage to houses than the EF4 tornado, but the damages have more probabilities to be found in EF1 or EF2 tornado than EF4 (Table 2 and Table 3). It means in this study houses in the area hit by EF1 or EF2 tornado may sometimes be damaged more seriously than those hit by EF4 according to the data we have.

Table 2. Parameter estimates in multinomial logistic regression model.

		B	Sig.	Exp(B)
1	Intercept	0.895	0.002	
	Building area	0.003	0.025	1.003
	[Building story number = 1]	−1.751	0.000	0.174
	[Building story number = 2]	0	.	.
	[Building structure = masonry-concrete]	0.151	0.412	1.163
	[Building structure = masonry]	0.903	0.000	2.466
	[Building structure = other hybrid]	0	.	.
	[Tornado strength estimation = EF1]	0.379	0.007	1.460
	[Tornado strength estimation = EF2]	2.298	0.000	9.950
	[Tornado strength estimation = EF3]	−0.487	0.000	0.614
	[Tornado strength estimation = EF4]	0	.	.
2	Intercept	0.465	0.148	
	Building area	0.005	0.002	1.005
	[Building story number = 1]	−1.024	0.000	0.359
	[Building story number = 2]	0	.	.
	[Building structure = masonry-concrete]	−1.277	0.000	0.279
	[Building structure = masonry]	−0.416	0.032	0.660
	[Building structure = other hybrid]	0	.	.
	[Tornado strength estimation = EF1]	0.395	0.035	1.484
	[Tornado strength estimation = EF2]	2.066	0.000	7.894
	[Tornado strength estimation = EF3]	0.116	0.516	1.123
	[Tornado strength estimation = EF4]	0	.	.

Table 3. Goodness of fit.

	Chi-Square	df	Sig.
Pearson	2842.742	2460	0.000
Deviance	2921.911	2460	0.000

4 CONCLUSION

This study analyzed the influencing factors of tornado-damaged houses in the 2016 Funing tornado in China. The field investigation and results from multinomial logistic regression model analysis showed that building area, building structure and number of stories are correlated with the classifications of the damage levels. This means the larger and the taller a house is, the more severely damaged it will be. In addition, in this tornado disaster the masonry houses were safer than the hybrid-structure houses. Finally, at our site it is hard to describe the correlation between tornado strength and the level of house damage.

There are still some limitations in this study that require further discussion. First, the classifications of the houses by investigators were sometimes subjective, which may not reflect the level of destruction. Second, the tornado strength in every village was estimated according to precedents instead of accurate calculation. These two disadvantages may affect the accuracy of the results. Third, this study only took tornado strength and some building factors into consideration, and there are some other factors that could be included in further study, for example the environmental factors such as water area and ratio of green space.

REFERENCES

Chen, J., Yang, H., Zhu, Y., Liu, J., Xuan, D., Jiang, A., and Jia, P. 2010. The survey and evaluation of tornadoes. *Journal of Natural Disasters*, 1999(04): 111–117.

Gao, R. 2018. *Tornado Characteristics Statistical Analysis and Post-Disaster Investigations in China*. Beijing Jiaotong University. (In Chinese).

Huang, D., Gao, G. Ye, D. and Xiao, C. 2017. Progress in tornado research and present situation of tornado warning operational system. *Science & Technology Review*, 2017(05):47–55. (In Chinese).

Huang, D., Zhao, S. and Gao, G. 2016. Disaster characteristics of tornadoes over China during the past 30 years. *Torrential Rain and Disasters*, 35(2):97–101. (In Chinese).

Meng, Z., Bai, L., Zhang, M., Wu, Z., Li, Z., Pu, M., Zheng, Y., Wang, X., Yao, D., Xue, M., Zhao, K., Li, Z., Peng, S. and Li, L. 2018. The Deadliest Tornado (EF4) in the Past 40 Years in China. *Weather and Forecasting*, 33(3):693–713. (In Chinese).

Shen, G. and Hwang, S. 2015. A spatial risk analysis of tornado-induced human injuries and fatalities in the USA. *Natural Hazards*, 77(2):1223–1242.

Zheng, F. and Xie, H. 2010. Progress in Tornado Researches in China in Recent 30 Years. *Meteorological Science and Technology*, 38(03):295–299. (In Chinese).

Zheng, Y., Zhu, W., Yao, D., Meng, Z., Xue, M., Zhao, K., Wu, Z., Wang, X. and Zheng, Y. 2016. *Mereorological Monthly*, 42(11):1289–1303. (In Chinese).

Risk Analysis Based on Data and Crisis Response Beyond Knowledge – Huang & Nivolianitou (eds)
© 2020 Taylor & Francis Group, London, ISBN 978-0-367-25146-8

Evaluation of ecological civilization construction level of prefecture-level cities in Guizhou based on grey relational projection method

Xia Feng
School of Big Data Application and Economics, Guizhou University of Finance and Economics, Guiyang, China
Guizhou Institution for Technology Innovation & Entrepreneurship Investment, Guiyang, China

Mu Zhang
School of Big Data Application and Economics, Guizhou University of Finance and Economics, Guiyang, China

ABSTRACT: Guizhou Province is currently in the stage of rapid economic development, but due to the adverse natural conditions and the dual pressures of economic development and environmental protection, it is particularly important to improve the level of ecological civilization construction in Guizhou prefecture-level cities. This paper introduces the grey relational projection method into the evaluation of the level of ecological civilization construction, and constructs the index of ecological civilization construction including Baidu Index. The system establishes the evaluation model of ecological civilization based on grey relational projection method, uses the analytic hierarchy process to determine the weight of indicators, compares the projection values of nine cities and cities, determines the level of ecological environment quality and the ranking of advantages and disadvantages of each city in Guizhou, evaluates the level of ecological civilization construction in Guizhou Province comprehensively, and puts forward the construction of ecological civilization in Guizhou Province in the future.

Keywords: Index system of ecological civilization construction, grey relational projection method, Baidu index, analytic hierarchy process, grey relational projection value

1 INTRODUCTION

The eighteenth National Congress of the Communist Party of China made the strategic decision of "vigorously promoting the construction of ecological civilization". The CPC Central Committee and the State Council have made a series of decisions. The report of the Eighteenth National Congress pointed out that "the construction of ecological civilization is a long-term plan concerning the well-being of the people and the future of the nation". The implementation of the document "National Ecological Civilization Construction Pilot Zone (Guizhou)" is to promote Guizhou Province to carry out the ecological civilization construction pilot zone. The plan proposes that we should take the construction of "Colorful Guizhou Park Province" as the overall goal, keep the two bottom lines of development and ecology. Guizhou Province has implemented the decision-making and deployment of the Central Committee of the Party and the State Council. However, the relevant system of ecological civilization construction needs to improve, the implementation efforts need to go deep into the grass-roots level. Therefore, this paper constructs a Baidu Index based on the data of the Internet users in Guizhou Province in the past five years through Baidu search for ecological civilization.

Wangqi (2018) and other provinces in China respectively analyzed the level of development of ecological civilization construction; Wangnan (2018) took Guizhou Province as

the research object, and put forward suggestions for the development of ecological civilization in Guizhou Province by constructing the index system of ecological civilization in Guizhou Province. The vast majority of literature will use environmental indicators, such as
forest coverage and environment. Big data drives everyone's development. Baidu advocates
the lifestyle of data decision-making, which makes more people realize the value of data,
reflects the attention of netizens in Guizhou Province to ecological civilization. Grey relational projection method is a multi-objective decision-making evaluation method from the
perspective of vector projection. At present, grey relational projection method has been
successfully applied in the fields of ecological environment evaluation, water quality evaluation, enterprise credit evaluation, ecological civilization evaluation and so on. In view of
this, this paper establishes an evaluation model of ecological civilization based on grey
relational projection, compares the magnitude of grey relational projection value, and
ranks the level of ecological civilization of prefecture-level cities in Guizhou, thus providing new ideas for the future development of ecological civilization construction in Guizhou
Province.

2 EVALUATION MODEL OF ECOLOGICAL CIVILIZATION BASED ON GREY RELATIONAL PROJECTION INTRODUCTION

The modeling steps are as follows:

Step 1: Construct Decision Matrix. Record $A=$ {Scheme 1,..., Scheme n}=$\{A_1, A_2, \ldots, A_n\}$,
A_0 is the optimum scheme, and record $V=$ {index 1, index 2,... Index m}=$\{V_1, V_2, \ldots, V_m\}$,
then the attribute value of the optimal scheme A_0 is Y_{0j}, and it satisfies:

When V_j belongs to benefit index, $Y_{0j} = max\left(Y_{1j}, Y_{2j}, \ldots, Y_{nj}\right), j = 1, 2, \ldots, m$

When V_j belongs to cost index, $Y_{0j} = min\left(Y_{1j}, Y_{2j}, \ldots, Y_{nj}\right), j = 1, 2, \ldots, m$

When V_j belongs to fixed index, Y_{0j} is the best stable value of the index, $j = 1, 2, \ldots, m$

The decision matrix Y of decision domain set A to factor index set V is as follows:

$$Y = \begin{pmatrix} Y_{01} & \cdots & Y_{0m} \\ \vdots & \ddots & \vdots \\ Y_{n1} & \cdots & Y_{nm} \end{pmatrix}, \text{ namely, } Y = \left(Y_{ij}\right)_{(n+1)\times m}, i = 0, 1, \ldots, n, j = 1, 2, \ldots, m$$

Step 2: Initialization Decision Matrix.

For benefit-oriented indicators, order

$$Y_{ij}^* = \frac{Yij - min\,Y_j}{max\,Y_j - min\,Y_j}, \quad i = 0, 1, \ldots, n, j = 1, 2, \ldots, m \tag{1}$$

For cost-based indicators, order

$$Y_{ij}^* = \frac{max\,Y_j - Yij}{max\,Y_j - min\,Y_j}, \quad i = 0, 1, \ldots, n, j = 1, 2, \ldots, m \tag{2}$$

For fixed indicators, order

$$Y_{ij}^* = 1 - \frac{\left|Y_{ij} - Y_j'\right|}{max\left|Y_{ij} - Y_j'\right|}, \quad i = 0, 1, \ldots, n, j = 1, 2, \ldots, m \tag{3}$$

Y_j' is the best stable value of j.

Step 3: Construct Grey Relational Degree Decision Matrix. The correlation between Y_{ij}^* and
Y_{0j}^* is expressed by $r_{ij}(i = 0, 1, \ldots, n, j = 1, 2, \ldots, m)$ The calculation is as follows:

$$r_{ij} = \frac{\lambda \min \min |Y_{0j}^* - Y_{ij}^*| + \lambda \max \max |Y_{0j}^* - Y_{ij}^*|}{|Y_{0j}^* - Y_{ij}^*| + \lambda \max \max |Y_{0j}^* - Y_{ij}^*|} \tag{4}$$

Taken as $\lambda = 0.5$, r_{ij} consists of $R = (r_{ij})_{(n+1)}$. Obviously,

$$R_{01} = R_{02} = \ldots = R_{0n} = 1.$$

Step 4: Determine Weight Coefficient of Factor Indicators.

(1) Establish hierarchical structure.
(2) Select two evaluation indexes, a_{ij} means comparing x_i with x_j, get $A = (a_{ij})_{n \times n}$
(3) From $AW = \lambda_{max} W$, get λ_{max} and the eigen value $W = (w_1, w_2, \ldots, w_n)^T$
(4) Define consistency indicators:

$$CI = \frac{\lambda_{max} - n}{n - 1} \tag{5}$$

When $CI = 0$, there's complete consistency; when CI is close to 0, there's better consistency.

Define the consistency ratio:

$$CR = CI/RI \tag{6}$$

When $CR < 0.1$, through consistency test.

(5) According to the above results, The final weight of each factor index can be obtained.

Step 5: Determine Weighted Grey Relational Decision Matrix. $W_k = (w_1, w_2, \ldots, w_m)^T$, R' is obtained by weighting R, then $R' = R \cdot W = (R_1', R_2', \ldots, R_m')$.
Step 6: Calculate grey relational projection value. If the projection angle of grey correlation between the decision-making scheme and the optimal scheme is θ_i, then:

$$\cos \theta_i = \sum_{j=1}^{m} r_{ij} w_j^2 / \sqrt{\sum_{j=1}^{m} w_j^2} \sqrt{\sum_{j=1}^{m} (r_{ij} w_j^2)^2}, \quad i = 1, 2, \ldots n \tag{7}$$

A_i on the optimal scheme is the grey relational projection value D_i:

$$D_i = \| A_i \| \cdot \cos \theta_i = \sum_{j=1}^{m} r_{ij} w_j^2 / \sqrt{\sum_{j=1}^{m} w_j^2}, \quad i = 1, 2, \ldots n \tag{8}$$

3 AN EMPIRICAL ANALYSIS ON THE CONSTRUCTION LEVEL OF ECOLOGICAL CIVILIZATION IN PREFECTURE-LEVEL CITIES OF GUIZHOU PROVINCE

3.1 *Index system and data source*

The index system is shown in Table 1. The sample period of this paper is selected from 2013 to 2017. The data of ecological civilization indicators mainly come from the statistical yearbook of 2014-2018 issued by the statistical bureaus of nine prefecture-level cities in Guizhou Province,

Table1. Ecological civilization construction level index system.

Target layer	Primary indicator	Secondary indicators
Index System of Ecological Civilization Construction	Eco economy A_1	Per capita gross domestic product A_{11}; Growth rate of added value of tertiary industry as a proportion of GDP A_{12}; Growth rate of per capita general public budget income A_{13}; Energy consumption per unit GDP A_{14}; Unit GDP Electricity Consumption A_{15}; General Public Budget Expenditure - Energy Conservation and Environmental Protection A_{16}
	Ecoenvironment A_2	Forest coverage A_{21}; Excellent rate of ambient air quality in urban area A_{22}; Achievement Rate of Water Quality in Centralized Drinking Water Sources above County Level A_{23}; Growth Rate of Fertilizer Application in Agriculture A_{24}; Comprehensive Utilization Rate of Industrial Solid Waste A_{25}; Total sulphur dioxide emissions A_{26}
	Ecolivability A_3	Per capita park green space area A_{31}; Harmless Treatment Rate of Municipal Domestic Waste A_{32}; Mean temperature A_{33}; Annual precipitation A_{34}; Per capita highway mileage A_{35}; Per capita housing area in urban areas A_{36}
	Eco culture A_4	General Public Budget Expenditure - Education Services A_{41}; General Public Budget Expenditure - Culture, Sports and Media A_{42}; Number of students in Higher Education A_{43}; Middle School Students A_{44}; Baidu Index PC Search Quantity A_{45}; Baidu Index Mobile Search Quantity A_{46}

the environmental bulletins and water resources bulletins issued by the environmental protection bureaus of all prefecture-level cities, and some indicators such as the per capita housing area of urban areas come from the Statistical Yearbook of Urban Construction in China.

3.2 *Empirical results and analysis*

(1) For nine prefecture-level cities, $A_0 =$ (74493, 17.5, 26.5, 0.62, 748, 18.61, 66.68, 100, 100, -13.6, 99.54, 3.1, 21.85, 100, 16.5, 1306, 83.88, 49.7, 129.79, 8.25, 441603, 681797, 59, 50)

(2) According to step 2, the decision matrix Y is normalized and obtained Y^*.

(3) According to step 3, the grey relational judgment matrix is obtained R.

(4) $W =$ (0.014, 0.019, 0.01, 0.036, 0.029, 0.038, 0.051, 0.063, 0.025, 0.052, 0.071, 0.092, 0.027, 0.037, 0.02, 0.072, 0.058, 0.075, 0.02, 0.027, 0.015, 0.052, 0.042, 0.055)

(5) $D_{2013} =$ (0.1259, 0.1023, 0.1322, 0.1133, 0.109, 0.1308, 0.1291, 0.1315, 0.1413), Similarly, projection values for 2014, 2015, 2016 and 2017 can be obtained. The projection values of different years in the same region are averaged, as shown in Table 2.

In Table 2, we can see that the average projection value of in Qiandongnan is the highest in the period of 2013-2017, with an average projection value of 0.1315, ranking first in Guizhou prefecture-level cities; the average projection value of Anshun City is 0.1308, followed by it; the construction level of ecological civilization in Liupanshui City is the lowest, with an average projection value of 0.1006.

From Figure 1, it is obvious that Construction of Ecological Civilization in Guizhou Province is relatively low and regional differences are large. Although the average projection value of five years shows that Qiandongnanis the highest, its fluctuation is relatively large; Anshun City fluctuates greatly, the main reason is that its carbon dioxide emissions and the growth rate of

51

Table 2. Gray correlation projection value and ranking of ecological civilization construction in prefecture-level cities of Guizhou.

	2013	2014	2015	2016	2017	mean value	ranking
Guiyang	0.1259	0.1276	0.1149	0.1259	0.1403	0.1269	4
Liupanshui	0.1023	0.0985	0.0933	0.1053	0.1038	0.1006	9
Zunyi	0.1322	0.1190	0.1086	0.1182	0.1306	0.1217	7
Anshun	0.1133	0.1390	0.1331	0.1389	0.1295	0.1308	2
Bijie	0.1090	0.1107	0.1093	0.1114	0.1041	0.1089	8
Tongren	0.1308	0.1188	0.1317	0.1111	0.1315	0.1248	6
Qiandongnan	0.1291	0.1369	0.1310	0.1381	0.1224	0.1315	1
Qianxinan	0.1315	0.1110	0.1287	0.1271	0.1320	0.1261	5
Qiannan	0.1413	0.1255	0.1215	0.1197	0.1344	0.1285	3

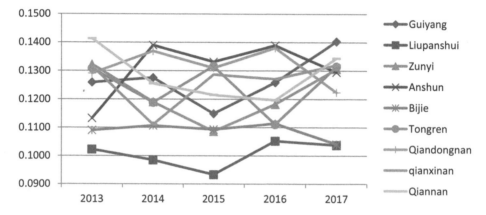

Figure1. Projection value polygons of different areas in the same year.

agricultural chemical fertilizer application have changed significantly; Qiannan shows that environmental protection has been neglected and the importance of ecological environment has been realized in 2017; Guiyang City, the construction of ecological civilization is at a medium level, but its level of development of ecological civilization is faster; Qianxinan fluctuates greatly in 2014 and 2015, and tends to be stable after 2016, which indicates that the construction level of ecological civilization has passed the initial stage.; Tongren City fluctuates greatly between 2015 and 2017, which indicates that it has realized the importance of environmental protection, but has not dealt with the relationship between ecological environment and economy well; Zunyi City's ecological civilization construction declines sharply from 2014 to 2015. Through the adjustment in 2016 and 2017, the level of ecological civilization construction is basically the same as that in 2013. The development of Bijie City is relatively stable, and the construction of its ecological civilization home demonstration site is relatively successful. Liupanshui City is relatively backward, and neglects the protection of the environment and focuses on the development of the economy.

4 CONCLUSIONS AND SUGGESTIONS

From the development of each prefecture-level city, Guiyang city's ecological and livelihood construction develops faster, but it should pay more attention to the protection of the natural environment; Liupanshui city develops more balanced, but it abounds in coal resources, which leads to waste of resources, so it should focus on reducing the rate of environmental pollution; The development of Zunyi is relatively unbalanced, the construction of people's livelihood and the protection of natural environment have changed greatly vertically; the growth rate of people's livelihood in Anshun is low in the whole province, which indicates that it has not invested much in people's livelihood in recent years, hoping to introduce a new model in social livelihood; Tongren and Bijie have slower ecological protection and lower development in ecological economy, but the people's livelihood is low. Guizhou province should speed up the construction of cultural industry, enhance the attention of residents to ecological civilization and people's livelihood, so as to improve the quality of life of residents.

ACKNOWLEDGMENTS

This research was financially supported by the Regional Project of National Natural Science Foundation of China (71861003) and the Second Batch Projects of Basic Research Program (Soft Science Category) in Guizhou Province in 2017 (Foundation of Guizhou-Science Cooperation [2017] 1516-1).

REFERENCES

Wang Qi, Miao Jingwen. 2018. Assessment of ecological civilization construction in China's provinces based on ecology. *Eco-economy*, 34 (09): 212–218.

Wang Nan, D. 2018. Evaluation and Development Strategy of Ecological Civilization Construction in Guizhou Province. *Tianjin Commercial University*.

DuYu, Liu Junchang. 2009. Research on evaluation index system of ecological civilization construction. *Scientific management research*, (3): 60–63.

Lin Zhen, Shuang Z.M. 2014, A comparative study on the evaluation index system of eco-civilization construction in provincial capitals: Guiyang, Hangzhou and Nanjing as examples. *Journal of Beijing University of Aeronautics and Astronautics (Social Science Edition)* 27 (05): 22–28.

Wang Ran, D. 2016. Establishment and Empirical Study on the Evaluation Index System of China's Provincial Ecological Civilization. *China University of Geosciences*.

Zhang Huan, Cheng Jinhua, Feng Yin, Chen Dan, Ni Lin, Sun Han. 2015. Evaluation index system and its application for the construction of ecological civilization in mega-cities: A case study of Wuhan. *Journal of Ecology*, 35 (02): 547–556.

Zhuo Shengjun, Wang Jian, Hu Shuheng. 2016. Measuring the level of ecological civilization construction based on the coupled model of AHP and entropy value — Taking Luan City as an example. *Journal of Anhui Architectural University* 24 (02): 59–64.

Xu Ye, Wei Ru, Jiao Yanfang. 2017. Construction of evaluation index system and comprehensive evaluation of urban ecological civilization construction: based on empirical analysis of 11 prefecture-level cities in Jiangxi Province. *Journal of Jiangxi Normal University (Natural Science Edition)* 41 (03): 265–270.

Wu Kaiya, Li Ruzhong, Chen Xiaojian. 2003. Grey relational projection model for regional ecological environment assessment. *Resources and environment of the Yangtze River Basin* (05): 473–478.

Zhou R.X, XuYaqi, Chen Qianqian. 2018. Water quality evaluation of Baiyan Lake in Hefei based on grey relational projection. *Journal of Zhejiang Institute of Water Resources and Hydropower* 30 (05): 50–54.

Zhang Mu, Zhou Zongfang. 2009. Enterprise Credit Evaluation Model Based on Grey Relational Projection Pursuit. *Statistics and Decision-making* (21): 35–37.

Zhu Yulin, Li Mingjie, Liu Huan. 2010. Comprehensive evaluation of urban ecological civilization degree based on grey relational degree: taking Changsha-Zhuzhou-Tan urban agglomeration as an example. *Journal of Central South Forestry University (Social Science Edition)* 4 (05): 77–80.

Fan D.C, Du Mingyue. 2017. Dynamic comprehensive evaluation of technological innovation capability of high-tech industry based on TOPSIS grey relational projection method - from the perspective of Beijing-Tianjin-Hebei integration. *Operational research and management* 26 (07): 154–163.

Zhang Mu, Zhou Zongfang. 2009. Enterprise credit evaluation model based on multi-objective programming and support vector machine. *China Soft Science* (04): 185–190.

Lu Feng, Cui Xiaohui. 2002. Grey Relational Projection Method for Multi objective Decision Making and Its Application. *Systems Engineering Theory and Practice* (01): 103–107.

LanQingxin, PengYiran, Feng Ke. 2013. Construction of evaluation index system and evaluation method for urban ecological civilization construction — Empirical analysis based on four cities of Guangzhou and Shenzhen in North China. *Research on financial and economic issues* (09): 98–106.

Risk Analysis Based on Data and Crisis Response Beyond Knowledge – Huang & Nivolianitou (eds)
© *2020 Taylor & Francis Group, London, ISBN 978-0-367-25146-8*

Research on China's meteorological disaster risk management system mechanism and associated countermeasures

Kaicheng Xing
Key Laboratory of Meteorological and Ecological Environment of Hebei Province, Shijiazhuang, China
Hebei Climate Center, Shijiazhuang, China

Daohong Chen & Shujun Guo*
Hebei Meteorological Bureau, Shijiazhuang, China

ABSTRACT: China's regional extreme weather and climate events are diverse and frequent. With the rapid development of the social economy, the vulnerability and exposure of meteorological disaster-bearing bodies have increased significantly, and the losses caused by meteorological disasters have also increased. This article analyzes the basic characteristics and development trends of regional extreme events such as drought, heavy precipitation, high temperatures and low temperatures in China. From the perspective of meteorological disaster risk management, studies related to Hebei and Hubei are used for in-depth analysis and the problems existing in meteorological disaster risk management system and mechanism are summarized. Based on the interrelationship between adaptive measures and meteorological disaster risks, this article proposes countermeasures for key components such as composite main body, stratified operation, classification implementation, and comprehensive support. This research work provides a very good reference for not only the overall framework and implementation mechanism of meteorological disaster risk management, but also for improving and perfecting the meteorological disaster risk management system and capability system and improving the meteorological disaster prevention capabilities.

Keywords: meteorological disaster, risk management, system mechanism, countermeasures

1 INTRODUCTION

On July 28, 2016, General Secretary Jinping Xi emphasized that disaster prevention, disaster mitigation and disaster relief are related to the safety of people and property, social harmony and stability. They are an important aspect of measuring the leadership of the ruling party, testing the government's execution, judging the country's mobilization, and embodying national cohesion. The new era requires a profound understanding of the basic national conditions of China's natural disasters, enhancing the sense of urgency and responsibility, and strengthening the management of meteorological disasters with more pragmatic and more effective measures, improving the system and mechanism, and optimizing disaster prevention, mitigation and relief.

---

*Corresponding author: gsjun888@sina.com

2   STATUS OF CHINA'S METEOROLOGICAL DISASTER RISK MANAGEMENT WORK

Under the background of global warming, extreme weather and climate events are frequent, and the risk of meteorological disasters is further aggravated. China's rapid and sustainable development has also placed higher demands on meteorological disaster risk management. In China, strengthening the risk management of meteorological disasters and alleviating the disaster damage is related to the stability of harmonious society, the construction of ecological civilization, and the security of public security. It is not only a scientific issue but a social, economic and even political issue.

Since 2008, the Chinese Meteorological Administration has promoted the construction of a meteorological disaster risk management business system and service practices, and has achieved remarkable results (Figure 1). So far, a database of historical meteorological disasters in 2,300 counties has been established, and national storms, floods, typhoons and frozen disasters risk zonation have been completed. Since 2012, the National Climate Center has carried out storm and flood disaster risk assessment and impact-based disaster early warning research, and pilot businesses have achieved initial results in the Hebei and Hubei Meteorological Bureau. The promotion of meteorological risk warning services for floods, urban waterlogging and geological disasters is progressing smoothly. A national, provincial, municipal and county meteorological disaster risk warning service business system was initially established, and the technical methods for risk census, risk zonation, disaster threshold determination, early warning, business inspection and benefit evaluation were formed.

3   METEOROLOGICAL DISASTER RISK MANAGEMENT BUSINESS PRACTICE

Hebei Meteorological Bureau and Hubei Provincial Meteorological Bureau have done significant work on meteorological disaster risk management and business technology innovation. The following is a brief introduction.

3.1   *Hebei province meteorological bureau related work*

3.1.1   Establish a new concept of disaster prevention and mitigation and improve the new meteorological disaster prevention and management system. Establish regular meteorological disaster prevention headquarters and public service agencies in provincial, municipal, and county governments and key townships. Clarify the power of affairs and the corresponding public financial security mechanisms, strengthen the performance management and accountability mechanism, mobilize social forces and volunteers to actively participate in cracking down on single-hazard species to prevent and reduce disasters.

3.1.2   Strengthen the construction of meteorological infrastructure and enhance the ability to support science and technology (Figure 2). Improve the six major systems: disaster monitoring, disaster warning, risk management, emergency response, legal disaster prevention and the disaster prevention demonstration system. Construct a system of meteorological services and support: meteorological big data resource pools; refined numerical weather predictions and severe weather monitoring and early warning systems; urban and rural refined public meteorological service production and release systems; meteorological popular science business systems and resource sharing and dissemination systems; as well as the National Aircraft Increasing Rain Science Experiment Shijiazhuang Base and the Jidong, Jixibei Aircraft Artificial Increasing Rain Support Base. Construct a system for operation condition monitoring, ecological monitoring and an evaluation system combining airborne equipment and satellite remote sensing, etc. Improve the technical support capability of meteorological disaster prevention, response to climate change and ecological civilization construction.

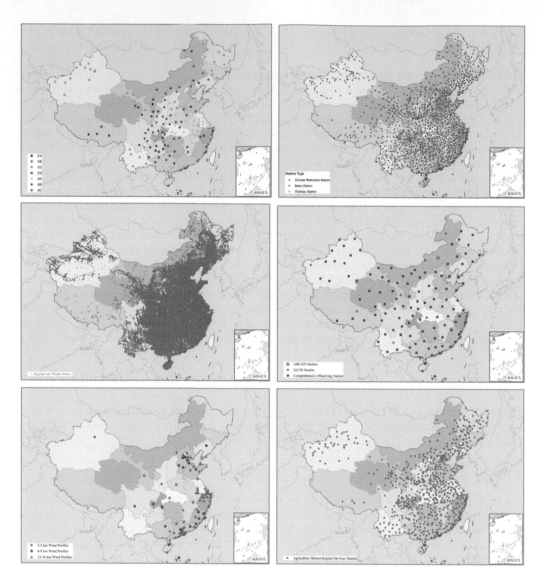

Figure 1. China meteorological observatory network layout (weather radar, national ground weather station, regional auto weather station, GRUAN station, wind profiler, agricultural meteorological services station).

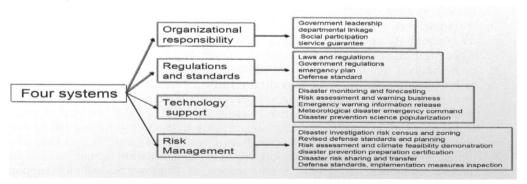

Figure 2. Hebei Province meteorological disaster comprehensive defense system.

3.1.3 Standardize the meteorological disaster prevention business and build a meteorological disaster prevention service system. Establish a meteorological disaster defense and command mobile decision service platform and a system to realize early warning information authority, network-wide, and with a unified release. Conduct comprehensive surveys and risk investigations of all meteorological disasters, and compile and publish detailed meteorological disaster risk zoning; The government leads to make a system of meteorological disaster prevention preparation certification for key units and personnel-intensive locations. Formulate meteorological disaster risk sharing, transfer, and government subsidy systems, and cooperate with insurance and reinsurance institutions to develop catastrophy insurance meteorological services. Formulate meteorological disaster prevention standards, measure the implementation, inspection and supervision system, and carry out performance inspection and disaster prevention and mitigation performance management assessment by the member units of the headquarters.

3.2 *Hubei provincial meteorological bureau related work*

3.2.1 Driven by technological innovation, build a meteorological disaster risk business system. Hubei Meteorological Bureau's self-developed business platform can realize real-time collection and data analysis of storms, floods, comprehensive display of various information, an integrated disaster report, disaster risk level warning information production, and automatic and precise targeting of one-click release to specific people. Various types of disaster information can verify the results of flood and flooding analysis and disaster loss assessment, and form a closed storm flood and risk warning service product chain.

3.2.2 Solve the risk warning and accuracy problem and improve the core competitiveness (Figure 3). In 2014, the flooded submerged model was put into operation, and the method of determining the critical rainfall of landslide based on soil erosion model was proposed. The critical rainfall of 24 watershed control stations reaching the fourth-level disaster standard could be derived, and the hydrological model was used to calculate the critical rainfall scheme of small and medium rivers. Realize rapid statistics, fine calculus and three-dimensional dynamic demonstration of rainstorms, surface rainfall, submerged area, and flash flood warning. The calculation speed is faster than that of foreign similar models, providing core technical support for accurate scientific early warnings of meteorological geological disasters, meteorological disaster management and services.

3.2.3 The defense gate moves forward and realizes the extension from "forecast" to "defense". From disaster collection, core technology research and development to meteorological disaster risk management, to achieve the integration of departments, resources, technology and management, form a meteorological disaster prevention

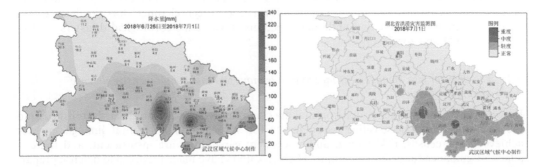

Figure 3.   Precipitation distribution and flood disaster monitoring map of Hubei Province from June 25 to July 1, 2018.

linkage mechanism with clear government leadership, clear division of duties, and reasonable allocation of resources. Based on defense planning and emergency plan preparation and weak grassroots unit disaster prevention, focusing on improving grassroots defense capabilities, and on defense key tasks and action plans, promoting the improvement of defense capabilities of the whole society and accurate prediction of meteorological disaster risk warning information and timely delivery, providing an important decision-making basis for emergency response and flood prevention work.

4   METEOROLOGICAL DISASTER RISK MANAGEMENT BUSINESS PRACTICE

4.1   *The legal system for meteorological disaster prevention is not sound and systematic*

China does not have a basic law on disaster prevention and mitigation for all types of natural disasters. The existing laws, regulations and rules are not specific, and there are no implementation rules and supervision mechanisms. Sound technical regulations, norms and standards systems have not yet been established, and there are a large number of gaps that are not scientific and reasonable and have insufficient binding force. The contingency plan is not operable, lacks comprehensive emergency drills, and the public's self-rescue and mutual rescue capability is insufficient. The awareness and ability of society's meteorological disaster prevention needs to be improved.

4.2   *The public service responsibility for disaster prevention and mitigation is not specific, the responsibility is not clear, and the guarantee is weak*

In some provinces the division of responsibility for disaster prevention and mitigation among local government departments is unscientific and unreasonable. The prevention, resistance, and rescue links of the same disaster are classified by different departments, and different disasters with the same source are also managed by different departments. Social responsibility lacks in important work positions, and the block-type disaster prevention system is not efficient [1]. The implementation system of grassroots government responsibilities is imperfect, the tasks of disaster prevention and mitigation of legal persons, social organizations and citizens are not clear, and the role of volunteers has not been fully exerted.

4.3   *The disaster prevention business system is incomplete and the core technical strength is not enough*

In many areas, the ability of monitoring, early warning and information release is insufficient, and the monitoring factors, scope and accuracy cannot meet the needs of fine-tuned forecasting, precision prevention and risk management. The business standards of different departments and fields are not uniform, and data cannot be shared in real time; The information dissemination system of the province, city and county connecting, standardizing and unifying is imperfect. A refined and time-sensitive information "green channel" has not been established, especially in remote and underdeveloped areas where meteorological disasters are frequent, it is a "blind zone".

4.4   *The disaster risk management mechanism is imperfect and comprehensive defense capability is weak*

In some areas, the emergency decision-making command system has not been established, and the scientific and time-sensitive requirements for vertical and horizontal linkage are not achieved. The census, risk zoning, and assessment capabilities are insufficient, and there is a lack of guidance for defense planning. High-risk units have made inadequate defense preparations, insufficient emergency prevention and disposal capabilities, and risk aversion mechanisms such as personnel evacuations and property transfer are not sound. The level of the agricultural insurance meteorological services is low, the coverage is not wide, and there is a lack of catastrophe risk transfer insurance. More attentions should be paid on the reduction of risks and losses. The cost of disaster reduction is large and the efficiency is not high.

5 COUNTERMEASURES FOR STRENGTHENING THE CONSTRUCTION OF DISASTER RISK MANAGEMENT MECHANISMS

5.1 *Improve the complex subject of meteorological disaster risk management and its cooperation mechanism*

Establish a complex subject linkage mechanism for meteorological disaster risk management. Strengthen the meteorological risk management mechanism of "government-led, departmental linkage and social participation", clarify the responsibilities and obligations of the government and various departments in the law, and confirm the composition, cooperation mechanism, and roles of the complex subject of disaster prevention [2], through public opinion guidance, media propaganda, and socialization (marketization) mechanisms, data sharing, emergency response, and effective interaction and integration among various departments. Promote the participation of social organizations, enterprises and citizens, and establish a cooperative mechanism of "complementary advantages, consensus-building, and collective action" to form a rule of law, society and public opinion, and a cultural environment that promotes the improvement of meteorological risk management.

5.2 *Strengthening the hierarchical operation and coordination mechanism of meteorological disaster risk management*

Adjust the management and technical layout of "upper and lower general" and establish a multi-level meteorological disaster risk management structure of "national, local and community" to adapt to the actual work of different levels. According to the level of disaster and the degree of impact and the actual situation of the disaster-body, in the various aspects of business technology support, early warning release, emergency response, and disaster management, the division of labor at each level is refined to achieve hierarchical operation. Achieve synergy between national and local planning, local governments at all levels, government and community organizations. Maximize the effectiveness of technologies, personnel, and resources at all levels, and ensure that local government can effectively carry out work based on local disaster levels and exposure and vulnerability of disaster-bearing bodies, and improve disaster prevention and risk management capabilities.

5.3 *Consolidate the classification implementation and linkage mechanism of meteorological disaster risk management*

China has a vast territory, diverse terrain and complex climate types. There are great differences in the types, impacts and disaster losses of meteorological disasters and disaster-bearing bodies. It is necessary to strengthen the investigation and zoning of meteorological disaster risks, and integrate the information data on various disaster-bearing bodies into the basic databases for disaster prevention. It would then be possible to establish a forecast and early warning business system based on meteorological disaster risks, so that the information on meteorological forecasting and early warning services and decision-making services are improved. According to the characteristics of the hazard factors and the law of transformation, research on the mechanism of transmission of different types of meteorological disasters and the linkage management of disaster prevention, include "drought-flooding", "drought-high temperature", extreme weather and climate events, secondary and derived meteorological disasters linkages, etc. According to the characteristics of meteorological disasters and key areas affected by localities, establish the evaluation criteria and countermeasure analysis systems for the differences in disaster risk management of local governments at all levels, and promote the improvement of meteorological disaster risk management.

5.4 *Effectively strengthen the comprehensive protection of meteorological disaster risk management and its accountability mechanism*

From the aspects of capital, science and technology, information, and law, we will improve the comprehensive protection of meteorological disaster risk management. Engineering and non-engineering disaster risk management measures highlight the key role of socialization and marketization in resource allocation to ensure that these measures can be drawn, used and won at critical times. Strengthen social supervision, and combine the principle of rigid accountability and flexible accountability to promote the gradual resolution of the basic issues, such as the subject, object, method and timing of the accountability of meteorological disaster risk management. While providing guarantees for the green and sustainable development of China's economy and society, we will protect national interests and the safety and well-being of people's lives and property, effectively adapt to and mitigate the adverse effects of climate change, and reduce the risks and losses of meteorological disasters.

In short, on the basis of improving institutional and mechanisms, strengthening basic research and applying basic research, and improving core business technologies, it is necessary to closely integrate the needs of ecological civilization construction according to the actual and trend of social and economic development in various regions. Based on the interrelationship between adaptive measures and meteorological disaster risks, the key links such as the complex subject, stratified operation, classification implementation and comprehensive support are taken as the entry point. It is important to establish and improve the overall framework and implementation mechanism of meteorological disaster risk management, expand and enhance the theoretical vision, and explanatory capacity of China's disaster prevention and analysis research, as well as the level of response to climate change.

ACKNOWLEDGMENTS

This study is supported by the National Key Research and Development Program of China (NO. 2018YFA0606302).

REFERENCES

China Meteorological Administration. 2015. Meteorological Disaster Risk Management Business Construction (2015–2016) Implementation Plan, 29–48.
Maosong Li, Shuangdi Pan. 2014. Meteorological Disaster Risk Management. Meteorological Press, 85–136.

Risk Analysis Based on Data and Crisis Response Beyond Knowledge – Huang & Nivolianitou (eds)
© 2020 Taylor & Francis Group, London, ISBN 978-0-367-25146-8

Quantitative analysis and risk assessment of terrorist attacks

Mei Hong* & Ren Zhang
Institute of Meteorology and Oceanography, National University of Defense Technology, Nanjing, China

Jingjing Ge
The troop of PLA, Nanjing, China

Ming Li & Liang Zhao
Institute of Meteorology and Oceanography, National University of Defense Technology, Nanjing, China

ABSTRACT: Deep analysis of the data related to terrorist attacks will help people to have deeper understanding of terrorism and provide valuable information supporting for counter-terrorism and anti-terrorism. Based on the global terrorism database, this paper introduces artificial intelligence theory, such as machine learning to conduct data mining and quantitative analysis of terrorist attack records. A quantitative classification model based on D-S evidence theory and adaptive Gauss Cloud Transform (AGCT) algorithm is constructed. Firstly, attribute dimensionality is reduced with expert knowledge and correlation analysis, and significant hazard assessment indicators are selected. Finally, based on the AGCT algorithm, the hazard index is classified to determine the hazard level of the event.

Keywords: data mining, terrorist attacks risk, AGCT algorithm

1 INTRODUCTION

Terrorism is a complex political and social phenomenon (Hu, 2001). Terrorist attacks have become one of the greatest threats to international peace and security. Moreover, it is a global focus of attention.

Both domestic and transnational terrorism pose challenges to governments and citizens. Only studying terrorist attacks, understanding the characteristics of terrorist organizations and mastering the motives of terrorist attacks can formulate effective counter-terrorism strategies. At present, Chinese scholars have done a lot of research work (Hao, 2002; Wang, 2002; Zhu, 2008; Li, 2009). However, the research mainly focuses on the qualitative description of terrorist attacks, lacking theoretical explanation basis. Moreover, the empirical research and the establishment of terrorist database are still blank. In contrast, foreign scholars have established a global terrorism database or tracking system (Rand Corporation, 2008; Lafree and Dugan, 2007; Start, 2012), and have made a lot of research achievements in the characteristics, root causes, prediction and evaluation models of terrorism and coping strategies (John et al., 2004; Clauset and Young, 2005; Ezell et al., 2010).

Unlike natural disasters, terrorist attacks have the characteristics of premeditation and intellectualization, changing with time and space. Therefore, in the face of terrorist attacks, besides strengthening the collection of intelligence work, we must also collect reliable historical data in order to obtain scientific conclusions and make wise decisions. At present, the international community has begun to attach importance to the establishment of terrorist attacks database, and the

---

*Corresponding author: flowerrainhm@126.com

establishment of the Global Terrorist Database (GTD). According to the record information of the GTD, with considering of the relevant factors affecting the harm of terrorist attacks, a quantitative classification model is established. The events in GTD are divided into five levels according to the degree of harm, and the risk levels of terrorist attacks are identified.

It can be seen from the analysis that the primary premise of the classification of the hazard assessment is to establish the index system of the hazard assessment of terrorist attacks, and the key to the construction of the system is the selection of the evaluation index. GTD records of every terrorist attack contain 126 attributes. On the one hand, attributes are interrelated, overlapping and influencing each other, which will lead to information redundancy. On the other hand, too many attributes will lead to a higher dimension of mathematical model, which is not conducive to modeling and analyzing. In this paper, we will study quantitative analysis of machine learning algorithm in terrorist incident records, and provide valuable information support for anti-terrorism.

2 ATTRIBUTE DIMENSION REDUCTIONS AND DATA QUANTIZATION

Records of terrorist attacks around the world in the Global Terrorist Database (GTD) are selected. (Notes on variables are excerpted from the database description document. The original text can be downloaded from http://www.start.umd.edu/gtd/)

Harm of terrorist attacks is related to many factors, such as casualties, number of abducted person, economic losses, weapons used, duration and so on, but there are correlations among different factors. If all the attributes listed in GTD are taken as evaluation indicators, the redundancy of index information will be too high. Based on expert knowledge, this paper synthesizes the relevant literature and screens out 10 influencing factors, including duration, occurrence area, selection criteria, success, attack type, weapon type, death number, injured number, property loss and kidnapped number, as an evaluation index of the harm of a terrorist attack. Among them, the occurrence areas of terrorist attacks mainly concentrate on cities, towns or urban areas.

Table 1. Hazardous assessment index system of terrorist attacks.

	Evaluation index	Index meaning	Quantized fraction
1	Duration	More than 24 hours	100
		Less than 24 hours	50
2	Occurrence area	City	100
		Town	80
		Non town	60
3	Attack standard	Satisfy three criteria at the same time	100
		Satisfy two criteria at the same time	80
		Meeting only one criterion	60
4	Success or not	Success	100
		fail	50
5	Type of attack	Three type	100
		Two type	80
		One type	60
6	Weapon type	Type 1-6	
		Type 7-13	
7	death toll	\	Raw Recorded Data
8	Number of injured persons	\	Raw Recorded Data
9	Number of abductees	\	Raw Recorded Data
10	property loss	Catastrophic	100
		Significant	80
		Smaller	60
		Unknown	40

Then Delphi method (Dewar and Friel, 2013) is used to quantify the qualitative indicators (such as duration, occurrence area and selection criteria) in the index system, and then data is to be standardized. The standard deviation standardization is adopted in this paper, and the specific calculation process is not described here. Table 1 shows the hazard assessment index system and quantitative scores.

3 QUANTITATIVE CLASSIFICATION: A CLASSIFICATION MODEL BASED ON ADAPTIVE GAUSS CLOUD TRANSFORM

The traditional hierarchical discretization method (Macri et al., 2003) is interval equal interval partition method, which is a hard partition either one or the other. It does not take into account the fuzziness and randomness of the index data and other uncertainties. It will easily lead to classification error of the index at the junction. Cloud model is a better uncertain knowledge representation model, which combines fuzziness and randomness based on classical probability theory and fuzzy mathematics. In this paper, adaptive Gauss Cloud Transform (AGCT) algorithm is used to softly divide the continuous index into discrete levels.

3.1 *Introduction of theory*

Adaptive Gauss Cloud Transform: Cloud model was first proposed by Academician Li Deyi in 1995. Since it was proposed, it has been successfully applied in intelligent fields, such as natural language processing, data mining and decision analysis. Academia defines cloud model as follows: suppose U is a quantitative domain expressed by precise numerical value and C is a qualitative concept on it. If the quantitative value $x \in U$ is a random realization of qualitative concepts, the determinacy of pairs $\mu(x) \in [0,1]$ is a random number with stable tendency (Liu, 2015):

$$\mu : U \rightarrow [0, 1], \ \forall x \in U, \ x \rightarrow \mu(x)$$

The distribution U in the universe x is called cloud, and each x is called a cloud drop, which is expressed as drop $(x, \mu(x))$. Clouds can be represented as a concept by three numerical characteristics, including expectation Ex, entropy En and hyperentropy He. It is $C(Ex, En, He)$.

Gauss mixture model (GMM) is an important method in probability statistics. It can transform the whole probability distribution function in the problem domain into the superposition of multiple Gauss distributions. Gauss Cloud Transform (GCT) combines cloud model with GMM to provide a method for discretizing continuous variables. Adaptive Gauss Cloud Transform (AGCT) is an improvement of GCT. It does not need to specify the number of concepts beforehand. Starting from the actual data samples, it can automatically form multiple conceptual clouds that conform to the human cognitive law, thus automatically dividing the data distribution of the problem domain into different concepts. AGCT takes into account the clustering principle of "strong intra-class relationship and weak inter-class relationship", and provides a method for the soft partition of continuous quantitative data.

Input: Data Sample Set X, Conceptual Ambiguity α
Output: m Gauss Cloud

Step 1: Statistically calculate the frequency distribution p of X and the number M of peaks as the initial value of the number of concepts;
Step 2: Data sets X are clustered into several Gauss clouds by the heuristic Gauss Cloud Transform (H-GCT).
Step 3: Adjust the number of concepts according to the ambiguity of each Gauss cloud.
Step 4: Cyclic Step 2 - 3 to generate a Gaussian cloud with ambiguity less than or equal to α.

3.2 *Model establishment and calculation results*

The steps of AGCT algorithm are specified as follows: using this algorithm to automatically generate cognitive levels, and then outputting the digital features of the corresponding concept cloud to design the cloud generator, converting the continuous quantitative data into discrete qualitative concepts, to achieve the classification of terrorist attack risk index.

The hazard index is divided into five levels. Figure 1 is the standard level cloud obtained from the hazard index softened. The discrete expression of the index level is shown in Table 2. The smaller the discrete value is, the higher the risk level is.

The hazard index of events listed in the problem is put into the cloud model, and the determination degree of each cloud level is calculated. According to the principle of maximum certainty, the risk levels of typical events are distinguished as shown in Table 3.

Based on the established quantitative classification model, hazards of terrorist attacks in 12 regions of the world in the past three years are evaluated. The hazard classification based on GIS platform is shown in Figure 2. According to the map, the risk of terrorist attacks in the Middle East, North Africa and South Asia belongs to the first level, with the highest level of risk; Western and Eastern Europe belongs to the second level; South-East Asia and sub-

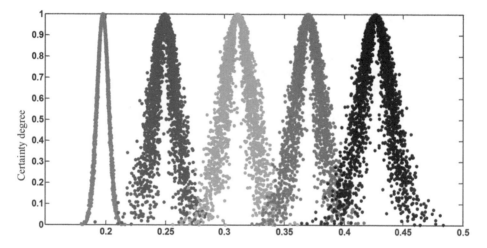

Figure 1.   Hierarchical nephogram of hazard index partition.

Table 2.   Cloud model representation for risk classification.

Indicators	Cloud Model Representation	Risk Level
Risk Index of Terrorist Attacks	$C1(0.1974, 0.0047, 0.0002)$	5
	$C2(0.2491, 0.0091, 0.0032)$	4
	$C3(0.3109, 0.0114, 0.0033)$	3
	$C4(0.3697, 0.0103, 0.0031)$	2
	$C5(0.4262, 0.0127, 0.0038)$	1

Table 3.   Risk levels of typical events.

Event Number	Hazard Index	Grade	Event Number	Hazard Index	Grade
200108110012	0.355	2	201411070002	0.205	5
200511180002	0.287	4	201412160005	0.181	5
200901170021	0.369	2	201705080012	0.234	4

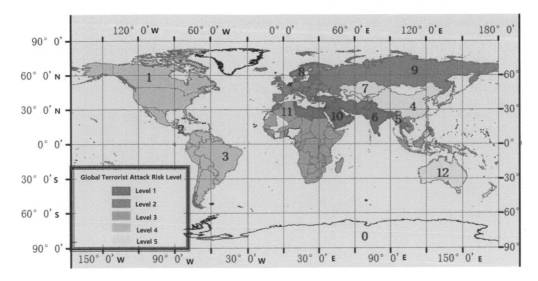

Figure 2. Global terrorist attack risk level division for the last three year.

Saharan Africa belong to the third level; America belongs to the fourth level; East Asia, Central Asia and Oceania belong to the fifth level, with the lowest level of risk.

4 CONCLUSIONS

Based on the quantitative and complete index data, this paper constructs the terrorist incident hazard index, quantifies and classifies the incident hazard with cloud model according to the fusion index. The adaptive Gauss cloud transform algorithm is used to quantify and classify the fusion indicators, which effectively handles the uncertainties such as the fuzziness and randomness of the index data and avoids the hard one or the other, with fuzziness of hierarchical boundaries taken into account.

ACKNOWLEDGMENTS

This study is supported by the Chinese National Natural Science Fund (No. 41875061; No. 41775165), the Chinese National Natural Science Fund (BK20161464) of Jiangsu Province and the Research Program of National Defense University of Science and Technology(ZK18-03-48).

REFERENCES

Clauset, A. and Young, M. 2005. Scale invariance in global terrorism. *Eprint Arxiv Physics*.
Dewar, J.A. and Friel J.A. 2013. Delphi method. *Encyclopedia of Operations Research & Management Science*: 406–408.
Garrick, B.J., Hall. J.E., Kilger, M. et al. 2004. Confronting the risks of terrorism: making the right decisions. *Reliability Engineering & System Safety* 86(2): 129–176.
Hao S., 2002. Ethno-separatism and terrorism. *Ethno-National Studies* 1(7).
Hu, L. 2001. *Terrorism and Countermeasures in Modern World*. Beijing: The Eastern Publishing Press.
Lafree, G. and Dugan, L. 2007. Introducing the global terrorism database. *Terrorism and Political Violence* 19(2): 181–204.
Li, W. 2009. *International Terrorism and Counter-terrorism Yearbook*. Beijing, Current Affairs Press.
Liu, Y.C. 2015. Adaptive concept abstraction method on multi-granularity—Gaussian cloud transformation. *Computer Engineering & Applications* 51(9): 1–8.
Macri, M., De, S. and Shephard, M.S. 2003. Hierarchical tree-based discretization for the method of finite spheres. *Computers & Structures* 81(8-11): 789–803.

Rand Corporation. 2008. Rand Database of Worldwide Terrorism Incidents. http://www.rand.org/nsrd/projects/terrorism-incidents/about.html

START, 2012. National Consortium for the Study of Terrorism and Responses to Terrorism (START). Global Terrorism Database [Data file]. Retrieved from http://www.start.umd.edu/gtd.

Wang, Y. 2002. *Roots of Terrorism*. Beijing: Social Sciences Academic Press.

Ezell, B.C., Bennett, S.P., Winterfeldt, D.V. et al. 2010. Probabilistic risk analysis and terrorism risk. *Risk Analysis* 30(4): 575–589.

Zhu, S. 2008. Theoretical analysis on terrorism motivation. *Journal of University of International Relations* 5: 1–5.

Risk Analysis Based on Data and Crisis Response Beyond Knowledge – Huang & Nivolianitou (eds)
© 2020 Taylor & Francis Group, London, ISBN 978-0-367-25146-8

Spatial and temporal patterns, change and risk analysis of meteorological disasters in East China

Jun Shi*, Linli Cui & Zhongping Shen
Shanghai Ecological Forecasting and Remote Sensing Center, Shanghai Meteorological Bureau, Shanghai, China
Shanghai Key Laboratory of Meteorology and Health, Shanghai, China

Hanwei Yang & Jianguo Tan
Shanghai Climate Center, Shanghai Meteorological Bureau, Shanghai, China

ABSTRACT: Based on the statistical data of meteorological disasters in six provinces and Shanghai city in East China, the overall patterns, spatial and temporal changes of meteorological disasters and the risks faced by different regions of East China were analyzed. The results indicated that rainstorm-induced flood (landslide and mud-rock flow) disaster affected the largest area and total failure area of crops, the largest population, and caused the greatest number of houses to collapse during 2004–2015. Strong convection weather (gale, hail, thunder, and lightning) disaster resulted in the highest number of deaths, and typhoon disaster caused the greatest direct economic losses. Over the past 12 years, there have been significant decreasing trends in the affected area and the total failure area of crops, the number of people affected and deaths by meteorological disasters in East China, while the direct economic loss caused by meteorological disasters showed no significant trends. The risk of meteorological disaster had obvious regional differences in East China. In the northern part of East China, the risk of drought and strong convection weather disaster was higher, but in the southern area there was a higher risk of rainstorm-induced flood disaster. In the eastern coastal areas of East China, the risk of typhoon disaster was higher. Disaster risk analysis has an important significance for disaster prevention and reduction, risk management and crisis response.

Keywords: meteorological disaster, spatial and temporal change, risk analysis, East China

1 INTRODUCTION

Meteorological disaster is one of the deadliest and costliest natural disasters in the world (Liu and Yan 2011; Guan et al. 2015). In recent years, many countries are experiencing more frequent meteorological disasters (such as floods, droughts, hurricanes, extreme heat or cold, snowstorms, etc.) as consequences of climate change and the intensification of human activities (Stott 2016; Witze 2018). These disasters seriously threaten people's lives, agricultural production, water resources and ecosystems (Lesk et al., 2016). China is one of the few countries in the world where various types of natural disasters occur frequently, causing severe losses. Wu et al. (2014) showed that meteorological disasters during 1994–2013 accounted for 55% of the deaths and 87% of the direct economic losses caused by natural hazards in China, and if weather-related secondary geological disasters, i.e., landslide and mud-rock flow, were included this would increase to 81% for deaths and 89% for economic losses. Thus, research on meteorological disasters is of great significance

---

*Corresponding author: sunrainlucky@qq.com

to human life and property and national economic construction and ecological security in China (Cui et al. 2017; 2018; Yu and Tang 2017).

Due to its location in the southeastern edge of Eurasian continent, East China is significantly affected by the East Asian monsoon, with prominent interdecadal climate changes and seasonal climate variations (Shi et al. 2018). The probability and intensity of extreme weather and climatic events such as high temperature, rainstorm and typhoon in East China will also increase with global warming (IPCC, 2013). In addition, East China is one of the most densely populated and economically concentrated regions in China. Meteorological disasters and the secondary disasters caused by various high-impact weather events have caused large economic losses and a wide range of social impacts and public concerns (Shi et al. 2012; 2016; He and Zhai 2015). Trend analysis of meteorological disasters is essential for disaster prevention and mitigation planning, and also for climate change adaptation (Wu et al. 2014). Therefore, in this study we analyzed the overall characteristics, changing trends and disaster risks faced by different regions of East China, with the support of the statistical data from various meteorological disasters in each province (city) of East China during 2004–2015, to provide reference for regional disaster prevention and reduction and sustainable development strategy (Huang 2012).

2 DATA AND METHODS

2.1 *Data*

Annual total losses of all meteorological disasters in each province (city) of East China, and annual losses from rainstorm-induced flood (landslide and mud-rock flow), drought, strong convection weather (gale, hail and thunder, and lightning), typhoon and low temperatures and snow disasters in each province (city) were used in this paper. These data came from the *Yearbook of Meteorological Disasters in China* (China Meteorological Administration 2005–2016), and data of disaster loss mainly included the affected area of crops, the area of total crop failure, affected population and death population, number of houses that collapsed and were damaged, the direct economic losses, etc. There were no data about the number of deaths caused by drought disaster.

2.2 *Methods*

In this study, East China includes Shandong, Jiangsu, Anhui, Zhejiang, Fujian and Jiangxi province and Shanghai city. The statistical data of meteorological disaster loss in the whole of East China from 2004 to 2015 were obtained by directly summarizing the corresponding disaster loss data of each province (city). The overall characteristics and changes of meteorological disasters in East China over 12 years were analyzed using statistical methods such as linear trend analysis, and the differences among different disaster types and provinces (cities) were compared and analyzed. According to the area of each province (city) in East China, the meteorological disasters per unit area of each province was calculated and was used as the risk index of meteorological disaster. The direct economic losses were analyzed according to the statistical value annually and were not adjusted by multiple price indexes according to the base year.

3 RESULTS

3.1 *The overall characteristics of meteorological disasters in East China*

From 2004 to 2015, meteorological disasters had affected an averaged area of 6.52 million hectares of crops per year in all provinces in East China, of which the area with total crop failure was 551,133 hectares per year. The affected population from meteorological disasters was 87.78 million people per year, with 398 deaths. Furthermore, meteorological disasters caused 259,000 houses to collapse and 726,000 houses were damaged, with a direct economic loss of 78 billion Yuan annually (Table 1). Rainstorm-induced flood (landslide and mud-rock flow) disaster had caused the largest affected area and total failure area of crops, and also the largest

Table 1. Mean annual loss from meteorological disasters in East China, 2004–2015.

Type of meteoro-logical disasters	Area of crops (10⁴ hm²)		Population		Number of houses (10⁴)		Direct eco-nomic loss (10¹⁰ Yuan)
	Affected area	Total failure area	Affected (10⁴)	Deaths)	Collapsed	Damaged	
Rainstorm-induced flood	227.8	27.9	2535.1	78.3	12.8	30.8	26.43
Drought	180.6	10.0	2154.2				7.61
Strong convection weather	58.5	4.8	1179.6	144.3	4.5	35.1	6.56
Typhoon	100.8	4.4	1800.3	115.6	6.9		28.87
Low temperatures and snow	78.4	6.4	1037.0	6.3	1.7	6.7	8.41
Other	6.1	1.6	71.6	53.6	0	0	0.11
Total	652.2	55.1	8777.8	398.1	25.9	72.6	77.99

number of people affected and houses collapsed during 2004–2015, accounting for 35%, 51%, 29% and 49% of the total losses from meteorological disasters, respectively. Strong convection weather disasters (gale, hail and thunder, and lightning) had resulted in the largest number of deaths and more houses damaged. Typhoon had caused the greatest direct economic losses and more deaths, and drought also caused more affected population and affected area of crops.

3.2 *The interannual variations of meteorological disasters in East China*

Over the past 12 years, there have been significant decreasing trends in the affected area and the total failure area of crops, at a rate of 391,500 hectares and 46,100 hectares per year, respectively (Figure 1a). Meanwhile, the number of people affected in East China and the number of deaths by meteorological disasters also decreased significantly at a rate of 5.28 million people per year and 63 people per year, respectively (Figure 1b). The direct economic loss caused by meteoro-logical disasters showed no significant trend, with greater economic loss of 123.75 billion Yuan in 2013 and a lower loss of 30.10 billion Yuan in 2014 (figure omitted).

During 2004–2015, the total affected area and the total failure area of crops caused by rain-storm-induced flood (landslide and mud-rock flow) disaster in East China had decreased signifi-cantly, at an annual rate of 105,600 and 27,900 hectares, respectively, but the trend change was not statistically significant (Figure 2a). The affected population and the deaths caused by rain-storm-induced flood disaster also decreased at a rate of 0.66 million people and 3.29 people per year, respectively, but the trend change was also not statistically significant (Figure 2b). The number of houses that collapsed and were damaged because of rainstorm-induced flood disaster

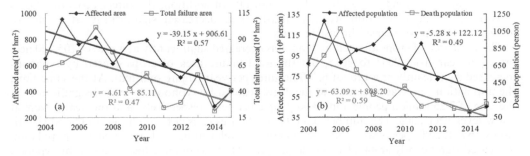

Figure 1. Trends in the losses caused by meteorological disasters in East China, 2004–2015. a) crops; b) population.

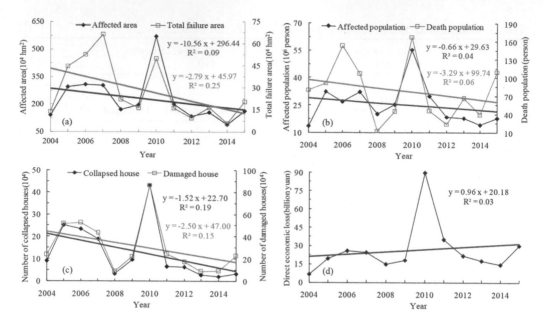

Figure 2.   Trends in the loss caused by rainstorm-induced flood (landslide and mud-rock flow) disasters in East China from 2004 to 2015. a) Crops; b) Population; c) Houses; d) Direct economic loss.

had fallen by 15,200 and 25,000 per year over the past 12 years, respectively, although the trend was not statistically significant (Figure 2c). The direct economic losses from rainstorm-induced flood disaster had increased slightly, although the trend was not statistically significant (Figure 2d).

The total affected area and the total failure area of crops caused by drought disaster in East China showed a decreasing trend at a rate of 6,600 and 1,700 hectares per year, respectively, during the period from 2004 to 2015, although the trend was not statistically significant (figure omitted). Meanwhile, the population affected by drought increased at a rate of 0.63 million people per year, and the population with poor drinking water decreased at the rate of 79,200 people per year, but both trends were not significant. Direct economic losses from drought disaster increased at a rate of 483 million Yuan per year over the past 12 years.

From 2004 to 2015, both the total affected area and the total failure area of crops caused by strong convection weather (gale, hail and thunder, and lightning) disaster in East China showed decreasing trends, at a rate of 72,100 hectares and 8,500 hectares per year, respectively (figure omitted). Meanwhile, the population that was affected and the deaths caused by strong convection weather disaster decreased significantly at the rate of 1.39 million people and 19.08 people per year, respectively. In the past 12 years, the number of houses that collapsed and were damaged by strong convection weathers decreased by 11,900 and 72,700, respectively, per year, and the trend was statistically significant. In addition, the direct economic losses caused by strong convection weather disaster had decreased significantly at a rate of 435 million Yuan per year in the last 12 years.

The total area of crops affected by typhoons decreased significantly at a rate of 105,100 hectares per year from 2004 to 2015, but the total failure area of crops caused by typhoons increased significantly at a rate of 6,300 hectares per year (figure omitted). At the same time, the population that was affected and the deaths caused by typhoon decreased significantly at the rate of 2.25 million people and 33.66 people per year, respectively, and the deaths were at a low level after 2008. The number of collapsed houses caused by typhoon decreased significantly by 18,300 per year, and the emergency relocation and resettlement population decreased significantly at 442,900 people per year. Direct economic losses caused by typhoon disaster had fluctuated greatly in the past 12 years, with the overall increasing rate of 84 million Yuan per year, but the trend change was not statistically significant.

During 2004–2015, both the total affected area and the total failure area of crops caused by low temperatures and snow disasters in East China showed a decreasing trend, at a rate of 76,600 hectares per year and 9,900 hectares per year, respectively (figure omitted). Meanwhile, the affected population and the deaths caused by low temperatures and snow disasters had decreased by 1.29 million people per year and 0.51 people per year, respectively, among which the trend of the affected population was statistically significant. The number of houses that collapsed and were damaged by low temperatures and snow disasters decreased at the rate of 27,000 and 10,500, respectively, in the past 12 years, but the trend change was not statistically significant. In addition, the direct economic losses caused by low temperatures and snow disasters had decreased by 567 million Yuan per year in East China, of which the most disastrous event occurred in 2008.

3.3 The risk of meteorological disasters in East China

According to the loss of meteorological disaster per unit area, the risk of meteorological disaster was preliminarily identified, and the results showed that the risk of meteorological disasters had obvious regional differences in East China (Table 2). As far as crops were concerned, Anhui, Jiangxi, Shandong and Jiangsu were generally at high risk of meteorological disasters, while Shanghai and Fujian were at low risk. In terms of the death population, those in Shanghai, Zhejiang and Fujian were at high risk of meteorological disasters, while those in Shandong had the lowest risk. Comparing the direct economic loss per unit area, Zhejiang province had the highest risk of meteorological disasters, followed by Fujian, Jiangxi, Anhui and Shandong, while Shanghai had the lowest risk of meteorological disasters.

For different types of meteorological disasters, the risk level in different regions of East China was also greatly different. In the northern part of East China, including Shandong, Jiangsu and Anhui, the risk of drought and severe convection weathers was higher, but in the southern part of East China, including southern Anhui, Jiangxi, Fujian and Zhejiang, there was a higher risk of rainstorm-induced flood disaster. In the eastern coastal areas of East China, including Fujian, Zhejiang and Shanghai, the risk of typhoon disaster was higher. The risk of low temperatures and snow disasters was higher in Shandong, Anhui, Jiangxi and Zhejiang.

4 CONCLUSION

In East China during 2004–2015, rainstorm-induced flood (landslide and mud-rock flow) disaster was the most important meteorological disaster, which caused the largest affected area and total failure area of crops and had largest affected population and the most collapsed houses. Strong convection weather (gale, hail and thunder, and lightning) disaster resulted in the largest number of deaths. Typhoon caused the greatest direct economic losses. Drought and low temperatures and snow disasters also had great influence on the social economy of East China. Over the past 12 years, there were significant decreasing trends in the affected

Table 2. Annual losses per 10,000 km^2 caused by meteorological disasters in each province (city) of East China, 2004–2015.

Province/ city	Area of crops (10^4 hm^2)		Population		Direct economic loss (10^9 Yuan)
	Affected area	Total failure area	Affected (10^4)	Deaths	
Shanghai	3.24	0.14	37.20	9.79	5.66
Jiangsu	9.37	0.59	94.93	4.69	5.94
Zhejiang	6.53	0.52	141.67	7.26	22.52
Anhui	10.87	1.08	154.54	4.73	8.06
Fujian	3.35	0.25	53.78	7.86	8.95
Jiangxi	7.19	0.81	100.93	4.96	8.47
Shandong	11.37	0.78	119.54	1.66	7.99

area and the total failure area of crops. Meanwhile, the number of people affected and deaths by meteorological disasters had also decreased significantly, but the direct economic loss caused by meteorological disasters showed no significant trend.

The risk of meteorological disaster had obvious regional differences in East China. Crops in Anhui, Jiangxi, Shandong and Jiangsu were at high risk of meteorological disasters, and people in Shanghai, Zhejiang and Fujian were at high risk according to the loss of meteorological disaster per unit area. For different types of meteorological disasters, the risk level in different regions of East China was also greatly different. In the northern part of East China, the risk of drought and severe convection weathers was higher, but in the southern part of East China, there was a higher risk of rainstorm-induced flood disaster. In the eastern coastal areas of East China, the risk of typhoon disaster was higher, and the risk of low temperatures and snow disasters was higher in Shandong, Anhui, Jiangxi and Zhejiang.

ACKNOWLEDGMENTS

The authors give thanks for funding provided by the National Natural Science Foundation of China (No. 41571044), the Open Research Fund Program of the State Key Laboratory of Subtropical Silviculture (No. KF2017-5), and by the Climate Change Special Fund of the China Meteorological Administration (No. CCSF201922).

REFERENCES

Chen, Y.F. and Gao, G. 2010. An analysis to losses caused by meteorological disasters in China during 1989–2008. *Meteorological Monthly* 36(2): 76–80.

China Meteorological Administration (CMA). 2005–2016. *Yearbook of Meteorological Disasters in China*. Beijing: China Meteorological Press.

Cui, L.L., Shi, J., Du, H.Q. and Wen, K.M. 2017. Characteristics and trends of climatic extremes in China during 1959–2014. *Journal of Tropical Meteorology* 22(4): 368–379.

Cui, L.L., Shi, J., Ma, Y. and Liu, X.C. 2018. Variations of the thermal growing season during the period 1961–2015 in northern China. *Journal of Arid Land* 10(2): 264–276.

Guan, Y.H., Zheng, F.L., Zhang, P. and Qin, C. 2015. Spatial and temporal changes of meteorological disasters in China during 1950–2013. *Natural Hazards* 75: 2607–2623.

He, Z.Y. and Zhai, G.F. 2015. Spatial effect on public risk perception of natural disaster: a comparative study in East Asia. *Journal of Risk Analysis and Crisis Response* 5(3): 161–168.

Huang, C.F. 2012. *Risk Analysis and Management of Natural Disaster*. Beijing: Science Press. (in Chinese).

IPCC. 2013. *Climate Change 2013: The Physical Science Basis*. Cambridge and New York: Cambridge University Press.

Lesk, C., Rowhani, P. and Ramankutty, N. 2016. Influence of extreme weather disasters on global crop production. *Nature* 529: 84–87.

Liu, L.F., Jiang, K.L. and Zhou, S.X. 2017. Analysis of Characteristics of meteorological disasters from 2011 to 2015 in China. *Journal of Hengyang Normal University* 38(3): 113–118.

Liu, T. and Yan, T.C. 2011. Main meteorological disasters in China and their economic losses. *Journal of Natural Disasters* 20(2): 90–95 (in Chinese).

Shi, J. and Cui, L. 2012. Characteristics of high impact weather and meteorological disaster in Shanghai, China. *Natural Hazards* 60: 951–969.

Shi, J., Wei, P.P., Cui, L.L. and Zhang, B.W. 2018. Spatio-temporal characteristics of extreme precipitation in East China from 1961 to 2015. *Meteorologische Zeitschrift* 27(5): 377–390.

Shi, J., Wen, K.M. and Cui, L. 2016. Patterns and trends of high-impact weather in China during 1959–2014. *Natural Hazards and Earth System Sciences* 16: 855–869.

Stott, P. 2016. How climate change affects extreme weather events. *Science* 352 (6293): 1517–1518.

Witze, A. 2018. Why extreme rains are gaining strength as the climate warms. *Nature* 563, 458–460.

Wu, J.D., Fu, Y., Zhang, J. and Li, N. 2014. Meteorological disaster trend analysis in China: 1949–2013. *Journal of Natural Resources* 29(9): 1520–1530.

Yu, X. and Tang, Y.D. 2017. A critical review on the economics of disasters. *Journal of Risk Analysis and Crisis Response* 7(1): 27–36.

Risk Analysis Based on Data and Crisis Response Beyond Knowledge – Huang & Nivolianitou (eds)
© 2020 Taylor & Francis Group, London, ISBN 978-0-367-25146-8

A new economic loss assessment system for urban severe rainfall and flooding disasters

Xianhua Wu*, Yun Kuai & Ji Guo
School of Economics and Management, Shanghai Maritime University, Shanghai, China
Collaborative Innovation Center on Climate and Meteorological Disasters, Nanjing University of Information Science & Technology, Nanjing, China

Guo Wei
Department of Mathematics & Computer Science, University of North Carolina at Pembroke, Pembroke, NC, USA

ABSTRACT: A new economic losses evaluation information system has been developed for monitoring severe rainfall and flooding disasters in cities. The data mining method, econometric regression model and input-output model are implemented in the system, on the basis of multi-source data such as hourly rainfall, geographical conditions, historical and real-time disaster information, socioeconomic data, and defense countermeasure. Combined with the weather forecast information, this system can report the real-time direct and indirect economic losses incurred by urban heavy rainfall and flooding disasters, automatically generate defense countermeasure reports for typical rainstorm and flooding points, and provide the spatial distribution of disasters. Finally, the system is conducive to improving the ability to manage disaster emergencies and eventually reducing the economic losses from the disaster.

Keywords: rainstorm, rainfall and flooding, disasters, economic loss evaluation, information system

1 INTRODUCTION

In recent years, with the acceleration of global warming and urbanization, together with the relatively backward planning of urban water systems and construction of underground pipe networks, some developing countries have frequently experienced adverse effects of heavy rains and flooding in urban areas (Huang. 2012; Huang and Huang. 2018). For instance, a heavy rainstorm in Beijing, China, on July 21 2012, killed 79 people, affected 1.602 million people and destroyed 10,660 houses, causing an economic loss of 11.64 billion yuan. Rainfall accompanies fatal flooding, leading to not only casualties but also huge economic losses, which has aroused wide spread concerns from all sectors of society. Therefore, it has become a difficult issue for the government and scholars, and the economic loss of such natural disasters as rainstorms should be quickly evaluated and effective countermeasures of emergency management should be advanced (Wu et al. 2016).

In mainland China, however, there is a lack of research on and methods for the comprehensive loss assessment of natural disasters, and a corresponding lack of evaluation systems and software. This seriously restricts the work of disaster emergency and risk management and is worthy of further studies and discussion. There are two innovations in this new economic loss assessment system. The first innovation is taking rainstorm and flooding points as the research objects, whose geographic scopes are determined by investigators. A survey questionnaire is then used to investigate the economic value of different types of properties. The second innovation is trying to integrate the data of different sources and different formats for evaluating economic losses and generating defense countermeasures. There are three kinds of data: socio-economic property value, precipitation data, and the text data of defense strategy.

2 CONSTRUCTION PLANS

2.1 *System architecture*

The overall system architecture diagram is illustrated in Figure 1, composed of data layer, service layer and application layer. The system uses the web development framework of MapGIS IGServer as the development platform, and combines with the logic process of specific business to build the application system.

(1) Data layer: This provides the system foundation for data support. The system data consists of two parts: the basic map data, and the corresponding business data.

(2) Service layer: Located on the data layer, this layer uses the IGServer MapGIS development platform as an application support, and accesses system data through the MapGIS data engine and relational database engine, respectively. The service layer provides GIS data services and functional services, releases management through the service management module and achieves business functions when called by the application system. With the REST services, the communication on the GIS platform and the GIS functions can be realized.

(3) Application layer: Based on data support and the GIS development platform, an economic loss evaluation system for urban rainstorm and flood disasters has been constructed, including social and economic data management, data management of historical disasters, data management of disaster countermeasures, economic loss assessment, generation of an assessment report, user management, system management and other functional modules.

This is illustrated in Figure 1.

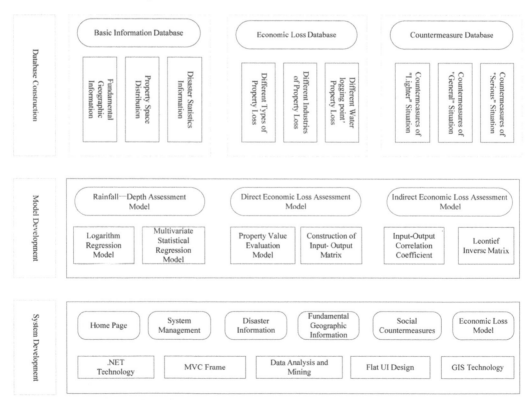

Figure 1.   Overall system architecture diagram.

2.2 Data organization

(1) Map data: This system is composed of two types of data format. The first is the base maps for the system, and directly uses the third-party free map data provided by the Map World Corporation. With web map technology, it provides users with smooth map roaming, seamless zoom, fast graded display and other online operating experiences. The second is vector data, which is stored in the form of MapGIS local data sources, including administrative divisions and a variety of infrastructure data. Through the layer control panel, it is possible to achieve free setting of the display and hiding all factors, and browsing both the spatial distribution and related attribute information of administrative divisions, places, roads, infrastructure, water systems, water conservancy facilities, and other surface features. Incorporating a variety of information from business databases and map data into the unified base map of geographic framework, the system can form a complete database of geographic frameworks, thus providing support for business analysis.

(2) Business data: This mainly consists of the information on historical rainstorm and flood disasters, the distribution of population density, the distribution of GDP density, community property value, individual buildings, defensive countermeasures, etc. As most of the data contain spatial information and attribute information, for a query in the application system the spatial location and attribute information are acquired from the business database, and then the spatial position information points are drawn on the client-side maps.

3 KEY TECHNOLOGIES

3.1 Database construction

Database is the foundation of system design. The first step is the construction of the basic database for the economic loss evaluation of rainstorm and flood disasters, whose data can be collected through investigation, outsourcing, remote sensing and other methods. The content covers the basic geographic information, disaster statistics, economic statistics and disaster countermeasures from the Longhua District in Shenzhen. At the same time, the basic data are systematically summarized, so via database recording and storage, the query, additions, calls, and statistical analysis can be achieved at any time.

3.2 Direct economic loss assessment of disasters

By combining census and sampling surveys, the property of urban rainstorm and flood disasters is divided into 15 categories, such as residential houses, office buildings, parking lots, commercial enterprises, industrial enterprises and public facilities. The corresponding tables are designed for different types of property and the data of socioeconomic losses are obtained through surveys. With the aid of mathematical statistics, the disaster loss rates of various types of objects subject to hazard effects in different situations are calculated according to the water depth, and the curve reflecting disaster loss rates is constructed. Finally, according to the survey data and the data in the official statistical yearbook, the economic losses of different types of properties affected by different water depths are calculated.

The calculation is divided into several steps as follows: first, estimate the total value of the properties; second, compute the losses at different water depths and obtain the loss rates by dividing the loss by the total loss; third, calculate the loss rates of various types of properties. Further detailed methods and the calculation process can be found in Wu et al. (2016).

3.3 Indirect economic loss assessment of disasters

The input-output model (IOM) is applied to the evaluation of indirect economic losses (Wu et al., 2016). The IOM, constructed based on input-output matrix, is put forward for the assessment of the indirect economic losses that a disaster has brought to different industries.

Its prominent feature lies in the ability to simulate and calculate the chain reaction and ripple effect that disasters have brought to economic disturbance. Describing the interaction between the buying and consuming economic flow of all the industrial sectors, it can assess the economic impact of disasters on one or more sectors through the changes in intermediate consumption demand. The calculation formula of this module is shown as follows:

$$\Delta Q = (I - A)^{-1} \Delta Y \qquad (1)$$

In the formula, $(I - A)^{-1}$ is the Leon Leontief inverse matrix calculated based on the Shenzhen input-output table (according to the RAS method), and represents the change in the total output of each department when the final output of a certain sector changes by a unit. ΔY expresses the change amount of the final output of the industry affected by rainstorm and flood disaster, while ΔQ indicates the change amount of the indirect output of the related industries.

It should be noted that in addition to the main research methods and models mentioned above, the construction of some sub models is also involved in the specific research process, which will not be further enumerated.

4 CASE ANALYSIS

4.1 "5.11" rainstorm and flooding disaster of Shenzhen in 2014

From 6:30 a.m. to 2:00 p.m. on May 11 2014, heavy rains occurred in Shenzhen City, including some torrential rain. At 1:00 p.m., rainstorm warning signal was issued by the Shenzhen Meteorological Observatory, as the precipitation reached 100 mm or more within 3 hours. Road traffic was almost paralyzed because of the severe flooding in a large area. "5.11 rainstorm disasters" caused different degrees of flood in the four street jurisdictions of Longhua New District. According to preliminary investigation, the total number of rainstorm and flooding points increased to 66 in the whole Longhua New District (22 in Guanlan, 10 in Longhua, 10 in Dalang and 14 in Minzhi).

Taking the worst-affected rainstorm and flooding points – Junzibu Community in the Guanlan Street – as an example, in the "5.11" storm disaster in 2014 the maximum sliding rainfall within 3 hours was 90.4 mm, with the average water depth of 607 mm and the longest flooding time of 7 h. The heavy rain trapped more than 300 residents; more than 100 cars and 400 shops flooded and more than 25 companies were affected. There were even three small landslides, which, luckily, did not cause any casualties. The direct property losses amounted to 760,000 yuan. The main reasons of the flood were prolonged rainfall, low-lying land, underdeveloped drainage facilities and delayed drainage.

4.2 Economic loss assessment of the "5.11" rainstorm disasters of Shenzhen in 2014

The economic loss assessment of the "5.11" rainstorm in Shenzhen in 2014 can be divided into two parts: direct economic loss assessment and indirect economic loss assessment. The spatial loss distribution of each flood point caused by the storm is shown in Figure 2. The point that suffered most was the Junzibu Community: The direct losses amounted to 760,000 yuan, indirect losses to 51,160,000 yuan, and the total loss to 51,920,000 yuan. The total economic loss was 89,360,000 yuan, with direct economic losses of 1,490,000 yuan and indirect economic losses of 87,860,000 yuan. From the perspective of the objects of hazard effect the residential-area losses were around 21,200 yuan, the business-district losses around 728,000 yuan, the industrial area losses around 760,000 yuan, and the office building losses around 9,000 yuan.

The five industries that suffered the most direct economic losses were wood processing and furniture manufacturing (576,000 yuan), wholesale and retail (84,675 yuan), accommodation and catering (51,700 yuan), real estate (19,400 yuan), and the garment and textile industry (17,600 yuan). The five industries with the most indirect economic losses were chemical

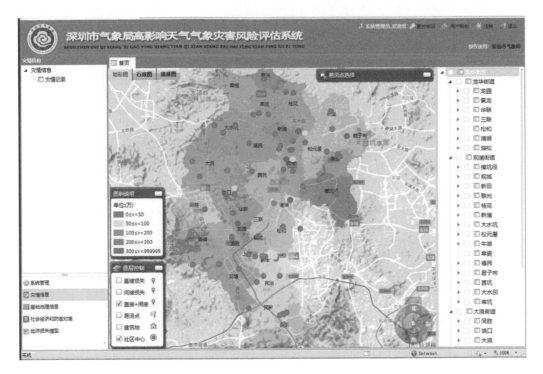

Figure 2.   Economic losses of all waterlogged points during the natural disaster on may 11 2014.

(7,123,700), electricity, heat production and supply (3,310,900 yuan), leasing and business services (3,218,500 yuan), farming (3,088,400 yuan), and the financial industry (2,914,000 yuan).

4.3   *Countermeasures of the "5.11" rainstorm and flooding disaster of Shenzhen in 2014*

Based on human-computer interaction and distributed data acquisition, this system can automatically generate not only the direct economic losses and the disaster-reduction efficiency of different types of properties as well as the indirect economic losses of different industries and regions, but also the disaster prevention report of typical rainstorm and flood points. The report includes six parts: basic information about the disaster, assessment ideas, direct economic losses, indirect economic losses, disaster description, and defensive countermeasures. The countermeasures are chosen from both macro and micro levels. At the macro level, attention is paid to the seriously damaged industries and their corresponding disaster prevention interventions. At the micro level, the classification of disaster losses is made based on the historical disaster information, and different countermeasures are put forward according to the characteristics of rainstorm and flooding points.

For the "5.11" rainstorm disaster in Shenzhen city in 2014, in terms of the rainstorm and flooding points of the Junzibu Community and based on the severity level of the disaster, the system has automatically extracted the defensive countermeasures, as shown in Figure 3.

5   CONCLUSIONS

Combined with weather forecasts, this system calculates instantly the direct economic losses of the rainstorm and flood disaster in the Longhua District of Shenzhen, obtains the indirect economic losses of the related industries, regions and departments, and generates automatically the economic loss evaluation and the countermeasure reports for typical rainstorm and flood

1. At the macro level, the countermeasures are as follows:

(1) Strengthen disaster prevention in the industrial zone, commercial district and residential area.

(2) Enhance disaster prevention in wood processing and furniture manufacturing, wholesale and retail trades, accommodation and catering, the real estate industry, textile and clothing industry, shoes, hats, leather and feather product industry.

2. At the micro level, the countermeasures are shown below:

(1) Relevant officials should be available on a 24-hour per day basis.

(2) Suspend classes in primary and secondary schools as well as kindergartens. The school should be equipped with professionals to ensure the children's safety in school.

(3) Teach students how to avoid danger on their way to and from school. Schools should protect the safety of students during school hours (including students on the school bus and boarding students).

(4) Inform outdoor people away from trees, billboards, overhead lines, towers and transformers and other areas of high-risk. Never touch the wire blown down by wind. Find a safe place to duck nearby.

(5) Personnel out on patrol should check the dangerous slopes, simple scaffoldings, flood-prone areas and precarious houses. Once any sign of disaster is seen, it should be promptly reported to sub-district office.

The dangerous places should be cordoned off.

(6) If the water in the water-logged ground is deep, personnel should throw a security cordon around the place or put up some signs warning people against entering those areas. Inform the water sector to remove the manhole covers. Place sandbags outside doors of shops situated in low-lying regions. Turn off the exposed power equipment.

Figure 3. Storm disaster prevention countermeasures that were used on may 11 2014.

points. All these can provide decision-making references for disaster prevention and the disaster emergency management of the municipal government and the Shenzhen Meteorological Bureau. At the same time, the interactive data collection of this system can facilitate the staff working for the communities, the streets and the municipal government to acquire data and evaluate disaster losses, saving labor and material costs for the staff at all levels of government. In addition, real-time updates of the database can be used for other disaster management. Finally, this system can be extended to the municipal and provincial ministry of environmental protection and civil affairs and other government departments as a norm of disaster economic loss assessment, to provide supplementary information for the natural disaster emergency management of Shenzhen and other similar cities. In this way, the risks of natural disasters can be drastically reduced, management costs lowered, and social welfare prioritized.

ACKNOWLEDGMENTS

This project was supported by Major Research Plan of National Social Science Foundation of China (18ZDA052), and the National Natural Science Foundation of China (91546117, 71373131, 41501555).

REFERENCES

Huang, C.F. 2012. *Risk Analysis and Management of Natural Disaster*. Beijing: Science Press. (in Chinese).

Huang, C.F., Huang, Y.D. 2018. An information diffusion technique to assess integrated hazard risks. *Environmental Research*. 161, 104–113. doi: 10.1016/j.envres.2017.10.037.

Wu, X.H., Zhou, L., Gao, G., et al, 2016. Urban flood depth-economic loss curves and their amendment based on resilience: evidence from Lizhong Town in Lixia River and Houbai Town in Jurong River of China. *Nat Hazards*, 82:1981–2000. doi: org/10.1007/s11069-016-2281-5.

Risk Analysis Based on Data and Crisis Response Beyond Knowledge – Huang & Nivolianitou (eds)
© 2020 Taylor & Francis Group, London, ISBN 978-0-367-25146-8

Assessing the joint dependence of air-water temperature and discharge-water temperature in the middle reach of Yangtze River using POME-copula

Yuwei Tao, Yuankun Wang* & Dong Wang*

Key Laboratory of Surficial Geochemistry, MOE, Department of Hydrosciences, School of Earth Sciences and Engineering, State Key Laboratory of Pollution Control and Resource Reuse, Nanjing University, Nanjing, China

ABSTRACT: Water temperature, a crucial environmental factor, has a direct impact on the ecological and biogeochemical processes in rivers. The water temperature regime in the middle reach of Yangtze River has been altered by intensified human activities, particularly the dam construction. In this study, the optimal moment-POME-copula framework is developed from a probabilistic perspective to examine the joint dependence structures of air-water temperature and discharge-water temperature in the middle reach of the Yangtze River. To assess the effect of TGR on water temperature, we compare the joint dependence in the pre-TGR (1983-2002) period with the post-TGR (2003-2014) period. The results show that dependence structures of air-water temperature and discharge-water temperature in the middle reach of Yangtze River have changed in different extent, which is mainly attributed to the impoundment of the TGR. In addition, it is found that marginal distributions of each hydrological variable derived from the POME-based method can well capture respective distribution patterns in contrast to traditional distributions. Thus the results of dependence structures appear to be more accurate based on well-fitted margins. We hope this study could provide a scientific reference for ecological operation of TGR facing biological conservation.

Keywords: dependence structure, water temperature, Three Gorges Reservoir

1 INTRODUCTION

As a key environmental factor for most aquatic organisms, water temperature plays a significant role in the overall health and function of aquatic ecosystems (Caissie, 2006). Distribution, growth, metabolism, food availability, migration, and reproduction of most riverine species are influenced by water temperature (Morrill et al., 2005). The thermal regimes of rivers is associated with climate changes, and human interference (e.g., dam construction, industrial pollution, deforestation etc.) (Chen et al., 2016). Water temperature has traditionally been strongly related to air temperature as a surrogate for net heat exchange. The volume of runoff also has been shown to affect water temperature. Besides, the dam construction as a severe form of human interventions in the natural river systems can alter the hydrological and thermal regimes, biodiversity of river species and water quality. Thermal stratification of reservoirs would result in the cooling or warming of water temperature downstream compared to conditions undisturbed by reservoir operations (Soja and Wiejaczka, 2014).

---

*Corresponding author: yuankunw@nju.edu.cn; wangdong@nju.edu.cn

Many varying sophisticated approaches have been developed to model different aspects of thermal behavior in space and in time and estimate the effect of dam constructions on water temperature. These have included empirical models that rely on statistical analysis to make predictions from weather data or information on catchment characteristics (Benyahya et al., 2007), the use of wavelet analysis to quantify the effect of multi-purpose dams in decreasing water temperature variability at multiple time scales from 1 to 8 days (Steel and Lange, 2007). Little attention has been paid to explore the dependence structures among water temperature and factors (e.g. air temperature and discharge) from a probabilistic perspective. Copulas provide a useful tool for exploring this type of problem. A copula is a multivariate probability distribution for which the marginal probability distribution of each variable is uniform; it describes the dependence structure among the marginal distributions of the individual variables (Liu et al., 2017). Copulas have been increasingly used in hydrometeorological multivariate applications including frequency analysis, dependence analysis and risk analysis regarding extreme conditions. To develop a flexible, unbiased statistical inference framework, the principle of maximum entropy (POME) is being increasingly coupled with copula (Liu et al., 2017). POME is to derive the marginal distribution for copula, best capturing the shape of probability density function.

With the operation of Three Gorges Reservoir (TGR), the largest electrical power station in the world, the dam-induced impact on the hydrology and ecology in the Yangtze River has aroused great concern. TGR has exerted flow regulation which altered discharge downstream. Past study revealed that streams affected by impoundment generally have weaker air-water temperature correlations (O'Driscoll and DeWalle, 2006). Therefore, the POME-copula method in this study is used to examine the dependence structures among water temperature, air temperature and discharge in the middle of Yangtze River and identify the effect of impoundment by comparing the dependence before and after the TGR. This study can inform science-based management of reservoir aimed at minimizing adverse ecological effects.

2 POME-COPULA METHOD

2.1 *Principle of maximum entropy (POME)*

For a continuous random variable X, its domain ranges from a to b and the entropy $H(X)$ can be expressed as (Shannon, 1948):

$$H(X) = - \int_a^b f(x) \log f(x) dx \tag{1}$$

where $f(x)$ is the probability density function (PDF) of X.

In accordance with principle of maximum entropy, the probability distribution is obtained by maximizing entropy subject to given data and constraints:

$$\int_a^b f(x) dx = 1 \tag{2}$$

$$\int_a^b g_i(x) f(x) dx = \mathrm{E}(g_i), (i = 1, 2, ..., m) \tag{3}$$

where $g_i(x)$ is the selected or specified function with respect to the properties of interest (e.g., moments); $E(g_i)$ is the expectation of $g_i(x)$; and m is the number of constraints. For a given set of constraints, a unique distribution can be defined. Therefore, finding proper constraints is critical (Singh, 2015) and leads to the mathematical optimization problem which can be solved using the method of Lagrange multipliers. The Lagrange function L can be expressed as:

$$L = -\int_a^b f(x)dx - (\lambda_0 - 1)[\int_a^b (f(x) - 1)dx)] - \sum_{i=1}^m \lambda_i[\int_a^b f(x)dx - E(g_i)] \qquad (4)$$

where $\lambda = [\lambda_0, \lambda_1, \ldots, \lambda_m]$ are the Lagrange multipliers. By differentiating L with respect to f and setting the derivative to zero, the maximum entropy probability density function can be derived as (Kapur and Kesavan, 1992):.

$$f(x) = \exp(-\lambda_0 - \lambda_1 g_1(x) - \lambda_2 g_2(x) - \ldots - \lambda_m g_m(x)) \qquad (5)$$

Using sample statistical moments as constraints has an advantage of reaching the universal PDF more accurately for given data. In general, $g_i(x)$ is set as x^i.

Accordingly, the cumulative distribution function (CDF) can be expressed as:

$$F(x) = \int_0^x f(x)dx \qquad (6)$$

In this study, the Lagrange multipliers with constraints of orders (denote m from 1 to 8) of moments are taken into account. Because more moments would lead to considerable computation-expense and less improvement in accuracy, the maximal order of moment is specified as 8. All the derived POME-based PDFs are compared with the Akaike information criterion (AIC) to select the optimal probability distributions for air temperature, water temperature and discharge. The best fitted distribution is the one which has the minimum AIC. AIC can be expressed as:

$$AIC = -2log(\text{ maximized likelihood for the model}) + 2(\text{numbers of fitted parameters}) \quad (7)$$

2.2 Copula

Copulas are functions that connect multivariate probability distributions to their one dimensional marginal probability distributions (Nelsen, 1999). For a continuous random vector (X, Y) with marginal distributions $F_X(x)$ and $F_Y(y)$, the joint cumulative distribution function (CDF) can be expressed as (Sklar, 1959):

$$P(X \le x, Y \le y) = C[F_X(x), F_Y(y); \theta] = C(u, v; \theta) \qquad (8)$$

where u and v are realizations of random variables $U = F_X(x)$ and $V = F_Y(y)$; θ is the copula parameter that tunes the dependence between margins.

Of all the copula families, the Elliptical and Archimedean copulas are widely applied in hydrological analysis. Because they can be easily constructed and present various properties of multivariate joint distributions, satisfying most actual application demands.

2.3 Assessment framework

Step 1. Obtain independent hydrometeorological series
The basic principle for applying the copula method is that series of hydrometeorological variables are supposed to be independent, while the strong autocorrelations of daily water temperature, air temperature, discharge cannot meet this requirement. Thus, we picked specific data from daily series based on the concentration period of water temperature, which refers to the date when higher water temperature appears in a specific period to form the independent sequence.

The basic principle for calculating the concentration period is based on vector analysis (Song et al., 2012). Each daily water temperature amount in a season is taken as the length of vector, whereas the corresponding date is taken as the vector direction. To define the direction of each day in a season, for example, January-March period is considered as a cycle (360°), and then one day is corresponding to $360°/92 = 3.91°$. Each daily water temperature is decomposed to two components in horizontal and vertical directions as Eq. (9) and (10):

$$T_x = \sum_{i=1}^{k} T_i \sin \theta_i \qquad (9)$$

$$T_y = \sum_{i=1}^{k} T_i \cos \theta_i \qquad (10)$$

where k is the number of days in one quarter; and T_x and T_y correspond to the horizontal and vertical component of water temperature. Thus, the concentration period of water temperature (D) can be expressed as:

$$D = \arctan(T_x/T_y)/(360/k) \qquad (11)$$

According to the concentration period of water temperature, the initial data series X = $\{x_1, x_2,..., x_n\}$, in the pre- and post-impoundment periods of water temperature, air temperature and discharge respectively are obtained for applying the following POME-copula method.

In order to test the independence of hydrometeorological series, the Pearson correlation test for serial dependence was employed (McCuen, 2002). The hypothesis for the Pearson test are H_0: $\rho = 0$; H_A: $\rho \neq 0$, in which ρ is the serial correlation coefficient of the population. Given a sequence x_i, $i = 1,2,..., n$, the statistic for testing the significance of the Pearson can be expressed as:

$$t = \frac{R_1}{\sqrt{(1 - R_1^2)/(n - 3)}} \qquad (12)$$

where R_1 is the sample correlation coefficient; n is the sample size; t is the statistic value of student t distribution with (n-3) degrees of freedom. The null hypothesis should be rejected if t is greater than $t_{v,\alpha}$ for a given significance level $\alpha = 0.05$ and one-tailed alternative hypothesis. As a result, all t values of each hydrometeorological series X are less than $t_{v,\alpha}$ values, indicating they are independent and is applicable to copula method.

Step 2. Marginal probability distribution inference using POME
(1) Normalize X to the unified interval [0,1]:

$$X' = \frac{X - X_{min}}{X_{max} - X_{min}} \qquad (13)$$

(2) Derive the maximum entropy distribution given constraints of different orders m (from 1 to 8) of moments;

$$g_i(x) = x^i \ (i = 1, 2, 3, ..., m) \qquad (14)$$

(3) Determine the optimal orders of moments by AIC;
(4) Obtain the optimal-moment based marginal PDF;

Step 3. Copula determination:
(1) Obtain CDF using Eq. (6) with the POME-based PDF;
(2) Identify the most appropriate copula by AIC;
(3) Evaluate the dependence of hydrometeorological variables based on the best copula model.

3 STUDY AREA AND DATA

The Three Gorges Reservoir (TGR) is located at the main stream of Yangtze River between Chongqing and Yichang. TGR began operation in 2003, and after three major impoundments, construction was completed in 2009. The water storage capacity is 39.3 km^3. TGR is constructed to make full use of the water resources for hydropower generation, irrigation and flood control.

The water temperature and discharge data from 1983 to 2014 at Yichang hydrological station in the middle of Yangtze River -44 km downstream of TGR, were provided by the Yangtze River Water Resources Commission. The data of air temperature from 1983 to 2014 at Yichang meteorological station were obtained from the National Meteorological Information Center.

4 RESULTS AND DISCUSSION

The dependence modelling is divided into 4 pairs: (1) seasonal air temperature (denoted as S1) and water temperature (denoted as S2) from 1983/1 to 2002/12; (2) seasonal air temperature (denoted as S3) and water temperature (denoted as S4) from 2003/1 to 2015/12; (3) seasonal discharge (denoted as S5) and water temperature (denoted as S6) from 1983/1 to 2002/12; (4) seasonal discharge (denoted as S7) and water temperature (denoted as S8) from 2003/1 to 2014/12.

The optimal order m of moments for each sequence was determined in the basis of the Akaike information criterion (AIC). Accordingly, the results for S1 to S8 are m = 8, 7, 1, 7, 4, 7, 1 and 7. The distributions from the optimal moment-POME method were compared to several commonly used marginal distributions, including Weibull, Normal, Lognormal, Gamma, Logistic and Exponential distribution. For S1 to S8, all of the commonly used marginal distributions have AIC values much greater than those of the optimal moment-POME distributions. Thus, the optimal moment-POME method performed better in determining marginal distributions than parametric methods.

Five candidate copulas, including meta-elliptical copula (Gaussian and Student t) and Archimedean copula (Clayton, Gumbel and Frank) were considered. The best copula model was selected based on the AIC criterion. The Frank copula yielded the smallest AIC value for S1-S2, S3-S4, S5-S6 and the Gaussian copula fitted best for S7-S8 (**Table 1**). Scatter plots of simulated data for each pair shown in **Figure 1** illustrated a clear shift in the patterns of dependence between pre- and post-impoundment.

The Frank copula is appropriate for capturing the positive correlation between air temperature and water temperature at Yichang both in the pre-TGR and post-TGR period. The dependence are symmetrical which means the dependence of air temperature and water temperature is the same if they are at the same level of high or low relative to their normal level. Tail dependence reflects the dependence in the extremes while both Frank and Gaussian copulas have no tail-dependence coefficients. After the impoundment of TGR, the strength of correlation between water temperature and air temperature diminishes. The weakening of the dependence may be caused by abnormal variations in water and air temperature, such as irregular fluctuations and periodic features. Also, the weakening of the relation has been attributed to the effects of the TGR (Cai et al., 2018).

Table 1. Summary of dependence structures of pairs.

Pairs	Copula	Parameter	Symmetry (Y/N)	Tail dependence (Y/N)
S1-S2	Frank	62.49	Y	N
S3-S4	Frank	44.25	Y	N
S5-S6	Frank	72.31	Y	N
S7-S8	Gaussian	0.97	Y	N

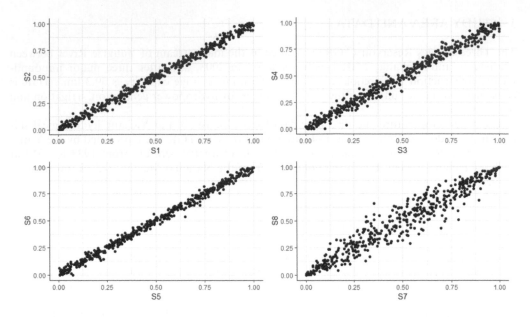

Figure 1. Scatter plots of simulated data for pairs.

The Gaussian copula better characterizes the positive correlation between discharge and water temperature at Yichang after impoundment, whereas the Frank copula provides the best characterization prior to impoundment. The relation exhibits symmetry with no tail dependence. Following impoundment, the relation weakened considerable, as indicated by the increased scatter, indicating a substantial disturbance of the relation between discharge and water temperature. It has been widely reported that the TGR has significantly altered the natural characteristic of river flow (Guo et al., 2018). The reservoir has apparently modified the relation between discharge and water temperature, suggesting that changes in the way in which flow is regulated by the TGR may could be a strategy to adjust water temperature.

5 CONCLUSION

In this study, the dependence among air temperature, discharge and water temperature were investigated by using the POME-copula method, which can capture detailed dependence patterns from probabilistic perspective. The results revealed: POME has captured optimal hydrological distributions compared with common used parametric distributions; the dependence structures of air-water temperature and discharge-water temperature were influenced noticeably by TGR, given that TGR has altered the discharge and water temperature.

ACKNOWLEDGMENTS

This study was supported by the National Key Research and Development Program of China (2017YFC1502704, 2016YFC0401501), and the National Natural Science Fund of China (51679118, 41571017, and 91647203), and Jiangsu Province"333 Project" (BRA2018060).

REFERENCES

Benyahya, L., Caissie, D., St-Hilaire, A., Ouarda, TBMJ., Bobee, B., 2007. A review of statistical water temperature models. *Canadian Water Resources Journal* 31: 179–192.

Cai, H.Y., Piccolroaz, S., Huang, J.Z., Liu, Z.Y., Liu, F., Toffolon, M., 2018. Quantifying the impact of the Three Gorges Dam on the thermal dynamics of the Yangtze River. *Environmental Research Letters* 13(0540165).

Caissie D (2006). The thermal regime of rivers: a review. *Freshwater Biology* 51: 1389–1406.

Chen, D.J., Hu, M.P., Guo, Y., Dahlgren, R.A., 2016. Changes in river water temperature between 1980 and 2012 in Yongan watershed, eastern China: magnitude, drivers and models. *Journal of Hydrology* 533:191–199.

Guo, L.C., Su, N., Zhu, C.Y., He, Q., 2018. How have the river discharges and sediment loads changed in the Changjiang River basin downstream of the Three Gorges Dam? *Journal of Hydrology* 560: 259–274.

Kapur, J.N., Kesavan, H.K., 1992. *Entropy optimization principles and their applications*. Academic Press.

Liu, D.F., Wang, D., Singh, V.P., Wang, Y.K., Wu, J.C., Wang, L.C., et al., 2017. Optimal moment determination in pome-copula based hydrometeorological dependence modelling. *Advances in Water Resources* 105: 28–50.

McCuen, R.H., 2002. *Modeling hydrologic change: statistical methods*. Taylor & Francis, Boca Raton.

Morrill, J.C., Bales, R.C., Conklin, M.H., 2005. Estimating stream temperature from air temperature: implications for future water quality. *Journal of Environment Engineering-ASCE* 131(1): 139–146.

Nelsen, R.B., 1999. *An introduction to copulas*. Springer, New York.

O'Driscoll, M.A. and DeWalle, D.R., 2006. Stream-air temperature relations to classify stream-ground water interactions in a karst setting, central Pennsylvania, USA. *Journal of Hydrology* 329: 140–153.

Shannon, C.E., 1948. *A mathematical theory of communication*. Bell Labs Tech.

Singh, V.P., 2015. *Entropy theory in hydrologic science and engineering*. McGraw-Hill, New York.

Sklar, M., 1959. *Fonctions de répartition à n dimensions et leurs marges*. Paris: Publications de l'Institut de Statistique de L'Université. 8, 229–231.

Soja, R., Wiejaczka, Ł., 2014. The impact of a reservoir on the physicochemical properties of water in a mountain river. *Water and Environment Journal* 28, 473–482.

Song C., Zhou X.D., Tang W., 2012. Evaluation indicators for assessing the influence of on downstream water temperature reservoirs. *Advances in Water Science* 23(3): 419–426. (in Chinese).

Steel, E.A. and Lange, I.A., 2007. Using wavelet analysis to detect changes in water temperature regimes at multiple scales: effects of multi-purpose dams in the Willamette River Basin. *River Research and Applications* 23: 351–359.

Risk Analysis Based on Data and Crisis Response Beyond Knowledge – Huang & Nivolianitou (eds)
© 2020 Taylor & Francis Group, London, ISBN 978-0-367-25146-8

Equity measurement of spatial layout for emergency shelters with consideration of the needs of vulnerable populations

Qiang Li
Faculty of Geographical Science, Beijing Normal University, Beijing, China

Jing Guo
Graduate School of Environmental studies, Nagoya University, Nagoya, Japan

Xiaoyue Tan & Jin Chen
Faculty of Geographical Science, Beijing Normal University, Beijing, China

ABSTRACT: As important facilities to ensure urban public safety, the spatial layout of emergency shelters should be able to meet the basic needs of urban residents for refuge. This paper presents a model for measuring the equity of spatial layout for emergency shelters based on the two-step floating catchment area (2SFCA) method. The model improves the searching methods for shelters and settlements, introduces the distance decay function and the vulnerable population parameters. The case study results using the improved 2SFCA in the Shichahai area shows that the model could indicate whether the shelter layout and population distribution are well-matched under the condition that the total scale of emergency shelters could meet all population needs.

Keywords: emergency shelter, equity, 2SFCA method, vulnerable population

1   INTRODUCTION

Emergency shelters provide residents with evacuation sites and temporary living, which can help reduce life and property losses in the case of a sudden disaster or incident. Examples of emergency shelters include green open spaces, squares, stadiums, schools, parking lots, etc. (Dai et al., 2010; Xu et al., 2012). It is one of the effective measures for cities to cope with sudden disasters and ensure public safety.

The layout balance of an emergency shelter highlights two points: (1) residents can quickly reach the shelter, which is convenience; (2) all residents with different behavioral abilities can reach the shelter, which is fairness. Therefore, in the context of increasingly complex urban spatial structures and an aging population, it is particularly important to pay more attention to the vulnerable population and meet the asylum needs of different groups when planning the construction of shelters. In addition to research on efficiency evaluation of construction cost and service provided (Chen et al., 2009), existing studies mainly focused on the service evaluation after the shelter is completed, including accessibility, layout rationality, suitability, overall services, etc. (Chen, 2014; Ji & Gao, 2014; Tao et al., 2014; Wang et al., 2014, 2015; Wei et al., 2015). However, the demographic characteristics have often been overlooked.

Accessibility refers to the ease of moving from one place to another, including origin, destination and the distance cost between the two points. The accessibility of an emergency shelter indicates the service available to residents within a certain distance. In order to include the emergency evacuation needs of vulnerable populations, this paper proposes an evacuation shelter accessibility model that comprehensively considers the scope of shelter service, the distribution of residents, the population structure, and the mobility characteristics of vulnerable populations.

2 MODEL SPECIFICATION

There are many models for measuring accessibility. The two-step floating catchment area (2SFCA) comprehensively considers the three core elements of accessibility and is widely used (Che, 2014; Tao et al., 2014; Tao & Cheng, 2016; Wang et al., 2015; Zhou et al., 2014).

2.1 Basic 2SFCA model

Step 1: For each evacuation shelter j, search for a settlement (k) within the range that is d_0 (distance) from j (service buffer of j), then calculate the ratio of the emergency shelter area to the settlement population as the service capacity (R_j) of the emergency shelter j:

$$R_j = \frac{S_j}{\sum_{k \in \{d_{kj} \le d_0\}} P_k} \tag{1}$$

where S_j refers the available area of the emergency shelter j, P_k is the population of the settlement k within the j service buffer, and d_{kj} is the distance from the settlement k to the emergency shelter j.

Step 2: For each settlement i, search the shelter (j) within the range of d_0 from i point (reachable range of i), then sum the service capacity R_j of all emergency shelters as the available service at the settlement i, which is the accessibility (A_i) for the residents of i.

$$A_i = \sum_{j \in \{d_{ij} \le d_0\}} R_j \tag{2}$$

2.2 Weakness of basic model

The 2SFCA method has some drawbacks that can be improved in the study of layout equity measurement of emergency shelters.

2.2.1 Limited to searching around the space centroid

When searching for the service buffer of the emergency shelter or the reachable range of the settlement, the space centroid is used as the starting point for searching. As shown in Figure 1, for an emergency shelter with a large area such as S_1, the service buffer only covers the area of its own land if searching from its space centroid, so that the service buffer is smaller than the real buffer, further resulting in smaller accessibility. Additionally, due to the irregular geometry of the settlement, the population centroid (P_1) of the L-shaped settlement at the lower left is located outside but falls in the service buffer of the shelter S_2. Apparently, S_2 does not actually cover the residents of the L-shaped settlement.

2.2.2 Regardless of distance decay

The 2SFCA method sets the reach ability in the range of the distance threshold but is unreachable outside the range. It does not consider the fact that reach ability decreases with distance in the same service range.

2.2.3 Lacks consideration of different behavioral abilities

The 2SFCA approach assumes that all populations have equal ability to reach facilities and lacks consideration of different demographic characteristics.

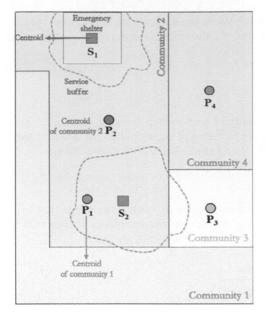

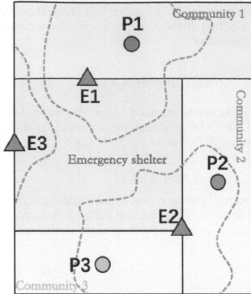

Figure 1.   Limitations of 2SFCA method.

Figure 2.   The service area for multiple entrance shelters.

2.3   *Model enhancement*

2.3.1   *Replacing centroid with the entrance of shelters*

A defect of the 2SFCA basic model is first, the entrance to the emergency shelter is used as the starting point of the search radius instead of the centroid. The service of the emergency shelter is considered as obtain able if any entrance is reachable. With regards to open spaces such as a green space where there is no fixed entrance, the connection point between the green space and the road network is viewed as an entrance. In the case of multi-entry parks used as shelters, residents can reach the site through different entrances, i.e., the shelter provides services to all residents from different entrances (Figure 2). Therefore, the service capabilities provided by each entrance to the shelter are considered equal:

$$R_j = R_j^{E1} = R_j^{E2} = \ldots = R_j^{Et} = \frac{S_j}{\sum P^{Et}} \tag{3}$$

where $R_j^{E1}, R_j^{E2}, \ldots, R_j^{Et}$ refer to the service capacity for entrance 1, 2,..., t of the emergency shelter j,P^{Et} is the population within the search buffer that is d_0 from the entrance t. Similarly, in the second step when searching for an emergency shelter within the range starting from the settlement i, the service capacity of the entrance R_j^{Et} is added to the accessibility result of i if the residents of settlement i could reach that entrance.

2.3.2   *Allocating population to the grid*

Secondly, the research area is divided into 50 $m \times$ 50 m grids using the Fishnet tool in GIS, and the population is allocated to each grid according to the floor area as the weight. The buildings in the study area are digitalized and the floor is identified for each building based on the 3D map. The settlement grid data are overlaid and the population p_m allocated to the grid m is calculated through equation (4):

$$p_m = \frac{A_m}{\sum_{m \in i} A_m} \times P_i \tag{4}$$

where A_m is the total floor area in the grid m, and P_i refers to the total population of settlement i.

2.3.3 Taking distance decay into account

To overcome the assumption that the reach ability in the range of the threshold is consistent and the out-of-range is unreachable, a distance decay function is added to the search buffer of the 2SFCA basic model. Combining the gravity function with the 2SFCA model and adding the exponential function to the search radius is widely used and beneficial to the accuracy improvement of the accessibility described.

2.3.4 Vulnerable population

In an emergency, a vulnerable population needs more time to reach the shelter than normal persons due to their limited mobility. It is necessary to set the proportion of the vulnerable population in the 2SFCA model for describing the actual asylum needs.

2.3.5 Enhanced 2SFCA model

Based on the improvements mentioned above, firstly, for each entrance of emergency shelter j, search for all population grids p_n within the radius d_0. Furthermore, the decay function $d_{nj}^{-\beta_n}$ is accounted for in the distance d_{nj} from the population grid to the emergency shelter. The service capacity R_j of the shelter j can be obtained as the following formula:

$$R_j = \frac{S_j}{\sum_{n \in \{d_{nj} \leq d_0\}} P_n d_{nj}^{-\beta_n}} \tag{5}$$

$$\beta_n = 1 + \alpha_n \tag{6}$$

$$\alpha_n = P_k'/P_k (n \in k) \tag{7}$$

where S_j is the area of the emergency shelter j; β_n the friction coefficient in the distance decay function expressed as a function of the proportion α_n of vulnerable population in the grid n; α_n is the ratio of the vulnerable population P_k' to the total population P_k of the settlement k. The higher the proportion of vulnerable population, the higher the friction coefficient of distance decay and the lower the accessibility.

Next, for each population grid m, search for emergency shelters within a range d_0 from m. Sum all service capacity R_j with a weighting function as the service accessibility A_m for m.

$$A_m = \sum_{j \in \{d_{mj} \leq d_0\}} R_j d_{mj}^{-\beta_m} \tag{8}$$

$$\beta_m = 1 + \alpha_m \tag{9}$$

$$\alpha_m = P_i'/P_i (m \in i) \tag{10}$$

where β_m, similar to β_n in equation (6), is the friction coefficient in the distance decay function that is based on the proportion α_m of the vulnerable population in the mesh m.

3 CASE STUDY

The Shichahai areais located in the old urban district of Beijing and has extensive land development and complex building space structures. Its population density is around 13,792 people/km^2, and the proportion of vulnerable people exceeds 1/3.

The school playgrounds, stadiums, parks, open green spaces and other facilities in the study area were first digitalized as emergency shelters. There are 25 available shelters with a total area of $3.6 \times 105\text{m}^2$. Based on the Open Street Map in ArcGIS and other network maps, the road network was completely digitalized for further calculation of the service buffer and OD matrix.

The Beijing Central City Earthquake and Emergency Evacuation Shelter (Outdoor) Planning Outline sets the service buffer of an emergency shelter at 500 m and residents need to walk to shelters within 5–15 min. Accordingly, 500 m is selected as the distance threshold d_0, and it is assumed that residents reach a shelter by walking; other factors such as road class and capacity are disregarded here.

4   RESULTS

4.1   *Layout equity of the basic 2SFCA model*

The 500 m buffer from the centroid of the emergency shelter and settlement are generated separately. The service capacity of each emergency shelter and accessibility of each settlement were calculated according to formulas (1) and (2), which was further interpolated to obtain the layout equity of the emergency shelter (Figure 3).

The average accessibility to the emergency shelter is 3.65 and the standard deviation is 11.17. The figures reveal that the spatial distribution differ by a large amount, indicating that overall, the emergency shelters can meet the evacuation demand, yet their spatial distribution are not equal. Some communities in the northwest have insufficient shelters. In contrast, large-scale shelters such as Beihai Park and Jingshan Park in the southeastern area offer evacuation service capacity that exceeds the corresponding demand, leading to wasted space resources.

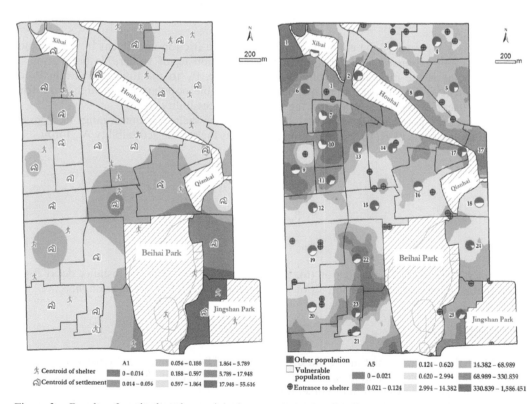

Figure 3.   Results of equity based on original 2SFCA method.

Figure 4.   Results of equity based on improved 2SFCA method.

4.2 Layout equity of enhanced 2SFCA model

Applying the enhanced 2SFCA model, the layout equity of the shelters evaluated on the grid scale is shown in Figure 4. Compared with Figure 3, the results break the barriers of the administrative boundaries, which is more practical. Moreover, the accessibility is higher around the entrance to the shelter. The gradient accessibility shows that a settlement which is farther away has a lower accessibility, which also avoids the uniformity in the threshold range of the basic model. Furthermore, the high proportion of vulnerable populations significantly reduces the accessibility. For example, the proportion of vulnerable population in communities 7 and 10 is as high as two-thirds, resulting in accessibility in these areas being much lower than the result in Figure 3. Conversely, the proportion of vulnerable population in community 6 is less than 1/10 and the accessibility is similar with that of Figure 3. This result reveals the impact of vulnerable populations on accessibility, indicating that considering the vulnerable populations is more instructive in designing the layout of shelters.

5 CONCLUSION AND DISCUSSION

An enhanced 2SFCA model was proposed to measure the spatial layout equity by improving the search method for emergency shelters and settlements, adding the distance decay function and the proportion of vulnerable populations. Through its application in the Shichahai area, the following main conclusions are obtained.

Although the total area of shelters in the study area can meet the needs of the entire population according to the basic 2SFCA model, the shelters' layouts do not match the population distribution. Some evacuation needs cannot be satisfied, or some shelters are not being fully utilized. The improved 2SFCA method fully considers the evacuation needs of the vulnerable population and can measure the layout equity of the shelter more accurately. Enhancing the search methods of shelters and settlements helps to meticulously describe the service buffer of the shelters and the reachable range of the population. Adding the distance decay function solves the problem of uniformity within the distance threshold while considering the proportion of vulnerable populations can accurately reflect the need for evacuation. Thus, the 2SFCA method has been improved to be more accurate and reasonable.

However, there are still some problems that need further study: (1) The model assumes that the residents reach the shelter using the shortest distance, regardless of road congestion or paralysis in an emergency. Parameters such as road class, width and capacity should be included in the model; (2) The research area should not be regarded as a closed area. The population in the area, especially the population near the edge of the area, may choose emergency shelters outside the area, thus reducing the evacuation needs inside the area.

ACKNOWLEDGMENTS

This project was supported by the National Key Research and Development Program of China (2017YFC1503004).

REFERENCES

Che, L.H. 2014. Evaluation of hospital layout based on the Gaussian 2SFCA accessibility mode. *Chinese Hospital Management* 34(2): 31–33.

Chen, Q.H. 2014. Land-use suitability evaluation on emergency shelters: A case study of Guangzhou. *Territory and Natural Resources Study* 6:12–17.

Chen, Z.F., Li, Q., Wang, Y. and Chen, J. 2009. Efficiency assessment of emergency shelter based on bounded variables DEA model. *Journal of Chinese Safety Science* 19(11): 152–158.

Dai, Q., Gao, Z.J. and Yang, H.P. 2010. A review of urban emergency shelter research. *Science of Technological Information* 6: 250–251.

Ji, J. and Gao, X. 2014. Evaluation method and empirical study on service quality of seismic emergency shelters. *Geographical Research* 33(11): 2105–2114.

Tao, Z.L. and Cheng, Y. 2016. Research progress of the two-step floating catchment area method and extensions. *Progressing of Geography Science* 35(5): 589–599.

Tao, Z.L. Cheng, Y. and Dai, T.Q. 2014.Measuring spatial accessibility to residential care facilities in Beijing. *Progressing of Geography Science* 33(5): 616–624.

Wang, J.B., Dai, S.Z. and Gou,A.P. 2014.Emergency service evaluation and planning measures of evacuation space. *Planners* 10: 104–109.

Wei, D., Tang, N. and Xu, S. 2015. Homeostatic principle-based research in reasonableness of the city's emergency shelter layout: A case of the central city in Xi'an. *Modern Urban Research* 5: 43–50.

Xu, L.P., Liu, Q.M. and Sun, J.J. 2012. Analysis and Optimization of Space Layout Characteristics of Anqing City Emergency Shelters Based on GIS. *Geometrics and Spatial Information Technology*, 35(2): 151–155.

Zhou, A.H. and Fu, X. 2013. Spatial distribution of the urban-emergency shelter in Beijing downtown areas. *Journal of Safety and Environment* 6:250–253.

Risk Analysis Based on Data and Crisis Response Beyond Knowledge – Huang & Nivolianitou (eds)
© 2020 Taylor & Francis Group, London, ISBN 978-0-367-25146-8

Assessment for current-carrying capacity of the power grid over South Hebei Province and its risk management

Hongyu Li, Kaicheng Xing* & Yuanyuan Jing
Key Laboratory of Meteorological and Ecological Environment of Hebei Province, Shijiazhuang, China
Climate Center of Hebei Province, Shijiazhuang, China

ABSTRACT: This study utilized the hourly measured data from 94 automatic meteorological sites over the southern part of Hebei to drive the wire energy balance equation for an analysis of carrying-current capacity. Based on the analysis of the sensitivity of allowable maximum carrying-current capacity to the variation of ambient wind speed and air temperature, the study determined the probability density distribution of ambient air temperature and wind speed, obtained the regional probability distribution of allowable maximum carrying-current capacity, and evaluated the transmission capacity at several safety operation levels. It shows that at the 95% safety operation level, the maximum carrying-current capacity of LGJ-400/35 transmission line in South Hebei can be set to 945, 774, 795 and 1012A in spring, summer, autumn and winter, respectively.

Keywords: Southern part of Hebei, overhead transmission line, carrying-current capacity, air temperature, wind speed

1 INTRODUCTION

As an important part of Beijing-Tianjin-Hebei region, southern Hebei Province has a strong demand for electric energy. In recent years, with the development of economy, the growth trend of industrial and civil electricity consumption is remarkable. However, southern Hebei is facing the aging and outdated situation of distribution facilities, and many other factors, such as land resources and environmental protection, limit the construction of transmission and distribution facilities to a certain extent. Therefore, on the basis of existing power grid facilities in southern Hebei, it is imperative to accurately assess the current carrying capacity of existing overhead conductors and tap the potential of power grid transmission to make full use of existing power grid resources.

The traditional load setting of southern Hebei power grid relies on the combination of fixed meteorological parameters (temperature 40°C from May to September, wind speed 0.5 m/s; temperature 25°C from October to April of next year, wind speed 0.5 m/s), and this combination of meteorological parameters is widely used in the setting of current carrying and transmission capacity of power grid. Such a strict and fixed combination of meteorological parameters may deviate from the actual meteorological conditions, affect the evaluation and further improvement of power transmission capacity, and increase the risk of power grid operation under the background of climate warming. Therefore, it is necessary to carry out scientific and reasonable calculation of carrying capacity and dynamic setting in combination with actual meteorological observation for the improvement of transmission capacity of power grid (Dunlop et al., 1979; Greenwood et al., 2014). Based on the sensitivity analysis of carrying capacity to meteorological parameters and the statistical analysis of wind speed and

---

*Corresponding author: xkc67@163.com

temperature at meteorological stations in southern Hebei, the allowable maximum carrying capacity of overhead conductors on seasonal scale is determined, and the operational risk is further assessed. On the premise of ensuring safety, we should maximize the current carrying potential of transmission equipment in Hebei South Power Grid, and try to provide scientific reference for the implementation of overhead conductor capacity-increasing technology in regional power grid.

2 DATA AND METHODS

2.1 *Data*

In order to grasp the complete diurnal variation of meteorological elements, 94 automatic meteorological stations in Baoding, Cangzhou, Xingtai, Hengshui, Shijiazhuang and Handan were used to observe hourly data. The meteorological elements involved were 1.5 m surface temperature and 10 m high wind speed. Ninety-four meteorological stations in southern Hebei have been equipped with automatic station observation equipment since 2003, and they will be installed one after another in 2010. The meteorological data used in automatic weather stations are the data from the automatic weather observation records to 2015. At the same time, the missing data are eliminated, and strict quality control is carried out.

In addition, the high temperature in summer is an important factor affecting the setting of overhead conductor carrying capacity. In order to understand the distribution of high temperature heat wave in southern Hebei on a long time scale, the daily maximum temperature data of 94 stations in southern Hebei from 1974 to 2015 were used to discuss the change of high temperature days.

2.2 *Study methods*

Understanding the three-dimensional probability density distribution of temperature and wind speed in southern Hebei can help to understand the combined distribution of these two meteorological parameters under actual conditions. In addition, the calculation of hourly maximum carrying capacity based on actual meteorological parameters and the analysis of its probability density distribution can help to determine the specific probability distribution of maximum carrying capacity under the constraints of actual meteorological parameters.

By analyzing the cumulative distribution over a certain allowable maximum carrying capacity, the allowable maximum carrying capacity can be further determined at different levels of safe operation. The conductor temperature can be retrieved from the current carrying capacity under a certain level of safe operation guarantee combined with the actual meteorological parameters. The probability density distribution of conductor temperature can help to evaluate the overheating condition of conductors caused by operation risk, and help the safe operation of power grid to control the operation risk of power grid.

3 RESULTS

3.1 *Analysis of carrying capacity variability*

According to the heat balance equation of the conductor surface, the temperature change will affect the radiation and convection heat dissipation terms. With the increase of temperature, the atmospheric long-wave radiation received on the conductor surface will increase, while the difference between the conductor temperature and the ambient temperature will decrease, which inhibits the convective diffusion of heat on the conductor surface. Therefore, the increase of temperature will weaken the maximum carrying capacity of the conductor. The increase of wind speed will enhance the convective diffusion on the surface of the conductor, thus indirectly enhancing the maximum carrying capacity of the conductor.

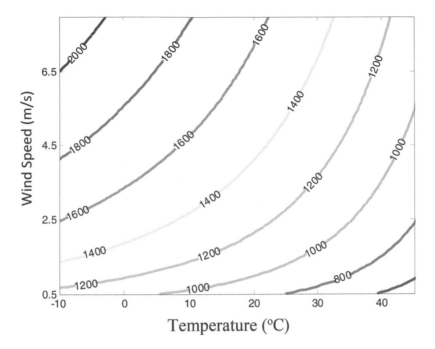

Figure 1. Two-dimensional distribution of maximum allowable carrying capacity of conductor under combination of ambient temperature and wind speed (the isoline number is the carrying capacity value).

It can be seen that the effect of temperature and wind speed on the maximum carrying capacity is reversed. In order to visually and quantitatively see the distribution of maximum carrying capacity of conductors corresponding to ambient temperature and wind speed, we give the two-dimensional distribution of maximum carrying capacity under the combination of atmospheric environment and wind speed (Figure 1). It can be seen from the graph that the conditions of low temperature and high wind speed correspond to the maximum carrying capacity of conductors, while the conditions of high temperature and low wind speed will greatly reduce the maximum carrying capacity. Specifically, when the ambient temperature is 20°C and the wind speed is 4 m/s, the maximum allowable carrying capacity is 1391A, while when the air temperature is 40°C and the wind speed is 1m/s, the maximum carrying capacity decreases to 717A. In the south of Hebei Province, in summer, the high-voltage system is often controlled by high temperature accompanied by low wind speed, which restrains the transmission capacity of conductors and increases the contradiction between industrial and residential power consumption and limited power transmission.

3.2 *Probability distribution of maximum carrying capacity*

The allowable maximum hourly carrying capacity of LGJ-400/35 conductor in southern Hebei power grid area is calculated by combining hourly automatic meteorological observation in southern Hebei. Traditionally, the current carrying capacity is set in two periods: summer half year (May-September) and winter half year (October-April). In this study, the current carrying capacity is set in four seasons. In order to compare the setting values of the maximum carrying capacity on different time scales, the probability density distribution of the maximum carrying capacity in summer and winter and four seasons is given (Figure 2).

The traditional carrying capacity is set at 590A and 799A in summer and winter respectively. After adjusting the carrying capacity on the seasonal scale, the maximum probability density in summer half year is 1051A, the probability density is 2.55 X 10^{-3}, the maximum probability density in winter half year is 1274A, and the probability density is 2.02 X 10^{-3}. The

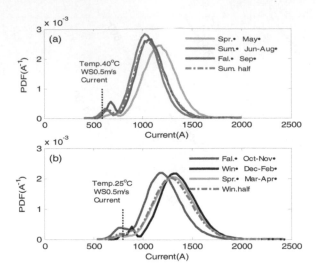

Figure 2. Probabilistic density distribution of maximum allowable carrying capacity of conductors in May-September (a) and April (b) of next year in Hebei South Power Network.

maximum load capacity of conductors shows obvious seasonal variation, with the lowest in summer and the highest in winter. From Figure 2, the maximum allowable carrying capacity calculated in different seasons will exceed the fixed value of traditional carrying capacity in most cases, which indicates that the traditional setting of carrying capacity is still conservative and can not give full play to overhead conductors. The setting of carrying capacity in season according to actual meteorological conditions is closer to the actual situation and scientific and reasonable, and can enhance the utilization rate of overhead conductors.

In addition, this paper calculates the cumulative probability distribution of the maximum carrying capacity exceeding a certain value in four seasons, and the corresponding cumulative probability can be seen as the safe operation probability of the overhead conductor in southern Hebei under this specific carrying capacity operating condition whose conductor temperature does not exceed 70°C. According to Figure 3, 1000-1500A is the range of carrying capacity with the largest probability density and drastic change of cumulative probability. The cumulative probability of the maximum allowable carrying capacity has obvious seasonal variation.

The spring and winter deviate to the direction of high carrying capacity obviously compared with summer and autumn, and both of them are far more than the traditional setting value of carrying capacity. In fact, the probability of meteorological parameters combination of extreme high temperature over 40°C in summer half year and small wind speed less than 0.5m/s is relatively small, and the meteorological conditions for setting reference of traditional carrying capacity are too harsh. If the overhead conductors in southern Hebei want to achieve 90% safety operation guarantee, the maximum carrying capacity in four seasons can be increased to 1017, 849, 903 and 1108A, respectively. To achieve 95% safety operation guarantee, the maximum carrying capacity in four seasons should be set at 945, 774, 795 and 1012 A, respectively. To achieve 97% safety operation guarantee, the maximum carrying capacity in four seasons should be divided into three parts. They are set at 884, 657, 731 and 898A.

The existence of grid operation risk will lead to the loss of conductor tension, which will impact the thermal aging process and safe operation of conductors. In order to further evaluate the operation risk of power grid, if the carrying capacity of overhead conductors in southern Hebei grid is taken as the fixed value of carrying capacity under 95% safe operation guarantee, then the temperature of overhead conductors can be calculated from actual ambient temperature and wind speed, and the thermal condition of conductors driven by fixed carrying capacity in different seasons can be obtained. Figure 4 (a) shows the probability

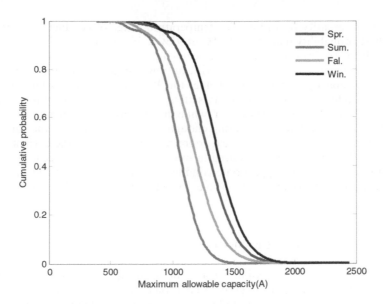

Figure 3. Cumulative probability distribution of the specific value of the maximum allowable carrying capacity in different seasons of Hebei South Power Network.

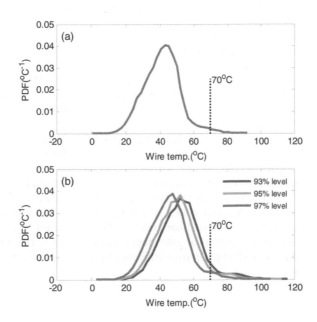

Figure 4. (a) Temperature probabilistic density distribution of overhead conductors in Southern Hebei Grid under traditional fixed carrier flow (b) Temperature probabilistic density distribution of overhead conductors at different safety levels.

density distribution of conductor temperature under the condition of traditional fixed current carrying capacity operation. The temperature range of the conductor is 0.7 to 92.3°C under the traditional current carrying capacity setting condition. The highest probability density conductor temperature is 43.0°C. A small part of the conductor temperature will exceed the maximum allowable value of 70°C. The safe operation level of the overhead conductor can meet 98.6% with the fixed value of the traditional current carrying capacity. In addition, Figure 4 (b) shows the probability density distribution of overhead conductor temperature at different levels of security operation. The temperature range of conductor is 5.9 to 112.6°C and the highest probability density is 49.7°C at 95% safe operation level. It can be seen that compared with the traditional fixed value of carrying capacity, although seasonally adjusted carrying capacity increases the operational risk of power grid to a certain extent, it enhances the transmission capacity of power grid. Under the 95% safe operation guarantee level, compared with the traditional carrying capacity setting, the carrying capacity increased by 184A (31%) in summer and 213A (27%) in winter, while the operational risk increased by only 3.6%. Therefore, under the condition of ensuring enough safe operation level, the current carrying capacity of overhead conductors has great room to improve under the current situation. In short, power grid operation managers need to weigh the economy and security of power grid operation, and determine the appropriate conductor carrying capacity and operational risk level according to the thermal extreme value and operational risk that the power grid can withstand.

4 CONCLUSIONS AND DISCUSSION

In order to fully tap the transmission capacity of overhead conductors in southern Hebei, based on the hourly actual meteorological observation data in southern Hebei and the analysis of key meteorological factors such as ambient temperature and wind speed, this paper drives the LGJ-400/35 conductor heat balance equation with the actual meteorological data, obtains the seasonal variation of the carrying capacity, and further compares and evaluates the different safe operation of the traditional carrying capacity setting. The following conclusions are drawn for the transmission capacity under probability:

(1) By investigating the sensitivity of maximum allowable carrying capacity of conductors to ambient temperature and wind speed, it is found that the higher ambient temperature is, the greater the negative rate of maximum carrying capacity varies with temperature, and the maximum carrying capacity is very sensitive to small wind speed. Moreover, the maximum carrying capacity variability is more sensitive to wind speed than temperature change, even more than an order of magnitude. Environmental wind speed is the primary meteorological factor affecting the setting of carrying capacity.

(2) Under 95% safe operation guarantee level, the seasonally adjusted current carrying capacity setting increases to a certain extent compared with the traditional fixed value of current carrying capacity, but greatly enhances the transmission capacity of the power grid. Compared with the traditional current carrying capacity setting, the summer current carrying capacity increases by 31% and the winter current carrying capacity increases by 27%.

(3) As far as Hebei South Power Network is concerned, there is still much room to improve the current carrying capacity of overhead conductors under a certain level of safe operation. Therefore, it is necessary to strengthen the assessment and research of regional meteorological spatial differences and climate change on overhead conductor transmission capacity and operational risk in the future.

REFERENCES

Dunlop, R.D., Gutman, R. and Marchenko, P.P. 1979. Analytical development of loadability characteristics for EHV and UHV transmission line. *IEEE Transactions on Power Apparatus and System*, PAS 98(2): 606–613.

Greenwood, D.M., Gentle, J.P. and Smyers, K. 2014. A comparison of real-time thermal rating systems in the U.S. and the U.K. *IEEE Transactions on Power Delivery*, 29(4): 1849–1858.

The analysis of data-based characteristics and risk on stable weather during the heating season in south-central Hebei Province

Yuanyuan Jing, Hongyu Li, Kaicheng Xing* & Jing Zhang

Key Laboratory of Meteorology and Ecological Environment of Hebei Province, Shijiazhuang, China
Hebei Climate Center, Shijiazhuang, China
Climate Center of Hebei Province, Shijiazhuang, China

ABSTRACT: The decrease in air-quality caused by stable weather during the heating season in south-central Hebei Province, brings a series of economic and social risks, such as an increase in the management of the atmospheric environment and a decrease of productivity and income. Based on the observational data from meteorological stations and ERA-Interim reanalysis data, the stable weather characteristics, which were divided into vertical and horizontal atmospheric movements, influence factors such as temperature inversion characteristics of the lower atmosphere and surface wind, in the heating season (November to March) in Hebei from 1981 to 2016. The results showed that an increase of frequency and intensity of the convergence of surface winds exacerbates the accumulation of pollutants in the horizontal direction, while an increase in the number of days, thickness and intensity of ground inversion limit the diffusion of pollutants in the vertical direction, which were unfavorable to the elimination of pollutants. Considering the weakness of the winter monsoon in the future, while the long-term likelihood of air pollution is influenced by stable weather, this paper analyzes the risk of exposure and vulnerability of humans and proposes corresponding protective measures and prevention suggestions.

Keywords: atmosphere, air pollution, risk, Hebei Province

1 INTRODUCTION

The atmospheric boundary layer is not only the main location for the diffusion and transmission of pollutants in the atmosphere, but also the main space for human production and living. Therefore, air pollution in this layer is closely related to human activities (Liu, 2005). When the source of pollution is strong, diffusion of atmospheric pollutants mainly depends on horizontal and vertical distribution of meteorological conditions. Stable weather usually refers to a low-level atmospheric feature with low wind speed near the ground and a stable atmosphere, which easily forms hazy weather. The vertical distribution of near-surface atmospheric temperatures, especially when there is an inversion layer of "upper warm and lower cold", will inhibit the occurrence of strong convections and seriously restrict the diffusion of pollutants in the vertical direction (Zheng and Shi, 2011). while, the convergence and divergence of the horizontal wind field near the ground, seriously affects the diffusion of pollutants in the horizontal direction, so that the pollutants accumulate in the lower layer, aggravating the pollution.

In recent years, with global climate change, rapid economic development and the acceleration of urbanization, a decrease in air-quality caused by stable weather during the heating season (November to March) in south-central Hebei Province brought a series of economic and social risks, such as an increase in management of the atmospheric environment and

---

*Corresponding author: Kaicheng Xing, (Email:xkc67@163.com)

a decrease in productivity and income (Xie et al., 2010; Gao and Tan, 2018). So, it is necessary to systematically analyze the characteristics of stable weather during the heating season in south-central Hebei. High-resolution reanalysis data (ERA-Interim) can be used for analysis (Gao and Hao, 2014), so vertical and horizontal atmospheric movement influence factors were analyzed, and the risk of exposure for and the vulnerability of humans, and the corresponding protective measures and prevention suggestions, were proposed.

2　DATA AND METHOD

The data was the third generation of reanalysis data of the European Medium Term Weather Forecasting Center (ECMWF), named ERA-Interim, taken four times per day (http://apps. ecmwf.int/datasets/). The spatial resolution was 0.125°*0.125°, and the vertical direction was 15 layers (1000~550 hPa) from 1981 to 2016. The observed data of meteorological stations from six stations in Hebei Province was taken four times per day from 1981 to 2016. The meteorological elements include temperature, air pressure and relative humidity.

The existence of an inversion layer, especially the ground inversion, mainly restrains the vertical movement of the atmosphere. The inversion days, thickness and intensity were analyzed when the diffusivity was strongest at 14:00 as follows. (1) Determine whether there is a ground inversion at a certain time (depending on the temperature difference between the upper and lower layers). (2) The inversion days were the calculation of inversion times, which were judged by the inversion conditions in a year, a month, a day. (3) The inversion thickness (H) refers to the thickness between the bottom and top of the inversion layer, according to the "Specifications for Upper-air Meteorological observation":

$$H = \frac{R_d}{G}\overline{T_v}(\ln p_{_b} - \ln p_t) \tag{1}$$

(4)　The inversion intensity (I) refers to the inverse increase of temperature (°C·(100 m)$^{-1}$) for every 100 m increase in the inversion layer, which depends on the inversion temperature (($t_t - t_b$)) and thickness.

The distribution of the horizontal wind field mainly restrains the horizontal movement of the atmosphere. According to the threshold, the frequency and intensity characteristics of wind convergence during the heating season are calculated by horizontal U and V wind fields. According to (Zhu, 2007), the formula of divergence is as follows, with unit 1, and positive for divergence, negative for convergence.

$$D = \frac{\Delta u}{\Delta x} + \frac{\Delta v}{\Delta y} \tag{2}$$

3　CHARACTERISTICS OF STABLE WEATHER DURING THE HEATING SEASON IN SOUTH-CENTRAL HEBEI

In order to analyze the convergence characteristics of the wind field around six cities (Shijiazhuang, Xingtai, Handan, Langfang, Baoding, and Hengshui) in south-central Hebei, we combined the calculated characteristics of ERA-Interim and the observational wind field, which defined the threshold of the convergence field.

Figure 1 showed that the convergence of wind field existed around Taihang Mountain and the south-central plain, which is also reflected in the calculation of ERA-Interim reanalysis data. Considering the area of convergence, and the accumulation on monthly convergence days, we defined -2E^{-5} as the threshold of the convergence field for subsequent calculations.

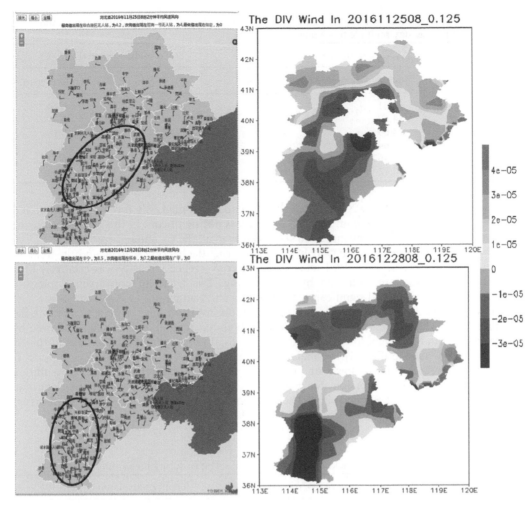

Figure 1. Comparison of convergence and divergence characteristics of wind between ERA-Interim and the observational wind field.

Table 1 shows the total convergence frequency (four times per day) in six cities during the heating season, which covers climate value (1981~2010), the past 10 years (2007~2016), and the past 5 years (2011~2016). The trend of convergence frequency was daily at 14:00 between 1981 and 2016. There are certain differences of convergence frequency between the six cities,

Table 1. The average and trend of frequency of convergence in six cities during the heating season, four times a day and at 14:00 on multiple time scales.

| City | Average of four times (%) | | | Trend of four times (%/10Y) | Trend at 14:00 (%/10Y) |
	1981~2010	2007~2016	2011~2016	1981~2016	1981~2016
Shijiazhuang	24.75	24.73	23.10	− 0.05	0.73
Xingtai	23.71	24.11	23.20	− 0.07	0.24
Handan	23.76	23.57	22.90	− 0.65	0.00
Langfang	17.89	19.41	18.64	0.45	0.08
Baoding	19.32	20.49	19.69	0.82	0.22
Hengshui	15.34	16.24	14.94	0.63	0.05

Table 2. The average and trending of total intensity of convergence in six Cities during the heating season, four times per day multiple time scales.

City	Average of divergence (E^{-5})		Trend of divergence (E^{-5}/10Y)	
	1981~2010	2007~2016	2011~2016	1981~2016
Shijiazhuang	−14.41	−14.59	−14.57	−0.14
Xingtai	−14.69	−14.70	−14.43	−0.04
Handan	−14.63	−14.78	−14.64	−0.13
Langfang	−12.64	−13.32	−13.25	−0.28
Baoding	−13.46	−13.88	−14.17	−0.31
Hengshui	−10.72	−10.98	−10.40	−0.08

which was relatively high for Xingtai and Shijiazhuang, and relatively low for Hengshui and Langfang. The trend of total convergence frequency is not significant in this period, while the trend of convergence frequency increased significantly in all the six cities.

Table 2 shows convergence intensity in six cities during the heating season, and the trend of convergence intensity between 1981 and 2016. The convergence intensity was relatively high for Xingtai and Handan, followed by Shijiazhuang. The trend of convergence intensity increased in this period, and significantly for Baoding and Langfang.

Table 3 shows the monthly mean and trend of total inversion days in six cities during the heating season. The inversion days were relatively high for Xingtai and Handan, followed by Shijiazhuang, and relatively low for Langfang. The total number of inversion days in the six cities increased, and significantly for Shijiazhuang and Xingtai.

Table 4 shows the monthly mean and trend of total inversion thickness in six cities during the heating season. The inversion thickness was relatively high for Baoding and Hengshui, and low for Xingtai and Shijiazhuang. The inversion thickness increased for all cities except Handan and Langfang, and significantly for Hengshui.

Table 3. The monthly mean and trend of inversion days in six cities during the heating season.

City	Monthly mean of inversion days (d)		Trend of inversion days (d/10Y)	
	1981~2010	2007~2016	2011~2016	1981~2016
Shijiazhuang	12.69	12.44	12.36	0.36
Xingtai	13.15	12.96	13.12	0.36
Handan	13.09	12.70	12.64	0.17
Langfang	9.32	9.00	9.00	0.01
Baoding	10.44	10.02	10.16	0.19
Hengshui	11.40	10.82	10.32	0.16

Table 4. The monthly mean and trend of inversion thickness in six cities during the heating season.

City	Monthly mean of inversion thickness (m)		Trend of inversion thickness (m/10Y)	
	1981~2010	2007~2016	2011~2016	1981~2016
Shijiazhuang	104.67	101.88	98.92	0.01
Xingtai	103.67	104.79	108.69	1.15
Handan	117.99	106.74	105.90	-3.83
Langfang	125.87	116.39	129.04	-1.07
Baoding	140.05	136.77	139.11	1.81
Hengshui	139.26	141.23	146.39	5.72

Table 5. The monthly mean and trend of inversion intensity in six cities during the heating season.

City	Monthly mean of inversion intensity (°C/100m)		Trend of inversion intensity (°C/100m·10Y))	
	1981~2010	2007~2016	2011~2016	1981~2016
Shijiazhuang	2.25	2.22	2.26	0.003
Xingtai	2.08	2.04	1.89	0.072
Handan	1.79	2.15	2.15	0.234
Langfang	1.32	1.22	1.19	−0.062
Baoding	1.42	1.42	1.58	0.024
Hengshui	1.11	1.07	1.11	0.074

Table 5 shows the monthly mean and trend of total inversion intensity in six cities during the heating season. The inversion intensity was relatively high for Shijiazhuang and Xingtai, and low for Hengshui. The inversion intensity increased for all cities except Langfang, and significantly on Handan. As with inversion days and thickness, the increase in inversion intensity suppresses the diffusion of pollutants.

According to (Jiang and Tian, 2013), the predicted weakness of the winter monsoon in the next few decades is conducive to the long-term stability of the weather. Therefore, we undertook a multi-faceted risk analysis of stable weather and accompanying hazy weather.

4 THE RISK CHARACTERISTICS OF STABLE WEATHER

4.1 *The risk for traffic safety*

In recent years, traffic safety has been severely affected by bad weather, resulting in frequent traffic accidents and disruptions. (Zhang, 2018) With frequent occurrences of hazy weather, the hidden dangers to road traffic safety can be caused by a deteriorated operating environment and wet roads, which seriously affects people's day to day lives. A series of problems, such as train delays, frequent accidents on the road, and the close of high-speed roads, have created a major threat to personal safety and property.

According to the characteristics of traffic systems and accident characteristics of hazy weather on highway, Liu et al. (2018) noted that when the speed and smog level increase, the probability of traffic accidents increases squarely. Most accidents in hazy weather are rear-end collisions, accounting for 61.46%. According to relevant statistics, accidents involving more than two vehicles accounted for 57.29% of collisions in hazy weather (Zhang, 2018).

4.2 *The risk to human health*

In recent years, due to rapid development of industry, high density population and the high concentration of pollutants, Beijing-Tianjin-Hebei and its surrounding areas have become one of the most serious areas of haze pollution in China, which has huge health risks to the population exposed to it.

Haze has many effects on health. Some pathogenic bacteria increase in hazy weather, and air pollutants can enhance this (Zheng, 2019), After the air pollutants enter the lungs, the particulate matter has a serious clogging effect, which decreases the ventilation, affecting the immune system and increasing the risk of cancer. People living long-term in cities with severe air pollution have a 10% – 15% higher risk of lung cancer than those in non-polluted cities.

4.3 The risk to industry

The hazy weather has brought new challenges to many industries such as agriculture, tourism and aquaculture. For the aquaculture industry, the unfavorable diffusion conditions caused by stable weather will aggravate the deterioration of living environment of animals, which is unfavorable to exchange fresh air, resulting in the decrease of immunity, even to the pestilence.

Beijing-Tianjin-Hebei region is rich in tourism resources and also has serious air pollution. Studies have shown that hazy weather has a strong influence on choice of tourist destination, and is positively correlated with the severity. In the hazy state, the weight of travel resources dedicated to the environment, traffic, and tourism factors is different from usual. The highest proportion of tourism experience is 0.40 and the resource environment is 0.32. Severe hazy weather will inevitably lead to loss of tourists, affecting tourism revenue, and also has an immeasurable negative impact on the city's tourism economic development. (Hao, 2017).

5 MEASURES AND SUGGESTIONS

In view of the climate conditions for pollutant diffusion during the heating season in south-central Hebei, people exposed to haze should educate themselves on prevention and take appropriate protective measures. For the general public, wearing effective isolation masks can greatly reduce dust particles entering the body, and because of the higher content of microbials in haze, it is recommended to reduce ventilation on hazy days. To avoid long-term outdoor aerobic exercise, take breathing healthy air as the premise of fitness and determine the optimal fitness mode. An appropriate supplement of vitamin D can improve resistance to a certain extent.

The control of haze has become China's strategic goal. First, we should base this on current social contradictions, learn from advanced technologies, and develop a low-carbon economy to use clean energy technologies. According to data from the *People's Daily Online*, in 2016 the proportion of clean energy power in the UK's electricity supply exceeded 50%, a significant increase from 25% in five years ago. Second, combined with dealing with the haze at the source (Pei and Shang, 2017), we should also prepare for the transition period of energy, and improve and strengthen the management and supervision of air pollution. At the same time, we should monitor the regional air-quality in all directions. With the help of advanced scientific and technological information, network-based supervision can optimize the development structure on urban spaces and improve urban traffic congestion problems (Gao and Tan, 2018). Finally, we should enhance people's environmental awareness and formulate practical and feasible environmental protection measures to return the blue sky to the people.

ACKNOWLEDGMENTS

This study is supported by the National Key Research and Development Program of China (NO. 2018YFA0606302).

REFERENCES

China Meteorological Administration. 2010. *Specifications for Upper-air Meteorological observation.* China Meteorological Press: Beijing, 1976, 71–83. (in Chinese).
Gao, F. and Tan, X. 2018. Evolution model and risk analysis of urban haze disaster chain. *Science & Technology Review.* 36(13): 73–81. (in Chinese).
Gao, L and Hao, L. 2014. Verification of ERA-Interim Reanalysis Data over China. *Journal of Subtropical Resources and Environment.* 9(2): 75–81. (in Chinese).
Hao, C.L. 2017. Analysis of tourist destination selection under haze. *Scientific and technological innovation Information.* 35: 4–5. (in Chinese).

Jiang, D.B and Tian, Z.P. 2013. East Asian monsoon change for the 21st century: Results of CMIP3 and CMIP5 models, *Chinese Science Bulletin*, 58(8): 707–716. (in Chinese).

Liu, D.S. 2005. Harmonious development of human and nature, *Arid Land Geography*. 28(2): 143–144. (in Chinese).

Liu, Z.Q., Wang, L., Zhang, A.H., et al. 2018. Study on Traffic Accidents Occurrence Mechanism in Haze Weather on the Highway. *Journal of Chongqing Institute of Technology*, 32(1): 43–49. (in Chinese).

Pei, G.F and Shang, W. 2017. The International Experience of Haze Management and Enlightenment. *Economic Relations and Trade*. 8: 17–20. (in Chinese).

Xie, P., Liu, X.Y., Liu, Z.R., et al. 2010. Impact of exposure to air pollutants on human health effects in Pearl River Delta. *China Environmental Science*. 30(7): 997–1003. (in Chinese).

Zhang, T.T. 2018. *Research on Road Traffic Flow Characteristics in Fog and Haze Weather*. Lanzhou Jiaotong University: Lanzhou, 1–5. (in Chinese).

Zhen, Q., Fang, Z.G., Wang, Y.Q., et al. 2019. Bacterial characteristics in atmospheric haze and potential impacts on human health. *Acta Ecologica Sinica*. 39(6) 1–10. (in Chinese).

Zheng, QF. and Shi, J. 2011. Temperature Inversion Characteristics of Lower Atmosphere over Shanghai. *Arid Meteorology*. 29(2): 195–200. (in Chinese).

Zhu, Q.G. 2007. Principles and methods of meteorology. China Meteorological Press: Beijing. (in Chinese).

Risk Analysis Based on Data and Crisis Response Beyond Knowledge – Huang & Nivolianitou (eds)
© 2020 Taylor & Francis Group, London, ISBN 978-0-367-25146-8

Difficulties and countermeasures of risk-based urban public safety assessments for high-impact weather

Kaicheng Xing
Key Laboratory of Meteorological and Ecological Environment of Hebei Province, Shijiazhuang, China
Hebei Climate Center, Shijiazhuang, China

Daohong Chen & Shujun Guo*
Hebei Meteorological Bureau, Shijiazhuang, China

ABSTRACT: Against the background of climate change, the occurrence and intensity of high-impact weather and climate events are increasing, and the scope and extent of public safety risks in high-impact weather cities are getting larger and deeper. Therefore, guaranteeing urban public safety means higher requirements for the space-time accuracy of high-influence weather forecasting and early warnings, and the pertinence and effectiveness of defensive measures. Risk-based urban public safety assessments for high-influence weather is the basis for solving these key problems. In this study, we explore the public safety incidents in high-impact weather cities in China in recent years, analyzing not only the complexity and urgency of the problems, but also the difficulties to be solved. Further, we enhance and strengthen the foundation of urban public safety risk assessments, implement an open and transparent assessment strategy, and upgrade unified administrative assessments to diversified assessments. Finally, we improve the current assessment system, enhance the scientific nature of risk assessments and offer suggestions for the application mechanism of the assessment results.

Keywords: high-influence weather, risk-based urban public safety, assessment difficulties, countermeasures

1 INTRODUCTION

Urbanization has led to changes in the underlying land surface, population increase, and increased emissions of atmospheric pollutants, making the urban climate characterized by urban heat islands, rain islands, low humidity, and low winds. In the context of global warming, extreme weather and climate events are frequent. High-impact weather poses serious risks to people's quality of life, safety of life and property, and normal operation of social and economic activities, especially as the scope of influence on public safety risks is growing, and the degree of influence is becoming deeper in large cities. In recent years, several high-impact weather events in large cities in China have affected public safety. These have received extensive attention from all walks of life and are key topics of online discussions. Defending against high-impact weather risks is an important part of urban public safety and has also received increasing attention. Adapting to the realistic needs of climate change in urban safety operation management, improving the time and space accuracy of forecasting and warning about extreme weather and climate events, and effectively carrying out risk-based urban public safety assessments for high-

---

*Corresponding author: hchongfu@126.com

impact weather, are all key to improving the pertinence and effectiveness of disaster prevention measures to safeguard the public.

2 CASES OF HIGH-IMPACT WEATHER THREATENING URBAN PUBLIC SAFETY

2.1 *Light snow caused the city traffic to be almost paralyzed in Beijing city*

On December 7, 2011, from 1 pm to 2 pm there was a light snowfall of only 1.8 mm in Beijing. At the beginning of snowfall, the atmosphere and surface temperature in the near-surface layer was higher, so snowflakes quickly absorbed latent heat and melt. The underlying surface temperature dropped rapidly and formed an ice surface, which prevented heat exchange between the road surface and the atmosphere, and its reflection further accelerated the temperature drop near the surface (Figure 1). This "falling snow into ice" subsequently caused tremendous social impact as hundreds of thousands of motor vehicles in Beijing's urban areas were unable to drive normally, causing almost complete traffic congestion. Millions of citizens were trapped in cars in peak rush hour, or faced a long journey trekking through the streets until late at night.

Beijing has a typical East Asian monsoon climate, and snowfall in winter is a common weather phenomenon. In the past, even if the snowfall was much deeper or the weather conditions were more serious than the "12.7" incident, the impact on Beijing's urban traffic was much less. The city's secondary disasters caused by high-impact weather are at special times (a large number of people return home during the weekend), in specific cities (in the rapidly developing metropolis the traffic volume has doubled), and on special road conditions (there are a large number of slopes formed by overpasses in the urban transportation network structure), etc. Non-meteorological factors (important environmental conditions) and other adverse weather conditions interact, and falling snow is a direct trigger for serious urban traffic.

The "high pressure reflow type" snowfall is a typical weather process in North China. At that time, the weather forecasting technical conditions were difficult, with weaker information, on a smaller scale and for a shorter duration [1]. Today, with the accuracy of time-space resolution of the numerical prediction models, the density of the meteorological observation network has been greatly improved and the ability to forecast and warn about such high-impact weather has been improved. However, if basic work such as risk-based high-impact weather urban public safety assessment is not well done, it is difficult to improve the temporal and spatial accuracy of forecasting and avoid repeating some events.

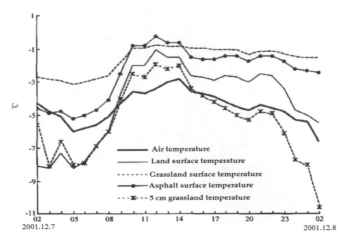

Figure 1.   The deceleration of surface temperature of different physical media over 24 hours, december 7 to december 8, 2001 [1].

2.2 Upstream mountain torrents the supply of tap water, which is interrupted for several days in Shijiazhuang city

From 3:00 pm on July 18 to 14:00 on July 21, 2016, Hebei Province experienced extraordinary heavy rain, with the largest range and the longest duration (63 hours) of rain since "63.8" (the August 1963 rainstorm). The accumulated rainfall on the Taihang Mountains in the west of Shijiazhuang, Xingtai and Handan City and the Yanshan Mountain in Qinglong County of Qinhuangdao City, exceeded 400 mm (Figure 2). There are 15 observation stations (in Zanhuang, Pingxiang, Jingjing, Yuanshi County of Shijiazhuang City, Lincheng County of Xingtai City, Cixian, Fengfeng County of Handan City, and Yixian County of Baoding City) that reported rainfall of more than 600 mm and the maximum precipitation process was 816.5 mm (in Zhangshiyan of Zanhuang County).

The continuous heavy rain caused floods in the central and southern Taihang Mountains, and the eastern Yanshan Mountains. The flood occurred once in more than twenty years and threatened 104,561,100 people in 152 counties (cities, districts) of 11 prefecture-level cities across the province, and according to incomplete statistics it caused direct economic losses of more than 50 billion yuan.

During the rainstorm, the precipitation in the Shijiazhuang urban area was more than 100 mm, which did not directly cause casualties and property losses. However, the Gangnan and Huangbizhuang Reservoirs (300,000 tons per day), which are responsible for 85% of the public water supply tasks in the main urban area of Shijiazhuang, carried a large amount of sediment and floating debris into the reservoir area due to the upstream flash floods. The amount of sediment and turbidity increased sharply, in addition to the technical accidents caused by human judgment and operational errors, and the treatment capacity of surface water plants almost haltd. The water supply decreased by 250,000 tons per day, directly affecting the urban residents' domestic water consumption and various economic activities. The Shijiazhuang Municipal Government initiated an emergency plan and issued timely official information to positively explain the work. Much work on repairing, dredging, cleaning and disinfection has been done. More than 200 vehicles were dispatched to provide water to citizens around the clock, giving priority to the basic living standards of residents. A temporary pipe network was laid to increase the amount of water from the Yangtze River from 100,000 m³/day to 350,000 m³/day. Through integrating the self-prepared wells of the

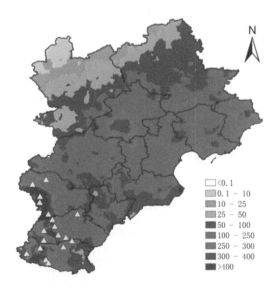

Figure 2. Beijing-Tianjin-Hebei process rainfall from July 18 to 21, 2016 (the yellow triangle marks the 18 severely affected counties).

enterprise into the urban water supply network, the water supply of the groundwater plant increased from 70,000 m^3/day to 100,000 m^3/day. By 16:30 on July 24, the urban public water supply capacity had gradually returned to normal.

During the rainstorm, both weather forecasting and rain notification services, as well as defense command and emergency responses, were accurate, timely, scientific and effective. However, there was a problem with the supply of tap water in the urban area of Shijiazhuang City, which was not fully taken into consideration This also provides a "live" case for the public safety impact assessment of high-impact weather-based cities in order to better deal with extreme weather and climate events.

3   RISK-BASED HIGH-IMPACT WEATHER URBAN PUBLIC SAFETY ASSESSMENT DIFFICULTIES

Risk-based high-impact weather urban public safety assessments have the general characteristics of urban public safety risk assessments, and have special background and actual requirements . Under the background of global warming, extreme weather and climate events are frequent, the urban population is highly dense and fast-moving, economic factors are highly accumulated, and political, cultural and international exchange activities are frequent, so large cities have become the hardest hit areas for high-impact weather public safety risks [2]. The frequent occurrence of major security incidents caused by high-impact weather in large cities has exposed many problems in public safety risk assessments. It is an inevitable choice to incorporate high-impact weather urban public safety assessments as the first line of defense and core analysis frameworks for risk management into government management functions [3].

3.1   *Basic requirements for urban public safety risk assessment*

The urban public safety risk assessment includes three parts: vulnerability assessment, hazard assessment and disaster reduction capability assessment. The focus is on assessing public safety areas such as natural disasters, public activities, accident disasters, and social security. From the risk assessment parameter system, the threshold setting of each parameter, the determination of the probability of occurrence of the risk, and the mapping of the risk matrix, a complete risk assessment system adapted to the pilot site is formed [4]. For each identified risk, it is necessary to comprehensively analyze the probability of occurrence of the risk and the severity of the consequences and compare the risk matrix map to assess the risk level and determine the value of the risk. The possibility of occurrence of risks, from low to high, is divided into low-level, medium-level, high-level, and extremely high-level. Various types of risk sources at various levels can be drawn on an electronic map to provide a basis for relevant departments to timely rectify and effectively respond to public security incidents. The risk assessments of the 2008 Beijing Olympic Games and the 2010 Shanghai World Expo [5] were fruitful. In October 2012, Shenzhen launched the city's public safety assessment, becoming the first region in China to carry out an urban public safety assessment. In June 2016, Guangzhou completed the "Guangzhou Urban Security Risk Assessment" [6], which was the first risk assessment work carried out in China for urban-level safety production.

3.2   *Characteristics of risk-based urban public safety assessment for high-impact weather*

Compared with the safety assessment of public activities, accident disasters, social security and other watersheds, the risk-based urban public safety assessment for high-impact weather is special and complex as the actual demand is more urgent and the level of demand is higher. The main conditions are as follows:

1)   More people can be affected by high-impact weather. The development of science and technology has meant urban residents have a wider range of activities, higher quality of

life and faster work pace. High-impact weather affects a wide geographical area, which can directly or indirectly affect the quality of life, safety of life and property, and social and economic activities of the masses, which can involve more people than other public security events.

2) More fields can be affected by high-impact weather. High-impact weather may pose a threat to urban infrastructure, such as urban transportation, water and electricity supply, and communications, which can influence more widespread fields than other public safety events.

3) More complex hazard factors can be invoked by high-impact weather. Meteorological disasters account for more than 70% of natural disasters, with many types, frequent occurrences, and rapid conversion of symbiosis. The hazard factors of high-impact weather have complex relationships with their location, intensity level, duration and environmental conditions, so it is more difficult to analyze and evaluate and has higher technical content.

4) The vulnerability of the disaster-bearing body is more difficult to grasp. The vulnerability of urban public safety hazard bodies is closely related to meteorological hazard factors, which may interact and change. This requires more timeliness and effectiveness of the emergency responses.

5) Meteorological hazard factors are predictable. With the improvement of meteorological science and technology, the predictability and forecasting and early warning capabilities of high-impact weather hazard factors are constantly improving. Compared with other security risk assessments, the results of risk-based urban public safety assessments for high-impact weather are more useful.

4   DIFFICULTIES OF RISK-BASED URBAN PUBLIC SAFETY ASSESSMENTS FOR HIGH-IMPACT WEATHER

4.1   *The principle of implementation of duties is not clear*

The subject of risk assessment is single, and the principle of implementing "government-led, professional assessment, and public participation" is not strict. The decision-maker "is both an athlete and a referee", the independence and objectivity of professional teams and institutions are not sufficient, the level of public participation is not deep and there are not many ways to participate. There is a shortage of professionals, and only a few particularly valuable results can enter into government decisions.

4.2   *Top-level design and institutional system are not sound*

Top-level design has the tendency of "emphasizing emergency management and ignoring risk assessment", and lacks understanding of the importance of risk-based urban public safety assessment for high-impact weather. It is not focused enough, and it is impossible to carry out unified planning and systematic design from the strategic level of urban public safety management. The laws, regulations and institutional systems are imperfect, lacking scientific and systematic theoretical support and guidance, and the standardization and systematization of mechanisms, principles, indicator systems, technical methods, models, basis, and procedures are low.

4.3   *The guidance results are not sufficiently informative*

The existing technical standards are mostly empirical judgments such as qualitative analysis or semi-quantitative analysis, there are few indicators for classification and quantification, and the accuracy of the evaluation results is insufficient. The technical route of risk assessment, the basic method of risk determination, the selection and setting of evaluation indicators, the weight measurement and model of evaluation indicators, and the low technical content of these aspects restricts the predictability and authority of the results.

The results of the risk assessment are the premise and basis for drafting an emergency plan, however, the reality is that this analysis and development of the risk assessment results are not enough. The establishment of mechanisms, the formulation of risk-reduction decisions and measures [7], and the effective control, mitigation of risks and dynamic tracking are difficult to implement, and the functional role of evaluation is difficult.

5 COUNTERMEASURES FOR STRENGTHENING RISK-BASED URBAN PUBLIC SAFETY ASSESSMENTS FOR HIGH-IMPACT WEATHER

In order to solve the above problems through systematic research, we propose the following countermeasures:

5.1 *Strengthen risk management awareness and build a "foundation" for public safety risk assessment of urban meteorological disasters*

Implement the principle of "safety first, prevention first", strengthen the assessment of theoretical guidance, construct a theoretical system of risk-based urban public safety assessment for high-impact weather with Chinese characteristics. Improve the legal and regulatory system and form a complete institutional framework. Accelerate the cultivation of professional talent and improve the level of specialization and technical support.

5.2 *Establish an effective evaluation mechanism and implement a diversified, open and transparent assessment structure*

Construct a benign collaborative assessment model, separating decision-making from evaluation functions, standardizinf institutionalized procedures, and guiding the public and professional organizations to participate in risk assessment and provide assessment services in an orderly manner. Improve the openness of risk assessment and improve the transparency of risk assessment. Strengthen the supervision of professional departments by government departments, ensuring the independence and objectivity of evaluation, and improving the credibility of professional evaluation agencies.

5.3 *Improve the evaluation technology system and enhance the scientific nature of risk assessment*

According to the complexity of high-impact weather risk assessments and urban spatial heterogeneity and comprehensiveness, adhere to the principle of combining quantitative analysis with qualitative analysis. Use comprehensive analysis methods, such as risk matrix analysis, analysis flow chart, mathematical model, scenario construction, etc. [8]. Make full use of new technologies such as GPRS, GIS, and database technologies to improve the accuracy and quality of risk assessment.

5.4 *Conduct dynamic tracking assessments to improve the timeliness of risk assessment*

Combine the current development of urban areas, and apply advanced research results of urban meteorological disasters, secondary disasters, and derivative disasters, using numerical weather prediction model products and the latest assimilation data. Continuously optimize and improve the results of risk assessments and effectively supports the government management, providing scientific and technological support for urban risk census and public safety risk management.

5.5 *Improve the application mechanism of assessment results and fully develop the role of risk assessment*

Focusing on the transformation of government functions and helping to innovate governance methods, transform process control into results orientation. Taking risk prevention as the core, the meteorological risk assessment is integrated into the urban public safety management and even the government's decision-making scientific, democratic, and legalized construction [9]. The risk assessment and the emergency response plan are closely linked, and one risk and one policy or multiple measures are taken together to achieve the treatment of both the symptoms and the root causes.

In short, strengthen the risk management of urban meteorological disasters, improve the system, technological innovations and application of results of public safety risk assessments for high-impact weather cities. This will help to explore a management system that is suitable for the safe operation of cities and improve the capacity of urban vulnerability management and risk management. It will contribute to forming a pattern of urban governance work between the management departments that are conducive to the safe operation of the city, such as information communication, resource sharing, rapid response, and strength integration. It will also help to reduce urban public safety risks of high-impact weather events in the context of global warming, to promote safe urban operations, and protect the lives and property of the people.

ACKNOWLEDGMENTS

This study is supported by the National Key Research and Development Program of China (NO. 2018YFA0606302).

REFERENCES

[1] Sun, J.S. et al. 2003. An analysis on serious city traffic trouble caused by light snow in Beijing. *Journal of Atmospheric Sciences* 27(6): 1058–1065.

[2] Brecken, P. et al. 2014. *Strategic Management of Emergencies: Risk Management and Risk Assessment.* Beijing: Central Compilation Press.

[3] Zhong, K.B. 2011. Internationalized metropolis risk management: challenges and experiences. *China Emergency Management* (4): 14–19.

[4] Xie, Z.C. 2011. *Emergency Management: Operational Models and Practices with Chinese Characteristics.* Beijing: Beijing Normal University Press.

[5] Rong, Z. 2012. Construction of urban public safety management system under the perspective of risk prevention and control—an empirical analysis based on Shanghai World Expo. *Theory Monthly* (4): 147–151.

[6] Sun, Y.W. et al. 2012. Application of risk assessment model in Sino-German disaster risk management pilot project in Baoan District, Shenzhen. *China Emergency Management* (2): 20–27.

[7] Zhang S.C. et al. 2015. Discussion on improvement of risk assessment and accident investigation. *Safety, Health and Environment* (12): 7–10, 27.

[8] Zhu, Z.W. et al. 2006. Analysis of China's regional public security evaluation and related factors. *China Public Administration* (1): 39–42.

[9] Rong, Z. 2014. *From Decentralization to Integration: Research on Public Risk Prevention and Control Mechanism in Megacities.* Shanghai: Shanghai People's Publishing House.

Risk Analysis Based on Data and Crisis Response Beyond Knowledge – Huang & Nivolianitou (eds)
© *2020 Taylor & Francis Group, London, ISBN 978-0-367-25146-8*

Study on ecological risk of mountain disasters in Nanling nature reserve

Qinghua Gong & Long Yang
Guangdong Open Laboratory of Geospatial Information Technology and Application, Guangzhou, China
Guangzhou Institute of Geography, Guangzhou, China

Yusi Li*
Xi'An University of Science and Technology, Xian, China

Junxiang Zhang
Tourism College, Huangshan University, Huangshan, China

ABSTRACT: Mountain disasters occur mostly in mountainous areas with steep slopes. Most mountainous areas are ecosystems with a relatively stable ecosystem structure and bear important regional ecological service functions. Mountain disasters destroy the components of the ecosystem of the affected area, which in turn affects the structure, function and health of the ecosystem. This study takes the forestry unit as the basic unit, mountain disasters as the risk source, and Nanling's various natural ecosystems as the risk receptors. Based on the mechanism of mountain disasters and the interaction mechanism between mountain disasters and ecosystems, the risk of mountain disasters, the vulnerability of risk receptors and the variability of ecosystem responses were evaluated in Nanling on the small-scale forestry scale.

Keywords: mountain disasters, ecological risk, nature reserve, land destruction

1 INTRODUCTION

Regional ecological risk assessment is an uncertain accident or disaster that may adversely affect a particular regional ecosystem and its components, functions and structures. Ecological risk assessment is the process of assessing the likelihood of adverse ecological impacts (Wang et al., 2014). In recent years, research on the impact of mountain disasters has focused on the loss of personnel, life, and property caused by disasters and changes in land use and paid less attention to the impact of disasters on ecosystems (CHEN et al., 2016; PENG et al., 2016). Mountain disasters frequently occur in mountainous areas with steep slopes. Most mountainous areas are ecosystems with relatively stable ecosystem structures and functions that often bear important regional ecological service functions. Mountainous disasters cause strong surface changes and ecological disturbances to destroy the components of the ecosystem of the affected area, thereby affecting the structure, function and health of the ecosystem (OU et al., 2008; WU et al., 2008; ZHANG et al., 2008). The Nan Ling mountain area has complex topographical and climatic conditions and has natural characteristics for nurturing mountain disasters. The Nan Ling mountain is a national nature reserve, an important ecological barrier and an ecological environment sensitive area in Guangdong province (DONG et al., 2012). The Nan Ling mountain ecosystem is threatened by mountain disasters such as landslides that are caused by geological environmental conditions and human activities, especially in the areas along the road where the mountain disaster is the most

*Corresponding author: 934350471@qq.com

advanced and the most serious. Under the influence of the unique climatic conditions of Nan Ling, the size of a single mountain disaster is small, but it is clustered in the area, and this cluster has a cumulative amplification effect which intensifies the risk of ecological damage of the disaster. This research focuses on the ecological risk assessment of mountain disasters, which is to identify mountain disasters as risk sources, taking all kinds of natural ecosystems in Nan Ling as risk receptors and describing and evaluating the possibility and harm degree of mountain disasters adversely affecting the structure and function of the Nan Ling mountain ecosystem on a regional scale. The ecological risk assessment and management of natural disasters has been recognized as an important part of the regional disaster prevention and mitigation strategy. The refined management of ecological risks is an important goal of ecological risk development (XU et al., 2011). The current risk assessment unit is mostly based on administrative units or grid units. For the Nan Ling area, the size of the single disaster unit is small, the spatial scale of the administrative unit is huge. Mountain disasters lack spatial coherence and are more difficult to control through ecological risk management. The small forestry class is under the conditions of comprehensive consideration of topography, ownership, tree species, and stand structure to determine basic zoning units for forest resource statistics and management. On the one hand, the small class of forestry can have a good consistency with the mechanism of disaster formation and destruction. On the other hand, it is also the management unit. Therefore, the selection of forestry small classes as the basic unit is the basis for ecological risk management.

This paper attempts to study the differences in mountain hazards, risk receptor vulnerability and ecosystem response based on the small-scale forestry unit, the mechanism of mountain disasters, and the interaction mechanism between mountain disasters and ecosystems. We form an ecological risk assessment system for mountain disasters. The risk of ecological damage caused by mountain disasters in the Nan Ling area is evaluated from the small-scale forestry scale, which provides a scientific basis for Nan Ling ecological security protection, disaster prevention and mitigation.

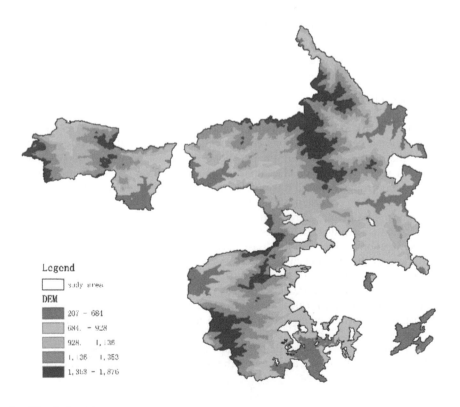

Figure 1. The study area.

2   STUDY AREA

The Nan Ling Range is the largest mountain in southern China and an important natural geographical boundary. It is the watershed between the Pearl River and the Yangtze River. It plays an important ecological role in South China. The study area is a typical subtropical climate with an average annual rainfall of 1705 mm. The horizontal zonal soil in the study area is red soil. Due to the large difference in the relative height of the mountain and the vertical distribution of hydrothermal conditions, the distribution of soil types is differentiated by height. Yellow soil is distributed below 1500 m, latent yellow soil distributed at 1500–1800 m, and above 1800 m is meadow soil. Due to its unique climatic conditions, Nan Ling mountain is an oasis with the largest remaining area in the world at the same latitude, with various types of forest vegetation including subtropical evergreen broad-leaved forest, coniferous and broad-leaved mixed forest, coniferous forest and hilltop lowland (LI et al., 2008; LIU et al;2009; WEI et al., 2011; CHEN et al., 2015; CHEN et al., 2008; ZHU et al., 2014; YANG et al., 2010; LUO, 2008).

3   MATERIALS AND METHOD

3.1   *Materials*

The data used in this paper mainly includes topographical maps, geological maps, land use status maps, geological hazard distribution, precipitation, vegetation types, and vegetation biomass statistics, which were obtained through field surveys and data collection. The field survey data includes: forest survey data, fixed plot survey data, small class in forestry attribute database, and geological disaster distribution. The terrain data is the DEM digital elevation data of the STM with a resolution of 30 m. The small class in forestry is used to calculate the slope, aspect, and terrain relief of each unit. Land use data is based on the database data on land use status. The geological data comes from the 1:2.5 million Chinese geological map vector files. Vegetation types and vegetation biomass are mainly derived from small class in forestry data and remote sensing data.

3.2   *Methods and indicator systems*

The ecological risk assessment of mountain disasters in this paper is to evaluate the risk of damage to ecosystem components, ecological structure and ecological function caused by the population or forest ecosystem vulnerability of the study area under the action of mountain disasters. The widely used ecological risk assessment method is the model PSR (Pressure-State -Response) model framework (NI et al., 2014). The model framework model has a very clear causal relationship. Natural disaster risk assessment based on the development of risk PSR causal chain model usually considers the risk source risk (H), risk receptor vulnerability (E), and Disaster Prevention Ability (D) in the process of natural disaster formation (ZHANG et al., 2007; SHANG et al., 2010; DU et al., 2016).

Ecological Risk (Risk) = Hazard (H) × Vulnerability (E) × Disaster Prevention Ability (D) (1)

This study takes the main mountain disaster as the risk source and takes the forest ecosystem of the Nan Ling Nature Reserve as the risk receptor, taking into account the risk of mountain disaster, the vulnerability of the forest ecosystem, and the relative ecology of forest ecosystem disaster prevention and resilience. The mountain hazard assessment is based on the formation mechanism of mountain hazards in the study area, and comprehensively calculates the possibility of various hazard factors affecting the occurrence of landslide disasters (LI, 2016; WANG, 2016). Vulnerability is a direct and potential loss of ecosystem damage caused by mountain disasters (Peng et al., 2014). This paper is based on the evaluation system of the forest ecosystem service function. The ecosystem disaster prevention capability is to determine the role of forest ecosystems in disaster prevention and mitigation through the mutual relationship between forest ecosystems and mountain disasters. This paper judges the disaster prevention capabilities of different vegetation ecosystems based on the size of roots of different vegetation types..

The primary task of disaster ecological risk zoning is to determine the scale and basic spatial unit of the zoning. Mountain disasters and ecosystem characteristics are mainly affected by precipitation conditions, topographic conditions, soil conditions, etc. (GONG et al., 2012). The formation process is controlled by natural zoning units and disasters in the study area. The scale is small because if the scale is too large it will seriously affect the risk management effect, and the risk management is bounded by the administrative unit. In the risk zoning process, on the one hand, there should be consistency of each factor within the unit and between the units, as much as possible. Differences, on the other hand, should choose appropriate spatial scales and facilitate ecological risk management. The small class in forestry is a basic division unit for forest management that is determined by a comprehensive consideration of natural conditions and management conditions. Small-scale forestry classes are more suitable for ecological risk assessment from the formation mechanism, spatial scale and management precision. In view of this, this paper chooses small class in forestry as the basic unit of ecological risk zoning. Based on the small-scale data of forestry surveys, the risk index, vulnerability index, and disaster prevention force are used as evaluation indicators to construct an ecological risk assessment system (LI et al., 2013; ZENG, 2010). After multiple indicators' comprehensive calculation, the ecological risk feature space is obtained. The distribution map, analysis and research on ecological risk characteristics are shown in Figure 2.

3.2.1 *Mountain disaster hazard*
The mountain disasters in the Nan Ling area are mainly landslide-type landslides and collapses, which are characterized by their scattered distribution, small-scale and poor stability. The rocks in the study area are mainly granite and the surface layer covers a thicker weathered layer. The weathered rock and soil form obvious zoning on the section, and the original rock structural surface remains in the strong weathering zone and the middle weathering zone, forming a relative geotechnical interface. Among the different weathering zones, the grain

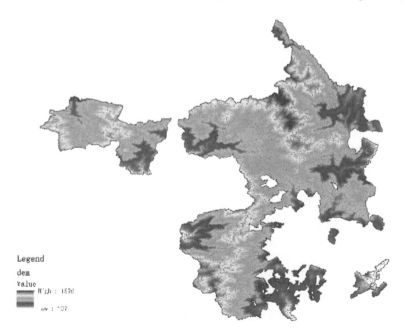

Figure 2.   The DEM of small class unit in forestry.

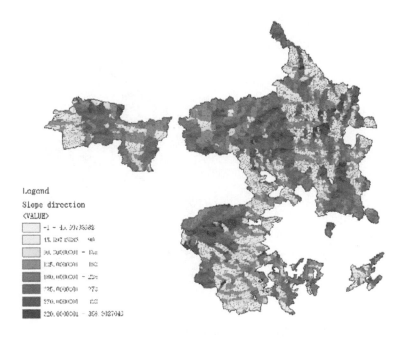

Figure 3.   The slope directions of small class unit in forestry.

composition of the rock and the porosity between the particles are also relatively different (Ross D J K et al., 2009; Toelle B E et al., 2011; Howarth R W et al., 2011). In the runoff caused by long-term rainfall, when the water penetrates downward, a relative groundwater runoff zone and a relative water-blocking interface (water accumulation zone) are formed. The continuous infiltration of rainfall replenishment water causes a temporary high

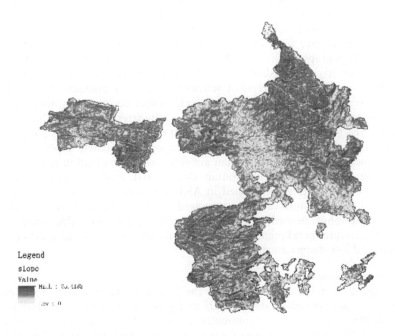

Figure 4.   The slope of small class unit in forestry.

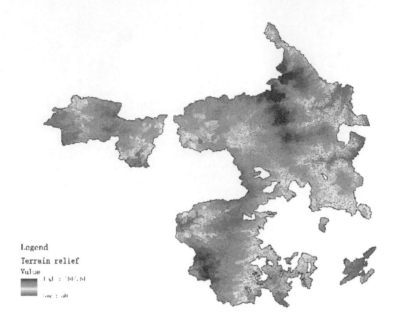

Figure 5.   The terrain relief of small class unit in forestry.

groundwater level in the loose rock layer of the slope and forms a sliding surface at the opposite water-blocking interface, which leads to the occurrence of a sloping surface landslide under the influence of geotechnical gravity. The risk of mountain disasters mainly considers the stability of the environment in which disasters occur and focuses on both internal and external factors. The mountain disasters in the Nan Ling area are mainly landslide-type landslides and collapses. The terrain relief, slope and elevation directly affect the landslide intensity and distribution (LIU et al., 2011). The formation lithology is the basis for triggering landslides. The slope runoff formed by long-term rainfall and the water infiltration causes the rock layer to be loose, and the landslide occurs under the action of soil gravity. The effect of slope runoff is represented by the distance from the main river. In addition to the intrinsic factors such as slope, elevation, topographic relief, and formation lithology, rainfall is one of the important external factors that trigger landslides. In addition, human activities are also important external factors that cause landslides. Human activities are mainly manifested in road cuts in the protected area, residential building and small hydropower construction, which have changed the original environmental geological conditions, and the stress balance of the rock (soil) body has been destroyed (Chang et al., 2013). Based on the above analysis, this paper will consider several factors such as multi-year average rainfall, topographic relief, slope, elevation, stratum lithology, distance from rivers, and human activities as the evaluation index system for the risk of mountain disasters. First, a mountain disaster risk analysis attribute table based on small class in forestry units is established in ARCGIS. The seven factors are weighted and summed up, and the maximum value is normalized to obtain the risk value of each small unit. The weight is determined according to the AHP method. The risk value range is 0–1. Then, according to the natural breakpoint method, the evaluation unit is divided into three levels of high, medium and low danger zones.

3.2.2 *Ecosystem vulnerability*
The direct impact of mountain disasters on the Nan Ling ecosystem is manifested in two aspects. First, the destruction of shrub layers and grass layers caused by mountain disasters causes the destruction of forest community levels, and the stability of the community declines. Second, the mountain disaster destroys the structural stability of the soil layer, causing the topsoil and vegetation to be destroyed, resulting in an increase in regional surface

exposure and an imbalance in the balance between topsoil and vegetation. The original terrain slope, slope length, etc. have also been changed, destroying the original balance. The topsoil structure is destroyed, and it is easy to cause soil erosion. In addition, changes in landscape elements caused by mountain disasters have caused changes in the landscape structure (GAO et al., 2011; PENG et al., 2015). The comprehensive reflection of soil erosion and the decline of ecosystem stability is the decline of ecosystem service function (CHANG et al., 2012). Therefore, this paper evaluates the vulnerability characteristics of an ecosystem from the perspective of the ecosystem service function, that is, when a mountain disaster occurs in a certain unit, then the unit will lose its ecosystem service function. Nan Ling's vegetation coverage is higher than 95%, which is a typical forest ecosystem. The forest ecosystem service function is the natural environmental conditions and effects formed by forest ecosystems and ecological processes to maintain human survival (HU, 2005). According to the characteristics of the Nan Ling Nature Reserve and the ecosystem in the area, the forest ecosystem services of the Nan Ling Nature Reserve include five aspects: biodiversity conservation, water conservation, soil and water conservation, carbon fixation and oxygen release, and nutrient maintenance. Different ecological service functions are calculated differently. This paper refers to the "Assessment of Forest Ecosystem Service Function" and determines the indicators and methods for evaluating the ecosystem service function of the Nan Ling Nature Reserve. Among them, biodiversity conservation considers both plant and animal protection aspects, and refers to the distribution and protection level of protected species in Nan Ling Nature Reserve to determine the level of biodiversity conservation function of each small unit. The importance of water conservation is to enhance soil infiltration, inhibit evaporation, and alleviate surface runoff. The water conservation function level calculates the water conservation capacity of each small unit according to the water balance equation. The total water conservation is mainly related to precipitation and evaporation. The surface vegetation cover type and unit area are closely related. Therefore, the importance of water conservation sources in the protected area depends on the evaluation indicators such as vegetation type, landform type and average annual precipitation. The soil retention function is measured by the amount of potential soil retention. The difference between the amount of soil erosion under the condition of surface damage and the amount of soil erosion under the condition of vegetation cover is calculated. The soil retention is mainly related to slope length, precipitation and vegetation, and highly related to soil type. Carbon fixation and nutrient maintenance functions depend mainly on vegetation type, forest age, net productivity, and biomass and unit area. Based on the above factors and combining them, select the dominant tree species, the priority tree species protection grade, the average annual precipitation, the slope length, the landform type, the soil type, the vegetation type, the tree age, the biomass and the small unit area as indicators for evaluating the vulnerability of the ecosystem. The 12 factors are weighted and summarized and normalized by the maximum value to obtain the vulnerability index of each small class. The weight is determined according to the AHP method. Then, according to the natural breakpoint method, the evaluation unit is divided into three levels of vulnerability level: high, medium and low.

3.2.3 *Ecological system disaster prevention capability*
By fixing soil transpiration and interception, changing underlying surface conditions, and regulating the slope hydrological cycle, plants can improve the formation conditions and scale of mountain disasters and effectively reduce the risk of disasters (2001–2015;1995–2014). This paper evaluates the disaster prevention capacity of forest ecosystems and integrates the soil-fixing capacity of vegetation and the hydrological regulation function of vegetation. The fixed soil capacity of forest ecosystems is manifested in the fact that the soil has a strong resistance to stress without the ability to resist tensile damage, while plant roots have strong tensile strength and are resistant to large tensile damage. The combination of the soil and root system allows the root-soil composite to have high strength composite properties. The function of soil conservation and slope protection of vegetation is mainly reflected by the effect of vegetation roots and rock and soil. The size of soil consolidation is mainly related to vegetation type, tree age and root biomass. The hydrological regulation function of vegetation is

mainly related to vegetation type and vegetation coverage. Therefore, the ecosystem disaster prevention index is measured by vegetation type, tree age and biomass. The three factors are weighted and summed up, the weight is determined according to the AHP method, and the maximum value is normalized to obtain the disaster prevention capability index of each small class, and then the evaluation unit is divided into high by the natural breakpoint method.

4 RESULTS

4.1 *Mountain disaster hazard*

According to the above-mentioned classification and statistical results of 2889 small class unit landslide hazard evaluation factors, a small class in forestry unit based attribute table is established in ARCGIS. Combined with the seven factors of the risk analysis, the weighted value of each small unit is calculated:

$$H_i = \sum WH_j P_{ij} \ (i = 1, \ 2..., \ 2889; \ j = 1, \ 2..., \ 7) \tag{2}$$

H_i is the risk of the i-th small class; WH_j is the weight of the j-th indicator; P_{ij} is the value of the j-th indicator of the i-th small class. When conducting the risk assessment of landslide hazards, it is necessary to determine the weight of each indicator in the risk assessment. This paper uses expert scoring to determine the weight of each factor. The risk of the i-th small class; WH_j - the weight of the j-th indicator; P_{ij} - the value of the j-th indicator of the i-th small class. When conducting the risk assessment of landslide hazards, it is necessary to determine the weight of each indicator in the risk assessment. This paper uses expert scoring to determine the weight of each factor, the results are shown in the Figure 6.

In ARCGIS, the risk of 2889 small class in forestry units was calculated using its map algebra function, with a risk range of 0–1. The small class in forestry units are then divided into three levels: high-risk zone, medium danger zone, and low danger zone at intervals of 0.2. The dangerous distribution of each small unit in the Nan Ling Nature Reserve is shown in Figure 3. According to the statistics of the hazard levels of 2889 small class in forestry units, there are 406 small class in forestry units with a low risk level, 740 small class in forestry units at medium risk level, and 1743 small class in forestry units located in high-risk areas. The risk of overall mountain disasters is high, which is consistent with the results of the classification of geological disasters formulated by the disaster investigation and geological disaster management departments. However, the results of this paper are more detailed and more spatially oriented.

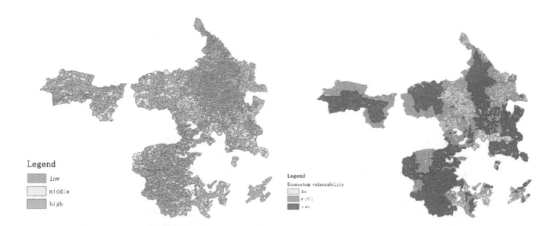

Legend
low
middle
high

Legend
Ecosystem vulnerability
low
middle
high

Figure 6. Mountain disaster hazard.

Figure 7. Ecosystem vulnerability.

120

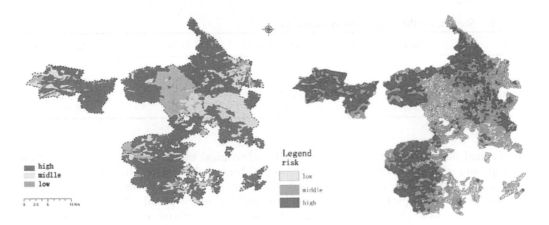

Figure 8. Ecological system disaster prevention capability. Figure 9. Ecological risk.

4.2 *Ecosystem vulnerability*

According to field surveys and remote sensing interpretation, there are five main types of forest land in the study area. Therefore, the five indicators of biodiversity conservation function index, water conservation function index, soil and water conservation function index, carbon fixation and oxygen release and nutrient function index were analyzed as factors for evaluating ecosystem vulnerability.

$$V_i = \sum WV_j Y_{ij} \ (i = 1, \ 2, \cdots 2889; \ j = 1, \ 2, \ 3, \ 4, \ 5) \tag{3}$$

where: V_j is the ecosystem vulnerability value of the j-th small class in forestry; WV_j is the weight of the j-th indicator; Y_{ij} is the value of the j-th indicator of the i-th small class in forestry.

The relevant data is normalized by the maximum value and substituted into equation (3). After map algebra calculation, the vulnerability value of the ecosystem in the nature reserve is between 0–1, and the small class in forestry is divided into high vulnerability area, moderate vulnerability area and low vulnerability area at intervals of 0.33. The levels are as shown. Areas with high vulnerability are concentrated in areas with high levels of biodiversity conservation in such small units, and the vulnerability of landslides with the same intensity is much greater than that of other units.

4.3 *Ecological system disaster prevention capability*

The function of soil conservation and slope protection of vegetation is mainly reflected by the effect of vegetation roots, rock and soil. The size of soil consolidation is mainly related to vegetation type, tree age and root biomass. The ecosystem's resilience index is measured by vegetation type, age, and biomass.

$$D_i = \sum WD_j L_{ij} \ (i = 1, \ 2, \cdots 2889; \ j = 1, \ 2, \ 3) \tag{4}$$

where: D_i is the value of ecological system disaster prevention capability. WD_j is the weight of the j-th indicator; L_{ij} is the value of the j-th indicator of the i-th small class.

The result of ecological system disaster prevention capability is shown in Figure 8. There are nine vegetation types in the study area, including subtropical evergreen broad-leaved forest, subtropical evergreen and deciduous broad-leaved mixed forest, subtropical evergreen coniferous forest, subtropical evergreen coniferous and broad-leaved mixed forest,

Table 1. Statistics of the proportions in different risks.

Level	Number of units	Area (km^2)	Unit ratio (%)	Area ratio (%)
Low risk	807	14.92	27.93	24.59
Middle risk	1126	24.40	38.98	40.21
High risk	956	21.36	33.09	35.20
Total	2889	60.67	100.00	100.00

subtropical bamboo forest, subtropical shrub, subtropical meadow, subtropical mountain swamp, and subtropical alpine lake vegetation. Subtropical evergreen broad-leaved forest, subtropical evergreen and deciduous broad-leaved mixed forest, subtropical evergreen coniferous forest, subtropical evergreen coniferous and broad-leaved mixed forest have better disaster protection ability. Subtropical bamboo forest, subtropical shrub, subtropical meadow, subtropical mountain swamps, and subtropical alpine lake have weak vegetation resilience.

4.4 *Ecological risk*

Through the attribute values of risk, vulnerability and resilience of GIS small class units, the comprehensive risk degree of each unit is calculated. According to the comprehensive risk degree, it is divided into three levels in ARCGIS, as shown in Figure. According to the statistics of the risk levels of 2889 small class units (Table 1), the degrees of disaster risk in the study area are different. The number of small class units in the three types of risk levels is relatively uniform, and the area at the medium risk level is relatively large. The medium-high-risk areas are mainly distributed in areas with medium elevations and medium vegetation distribution conditions. Areas with low risk are areas with low terrain, relatively low forest cover and biomass.

5 CONCLUSIONS AND DISCUSSION

In this paper, geological disasters are used as a risk source to analyze the ecological risk characteristics of Nanling, and the possible impacts of geological disasters on the ecological environment are expressed in quantitative form from the perspective of ecological service functions. The spatial distribution characteristics of risk have important practical guiding significance for regional ecological risk management.

The overall characteristics of ecological risk distribution are characterized by vertical distribution. The high-risk areas are mainly distributed in the area of 500–1000 m above sea level. The vegetation coverage is high, the terrain is large, the slope is large, and the mountain disaster risk is high. The ecosystem is also vulnerable. In areas with altitudes of less than 500 meters and greater than 1000 meters, vegetation coverage is relatively low, terrain fluctuations are relatively small, and disaster hazards and ecosystem vulnerability are also small.

An important feature of ecological risk assessment is that the spatial relationship between plaques plays an important role in determining the vulnerability The damage in one plaque can be transferred to other plaques, and the change of regional landscape patterns will inevitably lead to changes in the structure and function of regional ecosystems. Therefore, on the basis of relevant research, combined with the frequent occurrence characteristics of geological disasters in the study area, the ecological risk assessment of multi-risk sources based on landscape patterns is carried out; which has certain theoretical basis and reference significance for regional ecological risk assessment. This paper only considers the sources of risk such as geological disasters. There are other risk sources in the study area that affect the ecosystem, such as human activities, floods, soil erosion, etc., which can be more specifically and comprehensively analyzed in subsequent research.

ACKNOWLEDGMENTS

This work was supported by the project NSFC 41671506, 41977413,31770493), GDAS Special Project of Science and Technology Development (GDASCX-0701, 2017GDASCX-0101,2018 GDASCX-0101) and Guangdong Science and Technology Plan Project (2018B030324002).

REFERENCES

Chang Qing, Liu Dan, Liu Xiaowen. 2013. Ecological risk assessment and spatial prevention tactic of land destruction in mining city. *Journal of Agricultural Engineering* 29 (20): 245–254.

CHANG Qing, QIU Yao, XIE Miaomiao, PENG Jian. 2012. Theory and method of ecological risk assessment for mining areas based on the land destruction. *Acta Ecologica Sinica* 32 (16): 5164–5174.

CHEN Jun, WANG Hao, DAI Qiang. 2016. Risk assessment of landslide hazard caused by rainfall in Enshi city. *The Chinese Journal of Geological Hazard and Control* 27 (01): 15–21.

CHEN Renli, GONG Yuening, YANG Huai, XIE Guoguang, GU Maobin. 2015. Study on the diversity restoration of Papilionidae species in Nanling after a snow disaster. *Ecological Science*, 34 (02): 82–86.

CHEN Renli, HE Kejun, GONG Yuening, CHEN Zhenming, GU Maobin, CAI Weijing. 2008. Effect of the 2008 ice storm on the butterfly resources in Nanling area of China. *Ecological Science* 27 (06): 478–482.

Department of Land and Resources of Guangdong Province. *Prevention and cure of geological hazards plan of Guangdong Province*. 2001–2015.

Disaster Prevention and Mitigation Yearbook of Guangdong Committee. Disaster Prevention and Reduction Yearbook of Guangdong Provincial Guangdong: China Meteorological Press, 1995–2004.

DONG Anqiang, CHEN Lin, WANG Faguo, XING Fuwu. 2012. Study on vegetation of Nanling National Nature Reserve in Guangdong province. *Journal of Zhongkai Agrotechnical College* 25 (02):1–7.

DU Yueyue, PENG Jian, ZHAO Shiquan, HU Zhichao, WANG Yanglin. 2016. Ecological risk assessment of landslide disasters in mountainous areas of Southwest China: A case study in Dali Bai Autonomous Prefecture. *Acta Geographica Sinica* 71 (09):1544–1561.

GAO Bin, LI Xiaoyu, LI Zhigang, CHEN Wei, HE Xingyuan, QI Shanzhong. 2011. Assessment of ecological risk of coastal economic developing zone in Jinzhou Bay based on landscape pattern. *Acta Ecologica Sinica* 31 (12):3441–3450.

GONG Jie, ZHAO Caixia, WANG Heling, SUN Peng, XIE Yuchu, MENG Xingmin. 2012. Ecological Risk Assessment of Longnan Mountainous Area Based on Geological Disasters – A Case Study of Wudu. *Journal of Mountain Science* 30 (05):570–577.

Howarth RW, Ingraffea A, Engelder T. 2011. Natural gas: should fracking stop? Nature 477 (7364): 271–275.

Howarth RW, Santoro R, Ingraffea A. 2011. Methane and the greenhouse-gas footprint of natural gas f. rom shale formations. *Climatic Change* 106 (4): 679–690.

HU Yanlin. 2005. *The valuation of Tiantong forest ecosystem service functions based on GIS in Ningbo*. East China Normal University.

LI Xiang. 2016. *Ecological Risk Assessment within the condition forced by mountain disasters: A case of Longmen Mountain Town in Pengzhou City*. China West Normal University.

LI Xiehui, WANG Lei, MIAO Changhong, MIN Xiangpeng, LI Yating. 2013. Integrated Assessment of Ecological Risk for Flood and Drought Disasters in Henan Province. *Resources Science* 35 (11): 2308–2317.

LI Yide. 2008. Nanling Nature Reserves Caused by Low temperature, Frozen Rain and Snow Disasters– ecologically sensitive areas to be saved. *Scientia Silvae Sinicae* (06): 2–4.

LIU Ruihua, SUN Ning, TANG Guangliang. 2010. Analysis of Geological Environment and Causes of Landslides in Guangdong. *Tropical Geography* 01:13–17.

LIU Wei, WANG Chunlin, CHEN Xinguang, CHEN Huihua. 2013. Characteristics of heat resource in mountainous region of northern Guangdong, South China based on three-dimensional climate observation. *Chinese Journal of Applied Ecology* 24 (9): 2571–2580.

LUO Lv. 2008. Analysis of the characteristics of an early winter hail in northern Guangdong. *Rural Economy And Science-Technology* (02): 79–80, 84.

NI Xiaojiao, NAN Ying, ZHU Weihong, Choi Yunsoo, LIU Guoming, LIU Chen, YAO Kuo. 2014. Study on comprehensive assessment of ecological security in Changbai Mountain Region based on multi-hazard natural disasters risk. *Geographical Research* 33 (07): 1348–1360.

OU YANG Zhi-yun, XU Wei-Hua, WANG Xue-Zhi, WANG Wen-Jie, DONG Ren-Cai, ZHENG Hua, LI Di-Hua, LI Zhi-Qi, ZHANG Hong-Feng, ZHUANG Chang-Wei. 2008. Impact assessment of Wenchuan Earthquake on ecosystems. *Acta Ecologica Sinica* 28 (12): 5801–5809.

PENG Jian, DANG Weixiong, LIU Yanxu, ZONG Minl, i HU Xiaoxu. 2015. Review on landscape ecological risk assessment. *Acta Geographica Sinica* 70 (4): 664–677.

Peng Jian, Liu Yanxu, Pan Yajing, Zhao Zhiqiang, Song Zhiqing, Wang Yanglin. 2014. Study on the Correlation between Ecological Risk due to Natural Disaster and Landscape Pattern-Process: Review and Prospect. *Advances in Earth Science* 29 (10): 1186–1196.

PENG Ling, XU Suning, PENG Junhuan. 2016. Regional Landslide Risk Assessment Using Multi-Source Remote Sensing Data. *Journal of Jilin University* (Earth Science Edition) 46 (1): 175–186.

Ross DJK, Bustin RM. 2009. The importance of shale composition and pore structure upon gas storage potential of shale gas reservoirs. *Marine and Petroleum Geology* 26 (6): 916–927.

SHANG Zhi-hai, LIU Xi-lin. 2010. Assessment on Eco-environmental Risk and Losses for Natural Disasters: with the Disaster of Debris Flow Induced by "5·12" Wenchuan Earthquake as an Example. *China Safety Science Journal* 20 (09): 3–8.

Toelle BE, Alexander T, Baihly J, Boyer C, Clark B, Jochen V, Calvez JL, Lewis R, Miller CK, Thaeler J, Toelle BE. 2011. Shale gas revolution. *Oilfield Review* 23 (3): 40–55.

WANG Jiangwei, MENG Jijun. 2014. Ecological Risk Assessment and Management of Floods and Droughts in the Li River Basin. *Tropical Geography* 34 (3): 366–373.

WANG Sha. 2016. *Assessment of rainstorm hazard in Mountainous Ecological Scenic Spots – A Case Study of Huayang Ancient Town Resort in Qinling Mountains.* Shaanxi Normal University.

WEI Min. 2011. *Statistic Study on Geological Disasters Induced by Strong Precipitation in Guangdong.* Henan Polytechnic University.

WU Ning, LU Tao, LUO Peng, ZHU Dan. 2008. A review of the impacts of earthquake on mountain ecosystems: taking 5.12 Wenchuan Earthquake as an example. *Acta Ecologica Sinica* 28 (12): 5810–5819.

XU Xuegong, YAN Leisure, XU Lifen, LU Yaling, MA Luyi. 2011. Ecological Risk Assessment of Natural Disasters in China. *Acta Scientiarum* (Naturalium Universitatis Pekinensis) 47 (05): 901–908.

YANG Li, LV Jianbin. 2010. Influence of Snow and Ice Frozen Disaster in Nanling Mountain Area on Highway Cut Slope and Its Classification. *Guangdong Science & Technology* 19 (18): 119–121.

ZENG Yong. 2010. The regional ecological risk assessment of Hohhot City. *Acta Ecologica Sinica* 30 (3): 668–673.

ZHANG Chunmin, WANG Genxu. 2008. Impacts of Wenchuan Earthquake disasters on ecosystem and its spatial pattern: case study of Qingchuan, Pingwu and Maoxian counties. *Acta Ecologica Sinica* 28 (12): 5833–5841.

ZHANG Jiquan, LIANG Jingdan, ZHOU Daowei. 2007. Risk assessment of ecological disasters in Jilin Province based on GIS. *Chinese Journal of Applied Ecology* (08): 1765–1770.

ZHU Lirong. 2014. *Studies on the Forest Damage Restoration Response to Ice Storm, and Fast Recovery Methods in Nanling.* Zhongshan University.

Risk Analysis Based on Data and Crisis Response Beyond Knowledge – Huang & Nivolianitou (eds)
© 2020 Taylor & Francis Group, London, ISBN 978-0-367-25146-8

Temporal and spatial dynamic characteristics analysis of Inner Mongolia grassland fire

Duwala
Ecological and Agricultural Meteorology Center of Inner Mongolia Autonomous Region, Huhhot, China

Yushan*
School of Geographical Sciences, Inner Mongolia Normal University, Huhhot, China
School of Geographical Sciences, Northeast Normal University, Changchun, China

Guixiang Liu
Grassland Research Institute, Chinese Academy of Agricultural Sciences, Hohhot, China

Hongyan Zhang
School of Geographical Sciences, Northeast Normal University, Changchun, China

Huijuan Liu & Wunitu
Grassland Research Institute, Chinese Academy of Agricultural Sciences, Hohhot, China

ABSTRACT: This paper analyzed the temporal and spatial characteristics of grassland fires in Inner Mongolia from 2000 to 2016. It provides scientific basis for fire risk zoning and emergency management of Inner Mongolia grassland. The results show that Inner Mongolia grassland fires has regular distribution in time and space. During the 17 years from 2000 to 2016, Inner Mongolia grassland fires area showed a fluctuating and declining trend as a whole with time. The fire area in spring accounted for 67.55% of the total fire area. The fire area in autumn is only next to that in spring. Although the temperature is higher in summer, but the combustibles are in a strong growth period, and the moisture content of combustibles is high. In December, January and February, Inner Mongolia grassland area is covered with snow, and grassland fire is hardly happened in snow-covered area. Inner Mongolia grassland fires area is mainly distributed in the eastern and central Inner Mongolia grassland area. The total burning area of grassland in Hulun Buir, Xilin Gol League and Hinggan league accounted for 83.9% of the total burning area.

Keywords: Inner Mongolia, Prairie fir, temporal and spatial patterns, MODIS

1 INTRODUCTION

China's grasslands are mainly distributed in the arid and semi-arid regions of the north, with obvious temperate continental climate. The dry periods of grass cause the combustibles to ignite and form grassland fires (Duwala, 2012). Prairie fire is an inevitable interference factor in grassland systems. The precipitation in spring and autumn is small, the wind is large, the hay period is long, the air is dry, and the neighboring agricultural areas have the habit of burning straw in spring and autumn. It is easy to cause grassland fires. grassland fires not only cause human and animal casualties, but also burn houses, sheds and grassland vegetation. In addition, every year

*Corresponding author: dwlrsgis@163.com

the government will invest a lot of manpower and resources in the prevention and emergency rescue of grassland fires.

2 MATERIALS AND METHODS

2.1 *Study area*

The Inner Mongolia autonomous region is located in N37° 53 '~ N53° 20', E97° 10 '~ E126° 02' (Duwala 2012), most of the area lies between 40° ~ 50° north latitude. Located in the northern border of China, it is an important ecological defense line in the north of China. It covers an area of about 1.183 106 km^2, accounting for about one eighth of the country's land area and the third largest province in China (Wang, 2007). It borders eight provinces: Gansu in the west, Liaoning, Jilin and Heilongjiang in the east, and Hebei, Shanxi, Shaanxi and Ningxia Hui autonomous region in the south. It has the largest grassland and forest area in China, the largest animal husbandry base in China, and the key area for national ecological construction and protection.

The zonal soil types in Inner Mongolia autonomous region are distributed from east to west, including black soil, black calcium soil, chestnut calcium soil, brown calcium soil, gray desert soil, gray brown desert soil (Ge, 2007), and gray forest soil, gray brown soil, brown soil and brown soil developed in mountainous areas. There are non-zonal soil types such as meadow soil, swamp soil, saline soil, sandy soil and gravel soil. The dry humidity in the east-west climate of the Inner Mongolia autonomous region has obvious zonal distribution, which corresponds to the distribution characteristics of humid, semi-humid, semi-arid, arid and extremely arid climate. The grassland vegetation (Duwala, 2012) has obvious zonal distribution from northeast to southwest. The zonal grassland vegetation in the autonomous region (from east to west zonal grassland vegetation in turn is meadow, grassland, typical grassland and desert grassland) is the most widely distributed (Hu, 2016), with the largest area of 45 million hm^2, accounting for 57% of the autonomous region's grassland area. The second is desert, with an area of about 22.3 million hm^2, accounting for 28% of the grassland area of the autonomous region, and non-zonal vegetation meadows, swamps and other areas with an area of 11.55 million hm^2, accounting for 15% of the grassland area of the autonomous region. The vegetation types of warm temperate zone, middle temperate zone and cold temperate zone are distributed successively from south to north in Inner Mongolia, with obvious zonal distribution (Duwala, 2012).The precipitation resources in Inner Mongolia are insufficient, and the annual precipitation decreases from 450 mm in the east to 50 mm in the west. The main characteristics of precipitation in Inner Mongolia are uneven spatial and temporal distribution and low precipitation guarantee rate. The annual precipitation is the largest in the northeast of Hulun Buir City, with a value of 450 ~ 510 mm.The lowest annual precipitation is in the western Alxa League, below 100 mm. Precipitation in Inner Mongolia is mainly concentrated in summer, which accounts for 65 ~ 70% of annual precipitation. Winter precipitation is scarce, accounting for only 1 ~ 3% of annual precipitation. Precipitation in autumn and spring accounts for 15-18% and 12-15% of annual precipitation respectively.

2.2 *Data and methods*

2.2.1 *Data*

The NDVI data of this study was derived from the MODIS MOD13A1 data product with a time resolution of 16 d and a spatial resolution of 500×500 m. The area of the burned area is the MCD45A of the medium-resolution imaging spectrometer (MODIS sensor) provided by LPDAAC (Land processes distributed active archive center). This data is one of the MODIS land product 5 series, which is 500 m. The release of the monthly product form (Zhang and Qi, 2015). MCD45 (MODIS Burnt Areas Product), a MODIS standard fire-burning product from 2000 to 2016, was used to compare the results of the disturbance index algorithm in spatial distribution and range. The data is based on continuous multi-day reflectance data and angle information. The BRDF model is used to determine the difference between the time series change of the spectral reflectance on the target pixel and the normal condition by threshold value (Boschetti et al., 2010, You et al., 2013). The MCD45A data provides users with a variety of quality assessment information and a quality assessment score for each pixel that predicts the total fire area in the study area (Lina, 2017).

2.2.2 *Research method*
The fire point data is the level 3 products MOD14A1 (Terra) and MYD14A1 (Aqua) shared by the 6260 scene MODIS standard products from 2000 to 2015 (downloaded at: https://lads web.nascom.nasa.gov/data/), the product The spatial resolution is 1 Km and the time resolution is 1 day. First, the MODIS data is reprocessed. The fire mask data sets MOD14A1 (Terra) and MYD14A1 (Aqua) are divided into 10 levels, the extracted attribute values are 7, 8, and 9 pixels.

3 RESULTS AND ANALYSIS

The temporal and spatial distribution of grassland fires refers to the characteristic that the occurrence of grassland fires changes regularly with time (such as year, month, day) and space.

3.1 *Time distribution law*

3.1.1 *Inter-annual variability*
The occurrence of grassland fires is related to the climate difference in different years. In some years, the precipitation is less, the climate is dry, the temperature is high, the number of grassland fires is more, the fire loss is also serious; In some years, with more precipitation, wetter climate and lower temperature, grassland fires are relatively rare and losses are light.

Figure 1 shows the inter-annual fluctuation of burning area in Inner Mongolia grassland from 2000 to 2016. The total burning area of grassland fires during the 17 years from 2000 to 2016 was 58700 km², with an average annual burning area of 3452.94 km². The grassland fires

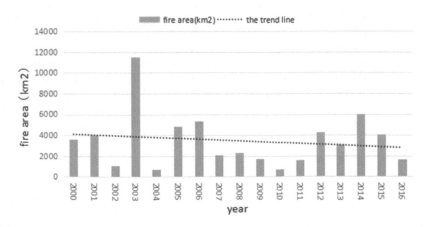

Figure 1.   Inter-annual fluctuations in the area of grassland burning in Inner Mongolia from 2000 to 2016.

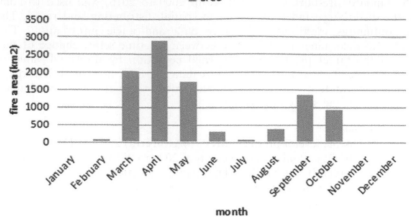

Figure 2. Distribution of the monthly fire area of Inner Mongolia grassland from 2000 to 2016.

area in 2000, 2001, 2003, 2005, 2006, 2012, 2014 and 2015 was larger than average. In 2003, the grassland fires area reached the maximum value in 17 years, and the grassland fires area was 11483 km². On the whole, the burning area of Inner Mongolia grassland fluctuated and decreased with time.

3.1.2 Month distribution

The number of grassland fires varies significantly in different months. Figure 2 shows the distribution of burning area in Inner Mongolia grassland from 2000 to 2016. It can be seen from the figure that the grassland fires area in March, April and may were respectively 2014.75 km², 2902 km² and 1734.5 km², and the fire area in spring accounted for 67.55% of the total fire area. In April alone, grassland fires accounted for about 1/3 of the total area. In April, the snow cover in the grassland region of Inner Mongolia had basically receded, the soil moisture was low, and the temperature gradually increased. In years with less precipitation, the combustible material in the grassland was in the most vulnerable period.

Another prairie fire season of the year occurs in September and October in the fall. It can be seen from Figure 2 that during the 17 years from 2000 to 2016, the grassland fires area in September was 1360 km², and that in October, the grassland fires area in Inner Mongolia was 938.5 km². From the end of August to the beginning of September, the yellow withered period of Inner Mongolia grassland gradually advances from northeast to southwest, and most of the yellow withered period of natural herbage is concentrated in September. Summer June, July and August although the temperature is higher, but the fuel is in the exuberant growth period, the fuel moisture content is high. Because the rainy season is also in the summer in the grassland area of Inner Mongolia, there are fewer grassland fires in summer. In December, January and February, Inner Mongolia grassland area is covered with snow, and grassland fires is hardly happened in snow-covered area.

3.2 Spatial distribution law

According to the spatial distribution of grassland burning area in Inner Mongolia from 2000 to 2016, as shown in Figure 3 and Table 1, the Inner Mongolia grassland burning area is mainly distributed in the eastern and central grassland areas of Inner Mongolia. In the 17 years, the burning area of grassland in Hulun Buir City was 10871.25 km²,

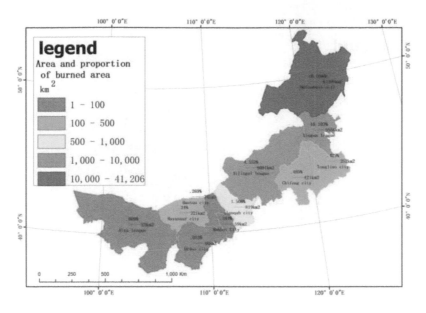

Figure 3. Spatial distribution of grassland fires in various cities of Inner Mongolia from 2000 to 2016.

Table 1. Spatial distribution of grassland fire area in various cities of Inner Mongolia from 2000 to 2016.

Union	Burning area (km²)	The percentage (%)
Hulun Buir	10871.3	56.90
Xilin Gol League	3397.4	17.80
Xingan league	1770.3	9.30
Tongliao city	996.0	5.20
Ulanqab city	558.9	2.90
Hohhot city	394.1	2.10
Alxa League	390.4	2.00
Chifeng city	380.0	2.00
Baotou city	356.1	1.90

accounting for 56.9% of the total burning area. During the 17 years, the grassland burning area of Xilin Gol Leagueand Hinggan League was 3397.38 km² and 1770.25 km² respectively, accounting for 17.8% and 9.3% of the total burning area. The total burning area of grassland in Hulun Buir, Xilin Gol League and Hinggan league accounted for 83.9% of the total burning area. Hulun Buir City, Old Barag banner and New Barag Right banner, Xilin Gol Leaguearea of east Ujimqin banner, west ujimqin banner, O ba ga banner and Xilinhot city. Arxancity and other areas of Hinggan league grassland fire prone areas.

In Hulun Buir City, Xilin Gol Leaguearea and Xingan League three prairie fire frequency area belongs to banner/county fire area of statistical results (Figure 4). Grassland fires in hulunbuir city are mostly concentrated in 7 counties, including Oroqen Autonomous Banner, Old Barag Banner, Moridawa daur autonomous banner,New Barag Right Banner, Ergun city, New Barag left Banner and Yakeshi city. The grassland fires in xilingol league are mainly distributed in the East Ujimqin banner. The grassland fires in Hinggan league mainly occurred in Zhalaitebanner.

129

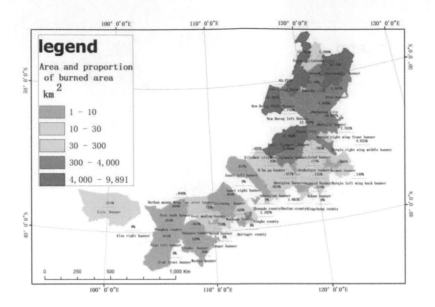

Figure 4.   spatial distribution of grassland fire in every banner/county of Inner Mongolia from 2000 to 2016.

4   CONCLUSION AND DISCUSSION

The spatial and temporal dynamics of grassland fires in Inner Mongolia are analyzed. The results are as follows:

(1) The annual dynamics of grassland fires in Inner Mongolia shows a fluctuating downward trend. During the 17 years from 2000 to 2016, the total burning area of grassland fires was 58700 km^2, with an average annual burning area of 3452.94 km^2. The grassland fires area in 2000, 2001, 2003, 2005, 2006, 2012, 2014 and 2015 was larger than average. In 2003, the grassland fires area reached the maximum value in 17 years, and the grassland fires area was 11483 km^2.

(2) Inner Mongolia grassland fires mainly concentrated in the spring, spring fire area accounted for 67.55% of the total fire area. Especially in April, grassland fire area accounts for about 1/3 of the total area. This is because in April, the snow cover in the grassland region of Inner Mongolia has basically receded, the soil moisture is low, the temperature increases gradually, and the combustible in the grassland is in the most vulnerable period when the precipitation is low. Another prairie fire season occurs in September and October in the fall. Because, the yellow withered period of most natural herbage in Inner Mongolia grassland area is concentrated in September, the moisture content of the combustible material decreases rapidly after yellow withered, and it is easier to ignite. Although the temperature is higher in summer, the combustible is in the prosperous growth period, and the moisture content of the combustible is high, and the precipitation in the Inner Mongolia grassland area is concentrated in summer, so it is not easy to happen fire. In winter, there is snow cover in the grassland area of Inner Mongolia.

(3) From the perspective of the spatial distribution of Inner Mongolia grassland fire, the Inner Mongolia grassland fire mainly distributed in the two grasslands of Hulun Buir and Xilin Gol in Inner Mongolia and the western part of the Arxancity.

ACKNOWLEDGMENTS

CASS-ASTIP-2016-IGR-04 and the National Natural Science Foundation of China (41761101) and the Inner Mongolia Science and Technology Project "Research on the Multi-scale Grassland

Fire Risk Assessment Technology of the Mongolian Plateau based on 3S Technology" and the Inner Mongolia Science and Technology Project "Arxan Forest Disaster Monitoring and Early Warning and Emergency Management System Research" were jointly funded.

REFERENCES

Duwala. 2012. *A Study of Grassland Fire Monitoring and Early Warning and Assessment Inner Mongolia.* Chinese Academy of Agricultural Sciences.

Wang, S.W. 2007. *Study on Woody Flora in Inner Mongolia.* Inner Mongolia Agricultural University.

Ge, Y.H. 2007. *The Database Construction for Pollen Morphology of Grassland and Desert Plants of Inner Mongolia.* Inner Mongolia University.

Hu, W. 2016. *Research and Application of MODIS Drought Monitoring Model Based on Cloud Background in Inner Mongolia.* Inner Mongolia Agricultural University.

Zhang, H.J. and Qi S.G. 2015. Spatio-temporal characteristics analysis of fires based on MODIS data in Henan Province from 2002 to 2012. *Journal of Hebei Normal University* (natural science edition) (4): 352–358.

Boschetti, L., Roy, D.P., Justice, C.O., et al. 2010. Global assessment of the temporal reporting accuracy and precision of the MODIS burned area product. *International Journal of Wildland Fire* 19(6): 705.

You, H., Liu, R.G, Zhu, S.Y and Liu, Y. 2013. Burned area detection in the Canadian Boreal forest using MODIS imagery. *Journal of Earth Information Science* 15(4): 597–603.

Li N. 2017. *Real-time Monitoring of Grassland Fire Based on Multi-Source Satellite Remote Sensing Data in China Mongolia Border Area.* Inner Mongolia Normal University.

Risk Analysis Based on Data and Crisis Response Beyond Knowledge – Huang & Nivolianitou (eds)
© 2020 Taylor & Francis Group, London, ISBN 978-0-367-25146-8

Impact factor selection and nonlinear assessment model for extreme disaster losses caused by typhoons

Hexiang Liu*, Qingjuan Xu & Jianwei Chen
School of Mathematical and Statistics Sciences, Nanning Normal University, Nanning, Guangxi, China

ABSTRACT: The linear and nonlinear dimensionality reductions of the 21 primary selection factors for the data of strong and super typhoons that landed in South China in the period of 1990–2016 are extracted through Alasso and combined weight method, and the linear and nonlinear characteristics of the disaster system are extracted in the low-dimensional subspace. The support vector machine (SVM) parameters are optimized, and an assessment model that combines the drag onfly algorithm and SVM (DA-SVM) for extreme disaster losses caused by typhoons is constructed through the dragonfly group optimization algorithm. The results show that the average absolute and relative errors of the proposed new nonlinear assessment model are 22.993 and 10.65%, respectively, and the root mean square error is 0.0315. A comparison of this evaluation result with those of the traditional SVM regression model calculation shows that the stability, fitting effect, and evaluation accuracy of the DA-SVM model based on Alasso and combined weights are significantly improved compared with the conventional SVM regression method in estimating the extreme disaster losses of typhoons in South China.

Keywords: Alasso, combined weight, dragonfly algorithm, SVM, typhoon in South China, extreme disaster.

1 INTRODUCTION

In the past few decades, variable selection methods and nonlinear intelligent computing techniques have been widely used in many disciplines and engineering fields (Jin et al., 2004; Zhao et al., 2014; Gong et al., 2015; Chen et al., 2016; Zen et al., 2017; Xu et al., 2018). The processing methods of various linear and nonlinear factors and the construction of various nonlinear mathematical models, especially the optimal processing of variable selection and the mathematical model of the construction, have become important ways to improve the model and improve the accuracy of assessment. For example, Jin Long et al. (2008) discovered that in the primaries of several meteorological forecasting factors, the conditional number calculation and analysis method is adopted. The combination of predictive factors with small complex collinearity is established to construct a forecasting model. A strong forecasting effect is also obtained specific to the hypothesis that the possible complex collinearity relationship will affect the forecast performance of the weather forecast equation in the stepwise regression forecasting modeling method commonly used in meteorological forecasting. Chen et al. (2011) used a typical correlation analysis method to separate linear variables from the two groups of tropical cyclone hazards and disasters. The correlation coefficient between the new variables is maximized, the linear combination coefficient of the hazard factors can be used as the weight coefficient of the set of variables, and the storm surge index is determined by historical disaster inversion methods. Jinkai et al. (2017)

---

*Corresponding author: hx_post@126.com

reanalyzed data from China's National Environmental Forecasting Center and Atmospheric Research Center. The variables are selected by the minimum absolute contraction and selection operator and stepwise regression technique, and the typhoon frequency prediction model based on arandom forest is established to achieve improved prediction results. Song et al. (2018) used the dragonfly group optimization to optimize the parameters of the support vector machine (SVM). They predicted the vibration of the cutter head by the optimized SVM. As a result, the prediction accuracy and generalization ability were greatly improved.

In recent years, many scholars have conducted fruitful research on the characteristics of typhoon disaster risks and constructed various linear or nonlinear mathematical models to assess disaster losses caused by typhoons. For example, Wei (2010) explored the highly uncertain non-linear relationship between the evaluation factor and the typhoon disaster. The initial weight of the BP neural network is optimized, and the PSO-BP neural network typhoon disaster quantitative assessment model is established through the particle swarm optimization algorithm based on the global stochastic optimization idea. Zhang et al. (2013) constructed a disaster loss prediction model and quantitatively expressed the law between the typhoon disaster damage and index factors through the typhoon disaster data of Hainan Province from 1992 to 2011, the fuzzy logic of the integrated T-S fuzzy neural network, and the performance of the neural network learning optimization. Li et al. (2014) indicated that a direct economic loss assessment model established by the fuzzy comprehensive evaluation method has a strong fitting effect through using 40 typhoon data and historical disaster data from 1984 to 2007 in the Zhejiang Province to select impact factors from four aspects: hazard factor, disaster-causing environment, disaster-bearing body, and disaster prevention and mitigation. Ugly et al. (2018) analyzed the spatial and temporal characteristics of the TC, the inter annual variation of the disaster situation, and the intensity characteristics of the disasters that affected Guangdong Province through the tropical cyclone data that affected the province from 1990 to 2015. The use of the disaster indicators to calculate the economic value of disaster losses improves the comprehensive disaster index model.

The above-mentioned studies demonstrate the application of linear and nonlinear intelligent computing techniques in risk analysis and disaster loss assessment of typhoon disasters. However, they focused less on the typhoon characteristics, the selection of disaster factor variables and the mathematical treatment methods in abnormal years, the mining of potential core information from limited samples, and the comprehensive study of assessment models of extreme disaster losses caused by typhoons. The current study attempts to fill this gap in literature. The important influencing factors of typhoon disasters are selected by the linear adaptive minimum absolute contraction and selection operator (Alasso) (Tibshirani, 1996, 2011) method. Meanwhile, the combined weighting method is used to realize the nonlinear dimensionality reduction of the original high-dimensional data, and the evaluation model of extreme disaster losses caused by typhoons based on the dragonfly algorithm (DA) is constructed to optimize the parameters of the SVM. This model is also applied to the assessment of extreme disaster losses caused by typhoons in South China to explore new ways for selecting impact factor selection techniques and assessment models for these losses.

2 LINEAR DIMENSION REDUCTION VARIABLE SELECTION METHOD

Multi-factor variables increase the difficulty in model construction and calculation while providing a large amount of information. Moreover, the correlation between multi-factor variables results in complex collinearity. In response to these problems, Tibshirani (1996, 2011) proposed the least absolute shrinkage and selection operator of a biased estimate. The Lasso method is effective at small calculation amounts, has a fast calculation speed, continuous parameter estimation, an dissuitable for high-dimensional data. An excellent estimation process should satisfy the nature of Oracle (Fan & Li, 2001). Unfortunately, the Lasso method does not occupy the consistency and asymptotic normality of the variable selection (Oracle nature).

Zou (2006) proved that Lasso can satisfy the nature of Oracle after the appropriate λ is selected and proposed an adaptive Lasso, which is an Alasso (Adaptive Lasso). The Alasso method can select variables and linear information of highly concentrated factors simultaneously to ensure data dimensionality reduction in parameter estimation. With the constraint that the sum of the absolute values of the regression coefficients is less than a constant, the sum of the squares of the residuals can be minimized to generate some regression coefficients that strictly equal to zero. The complex collinearity problem can be effectively solved, and the interpreted model can be obtained. Given that the problem is a convex optimization one, multiple local minimums need not be generated. The global optimal solution can be obtained quickly.

The mathematical expression of Alasso is given as follows:

$$\hat{\beta}_{\text{ALasso}}(\lambda) = \underset{\beta \in R^P}{\arg\min} \left\{ (y - X\beta)^T (y - X\beta) + \lambda \sum_{i=1}^{p} \frac{|\beta_i|}{|\hat{\beta}_{init,i}|} \right\}$$

where X is the independent variable, y is the response variable, β is the regression coefficient, and $\hat{\beta}_{init}$ is an initial estimator. In general, the parameter estimate obtained by least squares can be used as the initial estimate of β. However, in the case of high-dimensional data, $\hat{\beta}_{OLS}$ is unsatisfactory from the accuracy of model evaluation or the interpretable aspect of the model. Thus, the Lasso method is used to obtain the initial estimate of β.

Following the above-mentioned calculation method, the original multiple influence factors (independent variables) are compressed into a few variables that are highly correlated with the evaluation object (response variable) as a part of the model input.

3   COMBINED WEIGHTING METHOD FOR NONLINEAR DIMENSIONALITY REDUCTION

The combined weighting method combines the processing technique of nonlinear data with fuzzy mathematical theory to compress the data in a high-dimensional vector space to the optimized integration method in low-dimensional space research. The steps are described as follows:

A raw data matrix is given $X' = (x'_{ij})_{n \times m}$,

$$and \quad r_{ij} = \frac{x_{ij} - \min_{j} x_{ij}}{\max_{j} x_{ij} - \min_{j} x_{ij}} \quad (i = 1, 2, \cdots, n; j = 1, 2, \cdots, m).$$

The relative membership value r_{ij} is calculated, and the sample data x_{ij} are normalized to obtain the fuzzy evaluation matrix $R = (r_{ij})_n$ through (where $\max_{j} x_{ij}$ and $\min_{j} x_{ij}$ are the maximum and minimum values of the jth indicator). Then, the classification weights of the influence factors are determined as ai1 and the ranking weights ai2 by the particle swarm projection pursuit and fuzzy mathematics theory (Jinet al., 2003; Hexiang & Da-Lin 2012; Hexiang et al., 2013). The optimization algorithm is used to solve the value of the objective function as follows:

$$\min F = \sum_{j=1}^{m} \sum_{i=1}^{n} \left(\mu |a_{i1} - a_i| r_{ij} + (1 - \mu) |a_{i2} - a_i| r_{ij} \right)$$

$$s.t. \sum_{i=1}^{n} a_i = 1 且 a_i \geq 0, i = 1, 2, \cdots, n$$

where the classification coefficient is $\mu = 0.5$, that is, the weight of each classification is considered to have an equally important reference value.

The combined weight is obtained as $A = \{a_i, i = 1, 2, \cdots, n\}$.

Thus, the new variable matrix is $Z^* = R \cdot A = \left(z_{11}^*, z_{21}^*, \cdots, z_{n1}^*\right)^T$.

Using the combination weight optimization calculation method, the original large number of nonlinear factors can be compressed into a one-dimensional factor in relation to the evaluation object. The factor is used as another part of the model input.

4 DA-SVM ASSESSMENT MODEL

SVM is a small-sample machine learning method that was developed on the basis of statistical learning theory. This method defines the optimal linear hyper plane, introduces the kernel function, maps the sample space to a feature space of high-dimensional or even infinite dimension through nonlinear mapping, and obtains the decision function determined by only a few SVMs. Therefore, the method of linear learning machine can be applied in the feature space to solve the problem of highly nonlinear regression in the sample space. The problems of small sample, nonlinear, over-fitting, and local minima are also effectively solved.

DA is a new group intelligent optimization algorithm (Mirjalili S., 2016) that has a simple structure, easy implementation, excellent search performance, and strong robustness. The optimization of the parameters of the SVM can be realized by the search for food by a dragonfly group. As a result, an effective evaluation model is obtained.

In consideration of the characteristics of small samples, multiple indicators, and complex data of typhoon extreme disasters in South China, the advantages of Alasso and combined weight selection and DA combined with SVM (DA-SVM) are used to develop a DA-SVM nonlinear evaluation model based on Alasso and combined weights by referring to Wu et al. (2017) and Song et al. (2018). The specific algorithm steps are described as follows:

1. Read sample data, generate training and test sets, and normalize the sample data.
2. Determine the dragonfly population size N, the maximum number of iterations MIT, the range of values to be optimized (C, g, and ε), the spatial dimension D (where the spatial dimension corresponds to the number of parameters to be optimized, i.e., D = 3), inertia weight w, and neighborhood radius r.
3. Randomly generate an initial solution (i.e. dragonfly position) X of the individual, and randomly initialize the step vector ΔX.
4. Set t = 1 (the current iteration times).
5. Sequentially assign the position information of the dragonfly individual to C, g, and ε, and each individual corresponds to a set of model parameters.
6. Create an SVM regression model, train the training samples, and find the mean square error corresponding to each set of parameters as the fitness value of each dragonfly individual in the population.
7. Calculate the fitness values of all dragonfly individuals.
8. If t ≥ 2, then the greedy strategy is used to establish the ranking map of the dragonfly positions of the previous and current iterations, the excellent dragonfly individuals of the upper and lower generations are obtained, and the current optimal solution is recorded.
9. Update the food source position (current optimal solution) and the natural enemy position (current worst solution) using the Euclidean distance formula. Update the weights of the five dragonfly individual behaviors s, a, c, f, and e and inertia weight w. a and c are updated by a balancing strategy. s, f, and e are updated by a random value in the interval [0,1]. w is linearly decreasing with t in the interval [0.5,0.9].
10. Update the neighborhood radius. If a dragonfly has at least one neighbor dragonfly, then the step and position vectors are updated; otherwise, only the position vector is updated.
11. A linear combination of chaotic optimization of the dragonfly position is achieved by a combination strategy.

12. Check if the dragonfly position is within the boundary of the problem domain. If not, then correct it.
13. $t = t + 1$. If t is less than the maximum number of iterations MIT, then go to step 6.
14. Determine whether the termination condition is met. If it is satisfied, then end and output an optimal set of parameters and create an optimal SVM model; otherwise, return to step 4 to continue the iteration.

5   MODEL APPLICATION AND ANALYSIS

The South China coast (referring to Guangdong, Guangxi, and Hainan) is a region with frequent typhoon activity, a large number of occurrences, a serious impact, and a long and variable impact period in the entire year. The typhoon extreme disasters landing in South China involve not only many disaster-causing factors but also complex interactions. Moreover, the changes of extreme disasters are characterized by significant nonlinearity and time-varying features. Therefore, this study attempts to use a new nonlinear intelligent computing technology and method to carry out the risk analysis of typhoon extreme disasters landing in South China for comprehensively understanding the various linear and nonlinear characteristics of the typhoon extreme disaster factors in South China. In particular, the mathematical methods of selecting and processing important disaster factors are explored. A typhoon extreme disaster assessment model is then constructed.

5.1   *Sample data*

The experimental data of this study are taken from nine strong and super typhoons that landed in South China from 1990 to 2016. The assessment object (response variable) is direct economic loss (100 million yuan), and the data come from the Chinese Meteorological Disaster Dictionary from 1990 to 2000 (Guangdong, Hainan, Guangxi) (Wen, 2007), China Tropical Cyclone Yearbook from 2001 to 2016, and China Meteorological Disaster Yearbook. A total of 21 independent variables are considered in the primary selection, including the disaster-bearing body and disaster prevention and mitigation factors (local fiscal general budget revenue (100 million yuan), the whole society fixed assets investment (100 million yuan), construction industry housing completion area ($10,000 \text{ m}^2$), total output value of the construction industry (100 million yuan), total agricultural output value (100 million yuan), total sown area of crops (10,000 ha), economic density (10,000 yuan/km^2), population density (person/km^2), and telephone penetration rate (%)), per capita GDP (thousands per person), road mileage per unit area (km/10,000 km^2), railway operating mileage per unit area (km/10,000 km^2), number of doctors per thousand (persons), number of medical institutions per 10,000 people (number), number of college students per 10,000 students (persons), and number of full-time teachers in ordinary higher education institutions (10,000)). These data are derived from the China Statistical Yearbook from 1990 to 2016. Another dependent variable selected is the disaster source (center pressure (hp) at landing time, landing duration (h), maximum wind speed (m/s) at landing, number of rain days, and process (local) extreme rainfall value (mm)). The data come from the China Tropical Cyclone Yearbook from 1990 to 2016 and China Typhoon Network. Data from different sources are verified by repeated comparisons to ensure consistency.

5.2   *Assessment of extreme disaster losses caused by typhoons in South China*

For the 21 primary variables mentioned above, the Alasso method with variable selection is used. The cross-validation error of the calculation result is 0.3272, and the optimal adjustment parameter is 0.0021. The nine variables selected are the central air pressure (hp) at the time of landing, the maximum wind speed (m/s) at the time of landing, the process (local) rainfall extreme value (mm), the construction housing completion area ($10,000 \text{ m}^2$), the total planting area of crops (10,000 ha), population density (persons/km^2), the number of medical institutions per 10,000 people (number of students), and the number of college students per 10,000 students (persons). They are used as part of the model input. The 21 variables of the primary selection are successively decreased to one dimension by using the combined weight method to effectively mine the non-linear information of the evaluation object. The ten variables are used as model inputs, and the DA-SVM evaluation model A based on Alasso and combined weights is established. The RBF kernel function is used as the DA algorithm to optimize the kernel function of SVM. The maximum number of iterations in MIT is 200, and the dragonfly population size N is 50. The key steps in the dragonfly optimization SVM algorithm are the choice of the penalty factor C, the RBF kernel parameter g, and the insensitive loss coefficient ε. A small value of C indicates that the penalty for empirical error is small, the complexity of the learning machine is small, and the experience risk is large. When the value of C is infinite, the complexity of the algorithm increases. The selection of the RBF kernel parameter g determines the dimension of the feature subspace, which decides whether the regression plane can achieve the minimum empirical error. If the dimension of the feature subspace is high, then the resulting optimal regression plane is complex and the empirical risk is small, but the confidence range is large, and vice versa. The penalty factor C has a value range of [0.01, 1000], the RBF kernel parameter g to be optimized has a value range of [0.001, 100], and the insensitive loss coefficient ε has a value range of [0.0001, 0.5]. The equilibrium strategy is used to update the values of a and c. The updates of s, f, and e are taken from the random values in the interval [0, 1]. w is linearly decreasing with the decrease in the number of iterations t in the interval [0.5, 0.9].

The first seven groups of strong or super typhoons (approximately 80% of the total number of samples) from 1990 to 2016 are trained as training sets, and the remaining two groups (nearly 20% of the total number of samples) are used as test samples for evaluation. In the established model A, the number of independent samples is 1415, and 1522 are used for the evaluation test. The results are shown in Table 1. The table shows that that the average absolute error of the evaluation model A is 22.993, the average relative error is 10.65%, and the root mean square error (RMSE) is 0.0315.

5.3   *Performance analysis of DA-SVM model based on Alasso and combined weights*

The nine variables extracted from the Alasso method designed in Section 5.2 (excluding one variable obtained by successively reducing the dimension of the combined weights) are used as model inputs to establish the DA-SVM model B. The combined weighting method is used for the 21 primary variables mentioned in Section 5.2. The one variable mentioned before is used as the model input to establish the DA-SVM model C. The RBF kernel function, the maximum number of iterations, the size of the population, the range of the penalty factor, and the range of the parameters to be optimized in models B and C are the same as those in Section 5.2. The calculation results (Table 2) show that the absolute and relative errors of models B and C are 25.974 and 12.23%, and 65.059 and 26.45%, respectively, and the RMSEs are 0.0348 and 0.2352, respectively. These values are significantly greater than the values of 22.993, 10.65%, and 0.0315 of model A in Table 1.

Table 1.   Comparison of the actual value of typhoon extreme disaster loss in South China and the evaluation tests of model A to the observed one and their associated absolute and relative errors during the years of 1990–2016.

Number	Observed	Evaluation	Absolute error	Relative error
8 (1415)	176.4000	205.6700	-24.3170	13.79%
9 (1522)	288.0890	265.4110	21.6690	7.52%
Average	232.2445	237.5685	**22.9930**	**10.65%**
MSE		**0.0315**		

137

The Alasso method is used to mine the linear information of the evaluation factor, and then the weight reduction method is combined to extract the nonlinear information of the evaluation factor. DA-SVM effectively evaluates the extreme disaster losses caused by typhoons in South China.

To objectively analyze and evaluate the dimension reduction method of the variable and the evaluation performance of the DA-SVM model, the DA-SVM model A based on Alasso and combined weights is compared with the commonly used SVM regression model. Notably, the SVM regression model has been used to forecast the intensity and path of tropical cyclones in weather forecasting (Gu Jinrong et al. 2011; Teng Weiping, 2012). Using the ten variables calculated in Section 5.2 as the model input, the SVM regression model D is established. The results (Table 3) show that the average absolute error of model D is 82.1695, the average relative error is 34.89%, and the RMSE is 0.0405. The error is significantly larger than that of model A in Table 1.

The same variable as in model B is used as the model input to establish the SVM regression model E. The same variable as in model C is used as the model input to establish the SVM regression model F. The evaluation results of models E and F are shown in Table 4. On the basis of the comparison of Tables 2 and 4, the evaluation variables based on Alasso and combined weights are calculated, and the nonlinear DA-SVM mathematical model is constructed. In the case where the evaluation variables

Table 2. Comparison of the actual value of typhoon extreme disaster loss in South China and the evaluation tests of models B and C.

		B				C		
Number	Observed	Evaluation	Absolute error	Relative error	Evaluation	Absolute error	Relative error
8 (1415)	176.4000	208.7170	-29.2700	16.59%	211.6490	-35.2490	19.98%
9 (1522)	288.0890	266.4200	22.6780	7.87%	193.2200	94.8690	32.93%
Average	232.2445	235.5405	**25.9740**	**12.23%**	202.4345	**65.0590**	**26.45%**
MSE		0.0348			0.2352		

Table 3. Comparison of the actual value of typhoon extreme disaster loss in South China and the evaluation tests of models D (use of an SVM model).

Number	Observed	Evaluation	Absolute error	Relative error
8 (1415)	176.4000	209.1330	-85.3830	32.62%
9 (1522)	288.0890	261.7830	78.9560	37.75%
Average	232.2445	235.4580	**82.1695**	**34.89%**
MSE		**0.0405**		

Table 4. Comparison of the actual value of typhoon extreme disaster loss in South China and the evaluation tests of models E and F, in which the WNN algorithm is replaced by a linear regression prediction model.

		E				F		
Number	Observed	Evaluation	Absolute error	Relative error	Evaluation	Absolute error	Relative error
8(1415)	176.4000	213.632	-81.5290	46.22%	369.57	193.1700	109.51%
9(1522)	288.0890	257.929	74.4570	25.84%	193.22	94.8690	32.93%
Average	232.2445	235.7805	**77.9930**	**36.03%**	281.3950	**144.0195**	**71.22%**
MSE		**0.0527**			**1.0635**		

and the test samples are the same, the calculation accuracy is significantly improved compared with that of the linear SVM regression model.

6 CONCLUSIONS

In this study, a nonlinear DA-SVM model is constructed on the basis of Alasso and combined weights to assess the extreme disaster losses caused by typhoons in the coastal areas of South China. The Alasso method is used to select variables for 21 primary selection factors for achieving linear dimensionality reduction. The combined weighting method is used to perform nonlinear intelligent calculation on the 21 primary selected factors for achieving nonlinear dimensionality reduction and projecting high-dimensional data to low-dimensional space. Accordingly, the linear and nonlinear information of factor variables is fully exploited to provide useful information for evaluating the models. The DA is used to optimize the SVM parameters, and the DA-SVM nonlinear evaluation model based on Alasso and combined weights is constructed.

The constructed DA-SVM model is used to evaluate the extreme disaster losses caused by typhoons landing in South China. The results show that the average absolute and relative errors of the model are 22.993 and 10.65%, respectively, and the RMSE is 0.0315. A comparison of the evaluation results with those of the conventional SVM regression method indicates that the proposed nonlinear evaluation model has a significant improvement compared with the conventional SVM regression model method. The average mean square error of the proposed model is significantly decreased, which fully reflects its stability. This nonlinear evaluation method may be more suitable for characterizing the nonlinear variation characteristics of extreme disaster losses caused by typhoons than the conventional SVM regression method.

The DA-SVM model based on Alasso and combined weights is not perfect, especially when the disaster source of extreme typhoons involves some physical processes. Thus, the model needs further optimization in the future. In our future study, we may use physical variables such as vertical wind shear and vorticity as additional sources of disaster and input to the model to explore the impact of these variables on the aforementioned losses. This study also has important guiding significance for the construction of assessment models in other fields such as storm disasters, economics, and finance.

ACKNOWLEDGMENTS

This work was funded by China's NSFC Grants 41665006 and 11561009, China's Guangxi Natural Science Foundation 2018JJA110052.

REFERENCES

Chen Y.X., Liu H.X. Tan J.K. 2016. Probabilistic Neural Network Pre-Assessment Model Based on Isometric Feature Mapping Dimensional Reduction in Typhoon Disaster. *Journal of Catastrophology* 31(3): 20–25, 30.
Chen H.Y., Yan L.N., Lou W.P., Xu H.E., Yang S.F. 2011. On Assessment Indexes of the Strength of Comprehensive Impacts of Tropical Cyclone Disaster-Causing Factors. *Journal of Tropical Meteorology* 27(1): 139–144.
China Meterological Administration. *Tropical Cyclone Yearbook 1995–2014*. Beijing: Meteorological Publishing House.
China typhoon network.http://www.typhoon.gov.cn/,2017/07/14.
Chou J.M., Ban J.H., Dong W.J. et al. 2018. Characteristics analysis and assessment of economic damages caused by tropical cyclones in Guangdong Province. *Chinese Journal of Atmospheric Sciences(in Chinese)* 42 (2): 357–366.
Fan J., Li R. 2001. Variable selection via non concave penalized likelihood and its oracle properties. *Journal of the American Statistical Association* 96: 1348–1360.

Gong Z.W., Hu L. 2015. Influence factor analysis of typhoon disaster assessment. *Journal of Natural Disasters* 24(1): 203–213.

Gu J.R., Liu H.Q., Liu X.P., Lv Q.P. 2011. Application of Genetic Algorithm-Support Vector Machine model in tropical cyclone intensity forecast. *Marine Forecasts* 28(3): 8–14.

Jin J. L., Wei Y.M., Ding J. 2003. System evaluation model based on combined weights (in Chinese). *Mathematics in Practice and Theory* 33: 51–58.

Jin L. 2004. *Theory, method and application of neural network prediction modeling.*Beijing: Meteorological Press.

Jin L., Huang X.Y., Shi X.M. 2008. A study on impact of multicollinearity on stepwise regression prediction equation. *Acta MeteorologicaSinica* 66(4): 547–554.

Jin L., Kuang X.Y., Huang H.H. et al. 2005. Study on the over fitting of the artificial neural network forecasting model. *Acta Meteorologica Sinica* 19: 90–99.

Li G., Qiu X.F., Zhang M., Jin Y.J., Gong J.Y. 2014. Direct Economic Losses Assessment of Typhoon Disaster in Zhejiang Province. *Tropical Geography*34 (2): 178–183.

Lou W.P., Chen H.Y., Qiu X.F., Yang X.Z., Zhao H.J. 2010. Quantitative assessment of typhoon disaster based on particle swarm Optimization and BP neural network. *Journal of Natural Disasters* 19(4): 135–140.

Liu H.X., Zhang D.L. 2012. Analysis and prediction of hazard risks caused by tropical cyclones in southern China with fuzzy mathematical and grey models. *Applications in Mathematical Modelling* 36: 626–637.

Liu H.X., Zhang D.L., Chen J.W., Xu Q. J. 2013. Prediction of tropical cyclone frequency with a wavelet neural network model incorporating natural orthogonal expansion and combined weights. *Natural Hazards* 65: 63–78.

Mirjalili S. 2016. Dragonfly algorithm: a new meta-heuristic optimization technique for solving single-objective, discrete and multi-objective problems. *Neural Computing &Applications*27 (4): 1053–1073.

National Bureau of Statistics. *China Statistical Yearbook1995–2016.* Beijing: China Statistics Publishing House.

Song J.M., Li S.P., Zhou Y.Q., Zhong J.Q. 2018. Prediction Model of Cutter Vibration for Sugarcane Harvester Based on Support Vector Machine and Dragonfly Algorithm. *Journal of Agricultural Mechanization Research* 1: 20–28.

Tan J.K., Liu H.X., Li M.Y., Wang. J. 2018. A prediction scheme of tropical cyclone frequency based on lasso and random forest. *Theoretical and Applied Climatology* 133:973–983.

Tibshirani R. 1996. Regression Shrinkage and Selection via the Lasso. *Journal of the Royal Statistical Society* 58(1):267–288.

Tibshirani R. 2011. Regression shrinkage and selection via the lasso: A retrospective. *Journal of the Royal Statistical Society* 73(3):273–282.

Wen K.G. 2007. *China Meteorological disasters.* Beijing: Meteorological Publishing House.

Wu W.M., Wu W.Y., Lin Z.Y. et al. 2017. Dragonfly algorithm based on enhancing exchange of individuals' information. *Computer Engineering and Applications* 53(4): 10–14.

Xu L., Cai D.S. 2018. The Application of Improved PSO-SVM Model in Dam Deformation Prediction. *Journal of Rural Water Conservancy and Hydropower in China,* 3: 120–128.

Zeng J., Zhou J.J. 2017. Variable Selection for High-dimensional Data Model:Survey.*Journal of Applied Statistics and Management* 36(4): 678–694.

Zhang G.P., Zhang C.X., Xie Z. 2013. Typhoon Disaster Prediction Model Based on T-S Fuzzy Neural Network and Its Application-A Case Study of Hainan Island. *Journal of Catastrophology* 28(2): 86–89.

Zhao H.S., Jin L., Huang Y. et al. 2014. An objective prediction model for typhoon rainstorm using particle swarm optimization: neural network ensemble. *Natural Hazards* 72(2): 427–437.

Zhou Z.H. 2016. Machine Learning. *Beijing: Tsinghua University Publishing House,* 121–139.

Zou H. 2006. The adaptive lasso and its oracle properties. *Journal of the American Statistical Association* 101: 1418–1429.

Risk Analysis Based on Data and Crisis Response Beyond Knowledge – Huang & Nivolianitou (eds)
© *2020 Taylor & Francis Group, London, ISBN 978-0-367-25146-8*

Meteorological conditions analysis of road traffic accidents and risk management based on big data

Ming Yang & Kaicheng Xing

Key Laboratory of Meteorological and Ecological Environment of Hebei Province, Shijiazhuang, China
Climate Center of Hebei Province, Shijiazhuang, China

ABSTRACT: With the increase in car ownership, the frequency of and direct losses from traffic accidents increase year on year and social risks increase. There are lots of factors that lead to traffic accidents, among which adverse weather conditions are one of the important factors causing traffic accidents. This paper analyzes the correlation between meteorological conditions of road traffic accidents in Hebei during 2016 to 2018, the driving mechanism of meteorological conditions on the occurrence of traffic accidents, and the risk and management of traffic accidents. The results show that the traffic accidents in Hebei mainly occurred in autumn: the most accidents happened in November, followed by October and December. The frequency of traffic accidents is mainly affected by the meteorological conditions, including fog, rain and snow, and is higher in bad weather than that on sunny days. When the rainfall reaches heavy levels, the more rain there is, the more traffic accidents there are. Adverse weather conditions raise traffic risks mainly through affecting visibility and adhesion coefficient of road surface and vehicles, thus management measures for traffic accidents should be implemented dynamically in combination with this feature.

Keywords: big data, meteorology, traffic accident, risk management

1 INTRODUCTION

With the increase of car ownership, the frequency of traffic accidents also increases year by year, bringing huge losses to lives and property. Among the numerous background factors of traffic accidents, in addition to subjective factors such as improper driving and violation of traffic regulations, adverse weather conditions, especially rain, snow and fog weather conditions, have become the main cause of many traffic accidents (Lin et al. 2018).

Studies have shown that in the rain and snow the casualty rate of traffic accidents increases by 25% and the accident rate increases by 100. There are 5.86 vehicle collisions and scrapes per million vehicles on snow day: more than 13 times of the number of accidents on non-snow days (Lu et al. 2009). Schlosser (1977) studied the traffic accidents on the national highway of the Netherlands from 1965 to 1966. Among the total 36,364 accidents, on rainy days 2,360 accidents occurred on the highway and 5,243 accidents occurred on other roads. Therefore, the accident rate on rainy days is not low. There are also foreign studies on the impact of traffic-sensitive weather such as low visibility (especially dense fog), which show that there is a certain relationship between the frequency of fog and traffic accidents (Symons and Perry 1997; Eisenberg 2004; Keay and Simmonds 2005; Keay and Simmonds 2006; Koetse and Rietveld 2009). Many studies exist on the relationship between traffic accidents and adverse meteorological conditions in the domestic meteorological field (He et al. 2004; Xia et al. 2014; Wang et al. 2014), but most focus on a certain year and rarely involve risk management of traffic accidents, and there are even fewer studies that focus on Hebei province.

The distribution of traffic accidents and the relationship with meteorological conditions from 2016 to 2018 in Hebei province are analyzed in this paper. From the perspective of the impact of meteorological conditions on the traffic accidents, the risks and control measures of traffic accidents in Hebei province are discussed.

2 SPATIAL AND TEMPORAL DISTRIBUTION OF TRAFFIC ACCIDENTS IN HEBEI PROVINCE

The daily times of traffic accidents in 11 cities in Hebei province from 2016 to 2018 were collected and analyzed. According to the results, the most traffic accidents in Hebei province occurred in 2016, followed by 2018, and the least in 2017. The frequency of accidents in the second half of the year was slightly more than that in the first half.

In terms of the occurrence of accidents throughout the year, the percentage of accidents in the four seasons are similar: spring 24% (March to May); summer 24% (June to August); autumn 27% (September to November); and winter 25% (December, January and February). Autumn is the season with the most traffic accidents and spring has the least, and November is the month with the most accidents, followed by October and December.

3 RELATIONSHIP BETWEEN METEOROLOGICAL CONDITIONS AND TRAFFIC ACCIDENTS

The causes of traffic accidents are numerous, and many of these are accidents due to improper driving. In order to eliminate the influence of human factors and analyze the relationship between meteorological conditions and traffic accidents, the days with no fog, rain, snow and other adverse weather are recorded as sunny days. Compared with the daily traffic accidents on sunny days, the effects of only one adverse meteorological condition and multiple meteorological conditions on traffic accidents are discussed, respectively.

3.1 The influence of fog on traffic accidents

Fog, as one of the important causes of poor visibility, obstructs the sight of drivers and easily leads to traffic accidents. From 2016 to 2018, there were 500 foggy days in total in 11 cities of Hebei province, with an average of 103.1 daily traffic accidents. The total number of sunny days without fog, rain, snow and strong winds was 8915, with an average of 88.9 traffic accidents per day. It can be seen that compared with sunny days, foggy days are more likely to cause traffic accidents, and the average number of daily traffic accidents is higher.

Foggy days in Hebei province mainly occur in autumn and winter, accounting for about 43.6% and 37.6% of the total foggy days. The number of accidents occurred in fog days accounts for about 45.6% in autumn and 36.4% in winter of the total number of accidents in fog days.

According to the frequency statistics of the accidents because of fog, the average daily traffic accidents are 19~425, showing a single-peak distribution at 50 to100 times. That is to say, the most likely number of traffic accidents on fog days is 50 to 100.

3.2 The influence of rainfall on traffic accidents

From 2016 to 2018, there were a total of 1993 rainy days in 11 cities of Hebei province, with an average of 83.7 daily traffic accidents, which is lower than the average daily number of88.9 sunny days.

According to the National Standard of the People's Republic of China Grade of Precipitation (GB/T 28592-2012), the 24h rainfall is divided into different grades, and the number of rainy days and the average daily accidents are counted. The number of days of rainfall corresponding to different grades show a single-peak distribution, which increase first and then decrease, and the peak appears at the light rain level. That is, the number of days with light

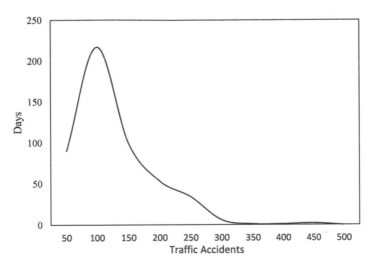

Figure 1. Traffic accidents on foggy days and the corresponding days from 2016 to 2018 in Hebei province.

rain accounted for the highest proportion in the total number of days of rainfall. The number of traffic accidents in light rain is lower than those on sunny days, and there is no significant increase. When the rainfall reached heavy, the average daily accidents exceeded those on the sunny days. The traffic accidents increase rapidly with the increase in the level of rainfall. In heavy rain, the average number of traffic accidents reached 109.2.

The daily rainfall above a heavy rain level was positively correlated with the number of traffic accidents, and the correlation coefficient was 0.17, which passed the significance test of 0.05. It can be seen that when the rainfall level is above heavy, the daily rainfall is significantly positively correlated with the daily traffic accidents. The higher the rainfall, the more traffic accidents.

3.3 *The influence of snow on traffic accidents*

The process of snowfall has a great impact on the visibility of drivers and the road, which will increase the road driving danger. A total of 255 days of snowfall have occurred in 11 cities in Hebei province in the past three years; of these 163 days only had snowfall without other

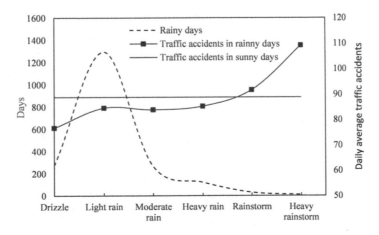

Figure 2. The variation of rainy days and traffic accidents with different rainfall grades.

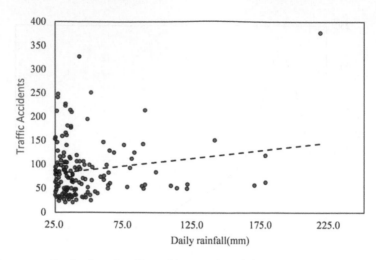

Figure 3.   The scatter distribution of traffic accidents and precipitation above heavy rain level.

adverse weather. These 163 days are recorded here as snow days. The daily average traffic accidents on snow days are 93, higher than the 88.9 on sunny days. The influence of snowy weather on different grades on traffic accidents varies. According to the *National Standard of the People's Republic of China Grade of Precipitation (GB/T 28592-2012)*, 24h snowfall is divided into different grades, and the number of snow days and average daily accidents of different grades are counted. According to the results, the snowfall days in Hebei province are mostly light snow days. The daily average traffic accidents increase first and then decrease, and also increase from moderate snow days to snowstorms. The first peak appears on the day of light snow. From the moderate snow to snowstorms, the more snowfall, the more traffic accidents, but the increasing rates are always changing. On the whole, the average daily traffic accidents increase as the snowfall level increases.

3.4   *The influence of multiple meteorological conditions on traffic accidents*

In addition to sunny days, fog days, rain days and snow days, there are also days with several adverse weather conditions at the same time. According to the statistical results of cities in

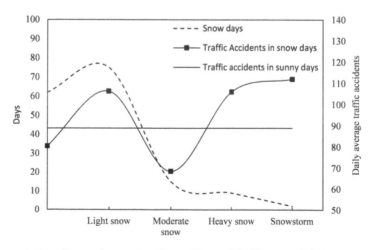

Figure 4.   The variation of snow days and traffic accidents with different rainfall grades.

Table 1. Daily average traffic accidents under different meteorological conditions.

Meteorological conditions	Foggy days	Rainy days	Snowy days	Two meteorological conditions	Three meteorological conditions	Sunny days
Daily average traffic accidents	103.1	83.7	93.0	93.8	140.4	88.9

Hebei province from 2016 to 2018, at most two or three of the above weather conditions occur at the same time in a single day. There were 330 days with two types of meteorological condition at the same time, and there were 93.8 daily traffic accidents on average. The occurrence of fog and rain at the same time accounts for the highest proportion at 43.3%, and there were 91.4 daily accidents on average, followed by the occurrence of wind and rain at the same time, accounting for 27.3%, with98.5 daily accidents. In addition, the occurrence of fog and snow at the same time happens on a small proportion of days, but there are 109.4 daily accidents on average.

The number of days in which the three meteorological conditions appear at the same time is 13, and there are140.4 daily traffic accidents on average. The number of days in which fog, rain and snow appear at the same time was eight (the most common combination), and the average daily traffic accidents was 185.5. The combination of three adverse weather conditions will increase traffic accidents and bring significant adverse effects to road traffic safety.

4 RISK ANALYSIS AND MANAGEMENT OF TRAFFIC ACCIDENTS IN HEBEI PROVINCE

4.1 *Risk analysis of traffic accidents*

In the recent three years of non-sunny days in Hebei province, traffic accidents in the days with only rain or fog account for the highest proportion, and the average daily traffic accidents accounted for 60.6% or 18.7% of all non-sunny days. It can be seen that fog and rainfall are the two most frequent meteorological conditions when only one bad weather condition occurs in a single day.

The main meteorological conditions affecting traffic accidents in Hebei province are different in different seasons. Rainfall mainly occurs in summer, fog mainly occurs in autumn and winter, and snow mainly occurs in winter. Traffic accidents happen most on foggy days and least on rainy days.

There are many hazy days in autumn and winter in Hebei province, which covers the ground in dust. Fog days also mainly occur in autumn and winter. The foggy moisture is mixed with dust, which reduces the adhesion coefficient between tires and the road. Heavy fog often causes heavy vehicle losses and casualties, leading to speed limits or closure of expressways and a delay in driving time (Bai et al. 2015).

The heavy rain in Hebei province has a significant impact on the traffic accident frequency. On the one hand, the strong rainfall makes the road slippery and causes a significant reduction of tire and road surface adhesion coefficient, affecting driving safety. On the other hand, the rain reduces visibility, the drivers' judgment of direction, and road conditions.

There are not many days in Hebei province when two meteorological conditions appear simultaneously, but the impact on traffic accidents is significant. For example, when fog and snow appear at the same time, the road adhesion coefficient is reduced, vehicles slip easily and the reduced visibility blocks the driver's line of sight, making them unable to accurately judge the current situation, leading to misjudgments and the occurrence of traffic accidents.

Although the days with three meteorological conditions are less in Hebei province, traffic accidents will increase sharply when three meteorological conditions occur at the same time.

In addition to the above points, risk factors of road traffic accidents may also include road sections, vehicle types, drivers and other factors. This paper only analyzes the risk and management issues related to meteorology.

4.2 *Risk management of traffic accidents*

The risks related to traffic accidents and meteorological conditions in Hebei province can be divided into the following three categories: traffic risk that affects visibility, such as fog; traffic risk brought by affecting the road surface and the vehicle adhesion coefficient, such as rain, snow, pooling after rain, ice after snow, etc.; the risk brought by the road adhesion coefficient, as well as the range of the driver's line of sight, such as heavy rain and heavy snow.

When there are traffic risks affected by visibility, the highway administration department should set up dynamic road signs and traffic signs at major roads and important junctions, and change settings according to the visibility. At present, many cities in Hebei province have liquid crystal variable signs on the road, which mostly give road instructions. When the visibility is low, the highway departments and related units can attract the attention of drivers by increasing the display font and changing the prompt content or speed limit value dynamically. Prompting drivers about current important road conditions can be effective in relieving drivers' tension. According to the survey, 70% of drivers felt very nervous as he entered a foggy area, 85% of the drivers felt fatigue while driving in the fog, 87.5% of drivers would change driving posture on foggy days, and therefore, it is easy to panic and cause traffic accidents when encountering an emergency (Luo and Wei 1999). When the road surface is wet and the adhesion coefficient is reduced, the traffic departments should keep in touch with the local meteorological departments and push the short-term and imminent forecast results in real time, to provide reference points for drivers on whether to continue driving or not. At the same time, according to the potential risk, the recommended safe driving speed and distance between two cars for different models should be given as soon as possible.

5 CONCLUSION AND DISCUSSION

By comparing the number of days and times of traffic accidents on non-sunny days and sunny days, this paper analyzes the average number of daily traffic accidents under different conditions. It briefly discusses the potential risks and management of traffic accidents in combination with the situation in Hebei province, and draws the following conclusions.

Autumn is the season with the most traffic accidents in Hebei province. November is the month with the most accidents, followed by October and December.

There are more traffic accidents on foggy days than on sunny days, and the probability of average traffic accidents is the highest between 50 and 100. In rainy days in Hebei province, the probability of light rain is the highest, and there are varying impacts on different grades of precipitation on traffic accidents. When the grade of rainfall is above heavy rain, the higher the rainfall, the more traffic accidents. The probability of light snow is the highest on snowfall days in Hebei province. Traffic accidents in light snow reach a peak value and from light snow to heavy snow, the traffic accidents decrease.

The traffic accidents with two meteorological conditions in a single day are higher than those on sunny days. The most traffic accidents occur in one day when fog and snow or heavy wind and rainfall occur at the same time. The number of days in which the three meteorological conditions occur simultaneously is small, but the daily average traffic accidents are significantly more then than in any other adverse weather.

The traffic accident risks in Hebei province caused by meteorological conditions mainly include reduced visibility, a reduced road adhesion coefficient, and the combination of reduced visibility and reduced road adhesion coefficient. According to the specific potential risk types and levels, the highway administration department and other relevant departments should dynamically adjust the control measures based on the real-time information of the meteorological department, design a safe driving speed, and reduce the possibility of traffic accidents.

ACKNOWLEDGMENTS

The authors wish to appreciate the assistance of colleagues of Hebei Provincial Meteorological Observatory and Cangzhou Meteorological Bureau.

REFERENCES

Bai, Y.Q., He, M.Q., Liu, J. and Qi, H.X. 2015. Study on the relationship between expressway traffic accidents and meteorological conditions. *Meteorological and Environmental Sciences* 38(02): 66–71.

Eisenberg, D. 2004. The mixed effects of precipitation on traffic crashes. *Accident Analysis & Prevention* 36(4): 637–647.

He, F.F., Fang, G.L., Wu, J.P. and Chen, J. 2004. Study on the relationship between adverse weather conditions and traffic accidents in Shanghai. *Quarterly Journal of Applied Meteorology* 2004(01): 126–128.

Keay, K and Simmonds, I. 2005. The association of rainfall and other weather variables with road traffic volume in Melbourne, Australia. *Accident Analysis & Prevention* 37(1): 109-124.

Keay K. and Simmonds I. 2006. Road accidents and rainfall in a large Australian city. *Accident Analysis & Prevention* 38(3): 445–454.

Koetse K.J. and Rietveld P. 2009. The impact of climate change and weather on transport: An overview of empirical findings. *Transport and the Environment* 14(3): 205–221.

Lin, Y., Li, Q., Zhang, K., Li, L., Qi, X., Lin, Z.G., Lin, S. and Zhang, Y.F. 2018. Study on the influence of meteorological conditions on traffic safety of expressway in Liaoning province. *Journal of Meteorology and Environment* 34(03):106–111.

Lu, T., Zhang, Y. and Gao, J.P. 2009. Research on safety management system of expressway operation under snow and ice weather. *Science and Technology Innovation Herald* 2009(01): 87.

Luo, Y. and Wei, L. 1999. Foggy weather and highway traffic safety. *CHINESE ERGONOMICS.* 1999(01): 4.

Symons L and Perry A. 1997. Predicting road hazards caused by rain, freezing rain and wet surfaces and the role of weather radar. *Meteorological Applications* 4(1): 17–21.

Xia, M.J., Cao, J. and Zhou, W.J. 2014. Analysis of meteorological conditions and road traffic accidents in Nanjing area. *Journal of the Meteorological Sciences* 34(03): 305–309.

Risk Analysis Based on Data and Crisis Response Beyond Knowledge – Huang & Nivolianitou (eds)
© 2020 Taylor & Francis Group, London, ISBN 978-0-367-25146-8

The combined effect of heavy rain and topography on the division of rainstorm and flood disaster risk in Hebei Province

Jing Zhang
Key Laboratory of Meteorology and Ecological Environment of Hebei Province, Shijiazhuang, China
Climate Center of Hebei Province, Shijiazhuang, Hebei, China

Lin Wu
Key Laboratory of Agricultural Water Resources, Hebei Laboratory of Agricultural Water-Saving, Center for Agricultural Resources Research, Institute of Genetics and Developmental Biology, CAS, Shijiazhuang, China
University of Chinese Academy of Sciences, Beijing, China

Rongfang Yang
Meteorological Service Center of Hebei Province, Shijiazhuang, China

Kaicheng Xing*
Climate Center of Hebei Province, Shijiazhuang, China

Xian Yang
Hebei Meteorological Information Center, Shijiazhuang, China

Jing Wei
Hebei Institute of Geological Survey, Shijiazhuang, China

Guidong Ma
Climate Center of Hebei Province, Shijiazhuang, China

ABSTRACT: Based on the natural disaster risk index and geography information system technology, four main factors of the risk of rainstorm and flood disasters in Hebei Province were studied. Considering the influence of topography, climate and social economy, the contribution of each major factor was determined and the flood disaster risk zoning map was drawn and analyzed. The results indicated that the regions at the highest risk of flood disaster were concentrated in the coastal areas of north-eastern Hebei Province from 1961 to 2008, mainly including Tangshan city and Qinhuangdao city. The zones at the second-highest risk of flood disaster were located in southeastern Hebei, including Cangzhou city, northern Hengshui city and eastern Xingtai city. The lowest and second-lowest flood risk zones were distributed in northwestern Hebei Province, including Zhangjiakou city and Chengde city, which was caused by low precipitation and low vulnerability.

Keywords: Risk assessment and zoning, flood, natural disaster risk index

1 INTRODUCTION

Under the background of global warming, extreme weather events and meteorological disasters have occurred frequently in China in recent decades (Renet al. 2005a,b; Kennedy, 2003). Many

---

*Corresponding author: 343617032@qq.com

scholars at home and abroad have carried out research on disaster risk zoning and evaluation. Blaikie, an American scholar, believed that the key to disaster reduction is reducing the vulnerability of the bearing body, and utilizing economic development and resources to increase the disaster resistance of the bearing body (Blaikie et al. 1994).Zhou et al. (2000) stated that flood risk zoning is an important part of flood assessment and management. On the basis of analyzing the catastrophic factors, they put forward the index model of disaster risk zoning based on GIS (Geographic Information System), and designed comprehensive zoning of flood risk in the Liaohe River basin (Zhou et al. 2000). Zhang et al. (2005) presented the integrated index and carried out flood disaster risk assessment and zoning study on the middle and lower reaches of the Liaohe River. Because of obvious differences in the risk of flood disasters in various areas, there are few studies on flood disaster zoning in Hebei Province regarding the influence of topography, climate and social economy. Therefore, carrying out flood disaster risk zoning in HebeiProvince can provide technical support for main functional area planning, flood risk management and control, and disaster reduction planning in Hebei Province.

This paper combines flood risk analysis method with GIS technology and the flood/waterlogging disaster risk index (FDRI) (Zhang and Li, 2007), assesses the risk level of flood and waterlogging disasters in Hebei Province and draws up a flood disaster risk zoning map of Hebei Province.

2 DATA AND METHODS

2.1 Site description

Hebei Province is located in the northeast of the North China Plain, the central area surrounding Beijing and Tianjin (36°03′~42°40′N, 113°27′~119°50′E) (Figure 1). The total area of Hebei Province is around 190,000 km^2 and it has a semi-humid semi-arid continental monsoon climate. The topography of Hebei Province is high in the west and low in the east, and the geomorphological type is complete. Affected by the East Asian summer monsoon, torrential rains in Hebei Province occur in the summer, accounting for around 89% of the total torrential rain in a year. The inter annual variation of rainstorm volume is large, and the regional difference of rainstorm makes the frequency and influence of flood disaster different all over Hebei Province. Serious flood disasters often occur in years with heavy rainstorms and concentrated rainstorms (Zhang et al. 2007).

Figure 1.   Municipal administrative boundaries of Hebei Province and Beijing and Tianjin.

2.2 Methods

The risk zoning of meteorological disasters refers to the quantitative analysis and evaluation of factors such as the environmental sensitivity of the disaster, the risk of disaster factors, the vulnerability of the disaster body, the ability of disaster prevention and reduction. In order to reflect the regional differences of the risk distribution of meteorological disasters, according to the size of the risk index, the risk zoning is divided into several levels. Considering that each evaluation factor has different effects on risk composition, after consulting experts on water conservancy, land, agriculture, meteorology, climate and other subjects, the risk of rainstorm and flood disasters involved in the factor weight coefficient are summarized in Figure 2. The risk index formula of rainstorm and flood disasters, *FDRI*, is represented as:

$$FDRI = (VE^{we})(VH^{wh})(VS^{ws})(10 - VR)^{wr} \qquad (1)$$

where the *FDRI* is a risk index of rainstorm and flood disasters, which is used to indicate the degree of risk. The greater the value, the greater the degree of disaster risk. The values of *VE*, *VH*, *VS* and *VR* represents the sensitivity of disaster pregnant environments, the risk of disaster-causing factors, the vulnerability of disaster-bearing bodies and the ability of disaster prevention and mitigation. *We*, *wh*, *ws* and *wr* are the weights of each evaluation factor.

2.3 Data

The meteorological data used in this paper were daily precipitation data of 81 meteorological stations in Hebei Province from 1961 to 2008. The disaster data were the general survey data of torrential rain, flood and waterlogging in Hebei Province from 1984 to 2008 (disaster-stricken population, disaster area, direct economic losses). The socioeconomic data were from *Hebei Statistical Yearbook*, published in 2008: the data of land area, total population at the end of the year, cultivated land area, gross national product (GDP), flood control and water-logging control area of county administrative region. Topographic data include DEM and water coefficient data from 1: 50000 GIS data in Hebei Province.

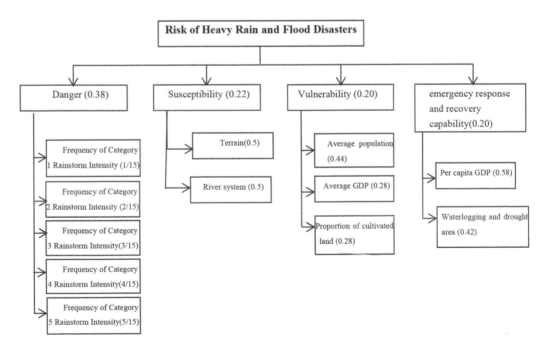

Figure 2.   Weight of risk zoning of heavy rain and flood in Hebei Province.

150

3 DATA AND METHODS

3.1 *Zoning of risk factors*

The greater the rainfall, the higher the risk hazard index of heavy rainfall and flood disaster-causing factors in Hebei Province. After weighting different scales of heavy rainfall, using the natural breakpoint method, five grades were determined. As can be seen from the danger zoning of the disaster-causing factors of rainstorm and flood disasters in Hebei Province in Figure 3, the lowest risk of rainstorm and flood is the northwest of Hebei Province, and the strip area in the east and south of the low-risk area is the sub-low rainstorm flood risk area. In general, the central and southern part of Hebei Province is a medium-risk area. The areas at high risk of disaster-causing factors of rainstorm and flood were located in the eastern part of Hebei Province, especially in the coastal areas of northeast Hebei Province.

3.2 *Zoning of sensitivity factor*

Astopography and water systems have similar effects on rainstorm and flood disasters in disaster pregnant environments, we integrated the opinions of many experts and assigned the weights of these two factors as 0.5. From the sensitivity zoning of disaster pregnant environments in Figure 4, the sensitivity of the Bashang area in northern Hebei Province to rainstorm and flood disasters is generally low, which is due to the low topographic index and water system index there. The areas with the highest sensitivity to rainstorm and flood disasters were

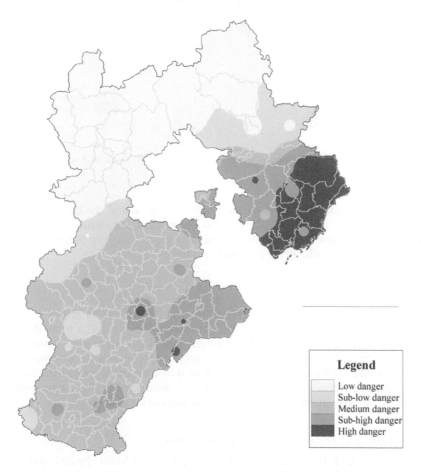

Figure 3.   Zoning map of risk factors of heavy rainfall and flood disasters in Hebei Province.

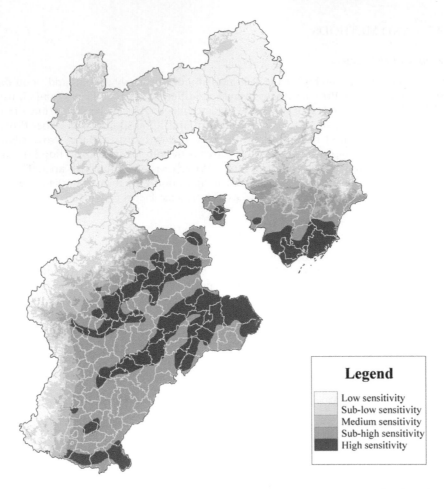

Legend

Low sensitivity
Sub-low sensitivity
Medium sensitivity
Sub-high sensitivity
High sensitivity

Figure 4. Zoning map of environmental sensitivity to heavy rain, flood and waterlogging disasters in Hebei Province.

located in the coastal areas of north-eastern Hebei Province and the eastern areas of south-central Hebei Province. This is mainly because these areas are located in plains or coastal areas, with low topography, and include the main rivers such as the South Canal, Yongding River, Ziya River and Luanhe River in Hebei Province.

3.3 *Zoning of vulnerability factor*

Figure 5showsthe vulnerability zoning of rainstorm and flood disasters in Hebei Province. It is clear that the north-east coastal zone with high population density and a more developed economy is the most vulnerable, followed by the pre-mountain plain area in south-central Hebei Province, mainly due to the large population density and high proportion of cultivated land in these areas. The moderately vulnerable areas are distributed in the eastern part of the secondary high vulnerability areas, and the low vulnerability areas are distributed in the northern part of Hebei Province, where the population is sparsely populated and the GDP density is low.

3.4 *Zoning of disaster prevention and mitigation capability*

The smaller the value of disaster prevention and response, the lower the ability of disaster prevention and resilience. From the distribution of disaster prevention and response in Figure 6,

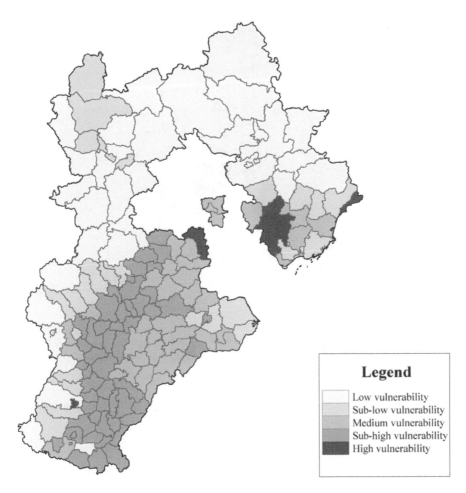

Figure 5. Vulnerability zoning map of disaster-bearing body of heavy rainfall and flood disasters in Hebei Province.

we can see that the wide area of flood control and high per capita GDP in the north-eastern coastal areas of Hebei Province meant it had the highest disaster prevention and response capacity. This was followed by the plain area in south-central Hebei Province and the eastern Taihang Mountain, which has a wide area of flood control and waterlogging. The areas with the lowest disaster prevention and resilience are mainly distributed in Zhangjiakou and Chengde in northern Hebei Province, mainly due to the smallest proportion of flood control and waterlogging area in these areas and the generally low per capita GDP.

3.5   Risk division of rainstorm and flood disasters

As can be seen from the risk zoning of rainstorm and flood disasters in Hebei Province in Figure 7, the areas with the highest risk of rainstorm and flood disasters in Hebei Province were mainly distributed in the coastal areas of north-eastern Hebei Province, because of the risk of disaster factors and the high sensitivity of the disaster environment in these areas. The sub-high and medium-risk areas of rainstorm and flood were located in the western part of the central and southern part of Hebei Province. The higher risk of rainstorm and flood is due to these areas being more prone to heavy rains under the influence of the windward slope topography of the southern foothills of Yanshan. The influence of the eastern and southern regions was significantly affected by the monsoon. The low risk of rainstorm and flood

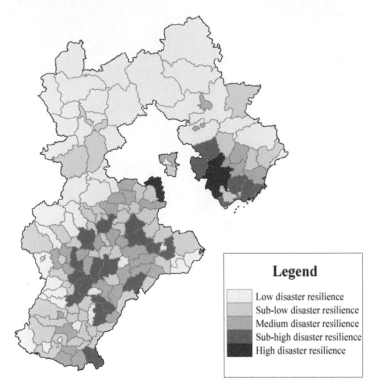

Figure 6. Zoning map of rainstorm flood and rainstorm resistance ability in Hebei Province.

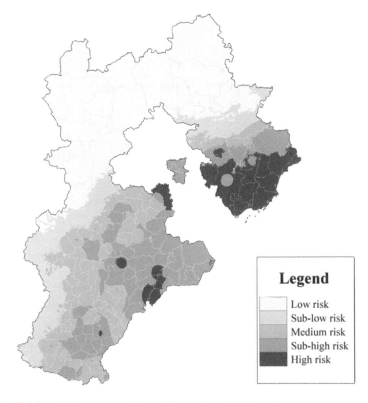

Figure 7. Risk division of rainstorm and flood disaster in Hebei Province.

disasters in the northern part of Hebei Province, especially in the northwest, is due to the particular risk of disaster factors in these areas and the lowest comprehensive vulnerability of the disaster complex, and the low sensitivity of the disaster pregnant environments.

3.6 *Verification*

With the data of rainstorm and flood disasters in Hebei Province in 1984–2008, the distribution map of the victims in Hebei Province by annual rainstorm and flood disaster was obtained by using the total index of affected population, affected area, direct economic loss and 2000 commodity retail price. Through the commodity price index in 2000, the annual average direct economic loss chart was calculated after the comparable price. Through comparison, it can be seen that the distribution of rainstorm and flood zoning in population, agriculture and economy is consistent with that after disaster statistics, and the results of the zoning are more reliable.

4 CONCLUSION

This article synthetically analyses and summarizes the information related to rainwater and flood disasters in geography, climate and economy of Hebei Province, in an intuitive and easy-to-understand way. This synthesizes the risk of factors causing rainstorm and flood disasters in Hebei Province, the sensitivity of geographical environment to the occurrence of floods, the vulnerability of disaster-bearing bodies and the ability of disaster prevention and mitigation. The risk zoning of flood and flood disasters in Hebei Province can provide a basis for flood disaster prevention planning in the region.

ACKNOWLEDGMENTS

This study is supported by the National Key Research and Development Program of China (NO. 2018YFA0606302).

REFERENCES

Ren, G.Y., Chu, Z.Y., Zhou,Y.Q., et al. 2005a.Recent advances in temperature change research in China. *Climate and Environment Research*10(4): 701–713.

Ren, G.Y., Guo J., Xu M.Z., et al. 2005b. Basic characteristics of surface climate change in China in recent 50 years. *Journal of Meteorology* 63(6): 942–956.

Kennedy, D. 2003. Breakthrough of the year. *Science* 302: 2033.

Blaikie, P., Kates, W., White, G.F., et al. 1994. *The Environments Hazard* (2nd eds.). New York: The Guildford Press.

Zhou, C.H., Wan, Q., Huang S.F., et al. 2000. Study on flood risk zoning based on GIS. *Journal of Geography*55(1): 15–24.

Zhang, H., Zhang, J.Q. and Han, J.S. 2005. Study on flood risk assessment and zoning based on GIS technology – Taking the middle and lower reaches of Liaohe River as an example. *Journal of Natural Disasters*14(6): 141–146.

Zhang, J.Q., Li, N. 2007. *Quantitative Methods for Risk Assessment and Management of Major Meteorological Disasters and Their Applications.* Beijing: Beijing Normal University Press.

Zang, J.S., Zhang, X.T., Zhao, X.P, et al. 2007. *China Meteorological Disaster Daily Code* (Hebei Volume). Beijing: Beijing Meteorological Publishing House.

Zhang, J.Q. 2004. Risk assessment of drought disaster in the maize-growing region of Songliao Plain, China. *Agriculture Ecosystem & Environment* 102(2): 133–153.

Risk Analysis Based on Data and Crisis Response Beyond Knowledge – Huang & Nivolianitou (eds)
© *2020 Taylor & Francis Group, London, ISBN 978-0-367-25146-8*

A peer-to-peer lending credit risk prediction method: Application of balanced relative margin machine

Beibei Niu*, Jinzheng Ren & Xiaotao Li
College of Economics and Management, China Agricultural University, Beijing, China

ABSTRACT: During the last few years, many classification methods have been applied to predicting peer-to-peer (P2P) lending credit risk. However, many models have given little consideration to data drifts and outliers. This paper examines the balanced relative margin machine (BRMM), and how it does in predicting default on a Chinese P2P lending platform, comparing its performance to the support vector machine (SVM) and logistic regression. In total, 55,596 P2P lending loans are examined. Discrimination performance for BRMM was found to be superior to SVM and logistic regression.

Keywords: P2P lending, Credit Risk, BRMM, SVM, Credit scoring

1 INTRODUCTION

The peer-to-peer (P2P) lending platform acts as a bridge where borrowers can borrow money directly from investors without the help of banks. P2P lending has received much attention in recent years due to its convenience, which offers quicker loan approvals than banks. It is emerging as an important supplement to the traditional financial institutions. Though P2P platforms have achieved rapid growth in many developing countries, platform credit risk management has been found to be weak still and an imperfect credit scoring system has led many P2P platforms to go bankrupt (Ma et al., 2018). Consequently, it is important for every P2P platform to seek an effective credit risk method to improve their risk management capabilities. Credit scoring is the most widely used risk management tool in the financial industry. P2P platform credit scoring methods are similar to those used by commercial banks. They both use demographic characteristics and credit history to build a data mining model in order to evaluate loan default probability (Emekter et al., 2015).

In recent years, a variety of data mining methods and statistical techniques have been tested in P2P lending data (Guo et al., 2016; Jiang et al., 2017). Compared with the risk management approach of commercial banks, there are still many problems with the P2P lending platform. First, the P2P lending platform copies the credit scoring approaches of commercial banks, although borrowers of P2P lending have greater differences than commercial bank customers (Aitken, 2017). Second, these methods poorly processing abnormal observations in the data-sets always. Most models do not make full use of the original data. Many outliers are deleted or replaced. Some important information will be ignored, so as to influence the prediction results. Alternatively, P2P lending credit scoring should make full use of outlier data and propose a highly targeted model to improve risk management capabilities.

Commercial banks use many different data mining algorithms to predict the probability of borrowers, such as LDA, linear regression, and logistic regression (Butaru et al., 2016). These methods often perform poorly when dealing with nonlinear relationships. Support vector machines (SVM) and neural networks (NN) can handle nonlinear relationships in credit risk

---

*Corresponding author: oubeigo121@163.com

assessment to improve prediction accuracy (Harris, 2015). Compared to NN, SVM can overcome many more problems, such as local extremum, dimensional disaster, over-fitting, etc. SVM has been applied successfully in credit scoring (Maldonado et al., 2017; Li et al., 2017; Maldonado et al., 2017). However, the SVM hyperplane pays no attention to the spread of the data. Shivaswamy and Jebara (2010) propose the relative margin machine (RMM) to overcome such shortcomings. A major limitation of the RMM method is the poorly processing of outliners at the outer margin. This limitation affects predictive accuracy. P2P-lending borrowers and bank borrowers are quite different. The internal differences between the customer groups are large. The P2P lending borrowers have more outliers than bank customers do. This fact will inevitably lead to the reduction of the accuracy of the model. It is important to search for a method to overcome the problem of huge outliers in loans. In order to tackle the limitation of the traditional SVM, Krell et al. (2014) propose using balanced relative margin machines (BRMM) to resolve it. BRMM defines the outer margin and the hyperplane in a new way. By using this definition, the outliers are included at the margin, which has significant advantages in datasets with greater amounts of abnormal data.

This paper proposes the BRMM method to improve the P2P platform prediction accuracy. We apply three default prediction models, including logistic regression, SVM, and BRMM to predict delinquency. The empirical research utilizes data from a Chinese P2P lending platform. Our study shows that BRMM provides a significant increase in prediction accuracy. This paper introduces BRMM to evaluate P2P-lending credit risk, to overcome the limitation existing in the current methods. It provides an effective method for the financial regulatory authorities and P2P platform to monitor, control, and manage risks.

2 METHODOLOGY

After the borrower receives the loan from the P2P lending platform, two statuses may arise: default and non-default. Corresponding to the state indicator $y_i \in \{1, -1\}$. Hand and Henley (1997) pointed out that credit scoring is a classification problem, wherein borrowers are classified as "good" or "bad." The bad ones are default borrowers. Default borrowers are assigned -1, while non-default ones are good and assigned 1. Every borrower contains n-dimensional characteristic indicator variables $x_i = (x_i^{(1)}, x_i^{(2)}, ..., x_i^{(n)})$. Each variable represents the relevant characteristics of the borrower, such as age, monthly income, loan amount, etc. Every sample point (x_i, y_i) contains characteristic indicator variable x_i and final state variable y_i. The platform predicted the borrower default or not based on the relevant characteristic indicator.

The maximum margin classification machine is the most fundamental SVM. It is also the basis of the advanced SVM. SVM defined by a separating hyperplane. The middle hyperplane separates two different types of points. The distance between the hyperplanes is the margin. The objective of the SVM is to find a hyperplane separating the different points.

The hyperplane should maximize the margin distance. Our training data consists of N pairs $S = \{(x_1, y_1), (x_2, y_2), ..., (x_n, y_n)$, with $x_i \in R^n$ and $y_i \in \{1, -1\}$. $f(x) = <w, x> + b$ is the decision function. w is the weighted variable. b is the bias. The classification rule induced by $f(x)$ is:

$$f(x) = sign[<w, x> + b] \tag{1}$$

Hence, the optimization problem is as follows:

$$\min_{w,b} \frac{1}{2} ||w||^2, \tag{2}$$
$$s.t. y_i(w \cdot x_i + b) \geq 1.$$

The above method can be used to solve the linear separable hyperplane. But in the general case, the data are not linearly separable. We use soft margin to overcome this limitation. The slack variable ξ_i is used to represent the distance between the classification error point to classification hyperplane. The hyperparameter C is the weight of the slack variable ξ_i, and present how we penalize slack variables. We write the optimization problem in equation (3).

$$\min_{w,b,\xi} \frac{1}{2}\|w\|^2 + C\sum_{i=1}^{n}\xi_i,$$
$$s.t. y_i(\langle w, x_i\rangle + b) \geq 1 - \xi_i,$$
$$\xi_i \geq 0. \tag{3}$$

SVM produces decision boundaries that separate data along its directions with large data spreads. RMM uses hyperparameter R to control the farthest distance between the data and the classification hyperplane. The initial function can, thus, be converted to the following:

$$\min_{w,b,\xi} \frac{1}{2}\|w\|^2 + C\sum_{i=1}^{n}\xi_i,$$
$$s.t. y_i(\langle w, x_i\rangle + b) \geq 1 - \xi_i,$$
$$\frac{1}{2}(\langle w, x_i\rangle + b)^2 \leq \frac{R^2}{2},$$
$$\xi_i \geq 0. \tag{4}$$

Inner margin borders are ± 1. We assume $R \geq 1$.

In RMM the margin is large only relative to the data spread. It fails to consider the outliers at the outer margin. A new method based on the RMM may, thus, be proposed, yielding the following:

$$\min_{w,w_0,\xi} \frac{1}{2}\|w\|^2 + C\sum_{i=1}^{n}\xi_i,$$
$$s.t. 1 - \xi_i \leq y_i(\langle w, x_i\rangle + b) \leq R + \xi_i,$$
$$\xi_i \geq 0. \tag{5}$$

BRMM defines outliers in the out margin and in the inner margin by adding qualifications. *Balanced* means treating the anomalies of the outer margin and the inner margin the same way. The hyperparameter R represents the maximum distance between the sample and the classification hyperplane. ξ_i is a slack variable. Constructing a Lagrangian function for the original BRMM, and finding the partial derivative of w, b, and ξ_i. The resulting Karush-Kuhn-Tucker (KTT) optimality condition is as follows:

$$\frac{\partial L(w, b, \xi, \alpha, \beta, \eta)}{\partial w} = w - \sum_{i=1}^{n}(\alpha_i - \eta_i)y_i x_i = 0, \tag{6}$$

$$\frac{\partial L(w, b, \xi, \alpha, \beta, \eta)}{\partial \xi_i} = C - \alpha_i - \eta_i - \beta_i = 0, \tag{7}$$

$$\frac{\partial L(w, b, \xi, \alpha, \beta, \eta)}{\partial b} = -\sum_{i=1}^{n}(\alpha_i - \eta_i)y_i = 0, \tag{8}$$

$$\alpha_i[y_i(\langle w, x_i \rangle + b) - 1 + \xi_i] = 0, \tag{9}$$

$$\eta_i[y_i(\langle w, x_i \rangle + b) - R - \xi_i] = 0, \tag{10}$$

$$\beta_i \xi_i = 0. \tag{11}$$

Substituting functions (6) (7) (8) (9) (10) (11) into the Lagrangian function and introducing the kernel function $K(x_i, x_j)$, we can get the following dual problem:

$$\max_{\alpha, \eta} -\frac{1}{2} \sum_{i,j=1}^{n} y_i y_j (\alpha_i - \eta_i)\left(\alpha_j - \eta_j\right) K\left(x_i, x_j\right) - \sum_{i=1}^{n} \eta_i R + \sum_{i=1}^{n} \alpha_i,$$
$$s.t. \sum_{i=1}^{n} (\alpha_i - \eta_i) y_i = 0, \tag{12}$$
$$0 \le \alpha_i + \eta_i \le C.$$

After solving dual problem (10), we can get the following optimal classification function:

$$f(x) = sign\left(\sum_{i=1}^{n} \alpha_i y_i K\left(x_i, x_j\right) + b\right) \tag{13}$$

The model has been built. We can evaluate the BRMM model based on the test data.

3   EXPERIMENTAL SET-UP

3.1   *Sample and data*

We use data from a Chinese P2P lending platform. It contains 55,596 loans. Out of this total, 7,183 (12.92%) failed and 48,413 did not (87.08%). Table 2 displays the research variable. The variables include the borrower's characteristics, credit card billing issues, web browsing behavior, credit card transaction record, loan period, etc. Default or not is the dependent variable.

The ratio of default and non-default data is 1:6.8. Model training of imbalanced dataset will affect prediction accuracy. Sampling techniques are thus used to select partial data from non-default sample datasets to reduce this imbalance. Before selection, the sample set was divided into a training set and a validation set. A random extraction of 10% of the dataset from the original sample, yielded 6,000 loans as a validation set. We balanced the remaining 49,456 loans. Taking

Table 1.   Variables used in the study.

Variable	Symbol	Variable	Symbol
Default or not	Y	Card monthly expenditure	X9
Gender	X1	Credit card number	X10
Occupation	X2	Credit card limit	X11
Marital status	X3	Number of cards used monthly	X12
Educational background	X4	Amount on cards used monthly	X13
Household registry type	X5	Number of defaults	X14
Monthly income	X6	Credit card available balance	X15
Personal non-wage income	X7	Cash advance limit	X16
Card expenditure monthly usage	X8	Monthly consumption proportion	X17

6,328 default loans as 1 unit, 6,473 non-default loans were extracted by random sampling, so that the ratio of default to non-default loans is 1:1. The training data set has 12,801 loans. In order to ensure the accuracy and stability of the prediction results, a total of fifty experiments were performed. The average classification accuracy was the final classification accuracy.

The dataset has missing data. Imputing continuous numerical variable's missing values using the average, with the majority being used to fill the missing data of binary discrete variables.

3.2 Benchmark models

In this section, we compare three types of credit scoring models: regularized logistic regression, SVM, and BRMM. SVM and BRMM have already been stated.

Although many data mining credit scoring models have been introduced, logistic regression is the most widely used in credit scoring practice because of its stability and interpretive clarity (Lessmann et al., 2015). The probability is expressed as the following logistic function:

$$probability = \frac{1}{1+e^{-z}} = \frac{1}{1+e^{-(\beta_0+\beta^T x)}} \tag{14}$$

The logistic regression results are represented by the conditional probability distribution, where x represents the borrower's characteristics, β is the weighted variable. β_0 is the bias. The maximum likelihood estimation method can be used for calculating the model's parameters. In order to perform better while out-of-sample, we use a regularized logistic regression.

3.3 Parameter selection

The values set of the kernel parameters in SVM is $\{2^{-5},2^{-4},2^{-3},2^{-2},2^{-1},2^0,2^1,2^2,2^3,2^4,2^5\}$, and the values set of C is $\{2^{-2},2^{-1},2^0,2^1,2^2,2^3,2^4,2^5,2^6,2^7\}$. The values set of the kernel parameters in BRMM is $\{2^{-3},2^{-2},2^{-1},2^0,2^1,2^2,2^3\}$, and the values set of C is $\{2^{-2},2^{-1},2^0,2^1,2^2\}$, the values set of R is $\{2^0,2^1,2^2,2^3\}$, the group with the highest average classification accuracy is the optimal parameter. The validity of the SVM model, the BRMM model, and the logistic regression was verified by using fifty random experiment accuracy averages. In the experiment, the training data set was used for training, and then the validation set was used to test the training model. The model parameter optimization was selected by the quadprog function in the Matlab 2016a toolbox. In order to ensure the comparability of the results, each cohort had the same grouping.

4 EXPERIMENTAL RESULTS

Table 2 compares the results of SVM, logistic regression and BRMM in training data and validation data. The results show that BRMM has the highest accuracy among the three methods.

Misjudging default loans as the non-defaulting customer will generate greater economic losses for the lenders, so estimating of defaulting loans is more important. Regardless of

Table 2. The Classification accuracy of SVM, logistic regression, and BRMM (hit ratio: %).

Methods	Training			Validation		
	Default	Non-default	Total	Default	Non-default	Total
SVM	73.96	82.21	79.03	72.64	79.89	77.18
Logistic	69.43	77.65	74.43	66.76	76.49	72.50
BRMM	76.71	85.47	82.07	74.48	82.06	79.21

whether it is in the training or validation set, the BRMM model has the highest accuracy for the default loans forecast. On the one hand, in the training set, BRMM is 7.28 and 2.75 percentage points higher than logistic regression and SVM. In the validation set, BRMM is 7.72 and 1.64 percentage points higher than logistic regression and SVM. On the other hand, the prediction accuracy of the non-default account of the three models in the training set or in the test set is higher than the default accuracy of the default account. BRMM have the highest accuracy, whether in terms of a default or non-default account.

5 CONCLUSION AND DISCUSSION

P2P lending is flourishing in China. It is a great supplement for the financial system, helping small enterprises and individuals obtain loans easily. Nevertheless, the weakness found in the risk management system among P2P platforms has been notorious. In order to enhance risk management skills, an accurate delinquency prediction approach is needed.

In this paper, BRMM was applied to predict whether P2P loans would default or not. We employed a large dataset from a Chinese P2P lending platform to test the validity of the BRMM method. The experimental results showed that BRMM outperformed the other methods. The results showed that BRMM has higher prediction accuracy and stability than SVM and logistic regression. BRMM can tackle the high-dimensional data in P2P lending. It can deal with the outliers on the outer margins by controlling the furthest distance from the data to the classification hyperplane.

From the results, we recommend BRMM as an alternative method for the P2P platform to build its credit scoring system. It is expected that BRMM may be used on other P2P platforms. For future work, we intend to use other P2P lending platform loan data to further test the model's validity.

ACKNOWLEDGMENTS

This project was supported by the National Natural Science Foundation of China (71103184).

REFERENCES

Aitken, R. 2017. 'All data is credit data': Constituting the unbanked. *Competition & Change* 21(4): 274–300.
Butaru, F., Chen, Q., Clark, B., Das, S., Lo, A.W., and Siddique, A. 2016. Risk and risk management in the credit card industry. *Journal of Banking & Finance* 72: 218–239.
Emekter, R., Tu, Y., Jirasakuldech, B., and Lu, M. 2015. Evaluating credit risk and loan performance in online Peer-to-Peer (P2P) lending. *Applied Economics* 47(1): 54–70.
Guo, Y., Zhou, W., Luo, C., Liu, C., and Xiong, H. 2016. Instance-based credit risk assessment for investment decisions in P2P lending. *European Journal of Operational Research* 249(2): 417–426.
Hand, D.J., and Henley, W.E. 1997. Statistical classification methods in consumer credit scoring: a review. *Journal of the Royal Statistical Society: Series A (Statistics in Society)* 160(3): 523–541.
Harris, T. 2015. Credit scoring using the clustered support vector machine. *Expert Systems with Applications* 42(2): 741–750.
Jiang, C., Wang, Z., Wang, R., and Ding, Y. 2017. Loan default prediction by combining soft information extracted from descriptive text in online peer-to-peer lending. *Annals of Operations Research* 266: 1–19.
Krell, M.M., Feess, D., and Straube, S. 2014. Balanced Relative Margin Machine - The missing piece between FDA and SVM classification. *Pattern Recognition Letters* 41(C): 43–52.
Lessmann, S., Baesens, B., Seow, H., and Thomas, L.C. 2015. Benchmarking state-of-the-art classification algorithms for credit scoring: An update of research. *European Journal of Operational Research* 247(1): 124–136.
Li, Z., Tian, Y., Li, K., Zhou, F., and Yang, W. 2017. Reject inference in credit scoring using Semi-supervised Support Vector Machines. *Expert Systems with Applications* 74: 105–114.

Ma, L., Zhao, X., Zhou, Z., and Liu, Y. 2018. A new aspect on P2P online lending default prediction using meta-level phone usage data in China. *Decision Support Systems* 111: 60–71.

Maldonado, S., Bravo, C., López, J., and Pérez, J. 2017. Integrated framework for profit-based feature selection and SVM classification in credit scoring. *Decision Support Systems* 104: 113–121.

Maldonado, S., Pérez, J., and Bravo, C. 2017. Cost-based feature selection for Support Vector Machines: An application in credit scoring. *European Journal of Operational Research* 261(2): 656–665.

Shivaswamy, P.K., and Jebara, T. 2010. Maximum Relative Margin and Data-Dependent Regularization. Journal of. *Machine Learning Research* 11: 747–788.

Risk Analysis Based on Data and Crisis Response Beyond Knowledge – Huang & Nivolianitou (eds)
© 2020 Taylor & Francis Group, London, ISBN 978-0-367-25146-8

Optimal catastrophic risk sharing for government boundary: A case study from egg price insurance in Beijing

Lei Xu, Bingbing Wang & Qinghui Geng
Agricultural Information Institute, Chinese Academy of Agricultural Sciences, Beijing, China

ABSTRACT: The establishment of a multi-level compensation mechanism for catastrophic risk is a critical issuefor the sustainable development of China's agricultural products price insurance. But the bottleneck problem which is to define the optimal boundary of government intervention is still unsolved. Therefore, this paper aims to provide a quantitative methodology of measuring the optimal catastrophic risk sharing ratio between government and market from the perspective of actuarial sciences and insurance economics. An empirical case showed that the payment rate of egg price insurance in Beijing city would top 958%, suffering the impact of a once-in-a-hundred-years catastrophic event.In this scenario, the optimal catastrophic risk sharing ratio is 172% between government and insurer, which means that the maximal indemnity of the insurance company is 172% of the premium in the month when the catastrophic event occurs, and the government begins to pay the rest of the indemnity over the 172% of the premium. Meanwhile, the government currently has a ceiling of just 261% of the premium on the rest of the indemnity payment in order to avoid subjecting itself to the "financial subsidies trap". Briefly, the indemnity responsibility undertaken by the government or the optimal boundary of government payment dropped into the interval from 172 % to 261% of the premium. The rest of risk and indemnity responsibility over the upper limit (261% of the premium) should be dispersedthrough the capital markets. This research is a profitable attempt to construct the mechanism of agricultural insurance catastrophic risk decentralization supported by the government.

Keywords: Catastrophic risk, Risk sharing, Government Boundary, Egg price insurance, Beijing

1 INTRODUCTION

Along with the continuous deepening of market-oriented reform and the acceleratinginternationalization process, various traditional and non-traditional risks are constantly protruding. The frequent and sharp price fluctuations of some agricultural products such as pork, eggs and vegetables has a huge effect on agricultural production and household consumption in China, and has become a hotspot of social concern, a key point of management decision-making and a difficulty of theoretical research (Xu Lei, 2018). Under this background, a great hope is placed by the agricultural sector on price insurance. It is an innovative risk management tool to disperse market risk, compensate economic losses and stabilize the agricultural market. Agricultural price insurance originated in the United States in the 1990s. In recent years, a number of "No 1 Central Document" proposed to "actively carry out pilot programs of agricultural price insurance". However, price insurance is aimed at the systematic price risk, which has similar price fluctuations trends and traits in different regions of the same product, the system aticness of price risk is likely to cause "catastrophe", it is difficult to achieve effective decentralization, and leaves insurance companies facing a large amount of operational risk. Judging from the price insurance pilot programs, the absence of "catastrophe" risk dispersion mechanisms of price insurance resulted in large-scale compensation occurring in many places, which poses a severe challenge to the sustainable development of price insurance. For example, the expiration loss ratio of hog price insurance in Anhua Agricultural Insurance Company is as high as 830.5%. Admittedly, in view of the huge compensation pressure faced by insurance companies, it is urgent that a "catastrophe" risk dispersion mechanism for agricultural price insurance is constructed. From the experience and lessons of local practice, the importance of government is self-evident, particularly the failure of the market mechanism in agricultural risk management has put forward an inherent demand for government involvement in agricultural risk management (Feng, 2004). In the formulation of government policy for "catastrophe" risk dispersion mechanism of agricultural price insurance, the first question to consider is what extent is appropriate? If the government undertakes excessive risk, it will inevitably put a heavy burden on finance, and "crowd out" the risk transfer function that the market should bear; but if the proportion shared by the government is too small, it will not be able to transfer and disperse the extreme risk effectively. This will affect the sustainability of agricultural price insurance. Although the agricultural price insurance subsidy pilot programs have basically reached the expected target, there are still many problems such as high fiscal costs, huge risk and high policy implementation costs that need further improvement (Huang et al., 2015). Determining the proper rate of government risk-taking may be one of the measures to solve these problems. In other words, the method to determine the starting point and upper limit of the government's financial commitment to the "catastrophe" risk compensation of agricultural price insurance becomes the key issue.

Up to now, there have been few clear and systematic literature research on the government's financial commitment boundary regarding the "catastrophe" risk compensation of agricultural price insurance, but related research has a long history and has gained many achievements. Existing research at home and abroad mainly focus on the use of financial market-level tools such as futures, options, and contract farming to avoid and disperse market price risks. Brag (1996) studied the important role of farm product options trading in avoiding agricultural market risks, Olivier Mahul (2002) studied the role of hedging financial products, futures and options in avoiding the market risks of crop yield insurance. Further, Blake K (2003) took cotton put option as an example to evaluate the function of options as a management tool of farm products price risk while Harwood et al. (1998) deemed that income insurance is cheaper and more effective than other tools in ensuring farmers' income. Moreover, Chad E. Hart et al. (2000) deemed that the design of agricultural income insurance is simple and flexible, is easy to be accepted by farmers and easy to generalize. Robert Dismukes (2006) proposed that income insurance can effectively manage the risks of farm product price fluctuations. MAFF (2001) deemed that sign price contract is an effective market risk management means while Turvey et al. (2000) explained how to use farm products selection correctly to effectively protect farm products price in the process of milk deep processing through empirical research. Wang et al. (2014) discussed the risk dispersion principle and feasibility of agricultural price insurance in China. Zhang (2015) raised the agricultural price risk dispersion model, that is agricultural price insurance plus futures and Ge et al. (2017) summarized the "insurance + futures" cotton pilot pattern in Shandong. Undoubtedly, these findings are of great value, but research on how to determine the starting point and upper limit of government financial commitment to "catastrophe" risk compensation responsibility of agricultural price insurance is still sorely lacking.

Under this background, this paper attempts to construct the optimal boundary model of government financial commitment to "catastrophe" risk compensation of egg price insurance. At the same time, taking Beijing's egg price insurance as an example makes for an empirical analysis and puts forward some corresponding policy suggestions.

2 THEORETICAL ANALYSIS FRAMEWORK

This paper targets the "catastrophe" risk of egg price insurance, attempts to put forward a risk dispersion mechanism of "insurance company + government + capital market", and divides the compensation amount (rate) of egg price insurance into three levels to be shared by each layer: the first layer is primary insurers, which is responsible for basic egg price insurance, the insurance company is in direct contact with farmers; the second layer is govern-ment, the government will partly share through the catastrophe risk reserve compensation beyond the insurance company's bearing capacity; the third layer is capital market, given the limited financial resources of the government, it is often unable to bear the responsibility of "catch-all", the risks beyond the bearing capacity of the government should be transferred to the capital market for resolution.

This paper focuses on the integrated risk-diversification mechanism of "insurance company + government + capital market", using the theory and method of modern insurance actuarial and financial engineering to determine the starting point and upper limit of the government's finan-cial commitment to the "catastrophe" risk compensation of agricultural price insurance: Build-ing the premium rate model of egg price insurance, to simulate and calculate the maximum loss ratio of egg price insurance faced by insurance companies when the egg market suffers the once-in-a-century "catastrophe" risk; regard the government as a risk sharer (similar to reinsur-ance), the insurance company as a risk distributer, referring to the optimal reinsurance pricing model and the co-insurance model, focus on the target of mean-variance, to explore the optimal boundary of government financial commitment to insurance compensations responsibility.

2.1 Constructing the premium rate determination model of egg price insurance

In this paper, the egg-feed ratio is used as the compensation indicator of egg price insur-ance. The guarantee level of egg-feed ratio (expected egg-feed ratio) set by insurance companies refers to the egg-feed compensation standard agreed by insurance companies and insured farmers during the signing of the insurance contract.This paper takes the break-even point of egg production, namely egg-feed ratio at 3.0 as the benchmark value of its guarantee level.

Supposing x_i $(i = 1,...n)$ is the time series of egg-feed ratio, we introduce the nonparametric kernel density estimation (KDE) and the Gaussian kernel density probability distribution function f(x) of the egg-feed ratio is expressed as follows:

$$f(x) = \frac{1}{nh\sqrt{2\pi}} * \sum_{i=1}^{n} \exp\left(-\frac{(x - x_i)^2}{2h^2}\right) \tag{1}$$

where n refers to sample capacity; h to bandwidth.

According to insurance actuarial principles, the pure premium rate of egg price insurance should be equal to the expected compensation of insurance companies, that is the expected loss E[loss] of the egg-feed ratio below 3.0. The calculation formula is as follows:

$$E[loss] = \text{Prob}(x < c\mu) * [c\mu - E(x|x < c\mu)]$$
$$= \int_0^{c\mu} f(x)dx \left[c\mu - \frac{\int_0^{c\mu} xf(x)dx}{\int_0^{c\mu} f(x)dx} \right] = \int_0^{c\mu} (c\mu - x)f(x)dx \tag{2}$$

The pure premium rate PR is:

$$\text{PR} = \frac{E[loss]}{c\mu} = \frac{\int_0^{c\mu} (c\mu - x)f(x)dx}{c\mu} \tag{3}$$

where c refers to guarantee level (this paper set it to 100%), it indicates the policyholder's risk preference degree and the expectation about the future to some extent, μ is the break-even point of egg-feed ratio, which is 3.0, $c\mu$ represents the guarantee level of expected egg-feed ratio chosen by farmers in the insurance contract (3.0 in this paper).

2.2 Constructing the egg price insurance loss ratio calculation model

Assuming that the farmer has purchased egg price insurance under a certain guarantee level (this paper sets the insurance period to one year), during this period,any reason other than the insurance contract exemption clause causing the actual egg-feed ratio to be lower than 3.0, can be regarded as a covered accident. Insurance companies should pay compensation according to the insurance clauses.

The formula for calculating compensation is as follows:

$$CA = max\left\{ \left(\frac{c\mu - x_i}{c\mu}\right), 0 \right\} \times IA \tag{4}$$

where CA refers to the amount of compensation paid by the insurance company; IA refers to the insurance amount; x_i refers to the actual egg-feed ratio; $c\mu$ is the expected egg-feed ratio which is 3.0.

Based on this, the loss ratio (the percentage of insurance companies' compensation and premium income) calculating formula of insurance companies is as follows:

$$LCR = \frac{CA}{IA * \alpha * PR} = \frac{max\left\{\frac{c\mu-x_i}{c\mu}, 0\right\}}{\alpha * PR} = \frac{max(LR, 0)}{\alpha * PR} \tag{5}$$

where LCR is the loss ratio; $LR = \frac{c\mu-x_i}{c\mu}$ is defined as the loss ratios of egg-feed ratio; PR is the pure premium rate to meet the essential operating expenses of insurance company; the actual premium is α times more than the pure premium (α is assumed to be 1.2 in this paper).

Then, this paper introduces the POT (Peak-Over Threshold) model of modern extreme values statistical theory. The fitting process is conducted on the tail distribution of LR to obtain its generalized pareto distribution (GPD). Based on this, the value at risk (VaR) method is used to calculate the maximum of LR under the situation that the market encounters aonce-in-a-century extreme risk, and then calculating the maximum loss ratio (LCR_{max}) that the insurance company may face at this time, namely the "catastrophe" risk faced by egg price insurance. The main calculation steps are as follows:

Firstly, constructingthe time series of LR that triggers insurance compensation:

$$y_{i\,(i=1,\dots n)} = LR = \frac{c\mu - x_i}{c\mu} \tag{6}$$

Step two, based on the time series of LR, using mean overrun function plot to determine the threshold u of GPD function in the POT model, and then fitting analysis to the sample which is above the threshold to obtain its GPD. The function F(y) is as follows:

$$F(y) = G_{\xi,\sigma}(z) = \begin{cases} 1 - \left(1 + \frac{\xi z}{\sigma}\right)^{-1/\xi}, \xi \neq 0 \\ 1 - exp\left(-\frac{z}{\sigma}\right), \xi = 1 \end{cases} \tag{7}$$

when $\xi \geq 0$, $\sigma > 0$, $z \geq 0$; $\xi < 0$, $0 \leq z \leq -\frac{\sigma}{\xi}$, ξ refers to shape parameter, and $\sigma > 0$ refers to scale parameter which determines the disappear speed in distribution tail, $z = y\text{-}u$.

Step three, calculating the value at risk (VaR) of egg-feed ratio loss rate (LR); VaR is the maximum expected loss at a given confidence level and within a certain holding period. Combined with the GPD function of LR, VaR can be expressed as the maximum possible deviation between the actual egg-feed ratio and expected egg-feed ratio (The biggest extreme risk of price falls). Given a confidence level p, take the inverse function of F(y) to get the VaR of LR's extreme risk, namely VaR_P^{POT}:

$$VaR_P^{POT} = \begin{cases} u + \frac{\sigma}{\xi}\left\{\left[\frac{n}{N_u}(1-p)\right]^{-\xi} - 1\right\}, \xi \neq 0 \\ u - \sigma In\left[\frac{n}{N_u}(1-p)\right], \ \xi = 0 \end{cases} \tag{8}$$

Step four, calculating the maximum loss ratio that insurance companies may face: when the market is in the extreme condition, the egg-feed ratio will reach its lowest point. Substituting the value of VaR_P^{POT} into Eq (5), we can obtain the maximum loss ratio LCR_{max} as follows:

$$LCR_{max} = \frac{max(LR, 0)}{\alpha PR} = \frac{max(VaR_P^{POT}, 0)}{\alpha PR} \tag{9}$$

2.3 Constructing the optimal boundary model of government financial commitment

2.3.1 The starting point of government financial commitment
This paper regards the government as risk sharer (similar to reinsurance), regards the insurance company as risk distributer, refers to the optimal reinsurance pricing model, focuses on the target of mean-variance. On the one hand, the averageloss ratio of the insurance company's egg price insurance can be close to an ideal value (80%) to ensure insured farmers can receive appropriate compensation, and the insurance company can also cover its operating costs i.e. the insurance company's mean gross profit margin is 20%. On the other hand, we should ensure that the fluctuation of egg price insurance claims (the variance of claim rate) of the insurance company is minimum, so as to realize a stable and healthy development of the egg price insurance business.

Using a linear programming technique, minimizing the product of the difference value between average and expected loss ratio of the insurance company's egg price insurance and the variance of insurance company's loss ratio to determine the starting point of the government financial commitment to loss caused by extreme risk, namely LCR_{opt} (the optimal shared proportion between government and insurance company), the formula is as follows:

$$LCR_{opt} = min\{|mean(LCR) - ECR| \times sd(LCR)\} \tag{10}$$

Here, mean (LCR) refers to the average loss ratio of the insurance company, ECR refers to the expected loss ratio of insurance company's insurance business (set as 80%), and $sd(LCR)$ refers to the variance of the insurance company's loss ratio.

2.3.2 The upper limit of government financial commitment
Assuming that the government's financial resources is limited and cannot fully undertake the compensation beyond the bearing capacity of insurance companies, this paper tries to propose an "integration insurance" model (government and insurance company as a joint insurance company or comboto pay compensation to farmers) to ensure the insured farmers can obtain sufficient compensation (the loss ratio is 100%). At this time, the mean value of the insurance company's gross margin is zero, the operating cost which is the 20% loss ratio generated by the insurance company carrying out the egg price insurance will be compensated by a financial fund, then the average loss ratio of the "integration insurance" is 120%, and the mean-variance method is used again to determine the LCR of the upper limit of government financial commitment, namely LCR_{opt}^{govmax}, the formula is as follows:

$$LCR_{opt}^{govmax} = min\{|mean(LCR) - ECR| \times sd(LCR)\} \tag{11}$$

Here, mean(LCR) refers to the average loss ratio of the insurance company; ECR refers to the expected loss ratio of the insurance company's insurance business (set to 120%); $sd(LCR)$ refers to the variance of the insurance company's loss ratio.

We can find that the boundary of government financial commitment to compensation is $[LCR_{opt}^{govmin} - LCR_{opt}^{govmax}]$. It should be noted that for the extreme case in the market, if the maximum loss ratio LCR_{max} faced by the insurance company's egg price insurance is higher than LCR_{opt} but lower than LCR_{opt}^{govmax}, the amount of government financial commitment can be counted according to $(LCR_{max} - LCR_{opt})$.

3 EMPIRICAL ANALYSIS

This paper takes Beijing's egg price insurance as an empirical analysis case, and uses the above model to calculate the optimal boundary of government financial commitment to the "catastrophe" risk compensation of Beijing's egg price insurance.

3.1 The premium rate determination of Beijing's egg price insurance

In this paper, the egg-feed ratio (egg price/price of compound feed for laying hens) is introduced into the price insurance as an important indicator to calculate the premium rate and determine the claim basis. To ensure the consistency of data statistics criteria, the egg price in Beijing and the price of compound feed for laying hens used in this paper are both selected from China's animal husbandry information network (www.caaa.cn), the time span is from January 2000 to July 2018 (see Figure 1). Looking at the egg-feed ratio, the maximum value is 4.03, the minimum value is 1.82, and the mean value is 3.11.

The determination of bandwidth is very critical when fitting on the Gaussian kernel density probability distribution function of Beijing's egg-feed ratio. A smaller h will make for a better result in the KDE fitting on sample, but the kernel density curve is not smooth. A bigger h will result in a smoother kernel density curve, but it will increase the deviation of KDE (Ye, 2003; Tan, 2003). This paper uses the modified thumb method to determine the optimal bandwidth h. The formula is: hopt = 0.9 * σ n - 1/5, in which σ is the smaller one between standard deviation and interquartile range divided by 1.34. According to this formula, the h is 0.1171.

Then, according to Formula (2) and Formula (3), MatlabR2017a is used to calculate the pure premium rate of Beijing's egg price insurance. The result is 3.74%, the actual premium rate is pure premium rate plus additional rate. The actual rate is set as 1.2 times of the net rate, namely the actual premium rate of Beijing's egg price insurance is 4.49%.

To ensure the premium calculation is more objective, this paper uses the egg futures price of the Dalian Commodity Exchange as a reference. Considering the huge trading volume of 1905 futures contract in December 2018, the average closing price of each trading day of the contract in that month (7.00 yuan/kg) is selected as the egg prices for calculating the premium.

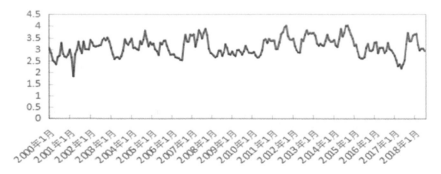

Figure 1.   The time series graph of egg-feed ratio in Beijing.

In the insurance contract, the total quantity of the insured is generally increased by integral multiples, such as 100 ton or 50 ton. For the convenience of calculation, the amount of insured is assumed to be only 1 ton, that is the insurance coverage of Beijing's egg price insurance is 7,000 yuan. The premium for a farmer to insure one ton of eggs is 314.30 yuan.

3.2 *Beijing egg price insurance loss ratio calculation*

According to the insurance contract, when actual egg-feed ratio is lower than the expected egg-feed ratio which is 3.0, the insurance company needs to pay to the insured farmers. In the calculation of the maximum loss ratio (catastrophe risk) of egg price insurance in Beijing, Formula (5) shows that in the case where the egg price insurance rate (4.49%) has been determined, the key point is how to effectively measure the maximum value of Beijing's LR (Figure 2).

This paper introduces the POT model from the modern extreme value statistical theory (EVT) to fit the GPD distribution of the *LR* in Beijing. The choice of threshold u in the GPD distribution is extremely important. An excessively low threshold results in the asymptotic behavior of the model being unsatisfied and produces a biased estimator. An excessively high threshold leads to fewer sample sizes being available and results in a higher variance of estimated parameters (Shi, 2006).

The common method to select the threshold is the mean residual life plot (Figure 3) It is not difficult to find that when the threshold is between 0 and 0.13, the graph is a curve and when the threshold is between 0.13 and 0.3, the graph is approximated as a straight line. Combined with parameter estimation changes of LR for different thresholds (Figure 4), the threshold is determined to be 0.13.

Then, using the maximum likelihood estimation method to estimate the parameter values of GPD distribution function of Beijing's *LR*, the GPD distribution function of Beijing's *LR* is as follows:

$$F(y) = 1 - \left(1 + \frac{0.05(y - 0.13)}{0.06}\right)^{-1/0.05} \tag{12}$$

In order to further verify the accuracy of the model, it is necessary to test the fitting effect. This paper uses the empirical probability plot (P-P plot), quantile plot (Q-Q plot), return level plot (return level plot), and the probability density plot (density plot) for analysis and

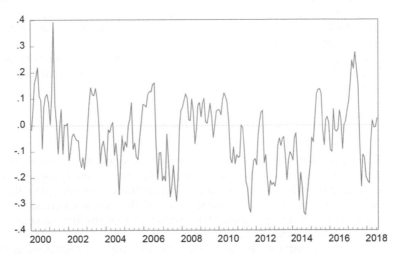

Figure 2. The time series graph of *LR* in Beijing.

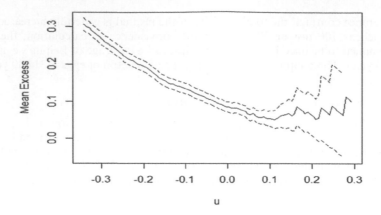

Figure 3. The mean residual life graph of *LR* in Beijing.

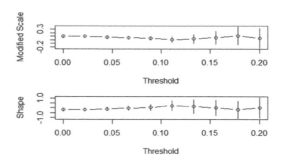

Figure 4. The parameter estimation graph of *LR* in Beijing under different threshold values.

comparison respectively. The results are shown in Figure 5. It is not difficult to find that all the points in the P-P plot and Q-Q plot are almost in a straight line, the return level plot is within the confidence level and the density curve in the density plot is consistent with the histogram. The four diagnostic diagrams indicate that the fitting effect of the GPD model of Beijing's LR is good, so the fitting effect of the sample is acceptable.

Based on the GPD distribution function of Beijing's LR, assuming that the market encounters a once-in-a-century extreme risk, because it is monthly data, the p should be taken $1-1/1200 = 99.92\%$. Putting it into Formula (8), the extremum value of Beijing's LR $\text{VaR}_p^{\text{POT}} = 0.43$ is obtained. This means that when encountering a once-in-a-century extreme risk, Beijing's egg-feed ratio will drop by 43% from the break-even point of 3.0 (the egg-feed ratio falls to 1.71). According to Formula (9), the maximum loss ratio LCR_{max} faced by insurance companies is as high as 958%.

3.3 *Determining the optimal boundary of Beijing municipal government's financial commitment to insurance compensation*

In terms of the starting point of the Beijing municipal government's financial commitment, focusing on the target of mean-variance, according to the programming model Formula (10) of the optimal sharing ratio between government and insurance company, the absolute value of the difference between the cumulative mean value of the insurance company's loss ratio and the target loss ratio (80%) is calculated. The product of this absolute value and the cumulative variance (Figure 6) is then obtained.

In Figure 6, taking loss ratio as the horizontal axis and objective function as the vertical axis, we can see that the target function reaches its minimum at 172%. This means that when

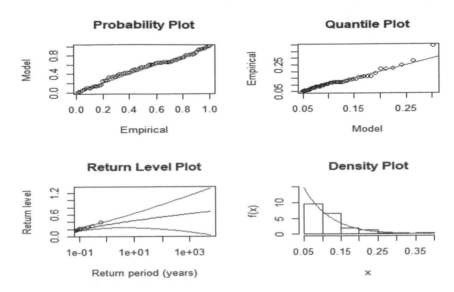

Figure 5. The GPD fitting diagnosis graph of Beijing's LR.

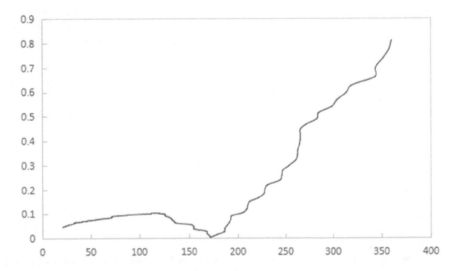

Figure 6. The optimal shared ratio of egg price insurance's compensation.
liability between government and insurance company

the loss ratio LCR$_{opt}$ is 172%, the insurance company's target loss ratio is close to 80%, and the fluctuation in loss ratio is at minimum. When the loss ratio of egg price insurance in Beijing reaches 172%, which is regarded as the starting point, the Beijing municipal government needs to intervene.

In terms of the upper limit of government financial commitment, assuming that the government's financial resources is limited, under the "Integration insurance" mode, to ensure farmers can get adequate compensation, the insurance company's compensation expense should equal its premium income (insurance company's loss ratio reach 100%) and the government compensates the insurance company for their operating costs (20% loss ratio). Therefore, the average loss ratio of "Integration insurance" is 120%. Similarly, according to Formula (11), the absolute value of difference between the cumulative mean value of the insurance company's loss ratio and the target loss ratio (120%) is calculated, then the product of this absolute value and the cumulative variance (Figure 7) is obtained.

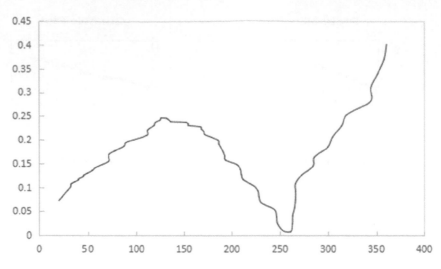

Figure 7. The upper limit of government financial commitment to egg price insurance.

In Figure 7, taking loss ratio as the horizontal axis and objective function as the vertical axis, we can see that the target function reaches its minimum at 261%, namely, when the loss ratio LCR_{opt}^{govmax} is 261%, the target loss ratio of "Integration insurance" is close to 120%, and the fluctuation of loss ratio is at minimum. The government begins to step in from a loss ratio 172% and ends its liability when the loss ratio reaches 261%, which is regarded as the upper limit of government financial commitment.

In this condition, the compensation for egg price insurance is 3010.99 yuan per ton, of which the insurance company needs to pay 540.60 yuan and the government needs to pay 279.73 yuan. The excess needs to be addressed by the capital markets, for example by issuing catastrophe bonds.

4   RESEARCH CONCLUSIONS AND POLICY RECOMMENDATIONS

Aiming at the "catastrophe" risk of egg price insurance, this paper tries to construct the optimal boundary model of government financial commitment. At the same time, by taking the Beijing egg price insurance as an example, it carries out an empirical analysis and draws the following basic conclusions:

Firstly, referring to the optimal reinsurance pricing model and the co-insurance model and focusing on the target of mean-variance. On the one hand, make sure the average loss ratio of the insurance company's egg price insurance is close to an ideal value (ensuring insured farmers can receive appropriate compensation and the insurance company can also cover its operating costs, keeping it from bankruptcy). On the other hand, ensuring the insurance company's egg price insurance has the smallest fluctuations of compensation (the variance of loss ratio) and using linear programming techniques to determine the starting point and upper limit of government financial commitment.

Secondly, the premium rate of Beijing's egg price insurance is 4.49% and the premium for farmers to insure a ton of eggs is 314.30 yuan. When encountering a once-in-a-century extreme risk, Beijing's egg-feed ratio will drop sharply to 1.71, down 43% from the break-even point of 3.0 At this moment, the loss ratio faced by the insurance company is as high as 958% and the payouts of egg price insurance is 3010.99 yuan per ton.

Thirdly, assuming that the government's financial resources are limited, the boundary of Beijing municipal government's financial commitment to the "catastrophe" risk compensation of agricultural price insurance is [172% ~ 261%]. When dealing with the once-in-a-century extreme risk in the egg market, the insurance company needs to pay 540.60 yuan and the government needs to pay 279.73 yuan. The rest of the compensation (2190.66 yuan) needs to be addressed by the capital market.

Combining the above conclusions, this paper puts forward the following policy recommendations: taking into consideration the current economic and social development in China, according to the principle of "financial support, risk sharing, limited compensation, continuous operation", we should pay attention to risk diversification and benefit balance on the risk-diversification chain, aim at the extreme risk of price fluctuations in agricultural markets, focus on the integrated risk management model of "insurance company + government + capital market", establish a stepwise decentralized system of "agricultural price insurance + government reserves for extreme price risk + price risk catastrophe bonds", and solve the difficulty of constructing an extreme risk dispersion system for farm products market price.

ACKNOWLEDGMENTS

This work was funded by Beijing Municipal Natural Science Foundation (Grant No. 9172022)

REFERENCES

Braga F. 1997. Over-the-Counter Derivatives and Price Management in a Post-Tri Partite Environment: The Case of Cattle and Hogs. *Canadian Journal of Agricultural Economics* 44(4): 207–223.

Dismukes, Rt. 2006. Managing Risk With Revenue Insurance, *FEATURE*, Originally published. Vol. 4, Issue 5.

Feng, G.S. 2004. *Research on Government Intervention in Agricultural Risk Management.* Zhejiang: Zhejiang University.

Ge, Y.B. and Cao, T.T. 2017. A New Model of Agricultural Product Price Risk Management: A Case Study Based on "Insurance + Futures" of Cotton. *Price Theory and Practice* (10): 119–121.

Hart, C.E., Hayes, D.J. and. Babcock, A. 2003. Insuring Eggs in Baskets.Publication is available online on the CARD. Website: www.card.iastate.edu.

Harwood, J., Somwaru, A. and Perry, J. 1998. The Potential for Revenue Insurance Policies in the South. *Journal of Agricultural and Applied Economics* (7): 47–61.

Huang, J.k., Wang, D. and Hu, J.L. 2015. Thoughts on Implementing the Target Price Policy of Agricultural Products–Based on the Analysis of Xinjiang Cotton Target Price Reform Pilot. *Rural Economy in China* (2): 10–18.

Huo, R. 2014. Crop Catastrophic Risk Sharing Ratio Analysis From Government Perspective: Taking Wheat Insurance in Henan Province For Example. *Chinese Academy of Agricultural Sciences.*

MAFF. 2001. Risk Management in Agriculture. A Discussion Document Prepared by the Economics and Statistics Group of the Ministry of Agriculture. *Fisheries and Food*, UK.

Mahul O. 2002. Hedging Price Risk in the Presence of Crop Yield and Revenue Insurance. *Zaragoza (Spain)* 3. Bennett. Evaluation of Cotton Put Options as a Price Risk Management Tool. *The Journal of Cotton Science* (7): 39–44.

Shi, D.J. 2006. *Practical Extremum Statistical Method.* Tianjin: Tianjin Science and Technology Press.

Tan, Y.P. 2003. Application of Nonparametric Density Estimation in Individual Loss Distribution. *Statistical Research* 8: 40–44.

Turvey, C.G and Romain, R. 2000. Using U.S.BFP/class III futures contracts in Risk Reduction Strategies for Subclasses 5a and 5b Milk for Further Processors. *Canadian Journal of Agricultural Economics* (4): 475–498. Wang, K. 2008. A Study on the Influence of Flexible Crop Yield Distribution on Crop Insurance Premium Rate. *Chinese Academy of Agricultural Sciences.*

Wang, K., Zhang, Q., Xiao, Y.G., Wang B.W., Zhao, S.J. and Zhao, J.Y. 2014. Feasibility of Price Index Insurance for Agricultural Products. *Insurance Research* (1): 40–45.

Xu, L. and Wang, B. 2018. Study on Extreme Risk Evaluation of Pork Price Fluctuation in Beijing City. *8th Annual Meeting of Risk Analysis Council of China Association for Disaster Prevention (RAC 2018).* Atlantis Press.

Ye, A.Z. 2003. *Nonparametric Econometrics.* Tianjin: Nankai University Press.

Zhang, Q., Wang B.W. and Wang, K. 2015. Feasibility Analysis and Key Points of Scheme Design of Pig Price Insurance in China. *Insurance Research* (1): 54–61.

Zheng, C.L., Zhou, X.Y., Zhang, Y.Y. and Chen, K. 2018. Design of Egg Futures Price Insurance and Farmers' Optimal Insurance Choice: A Case Study of Hubei Province. *Insurance Research* (10): 51–64.

Risk Analysis Based on Data and Crisis Response Beyond Knowledge – Huang & Nivolianitou (eds)
© 2020 Taylor & Francis Group, London, ISBN 978-0-367-25146-8

Study on integrated loss assessment of earthquake-flood disaster in Yunnan Province based on dynamic spatial panel

Shanshan Du, Ye Xue*, Xiaodong Ji, Xueyu Xie, Chongyi Xue & Huiwen Wang
School of Economics and Management, Taiyuan University of Technology, Taiyuan, China

ABSTRACT: Based on the theory of a regional disaster system and data from the Yunnan region from 1991 to 2015, this article constructs an index system for assessing the integrated loss of earthquake-flood disaster, which is the hazard and the vulnerability of hazard-bearing bodies. A dynamic spatial panel model for integrated loss assessment of earthquake-flood disaster is established. The results show that there are significant positive spatial spillover effects and time lag effects in the integrated loss of earthquake-flood disaster, and the coefficient of integrated hazard is estimated to be 0.7183, which is 3.7 times the integrated vulnerability index. It has a greater effect on the integrated loss. From the determination coefficient R2, logarithmic likelihood value log$L$ and the t value of coefficient significance test, the model is used to explain the loss change, and the integrated loss level of the whole earthquake-flood disaster in Yunnan is higher. The assessment results have important theoretical and practical significance for integrated disaster prevention and mitigation.

Keywords: earthquake, flood, integrated loss assessment, dynamic spatial panel model

1 INTRODUCTION

Earthquakes, floods and other natural disasters occur frequently, which not only causes serious economic losses, but also heavy casualties (Wu, 2018), which is a major obstacle to the realization of social and economic sustainable development in China. Therefore, it is necessary to assess effectively the losses caused by earthquakes, floods and so on, in order to take timely and targeted measures to reduce their impact.

At present, scholars mostly assess the losses caused by natural disasters from two angles before and after disasters: pre-disaster loss assessment and post-disaster loss assessment. In studies on pre-assessment of pre-disaster losses, foreign scholars (Tao et al., 2009; Xu et al., 2015) have used the methods of loss curve, cost-benefit analysis and fuzzy integrated assessment to model a variety of factors and disaster losses, and pre-evaluate the degree of possible losses caused by various disasters. Similar to foreign research, domestic scholars (Wang et al., 2018; Wen et al., 2018) have also used loss curve, cost-benefit analysis and fuzzy integrated assessment methods to pre-evaluate the pre-disaster loss through a variety of factors. In studies on the post-disaster loss assessment research, foreign scholars (Nuku et al.,1995; Profeti et al., 2015) have selected different indicators for disaster loss assessment according to the characteristics of flood, earthquake or meteorology and other disasters. Domestic scholars (Wang, 2012) have also used a variety of indicators, such as economic losses, social impact and affected population to assess the losses caused by different natural disasters.

From the point of view of index selection and assessment methods, the above studies have not been able to analyze systematically the causes of disaster losses, and the loss assessment results obtained by many previous studies may have a large deviation from the actual results.

---

*Corresponding author: xueye0412@126.com

At present, the methods used do not take into account the impact of disasters in the surrounding areas on the region and the impact of previous disasters on the current period. The effectiveness of the evaluation model may be low, and the research object is focused on a single type of natural disaster. Therefore, based on the theory of regional disaster system, this study designed an assessment index system from the two angles of the hazard and the vulnerability of hazard-bearing bodies, and constructed a dynamic spatial panel assessment model for integrated loss of earthquake-flood disaster.

2 MATERIALS AND METHODS

2.1 Selected study area

Yunnan is located in the southwest border of China, with a total area of about 390,000 square kilometers, accounting for 4.11% of the country's area. Due to geographical and natural factors, frequent earthquakes and floods occur in Yunnan. Through the statistics of the occurrence of earthquakes and floods in the province, it can be seen that the spatial distribution characteristics of earthquakes and floods are remarkable. Lijiang City, Chuxiong Prefecture and Pu'er City suffer more serious earthquake and flood disasters, especially frequent flood disasters. Therefore, Lijiang City, Chuxiong Prefecture and Pu'er City, which are rich in earthquake and flood disaster data and have a wide coverage of occurrence time (1991–2015), are used for the study area.

2.2 Assessment index system and data sources

Earthquakes and flood are natural disasters that cause loss to humans and property, and are the result of the interaction between humans and nature. Therefore, there are two aspects of social and economic attributes and natural attributes. There are two necessary conditions for the formation of earthquake and flood disasters: factors causing earthquake and flood disasters (the hazard), and the impact of disasters on humans or social property (hazard-bearing bodies). That is to say, the integrated loss of earthquake-flood disaster is caused by the action of the hazard of earthquake and flood disaster on the hazard-bearing bodies.

Therefore, based on the theory of regional disaster system, this study constructed the integrated loss assessment index system of earthquake-flood disaster from the point of view of the hazard and the vulnerability of hazard-bearing bodies. The relationship among the three can be reflected by functional expressions in the following form:

$$D = f(H, V) \tag{1}$$

D represents the integrated loss of earthquake-flood disaster in Yunnan area, H and V indicate the hazard and the vulnerability of hazard-bearing bodies respectively.

The establishment of the assessment index system mainly uses the frequency statistics method to consult the research results of the national natural disaster loss assessment, selects the indicators with high frequency of use and according to the principles of the construction of the index system. The index system of integrated loss assessment of earthquake-flood disaster is then constructed. The first index is the integrated loss of earthquake-flood disaster in Yunnan area, and the second includes the hazard and the vulnerability of hazard-bearing bodies. The third is a specific index used to characterize the characteristics of the second-level index, and indicates the standard unit of each index. The indicator system constructed is shown in Table 1.

2.3 Dynamic spatial panel model

The occurrence of earthquake and flood disaster is not only related to a change of time series, but also a spatial process. The sample data of earthquake and flood disaster loss must contain time series and cross section data. The spatial effect of the integrated loss of earthquake-flood

Table 1. Assessment index system of integrated loss of earthquake-flood disaster.

The first index	The second index	The third index	Unit	Raw data source
Integrated loss of earthquake-flood disaster in Yunnan Province (D)	The hazard (H)	Magnitude (N_1)	Richter magnitude	Yunnan disaster reduction yearbook
		Rainfall (N_2)	mm	Yunnan disaster reduction yearbook
	The vulnerability of hazard-bearing bodies (V)	Grain planting density (S_1)	Ha/Km^2	Yunnan statistical yearbook
		Population density (S_2)	People/K^2	Yunnan statistical yearbook
		Ground average GDP (S_3)	10^4 Yuan/K^2	Yunnan statistical yearbook
		Per capita income level of farmers (S_4)	Yuan	Yunnan statistical yearbook

disaster in Yunnan is not only reflected in the spatial correlation between the integrated loss of earthquake-flood disaster in this area and the integrated loss of earthquake-flood disaster in adjacent areas: it is also reflected in the influence of the hazard and the vulnerability of hazard-bearing bodies on the integrated loss of earthquake-flood disaster in adjacent areas. From the time point of view, it is manifested as the dynamic continuity of the impact of the loss, especially the impact of the previous period on the current period. However, most of the existing studies related to regional earthquake and flood disaster loss assessment ignore the impact of spatial correlation and time lag, and further studies should explore the accuracy of the research results.

Therefore, according to the characteristics that the integrated loss of regional earthquake-flood disaster has spatial effect and time lag, this study explores the following aspects. Suppose i represents the i region and t represents the t time.

(1) The integrated loss of earthquake-flood disaster in this area affects the integrated loss of earthquake-flood disaster in adjacent areas through the spatial correlation between the hazard and the vulnerability of hazard-bearing bodies, which is expressed by W_iD_t.

(2) The influence of the integrated loss of the previous earthquake-flood disaster in this area on the integrated loss of the current earthquake-flood disaster in this area, as a result of the spatial correlation acting on the integrated loss of the earthquake-flood disaster in the adjacent area, which is expressed by WD_{t-1}.

(3) The influence of the previous period in this area on the integrated loss of earthquake-flood disaster in the current period, which is expressed by D_{it-1}.

(4) The hazard and the vulnerability of hazard-bearing bodies in this area, expressed by W_iH_t and W_iV_{t-1} respectively, because of the spatial correlation affecting the integrated loss of earthquake-flood disaster in adjacent areas.

(5) The influence of the previous hazard and the vulnerability of the hazard-bearing bodies on the integrated loss of earthquake-flood disaster in the current period, expressed by H_{t-1} and V_{t-1} respectively.

(6) The previous hazard and the vulnerability of the hazard-bearing bodies in this area play an important role in the integrated loss of earthquake-flood disaster in this area, and the integrated loss of earthquake-flood disaster in adjacent areas is affected by the spatial correlation, which are expressed by W_iH_{t-1} and W_iV_{t-1} respectively.

$$W_i = (W_{i1}, W_{i2}, ..., W_{iN}), W_iD_t = W_{i1}D_{1t} + W_{i2}D_{2t} + ... + W_{iN}D_{Nt} \qquad (2)$$

Equation (2) represents the spatial interaction effect of disaster loss, that is, the impact of disaster loss in the surrounding area on the disaster loss in the area. W_{i1} is generally 0, indicating that there is no spatial interaction between the regional disaster loss and itself.

Therefore, the selected empirical model should be able to estimate and test the time series and cross-sectional data, so the dynamic spatial panel model is the most suitable.

Considering the study of the integrated loss of regional earthquake-flood disaster in the above aspects, Equation (1) is concretized into Equation (3) in combination Equation (2), and the random error term is expressed in detail. The general dynamic spatial panel data measurement model of integrated loss of earthquake-flood disaster in Yunnan area is constructed in this article, as shown in Equation (3).

$$D_{it} = \tau D_{it-1} + \delta W_i D_t + \eta W_i D_{t-1} + \beta_{11} H_{it} + \beta_{21} W_i H_t + \beta_{31} H_{it-1} + \beta_{41} W_i H_{t-1} + \beta_{12} V_{it} + \beta_{22} W_i V_t + \beta_{32} V_{it-1} + \beta_{42} W_i V_{t-1} + vit, \ \nu_{it} = \rho \nu_{it-1} + \lambda W_i \nu_t + \mu_i + \delta_t + \varepsilon_{it} \tag{3}$$

In the equation, β_{11} is the coefficient of the hazard. β_{21}, β_{31} and β_{41} are the spatial lag term, time lag term and spatio-temporal lag term coefficient of the hazard, respectively. β_{12} is the coefficient of the vulnerability of the hazard-bearing bodies. β_{22}, β_{32} and β_{42} are the spatial lag term, time lag term and spatio-temporal lag term coefficient of vulnerability of hazard-bearing bodies, respectively. μ and ν_t are spatial fixation effect and time fixation effect respectively. τ, δ and η represent the time lag term, spatial lag term and spatio-temporal lag term coefficient of the integrated loss of regional earthquake-flood disaster, respectively.

In this article, we mainly consider the geographical weight matrix, which is also spatially related at a longer distance, to investigate the spatial effect of the integrated loss of earthquake-flood disaster, that is, the rate of attenuation with geographical distance. Equation (4) shows the specific form of the distance space weight matrix (W):

$$W_{ii'} = \begin{cases} 1/d_{ii'}^2, & i \neq i' \\ 0 & i = i' \end{cases} \tag{4}$$

Where $d_{ii'}$ is the distance between the geographical center of area i and area i'.

3 INTEGRATED LOSS ASSESSMENT OF EARTHQUAKE-FLOOD DISASTER IN YUNNAN PROVINCE

3.1 *Dynamic spatial panel model determination*

In order to select the specific dynamic spatial panel model form to carry on the empirical research to the earthquake-flood disaster integrated loss assessment in Yunnan area, it is necessary to carry on the correlation test to Equation (3).

First, we tested test whether there is a spatial effect in the model. According to the test results (Table 4), the *lnD* global spatial autocorrelation index *Moran'I* = 0.090, the original assumption of "no spatial autocorrelation" is rejected at a significant level of 1%. In addition, the spatial autocorrelation test was carried out on the influencing factors. It was known from Table 2 that at the significant level of 1% and 5%, the *Moran'I* values of *lnH* and *lnV* passed the significance test and were positive, indicating that these two indexes were also spatially dependent. Therefore, spatial factors should be included in the integrated loss assessment model of earthquake-flood disaster.

Table 2. Spatial correlation test.

Indicators	*Moran'I*	Probability value
lnD	0.090	0.000
lnH	0.058	0.000
lnV	0.123	0.008

Note: the spatial weight matrix W in the test is the geographical distance weight matrix.

177

Second, the mixed least square regression was used to determine the specific form of spatial effect in the model. As can be seen from Table 3, both LMlag and LMerr passed the 1% significance test, that is, rejecting the original assumptions of "no spatial lag effect" and "no spatial error effect", indicating that both SAR and SEM models were applicable. However, in the robust LM test, LMlag is significant at 1%, while the robust LMerr is not significant, indicating that the spatial lag model was superior to the spatial error model.

Third, the model tested whether there is heterogeneity of time or cross section, as well as fixed effect and random effect. Table 4 shows that the Wald test of time fixed effect and cross section fixed effect has not passed the significance test, indicating that the original hypothesis is accepted. The F and χ^2 statistics, which retest the time effect, accept the original hypothesis at the significant level of 1%, which indicates that the time effect exists but the cross section effect does not exist, and the time effect model is more suitable. At the same time, this study used the Hausman method to test the fixed effect and random effect of panel data, which did not pass the significance test, indicating that the original hypothesis was true, so the random effect model was more consistent.

Finally, after the preliminary estimation of the model, the spatial lag terms of the hazard and the vulnerability of hazard-bearing bodies were not significant, so were eliminated in the final evaluation model.

In summary, the dynamic spatial lag stochastic model was finally selected to study the integrated loss of earthquake-flood disaster in Yunnan area, as in Equation (5):

$$lnD_{it} = a + \tau lnD_{it-1} + \delta W_i lnD_t + \beta_1 lnH_{it} + \beta_2 lnV_{it} + \mu_i + \nu_t + \varepsilon \quad (5)$$

D is the integrated loss variable of earthquake-flood disaster in Yunnan area. a is a constant term. H and V are the variables of hazard and the vulnerability variables of hazard-bearing bodies, respectively. ε is a random error term. μ_i and ν_t are spatial fixation effect and time fixation effect, respectively. β_1 and β_2 are the coefficients of the hazard and the vulnerability of hazard-bearing bodies, respectively. τ represents the coefficient of the time lag term for the integrated loss of regional earthquake-flood disaster. δ is the coefficient of spatial lag term for the integrated loss of regional earthquake-flood disaster. W is the set spatial weight matrix.

Table 3. Spatial effect test of integrated loss of earthquake-flood disaster in Yunnan area.

Test method	Hypothetical	Test value	Probability value
LMlag	No spatial lag	58.3925	0.0000
Robust LMlag		60.9678	0.0015
LMerr	No spatial error	18.3193	0.0000
Robust LMerr		21.2635	0.0700

Table 4. Model setting test results.

Test object	Test method	Statistical value	P value
Section fixed effect model	Wald	$F = 76.25$	0.257
Time fixed effect model	Wald	$F = 5.48$	0.135
Time effect	Wald	$F = 3.57$	0.0001
	LR	$\chi^2 = 48.32$	0.0000
Fixed effect	Hausman	$\chi^2 = 12.00$	0.1294

3.2 *Estimation result analysis*

Using Stata software, the variable parameters of Equation (5) were estimated, and the estimation results of dynamic spatial panel model (Table 5) and the assessment model (Equation (6)) were obtained.

The dynamic spatial panel model of integrated loss of earthquake-flood disaster in Yunnan is as follows:

$$lnD_{it} = 0.4502lnD_{it-1} + 0.0625W_i lnD_t + 0.7183lnH_{it} + 0.1916lnV_{it} + \mu_i + \nu_t + \varepsilon \quad (6)$$

Table 5 shows the following:

(1) The interpretation ability of the model is strong. From the point of view of goodness of fit, R^2 is 0.8994 – more than 0.8 – indicating that the dynamic spatial panel model has a better degree of fitting. From the logarithmic likelihood function value, the *LogL* value is 268.7936, which further indicates that the fitting data of the model is better, and the explanatory variables and the explained variables have a strong linear relationship. From the estimated results of the model, the coefficients of all explanatory variables are significant at 1% or 5%. Therefore, according to the sample data studied in this paper, the spatial panel model considering spatial spill over effect has a strong ability to explain the integrated loss of earthquake-flood disaster.

(2) There is significant spatial spillover in the integrated loss D of earthquake-flood disaster in Yunnan area.

The spatial autoregressive coefficient δ of the model is 0.0625, and it is significant at 1%, which indicates that there is a significant positive spatial dependence on the integrated loss of earthquake-flood disaster in Yunnan. When the integrated loss of earthquake-flood disaster in adjacent areas is increased by 1%, the integrated loss of earthquake-flood disaster in this area will be increased by 0.0625%, on average. This shows that the increase or decrease of the integrated loss of earthquake-flood disaster in the adjacent area will lead to the change of the integrated loss in the same direction because of the relative geographical position between the regions.

(3) There is a significant time lag in the integrated loss D of earthquake-flood disaster in Yunnan area.

The $\tau = 0.4502$ of the model is significant at 1%, indicating that there is a significant time-dependent integrated loss of earthquake-flood disaster. Once the integrated loss of earthquake-flood disaster is formed, it will not only have a great adverse impact on the current period, but also have a strong positive impact on the level of disaster loss in the future.

(4) Compared with the vulnerability of hazard-bearing bodies, the positive relationship between the hazard and the integrated loss of earthquake-flood disaster is more obvious.

In the model, β_{11} is 0.7183, which is significant at 5% and indicates that H has a significant positive promoting effect on D. At the same time, β_{12} is 0.1959, which is significant at the

Table 5. Estimation results of dynamic panel model based on geographical distance spatial weight.

Variable	Equation (5) (Dynamic spatial panel)	
	Coefficient	*t* value
WlnD	0.0625***	4.2940
lnD(t − 1)	0.4502***	4.3518
lnH	0.7183***	5.1792
lnV	0.1916*	1.4731
R^2	0.8994	
LogL	268.7936	

Note: *, * and * are significant at 1%, 5% and 10% respectively.

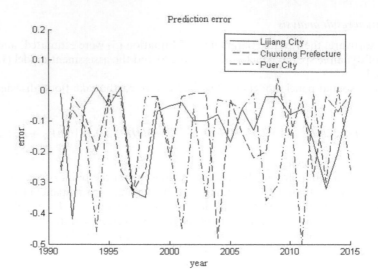

Figure 1.   Mean square error of loss assessment.

significant level of 5%, indicating that V also has a positive impact on D, that is, the vulnerability of the hazard-bearing bodies will promote the increase of the integrated loss of earthquake-flood disaster.

3.3   *Comparison of integrated loss assessment results of earthquake-flood disaster in Yunnan Province*

The integrated loss of earthquake-flood disaster was assessed to verify further the consistency between the assessment results of the model and the actual situation, using the earthquake and flood disaster data of Lijiang City, Chuxiong Prefecture and Pu'er City from 1991 to 2015, based on Equation (6). Considering that the model loss assessment is an estimate of the future uncertain loss, 0 is used to represent the value of the time lag term of the integrated loss of earthquake-flood disaster in 1991. If the actual value of the integrated loss of earthquake-flood disaster is 0, this indicates that there is no earthquake or flood disaster. Finally, the mean square error is used to measure the accuracy of the assessment results. The results are shown in Figure 1.

Figure 1 shows that the prediction error of the overall loss assessment results in Yunnan is small, and the prediction effect is good. Among them, the mean square errors of Lijiang City, Chuxiong Prefecture and Pu'er City are 0.03, 0.04 and 0.06, respectively, the root mean square errors are 0.16, 0.19 and 0.24, respectively, and the average absolute errors are 0.12, 0.13 and 0.17, respectively. The larger the mean square error, the lower the accuracy of the model assessment results, and the more inconsistent with the actual situation. In summary, the assessment results of dynamic spatial panel model are close to the objective reality.

4   CONCLUSIONS

In this study, based on the theory of regional disaster system, an integrated loss assessment index system for earthquake-flood disaster was designed using the hazard and the vulnerability of hazard-bearing bodies. Based on the spatial econometric theory, the assessment model of integrated loss of earthquake-flood disaster in Yunnan area was constructed. The results show that there is significant positive spatial spillover effect and significant time lag in the integrated loss of earthquake-flood disaster in Yunnan Province, and the integrated loss level of earthquake-flood disaster in Lijiang City, Chuxiong Prefecture and Pu'er City is relatively high. Through the mean square error analysis of the integrated loss assessment results of earthquake-flood disaster in Yunnan area, the loss value assessed by the dynamic spatial panel model is close to the objective reality.

ACKNOWLEDGMENTS

This project was supported by the Program for the Philosophy and Social Sciences Research of Higher Learning Institutions of Shanxi (PSSR) (No.2017314) and the Program for the Soft Science of Shanxi Province (No.2017041025-2).

REFERENCES

WU, J.D. 2018. Critical review on theory and method of natural disaster losses estimation: an economic perspective. *Journal of Natural Disasters* 27(3): 190–198.

Tao, W. and Kun, Q. 2009. *Cloud Model Method in Disaster Loss Assessment*. IEEE Computer Society. 2009 International Forum on Information Technology and Applications. Chengdu: IEEE, pp. 673–676.

Xu, J., An, J. and Nie, G. 2015. A dasymetric data supported earthquake disaster loss quick assessment method for emergency response in China. *Natural Hazards & Earth System Sciences Discussions* 3(2): 1473–1510.

Wen, Q.P., Yuehua Zhou, Y.H. and Huo, Z.G. 2018. Quantitative Assessment on Vulnerability of Storm Flood Disasters in Hubei Province. *Chinese Journal of Agrometeorology* 39(08): 547–557.

Tiantian Wang, T.T. and Liu, Q. 2018. The assessment of storm surge disaster loss based on BAS-BP model. *Marine Environmental Science* 37(03): 457–463.

Profeti, G. and Macintosh, H. 2015. Flood management through LANDSAT TM and ERS SAR data: a case study. *Hydrological Processes* 11(10): 1397–1408.

Nuku, K.K. and Rangi, R.T. 2015. Protecting vulnerable communities. *Nursing New Zealand* 21(3): 45.

Wang, H.Y., Liu, X.J. and Li, Z.W. 2012. Reviews of estimation on earthquake-disaster direct economic losses for industrial enterprises. *Ecological Economy* 2: 306–309.

Risk Analysis Based on Data and Crisis Response Beyond Knowledge – Huang & Nivolianitou (eds)
© 2020 Taylor & Francis Group, London, ISBN 978-0-367-25146-8

Assessment of financial agglomeration level of prefecture-level cities in Guizhou based on Internet Big Data

Xiaonan Huang

School of Big Data Application and Economics, Guizhou University of Finance and Economics, Guiyang, China

Guizhou Institution for Technology Innovation & Entrepreneurship Investment, Guizhou University of Finance and Economics, Guiyang, China

Mu Zhang

School of Big Data Application and Economics, Guizhou University of Finance and Economics, Guiyang, China

ABSTRACT: n order to objectively evaluate the level of financial agglomeration in Guizhou Province, this paper constructs an evaluation index system of financial agglomeration level from four aspects: financial industry, banking industry, securities industry and insurance industry. Baidu index derived from the keywords of "finance", "bank", "securities" and "insurance" is included in the evaluation index system. Based on the hesitant fuzzy linguistic set theory and PROMETHEE multi-attribute decision-making method, the evaluation results of financial agglomeration level of nine cities in Guizhou Province are obtained. The empirical results show that Baidu Index can improve the scientific evaluation of financial agglomeration to a certain extent. The introduction of Baidu Index evaluation index system is more in line with the development of the era of Internet big data. At the same time, hesitant fuzzy language set can effectively solve the decision-making problems in complex information environment. The application of hesitant fuzzy language PROMETHEE method in the evaluation of financial agglomeration level has a strong accuracy.

Keywords: financial agglomeration, fuzzy linguistic, Baidu Index, Guizhou Province

1 INTRODUCTION

With the rapid development of Internet technology, search engines are widely used among netizens. Traditional index statistics cannot fully reflect all aspects of a research problem. Search data is the effective subversion of traditional data statistics. It has gradually become one of the methods to study big data. Moreover, the hesitant fuzzy linguistic set can effectively express the uncertainty in decision-making. Compared with the single semantic term of traditional semantic set, hesitant fuzzy linguistic set allows the use of multiple semantic terms to express decision preferences and better meet the actual decision-making needs. At present, hesitant vague language has been widely used in brand evaluation, disaster risk assessment, service provider selection and so on.

At present, Foreign scholars mainly study financial agglomeration from the connotation of financial agglomeration, the formation mechanism of financial agglomeration, the evolution of financial agglomeration, and the effect of financial agglomeration. Domestic scholars also elaborate on the above aspects from different angles. Huang Jieyu (2011) thinks that the internal motivation of financial agglomeration includes five factors: industrial agglomeration, finance as the core and leading factor of economy, externality, flow of information and scale economy. Zhang Fan (2016) analyses the changes of financial industry agglomeration effect in China from two perspectives of financial resources and financial scale, and thinks that financial industry agglomeration effect has obvious regional characteristics. Ru Lefeng et al. (2014) constructed the index system of financial level measurement from four aspects of financial background, financial scale, financial density and financial density, and used factor analysis to evaluate the level of financial agglomeration in 286 central cities above prefecture level in 2010.

Therefore, this paper will take into account the fuzziness of the evaluation process and the development of big data information, follow the connotation of financial agglomeration and the principle of index selection, construct the index system of financial agglomeration level, and add Baidu index as Internet data into the evaluation index system of financial agglomeration. On this basis, the paper evaluates and analyses the level of financial agglomeration in nine cities and municipalities of Guizhou Province by using the hesitant fuzzy language PROMETHEE method.

2 PROMETHEE METHOD BASED ON HESITATIVE FUZZY LANGUAGE

As a common form of expression for decision makers, hesitant fuzzy language can accurately and clearly express the preference information of decision makers in uncertain environments. The calculation steps of PROMETHEE method based on hesitant fuzzy language are given below.

Step 1: Define a multi-attribute decision-making problem according to the decision-making needs: Determine the solution set A={$A_1, A_2, \ldots A_m$}, a set of attributes consisting of n attributes X={$X_1, X_2, \ldots X_n$}.

Step 2: Qualitative evaluation of the performance of each scheme A_i under each attribute X_j is given by using linguistic expressions for the above decision-making problems. At the same time, in order to correctly calculate the two hesitant fuzzy linguistic numbers, new linguistic terms are added to make each hesitant fuzzy linguistic number contain the same number of linguistic terms.

Step 3: Calculate the weight of each attribute, and compose the set of attribute weights W={$W_1, W_2, \ldots W_n$}, where $0 \leq W_j \leq 1$, and $\sum_{j=1}^{n} W_j = 1$. This paper introduces the method of entropy weight to calculate attribute weight. Entropy weight method can be roughly divided into the following two steps to determine attribute weight:

(1) For the multi-attribute decision-making matrix $(h_{ij})_n$ of n attributes of m schemes, the entropy value of the index J is calculated.

$$e_j = -\frac{1}{lnn} \sum_{i=1}^{n} (\varepsilon ln\varepsilon), \sum_{i=1}^{n} h_{ij} \neq 0, \ j = 1, 2, \ldots, m \tag{1}$$

Among them, $\varepsilon = h_{ij} \left(\sum_{i=1}^{n} h_{ij} \right)^{-1}$, when $\varepsilon = 0$, $ln \ \varepsilon = 0$.

(2) According to the result of calculating the entropy value, the weight of each attribute is calculated.

$$w_j = 1 - e_j / \sum_{j=1}^{n} (1 - e_j), \ j = 1, 2, \ldots, m \tag{2}$$

Step 4: Determine the priority function. Six classical priority functions are given in reference [13]. However, in the context of hesitant fuzzy language, the two schemes are described in the form of hesitant fuzzy language and cannot be directly calculated, thus the six priority functions mentioned above cannot be directly used. Therefore, this paper draws on the improved priority function proposed by Liao Huchang et al. The improved priority function is expressed as follows:

$$P_j(a_i, \ a_k) = \begin{cases} 0, & d_j(a_i, a_k) \leq 0 \\ \frac{d_j(a_i, a_k)}{\theta d_j\left(A_j^+, A_j^-\right)} & 0 < d_j(a_i, a_k) \leq \theta d_j\left(A_j^+, A_j^-\right) \\ 1, & d_j(a_i, a_k) > \theta d_j\left(A_j^+, A_j^-\right) \end{cases} \tag{3}$$

where d_j (a_i, a_k) $= \sigma_s^{ij}-\sigma_s^{kj}(i,k = 1,2,3,\ldots,n)$. σ_s^{ij} is the sum of all hesitant fuzzy linguistic numbers in a set of hesitant fuzzy languages. In addition, $0<\theta<1$, the decision maker can choose θ value independently.

Step 5: Calculate the priority index $\prod(A_i, A_j)$. Priority index indicates the degree to which scheme A_i is superior to other scheme A_j:

$$\prod(A_i, A_j) = \sum_{k=1}^{n} W_k P_k(A_i, A_j) \tag{4}$$

Step 6: Calculate the outflow $\Phi^+(A_i)$, inflow $\Phi^-(A_i)$ and net flow $\Phi(A_i)$ of each scheme A_i:

$$\Phi^+(A_i) = \sum_{A_j \varepsilon A} \prod(A_i, A_j) \tag{5}$$

$$\Phi^-(A_i) = \sum_{A_j \varepsilon A} \prod(A_j, A_i) \tag{6}$$

$$\Phi(A_i) = \Phi^+(A_i) - \Phi^-(A_i) \tag{7}$$

3 AN EMPIRICAL ANALYSIS OF FINANCIAL AGGLOMERATION LEVEL OF PREFECTURAL CITIES IN GUIZHOU

3.1 *Index system and data source*

This paper follows the principles of representativeness, objectivity, pertinence and comparability when constructing the index system of financial agglomeration level. It not only emphasizes the agglomeration characteristics of financial industry, but also considers various factors from a systematic point of view to truly reflect the level of regional financial agglomeration. On the basis of the previous studies, the index system of financial agglomeration level, which includes four first-level indicators and 17 second-level indicators, is constructed by adding Baidu Index from four aspects: financial industry, banking industry, securities industry and insurance industry, as shown in Table 1.

The Baidu Index used in this paper is the data of Internet users' attention to keyword search in each year. The four keywords of "finance", "bank", "securities" and "insurance" are input into the Baidu Index to obtain the average of keyword search frequency in nine cities of Guizhou Province each year, so as to form the data base of this study.

In addition to Baidu Index, the other indicators of financial agglomeration level in this paper are from the Statistical Bulletin of National Economic and Social Development of Nine Cities in Guizhou Province from 2013 to 2017, Statistical Yearbook of Nine Cities in Guizhou Province from 2014 to 2018, website of Guizhou Statistical Bureau and RESSET financial research database.

3.2 *Empirical results and analysis*

(1) Define multi-attribute decision making problems. From Table 1, we can see that the set of schemes determined is: A= {Guiyang City, Liupanshui City, Zunyi City, Anshun City, Bijie City, Tongren City, Southwest Guizhou, Southeast Guizhou, Southwest Guizhou Prefecture, Attribute Set X={$X_1, X_2, X_3, \ldots, X_{17}$}. Meanwhile, the language terminology set S of the 17 attributes mentioned above can be expressed as S={S_0 = extreme difference, S_1 = poor, S_2 = slightly poor, S_3 = medium, S_4 = slightly good, S_5 = good, S_6 = excellent}.

(2) According to the subjective evaluation of financial agglomeration of nine cities in Guizhou by decision-making experts, the qualitative evaluation data of nine cities in Guizhou under 17 attributes are obtained. And the evaluation matrix H_S of hesitant fuzzy languages is

Table 1. Index system for measuring financial agglomeration level.

Target layer	First level index	Second level index	Variable
Level of financial agglomeration	Finance	Gross financial industry output (RMB 100 million)	X_1
		Location Entropy of Financial Industry	X_2
		Financial Baidu Index	X_3
	Banking	Deposit balance of financial institutions at the end of the year (RMB 100 million)	X_4
		Loan balance of financial institutions at the end of the year (RMB 100 million)	X_5
		Residents' savings deposit balance (RMB 100 million)	X_6
		Location Entropy of Banking Industry	X_7
		Bank Baidu Index	X_8
	Securities business	Number of Listed Companies	X_9
		Total stock market value (RMB 100 million)	X_{10}
		Location Entropy of Securities Industry	X_{11}
		Baidu Securities Index	X_{12}
	Insurance industry	Annual premium income (RMB 100 million)	X_{13}
		Insurance density (%)	X_{14}
		Depth of insurance (%)	X_{15}
		Location Entropy of Insurance Industry	X_{16}
		Insurance Baidu Index	X_{17}

constructed (The following calculation process takes 2013 data as an example). In order to operate accurately, new linguistic terms need to be added so that each hesitant fuzzy language number contains the same number of linguistic terms.

(3) Calculate attribute weights. In this paper, the entropy weight method is used to calculate the attribute weight. First, the weight of each attribute is calculated according to formula (1) and formula (2). Considering the comparability of data, this paper adds the five-year attribute weights together to get an average attribute weights, which is used as the attribute weights to calculate the annual financial agglomeration level of prefecture-level cities in Guizhou. So the set of attribute weights is W={0.09171, 0.036433, 0.084498, 0.096437, 0.053392, 0.01091, 0.086375, 0.091876, 0.059646, 0.059674, 0.037594, 0.021871, 0.019661, 0.061762, 0.055871, 0.078321, 0.053967}.

(4) Calculate inflow, outflow and net flow. Since these 17 attributes are all profit-oriented attributes, the positive and negative ideal solutions of the hesitating fuzzy language are determined as A^+={S_6, S_6, S_6, S_6, S_6, S_6, S_6, S_6, S_6, S_6, S_6, S_6, S_6, S_6, S_6, S_6, S_6}and A^-={S_0, S_0, S_0, S_0, S_0, S_0, S_0, S_0, S_0, S_0, S_0, S_0, S_0, S_0, S_0, S_0, S_0}. Under the income attribute X_j, the priority index is calculated according to formula (4). Then the inflow, outflow and net flow of nine cities in Guizhou Province can be obtained by formula (5), formula (6) and formula (7). The results of the inflow, outflow and net flow of nine cities in Guizhou Province in 2013 are shown in Table 2.

Similarly, we can calculate the inflow, outflow and net flow of nine cities in Guizhou Province in 2014, 2015, 2016 and 2017. The specific results are shown in Table 3. Based on the data in Table 3, the trend maps of net flow in nine cities of Guizhou Province from 2013 to 2017 are drawn, as shown in Figure 1.

Because the net flow value of nine cities and counties in each year calculated above is not absolute value, we can not simply compare the annual trend changes through the above results. This paper uses the same calculation method to calculate the priority index between different years of a single city, and then calculates the inflow and outflow, so as to get the net flow of different years of a single city as the research object. The specific results are shown in Figure 2.

From Table 3 and Figure 1, we can see that the first two places in the ranking of annual net flow from 2013 to 2017 are Guiyang City and Zunyi City, which are far more than the other seven cities. This shows that Guiyang City and Zunyi City are the closest to the ideal

Table 2. Results of inflow, outflow and net flow in nine prefectures of Guizhou Province in 2013.

Prefectures	Inflow	Outflow	Net flow
Guiyang	6.676516	0.243435	6.433081
Liupanshui	1.052764	1.617489	-0.564725
Zunyi	4.478786	0.546521	3.932265
Anshun	0.881863	1.883039	-1.001175
Bijie	0.359823	2.065420	-1.705597
Tongren	0.165825	2.177848	-2.012023
Qianxinan	0.067525	2.261729	-2.194204
Qiandongnan	0.436156	1.928566	-1.492410
Qiannan	0.455793	1.851003	-1.395210

Table 3. Results of net flow in nine cities of Guizhou Province from 2013 to 2017.

Prefectures	2013	2014	2015	2016	2017
Guiyang	6.433081	6.394074	6.256840	6.242496	5.932867
Liupanshui	-0.564725	-0.643036	-1.436936	-1.494744	-1.643824
Zunyi	3.932265	4.039012	3.842962	3.544785	3.545347
Anshun	-1.001175	-0.984385	-0.709096	-0.984662	-1.492438
Bijie	-1.705597	-1.944864	-1.647272	-1.196513	-0.804450
Tongren	-2.012023	-1.800678	-1.736526	-1.797468	-1.629730
Qianxinan	-2.194204	-2.237299	-2.300127	-2.436923	-2.200559
Qiandongnan	-1.492410	-1.729410	-1.557074	-1.499323	-1.455578
Qiannan	-1.395210	-1.093415	-0.712771	-0.377647	-0.251634

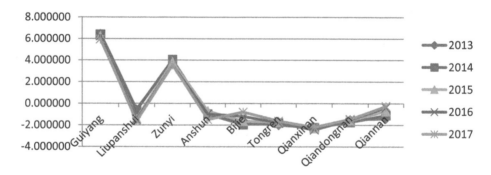

Figure 1. Trend map of net flow in nine cities of Guizhou Province from 2013 to 2017.

scheme, belonging to the high concentration area of financial industry. Tongren, southeastern Guizhou and southwestern Guizhou all have relatively low degree of financial agglomeration. The net flow value is between - 1 and - 3, which belongs to the low level of financial agglomeration. In addition, we can find from Table 1 that there are many differences between the highest and lowest net flows of financial agglomeration in Guizhou Province each year, which shows that there are serious imbalances in the development of regional financial industry.

From Figure 2, we can see that almost all prefectures and municipalities have increased year by year with the passage of time. From 2013 to 2017, the financial agglomeration degree of Guizhou Province shows an overall trend of growth. Among them, the net flow of Qiannan in 2013 was -1.35, and in 2017 was 1.26. The five-year variation difference of net flow was the highest.

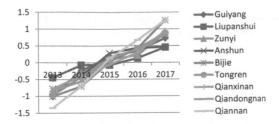

Figure 2. Trend map of net flow change in nine cities of Guizhou Province from 2013 to 2017.

Moreover, Tongren City, Bijie City and Qiandongnan Prefecture are also in the forefront of the province's five-year net flow change value.

4 CONCLUSION AND DISCUSSION

Based on the empirical results, we can draw the following conclusions: (1) Spatially speaking, Guiyang City and Zunyi City are the areas with high level of financial agglomeration, Tongren City and Southwest Guizhou are the areas with low level of financial agglomeration, and there are serious regional imbalances in financial agglomeration in Guizhou Province. (2) From the time point of view, the level of financial agglomeration of almost all prefectures and municipalities is increasing year by year. At the same time, the lower level of financial agglomeration of prefectures and municipalities has a strong momentum for development, there is a lot of room for development, the government should take measures to make full use of various advantages to establish financial agglomeration.

In view of the above conclusions, this paper puts forward the following policy recommendations: Firstly, the government should strive to achieve a balanced development in the financial industry region. Relatively backward financial sub-industries need to arouse the government's great attention, give appropriate tilt in policy and resources, promote the accelerated development of short-board industries. Secondly, the government needs to support the development of financial industry in backward areas in many ways. We should speed up the infrastructure construction of financial agglomeration areas, strengthen regional exchanges and cooperation, and enhance the level of financial agglomeration from various aspects and perspectives. Finally, in order to keep up with the development of the big data era, we should closely integrate the big data with the financial industry through the construction of the national big data (Guizhou) comprehensive pilot area, and promote the development of the Internet, so as to realize the rapid development of the big data financial industry.

ACKNOWLEDGMENTS

This research was financially supported by the Regional Project of National Natural Science Foundation of China (71861003) and the Second Batch Projects of Basic Research Program (Soft Science Category) in Guizhou Province in 2017 (Foundation of Guizhou-Science Cooperation [2017] 1516-1).

REFERENCES

Vernon R. 1966. International Investment and International Trade in the Product Cycle. *Quarterly Journal of Economics* 80(3):197-207.
Milway J., Martin R. 2007. Assessing Toronto's Financial Service Cluster. *Presentation to Toronto Board of Trade* 78-86.

Dow S.C. 1994. The Staged of Banking Development and Spatial Evolution of Financial Systems. *Money and the Space Economy*. John Wiley & Sons 56-64.

Marius B., Nicole A.M. 2008. Sectoral Agglomeration Economies in a Panel of European Regions. *Regional Science and Urban Economics* 38:360-376.

Huang Jieyu. 2011. Analyses on the Impelling Force of Financial Agglomeration. *Journal of Industrial Technological Economics* 03: 129-136.

Zhang Fan. 2016. The cluster effect of China's financial industry and its temporal-spatial evolution. *Science Research Management* 37: 417-425.

Ru Le-feng, Miao Chang-hong and Wang Hai-jiang. 2014. Research on the Level and Spatial Pattern of Financial Agglomeration on Central Cities in China. *Economic Geography* 34(02): 58–66.

Deng Wei. 2015. Analysis on the Spatial Distribution and Influencing Factors of China's Financial Industry. *Statistics & Decision* 21: 138-142.

Liao Huchang, Yang Zhu, Xu Zeshui and Gu Xin. 2018. A hesitant fuzzy linguistic PROMETHEE method and its application in Sichuan liquor brand evaluation. *Control and Decision*: 1–10.

Xu Xukan, and Wang Jing. 2019. Risk Assessment of Urban Flood Disaster Based on Hesitative Fuzzy Set. *Statistics & Decision* 35(05): 51–55.

Liu Jun. 2018. Selection of cloud computing service provider based on hesitant fuzzy group decision making model. *Computer Engineering and Applications* 54(23): 109-114.

Risk Analysis Based on Data and Crisis Response Beyond Knowledge – Huang & Nivolianitou (eds)
© 2020 Taylor & Francis Group, London, ISBN 978-0-367-25146-8

Analysis on temporal and spatial variation of drought in Northeast China based on SPEI and DSI

Weidan Wang, Li Sun* & Zhiyuan Pei
Key Laboratory of Cultivated Land Use, Ministry of Agriculture and Rural Affairs, P.R. China
Chinese Academy of Agricultural Engineering & Planning Design, Beijing, China

ABSTRACT: With climate warming and increased occurrence of extreme weather and climate events, drought is more frequent in China. The Standardized Precipitation Evapotranspiration Index (SPEI), combining the advantage of Standardized Precipitation Index (SPI) and Palmer Drought Severity Index (PDSI), is computed at different time scales in Northeast China, based on monthly meteorological data from the China surface climatological daily data set provided by the National Meteorological Information Center. Remote sensing is another method used to monitor drought conditions on a regional scale, especially in the areas with few meteorological stations. The Drought Severity Index (DSI) was selected to analyze drought conditions on a regional scale, and the correlation between SPEI and DSI was calculated. The results showed that between 1968 and 2017 there were drought trends in Northeast China, except in winter, and were more obvious in autumn. From 1975 to 1987 there was a trend of drought but this was not significant, and in the mid-1980s, there were signs of wetting in Northeast China but disappeared quickly. Before that, inter-annual drought fluctuated slightly and after that, the region was referred to as dry. After 2009, dry and wet conditions fluctuated again, but more extremely. Spatial distribution of drought in Northeast China was heterogeneous and complex. The western region was the most seriously affected area, with the highest drought frequency. Correlation analysis between DSI and SPEI showed that positive relationships existed. Combing this stationary index SPEI and remote-sensing index DSI may improve the accuracy of drought monitoring. The results of this study can provide a scientific basis for early drought prediction and risk management of water resources and agricultural production in Northeast China.

Keywords: Standardized Precipitation Evapotranspiration Index, Drought Severity Index, drought, temporal and spatial variation

1 INTRODUCTION

As one of the most frequent disasters, drought has great influence on agricultural production, the natural ecosystem and socioeconomic development. Analysis on temporal and spatial variation of drought can provide a basis for early warnings and the risk management of water resources and agricultural production. The Palmer Drought Severity Index (PDSI), Standardized Precipitation Index and (SPI) Standardized Precipitation Evapotranspiration Index (SPEI) are widely used indexes based on situ data. SPEI (Vicente-Serrano, 2010) combines the sensitivity of PDSI (Palmer, 1965) to changes in evaporation demand (caused by temperature fluctuations and trends) with the simplicity of calculation and the multi-temporal nature of the SPI

---

*Corresponding author: wangwd52@mail.bnu.edu.cn

(McKee, 1993). SPEI is used to analyze temporal-spatial variation, characteristic and trends of drought (Cai et al, 2017; Chen et al, 2017; Li et al, 2017; Shen et al, 2017; Wei et al, 2018). Some scholars have discussed the applicability of SPEI to specific areas and indicated that in most parts of China, the precipitation minus potential evapotranspiration series match the presumed Log-logistic distribution (Wang et al, 2014); and the SPEI is suitable for quantifying droughts in Northeast China (Shen et al., 2017). But station-based drought index of meteorological variables cannot completely reflect drought characteristics.

By combining Normalized Difference Vegetation Index (NDVI) and the rate of evapotranspiration (ET) and potential evapotranspiration (PET), Mu et al. (2013) proposed a new drought index, the Drought Severity Index (DSI), which can monitor agricultural drought on a global scale. Zhang et al. (2019) determined the relationship between the crop yield and drought-affected areas monitored by DSI, and provided a theoretical basis for the application of DSI to agricultural monitoring across China. The uncertainties of remote-sensing data may result in a false DSI drought detection signal. This study used SPEI to investigate spatial-temporal evolution characteristics of regional drought in Northeast China, and the correlation of SPEI and DSI was explored. This may provide some reference for drought monitoring, early warnings and agricultural production in major grain producing areas.

2 STUDY AREA AND DATA

2.1 *Study area*

Historically, severe droughts have frequently occurred in Northeast China. This zone has a vast territory, and an area of about 845,300 km^2. It has a temperate monsoon climate, the annual rainfall averaged for available weather stations in the region is about 600 mm, and the annual average temperature is 5.4°C. Droughts occurred frequently in the study area, especially entering the 21st century, and there are few years without drought.

2.2 *Remote-sensing data*

To simplify the study, we directly adopted the DSI product provided by The Numerical Terradynamic Simulation Group (NTSG) at the University of Montana. The DSI dataset is available at http://files.ntsg.umt.edu/data/NTSG_Products/DSI/, with a spatial resolution of 0.05 degrees, at annual time intervals over 2000–2011, which is computed based on the vegetation greenness products and evapotranspiration/latent heat flux products.

Figure 1.   Distribution of meteorological stations in Northeast China.

2.3 Situ data

We collected the situ data of meteorological variables during 1968 – 2017 in the China surface climatological daily data set provided by the National Meteorological Information Center. The drought-affected and drought-damaged areas were collected from the *China Drought and Flood Disaster Bulletin* and the *China Meteorological Disaster Yearbook*.

3 METHODOLOGY

3.1 Computation of SPEI

First, the monthly potential evapotranspiration was computed based on the Thornthwaite method, which is the simplest approach for calculating PET (Thornthwaite, 1948),

$$PET = 16K\left(\frac{10T}{I}\right)^m \tag{1}$$

Second, the difference between the precipitation (P) and PET is calculated:

$$D = P - PET \tag{2}$$

The calculated D values are aggregated at different time scales, following the same procedure as that for the SPI.

Third, the Log-logistic distribution is selected to fit the D series. The probability density function of a three parameter Log-logistic distributed variable is expressed as follows:

$$f(x) = \frac{\beta}{\alpha}\left(\frac{x-\gamma}{\alpha}\right)^{\beta-1}\left(1+\left(\frac{x-\gamma}{\alpha}\right)\right)^{-2} \tag{3}$$

The probability distribution function of the D series according to the Log-logistic distribution is given by this equation:

$$F(x) = \left[1+\left(\frac{\alpha}{x-\gamma}\right)\right]^{-1} \tag{4}$$

where α, β and γ are scale, shape and origin parameters, respectively, for D values in the range $(\gamma > D > \infty)$.

Then, with F(x) the SPEI can easily be obtained as the standardized values of F(x).

$$SPEI = W - \frac{C_0 + C_1 W + C_2 W^2}{1 + d_1 W + d_2 W^2 + d_3 W^3} \tag{5}$$

Where, $W = \sqrt{-2\ln(P)}$ for $P \leq 0.5$, and P is the probability of exceeding a determined D value, $P = 1 - F(x)$. If $P0.5$, P is replaced by $1-P$ and the sign of the resultant SPEI is reversed. The constants are: $C_0 = 2.515517$, $C_1 = 0.802853$, $C_2 = 0.010328$, $d_1 = 1.432788$, $d_2 = 0.189269$, $d_3 = 0.001308$.

This paper computes the SPEI of 89 stations from 1968 to 2017 in multi-scales to analyze the characteristics of temporal and spatial evolution of seasonal and inter-annual drought in Northeast China, and drought grade according to the criteria in Table 1.

Table 1. Drought intensity based on categories by SPEI.

Grade	SPEI	Category
1	$-0.5 \leq$ SPEI	No drought
2	$-1.0 <$ SPEI≤ -0.5	Mild drought
3	$-1.5 <$ SPEI≤ -1.0	Moderate drought
4	$-2.0 <$ SPEI≤ -1.5	Severe drought
5	SPEI≤ -2.0	Extreme drought

3.2 Mann-Kendall trend test

The Mann-Kendall (MK) trend test method (Mann 1945; Kendall 1975) is a nonparametric statistical test recommended and widely used by the World Meteorological Organization (WMO). The advantages of this method are not requiring samples to obey a certain distribution and not being disturbed by a small number of outliers. It has a high degree of quantification, a wide range of detection and convenient calculation, and is more suitable for sequential variables and type variables.

In the test curve, if the UFk line changes within the critical boundary, it indicates that the trend and mutation of the change curve are not obvious, and the value of UFk is greater than zero, indicating that the sequence shows an upward trend, and vice versa. When it exceeds the critical line, it indicates a clear upward or downward trend.

3.3 Correlation analysis

The Pearson correlation coefficients between yearly situ-SPEI and the DSI of the corresponding position in the specific year are analyzed in the study area.

4 RESULTS AND DISCUSSION

4.1 The temporal characteristics of drought

4.1.1 Inter-annual variation characteristics

Figure 2 demonstrates the change of SPEI across China at the annual scale base on mean of monthly SPEI from all stations during 1968–2017. It has experienced four different periods: (1) Before 1975 the climate was slightly damp but fluctuated; (2) From 1975 to 1987 there was

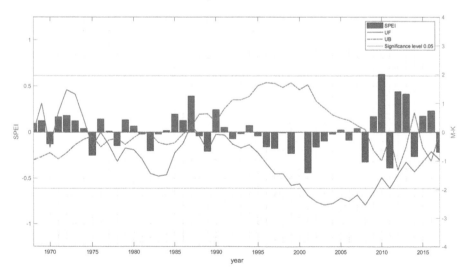

Figure 2. Inter-annual trends based on SPEI in Northeast China.

a trend of drought but this was not significant. It is worth mentioning that, from 1984 the trend of drought alleviated and there were signs of humidification but disappeared quickly. (3) From 1987, the trend of drought aggravated. By 2000, the aridification reached a significant level of 0.05. (4) There was a insignificant trend of drought from 2009 to 2017, but dry and wet conditions began fluctuating again, more extremely. As a whole, there are drought trends in Northeast China.

An MK trend test gives $\beta = -0.0025$, $Z = -0.95$, which is smaller than 1.96 so the inter-annual drought trend exists but is not significant.

4.1.2 *Seasonal variation characteristics*

An MK trend test gives $\beta = -0.003$, $Z = -0.636$ in spring, and $\beta < 0$ indicates the SPEI is descending, which means there is an existing drought trend in spring but $|Z| < 1.96$ means the trend is not significant. The same situation applies for summer. In autumn, the drought trend is obvious. But the data in winter shows that it is getting wet but not significantly.

4.2 *Spatial distribution characteristics of drought*

The spatial difference of drought frequency is shown in Figure 3. According to Table 1, the drought frequency distribution is plotted once SPEI < -0.5. This gives the frequency of drought as 31.8–36.3%, mild drought as 12.87–21.66%, moderate drought as 8.67–15.33%, severe drought as 2.67–6.33%, and extreme drought as 0–1.49%. The more severe the drought, the lower its frequency. From the perspective of space patterns, the regional difference of drought frequency is obvious. The western region was the most seriously affected area, with the highest drought frequency.

4.3 *Correlation analysis between SPEI and DSI*

From Table 3. we find that, except for 2003, 2006, 2009, SPEI-12 and DSI is positively related, and for most years the P-values for testing the hypothesis of no correlation is small, then the correlation is significant. But this does not pass the significance test every year. For example, the correlation coefficient in 2003 is greater than zero, but P-value is too large. This may be because the values identified from DSI (area data) corresponding to the stations (point data)

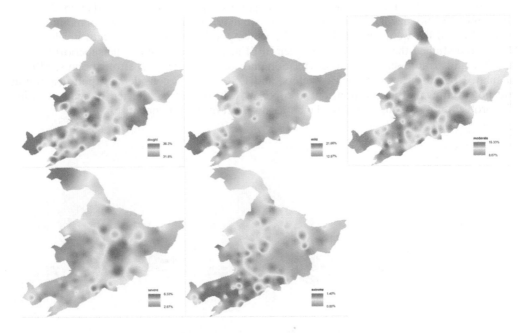

Figure 3.   Frequency of drought in each grade based on SPEI.

193

Table 2. The result of MK trend test based on SPEI.

Time	β	Z	Trend	Significance (0.05)
Year	-0.003	-0.954	Descending	Not significant
Spring	-0.003	-0.636	Descending	Not significant
Summer	-0.003	-0.87	Descending	Not significant
Autumn	-0.008	-2.108	Descending	Significant
Winter	0.009	2.058	Rising	Significant

Table 3. The result of correlation analysis between SPEI and DSI.

Year	SPEI-12/DSI (R)	SPEI-12/DSI (P)	Year	SPEI-12/DSI (R)	SPEI-12/DSI (P)
2000	0.3756	0.0007	2006	-0.1284	0.2415
2001	0.2965	0.0059	2007	0.2556	0.0182
2002	0.5721	1.07e-08	2008	0.3880	0.0002
2003	-0.2182	0.0448	2009	-0.0134	0.9033
2004	0.1860	0.0882	2010	0.4799	3.37e-06
2005	0.1807	0.0980	2011	0.2540	0.0190

are in line with large spatial uncertainties. Here SPEI-12 in December is used to contrast with annual DSI, not considering the timescale of drought. SPEI is calculated based on readily available monthly precipitation and temperature. Nevertheless, DSI refers to ET/PET and NDVI, not only the meteorological factors, and PET in DSI refers to the Penman–Monteith equation (Monteith 1965), instead of the Thornthwaite equation (Thornthwaite 1948) in SPEI.

Generally, the situ data is more accurate, but because the site distribution is limited, it is difficult to get relatively accurate global data for the whole area. The situ-SPEI and the annual DSI in 2003, 2006, and 2009 is shown in Figure 4. According to the *China Flood and Drought Disaster Bulletin*, there is a severe spring drought in Northeast China in 2003, especially in Heilongjiang Province. DSI can identify the drought by year, however, SPEI is not good enough. In 2006, summer drought occurred southwest of this area, but the data of DSI is not obvious, nevertheless it was reflected in SPEI to some degree. The drought was more serious in 2009 than in 2003, but DSI cannot identify this. The combination of the two may improve the reliability of monitoring.

Based on SPEI, we analyzed the spatial and temporal variation of drought in Northeast China, and studied the correlativity between SPEI and DSI to combine the superiority of in-situ data and remote-sensing data. Correlation may not be particularly strong because SPEI is calculated based on meteorological factors, and DSI not only considers evapotranspiration and potential evapotranspiration, but also NDVI, which may be affected by many factors. The problem is also complicated by the buffering effect of the soil moisture.

Figure 4. Drought distribution of Northeast China monitored by SPEI and DSI.

5 CONCLUSION

Based on meteorological data from 1961 to 2017 at 89 stations, SPEI was used to quantitatively analyze the spatiotemporal variation characteristics of drought, and the correlation between SPEI and TVDI.

Dry and wet conditions were experienced in four different periods: (1) Before 1975, the climate was slightly damp but fluctuated. (2)From 1975 to 1987, there was a trend of drought but not significantly. Since 1984 the trend of drought has mitigated and there were signs of wetting. (3)But since 1987, the trend of drought has aggravated again. From 2000 to 2009, it reached a significance level of 0.05. (4) The trend of drought still exists since 2009, but dry and wet conditions are fluctuating again, more extremely. As a whole, there are trends of drought in Northeast China.

There is existing drought trend in spring and summer, but the trend is not significant. In autumn, the drought trend is significant. The data in winter shows that it is wet but not significantly. Spatially, the frequency is drought 31.8–36.3%, mild drought 12.8–21.66%, moderate drought 8.67–15.33%, severe drought 2.67–6.33%, and extreme drought 0–1.49%. The more severe the drought, the lower its frequency. From the perspective of space patterns, the regional difference of drought frequency is obvious. On the whole, the western region was the most seriously affected area, with the highest drought frequency.

The relation between SPEI-12 and DSI is positive for most years, but does not pass the significance test every year. This is only a preliminary exploration. To synthesize the advantages of the two methods in appropriate time scale and space scope, further research is still needed.

ACKNOWLEDGMENTS

This project was supported by National Key Research and Development Program of China (2016YFB0501505).

REFERENCES

Chen S.D., Zhang L.P., Tang R.X., Yang K., Huang Y.Q. 2017. Analysis on temporal and spatial variation of drought in Henan Province based on SPEI and TVDI[J]. *Transactions of the Chinese Society of Agricultural Engineering* 2017, 33(24): 126–132.

Cai, S.Y., Zuo, D.P., Xu, Z.X., Yang, X.J. 2017. Spatiotemporal characteristics of drought in Northeast China based on SPEI. *South-to-North Water Transfers and Water Science & Technology* 15(5): 15–21.

Kendall, M.G. 1975. *Rank Correlation Methods*. Griffin, London, UK.

Mann, H.B. 1945. Nonparametric tests against trend. *Econometrica* 13, 245–259.

McKee, T.B., Doesken, N.J. and Kliest, J. The Relationship of Drought Frequency and Duration to Time Scales. *Proceedings of the 8th Conference on Applied Climatology*, Anaheim, 17-22 January 1993, 179–184.

Mu, Q.Z., Zhao, M.S., Kimball, J.S., McDowell, N.G., Running, S.W. 2013. A remotely sensed global terrestrial drought severity index. *Bulletin of the American Meteorological Society* 94(1): 83–98.

Li, X.X., Ju, H., Liu, Q., Li, Y.C., Qin, X.C. 2017. Analysis of drought characters based on the SPEI-PM index in Huang-Huai-Hai Plain. *Acta Ecologica Sinica* 37(6): 2054–2066.

Palmer W. *Meteorological drought*. U.S. Department of Commerce Weather Bureau Research Paper, 1965.

Shen, G,Q., Zheng, H.F., Lei, Z.F. 2017. Applicability analysis of SPEI for drought research in Northeast China. *Acta Ecologica Sinica* 37(11): 3787–3795.

Vicente-Serrano, S.M., Begueria, S., Lopez-Moreno, J.I. 2010. A multiscalar drought index sensitive to global warming: The standardized precipitation evapotranspiration index. *Journal of Climate* 23(7): 1696–1718.

Wang, L., Chen, W. 2014. Applicability analysis of standardized precipitation evapotranspiration index in drought monitoring in China. *Plateau Meteorology* 33(2): 423–431.

Wei, X., Hu, Q., Ma, X., Zheng, S., Tang, X., Zhang, Y., Pan, X., He, Q. 2018. Temporal-spatial Variation Characteristics of Drought in Summer Maize Growing Season in North China Plain Based on SPEI. *Journal of Arid Meteorology* 36(4): 554–560.

Zhang, Q., Yu, H., Sun, P., Singh, V. P., Shi, P. 2011. Multisource data based agricultural drought monitoring and agricultural loss in China. *Global and Planetary Change* 172(1): 298–306.

Risk Analysis Based on Data and Crisis Response Beyond Knowledge – Huang & Nivolianitou (eds)
© 2020 Taylor & Francis Group, London, ISBN 978-0-367-25146-8

A Python approach to collecting disaster data for the Internet of Intelligences

Wen Tian & Chongfu Huang*

Key Laboratory of Environmental Change and Natural Disaster, Ministry of Education, Beijing Normal University, Beijing, China
State Key Laboratory of Earth Surface Processes and Resources Ecology, Beijing Normal University, Beijing, China
Academy of Disaster Reduction and Emergency Management, Faculty of Geographical Science, Beijing Normal University, Beijing, China

ABSTRACT: In this paper, we suggest a Python approach to collecting disaster data for the Internet of Intelligences (IOI) for emergency rescue. Information including specified keywords was collected from Weibo and WeChat, as social media, and the web spider crawler was written in the Python programming language. By mining data from the information, we provided an improved TextRank algorithm based on attribute vector for data mining, which can obtain the disaster narratives to help decision-makers understand disaster situations. The final data results were visualized through the risk map display module of the IOI. Finally, we selected the 7/11 flood in Santai County, China, on July 11, 2018 as an example for applying the disaster data collection module in the IOI.

Keywords: Internet of Intelligences, social media, Python, TextRank, Santai County

1 INTRODUCTION

The Internet of Intelligences (IOI) is a platform for risk analysis and management (Huang, 2016). One of mathematical models in IOI, based on geospatial information diffusion technology, can supplement incomplete geospatial data to make it complete (Huang, 2019). Only with complete disaster data can decision-makers design a reasonable emergency rescue plan. The geospatial diffusion model fills disaster data in gap geographic units by using disaster data in observed units and the background data of the observed and gap units.

In this article we will explore an approach, embodied in IOI, to online collect the disaster data in the units where people have scattered disaster data and can access the Internet to share this on social networks.

Python, as a web application development language, was a potential tool for accomplishing our task. Python not only has a rich technical library and powerful computing capacity, but also has many advantages (it is object-oriented, open source, portability, etc.), which makes it the first choice for designing the disaster data collection module for the IOI.

Most of the disaster data are hidden in webpages. A web spider data crawler is computer software technology that can capture the data from a webpage. This crawler technology was developed in 1993. The traditional crawler crawls all the webpages related to the seed URL. In 2001, IBM designed distributed crawlers that could implement web page storage (Edwards et al., 2001). With the development of network technology, the performance of crawler technology has gradually improved, and some new methods have

---

*Corresponding author: hchongfu@bnu.edu.cn

appeared, including multi-threaded crawlers (Harthet et al., 2006), API crawlers (Cao et al., 2017), user-guided social media crawler methods (Fredrik et al., 2017), etc. However, along with the advancement of crawler technology, there is also anti-crawler technology. Therefore, this study investigated how to design high-performance crawler code and deal with anti-crawler technology.

The disaster data obtained through the data crawler are mostly unstructured text data that need to be further mined. Existing approaches to mining text data, such as NLP (Natural Language Processing), IR (Information Retrieval) and AS (Automatic Summarization), are most applied by counting the word frequency, analyzing sentiments, or extracting keywords, rather than providing narratives. If the data content can be mined for the narratives containing the disaster attribute information, it will help the decision-makers to know more about the disaster situation. The TextRank algorithm proposed by Mihalcea is based on the PageRank algorithm and is widely used in key sentence extraction (Mihalcea et al., 2004). Many scholars at home and abroad have improved this algorithm depending on their own research needs (Hamid, 2016; Yu et al., 2016; Ahmad et al., 2017; Pu et al., 2017) with good results. Thus, study paper attempted to improve the TextRank algorithm to mine disaster attribute information in text data.

Social media refers to online communication between users, often in blogs, content communities, and social networking sites, it has been crawled over the years to collect disaster-related multimedia content (Rogstadius et al., 2013; Said et al., 2019). Online social media provides new sources for the collection of disaster data. China has the most online social media users in the world (Xue, 2016). When a disaster occurs, many social media users try to locate the most up-to-date information about that event, and survivors also use social media sites to keep in touch with the world and share disaster messages on the sites (Gao et al., 2011). Therefore, social media data will become an important data source for IOI.

This article will focus on these four aspects. (1) Design a Python-based crawler code to collect disaster data from social media webpages. (2) Improve the TextRank algorithm to mine the disaster text data. (3) Call and visualize the data results through the IOI. (4) Taking the 7/11 flood in Santai County, Sichuan Province, China, as an example, apply the disaster data collection module of IOI to the county scale, to support accurate rescue operations.

2 INTERNET OF INTELLIGENCES

After natural disasters occur in China, emergency rescue work is arranged by the government emergency agency. With the development of network technology, it is possible to support emergency rescues through the online network systems. Most existing online disaster mitigation systems are only for a single disaster (Lin, 2018) or a certain region (Wang et al., 2018), so it is necessary to implement a comprehensive disaster mitigation system. The concept of IOI was proposed by Huang in 2011 (Huang, 2011) and related technology has been developed. Ten IOI platforms have been developed (Huang, 2017) for online risk assessment, including a risk radar driven by IOI, a marine environmental risk management, and risk timeliness evaluation, etc.

IOI relies on the background data to calculate the disaster data based on the geospatial information diffusion technology, which can realize the accuracy and completeness of results. (Huang, 2019). So, we need to collect more background data, and online social media data will be an important data source for IOI.

The technical framework of IOI uses the PHP language and the Yii framework for system development, the MySQL database for data management, and visualizes the data result through WebGIS. Supported by this technical path, the disaster data collection module in IOI was designed mainly based on the Python language. This module also can be executed by PHP, and then the data will be transferred by MySQL. The design of the disaster data collection module in IOI will be introduced in the following chapter.

3 DISASTER DATA COLLECTION MODULE IN IOI

3.1 *Data source selection and data management*

Social media has a variety of communication forms and operating modes. Among them, Weibo and WeChat have become the platforms of mass socialization in China (Tan et al., 2017). Compared with other types of social media, Weibo and WeChat public accounts have more prominent media attributes which has made them the largest public information publishing platforms in China today. They have a huge influence on the current network public opinion in China, especially after a public event. Therefore, in this article, we choose Weibo and WeChat public accounts as the source of disaster data.

For the management of disaster data, we use a software named SQLyog under the WAMP-SERVER environment to connect the MySQL database, and use MySQLdb library in Python to transfer the disaster data. The database table design is shown in Table 1.

3.2 *Data crawler*

The principle method of a data crawler is to make a request to the server through the simulation of a browser by programming and send the Uniform Resource Locator (URL) as the content of the Hyper Text Transfer Protocol (HTTP) request to the server. Though a data crawler we can ascertain the response resource of the server.

A data crawler has two patterns: API-based (Application Programming Interface) and HTML-based (Hyper Text Markup Language). API-based sometimes crawlers have errors, such as the limitation of the number of crawls per day, or some social media sites don't provide API. Thus, in this article, we choose an HTML-based crawler. HTML-based crawlers must solve three key problems: (1) selecting the webpages, that is, how to find the target webpage; (2) parsing the webpages, that is, how to find the node containing the information from the webpage and obtain the path; (3) responding to the anti-crawler, that is, how to deal with the prevention set by webpage. For the first two problems, Python has some mature libraries, such as Request and Selenium, which can be used to design code by calling the methods in the libraries. Among them, the Selenium library can link the search engine and enter keywords for crawling, which can make the information more targeted. For the third problem, as most anti-crawler technologies judge whether or not it is a crawler by the access frequency of the IP address, this paper used the proxy pool of IP technology to respond.

In addition, we mainly used keyword matching technology to clean and filter the crawled data. Given the limitation about the amount of crawled data, we further filtered the unrelated disaster data by manual browsing to improve the correlation of the data.

3.3 *Data mining*

The TextRank algorithm is based on the improvement of PageRank: the link relationship between nodes constitutes an undirected network to calculate the similarity of the distance.

Table 1. The design of database tables.

Platform	Number	Field	Description	Type
Weibo	1	name	User's Nickname	Varchar (50)
	2	time	Posting time	Varchar (50)
	4	txt	Content	Longtext
WeChat Public Number	1	title	Article title	Varchar (50)
	2	time	Posting time	Varchar (50)
	3	source	Source	Varchar (50)
	4	txt	Content	Longtext

This study improved the algorithm by adding some attribute vector which can be constructed by the township name in the disaster occurrence area. The attribute vector was recorded as P_i.

For the crawled text data, W, which includes n sentences, by calling the Jieba library in Python for word segmentation, we get the word vector and the eigenvectors $S_n = \{K_1, K_2, K_i\}$ of each sentence. Through the eigenvectors we can form the eigenvector matrix D of the text. According to the eigenvectors of each sentence, the similarity between sentences is calculated by Euclidean distance, and then the similarity matrix Q is obtained. For each term K_{ni} in D, we can use the TF-IDF (term frequency–inverse document frequency) algorithm to evaluate word frequency and get the first n word vectors, which can construct an eigenvector F_n. Through assigning a value in range $[(0,1)]$ to each term in F_n by determining if F_n is similar to the word vector in P_i, we can get the weight vector W_n. Finally, we can get the vector A through the multiplication of Q_i and W_n. Vector A can be used to select key sentence to construct narratives.

3.4 *Interaction with IOI for disaster mitigation*

By encapsulating the above code as data collection module in IOI, it is possible to implement real-time data collection when a disaster occurs by triggering this through keywords. By connecting the MySQL database, data collection results can be visualized through the risk map display module (Figure 2), which will support decision-makers to intuitively grasp the disaster situation.

4 APPLICATION

4.1 *Data crawler*

This study used the 7/11 flood in Santai County, Sichuan Province as an example for data collection. The 7/11 flood was caused by seasonal rainfall. According to reports, from July 9 to 11, 2018, there was large-scale rainfall in the middle and upper reaches of the Fujiang river and the Kaijiang River. Rescue was the primary task, and therefore the target of the disaster data collection was obtaining the distribution of the people affected by the flood. According to the above analysis, this paper selected the data collection period from July 9 to 11, 2018 and "the flood in Santai County" as keywords. Finally, by filtering and cleaning, we found 210 pieces of disaster data, include 135 pieces of Weibo data, and 75 pieces of WeChat public accounts data.

4.2 *Data mining*

Narratives and keywords were extracted based on the improved TextRank algorithm. According to the results, we selected three indicators: the name of affected township, the number of affected people, and the disaster situation. Some results are shown in Table 2.

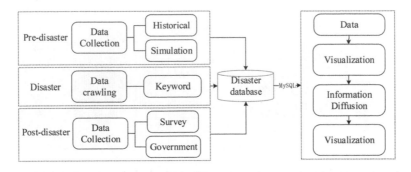

Figure 1. Interaction with IOI for disaster mitigation.

Table 2. Extraction in multi-document.

Town	Victims	Situation	Time
Yongming	About 400	The Yongming Town, on Jiangxin island, Santai County has over 400 people.	July 11
Longshu	Over 2	Rainstorm in the upper reaches of the Zijiang River caused the water level to soar.	July 10
Laoma	Over 10	The water level rose to the street level of the town. The water depth in Hejiaqiao was over 1.6 meters. The villages of Ganjiang, Yujiu and Xiao-wei were seriously affected. The farmland and houses are flooded.	July 11
Licheng	About 400	Licheng village was seriously affected.	July 11
Luxi	None	Ophiopogonis of Pengjia Lane was submerged, and the flood had flooded the top of the corn. About 100 Hectares of Ophiopogonis land is affected in groups 2, 4, 5, 6, 7, 8 of two villages.	July 11
Huayuan	None	The farmland in Yucheng Village was destroyed by floods, and five groups were affected by disasters.	July 11
Liuying	None	The Ophiopogonis of Yangjiatun Village was affected.	July 11
Tongchuan	None	Due to the influence of the upstream flood, about 100 Hectares of land over 20 farm houses were flooded in Liulin Village.	July 11

4.3 *Data visualization*

The disaster situation of the 7/11 flood is showed in Figure 2 according to the risk map display module in IOI. (a) The data collected by a Python approach. (b) The actual data published by the local government on July 12.

In these risk maps, red indicates the severely affected areas, brown indicates that the disaster is moderate; light blue indicates that the disaster level is light. Both (a) and (b) show that the disaster-stricken areas are mainly distributed on the banks of the Fujiang River and the Zijiang River.

In the Python approach, the red areas include Yongming Township, Licheng Township, Laoma Township, Luxi Township and Huayuan Township, and the brown areas include Longshu Township, Liuying Township, Zhengsheng Township, Lixin Township and Ling Township. Comparing the results of the Python approach with the actual data published by the local government shows that the affected-area is basically the same; the only difference is that the first method shows Laoma Township as severely affected, but actually it was Liuying Township. The slight deviation regarding property loss is because the Python approach time setting was is from July 9 to 11, but the actual disaster was aggravated by a dam break and flood discharge after July 11.

(a)   (b)

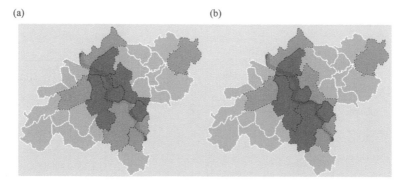

Figure 2. Risk map for Santai County. (a) Python approach; (b) Local government data.

5 CONCLUSION

In this paper, IOI and Python-based data mining technology are combined to explore the real-time collection of disaster data. We used the proxy pool of IP method to solve the anti-crawler problem, which improved the speed and effectiveness of the crawler code, and we improved the TextRank algorithm to collect 210 pieces of text data. According to the disaster data, most of the affected areas in Santai County were concentrated along the Fujiang River and the Zijiang River. Among them, Yongming Township, Licheng Township, Laoma Township, Luxi Township and Garden Township were the most affected areas. Comparing the results of the Python approach with the actual data published by the local government shows that the affected areas are basically the same. The few deviations show there is a gap unit in some regions, so we need geospatial information diffusion technology to improve this. In the future, these two problems will be used to focus on improving data accuracy and reliability; and improving the intelligence of the data collection module.

ACKNOWLEDGMENTS

This project was supported by the National Key Research and Development Plan (2017YFC1502902).

REFERENCES

Ahmad A., Mardhani R. 2017. Document Summarization using TextRank and Semantic Network. *I.J. Intelligent Systems and Applications* 11: 26–33.

Cao, Y.B., Mao, Z.J. 2017. Temporal and spatial characteristics of the Jiuzhaigou 7.0 earthquake based on microblog data mining. *China Earthquake* 33(4): 613–625.

Edwards, J., Mccurley, K., & Tomlin, J. 2001. An Adaptive Model for Optimizing Performance of an Incremental Web Crawler. *International Conference on World Wide Web*. ACM.

Fredrik, E., Bródka Piotr, Martin, B., Henric, J. 2017. Do we really need to catch them all? A new user-guided social media crawling method. *Entropy* 19(12): 686–701.

Gao, H., Barbier, G., & Goolsby, R. 2011. Harnessing the crowdsourcing power of social media for disaster relief. *IEEE Intelligent Systems* 26(3): 10–14.

Hamid, F. 2016. Evaluation techniques and graph-based algorithms for automatic summarization and keyphrase extraction. *Unt Theses & Dissertations*.

Harth A., Umbrich J., Decker S. 2006. MultiCrawler: A Pipelined Architecture for Crawling and Indexing Semantic Web Data. In: Cruz I. et al. (eds) *The Semantic Web - ISWC 2006*. ISWC 2006. Lecture Notes in Computer Science, vol 4273. Springer, Berlin, Heidelberg.

Huang, C.F. 2011. The intelligent network of risk analysis online services. *Journal of Risk Analysis and Crisis Response* 1(2): 110–117.

Huang, C.F. 2015. Internet of intelligences can be a platform for risk analysis and management. *Human and Ecological Risk Assessment* 21(5): 1395–1409.

Huang, C.F. 2017. The development of the principle of intelligent networking and its core technology. *Journal of Risk Analysis and Crisis Response* 7(3): 146–155.

Huang C.F. 2019. Geospatial Information Diffusion Technology Supporting by Background Data. *Journal of Risk Analysis and Crisis Response* 9(1): 2–10.

Lin C. 2018. Construction of a material information management platform for earthquake emergency rescue support and decision-making. *Doctoral dissertation, Institute of Earthquake Research, China Earthquake Administration*.

Mihalcea, R., Tarau, P. 2004. *Textrank: bringing order into texts*. Association for Computational Linguistics, 404–411.

Pu, M., Zhou, F., Zhou, J.J., Yan, X., and Zhou, L.J. 2017. Extraction of news key event topic sentences based on weighted Textrank. *Computer Engineering* 43(8): 219–224.

Rogstadius, J., Vukovic, M., Teixeira, C., Kostakos, V., Laredo, J. 2013. Crisis tracker: crowdsourced social media curation for disaster awareness. *IBM Journal of Research and Development* 57(5).

Said, N., Ahmad, K., Regular, M., Pogorelov, K., Hassan, L., & Ahmad, N., et al. 2019. Natural disasters detection in social media and satellite imagery: a survey.

Tan, T., Zhang, Z.J. 2017. The status quo, development and trend of social media in China. *Friends of the Editor* (1), 20–25.

Wang, W., Jin, S.S., Guo, W., Li, C.Y. 2018. Decision Support Platform for Disaster Prevention, Disaster Mitigation and Relief. *Popular Science* (5): 16–17.

Xue, X.J. 2016. Self-media art communication of China's national image. *Media* (6): 82–84.

Yu, S.S., Su, J.D., Li, P.F. 2016. Improved TextRank-based Method for Automatic Summarization. *Computer Science* 43(6): 240–247.

Risk Analysis Based on Data and Crisis Response Beyond Knowledge – Huang & Nivolianitou (eds)
© 2020 Taylor & Francis Group, London, ISBN 978-0-367-25146-8

The impact of ground motion prediction equations on estimating insurance losses in the EQC at earthquake catastrophe model

Zhenghui Xiong & Zhijun Dai
Institute of Geophysics, China Earthquake Administration, Beijing, China
China Earthquake Risk and Insurance Laboratory, Beijing, China

Xiaojun Li*
College of Civil Engineering, Beijing University of Technology, Beijing, China

ABSTRACT: Based on the EQC at catastrophe model proposed by this article, the effects of different GMPEs of *Seismic Ground Motion Parameters Zonation Map of China* (GB18306, version 2001/2015) are discussed on calculating pure premium rates and occurrence exceedance probability. The results show that GMPEs determine the distribution area of peak ground motion, thereby affecting the amounts and damage extent of insured exposures. Taking a single point, the loss calculated by the new 2015 GMPE is up to 3.7 times larger than the one from 2001 at the return period of 1-in-500. The pure premium rates of typical frame-shear wall structures are given as a color map of Beijing city, offering an intuitive view of seismic risk.

Keywords: seismic risk, catastrophe model, earthquake insurance, GMPE, Beijing

1 INTRODUCTION

Catastrophe modeling is a significant tool for assessing insurance losses in horrible disasters such as earthquakes, floods and hurricanes. As an effective means of risk management, insurance plays an important role in mitigating natural disaster losses and helping recovery and reconstruction post-disaster. Insurance and reinsurance operations commonly rely on large sets of historical loss experience data to conduct their business, set premiums, and to conceive their risk management strategy. However, for operations concerned with less frequent perils but huge losses such as earthquakes, the data historically available is insufficient to obtain reliable risk statistics (Franco, 2012). The industry makes the use of numerical models that, based on the current understanding of the mechanisms underlying environmental hazards and supported by existing data, estimate the probability of occurrence as well as the consequences of catastrophic events in terms of monetary loss (Grossi & Kunreuther, 2005). These models are known as catastrophe models.

According to the commonly used risk framework (Nakada et al., 1999; Porter, 2003; Calviet al., 2014; Yoshikawa & Goda, 2014), catastrophe models involve four basic components: stochastic module, hazard module, vulnerability module and financial analysis module. Ground Motion Prediction Equations (GMPEs) are crucial for catastrophe models to assess seismic hazard, as they provide the shaking (strong ground motion) that may occur at a site if an earthquake of a certain magnitude occurs at a nearby location. Through generating a hazard footprint for earthquakes by the GMPE, the intensity measure of all modeled exposures that are located within its extent can be obtained. Obviously, the GMPEs determine the distribution areas of ground motion such as peak ground acceleration (PGA). That affects the amounts and damage extent of insured exposures.

---

*Corresponding author: beerli@vip.sina.com

In this paper, we develop an earthquake catastrophe model by integrating the achievements of the *Seismic Hazard, Risk and Policy Research for Earthquake Insurance* (SHRPREI) project proposed by the Institute of Geophysics, China Earthquake Administration (IG-CEA).The EQC at aims to calculate the insurance losses such as Average Annual Loss (AAL)or pure premium expected from earthquakes. Based on this catastrophe model, the effects on estimating pure premium and occurrence exceedance probability (OEP) are discussed using GMPEs from different *Seismic Ground Motion Parameters Zonation Map of China* (GB18306, version 2001/2015) on exposures in Beijing.

2 METHODOLOGY OF EARTHQUAKE CATASTROPHE MODEL

Prior to the advent of catastrophe models, the insurance industry's usual approach was to estimate the maximum percentage of total insured value in an area that might suffer loss from a realistic earthquake event (Winspearet al., 2012). Instead of using his torical loss experience to determine the insurance rate, the estimation methodology of catastrophe modeling has been developed to the point that losses from destructive earthquakes can be quantified with reasonable accuracy using scientific modeling techniques (Dong, 2002). Catastrophe models have been widely used in the insurance and reinsurance industries for over twenty years for calculating the amount of insured loss expected from natural catastrophes (such as earthquakes or typhoons) at an annual probability of exceedance (e.g. 1/500). Commercial models (proprietary) supplied by companies like Risk Management Solutions (RMS) Inc., AIR WorldwideCorp. and Core Logic Inc. are presently used globally. There are also multiple publicly accessible catastrophe models focusing on seismic risk estimation and earthquake loss calculation. Among these models are the open-source and well-known ones such as Hazard-U.S. (HAZUS) and Global Earthquake Model (GEM).

These catastrophe models make use of a relatively standard frame work. Taking the simplified IRAS Earthquake Model of RMS Inc. as an example, there are regularly four major components of analytical modules in a catastrophe model: stochastic module, hazard module, vulnerability module and financial analysis module (Nakada et al., 1999). It begins with defining a set of potential catastrophic events, along with their relative likelihood, and then assessing the hazard generated by each event. The model creates an estimate of the damage that could occur to the exposure, which is then allocated between property owners, insurers, and reinsurers to quantify the financial loss.

- **Stochastic Module** creates a set of representative events and associated probabilities. It is a collection of modeled events that represent the range of what could occur: magnitude-from small to large, location of the epicenter in different areas, and with a range of physical parameters.The stochasticsetis simulated and non-historical, and createsafull range of potential events. For earthquakes, the potential seismic sources and seismicity model are considered in generating the stochastic events.
- **Hazard Module** calculates the level of hazard of each event at all sites. The amplitude and character of ground motion produced by an earthquake rupture is dependent on the type of earthquake, its magnitude and depth, but those ground motions will gradually die out or attenuate with distance from the event. Earthquake models rely on GMPEs derived from historical measurements to estimate the extent of shaking with distance from a given event. The site conditions impact the relative shaking, which means soft soils and mud will generally amplify shaking relative to rock. Once the stochastic events are defined and the hazard generated by each event has been assessed, the next step is to understand which exposures are affected and the exposure characteristics of each risk that determine damage.
- **Vulnerability Module** calculates the average damage and associated variability. Damage is calculated through the vulnerability module, which estimates the degree to which a specific building or location is affected by the intensity measure (IM) of a physical phenomenon. For instance, how would a given location expect to perform based on the peak ground

motion at that site? It comes to the mean damage ratio (MDR), or the average loss of a group of locations divided by their values. A vulnerability function models the MDR, which equals average loss divided by replacement value. It also considers the potential uncertainty in damage. Individual structures will perform differently due to variations in their construction and quality, which is reflected in a range of possible loss outcomes. There are multiple vulnerability functions embedded into a catastrophe model which are differentiated on the basis of the primary characteristics.

- **Financial Analysis Module** calculates the financial impact for all participants and generate metrics. It takes the damage from the vulnerability module and calculates the losses. The module then allocates the loss to the various stakeholders. In this process, it creates an event loss table (ELT) or a year loss table (YLT) depending on whether an analytical or simulation approach is used. From this information, probabilistic outputs such as return period losses are generated. The last step in the catastrophe modeling framework is to evaluate the outputs from the model and then apply those results to real life business decisions.

We establish an EQC at catastrophe model by referring to the typical frameworks above. The EQC at integrates the latestachievements of project SHRPREI and GB18306, which were both accomplished by the research team of IG-CEA. For the Stochastic Module, we generate a set of 100,000-year earthquakeevents with magnitudes larger than 4.7, which are simulated by the Monte Carlo method using these is micactivity information from GB18306-2015 (Xuet al., 2018). The distribution of this stochastic setcovers all of mainland China, as well as buffers near the borders. Each event contains several attributes such as magnitude, epicenter location (longitude and latitude), simulated date, fault strike angle and GMPE zone.For the Hazard Module, we offer different GMPEs (Huo & Hu, 1992; Wang et al., 2000; Yu, 2013) to calculate the ground motion parameters (e.g. PGA) of each event at all sites. Due to the lack of sufficient earthquake records, catastrophe models always apply the ellipsoid attenuation modelto estimate the ground motion in China. The EQC at allows selecting GMPEs available in the system or uploading a custom GMPE with a fixed format. For the Vulnerability Module, we propose an insurance-oriented building classification with the factors of construction class, occupancy type, height, build year and seismic fortification level. Therefore, each vulnerability function will be defined by the five primary characteristics for existing buildings. Then a database with various vulnerability curves is designed and completed to support the EQC at. For the Financial Analysis Module, users can upload their insurance policies or portfolios with a template, and download the results sheets including ELT and OEP.

Figure 1 is a simple schematic to help understand the general workflow of calculating insurance losses in the EQC at. Once the coverage of insurance policies are determined, the EQC at will figure out the possible events of destructive earthquakes from the stochastic set. Taking event No.1 as an example, the model then generates the hazard footprint with the GMPE by the attributes of the event, and matches the ground motion (intensity or PGA) with every insured building by location. Based on the building characteristics, EQC at will search for the matching vulnerability curves from the database, and find out the mean damage ratio (MDR) corresponding to the intensity or PGA. For instance, the MDRs of the insured buildings No.1, 2, and M are separately 15%, 85% and 0. The ground up loss can then be obtained by multiplying the value of the building by the MDR, e.g. the ground up loss of building No.1 is 600,000 (4,000,000*15%). In combination with the insurance conditions, the net loss of each building equals the ground up loss minus the deductibles. Commonly it is no more than the total insured value (limit). We can then obtain the event loss table (ELT) by accumulating all building losses (net loss or ground up loss) for each event. For an insured building, its AAL or pure premium can be calculated with Equation 1 below:

$$E = \sum_{j=1}^{N} P_j L_j \qquad (1)$$

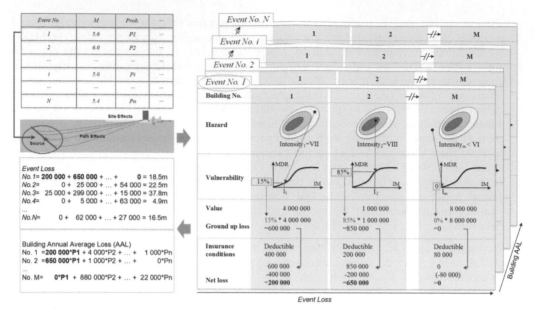

Figure 1.   Flow of calculating insurance losses in EQC at.

Where P_j = occurrence probability of the given event j, L_j = net loss of the given event j. Since the EQC at's stochastic set is generated by the Monte Carlo simulation where each event just occurs once in a 100,000 years, this results in P_j = 1/100,000.

3   GMPE OF SEISMIC GROUND MOTION PARAMETERS ZONATION MAP OF CHINA (GB18306)

Ground Motion Prediction Equations (GMPEs) are frequently also referred to as attenuation relationships, empirical ground-motion models, or just ground-motion models. They are a critical element of any probabilistic seismic hazard or risk analysis (Stafford, 2013). A GMPE is essentially a function that takes a number of input parameters that relate to the properties of an earthquake and its spatial relationship with a site and defines the distribution of ground-motion values that is to be expected for the considered scenario.

The generalized GMPE model for a rock site (Vs 30 ≥ 500m/s) is described as:

$$Ln\,Y(M, R) = c1 + c2M + c3M_2 + c4Ln(R + c5\ exp\ (c6M)) \tag{2}$$

where Y is the ground-motion parameter (peak acceleration or response spectral value), M is the magnitude, R is the distance measure, c3 = 0 but c6 ≠ 0. A single random variable method is used in the regression analysis in which the variable is $Ln\ Y$ and the coefficients of c1, c2, c4, c5, and c6 are determined by using the least root mean square method (Yu, 2013).

Due to the lack of seismic data in most parts of China, complex GMPEs with multiple parameters are difficult to apply. The simplified GMPE model above (Equation 2) has been widely used in earthquake engineering (Huo & Hu, 1992; Wang et al., 2013). The advantage of this model is that it not only has a simple form but also reflects magnitude saturation. The most typical application of this GMPE is calculating the hazard foot print for probabilistic seismic hazard assessment (PSHA) in the GB18306.

The *Seismic Ground Motion Parameters Zonation Map of China*, GB 18306, has been evolving and published in five versions (5th = 2015, 4th = 2001) so far. The latest one was released in 2015 and is called the 5th generation map, while the one released in 2001 is known as the

4th generation. The advances of the new GMPE in GB18306-2015 focus on the following four aspects (Yu, 2013),: (1) More abundant data is collected. The strong-motion observation network has developed rapidly during the last twenty years in China and numerous stations with reliable accelerometers have been built. Therefore, a great number of acceleration recordings were captured during large earthquakes. For instance, the 2008 Wenchuan Earthquake (Ms 8.0) produced more than 20,000 records in the main shock and aftershock. Plentiful seismic data can well reflect localized seismic characteristics and geological conditions. (2) A methodology representing ground motion saturation is used. The new GMPE's coefficients are obtained by fitting separately in two parts of the whole data with a threshold magnitude of M 6.5.It fully considers the distance-saturation and magnitude-saturation in a near field with large magnitudes.(3) A more robust step-regression approach is applied. Regressing the source and path parameters (M, R) relatively helps reduce the uncertainty within and between events. The results are also unbiased and stable (Xiao & Yu, 2010). (4) Anew attenuation zoning scheme is utilized. The 5th GMPE takes four zones according to the different levels of seismic activity compared to just two in the 4th. The more specific zoning can improve the accuracy of regionalized ground motion predictions.

Figure 2 shows the comparison of predicted ground motions on a rock site using GMPEs from different GB 18306 versions in northern China. For a large earthquake (M = 8), the predicted PGA using the 5th GMPE is slightly lower than the 4th GMPE near the source, while the differences become obvious when the distance is over 20km. For smaller earthquakes (M = 5, 6, 7), the predictions with the 5th GMPE are much higher withina20km distance and become a bit lower when the distance is over 50 km. One reasonable explanation is that the application of the saturation-considered model and abundant strong motion data in a near source for the new GMPE increases the PGA value (except for M = 8) in the near field but quickens the attenuation rate in the far field.

4 EXAMPLE ANALYSIS

Beijing is the capital of the People's Republic of China with more than 21.54 million inhabitants in 2018. It is the political and cultural center of the country, and a world-famous ancient capital and modern international city. Beijing ranges across the North China Plain seismic belt and Fenwei seismic belt. It lies on a medium-strong seismic hazard zone where earthquakes associated with different active seismic faults can generate massive damage and disruptions on its

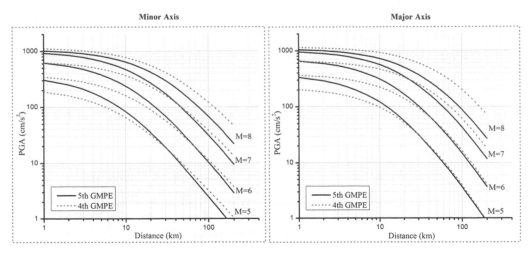

Figure 2. Comparison of predicted ground motionson rock site using GMPEs from different GB 18306 versions (5th = 2015, 4th = 2001) in northern China.

infrastructure. The urban areas of the city are divided into 16 districts (counties), each of them with important differences from social, economic and infrastructural perspectives.

As a megacity with national key fortification requirements, Beijing is extremely sensitive and vulnerable to earthquake disasters. Recently two earthquakes of magnitudes 2.9 and 3.0 respectively occurred on April 7th and 14th of 2019 successively in suburban Beijing. Since both earthquake epicenters were very close to the city center (with in 30 km) and the depth of the hypocenters was quite shallow (with in 20 km), numerous citizens said that they felt the shocks on WeChat and Weibo. Even though seismologists have claimed that these were normal seismic activity, seismic safety remains a cloud on the horizon for Beijing residents.

The Occurrence Exceedance Probability (OEP) is the probability that at least one loss exceeds the specified loss amount or rate. The OEP is the distribution of the largest loss in the period and is based on the theory of order statistics (Grossi & Kunreuther, 2005). Industry professionals commonly refer to the curves in making decisions regarding the risk of a given exposure. Using the well-developed business environment with massive office buildings in Beijing, we choose a typical office building of frame-shear wall structure as the input building class to calculate its OEP curve via the EQC at. The vulnerability curve of this building type was obtained from Zhanget al. (2018). Figure 3 shows the OEP curves of the Financial Street Center in the Xicheng District of Beijing with GMPEs from different GB 18306 versions (5th = 2015, 4th = 2001).The OEP with 5th GMPE is far greater than the other. For the return period of 500years, the loss ratio calculated with the 5th GMPE is 2.52%, which is nearly 3.7 times larger than the 0.68% with the 4th GMPE. The main reason is that earthquakes within the 500-year return period are usually smaller than magnitude 7.0. For these earthquakes, the predicted PGAs with the 5th GMPE are obvious larger than the values with the 4th GMPE (Figure 2). In addition, the number of small earthquakes is much higher than large earthquakes. The cumulative effect on amount and amplitude expands the difference.

The average annual loss is also a common statistic used when analyzing model results. The AAL represents the expected value of the aggregate loss distribution, or the premium needed to cover losses from a disaster over time. It's believed that the newest GMPE (GB18306-2015) represents a better, more reliable technique and available data. In order to estimate a more accurate AAL, we choose the 5th GMPE to calculate the distribution map of pure premium rates on a typical office building of frame-shear wall structure in Beijing (Figure 4). The color

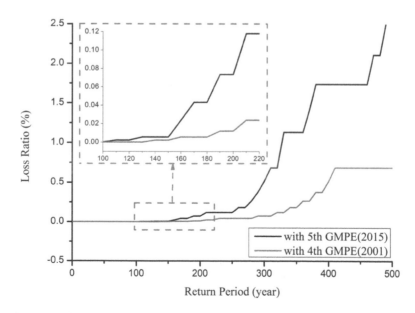

Figure 3.   Occurrence exceedance probability (OEP) curves of Financial Street Centerin Xicheng District, Beijing with GMPEs from differentGB18306versions (5th = 2015, 4th = 2001).

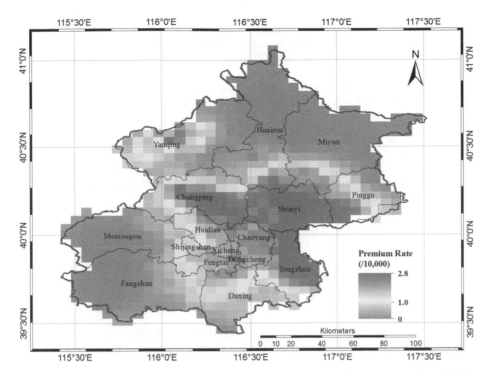

Figure 4. Pure premium of typical office building (frame-shear wall structure) with GMPE from GB18306-2015.

map shows that the red areas mean higher premium rates while the green zone represents lower rates. In other words, the changes in the color map from green to dark red expresses the seismic risk level. Insurers and reinsurers can benefit from the zone map of pure premiums to identify the earthquake risk of portfolio loss.

Most areas of Tongzhou, Shunyi and Changping districts are located in the high premium rate zone. The major urban districts consisting of Dongcheng, Xicheng, Haidian, Chaoyang, Fengtai and Shijingshan are at moderate premium rate. Most of the rural districts are at a relatively lower rate except for the northwest of Yanqing. As the Beijing government offices are moving from Dongcheng to Tongzhou, the premium rates of new office buildings turn into a higher one. it seems that the Beijing government is confronted with allocating more money to the new office building sin order to manage the increasing seismic risk. Whether the money is used for paying the higher premium rate or strengthening the buildings, the potential earthquake disaster in Tongzhou should not be ignored.

5   CONCLUSION AND DISCUSSION

A catastrophe model based on physical mechanisms helps insurers and reinsurers to better understand the seismic risk by providing outputs like ELT, OEP, AAL, etc. It is believed that a complete catastrophe model should contain the functions of estimating the loss for buildings/structures, contents and time element (business interruption). This is the direction that the EQC at strives to accomplish.

GMPEs have a great influence on the distribution areas of PGA, thereby affecting the amount and damage extent of exposures. The OEP of a single point is sensitive to the differences between the GMPEs, even though they have a similar form. In the case of bigger coverage the differences tend to be smaller. As more strong-motion data become available, GMPEs can be made region specific to reflect differences in the earthquake source or differences in the site categories. The regional GMPEs can effectively reduce the total uncertainty in a typical risk analysis which is commonly dominated by the uncertainty associated with ground-motion prediction.

The pure premium rate ranges between 0.01 to 2.8 per 10,000 for a typical office building with frame-shear wall structure in Beijing. The new government offices in Tongzhou are predicted to lie in a place with a relatively higher seismic risk compared to the core urban areas.

ACKNOWLEDGMENTS

This project was supported by the National Key R & D Program of China (2017YFC1500400, 2018YFC1504600), the Special Fund of Institute of Geophysics, China Earthquake Administration (DQJB17C03, DQJB17C05, DQJB17C06).

REFERENCES

Franco, G. 2012. Optimal Selection of Simulated Years of Catastrophe Activity for Improved Efficiency in Insurance Risk Management. *In 2012 IEEE Conference on Computational Intelligence for Financial Engineering & Economics (CIFEr)* 1–7.

Grossi, P. & Kunreuther, H. 2005. Catastrophe Modelling: A New Approach to Managing Risk. New York: Springer.

Nakada, P., Shah, H., Koyluoglu, H.U., & Collignon, O. 1999. P&C RAROC: A Catalyst for Improved Capital Management in the Property and Casualty Insurance Industry. *The Journal of Risk Finance* 1 (1): 52–69.

Porter, K.A. 2003. An Overview of PEER's Performance-Based Earthquake Engineering Methodology. 9th international conference on applications of probability and statistics in engineering, San Francisco, CA.

Calvi, G.M., Sullivan, T.J., & Welch, D.P. 2014. *A Seismic Performance Classification Framework to Provide Increased Seismic Resilience* 361–400.

Yoshikawa, H., Goda, K. 2014. Financial Seismic Risk Analysis of Building Portfolios. *Natural Hazards Review* 15(2):112–120.

Winspear, N., Musulin, R., & Sharma, M. 2012. Earthquake catastrophe models in disaster response planning, risk mitigation and financing in developing countries in Asia. *Geological Society, London, Special Publication*, 361(1): 139–150.

Dong, W.M. 2002. Engineering Models for Catastrophe Risk and Their Application to Insurance. *Earthquake Engineering and Engineering Vibration*, 1(1): 145–151.

Xu, W.J., Gao, M.T, & Zuo, H.Q. 2018. *Simulation of Seismic Catalog for Earthquake Catastrophe Model Based on the Monte Carlo Method*. The 15th International Symposium on Structural Engineering (ISSE-15), Hangzhou, China.

Wang, Y.S., Li, X.J., & Zhou, Z.H. 2013. Research on attenuation relationships for horizontal strong ground motions in Sichuan-Yunnan region. *Acta SeismologicaSinica* 35(2): 238–249. (In Chinese).

Xiao, L., & Yu Y.X. 2010. A new step regression approach for fitting groundmotion data with attenuation relation. *Acta SeismologicaSinica* 32(6): 725–732. (In Chinese).

Huo, J.R., & Hu Y.X. 1992. Study on attenuation laws of ground motion parameters. *Earthquake Engineering and Engineering Vibration* 12(2): 1–11. (In Chinese).

Wang, S.Y., Yu Y.X., Gao, A.J., & Yan, X.J. 2000. Development of Attenuation Relations for Ground Motion in China. *Earthquake Research in China* 16(2):99–106. (In Chinese).

Yu, Y.X., Li, S.Y., & Xiao, L. 2013. Development of Ground Motion Attenuation Relations for the New Seismic Hazard Map of China. *Technology for Earthquake Disaster Prevention* 8(1): 24–33. (In Chinese).

Wang, Y.S., Li, X.J., & Zhou, Z.H. 2013. Research on attenuation relationships for horizontal strong ground motions in Sichuan-Yunnan region. *Acta Seismologica Sinica* 35(2): 238–249. (In Chinese).

Stafford, P.J. 2013. Uncertainties in ground motion prediction in probabilistic seismic hazard analysis (PSHA) of civil infrastructure. *Handbook of Seismic Risk Analysis and Management of Civil Infrastructure Systems*, 29–56.

Zhang, L.X., Li, M.D., Liu, J.P., & Zhu, B.J. 2018. Seismic fragility analysis of frame-shear wall structures considering structural parameter uncertainty. *Journal of Natural Disasters*, 27(04): 112–118. (In Chinese).

Risk Analysis Based on Data and Crisis Response Beyond Knowledge – Huang & Nivolianitou (eds)
© 2020 Taylor & Francis Group, London, ISBN 978-0-367-25146-8

Mass notification for public safety: Current status and technical challenges

Christodoulos Asiminidis* & Sotirios Kontogiannis
Laboratory Team of Distributed Microcomputer Systems, Department of Mathematics, University of Ioannina, Ioannina, Greece

George Kokkonis
Department of Business Administration, Technological Educational Institute of Western Macedonia, Kastoria, Greece

Nikolaos Zinas
Tekmon LTD, Ioannina, Greece

Charis Papadopoulos
Department of Mathematics, University of Ioannina, Ioannina, Greece

ABSTRACT: Critical and emergency incidents often lead to asset, infrastructure and human loss, all major concerns for companies, organizations and governments. Early prediction, emergency management plans and immediate dissemination of information can mitigate such events from occurring. To this end, early warning systems have been deployed. Early warning systems comprise of sensors, event detection and decision subsystems that work together to forecast the occurrence of possible incidents. These forecasts are then processed by a mass notification system for the dissemination of public alert guidelines and coordinated emergency group actions. While these systems reduce risks and enable authorities' coordination, their deployment often fails to offer timely alerts specifically in case of incidents affecting a massive number of participants. This study presents existing notification systems, their minimum requirements and technical challenges for large scale deployments. Furthermore it proposes a high level architecture and required features for a mass notification system which focuses on public safety and protection of life and property on a large scale.

Keywords: Incident response, early warning systems, mass notification systems, systems evaluation, public safety

1   INTRODUCTION

The increasing number of the emergency events, in conjunction with the innovation and diffusion of technology has led to the development of notification systems to alert groups of people for a critical incident call that might put themselves and/or their own property in danger. Notification Systems are divided into four categories: The first one is natural disasters such as earthquakes, fires, critical climate change, storms, tsunamis etc. The second one is critical infrastructures which may include airports, stadiums, highways, social events etc. The third one involves buildings and neighbors' safety systems, which are notification systems can be installed into establishments for thief, burglary prevention, fire incidents confrontation or common resource monitoring such as

---

*Corresponding author: chasiminidis@cs.uoi.gr

natural gas. The fourth and the last one is an hybrid category that focuses on several targeted resources and include Incidents of Compromise configuration capabilities, risk assessment categories setup, risk management indicators and incident learning capabilities. That is AI capable algorithms which take feedback from the incidents dissemination process in order to automate or augment the issue of alerts or offer suggestions via bot services.

2 RELATED WORK

There is several systems emergency notification software available at the market that covers the needs of the previous section categories. In this related work the authors focus on some Incident response systems that favor public safety and the different capabilities offered by these systems.

Sikder et al. have developed an application called Smart Disaster Notification System which its main purpose is to alert people before the natural disaster and give information about the shortest path to the closest shelter. Their application, based on the user's locations, receives data about the weather condition. Then, the system estimates the probability of the disaster and sends back a smart disaster notification which depends on the weather data received. The application works very well for the users who have subscribed by giving their name, mobile number and region. People who have not registered will not get notified. Disadvantageous is also the fact that the user must carry an Android smart phone.

Malizia et al. suggested an emergency notification system for a natural disaster which interacts non-conventionally and it is tailored to vulnerable groups of people such as the ones who are disabled, arthritis and limited motor capacities, deaf and blind. Particularly, vulnerable groups should be subscribed to an Emergency Notifications System so they can get the mobile device, e.g. Personal Digital Assistant depending on the user's ability. Specifically, the Simple Emergency Alerts fo[u]r All ontology they built into the mobile device receives alerts in case of emergency by vibrating which is for all the groups of people, but also can give information to the user with eye moving using a camera especially for the disabled people and the ones who are able to see.

Hossain et al. developed a crowd sourced framework for neighbor assisted medical emergency system. The system they presented called Neighbor Assisted Medical Emergency System (NAMES) has as its main purpose to reduce the response time of the ambulance to the emergency event and medical operational cost. The patient can either call an emergency number or from the pendent/wrist belt and notification is sent to the hospital. At that point, the system examines the nature of the emergency. In the worst case, that the emergency is severe then the system directly calls the Medical Emergency Team and then the system finished all the procedures. In any other case, the closest neighbors are called and then the system waits for their response.

Call-Em-All has developed an Emergency notification system that uses mass texting XMPP protocols for mass texting, text to speech conversion capabilities as well as voice broadcasting for up to 10,000 contacts instantly. Call-Em-All does not require a client mobile phone application to operate. Informcast has created alert software for mass notifications using SMS, e-mail, calls and custom application dialogs, custom mobile phone applications for organizations and companies, as well as panic buttons and IoT integration.

Fireeye has released MIR 2.0 a web based incident response system, a highly configurable system with Indicators of Compromise (IOC), IOC scanning and incident analysis, dissemination and management via a series of wizard steps. This system offer voice call and e-mail capabilities towards organization groups and thus may extend to a mass notification use.

Finally, Tekmon offers a secure web-based mass notification system with automated IOC configuration that uses SMS, voice calls and e-mail capabilities. Tekmon's system ensures immediate personalized notification and does not require a mobile phone application. The system's advanced escalation logic ensures that all recipients will receive the alert gradually by changing its transmission technologies if the alert is not received.

3 MASS NOTIFICATION FRAMEWORK

A mass notification system is a system that propagates incident alerts to the masses. A mass notification process is part of different systems cooperation. This process is illustrated at Figure 1. At first an Incident response system is required that monitors different types of indicators. Then an adaptation or risk-assessment policy or even a filtering process will lead to metrics or key point indicators than in turn will be used as input on a risk assessment methodology, plan or process. The methodology, adaptation or automation of this process is usually supervised by the risk partners or stakeholders.

The final stakeholders decision according to the assessment plan will activate the Mass notification system (if autonomously implemented) or the Mass notification module (if part of the Incident response system), which in turn will initiate the process of alerts dissemination to the public. Incident response alerts and notifications data transmissions through organization groups involve MQTT, XMPP or custom TCP - UTP based services and protocols and client mobile applications. However, in Mass Notification systems for public safety purposes the use of custom services and dedicated applications shadows the risk of not being notified, due to digital divide or technological discrepancies and are utilizing alert channels of well-known services: SMS, voice message and email, passing through telecommunication aggregators.

The alerts flow of a Mass notification framework is as follows: Sensors transmit data via a concentrator to a cloud data logging service. Incident response logic enforces data classification, sub-categorizes the incidents and adapts to the appropriate response according to assessment planning. Then, with the use of MQTT, XMPP or TCP/UDP protocols the appropriate organization's manager or stakeholders get notified about each event. Stakeholders come across with a decision if they are going to inflict an alert to a group of people if they or their property is at stake. If so, they send the corresponding alerts via HTTP protocol to the corresponding Telekom subsystem. The Telekom subsystems in turn are passing the alert messages to people throughout the communication providers' infrastructure, either as SMS, or as voice messages or as e-mail.

There are two major types of Mass notification system-Telecom provider policies: In the first policy the Telekom provider has minimum queuing capabilities but acknowledges to the Mass notification service if the user has successfully received the alert, allowing the use of retrans-

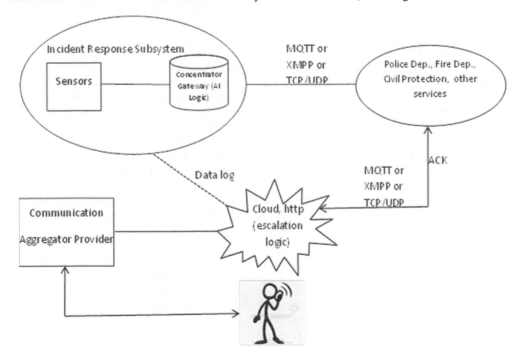

Figure 1.   Mass notification framework for systems implementing public safety alert systems.

missions and escalation logic and in the second policy the provider batch stores the requests and then is alone responsible for the alerts delivery offering 99.9% SLA's for batch storage only, but no guaranties for end user deliveries. In the first case, the mass notification subsystem may use an escalation logic technique by which if the alert message is not received it can be retransmitted or forced into another transmission type. This of course requires extra timeout time in order to guarantee alerts reception.

In this paper, authors are examining the response capabilities of a mass notification subsystem for public safety purposes using the three different alerting mechanisms through SLAs with a communication aggregator provider that offers SMS, voice message and e-mail deliveries. Considering the wildfires in the Attica region, Athens that happened in July 2018, authors are also taking into consideration the parameters in order to send a massive number of messages in less than 45 minutes as a considerable maximum amount of time needed.

Authors' experimentation with the use of simulation tools tries to indicate the mean total time required for a mass notification service that issues an alert involving public safety. In the experimental scenarios I and II, the mass notification system, placed on the cloud, communicates (Figure 1) to one or several Telecommunication aggregator services via HTTP in order to give out user personal information (email or telephone number) and the alert content respectively. This simulation is divided into two parts: Scenario I involves a number of HTTP requests from the Mass notification service to the Telecommunication providers' web interfaces. In this scenario an estimation of the time required for all this information to be propagated is calculated. Scenario II involves the total time estimation required for one Telecommunication aggregator to successfully serve using only e-mail or SMS or voice calls (with no escalation logic) respectively, a vast number of alerts out to the public. This estimation has been focused on information given from the Telecom aggregators in Greece and the contract services-guarantees they can offer to third party companies.

3.1 Alerts time estimation from the MNS to the Telecom aggregators

In scenario I, authors used OPNET to simulate http traffic from a ip32 cloud source to the aggregator service using http requests. The link capacity between the cloud client sending the http requests and the aggregator service was set to 10Mbit/s with network latency 45-70ms, network jitter 5-30ms and minimum packet loss. Each HTTP request has been processed by the Telecom aggregator service and put to the aggregator queue with an average processing delay per request of 150ms-250ms. Each HTTP request has been processed as part of a batch by the Telekom aggregator and an appropriate HTTP PUSH ACK message notification is sent back to the Mass Notification Service (MNS). In case of request or socket timeout another HTTP request can be issued with the same payload, until all HTTP requests are acknowledged (SLA close to 99,9%). Authors assume that the Telekom aggregator interface is similar to the ones such as: Twilio or Tropo. Each HTTP request payload contains a total of 16 users in JSON encoding and the user fields are ID, Type and short alert text (total of 64-128Bytes of payload per user).

Figure 2 below presents the average estimated total time in minutes depending on four different types of server architectures applied at both mass notification and Telekom aggregator service sides (parallel service threads running concurrently 1 thread/core-1 request/core).

This means that for a maximum service utilization of 64 parallel HTTP requests, the total time required for all alert messages to be queued at the Telecom provider side is less than 1minute, for

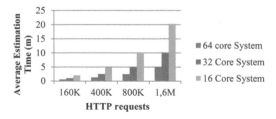

Figure 2.   Average estimation time (m)/HTTP requests.

214

Table 1. Estimated total time for sending and contracting capacity using SMS and voice calls.

No. of Alerts	Estimated time SMS delivery (min)	Estimated time voice calls delivery (min)
160K	25 min	40 min
400K	62.5 min	100min
800K	125 min	200min
1,6M	250 min	400min

160K alerts and 5minutes for 1,6Million alerts. Total time estimation of Telekom provider SMS, e-mail or voice call alerts propagation to the end users follows.

3.2 *Time estimation using e-mail, SMS and voice calls respectively*

In this scenario, authors examined the maximum capacity that the Telekom aggregator can send and the maximum SLA contract the aggregator can offer to third parties in Greece for E- mails, SMS and Voice calls. The maximum capacity for e-mails per communication aggregator provider depends on the number of static IPs that the third party uses (hires from the Telekom aggregator). The Telecom aggregator maximum capacity is 10.000 E-mails/min, however, this number cannot be reached by third party SLA's. That is, for a number of 4 hired static IPs the maximum e-mail throughput that can be achieved depends on e-mail size. For e-mails of minimum content (less than 512Bytes) the maximum throughput is no more than 2100/min. Nevertheless, if 16-24 static IP's are hired then MNS can reach up to the 10.000 e-mails/min throughput, however the cost of such an IP pool is prohibitive.

In the case of e-mails, the maximum Telecom aggregator throughput is used. For 160K, 400K, 800K and 1,6M the estimated time for one aggregator service to send 16 minutes, 40 minutes, 80 minutes and160 minutes accordingly.

Regarding SMS, the maximum capacity of the SMS that an aggregator in Greece can contract is 12000 SMS/min, if the SMS number if alphanumeric. However, in case of numeric numbers in the range of 697-699, the maximum capacity that can contract is 1800 SMS/min. Using as a third party company both the four Telekom providers in Greece with an SLA agreement at their maximum offered SMS SLA capacities the total of 6400 SMS/min can be reached, the estimated time for SMS delivery time using both the four Telecom aggregators in minutes is presented at Table 1.

Specifically, for 160K, it takes 25 minutes for the total estimated SMS sending time. For 400K, it requires 62.5 minutes (more than the limit of 45min set by the authors). This means that the current infrastructure in Greece cannot sustain a Mass notification service via SMS of more than 300K SMS alerts.

The maximum capacity of the Voice Calls that the aggregator can send is more than 15,000Voice Calls/min/cell (In case of concerts or events with multiple antenna coverage points), however, the maximum capacity that can offer SLA's to third parties is 160 Voice Calls/min per two phone numbers (dedicated circuits). Table 1 shows the total estimated time for voice calls. In this scenario authors used an hypothetic aggregating set of providers offering SLA agreements for a total of 25 phone numbers, sending an average of 4000 Voice calls/min. In particular, for 160K voice calls, the aggregator needs 40 minutes to send all of the voices call alerts. For 400K, the estimated total time is close to 100minutes. That is, the limit for sending voice calls in the expected interval of 45minutes is set by authors to 180K voice call alerts.

4 CONCLUSIONS

In this paper, authors present existing incident response systems with embedded mass notification capabilities and analyze the framework and interconnection between incident response and mass notification on a technological ground. Authors classified and described the architecture of a mass notification system and tested, through simulations, the estimated total time for sending critical events out to the public.

According to the authors simulations and study based on data from Greece, only for a number of 180,000 voice call alerts existing Telekom infrastructure can offer SLA services, since it reaches the maximum hypothetical time of imminent mass notification set by the authors at 45 minutes. For the same interval Telekom infrastructure can deliver up to 300,000 SMS alerts. E-mails from the other hand offer the fastest way of notification, despite the fact that the e-mail service is not designed for interactive and real-time response.

As a main conclusion authors pinpoint that the classical mass notification ways for public no-tice must be circumvent with better performing TCP/IP or UDP based protocols as: MQTT, XMPP or even CoAP, already used by the incident response industry, social media and the IoT. This of course requires further public training and educational efforts towards the public in order to confront digital divide phenomena.

ACKNOWLEDGEMENTS

This research has been co-financed by the European Union and Greek national funds through the Operational Program Competitiveness, Entrepreneurship and Innovation, under the call RESEARCH – CREATE – INNOVATE (project code: T1EDK-02374).

REFERENCES

CISCO Inc. 2013. *Tropo Framework and interfaces for speech recognition, text to speech, voice and text messaging.* Available at: https://www.tropo.com/ (accessed December 1, 2018).

Fielding R., Gettys J., Mogul J., Frystyk H., Masinter L., Leach P. and Berners-Lee T. 1999. Hypertext transfer protocol–http/1.1. *RFC Editor, United States,* DOI: 10.17487/RFC2616

FireEye Company 2011 *MIR 2.0 Incident response system.* Available at: https://www.fireeye.com (accessed November 25, 2018).

Gounopoulos E., Kokkonis G., Valsamidis S. and Kontogiannis S. 2018. Digital Divide in Greece-A quantitive examination of Internet non-use *Springer Procedia in Business and Economics,* Economy, Finance and Business in Southeastern *and Central* Europe, *Springer* ISBN 978-3-319-70376-3, pp.889–903.

Herrmann, Nguyen, Barclay, Call-Em-All 2005 Emergency Notification System. Available at: https://www.call-em-all.com/ (accessed October 28, 2018).

Hossain A., Mirza F., AsifNaeem A. and Gutierrez J. 2017. A crowd sourced framework for neighbor assisted medical emergency system *27th International Telecommunication Networks and Applications Conference (ITNAC),* DOI: 10.1109/ATNAC.2017.8215436

Kokkonis G., Chatzimparmpas A. and Kontogiannis S. 2018, Middleware IoT protocols per-formance evaluation for carrying out clustered data. *In Proc of 3rd SEEDA-CECNSM conference, Kastoria Greece, IEEE,* DOI: 10.23919/SEEDA-CECNSM.2018.8544939

Malizia A., Onorati T., Bellucci A., Diaz P. and Aedo I. 2009. Interactive Accessible Notifications for Emergency Notification Systems *5th International Conference on Universal Access in Human-Computer Interaction,* DOI: 10.1007/978-3-642-02713-0_41

MQTT.org. 2013. *Message Queuing telemetry transport.* Available at: http://mqtt.org/ (accessed November 28, 2018).

Riverbed Technology. 2013. *OPNET-Riverbed IT Guru Academic edition Network simulator-modeler.* Available at: https://www.riverbed.com/gb/products/steelcentral/opnet.html (accessed November 10, 2018).

Sikder F., Halder S., Hasan T., Uddin J. and Baowaly M. 2017. Smart Disaster Notification System *Proceedings of the 2017 4th International Conference on Advances in Electrical* Engineering *(ICAEE),* DOI: 10.1109/ICAEE.2017.8255438

SingleWireCompany Software. 2012. *Informacast Mass notification system.* Available at: https://www.sin glewire.com/informacast (accessed December 5 2018).

TwilioInc. 2014. *Framework and cloud communications platform for building SMS, Voice and messaging applications, video and authentication.* Available at: https://www.twilio.com/ (accessed December 1, 2018).

Xin J. and Huang C. 2013. Fire risk analysis of residential buildings based on scenario clusters and its application in fire risk management *Fire Safety Journal* 62(1): 72–78.

XMPP.org 2013. Available at: https://xmpp.org/ (accessed December 7 2018).

Zinas N. 2014. *Tactical Mass Notification System.* Available at: https://www.tekmon.gr/ (accessed January 5, 2019).

3-D simulation and visualization of diversion flood risk in small and medium-sized flood storage polder: A case study of Linan flood storage polder in Dongting Lake area, China

Dehua Mao
College of Resources and Environmental Sciences, Hunan Normal University, Changsha, China

Chang Feng*
College of City and Tourism, Hengyang Normal University, Hengyang, China

Ruizhi Guo
College of Mathematics and Statistics, Hunan Normal University, Changsha, China

ABSTRACT: A system was developed for 3-D dynamic simulation and visualization of diversion flood routing in the Linan flood storage polder. First, C# was used to program diversion flood routing system for the Linan flood storage polder in the Visual Studio platform, which is based on the ArcGIS engine. A given water level inundation programming thought were proposed to simulate diversion flood routing, considering the specificity of the flood storage polder's diversion flood. Second, the inundated simulation results were compared to numerical simulation results in similar studies. These results show that this system can not only reasonably simulate diversion flood routing in the Linan flood storage polder, but also dynamically and visually demonstrate the inundation process and self-defined flood elements, such as water level, inundated range and water volume at any given diversion flood time. Finally, the dynamic flood risk assessment can be realized by coupling these dynamic flood elements with the socio-economic and resource distribution of the flood storage polder. This research shows that the thought of programming is more suitable for small- and medium-sized flood storage polders, which provides a scientific basis for the flood control operation and risk management.

Keywords: ArcGIS engine, 3-D dynamic simulation and visualization, flood routing, diversion flood risk, Linan flood storage polder, secondary development

1 INTRODUCTION

Flood storage areas in the Dongting Lake are an important part of the flood control project in the middle and lower Yangtze River. According to the extraordinary flood emergency plan in the Yangtze River and the flood control plan in the Three Gorges reservoir, flood storage areas of Dongting Lake are used to store excess water to ensure the safety of key towns and traffic trunks when the Yangtze River and the Dongting Lake area is flooding simultaneously. At home and abroad, most of the research on flood simulation is for floods in cities and rivers (Jiang et al., 2009) based on 1-D and 2-D simulations (Meire et al., 2010; Shahapure et al., 2010). There are few studies on 3-D simulation (Li et al., 2005) of diversion floods in flood storage areas. In this paper, the Linan flood storage polder, one of the 24 flood storage areas in the Dongting Lake area of China, was used as the research area to simulate the diversion flood storage and diversion flood process, and then the water level, inundated range and

*Corresponding author: fengchang8802@163.com

water volume were analyzed. This provides a scientific basis for diversion flood control and planning in flood storage areas. Consequently, a system for 3-D simulation and visualization of diversion floods in the Linan flood storage polder was developed using the C# language under the ArcGIS engine and the Visual Studio platform.

2 RESEARCH AREA

The Linan flood storage polder is located in the lower reaches of the Li River, in the Dongting Lake area. It is built on the west side of the mountain and on the other three sides is surrounded by water, and its elevation is roughly between 35.8 and 38.2 m. The first-line flood-wall is 24.4 km (10.5 km in the north of Li River and 13.9 km in the southeast of Dao River). The elevation at the top of the floodwall is between 45.7 and 48.3 m, and the surface width is 3.5 to 8 m (Mao, 2009).

Moreover, there is a flood gate in the Linan flood storage polder, which has 9 holes, and is 113 m long and 38.66 m high at the bottom. The designed flood level is 45.67 m (at Liujiahe Hydrologic Station), the designed flood storage capacity is 200 million m^3, and the designed flood process is divided into 24 hours. To reduce the flood peak, the operation of the flood gate must be strictly obeyed (Water Resources department in Hunan province). Therefore, the process and mode of opening the gate must be strictly implemented according to the dispatching and operation plan of the flood control gate of the Linan flood storage polder (2005).

3 PRE-PROCESSING OF IMAGE DATA

3.1 *Processing of image layer data of the Linan flood storage polder*

The system of 3-D simulation and visualization of diversion flooding in the Linan flood storage polder is built using the ArcGIS Engine's "Analyst 3D", "Carto" and other program packages. Using the main controls such as "axSceneControl" and "axTOCControl", the methods and properties of the ArcGIS engine's interface are used to implement some functions in ArcScene. The core of the Analyst 3D expansion module is the ArcScene application, which is designed to efficiently manage 3D data and perform 3D analysis (Qiu, 2010) and mainly uses Tin data for 3D display. According to the map data of the Linan flood storage polder, the method of creating a Tin with vector data is used to generate the Tin surface. First, we imported the "1:1 million" database data (*.dwg) into Arcmap. Secondly, the conversion program "CreateTinfromFeatures" under the 3D Analyst tool and its appropriate "Feature" was selected to generate the Tin surface of the Linan flood storage polder. At that point, the effect of the 3D presentation was achieved in the system.

In addition, under the 3-D inundated simulation of a given water level programming idea, another image layer is needed to dynamically demonstrate flooding by the superimpose method. Accordingly, we have created a DEM layer called the "water depth layer" and have selected the method of creating a raster surface from the Tin to create a raster layer according to the obtained Tin map of the Linan flood storage polder. Hence, the specific process is listed as follows: we used "Convert Tin to Raster" in 3D Analyst to convert the data, which converted the elevation field of the Tin layer to the raster layer, and then we selected the cell size "5" and the elevation factor (z factor) "1" to convert the raster layer.

3.2 *Processing of remote sensing image data*

The calibration of the remote sensing image in the Linan flood storage polder superimposes the Tin data for the 3D display. The processing of remote sensing image data generally includes the projected coordinate system, the geographic coordinate system, the rotational transfer of the projection center, and the calibration of coordinates. Accordingly, the remote sensing image projection, the transformation of the geographic coordinate system, and the calibration of the projection center or coordinates can be performed by referring to the Tin map of the Linan flood storage polder. First, we used the "Define Projection" in the "Data Management tools" to perform remote sensing image transformation and transform the coordinate system through the Tin map of the Linan flood storage polder generated by "Coordinate System import from". Second, we used the "Georeferencing" tool to set up six "Control points" and the Link Table to calibrate the projection center and coordinates.

4 THE FUNCTION OF THE SYSTEM OF 3-D SIMULATION AND VISUALIZATION OF DIVERSION FLOODING IN THE LINAN FLOOD STORAGE POLDER

The system for 3-D simulation and visualization of diversion flooding in the Linan flood storage polder includes basic functions (map browsing and map loading, etc.), image data rendering (elevation conversion and color rendering, etc.), diversion flooding simulation, diversion flooding analysis, and the interface shown in Figure 1. In the system's interface, the "Scene-Control" is used as the main display control, the "Arcscene" is used to 3D display in ArcGIS, the "TOCControl" and "ToolBarControl" are used as the main view browsing and display control, and this program code is written in C#.

Tin R1 = Tin layer rendering; Tin R2 = Tin monochrome rendering; Tin conversion = Tin elevation conversion; RS image overlay Tin = Remote sensing image overlay Tin data; Water depth render = Raster water depth rendering; Flood simulation = Flood storage simulation; Flood area = Flood submerged area; Flood volume = Flood submerged volume; Stop S = Stop simulation.

4.1 Image data processing and rendering – Part 1

(1) Setting the elevation conversion factor. The elevation conversion coefficient is a constant to convert the unit of the horizontal coordinate system into the elevation coordinate system. The terrain of the Linan flood storage polder is flat, and the elevation difference is not obvious, thereby it is necessary to convert the elevation factor of the Tin map for 3-D display. This system sets the elevation conversion factor for the Linan flood storage polder by "I3Dproperties", "IlayerExtensions", "Apply3Dproperties" and other interfaces.

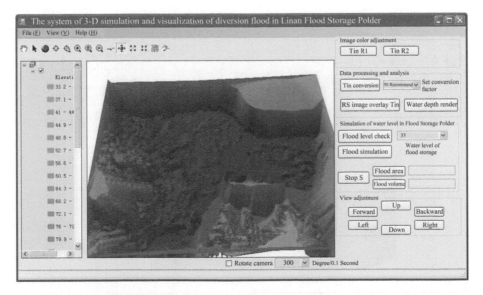

Figure 1. The system interface of 3-D simulation and visualization of diversion flood in Linan flood storage polder.

(2) Image color rendering. When ArcScene displays image data, it is always displayed in a fixed color. In order to reflect the surface features and elevation, color rendering is required. This system has designed the Tin layer rendering and Tin monochrome rendering where color rendering uses the RGB color system.

(i) Tin monochrome rendering uses methods and properties in the interfaces of the "Itin-Renderer", "ItinLayer", "ItinSingleSymbolRenderer and "IsimpleFill Symbol" to build a function "TinRenderer" to insert the rendering model. Then this system uses the "Itin-Layer", "Icolor" interface to build the "getrgb" function and to set the image color.

(ii) Tin layer rendering uses methods and properties of the "ItinLayer", "ItinRenderer", "ItinAdvanced", "ItinColorRampRenderer", "IalgorithmicColorRamp" and "Isimple-Fill Symbol" to construct a function "TinGradient Renderer" that inserts the layered rendering model. Then this system uses "ItinLayer", "Icolor" (instantiate "Fromco-lor", "Tocolor" respectively) to render Tin layer.

(3) Superimpose Tin map and remote sensing data. The remote sensing data can truly reflect things on the surface, while the Tin map can display the 3-D elevation of the surface. Therefore, the superposition of both is able to create virtual reality, and the higher the resolution of the remote sensing data, the more realistic to reflect the features in the ground. This system mainly displays the surface elevation by the superimposition of the Tin map and remote sensing image data, which uses methods and properties of "Itin-Layer", "IrasterLayer", "ItinAdvanced", "Isurface", "IlayerExtensions", "I3Dproperties" and other interfaces.

4.2 *The 3-D simulation and visualization of diversion floods in the Linan flood storage polder – Part 2*

The 3-D simulation and visualization of diversion flooding is the core of the whole system (Figure 2 and 3). The diversion flooding simulation is based on a given water level (Feng, 2013). Therefore, it mainly simulates the entire flood storage process of the Linan flood storage polder after opening the flood diversion gate to reach the guaranteed water level.

The code programming of the flood submerged simulation by a given water level was selected after studying the topographical features and flood process of the Linan flood storage polder. The specific considerations are listed as follows. First, the hydrodynamic basis of submerged diversion flooding is the water level difference between flood diversion area

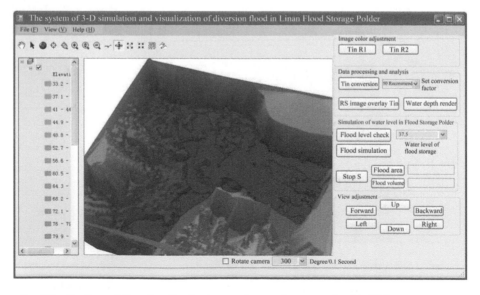

Figure 2.   Visualization of diversion flood submerged area when water level is 37.5.

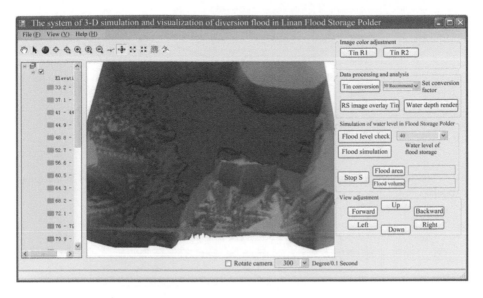

Figure 3.   Visualization of diversion flood submerged area when water level is 40.

and flood source area, and the balance of water level between both is the inevitable result of flood diversion (Feng et al., 2013). Second, the terrain of the Linan flood storage polder is relatively flat and simple (the ground elevation is 36.0~39.0m), the bottom of this flood storage polder is 33.2m, and the difference between the guaranteed water level at the time of the flood storage is 13.1 m. Third, there is a diversion flood gate in the Linan flood storage polder, and the time and process of opening the gate are strictly implemented according to the dispatching and operation plan of the flood control gate of the Linan flood storage polder (2005). The 9-hole gate will open step-by-step at 0, 8, 10, 12, 14, and 16 hours, and flood flow through the gate will increase from 433 m^3/s to 2551 m^3/s. Therefore, the designed diversion flood process takes 24 hours, which is obviously different from the hydrodynamic characteristics of sudden dam break (Luka et al., 2007). Fourth, the fitting of the water level of flood simulation to the measured water level of actual flood is proportional to the time of the diversion flood process. According to the 10 sections of the 1-D flow model established by the relevant research (Hunan Hydropower Survey and Design Research Institute), the flood is sumbergd to the Linan flood storage polder after 4 hours, and the water level is the same throughout, with the average water depth around 0.397 m. We compared the results of the system simulation with the water level data of the dispatching and operation plan of the flood control gate of the Linan flood storage polder (2005) and the fit was good.

In summary, it is possible to obtain flood data, such as the flood area, flood storage volume and water level using the code programming of the flood submerged simulation by a given water level. This method has more stability in a small- and medium-sized flood storage.

4.3   *Flood submerged analysis – Part 3*

The flood submerged analysis of the system of 3-D simulation and visualization of diversion flood in the Linan flood storage polder, can find the submerged area and submerged volume of the diversion flood by input of water level. Therefore, this study has analyzed the relationship between the submerged volume and submerged elevation, and the submerged area and submerged elevation, according to the simulation results (Figure 4 and 5). The flood submerged analysis part mainly uses the interfaces "ItinAdvanced" and "Isurface".

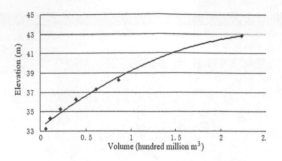

Figure 4.   The submerged Volume and submerged elevation graph of diversion flood in the Linan flood storage polder.

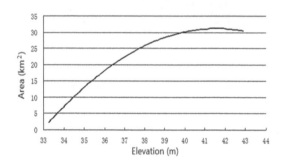

Figure 5.   The submerged area and submerged elevation graph of diversion flood in the Linan flood storage polder.

5   CONCLUSION AND DISCUSSION

The research uses C# to establish the system of 3-D simulation and visualization of diversion flooding in the Linan flood storage polder by combination of the ArcGIS engine and Visual Studio software. Given the characteristics of the diversion flood, a code programming method of a given water level was selected. The results of the system show that this code programming method is adapted to the small- and medium-sized flood storage polder, which can provide a reference for flood control planning and diversion flood simulation in the small- and medium-sized flood storage polder, such as Linan. However, in areas where the river slope is large (e.g. the middle and upper reaches of the Yangtze River, China), this simulation method of a given water level may cause distortion, and further research is needed for subsequent research.

ACKNOWLEDGMENTS

This project was supported by the Research Project of Hunan Provincial Water Resources Department (Grant No. XSKJ 2018179-09) and First-class Disciplines (Geography) in Hunan Normal University (Grant No.810006-1213).

REFERENCES

Jiang, J., Wu, L.D., Xu, J.B. 2009. Digital earth-based flood routing model visualization. *Computer Engineering and Applications* 45(36): 1–4.
Shahapure, S.S., Eldho, T.I., Rao, E.P. 2010. Coastal Urban Flood Simulation Using FEM, GIS and Remote Sensing. *Water Resource Management* 24(13): 3615–3640.

Meire, D., Doncker, L.F., Declercq, F. 2010. Modelling river-floodplain interaction during flood propagation. *Nature Hazards* 55(1): 111–121.

Li, Y., Fan Z.W., Wu S.Q., Wu X.F. 2005. Numerical simulation and 3-D visualization of flood propagation in large-scale detention basins. *Journal of Hydraulic Engineering* 36(10): 1158–1164.

Mao, D.H. 2009. *Theoretical Methods and Applied Research on Comprehensive Risk Analysis of Flood Disasters*. China Water Power Press: China: Beijing.

Qiu, H.G. 2010. ArcGIS Engine development from entry to mastery. *China Posts and Telecom Press*, China: Beijing.

Feng, C., Mao, D.H., Li, Z.L., Hu, G.W. 2013. Numerical simulation analysis of diversion flood routing in Linan Flood Storage Polder. *Water Resources and Power* 31(7): 44–47.

Feng, C. 2013. The simulation, visualization and control of diversion flood routing in the Linan Flood Storage Ploder. *Doctoral Thesis of Hunan Normal University*, China: Changsha.

Luka, S., Senka, M., Danko, H. 2007. Tribalj dam-break and flood wave propagation. Annali dell' Università di *Ferrara* 53: 405–415.

Human health risk assessment via groundwater DNAPLs migration model

Yue Pan, Xiankui Zeng*, Jichun Wu & Dong Wang
School of Earth Sciences and Engineering, Nanjing University, Nanjing, China

ABSTRACT: With the rapid progress of industrialization, the groundwater contamination has become a tricky environmental issue. In the past, the groundwater contamination issues usually focus on the soluble pollutants. Recently, the insoluble pollutants have attracted the attention of the public. The insoluble aqueous pollutants are also called non-aqueous phase liquids (NAPLs). When the pollutant density is over 1g/cm^3 it is dense non-aqueous phase liquids (DNAPLs). Part of the DNAPLs are toxic and pose a great threat to the ecological environment and human health. In this paper, we propose a human health risk assessment method based on the numerical simulation process and parameter uncertainty analysis. Based on a sandbox experiment, the DNAPL migration process is simulated by TOUGH program. In addition, the modeling uncertainty is considered explicitly in this study. The unreliable model parameters would cause inaccurate results. In order to reduce parameter uncertainty, we calibrate the unknown model parameter in Bayesian theory. The Markov chain Monte Carlo simulation is applied to reduce the uncertainty of parameters. The human health risk is illustrated in the distribution of a WTO metric – Maximum Concentration Level. Compared with the results assessed by the specific parameters, the distribution of risk could provide more flexible and reliable information.

Keywords: human health risk, dense non-aqueous phase liquids, Markov chain Monte Carlo simulation, Maximum Concentration Level

1 INTRODUCTION

With rapid industrialization and excessive agricultural activities, groundwater pollution has become an important environmental issue that threatens ecosystems and human health. In addition to the conventional pollutants, such as heavy metals, nitrogen, chlorine and so on, various organic pollutants with complex physical and chemical properties have gradually attracted the attention of the society. This type of organic pollutants is denser than water and insoluble in water. It is dense non-aqueous phase liquid (DNAPL). Because the physical and chemical properties of such pollutants are more complicated than soluble pollutants, it increases the difficulty of monitoring and controlling groundwater pollution (Haley et al. 1991). In order to control the threaten of DNAPL in groundwater to human health, the health risk to DNAPLs has become an important part of groundwater environmental management.

Human health risk assessment refers to the evaluation of the probability that an individual may be affected by health in a potentially polluted environment (Len et al. 2002). The assessment contains identifying potential sources of pollution, estimating the concentration of pollutants in contact with an individual, and quantifying the health risks of exposure to pollution. The commonly used assessment model is RAGS

---

*Corresponding author: xiankuizeng@nju.edu.cn

(risk assessment guidance for superfund) model proposed by the US Environmental Protection Agency (EPA 2009). The process can be divided into four steps: (1) risk identification, (2) dose response, (3) exposure evaluation, and (4) risk characterization. The focus of human health risk assessment is on accurately calculating the concentration of pollutants. Conventional methods for obtaining pollutant concentrations are field sampling and laboratory analysis (Teh et al. 2016). But in fact, the impact of pollutants on the human health is continuous. As time goes by, the temporal and spatial distribution of pollutant concentrations will change significantly. Therefore, applying numerical simulation techniques and obtaining dynamically changing concentration values for human health risk assessment could be very necessary.

Due to the complexity and variability of groundwater systems, the migration process of pollutants is affected by many factors, such as flow rate, aquifer properties, and pollutant degradation (Mclaren et al. 2012). Therefore, when using the simulation model to describe the migration process of pollutants, the simulation results (such as the concentration of pollutants) are usually affected by these factors and become uncertain. In this work, human health risk assessment is carried out by the indoor sandbox DNAPL migration experiment. The pollutant is tetrachloroethylene (PCE) and the TOUGH program is used to simulate the migration process. The PCE is a common DNAPL pollutant used to make dry cleaning agents, pesticides, metal degreaser, etc. It has a density of 1.62 g/cm^3, a viscosity of 0.9 cp, and a solubility of 150 mg/L. The PCE pollutant has a stimulating effect on human dermal and respiratory tract. Moreover, it can lead to impaired liver and kidney function, even death (Zhou et al. 2005). In domestic wastewater, agricultural wastewater, industrial wastewater, PCE pollutants are commonly detected in the world. In a site of Reno, Nevada, the PCE concentration in groundwater is higher than 36 mg/kg (Kropf et al. 2003). Moreover, the concentration is as high as 40000 mg/kg in a site of Jacksonville, Florida and the pollution source is the dry cleaning facility wastewater (Strbak 2000). However, the hydrogeological information is limited in real cases, especially the source architecture, assessing the impact of model uncertainty on human health risk becomes infeasible. As a result, we apply a sandbox experiment with the known pollution source architecture and simplified hydrogeological conditions to assess human health risk. The key parameters (e.g. permeability) of the PCE migration model are analyzed by the Markov chain Monte Carlo (MCMC) method. PCE saturation observation data are used for parameter inversion to obtain its posterior distribution. And then conduct human health risk assessment based on the uncertainty of model parameters.

2 METHODOLOGY

2.1 Human health risk assessment

In this work, the RAGS model is applied to human health risk assessment. PCE is a toxic organic substance and is paralysis on the central nervous system of human being. We consider the index Maximum Concentration Level (MCL) to assess human health risk. The acceptable concentration of PCE is 4.0×10^{-2} mg/L according to WHO's drinking water standards (WHO 2011). If MCL is higher than 4.0×10^{-2} mg/L, we believe that the water is not drinkable and human health is in risk. We calculate the probability density distribution of the concentration and determine the probability of MCL higher than 4.0×10^{-2} mg/L.

2.2 DNAPL migration model

DNAPL is an insoluble aqueous. When DNAPL is present in groundwater, the entire groundwater system can be considered as a three-phase(α) three-component(χ) system, aqueous phase, DNAPL phase, gas phase, and water component, DNAPL component, air component. The balance control equation is based on Darcy law (Abriola and Pinder 1985):

$$\frac{\partial}{\partial t}\left(\rho^{\alpha} v_{\alpha} m_{\chi}^{a}\right) + \nabla\left(\rho^{\alpha} v_{\alpha} v^{a} m_{\chi}^{a}\right) - \nabla J_{\chi}^{a} - \rho^{\alpha} v_{\alpha} s_{\chi}^{a} = \rho^{\alpha} v_{\alpha}\left(e_{\chi}^{a} + I_{\chi}^{a}\right) \qquad (1)$$

where ρ^{α} is the density of α phase, v_{α} is the fraction volume occupied by α phase, m_{χ}^{a} is the mass fraction of χ component in α phase, v^{α} is the mass average of α phase velocity, J_{χ}^{α} is the nonconvective flux of χ component in α phase, s_{χ}^{α} is the external supply of χ component in α phase, e_{χ}^{α} is the exchange mass of χ component for phase change, I_{χ}^{α} is the exchange mass of χ component for interphase diffusion.

2.3 Markov chain Monte Carlo simulation

For a nonlinear equation (such as the DNAPL migration equation), the observed data y can be expressed as

$$y = f(\vartheta) + \varepsilon \qquad (2)$$

where $f(\vartheta)$ is the simulation model, ϑ is the input parameter, ε is the residual error.

In order to determine the input parameters, we should calculate the posterior distribution of them (Hill 1974):

$$p(\vartheta|y) = \frac{p(y|\vartheta)p(\vartheta)}{\int p(y|\vartheta)p(\vartheta)d\vartheta} \qquad (3)$$

where $p(\vartheta|y)$ is the posterior distribution of ϑ, $p(y|\vartheta)$ is the joint likelihood function between y and ϑ, $p(\vartheta)$ is the prior distribution of ϑ.

The basic idea of Markov chain Monte Carlo simulation can be summarized as follows: (1) According to expert knowledge or experience, set the prior information of model parameters ϑ; (2) Define the likelihood function to describe the degree of fitting between the model simulation value $f(\vartheta)$ and the observations y; (3) selecting a prior algorithm (e.g., DREAM algorithm) for the sampling the model parameter, then construct a stable distributed Markov chain to fully search the probability distribution space of the objective function $p(\vartheta)$, and the posterior probability distribution $p(\vartheta|y)$ is obtained; (4) Obtain the posterior probability distribution of the model output according to the posterior probability distribution of the model input parameters.

3 SANDBOX EXPERIMENT

The size of the two-dimensional sand box is 0.40 m in height (x), 0.55 m in width (y), and 0.018 m in thickness (z). The quartz sand with a particle size of about 0.5 mm is uniformly filled in the flask, and five lenticles with a particle size of about 0.2 mm are placed at the specified positions. The fixed flow boundary is set at y = 0 and y = 0.55 m, and the flow velocity is 1.5 m/d controlled by a peristaltic pump. Beside the right boundary we set a water source well which is significantly for drinking. The pollutant injection point is at x = 0.35 m, y = 0.27 m, z = 0.009 m, as shown in Figure 1.

At the beginning of the simulation, the injection of PCE has been stopped, and the PCE distribution in the sandbox is shown in Figure 2. The water source well takes the midpoint of the right boundary of the sandbox and we evaluate the risk from the PCE concentration in it. In the numerical DNAPL migration simulation, parameters such as medium permeability, porosity, and reciprocal of DNAPL inlet pressure have significant effects on DNAPL distribution. Among these factors, the sensitivity of medium permeability is higher than the others.

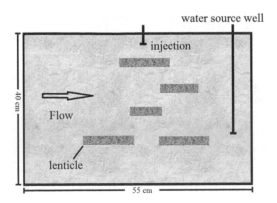

Figure 1. The diagram of sandbox.

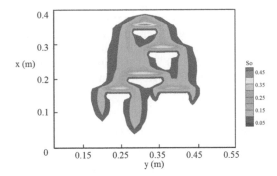

Figure 2. The distribution of PCE saturation.

The common simulation is based on the empirical permeability, then the results would obtain the uncertainty. This case assumes that the permeability of the quartz sand medium is an unknown variable (e.g., k_b represents the background medium and k_l represents the lenticle), and the PCE migration numerical model is established on this assumption. The light transmission method is used to monitor the saturation distribution of PCE in the sandbox, and a digital image was obtained. The digital image was processed by Tecplot 360 EX (2015 R1) software.

4   RESULTS AND DISCUSSION

In this section, we identify the permeability of quartz sand medium to compare the influence of parameter uncertainty on human health risk assessment. The probability density distribution of permeability is used to simulate PCE migration for risk assessment. Moreover, the permeability is set fixed to a reference value for risk assessment again. Then the two results are compared.

4.1   *Analysis of parameter uncertainty*

The DREAM algorithm is used to search the probability space of model parameters. Four parallel Markov chains are collected and the length of each chain is set to 20,000. The burn-in period is 12,000 and after burn-in period is 8,000. The posterior probability density function is shown as Figure 3.

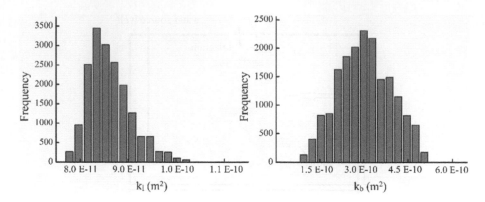

Figure 3. The frequency of the permeability.

Obviously, the posterior distributions of the parameters are converged to narrow ranges, indicating that the permeability is highly sensitive to the observed value (saturation). The prior distribution range of k_b is $[1.0 \times 10^{-10}, 6.0 \times 10^{-10}]$, the identified posterior distribution range is imploded to $[1.5 \times 10^{-10}, 4.5 \times 10^{-10}]$. The prior distribution range of k_l is $[3.0 \times 10^{-11}, 1.0 \times 10^{-10}]$, and the identified posterior distribution range is to $[8.0 \times 10^{-11}, 1.0 \times 10^{-10}]$. Both k_b and k_l obey the Gaussian distribution.

4.2 Human health risk assessment on parameter uncertainty

Based on the posterior parameter frequency distribution, the cumulative distribution of the MCL is obtained. As shown in Figure 4, the frequency of MCL also shows obvious Gaussian distribution characteristics. In the WHO's drinking water standards, the risk is exist when MCL is higher than 4.0×10^{-2} mg/L. The plus sign area represents MCL higher than 4.0×10^{-2} mg/L and it illustrates that there is a more than 30% probability of risk threaten to human health. In order to compare the influence of parameter uncertainty on the risk assessment results, the parameters k_b and k_l are fixed at the empirical value ($k_b = 2.5 \times 10^{-10}$, $k_l = 3.2 \times 10^{-11}$) (Schroth et al. 1996, Zheng et al. 2015), and the simulation results in the empirical value of MCL is 3.53×10^{-2} mg/L, this value has a lower probability of appearing in the MCL distribution, and it indicates that there is no risk on human health. Therefore, it is improper to set the parameters of the pollutant migration model to the determined values and ignore the uncertainty of the parameters for human health risk assessment.

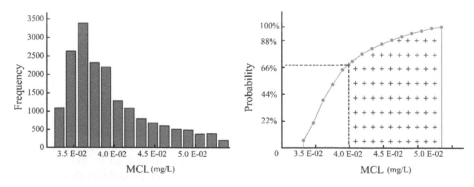

Figure 4. The cumulative probability of MCL.

5 CONCLUSION

Human health risk assessment is an important basic work for the restoration of polluted sites. In order to make the assessment results more authentic, it is usually necessary to establish a migration model of pollutants. This study assesses human health risk of exposure to DNAPL and quantified the risk using the concentration index MCL. Based on the sandbox experiment, the influence of the parameter uncertainty of the PCE migration model on health risk assessment results was compared by Bayesian parameter uncertainty analysis method. The results show that human health risk assessment obtained with the determined model parameter values are unprecise. Moreover, it may even lead to a wrong estimate of risk, thus delaying the proposal of control measurements.

ACKNOWLEDGMENTS

This study was supported by The National Key Research and Development Program of China (2016YFC0402802), the National Natural Science Foundation of China (41672233, 41501570).

REFERENCES

Abriola, L.M. and Pinder, G.F. 1985. A Multiphase Approach to the Modeling of Porous-Media Contamination by Organic-Compounds .1. Equation Development. *Water Resources Research* 21(1), 11–18.

EPA, U.S. 2009. Risk Assessment Guidance for Superfund Volume I: Human Health Evaluation Manual (Part F, Supplemental Guidance for Inhalation Risk Assessment). EPA, U.S. (ed), Washington DC.

Haley, J.L., Hanson, B., Enfield, C. and Glass, J. 1991. Evaluating the Effectiveness of Ground-Water Extraction Systems. *Ground Water Monitoring and Remediation* 11(1), 119–124.

Hill, B.M. 1974. Bayesian Inference in Statistical Analysis Technometrics 16(3), 478–479.

Kropf, C.A., Benedict, J. and Berg, J.H. 2003. Widespread PCE Contamination: Characterization and Source Investigation to Protect Municipal Wells, U.S.A.

Len, R., Keith, S., Paul, S., Ken, H., Patricia, K., Gevan, M. and Beth, L. 2002. SOURCES, PATHWAYS, AND RELATIVE RISKSOF CONTAMINANTS IN SURFACE WATER ANDGROUNDWATER: A PERSPECTIVE PREPARED FOR THE WALKERTON INQUIRY. *Journal of Toxicology and Environmental Health Part A*, 65(1), 142.

Mclaren, R.G., Sudicky, E.A. and Park, Y.J. 2012. Numerical simulation of DNAPL emissions and remediation in a fractured dolomitic aquifer. *Journal of Contaminant Hydrology* (136–137), 16.

Schroth, M.H., Ahearn, S.J., Selker, J.S. and Istok, J.D. 1996. Characterization of miller-similar silica sands for laboratory hydrologic studies. *Soil Science Society of America Journal* 60(5), 1331–1339.

Strbak, L. 2000. In situ flushing with surfactants and co-solvents. National Network for Environmental Management Studies, Washington.

Teh, T., Norulaini, N.A.R.N., Shahadat, M., Wong, Y. and Omar, A.K.M. 2016. Risk Assessment of Metal Contamination in Soil and Groundwater in Asia: A Review of Recent Trends as well as Existing Environmental Laws and Regulations. *Pedosphere* 26(4), 431–450.

WHO 2011. Guidelines for Drinking-Water Quality Fourth Edition. World Health Organization.

Zheng, F., Gao, Y., Sun, Y., Shi, X., Xu, H. and Wu, J. 2015. Influence of flow velocity and spatial heterogeneity on DNAPL migration in porous media: insights from laboratory experiments and numerical modelling. *Hydrogeology Journal* 23(8), 1703–1718.

Zhou, C.F., Ding, R. and Shen, T. 2005. Cytotoxicity of tricolorethylene, perchioroethtlene and dichloroethylene on human keratinocytes. *Chinese Journal of Industry and Medicine* 18(2), 4.

Gephi network analysis method for determining assessment indexes of extreme high temperature weather risk about electric power system—a case study of Beijing City

Chunming Shen, Wei Zhu*, Liping Xu, Xiru Tang, Jianchun Zheng & Shunquan Yuan
Beijing Research Center of Urban System Engineering, Beijing, China
Beijing Academy of Science and Technology, Beijing, China

ABSTRACT: The power system of Beijing is a complex giant system. The risk of peak power load supply under the extreme high temperature weather was prominent, and there was few research about the scientific and reasonable method to determine the assessment indexes on the extreme weather risk. The risk-causing mechanism of power system under the influence of extreme temperature was analyzed. Based on the analysis of energy flow in life cycle, the network structure model of Beijing power system was constructed from the aspects of power production and transformation, power transmission and distribution, and power terminal consumption. The network structure weight of each link factor was quantified by the Gephi software using graph theory analysis method, and the extreme weather risk assessment indexes of power system were determined. The analysis showed local power, external power, urban transmission and distribution network and tertiary industry consumption were at the higher level in the importance of system structure. Inner Mongolia was the most influential source of external power for Beijing, and natural gas was the most influential basic energy for local power supply. The Anding 500 kV hubs have the greatest degree of correlation among the key stations affecting the power transmission and distribution of Beijing.

Keywords: extreme weather risk; risk assessment index determination; urban power system; Gephi network analysis.

1 INTRODUCTION

Extreme high temperature has become one of the most serious extreme climate disasters in the world (Meehl and Tebaldi, 2004; Russo *et al*, 2017). As the main region of climate change risk, the cities are facing unprecedented severe challenges (Willems *et al*, 2013). Electric power system is the necessary basic guarantee for the safe and stable operation of modern cities. Especially in the future, a large increase in the proportion of energy demand in the terminal sector would become a rigid condition for achieving global temperature rise and carbon emission reduction targets (Kang and Eltahir, 2018). The reliability of electric power system is becoming more and more critical to the operation and development of cities. In recent years, due to the urbanization expansion, the frequent occurrence of extreme high temperature events and the urban heat island effect (Liu *et al*, 2007), the urban power load in China has been rising continuously, with significant difference between peak and valley charges. It increases the pressure of power system regulation and the risk of urban power grid operation. Especially in Beijing,

---

*Corresponding author: Wei Zhu, doctor, research fellow. E-mail: zhuweianquan@126.com.

which relies closely on the North China Power Grid and has about 70% of the power transferred from outside to the typical large-scale receiving-end power grid, there was little room for the improvement of the local power supply regulation capacity and the inherent vulnerability of the power system was relatively large. With the rapid growth of electricity consumption in the tertiary industry and cities, the proportion of retail users such as shopping malls, office buildings and residential buildings has increased significantly by nearly 70% (Beijing, 2017), especially in the peak power consumption. When the adequacy was low, the difficulty and risk of power grid regulation were aggravated.

In this paper, based on the energy flow network diagram of urban power system, the risk-causing mechanism of power system under the influence of extreme temperature was analyzed. Gephi software was used to quantify the network structure weight of each link factor by graph theory method, and determined the key indicators of extreme high temperature risk assessment of power supply system.

2 ENERGY FLOW DIAGRAM ANALYSIS OF BEIJING ELECTRIC POWER SYSTEM

Referring to the U.S. Energy Flow Map drawn by the National Laboratory of the United States of Lawrence Livermore, a simplified version of the Beijing Power System Energy Flow Map in 2016 was drawn from left to right, which was based on the Beijing Energy Balance Table in 2016 of China Energy Statistics Yearbook 2017, and with the main line of primary energy supply, power conversion, power transmission and distribution to terminal power consumption, as shown in Figure 1. There were many kinds of primary energy used for power

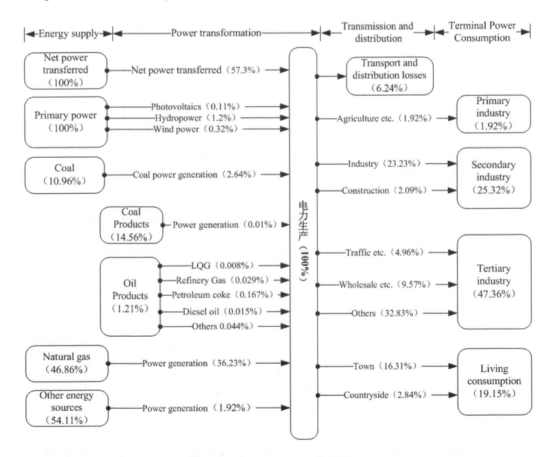

Figure 1.   Energy flow map of Beijing electric power system in 2016.

conversion in Beijing, including raw coal, natural gas, water, wind, solar energy, oil and other energy sources. However, the proportion of various energy supply varies greatly, and the focus of terminal power consumption was prominent.

In 2016, raw coal, natural gas and other energy used for converted power generation in Beijing accounted for 10.96%, 46.56% and 54.11% of the total primary energy, respectively. The proportion of converted energy to total electricity production in Beijing was about 60% as shown in Figure 2 (a). The primary energy used for power conversion was mainly natural gas, and the respective proportion of raw coal, primary power and other energy sources for power generation. All of them were less than 3%, while coal products and petroleum products account for less than 0.5% of their respective electricity generation. Electric power terminal consumption mainly concentrated on the tertiary industry, secondary industry and living consumption, accounting for 47.36%, 25.32% and 19.15% of the total electric power consumption, of which 91.75% was industrial electricity consumption and 85.17% was urban domestic electricity consumption. Therefore, the tertiary industry, industry and urban living consumption are the main factors affecting electricity consumption.

From the point of view of power supply and demand balance, the disaster chain of power system under the influence of extreme high temperature weather was analyzed. As shown in Figure 3, under extreme high temperature weather events, air temperature was at an extremely high level, and human comfort, human health, industrial equipment and production and operation environment would be seriously negatively affected. In order to meet the needs of human body and industrial production environment, power refrigeration were needed. Maintaining the reasonable air temperature in the region would increase the demand for refrigeration load on the basis of the original power system load operation. At the same time, extreme high temperature would also

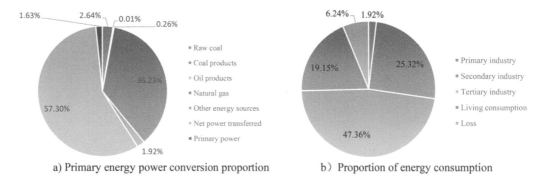

a) Primary energy power conversion proportion   b）Proportion of energy consumption

Figure 2.   Power production conversion and consumption proportion of Beijing in 2016.

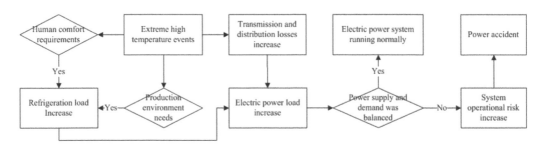

Figure 3.   Diagram of risk-induced disaster Chain of power system under extreme high temperature weather.

lead to an increase in power transmission and distribution losses (Zhao et al, 2017), so the power load of the power system would rise. When the power supply of power system could meet the demand of power load, the power system would operate normally; when the power supply could not meet the demand of power load, the system operation risk would increase in a power law function, and it would easily lead to power system accidents. Therefore, driven by extreme high temperature weather event factors, the increase of refrigeration load and insufficient power supply were the main factors leading to system instability and disaster, and were also the main content of risk assessment indicators for extreme high temperature events in power systems.

3 GEPHI VISUALIZATION PROCESSING AND KEY POINT DETERMINATION OF SYSTEM STRUCTURE

Power load and power supply are the important contents of risk assessment index of extreme high temperature events in power system. The basic event factors affecting power load are mainly terminal consumer industries, and the basic events affecting power supply mainly include power supply and transmission and distribution network. Based on the energy flow chart of Beijing electric power system in 2016 and the Energy Development Plan of Beijing during the 13th Five-Year Plan, the network structure of Beijing electric power transmission and distribution network is simplified, and the network structure chart of Beijing electric power system is constructed by Gephi visual processing software from the perspective of the whole life cycle. The layout is shown in Figure 4 after the change of huiyifan. The network structure parameters are shown in Table 1. Figure 4 (a) is edge directed network with different weights, and Figure 4 (b) is edge directed network with the same weights. Obviously, the network of Figure 4 (b) is more complex.

Based on the energy flow diagram of Beijing power system in 2016, the visualization processing of network diagram was shown in Figure 4 (a). The nodes in the network diagram represent the main links of power production, transmission and consumption, while representing the proportion of power load transmission. According to the value of node degree and edge weight in network graph, the partition matrix of node importance level is established, as shown in Table 2. Statistical analysis of the importance of each node in Figure 4 (a) shows that local power, external power, urban transmission and distribution network and tertiary industry consumption are of high importance, natural gas power generation is of high importance, and industrial consumption and industrial consumption in the second year are of medium importance. Because of the high proportion of Beijing's external power transfer, the network visualization processing of Beijing's external power sources, external power transmission channels and urban transmission and distribution stations is shown in Figure 4 (b). Compared with Shanxi, Hebei and Tianjin, the Inner Mongolia Autonomous Region of the external power sources has the highest output and the largest total weight of transmission and distribution capacity. Therefore, Inner Mongolia has the largest output in the external power supply areas. Among the urban power transmission and distribution stations, the degree of

Table 1. Statistics of network structure parameters of Beijing electric power system.

Category	Average degree	Average Weighting Degree	Network diameter	Graph density	Average path length	Average clustering coefficient
Structural chart of electric power system	0.958	1.401	5	0.042	3.155	0
External power regulation system structure	1.647	1.647	4	0.05	2.108	0.013

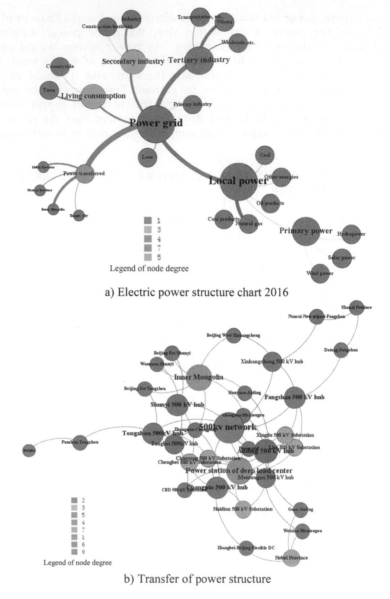

a) Electric power structure chart 2016

b) Transfer of power structure

Figure 4.   Gephi visual network map of power system structure in Beijing.

Table 2.   Matrix table of node importance in Beijing power system.

Importance level			Edge Weight (Maximum)				
			Lower	Low	Medium	High	Highe
			(0,0.5%]	(0.5%,5%]	(5%,15%]	(15%,30%]	(30%,60%]
Degree	Lower	1	Lower	Lower	Low	Medium	High
	Low	2, 3	Lower	Low	Medium	Medium	High
	Medium	4	Low	Low	Medium	High	High
	High	5, 6	Low	Medium	Medium	High	Higher
	Higher	7	Medium	Medium	High	Higher	Higher

234

the Anshan 500 kV hub node is 7, and that of Fangshan, Tongzhou, Shunyi and Changping 500 kV hubs is 5. Therefore, Anshan, Fangshan, Tongzhou, Shunyi and Changping 500 kV hubs were the key stations affecting Beijing urban power transmission and distribution, and the Anshan 500 kV hub has the greatest correlation.

4 CONCLUSION

Based on Gephi software visualization processing, this paper divides the importance level of influencing factors of Beijing's power supply. The research showed that the power supply in Beijing was extremely unbalanced, and the power load and power supply were the important contents of the risk assessment index of extreme high temperature events in power system. Local power, external power, urban transmission and distribution network and tertiary industry consumption were at the higher level in the importance of system structure. Inner Mongolia was the most influential source of external power for Beijing, and natural gas was the most influential basic energy for local power supply. The 500 kV hubs in Anding, Fangshan, Tongzhou, Shunyi and Changping are the key stations affecting the power transmission and distribution in Beijing. Among them, Anding 500 kV hubs have the greatest degree of correlation.

ACKNOWLEDGMENTS

This project was supported by the National Key R&D Program of China (2018YFF0213801-006) and China Postdoctoral Science Foundation (2018M63137) and Beijing Postdoctoral Working Foundation (2018-22-050) and Budding Project of Beijing Academy of Science and Technology.

REFERENCES

Beijing Statistical Bureau. 2017. Beijing Statistical Yearbook. Available at: http://tjj.beijing.gov.cn/nj/main/2017-tjnj/zk/indexch.htm.
Kang S., and Eltahir E.A.B. 2018. North China Plain threatened by deadly heatwaves due to climate change and irrigation. *Nature Communications* 9: 2894.
Lawrence Livermore National Laboratory. Energy flow chart. Available at: https://publicaffairs.llnl.gov/news/energy/energy.html.
Liu W., Ji C., Zhong J., et al. 2007. Temporal characteristics of the Beijing urban heat island. *Theoretical and Applied Climatology* 7: 213–221.
Meehl G.A. and Tebaldi C. 2004. More intense, more frequent, and longer lasting heatwaves in the 21st century. *Science* 305: 994–997.
Russo S., Sillmann J. and Sterl A. 2017. Humid heat waves at different warming levels. *Scientific Reports* 7: 7477.
Willems P., Arnbjerg—Nielsen K., Olssond J., et al. 2013. Climate change impact assessment on urban rainfall extremes and urban drainage: Methods and short comings. *Atmospheric Research* 103: 106–118.
Zhao, N., Shi Y.H., Li N.J., and Shen J.H. 2017. The relationship of temperature humidity index and meteorology sensitive power load in Beijing. *Electric Power* 50 (2): 175–180.

Risk factors analysis for falls among the urban elderly in Beijing, China

Qiujie Zhang
Beijing Academy of Science and Technology, Bejing, China

Wei Zhu
Beijing Research Center of Urban Systems Engineering, Bejing, China

Lugang Guo
Beijing Academy of Science and Technology, Bejing, China

ABSTRACT: Falls are the leading cause of fatal and non-fatal injuries among Chinese people aged 65 or older. In this paper, we conducted a survey of 1,059 seniors aged 60+ in seven urban communities in Beijing and analyzed the risk factors of their falls with logic regression. For the first time, we integrated family support as a variable with physical, psychological and environmental variables among risk factors for senior falls. Results revealed that family support had a remarkable influence on the respondents' falls within the previous 2 years and a negative family relationship is a risk factor for falls among the elderly. Some of the psychological factors had a remarkable influence on falls by the respondents within the previous 2 years. Anxiety over falling was a risk factor for falls among the elderly. Some of the environmental factors had significant influence on falls by the respondents within the previous 2 years. Seniors with anti-slip handles in washrooms would be 5.1% less likely to fall than those without handles. It was concluded that the above-mentioned risk factors should be taken into consideration for intervention programs against falls by seniors. Policies for seniors should consider the perspective of families instead of individuals, and measures contributing to family support should be included in intervention programs to facilitate positive family relationships for seniors.

Keywords: risk factors, falls, injuries, family support

1   INTRODUCTION

Falls are the leading cause of fatal and non-fatal injuries among Chinese people aged 65 or older. The Causes of Death Statistics 2006 released by the National Disease Surveillance System shows that, in China, mortality rates aged 65+ due to falls were 49.56 deaths per 100,000 for men and 52.80 deaths per 100,000 for women. A higher incidence of falls and often more serious health consequences are observed among seniors. Therefore, the identification of risk factors for falls among the elderly and that of preventive interventions are a key priority in reducing senior falls and accidents.

Academics at home and abroad have a number of research findings on risk factors for falls by senior citizens. First, the risk factors could be physical. Older people tend to sustain more falls (Masud and Morris, 2001). Women are more likely to fall than men and the difference becomes more distinct among older seniors (Li and Wany, 2001). Chronic diseases such as cardiovascular diseases, hypertension and vertigo can increase risk of falling (Lawlor, Patel and Ebrahim, 2003). Coleman (2004), Kulmala (2008) and Macedo (2012) suggested that vision impairment may lead to higher risks of falling. Elderly people who have already fallen once are more likely to fall again (Stalenhoef et al., 2002). Second, the risk factors could be psychological. Seniors living with fear of falling are more prone to falling than those who are not scared (Delbaere et al., 2004). Third, the risk factors could be sociological. Wilson et al. (2006) discovered the relevance of education to falling, and people with high school, college or higher education tend to fall less than those with lower levels of education. Marital status also has influence on falls by senior citizens. Based on the statistics of falls by senior citizens in Jiangsu Province, Xiang et al. (2010) concluded that widowed senior citizens are a highly at-risk population. Senior citizens living alone are more likely to fall compared to those living with others. Fourth, the risk factors could be environmental. Both domestic and international researchers found slippery and uneven surfaces and poor lighting could be risk factors for falls among the elderly.

However, few of the domestic findings mentioned the important factor of family support network and no significant correlation has been discovered. As a basic social network, family is an important way to provide economic and emotional support. In view of this, the author made an analysis of 1,059 senior citizens aged 60+ in seven urban communities in Beijing and placed family support as a variable among risk factors for falls by senior citizens, in addition to physical, psychological and environmental factors.

2 RESPONDENTS AND METHODS

A stratified cluster sampling technique was applied and the respondents were taken randomly from the residence registration records provided by seven community administrations in Xicheng District, Beijing. Inclusion criteria were senior citizens aged 60+ and informed consent on a voluntary basis. Exclusion criteria was cognitive impairment.

Guided by community staff, trained investigators collected statistics with unified questionnaires for screening and gathering an overview of falls, as well as interviews with the respondents. For senior citizens unable to provide information, their carers were interviewed instead.

Stata 12.0 is used to model the descriptive statistics for logistic regression analysis.

3 SAMPLE DESCRIPTION

1,059 questionnaires were sent out and there were 998 valid ones collected. More women (61.42% of the sample) were included in the survey than men (38.58%), and in terms of age distribution, the sample was well-balanced: age 60–64, 30.06% of the total; age 65–69, 22.24% of the total; age 70–74, 15.33% of the total; age 75–79, 15.13% of the total; and age 80+, 17.23% of the total. In terms of educational background, the respondents had, in general, received a decent education: primary education or below, 17.23% of the total; middle school education, 28.66% of the total; high school or secondary vocational education, 25.15% of the total; college education or higher, 28.96% of the total.

As shown in Table 1, 79.26% of the respondents were married and had living spouses while 20.74% were divorced, widowed or unmarried. Most of the respondents (84.67%) lived with their family, while 15.33% lived alone. As for family relationships, 90.68% of the respondents enjoy positive ties with their families and only 9.32% reported negative family relationship.

As shown in Table 2, falls by the respondents within the previous two years were not uncommon at 170 (17.03%).

4 ANALYSIS OF RISK FACTORS FOR FALLS AMONG THE ELDERLY

Stata 12.0 and logistic regression model are used to handle data collected in the survey in order to have quantitative analysis of how and by how much different risk factors would influence the senior falls.

Table 1. Sample distribution.

Variables	Type	Number (people)	Percentage (%)
Sex	Male	613	61.42
	Female	385	38.58
Age	60 to 64	300	30.06
	65 to 69	222	22.24
	70 to 74	153	15.33
	75 to 79	151	15.13
	80 or older	172	17.23
Educational background	Primary School	172	17.23
	Middle school	286	28.66
	High school or secondary vocational school	251	25.15
	College education or higher	289	28.96
Marital status	Married (with Living Spouses)	791	79.26
	Divorced/widowed/unmarried	207	20.74
Living arrangements	Not living alone	845	84.67
	Living alone	153	15.33
Family Relationship	Positive	905	90.68
	Negative	93	9.32

Table 2. Fall incidence of the interviewed elderly within recent two years.

Variables	Type	Number (people)	Percentage (%)
Fall incidence within the previous two years	Yes	170	17.03
	No	828	82.97
TOTAL		998	100

The variable in the measurement model is whether or not the respondents have ever fallen within the previous two years (in the equation as fall). Sociological, physical, psychological and environmental factors as well as family support are inserted into the equation as independent variables (see Equation 1).

$$Ln\left(\frac{P_{fall}}{1 - P_{fall}}\right) = \beta_0 + \beta_i Indi_i + \beta_j Phy_j + \beta_k Psy_k + \beta_l Env_l + \beta_m Fami_m + \varepsilon \quad (1)$$

As shown in Table 3, fall as a dependent variable reflects whether or not the respondents have fallen within the previous two years and Yes and No are coded as 1 and 0, respectively. The independent variables include $Indi_i$, the sociological information of the respondents, Phy_j, the physical variables for fall incidence, Psy_k, the psychological variables for fall incidence, Env_l, the environmental variables for fall incidence, and $Fami_m$, the family support variables for fall incidence.

5 CALCULATION OF REGRESSION COEFFICIENT AND DESCRIPTION OF MODEL RESULTS

Logistic regression was conducted with Equation (1) to identify the influence coefficient of each independent variable on the dependent variable. See Table 4 for results: the variables

Table 3. Description of variables.

Variables		Description	
Family support variable	$Fami_1$: Dummy variable $Fami_2$: Dummy variable $Fami_3$: Dummy variable	Single or not Positive or negative family relationship Living alone or not	Single = 1 Negative = 1 Living alone = 1
Psychological variable	Phy_1: Dummy variable	Anxious for falling or not	Anxious=1
Environmental variable	Env_1: Dummy variable Env_2: Dummy variable Env_3: Dummy variable Env_4: Dummy variable Phy_1: Categorical Variable, each category is taken as a dummy variable	Anti-slip handle in the washroom or not Stairs well-lighted or not Roads smooth and dry or not Living on the appropriate floor or not Age (respondents aged 60–65 are taken as the reference group)	No = 1 No = 1 No = 1 No = 1 65–69 70–74 75–79 80+
Physical variable	Phy^2: Dummy variable Phy_3: Dummy variable Phy_4: Dummy variable Phy_5: Dummy variable Phy_6: Dummy variable Phy_7: Dummy variable Phy_8: Dummy variable Phy_9: Dummy variable Phy_{10}: Categorical variable each category is taken as a dummy variable	Sex (male respondents are taken as the reference group) Hypertension Hypotension Have heart disease Incidence of stroke before Incidence of vertigo before Frequent urination, urgent urination, urinary incontinence Diabetes Vision performance (capability to see things or people within 4 meters clearly)	Female Yes = 1 Yes = 1 Yes = 1 Yes = 1 Yes = 1 Yes = 1 Yes = 1 Visible Barely visible Invisible
Dependent Variable	*falling*: dummy variable	Fall incidence in recent 2 years	Yes = 1

with $p \leq 0.05$ are considered as remarkably relevant and those with $p > 0.05$ are unremarkably relevant. An analysis of Table 4 elicited the following conclusions:

If the impact of sociological, physical, psychological and the environmental factors are excluded, whether the respondents have positive family relationships would have a remarkable influence on the fall incidence within the previous two years, i.e. seniors with negative family relationships would be 10.1% more likely to fall than those with positive family relationships. Neither living alone nor marital status had a remarkable influence on fall incidence within the recent 2 years.

If the impact of sociological, physical, environmental and family support factors are excluded, the psychological factor (such as anxiety over falling) would have a remarkable influence on the fall incidence within the previous two years. Compared with those who do not worry about falling, seniors worrying about falling would be 7.7% more likely to fall within the recent 2 years.

If the impact of sociological, physical, psychological and family support factors are excluded, the environmental factor would have remarkable influence on fall incidence within the previous two years.

For instance, seniors with anti-slip handles in the washrooms would be 5.1% less likely to fall than those without. However, neither a smooth and dry road surface in the neighborhood nor living on the appropriate floor has a remarkable influence on falls by senior citizens.

If the impact of sociological, psychological, environmental and family support factors excluded, age of the respondents would have a remarkable influence on fall incidence within

Table 4. Regression results of variables on falling in the previous two years.

Variables		OR	dy/dx	p > |z|
Married		0.886	-0.015	0.630
Negative family relationship		1.969	0.101	0.032
Living alone		1.127	0.015	0.650
Anxiety over falling		1.956	0.077	0.001
Anti-slip handles in the washing room		0.650	-0.051	0.024
Well-lit stairs		1.428	0.041	0.088
Smooth and dry road surface in the neighborhood		1.042	0.005	0.857
Living on the appropriate floor and easy access to the floor		0.754	-0.037	0.165
Age (respondents aged 60–65 are taken as the reference group)	65–69	1.917	0.091	0.021
	70–74	1.389	0.044	0.306
	75–79	1.148	0.018	0.668
	80+	1.922	0.094	0.056
Sex (male respondents are taken as the reference group)	Female	0.789	-0.029	0.223
Suffering from hypertension		1.066	0.008	0.729
Suffering from hypotension		0.618	-0.051	0.369
Suffering from heart disease		1.173	0.020	0.504
Incidence of stroke before		0.492	-0.068	0.424
Incidence of vertigo before		1.531	0.060	0.248
Frequent urination, urgent urination, urinary incontinence		1.461	0.053	0.409
Diabetes		1.220	0.026	0.409
Vision performance (ability to see things or people clearly within 4 m)	Visible	0.979	-0.003	0.925
	Barely visible	1.487	0.054	0.128
	Invisible	1.548	0.063	0.750

the previous two years. Those aged 65 to 69 are 2.1% more likely to fall than those aged 60 to 64. No remarkable relevance has been discovered among other age groups. Sex has no remarkable influence on falling among the elderly. Neither chronic diseases such as hypertension, hypotension, heart disease, diabetes, vertigo and stroke, nor vision, has a remarkable influence.

6 CONCLUSION AND DISCUSSION

From the analyses above it could be concluded that the incidence rate of falls among the elderly is 17.03%, which is quite close to previous domestic empirical research findings. Maolong and Yuetao (2014) performed a meta-analysis on 23 research findings from 2001 to 2004, and used a random effect model to obtain a combined incidence rate of falls by senior citizens at 18.3% (95% CI from 15.7% to 20.8%).

The family support factor has a remarkable influence on the respondents' falls within the previous two years, and negative family relationship is a risk factor for falls among the elderly, in agreement with Zhaohui et al. (2006). Living alone and marital status have no remarkable influence on falls by senior citizens within the previous two years, which is inconsistent with the conclusions of Xiang et al. (2010) or Qinghua et al. (2014), and further research is needed.

The psychological factor has a remarkable influence on falls by senior citizens within the previous two years. Anxiety over falling is a risk factor for falls by senior citizens, which is consistent with the conclusion of Delbaere et al. (2004).

The environmental factor has a remarkable influence on falls by senior citizens within the previous two years. Senior citizens with anti-slip handles in the washrooms would be 5.1% less likely to fall than those without, supporting some of the conclusions of Kallin et al. (2004), Dehua et al. (2013) and Zhaohui et al. (2006). Neither a smooth and dry road surface in the neighborhood nor living on the appropriate floor has a remarkable influence on falls by senior citizens.

If the impact of sociological, psychological, environmental and family support factors are excluded, the age of the respondents would have a remarkable influence on fall incidence within the previous two years. Those aged 65–69 are 2.1% more likely to fall than those aged 60–64, which is consistent with the conclusion of Masud et al. (2001). Sex, chronic diseases and vision have no remarkable influence on falls, which is inconsistent with current research findings, and further research is needed.

Based on the conclusions above, the following suggestions can be made:

(I)   The above-mentioned risk factors should be taken into consideration when relevant authorities plan intervention programs for falls by senior citizens. Policies for senior citizens should be based on families instead of individuals, and measures contributing to family support should be included in intervention programs to facilitate positive family relationships for senior citizens.

(II) In view of the anxiety of the elderly over falls, public information on fall prevention should be intensified to help the elderly have more understanding of and develop more capabilities against falls.

(III) In view of hidden dangers in the home, public information danger screening should be intensified to teach senior citizens and their carers how to identify and remove possible dangers in the household environment. For seniors living alone and those at advanced old age, community administration staff, social workers and volunteers should provide regular on-the-spot screening services to highlight hidden dangers.

(IV) For senior citizens aged 60+, PPP programs, government procurement programs or government-subsidized programs for refurnishing residences should be developed to offer safer living environments.

ACKNOWLEDGMENTS

This project was supported by Soft Science Research Project of Hebei, China (17456115).

REFERENCES

Coleman A.L, Stone K, Ewing S.K et al. 2004. Higher risk of multiple falls among elderly women who lose visual acuity. *Ophthalmology* 111: 857–862.

Delbaere K., Crombez G., Vanderstraeten G., et al. 2004. Fear-related avoidance of activities, falls and physical frailty: a prospective community-based cohort study. *Age Ageing* 33(4): 368–373.

Kallin K., Jensen J., Olsson L.L., et al. 2004. Why the elderly fall in residential care facilities and suggested remedies. *Journal of Family Practice* 53: 41–52.

Kulmala J., Era P., Parssinen O. et al. 2008. Lowered vision as a risk factor for injurious accidents in older people. *Aging Clinic and Experiment Research* 20: 25–30.

Lawlor D.A., Patel R., Ebrahim S. 2003. Association between falls in elderly women and chronic diseases and drug HSe: cross-sectional study. *BMJ* 327(7417): 712–717.

Li L.T., Wang S.Y. 2001. Disease burden and risk factors of the elderly falling. *Chinese Journal of Epidemiology* 22(4): 262–264.

Macedo B.G., Pereira L.S.M., Rocha F.L. et al. 2012. Association between functional vision, balance and fear of falling in older adults with cataracts. *Revista Brasileira De Geriatria E Gerontologia* 15(2): 265–274.

Masud T, Morris R.O. 2001. Epidemiology of falls. *Age Ageing* 30(4): 3–7.

Stalenhoef P.A, Diederiks J.P, Knotmerus J.A et al. 2002. A risk model for the prediction of recurrent falls in community dwelling elderly: a prospective cohort study. *Journal of Clinic Epidemiology* 55(11): 1088–1094.

Stewart M, Davidson K, Meade D et al. 2000. Myocardial infarction: survivors' and spouses' stress, coping, and support. *Journal of Advanced Nursing* 31(6): 1351–1360.

Wilson R.T, Chase G.A, Chrischilles E.A. et al. 2006. Hip fracture risk among community-dwelling elderly people in the United States: a prospective study of physical, cognitive, and socioeconomic indicators. *American Journal of Public Health* 96(7): 1210–1218.

Yu X, Xue C.B, Hu Y, et al. 2010. Analysis of falls and risk factors among older residents in Jiangsu Province. *Chinese Journal of Disease Control Prevention* 14(10): 939–941.

Risk Analysis Based on Data and Crisis Response Beyond Knowledge – Huang & Nivolianitou (eds)
© *2020 Taylor & Francis Group, London, ISBN 978-0-367-25146-8*

Global heat wave hazards, considering the humidity effects under greenhouse gas emission scenarios

Ning Li*, Xi Chen & Wei Gu
Key Laboratory of Environmental Change and Natural Disaster, MOE, Beijing Normal University, Beijing, China
Academy of Disaster Reduction and Emergency Management, Ministry of Civil Affairs & Ministry of Education, Beijing, China

ABSTRACT: In this paper, the wet-bulb globe temperature (W), considering both temperature and humidity effects, was used as a heat index to quantify the number of annual total heat wave days (HWDs). Changes in surface air temperature, relative humidity, W, and the total number of HWDs in a year were analyzed using multi-model simulations for the reference period (1986–2005) and different greenhouse gas emission scenarios. Our results suggest that HWDs per year will continue to increase throughout the 21st century, and by 2086–2095, approximately one-half of the land area of the earth could be exposed to heat waves for more than 100 days in a year, under RCP 8.5. At the same air temperature, there should be more focus on tropical countries exposed to the greatest increases in HWDs in the future.

Keywords: climate change, humidity effects, heat wave hazard, RCP scenario

1 INTRODUCTION

Studies on heat waves often focus on surface air temperature as the main variable (Cowan et al. 2014). However, in addition to extreme high temperatures, surface relative humidity is also an important factor in defining heat waves as it is directly related to body heat exchange (Kovats and Hajat 2008). If the environmental relative humidity is high, it will cause heat accumulation in the body (Budd 2008) and both morbidity and mortality will increase. Thus, a variety of heat stress indices like humidex (Barnett et al. 2010), apparent temperature (Russo et al. 2017), and wet-bulb globe temperature, take both temperature and humidity into consideration.

Wet-bulb globe temperature (WBGT) is the most common index including both temperature and humidity effects to quantify heat stress and has a long history of use. The heat stress index WBGT (°C) is a combination of the black globe temperature, the natural wet bulb temperature, and the air temperature, to take into account the effect of heat absorption from the sun and evaporative cooling, which is strongly related to air humidity. As ambient WBGT approaches human body skin temperature, related to air humidity, it becomes difficult for the body to cool itself down.

Recent studies have pointed to a growing concern about the increasing heat stress in the 21st century, considering humidity effects as well as extreme temperatures. By applying 35°C as a threshold for human adaptability, Pal and Eltahir (2015) predicted that extremes of wet-bulb temperature in the region around the Arabian Gulf are likely to approach and exceed this threshold under greenhouse gas emission scenarios. Russo et al. (2017) quantified humid

---

* Corresponding author: ningli@bnu.edu.cn

heat wave hazard at different levels of global warming using the apparent temperature, showing that humidity can amplify the magnitude and apparent temperature peak of heat waves. Kang and Eltahir (2018) stated that the North China Plain is likely to experience deadly heat waves with wet-bulb temperature exceeding the threshold. Lin et al. (2017) determined trends of heat wave variation and stress thresholds in three major cities of Taiwan based on WBGT, and suggested that the heat stress in all three cities will either exceed or approach the danger level (WBGT ≥ 31°C) by the end of this century.

Quantifying the spatial pattern and temporal changes of heat wave hazard, considering both temperature and humidity effects on a global scale, is of significant importance that can provide useful information for further global heat risk assessment. Therefore, this study focuses on the spatial extent and temporal variation of heat stress as well as air temperature and relative humidity, and WBGT was utilized as a heat index and the number of annual total heat wave days (HWDs).

2 DATA AND METHOD

In this study, daily data were used of mean surface air temperature and near-surface relative humidity from multi-model simulations for the Coupled Model Inter-comparison Project Phase 5 (CMIP5), which is available from the Earth System Grid Federation (ESGF) (https://esgf-index1.ceda.ac.uk/search/cmip5-ceda/). CMIP5 productions are based on general circulation model (GCM) estimates of present and future climate in response to increases in active atmospheric constituents (Leng et al. 2016). We used one run (rli1p1) for each model built and all the data were interpolated to a common 0.5° grid cell size using a bilinear function.

WBGT was used as a heat index following previous studies on long-term changes in heat stress (Knutson and Ploshay 2016) and the number of total HWDs in a year was employed to quantify the heat hazard (Dong et al. 2015). Eq. (1) was used to evaluate the heat stress outdoors where solar radiation is present, and Eq. (2) was used to evaluate the heat stress indoors, or outdoors without solar radiation:

$$WBGT = 0.7T_{nwb} + 0.2T_g + 0.1T_a \tag{1}$$

$$WBGT = 0.7T_{nwb} + 0.3T_g \tag{2}$$

where T_{nwb} is wet-bulb temperature °C, T_g is blackglobe temperature (°C), and T_a is air temperature (°C). Considering the black globe temperature is difficult to obtain, we used the 'simplified WBGT' (hereafter denoted simply as 'W', as in Eq. (3)) which was developed by the American College of Sports Medicine (ACSM 1984). W depends only on air temperature and relative humidity, and represents heat stress for average daytime conditions outdoors:

$$W = 0.567T_a + 0.393e + 3.94 \tag{3}$$

$$e = (RH/100) \times 6.105\exp\left(\frac{17.27T_a}{237.7 + T_a}\right) \tag{4}$$

Where e is water vapor pressure (hPa), which was calculated on the basis of daily air temperature and relative humidity using Eq. (4). RHis relative humidity (%).

We first calculated daily W during the reference period (1986–2005) and the future period (2006–2095) using daily simulations from 19 CMIP5 GCMs under Representative Concentration Pathways (RCP) 2.6, 4.5 and 8.5 on each grid. In climatology, the occurrence of at least three consecutive hot days has been defined as a heat wave (Perkins and Alexander 2012). A heat wave is considered to be daily W exceeding a given region's threshold (the local

95th percentile value for daily W over the reference period) for at least three consecutive days. The HWDs, used to quantify heat wave hazard, is defined as the total days of heat waves in a year. In addition, we selected future periods 2076–2095 to identify the long-term future changes of HWDs over global land. For both the reference period and the long-term future periods we used 20-year averages of HWDs in order to minimize inter-annual variations.

3 RESULTS

3.1 *Future changes in air temperature, relative humidity and W*

Models suggest that the changes in annual mean W under the three RCP scenarios are overall smaller, more spatially uniform, and thus have less inter-model variations than those in annual mean air temperature (Figure 1). For the RCP 8.5 scenario, the median change in air temperature is nearly 1°C higher than that in W (Figure 1a, b). This is because GCMs that simulate stronger warming also tend to show larger decreases in relative humidity, generating a cancelation effect between air temperature and water vapor. The result is consistent with previous studies that found the spread of changes in WBGT was substantially smaller than that in air temperature from both observations and models (Coffel et al. 2018).

3.2 *Future changes in HWDs*

Figure 2 displays the spatial pattern of the heat stress expansion over the global land. We find that the number of HWDs per year under RCP 8.5 during future periods 2076–2095 have more spatial heterogeneity and more variable than that of RCP 4.5 and RCP 2.6 during these three future periods (Figure 2). In addition, the heat waves emerge more prominently and frequently at low latitudes relative to middle latitudes and high latitudes under all scenarios. It shows that approximately half the land area of the earth could be exposed to heat waves for more than 100 days in a year under high future emissions (RCP 8.5).

Between 2076 and 2095, the average number of annual total HWDs in the tropical areas, including northern South America, central Africa, Southeast Asia, and some Pacific island countries, will even exceed 350 under RCP 8.5, much more than that of higher latitudes. It is noted that some of these areas most susceptible to heat waves are also the most densely populated and have substantial population growth, resulting in large increases in the number of people exposed to dangerous heat and humidity combinations (Coffel et al. 2018).

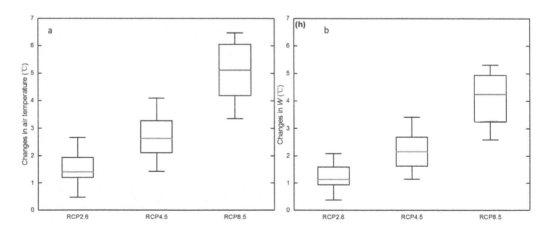

Figure 1. Changes in annual mean air temperature (left) and annual mean W (right) during 2076–2095 relative to 1986–2005 under scenarios of RCP 2.6, RCP 4.5 and RCP 8.5. The corresponding range of projected spatially averaged increases in annual mean temperature over land was based on the results of 19 CMIP5 GCMs.

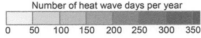

Number of heat wave days per year

0   50   100 150 200 250 300 350

Figure 2.   Spatial distribution of the average number of HWDs during (a) the reference period and (b) the three future periods 2076–2095 under RCP 8.5.

4   CONCLUSION AND DISCUSSION

Our study finds that the spatially average number of HWDs in a year will continue to increase throughout the 21st century under these emission scenarios, and the amount will be much greater under higheremission scenarios. Our analysis indicates that the tropical areas will be exposed to heat waves for almost the entire year under RCP 8.5 in the late 21st century, compared to less 50 HWDs in the reference period. During the period of 2086–2095, approximately half the land area of the earth could be exposed to heat waves for more than 100 days in a year under high future emissions (RCP 8.5).

Many earlier studies have shown that heat wave stress is projected to increase over all land regions along with rising temperatures. This paper shows that tropical humid areas will be disproportionately exposed to more HWDs, because these areas have both year-round warm temperature and higher relative humidity. These results show that HWDs has a better indication of heat waves.

ACKNOWLEDGMENTS

This work was supported by the National Key Research Program China (No.2016YFA0602403), National Natural Science Foundation of China (No. 41775103).

REFERENCES

ACSM. 1984. Prevention of thermal injuries during distance running: Position stand. *Med J Aust* 141: 876–879.

Barnett, A.G., Tong, S., Clements, A.C.A. 2010. What measure of temperature is the best predictor of mortality? *Environ Res* 110: 604–611.

Coffel, E.D., Horton, R.M., Sherbinin, A.D. 2018. Temperature and humidity-based projections of a rapid rise in global heat stress exposure during the 21st century. *Environ Res Lett* 13:1.

Cowan, T., Purich, A., Perkins, S., Pezza, A., Boschat, G., Sadler, K. 2014. More frequent, longer, and hotter heat waves for Australia in the twenty-first century. *J Clim* 27: 5851–5871.

Kang, S., Eltahir, E.A.B. 2018. North China Plain threatened by deadly heat waves due to climate change and irrigation. *Nat Commun* 9: 2894.

Kovats, R.S., Hajat, S. 2008. Heat stress and public health: a critical review. *Annu Rev Publ Health* 29: 4155.

Leng, G.Y., Tang, Q.H., Huang, S.Z., Zhang, X.J. 2016. Extreme hot summers in China in the CMIP5 climate models. *Clim Change* 135(3–4): 669–681.

Lin, C.Y., Chien, Y.Y., Su, C.J., Kueh, M.T., Lung, S.C. 2017. Extreme variability of heat wave and projection of warming scenario in Taiwan. *Clim Change* 145: 305–320.

Meehl, G.A., Tebaldi, C. 2004. More intense, more frequent, and longer lasting heat waves in the 21st century. *Science* 305(5686): 994–997.

Perkins, S.E., Alexander, L.V, 2012. On the measurement of heat waves. *J Climate* 26(13): 4500–4517.

Russo, S., Sillmann, J., Sterl, A. 2017. Humid heat waves at different warming levels. *Sci Rep* 7: 7477.

Risk Analysis Based on Data and Crisis Response Beyond Knowledge – Huang & Nivolianitou (eds)
© *2020 Taylor & Francis Group, London, ISBN 978-0-367-25146-8*

Nonstationary risk analysis of extreme rainfall events

Pengcheng Xu
School of Geographic and Oceanographic Science, Nanjing University, China

Dong Wang*
Key Laboratory of Surficial Geochemistry, Ministry of Education, Department of Hydrosciences,
School of Earth Sciences and Engineering, State Key Laboratory of Pollution Control and Resources
Reuse, Nanjing University, China

ABSTRACT: Risk analysis of extreme hydrological events should be emphasized critically in water resources management. Due to the obviously changing environment and urbanization, the stationarity assumption of extreme events for performing the hydrological frequency analysis to make a systematic risk assessment is challenging and problematic. We investigate nonstationarity and trends in the annual precipitation extremes in six selected gauges in Haihe River Basin of China. This paper is aimed at make a deep comparison of three models in terms of trend analysis, goodness of fit, extreme rainfall quantiles corresponding to different return periods. This results denote the importance of considering nonstationarity when assessing the return period and hydrological risk. It is concluded that nonstationary risk analysis of annual maximum rainfall series can be necessary to water resources management coping with climate change.

Keywords: extreme hydrological events, nonstationary risk analysis, return period, climate change

1 INTRODUCTION

Under the background of the rapid progress of global climate change and urbanization, the heavy rains-induced floods all over the world have risen in recent decades which is gradually becoming a major obstacle to the sustainable development of social economy (Donat et al., 2016; Ali and Mishra, 2018). Hydrological extreme events, such as storms, droughts and floods, have a great impact on agriculture, urban construction, and ecosystems. Risk analysis of these extreme events is essential and fundamental for the design of hydraulic infrastructures and water resources management.

Traditional hydrological frequency analysis is based on stationary assumptions, which means that environmental control factors such as climatic factors and land use rate have a constant mechanism or pattern that affects hydrological variables in the past, present and future). The feasibility of hydrological frequency and risk analysis based on stationary assumptions is challenged under the multiple effects of climate change, urbanization and heat island effects. In recent years, nonstationary hydrological frequency analysis has become a research hotspot in hydrology under the drive of climate change (Chen and Sun, 2017; Call et al., 2017).

Therefore, great efforts have been made to incorporate the influences of non-independence and nonstationarity of hydrological extreme events. Nonstationarity may be attributed to climate variability or different human influences (e.g. urbanization, regulation of rivers, construction of dikes, deforestation, etc.). The approaches of nonstationary frequency analysis

---

*Corresponding author: wangdong@nju.edu.cn

proposed in the literature involve parameters of a chosen distribution that are dependent on time. Delgado et al. (2010) made a deep test of the GEV model with time-varying location and scale parameter for the Mekong River. The results indicated rising probability of extreme floods can be detected from the increase of the scale parameter. Chen and Sun (2017) conducted a trend analysis of the CMIP5 (Coupled Model Intercomparison Project Phase 5) climate model dataset to verify the impact of human activities on China's daily extreme precipitation increase. Ganguli and Coulibaly (2017)investigate nonstationarity and trends in the short-duration precipitation extremes in selected urbanized locations in Southern Ontario, Canada, and evaluate the potential of nonstationary intensity–duration–frequency (IDF) curves.

Since the number of extreme rainfall events hasbeen observed over recent decades in Haihe River Basin of China, especially in the city of Beijing, the accuracy of existing estimation methods has diminished to some extent. Therefore, the need for a nonstationary approach of extreme rainfal lrisk analyses has arisen. In fact, a few studies in the past have shown that Gumbel distribution or extreme value type I fits poorly to the historical rainfall extremes (Coulibaly et al., 2015). Therefore, in the present study, we perform risk analysis of extreme precipitation using a generalized extreme value (GEV) distribution model.

2 METHODS

The methods used in this paper including the trend analysis, GEV distribution parameter estimation corresponding to nonstationary model and the calculation of hydrological risk. The details of these methods can be found in these reference (Condon et al., 2015; Gilleland and Katz, 2016).

Let $F(X)$ be the cumulative probability distribution function (CDF) of the quantity of interest, in this study, maximum daily precipitation in a year (Im). The GEV distribution is composed of three parameters, the location, the scale, and the shape of thedistribution, which describes the center of the distribution,the deviation around the mean, and the shape or the tail ofthe distribution respectively. The cumulative distribution function of stationary GEV model can be derived as follows:

$$F(x) = \begin{cases} \exp\left\{-\left[1 + \kappa\left(\frac{x-\mu}{\sigma}\right)\right]_+^{-\frac{1}{\kappa}}\right\} if \kappa \neq 0 \\ \exp\left\{-exp\left(-\frac{x-\mu}{\sigma}\right)_+\right\} if \kappa \to 0 \end{cases} \tag{1}$$

where z_+=max{y,0} and
$x \in [(\mu - \sigma)/\kappa, +\infty)$ when $\kappa > 0$,
$x \in (-\infty, (\mu - \sigma)/\kappa)$ when $\kappa < 0$, and
$x \in (-\infty, +\infty)$ when $\kappa = 0$.

μ is the location parameter, σ is a scale parameter and κ is a shape parameter. In this study, two kinds of nonstationary GEV models (GEVns-1 and GEVns-2) are assumed with shape parameter being constant. GEVns-1 model considers the time-varying characteristics from location parameter only while GEVns-2 model incorporates the time varying features into both location and scale parameter. These two nonstationary models is to regard the significant trends as a linear function of time (in years):

$$\mu(t) = \mu_o + \mu_1 t \tag{2}$$

$$\sigma(t) = \exp(\sigma_o + \sigma_1 t) \tag{3}$$

where the scale parameter is always positive throughout, it is usually calculated on the basis of a log link function.

In this study, the Bayesian method through the Markov chain Monte Carlo (MCMC) approach (Cheng et al., 2014) is performed to estimate the nonstationary GEV model. Simultaneously, the Deviance Information Criterion (DIC) and Bayes factors (BF) for different stationary and nonstationary models are calculated to select the best frequency model. The minimum value of DIC recommends the best model while BF smaller than 1 indicates the best fitting.

Conventionally, the T-year return level for certain daily precipitation x_T is equal to the (1-1/T)-th quantile of marginal distribution of Im (The probability distribution is the same for all years in a stationary situation). Equivalently, on average, one out of T years has at least one daily rainfall that exceeds x_T, so that $T(1 - F(x_T)) = 1$. That is to say that the probability of the annual maximum daily rainfall exceeds x_T is 1/T.

So, hydrological stationary risk can be defined as follows:

$$R_s = 1 - F(x_T)^n = 1 - (1 - 1/T)^n \tag{4}$$

Considering the circumstances of time-varying exceedance probabilities, the probability of the annual maximum daily rainfall exceeding x_T in each year is different. So here we use $F_t(x_T)$ to represent the probability of daily rainfall exceeding design level x_T at t-th year. So the hydrological risk for the nonstationary case is:

$$R_{ns} = 1 - \prod_{t=1}^{n} F_t(x_T) \tag{5}$$

3   CASE STUDY

The related daily precipitation of 1958-2017 from six gauges of the Haihe River Basin of China observed were analyzed through the proposed nonstationary model (**Figure 1**).

3.1   *Detection of nonstationarity*

Before the establishment of the nonstationary frequency analysis model, it is essential to examine nonstationarities of AMP (Im). A series of statistical tests (i.e. a Ljung-Box test, Man-Kendall test and Pettitt test) are performed in this section to detect the presence of nonstationary trends and abrupt shifts in the extreme precipitation time series. A more

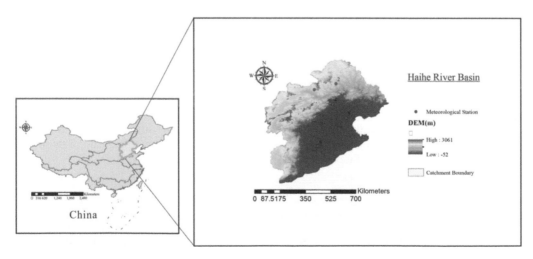

Figure 1.   Selected meteorological stations in Heihe River Basin.

Table 1. Detection of trends and nonstationarity in Heihe River Basin.

Station No.	L-jungBox Test p.value	MK test p.value	Z statistics	Pettitt Test p.value
1	0.798	0.674	0.421	0.708
2	0.462	0.132	-1.649*	0.091
3	0.575	0.051	-1.963**	0.012
4	0.971	0.054	-1.926*	0.089
5	0.747	0.214	-1.244	0.678
6	0.923	0.067	-1.831*	0.024

comprehensive and rigorous evaluation of the over trend in the time series can be addressed objectively according to various tests. Table 1 shows the detailed results of the nonstationary detecting tests.Firstly, time series of Im for all 6 stations can pass the Ljung-Box test with 20 lags (p.value > 0.05 in **Table 1**).

In the figure, the extreme observations are mutually independent with no serial autocorrelation, so it is credible to perform the standardized Mann-Kendall test to estimate the statistical significance of trend without any modification. As shown in **Table 1**, co-currences of significant trends, change points and nonsationarities for the rainfall extremes can be detected in multiple stations (station 2, 3, 4 and 6). Station 1 exhibits shows insignificant decreasing trend. On the other hand, station 5 shows a weak decreasing trend. Above tests taken together detect the presence of nonstationarity in rainfall extremes across three out of 6 sites. According to Porporato and Ridolfi (1998), an insignificant trend can also have a significant impact on the results of probability analysis. Hence even if precipitation extremes in certain station exhibit statistically insignificant trends, we assess the performance of both nonstationary and stationary models for all sites in the following section.

3.2 Establishment of nonstationary GEV model

In the second phase of this study, stationary and nonstationary GEV model are established according to Equation (1)-(3). The results of the fitted model parameters through MCMC approach can be found in **Table 2**. As shown in Table 2, best fitted GEV model for station 1 is

Table 2. Best fitted GEV model for six stations.

No.	Model	μ	σ	κ	DIC	BF
1	**GEV$_s$**	**50.64**	**18.13**	**0.012**	**1593.71**	-
	GEV$_{ns}$-1	46.69+0.0013t	17.81	0.035	1597.58	1.0011
	GEV$_{ns}$-2	51.7-0.00062t	exp(2.96-0.000037t)	0.019	1596.53	1.004
2	GEV$_s$	40.71	14.01	0.053	1497.533	-
	GEV$_{ns}$-1	42.02+0.00026t	14.30	0.099	1493.47	0.997
	GEV$_{ns}$-2	**42.90-0.00031t**	**exp(2.39+0.00012t)**	**0.082**	**1492.82**	**0.996**
3	GEV$_s$	34.31	11.00	0.13	1455.2	-
	GEV$_{ns}$-1	**37.51-0.00067t**	**11.38**	**0.10**	**1448.41**	**0.995**
	GEV$_{ns}$-2	41.76-0.0030t	exp(2.37+0.000019t)	0.11	1449.82	0.996
4	GEV$_s$	56.36	22.29	0.34	1751.59	-
	GEV$_{ns}$-1	65.73-0.0035t	23.17	0.31	1749.81	0.9995
	GEV$_{ns}$-2	**60.33-0.00055t**	**exp(3.10+0.000026t)**	**0.31**	**1749.48**	**0.9988**
5	GEV$_s$	67.46	26.37	0.072	1752.36	
	GEV$_{ns}$-1	77.72-0.0041t	26.88	0.047	1750.30	0.9986
	GEV$_{ns}$-2	**70.19-0.00072t**	**exp(3.66-0.00019t)**	**0.067**	**1749.53**	**0.9979**
6	GEV$_s$	63.02	29.43	0.033	1767.72	-
	GEV$_{ns}$-1	69.26-0.0053t	27.93	0.059	1767.24	1.0001
	GEV$_{ns}$-2	**57.05+0.00083t**	**exp(3.45-0.000071t)**	**0.072**	**1766.92**	**0.9999**

the stationary model (GEV$_s$) with smallest *DIC* value. Meanwhile Bayes factors (*BF*) for-GEVns-1 and GEVns-2 is larger than 1 which help recommends the stationary model is better than two nonstationary models. For most cases, the GEVns-2 is most likely distribution model for four stations which means incorporation of the time trend into the scale parameter, in addition to the location parameter, can help yield a significant improvement over the nonstationary model with time trend only in the location parameter.

4 CONCLUSION

The present paper has conducted a nonstationary risk analysis model through the GEV distribution function. The best fitted GEV model was selected from the stationary GEV and two nonstaionary models based on the smallest *DIC* value and smaller *BF*. The Bayesian method through the Markov chain Monte Carlo (MCMC) approach is proposed to estimate the GEV parameters. And the proposed risk analysis model was implemented in the six gauging stations of Heihai River Basin of China. Comparison of return period and risk calculation between stationary and nonstationary conditions recommend that use time-varying risk analysis model to assist decision making for the management of water resources exacerbated by climate change is more feasible than the traditional stationary model which can apply to the many similar situations around theworld.

ACKNOWLEDGEMENTS

This study was supported by the National Key Research and Development Program of China (2016YFC0401501,2017YFC1502704), and the National Natural Science Fund of China (41571017, 51679118 and 91647203), and Jiangsu Province"333 Project" (BRA2018060).

REFERENCES

Ali, H., and Mishra V. 2018. Increase in subdaily precipitation extremes in India under 1.5 and 2.0°C warming worlds. *Geophysical Research Letters*, *45*, 6972–6982.

Call, B.C., Belmont, P. Schmidt, J.C., and Wilcock, P.R. 2017. Changes in floodplain inundation under nonstationary hydrology for an adjustable, alluvial river channel. *Water Resources Research, 53(5)*, 3811–3834.

Chen, H., and Sun J. 2017. Contribution of human influence to increased daily precipitation extremes over china. *Geophysical Research Letters, 44(5)*, 2436–2444.

Cheng, L., AghaKouchak, A., Gilleland, E., and Katz, R.W., 2014. Non-stationary extreme value analysis in a changing climate. *Climatic Change*, 127, 353–369.

Condon, L.E., Gangopadhyay, S., and Pruitt, T. 2015. Climate change and non-stationaryflood risk for the upper Truckee River basin. *Hydrology and Earth System Sciences*, 19, 159–175.

Coulibaly, P., Burn, D., Switzman, H., Henderson, J., and Fausto, E. 2015. A comparison of future IDF curves for Southern Ontario, Technical Report, McMaster University, Hamilton, available at: https://climateconnections.ca/wp-content/uploads/2014/01/IDF-Comparison-Report-and-Addendum.pdf (last access: 9December 2016).

Delgado, J.M., Apel, H., and Merz, B. 2010. Flood trends andvariability in the Mekong river. *Hydrology and Earth System Sciences*, 14, 3, 407–418.

Donat, M.G., Lowry, A.L., Alexander, L.V., O'Gorman, P.A., and Maher, N. 2016. More extreme precipitation in the world's dry and wetregions. *Nature Climate Change*, 6, 508–513.

Ganguli, P., and Coulibaly, P. 2017. Does nonstationarity in rainfall requires nonstationary intensity-duration-frequency curves? *Hydrology and Earth System Sciences*, 21, 6461–6483.

Risk Analysis Based on Data and Crisis Response Beyond Knowledge – Huang & Nivolianitou (eds)
© 2020 Taylor & Francis Group, London, ISBN 978-0-367-25146-8

Flood risk analysis based on the copula method

Pengcheng Xu
School of Geographic and Oceanographic Science, Nanjing University, China

Dong Wang*
Key Laboratory of Surficial Geochemistry, Ministry of Education, Department of Hydrosciences,
School of Earth Sciences and Engineering, State Key Laboratory of Pollution Control and Resources
Reuse, Nanjing University, China

ABSTRACT: Due to climate change and increasing urbanization, heavy rains-induced floods
have occurred more frequently all over the world in recent decades, which is becoming a major
deterrent to the sustainable development of social economy. We try to detect risk and return
period from both univariate and bivariate cases according to the annual maximum daily streamflow observed at two selected gauges in Yangtze River Basin of China. In this study, copulabased approach is developed for frequency analysis of flood events. The GEV distribution parameter and copula parameter are both estimated through the Maximum Likelihood Estimation
method (MLE). Results show that Kendall's return period can be achieved in the same joint
distribution probability level corresponding to the same dangerous area, compared to the traditional return period method (AND/OR method). The definition of Kendall's return period is
more reasonable and can effectively avoid the mistakes of assessing flood events.

Keywords: extreme hydrological events, copula, Kendall's return period, bivariate
modelling

1   INTRODUCTION

Under the background of the rapid progress of global climate change and urbanization, the
heavy rains-induced floods all over the world have risen in recent decades which is gradually
becoming a major obstacle to the sustainable development of social economy (Donat et al.,
2016; Ali and Mishra, 2018). Hydrological extreme events, such as storms, droughts and
floods, have a great impact on agriculture, urban construction, and ecosystems. Risk analysis
of these extreme events is essential and fundamental for the design of hydraulic infrastructures
and water resources management.

One of the challenges facing the field of hydrology is gaining a better understanding of
flood regimes. To do this, flood frequency analysis (FFA) is most commonly used by engineers and hydrologists worldwide and basically consists of estimating flood peak quantiles for
a set of non-exceedance probabilities. Generally current methods of FFA assume that the
flood series data are independent and identically distributed or, in other words, they are free
of trends and abrupt changes. Univariate frequency analysis has been widely used to assess
the various flood characteristics (Cancelliere and Salas, 2004; Deo et al., 2017). Since one
characteristic, based on traditional univariate frequency analysis, is insufficient for
a comprehensive assessment of droughts, it is important to simultaneously consider all flood
characteristics using a joint probability distribution of flood variables. The extreme events

*Corresponding author: wangdong@nju.edu.cn

have varying characteristics. For example, characteristics of extreme rainfall are amount, duration, number of events in a year, and inter-arrival time; droughts are characterized by duration, severity, areal extent, and inter-arrival time; and floods are characterized by peak, volume, duration, and inter-arrival time. These characteristics are usually dependent variables. Since the joint behavior of variables has pivotal implications in practical applications, the construction of joint probability distributions is important for planning, design, and management of water resources and environmental systems. Copulas, with the advantage of modeling the dependence structure independently of the marginal distributions of variables, are now becoming a standard tool for multivariate frequency analysis of rainfall storms (Kao and Govindaraju, 2008;); droughts (De Michele et al., 2013); floods; and other hydrological events (Bárdossy and Pegram, 2009).Bivariate copulas is able to investigate the cross-dependence between meteorological and hydrological droughts.

Therefore, in the present study, we perform risk analysis of annual daily stream flow using a generalized extreme value (GEV) distribution combined with copula model. The objectives of this paper can be concluded as follows: (1) marginal probability distribution and bivariate probability distribution modeling; (2) risk analysis through the Kendall's and traditional return period methods; (3) comparison of design values series corresponding to various return period methods.

2   METHODS

The methods used in this paper including GEV distribution parameter estimation and copula parameter estimation. The details of these methods can be found in these reference (Gilleland and Katz, 2016; Grobmab, 2007).

Let $F(X)$ be the cumulative probability distribution function (CDF) of the quantity of interest, in this study, maximum daily stream flow in a year (F_1 and F_2). The generalized extreme value (GEV) distribution model is composed of three parameters, the location, the scale, and the shape of the distribution, which describes the center of the distribution, the deviation around the mean, and the shape or the tail of the distribution respectively. The cumulative distribution function of stationary GEV model can be derived as follows:

$$
F(x) = \begin{cases} \exp\left\{-\left[1 + \kappa\left(\frac{x-\mu}{\sigma}\right)\right]_+^{-\frac{1}{\kappa}}\right\} \text{if } \kappa \neq 0 \\ \exp\left\{-\exp\left(-\frac{x-\mu}{\sigma}\right)_+\right\} \text{if } \kappa \to 0 \end{cases}
\tag{1}
$$

where $z_+ = \max\{y, 0\}$ and
   $x \in [(\mu - \sigma)/\kappa, +\infty)$ when $\kappa > 0$,
   $x \in (-\infty, (\mu - \sigma)/\kappa]$ when $\kappa < 0$, and
   $x \in (-\infty, +\infty)$ when $\kappa = 0$.
μ is the location parameter, σ is a scale parameter and κ is a shape parameter.

And the copula model can be presented as follows:

Let $[X_1, X_2]$ be a bivariate discrete random vector with joint probability density function $f(x_1, x_2)$ and one-dimensional marginal density functions, respectively, as $f_{X_1}(x_1), f_{X_2}(x_2)$ for X_1 and X_2. The joint probability distribution was determined using the copula method which is based on Sklar's theorem. Accordingly, a 2-dimensional copula C can be expressed as (Grobmab, 2007):

$$
F(x_1, x_2) = C(F_{X_1}(x_1), F_{X_2}(x_2); \theta),
\tag{2}
$$

where F denotes the bivariate cumulative distribution function (cdf), F_{X_i} is the marginal distribution of X_i, and θ is the copula parameter. The copula estimation can be summarized as follows.

The marginal empirical distribution function which is the pseudo-observation can be defined as:

$$\hat{F}_i(x) = \frac{1}{n+1} \sum_{t=1}^{n} 1(X_{it} \le x), \forall i = 1, 2 \tag{3}$$

Then, the maximum pseudo-likelihood estimator (MLE) is given by (Chenand Fan, 2006):

$$\hat{\theta} = \arg\max_{\theta \in \Theta} \left\{ \frac{1}{n} \sum_{t=1}^{n} \log c\left[\hat{F}_1(x_{1t}), \hat{F}_2(x_{2t}); \theta\right] \right\} \tag{4}$$

From the perspective of bivariate case, the joint return period (JRP) of extreme rainfall events can be calculated through three methods in stationary situation (Salvadori et al., 2011). They are AND method corresponding to the probability of $P(X \ge x \cap Y \ge y)$, OR method corresponding to $P(X \ge x \cup Y \ge y)$,and Kendall return period method. Details of Kendall return period can be referred to Salvadori and De Michele (2004). They can be calculated as follows:

$$T_{and} = \frac{1}{P((X \ge x \cap Y \ge y))} = \frac{1}{1 - F_X(x) - F_Y(y) + C[F_X(x), F_Y(y)]} \tag{5}$$

$$T_{or} = \frac{1}{P((X \ge x \cup Y \ge y))} = \frac{1}{1 - C[F_X(x), F_Y(y)]} \tag{6}$$

$$T_{ken} = \frac{1}{P\{C[F_X(x), F_Y(y)] \ge p_{ken}\}} = \frac{1}{1 - K_c(p_{ken})} \tag{7}$$

where $K_c(\cdot)$is the Kendall distribution function.

3   CASE STUDY

The related daily stream flow of 1952-2012 from two gauges of the Yangtze River Basin of China observed at Hankou and Yichang gauge. The annual daily flow is extracted from the above dataset which can be regarded as two flood series (F_1 and F_2 in the following sections).

3.1   *Modeling the marginal and bivariate distributions*

In the first phase of this study, GEV model for marginal distribution are established according to Equation (1). The results of the fitted model parameters through maximum likelihood estimation (MLE) approach can be found in **Table 1**. Meanwhile, the MLE method is compared t0 the generalized maximum likelihood method (GMLE) proposed by Martinsand Stedinger (2001). Here, the minimum *AIC* criterion is adopted to compare these two estimation methods. For flood series F_1, the location parameter is 50.26 while the scale and the shape parameter are 9.05 and -0.27 respectively. And the MLE method can provide the smaller *AIC* value than that provided by GMLE method. On the other hand, the results shown in Table 1 also demonstrated that MLE approach for smaller samples provides more accurate parameter estimates, i.e. narrower confidence intervals than GMLE approach. For flood series F_2, the location parameter is 45.56 while the scale and the shape parameter are 9.10 and -0.29 respectively. And in Table 1, comparison of the density of selected GEV model and empirical cumulative distribution function (CDF) also recommends that the GEV model is more appropriate for F_2 than that for F_1.

Elliptical and Archimedean copulas which can reflect several desirable properties with simple computation of measures of dependence pattern have been two potential kinds of copulas in hydrological practical application. In the second phase of this study, elliptical copulas,

Table 1. Parameters of GEV model for two flood series based on MLE and GMLE method.

Series	EM	μ	σ	κ	AIC
F_1	MLE	**50.26**	**9.05**	**−0.27**	**455.84**
		[47.77,52.84]	**[7.30,10.80]**	**[−0.42,−0.11]**	
	GMLE	50.31	9.33	−0.29	468.25
		[47.79,52.84]	[7.42,11.23]	[−0.44,−0.14]	
F_2	MLE	**45.56**	**9.10**	**−0.29**	**454.61**
		[43.07,48.04]	**[7.37,10.84]**	**[−0.44,−0.14]**	
	GMLE	45.57	9.23	−0.30	489.59
		[43.06,48.08]	[7.42,11.06]	[−0.45,−0.16]	

Table 2. Copula modeling over the dependence pattern between two flood series.

copula	θ	RMSE	Sn	$\widehat{xv_n}$	LL
Clayton	**1.2971**	**0.018**	**0.0203**	**17.43**	**14.58**
Gumbel	1.4598	0.028	0.0491	2.73	7.02
Frank	3.2052	0.025	0.0417	6.27	7.18
normal	0.5450	0.023	0.0330	6.14	9.35
t	0.5452	0.024	0.0320	9.43	9.99

Student t and Normal copula, as well as three widely-used types of Archimedean copulas, Clayton, Gumbel and Frank copula were selected as candidate copulas to model the dependence structures of hydrometeorological variables. The detailed results of goodness-of-fit are shown in **Table 2**. The best fitted copula (bold face in Table 2) achieved the minimum RMSE, the minimum Cramér-von Mises statistic (S_n), the maximum log-likelihood value (LL), and the minimum cross-validation value ($\widehat{xv_n}$). According to the results, Clayton copula is selected as the best fitted copula to model the dependence structure between these two flood series (F_1 and F_2). And the copula parameter is estimated through maximum likelihood estimation method. In conclusion, the MLE approach is reliable for estimate the parameter both from marginal and bivariate sides.

3.2 *Risk analysis for flood series*

After the best fitted marginal and copula distribution model are achieved, the risk analysis can be conducted through AND, OR and KEN methods. In this study, the return period analysis is adopted as the entry point of risk analysis. The return periods of T_{and}, T_{or} and T_{ken} ranging from 2 to 200 years were calculated according to Equations (5–7).Results of return period are shown in Table 3, Table 4. As shown in Table 3, T_{or} is always smaller than the return period under univariate cases while T_{and} and T_{ken} are larger than that under univariate cases. The T_{ken} value is between T_{and} andT_{or}, which is the most reproducible with the design return period of the sample. The rationality of Kendall's return period method is validated further.

The design value of flood quantiles corresponding to different return periods can be obtained according to the specific marginal and copula distribution function (**Table 4**). Generally, the design value under the T_{or} criterion is higher than that under T_{and} and T_{ken} criterion while the design value under the T_{ken} criterion is smaller than that under T_{and}. That is becauseT_{or} criterion expands the scope of dangerous events, leading to overestimation of hazardous areas. Similarly, the reason for the large value of T_{and} is due to the narrowing of the

Table 3. Three joint return periods for these two flood series (F_1 and F_2).

T_{F_1}	T_{F_2}	$T_{or}(F_1, F_2)$	$T_{and}(F_1, F_2)$	$T_{ken}(F_1, F_2)$
200	200	100.49	17525.90	6135.00
100	100	50.58	4409.70	2941.18
50	50	25.57	1116.53	416.67
20	20	10.57	185.41	102.04
10	10	3.18	49.17	29.24
5	5	3.06	13.70	8.43
2	2	1.54	2.86	2.25

Table 4. Design values of two flood series corresponding to various return levels.

T/a	F_1				F_2			
	$T(id115)$	T_{or}	T_{and}	T_{ken}	$T(F_2)$	T_{or}	T_{and}	T_{ken}
2	53.42	57.40	50.12	52.18	48.72	52.23	45.76	48.20
5	61.44	65.41	56.68	58.98	56.62	59.70	52.66	54.24
10	65.56	70.99	60.92	62.62	60.58	62.00	54.82	56.83
20	68.81	73.54	62.50	65.65	63.65	64.879	57.84	59.27
50	72.19	76.09	65.78	67.84	66.78	67.754	61.14	61.72
100	74.22	75.97	67.84	76.46	68.63	70.629	62.29	61.43
200	75.90	79.13	69.54	70.02	70.13	70.773	63.58	65.74

range of dangerous events. To some extent, K end all's return period overcomes the shortcomings of the traditional recurrence period to overestimate the dangerous area by reasonably dividing the dangerous area.

A difference analysis on the design values of **Table 4** is performed and the results are shown In terms of the quantiles of flood series F_1 under different return periods, the maximum error (7.74%) of T_{ken} occurs at the return level of 200 years which is smaller than that of T_{and} and T_{or}. Similarly, for flood series F_2, the error provided by T_{ken} is smaller than the error by T_{and} and T_{or}. This indicates that design value under K end all's return period is closer to design value under univariate case than the traditional return period.

4 CONCLUSIONS

The present paper has conducted a risk analysis model through the GEV distribution function and copula model. The parameters of GEV and copula model are estiamated through the MLE appraoch. And the proposed risk analysis model considering the was implemented in the two gauging stations of Yangtze River Basin of China. Results show that Kendall's return period can be achieved in the same joint distribution probability level corresponding to the same dangerous area, compared to the traditional return period method (AND and OR method). The definition of Kendall's return period is more reasonable and can effectively avoid the mistakes of assessing flood events, which is more conducive to guiding the risk management of the tide and regarded as a reference for the prevention and control of urban innudation.

ACKNOWLEDGEMENTS

This study was supported by the National Key Research and Development Program of China (2017YFC1502704, 2016YFC0401501), and the National Natural Science Fund of China (41571017, 51679118 and 91647203), and Jiangsu Province" 333 Project" (BRA2018060).

REFERENCES

Ali, H., and Mishra V. 2018. Increase in subdaily precipitation extremes in India under 1.5 and 2.0 °C warming worlds. *Geophysical Research Letters*, *45*, 6972–6982.

Bárdossy, A., and Pegram, G.G.S., 2009. Copula based multisite model for daily precipitationsimulation. Hydrology Earth System Sciences, 13(12), 2299–2314.

Cancelliere, A., Salas, J.D., 2004. Drought length properties for periodic-stochastic hydrologic data. *Water Resources Research*, 40(2), 389–391.

Deo, R.C., Byun, H.R., Adamowski, J.F., Begum, K., 2017. Application of effective drought index for quantification of meteorological drought events: a case study in australia. *Theoretical and Applied Climatology*, 128(1-2), 359–379.

Donat, M.G., Lowry, A.L., Alexander, L.V., O'Gorman, P.A., and Maher, N. 2016. More extreme precipitation in the world's dry and wetregions. *Nature Climate Change*, 6, 508–513.

Gräler, B., van den Berg, M.J., Vandenberghe, S., Petroselli, A., Grimaldi, S., De Baets, B., Verhoest, N.E.C., 2013. Multivariate return periods in hydrology: a critical and practical review focusing on synthetic design hydrograph estimation. *Hydrology Earth System Sciences*, 17, 1281–1296.

Gilleland, E. and Katz, R.W. 2016. Extremes 2.0: an extremevalue analysis package in. *Journal of Statistical Software*, 72, 1–39.

Grobmab, T., 2007. Copula and tail dependence. Berlin: Diploma thesis, Humboldt-University.

Risk Analysis Based on Data and Crisis Response Beyond Knowledge – Huang & Nivolianitou (eds)
© *2020 Taylor & Francis Group, London, ISBN 978-0-367-25146-8*

An approach to measure interaction in precipitation system under changing environment

Wenqi Wang, Dong Wang & Yuankun Wang
Department of Hydrosciences, School of Earth Sciences and Engineering, Nanjing University, Nanjing, China

ABSTRACT: Precipitation system is a common complex system in nature. The interaction and dependent structure of the meteorological factors in the system are of great significance to the establishment of hydrological model. In this study, we propose a framework coupling entropy and copula for analyzing the process correlation of meteorological factors in precipitation system. Based on long time series of daily observational data (1985–2015), we study process interaction between meteorological factors and make better inference for complex system network behavior shifts under changing environment.

Keywords: complex system, entropy, Copula, mutual information, copula entropy, system interaction

1 INTRODUCTION

How to define a system is a problem faced routinely in science and engineering, with solutions developed from our understanding of the processes inherent, to assessing the underlying structure based on observational evidence alone (Goodwell and Kumar, 2017). Information theoretic measures are especially suitable for characterizing and quantifying the complex nonlinear relationship between variables in the system. Also for systems with inconspicuous physical relationship, information theoretic measures can make full use of observational data for detecting connectivity between the variables and to specify the strength of association. Copulas can establish dependent structures between two or more variables independent of different marginal distributions of the individual variables, especially for capturing asymmetric nonlinear dependent structures.

Here we propose a framework coupling information theory and copula for analyzing the process correlation of meteorological factors in precipitation system. We study time series of different meteorological variables, e.g., temperature, air pressure, wind speed, humidity, vapor pressure, sunshine duration and precipitation to quantify dependent and connective structure, and their links and directional information flow intensity under changing environment.

Taihu Lake basin is one of the highly developed areas in the Yangtze River Delta region in China. The west region of Taihu Lake basin is a typical area influenced by subtropical monsoon climate, and rainfall characteristics study of this area is particularly valuable and important for reasonable utilization and planning of water resources, as well as efficient hydrological forecasting under changing environment. The study of precipitation system behavior can also be valuable for better characterization and understanding of impacts on interaction between meteorological factors and holistic precipitation network brought by changing environment.

Our approach presents a way to study and characterize interactions between meteorological factors for precipitation system from a new perspective, and provides a new method for indicating interactions in complex natural system with little preliminary assumption and knowledge.

2 METHODS

2.1 *Information theoretic measure*

Information theory is proposed and developed by Shannon (1948). For a random variable X with the probability density function (PDF) f(x) defined on the interval [a, b], the marginal entropy of the random variable X can be defined as:

$$H(X) = H(p_1, p_2, p_3, \ldots, p_N) = -\sum_{i=1}^{N} p_i \log p_i \tag{1}$$

where $p_i (0 \leq p_i \leq 1)$ is the probability of occurrence $x_i (i = 1, 2, \ldots, N)$ and $\sum p_i = 1$.
For bivariate case, the joint entropy between them can be defined as:

$$H(X, Y) = -\sum_{i=1}^{N} \sum_{j=1}^{M} p_{ij} \log p_{ij} \tag{2}$$

where the joint probability of x_i and y_j denoted as $p(x_i, y_j) = p_{ij}, i = 1, 2, \ldots, N; j = 1, 2, \ldots, M$.

2.2 *Copulas*

Copula provides a flexible way for representing the dependence between random variables, which is guaranteed by Sklar's Theorem (Sklar, 1959). According to Sklar's Theorem, let H be a joint cumulative distribution function with marginal univariate distributions F and G, then there exists a copula C so that $H(x, y) = C[F(x), G(y)]$ (Nelsen, 1999). This copula C is unique if F and G are continuous. For bivariate random vector (X, Y) with marginal functions as $F_X(x)$ and $F_Y(y)$, the joint distribution function can be expressed with copula C as:

$$P(X \leq x, Y \leq y) = C(F_X(x), F_Y(y); \theta) \tag{3}$$

where $U = F_X(x), V = F_Y(y)$ with U, V~U (0,1) and $(U, V)F$~$UV \equiv C$. The multivariate case of the copula function with margins $\{F_i, i = 1, \ldots, N\}$ is defined as:

$$P(\mathbf{x}) = C(F_1(x_1), \ldots, F_N(x_N)) \tag{4}$$

2.3 *Mutual information and copula entropy*

Mutual information describes the amount of information transmission or shared between two random variables, which is represented as:

$$I(X, Y) = \sum_{i=1}^{N} \sum_{j=1}^{M} p_{ij} \log \left(\frac{p_{ij}}{p_i p_j} \right) \tag{5}$$

where p_i, p_j is the probability of occurrence x_i, y_j $(i = 1, 2, \ldots, N; j = 1, 2, \ldots, M)$.
Considering copula entropy, let $\mathbf{x} \in R^N$ be random variables with marginal functions $u = [F_1, \ldots, F_N]$ and copula density c($\mathbf{u}$). Then copula entropy of x is defined as:

$$H_c(x) = -\int_u c(\mathbf{u}) \log c(\mathbf{u}) d\mathbf{u} \tag{6}$$

where $c(\mathbf{u}) = \dfrac{d^N C(\mathbf{u})}{du_1 du_2 \ldots du_N}$.

The entropy of the copula density function (termed as the copula entropy or copula information) is defined as a measure of the dependence uncertainty represented by the copula function.

The copula entropy can be used as a measure of nonlinear dependence, for which the mutual information (MI) is shown to be equivalent to the negative copula entropy (Ma and Sun, 2011).

3 STUDY AREA AND DATA

3.1 *Study area*

Taihu Lake basin is one of the highly developed areas in the Yangtze River Delta region in China. The terrain of the basin is high in the west and low in the east. Meanwhile, the landscape is also low in the center and highin the surrounding area. Mountainous and hilly areas are mainly distributed in the western part of the basin. The area of the western hilly areas is 7338 km^2, accounting for about 20% of the total area of the basin. From north to south, there are Maoshan Mountains, Jieling Mountains and Tianmu Mountains in the western hilly region, which belong to Jiangsu Province and Zhejiang Province. Maoshan Mountains and the northern area of Jieling Mountains are in Jiangsu Province,and the southern area of Jieling Mountains and Tianmu Mountains are in Zhejiang Province.

Taihu Lake Basin belongs to the middle subtropical monsoon climate area, with distinct seasons, mild climate and abundant rainfall.Most of the precipitation is concentrated from May to September. The regional distribution of annual precipitation is affected by climate, water vapor sources and topography. Our study area is located in the west region of Taihu Lake basin. The selected weather station (Liyang station) is marked as the purple point in the map (Figure 1).

3.2 *Data*

Generally, precipitation process is affected by atmospheric circulation and characteristic field of atmospheric physical parameters. Atmospheric circulation can be reflected by air pressure condition (air pressure), and characteristic field of atmospheric physical parameters can be reflected by vapor pressure condition (vapor), dynamic condition (wind speed) and thermodynamic condition (temperature).

Therefore, we choose five meteorological variables and daily observation records of Liyang weather station from 1985 to 2015. Specifically, five meteorological variables include one core

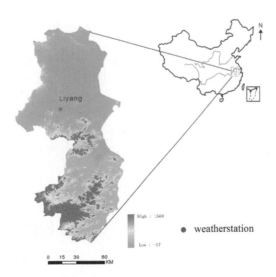

Figure 1. West of Taihu Lake and the selected weather station.

Table 1. Five meteorological variables used in the study.

Variable	abbreviation	units
precipitation	PPT	0.1mm
wind speed	WS	0.1m/s
air pressure	AP	0.1hPa
vapor pressure	VP	0.1hPa
air temperature	AT	0.1°C

variable, i.e., precipitation, and four related variables:wind speed, air pressure, vapor pressure and air temperature. The abbreviation of five variables and corresponding units are shown in Table 1.

4 RESULTS AND DISCUSSIONS

4.1 *Temporal variation of the five variables*

To re-analysis the temporal variation trend of the precipitation system, we first provide their variation curves between 1985 and 2015 (Figure 2). AP, AT and VP showed similar variation patterns, while WS and PPT showed another different variation trend.

4.2 *Measuring interaction by mutual information*

The structure of network system of interactions can be simply described in Figure 3, where lines with arrows represent links, connectivity and interaction, and circles represent nodes, which are time-series variables.

We first compute mutual information between PPT related factors and PPT and within PPT related factors, which means the two cases in Figure 4. Mutual information simply calculated from a kernel method is first provided in Table 2. Two periods (1985–1999 and 2000–2015) are compared for different pairs. It can be observed that the mutual information in these two periods are similar with small changes.

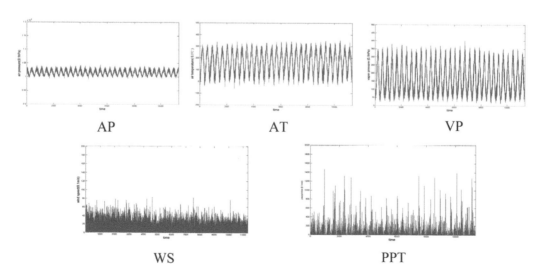

Figure 2. Temporal variation between 1985 and 2015 for five variables (AP, AT, VP, WS, PPT).

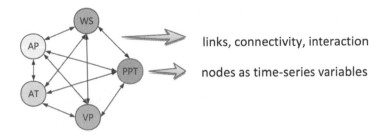

Figure 3. Diagram for the interactive precipitation system.

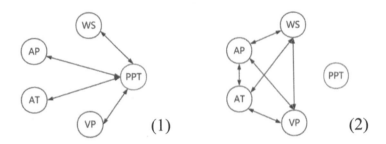

(1)

(2)

Figure 4. Interaction between five variables (1) PPT and other factors; (2) within PPT related factors.

Table 2. Mutual information between five meteorological variables from a kernel method.

Mutual Information (Kernel MI)			
1985–1999		2000–2015	
WS_PPT	0.2072	WS_PPT	0.1405
AP_PPT	0.0762	AP_PPT	0.0664
AT_PPT	0.0836	AT_PPT	0.0767
VP_PPT	0.0974	VP_PPT	0.0921
WS_AP	0.049	WS_AP	0.0483
WS_AT	0.0474	WS_AT	0.0491
WS_VP	0.0506	WS_VP	0.0533
AP_VP	0.7784	AP_VP	0.7397
AP_AT	0.7877	AP_AT	0.7944
VP_AT	1.4082	VP_AT	1.2171

4.3 *Measuring interaction by copula entropy*

Further, we will use copula entropy for computing the dependence. For clarity, here we just choose two cases as examples. Example 1 (eg1) isbetween wind speed (V5) and vapor pressure (V17). Example 2 (eg2) is between vapor pressure (V17) and precipitation (V20). The dependence structure is shown in Figure 5.

Three Copulas, Gumbel Copula, Frank Copula and Clayton Copula, are chosen as candidate Copula functions for Copula entropy estimation. First, we choose the best Copulabased on cross validation results. Then mutual information is estimated indirectly from copula entropy. Copula fitting results and estimated mutual information for two selected pairs are shown in Table 3. From these two cases, we can preliminarily conclude that the link between VP and PPT is stronger than the link between WS and VP from copula-entropy perspective. Furthermore, temporal evolving trend is made to see impact of changing environment on the

261

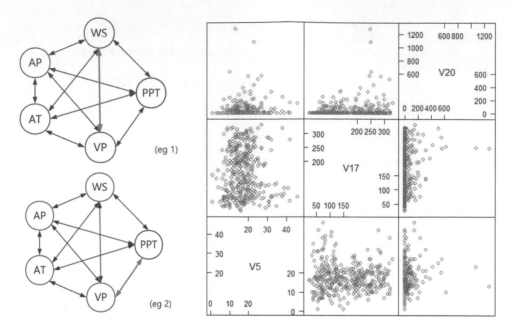

Figure 5. Dependence structure between two variables (eg1: WS-VP; eg2:VP-PPT).

Table 3. Copula entropy between variable pairs.

Pair	Copula	parameter	p-value	Copula entropy	Mutual information
WS-VP	Clayton	0.037	0.052	-0.005	0.005
VP-PPT	Gumbel	1.219	0.916	-0.021	0.021

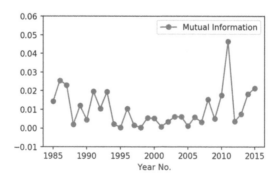

Figure 6. Temporal variation of mutual information between VP and PPT.

dependencies within the precipitation system (Figure 6). The results can be used to analyze and characterize process connectivity in a precipitation system at varying times and changing environment.

5   CONCLUSION AND OUTLOOK

Our approach presents an easy way to study and characterize interactions between meteorological factors for precipitation system from a new perspective, and provides a method for

262

indicating interactions in complex natural system with little preliminary assumption and knowledge. The results would be beneficial for detecting network behavior shifts and bring inspiration for further analysis on regional rainfall characteristics.

The choice of copula family is limited or even not the most suitable, so how to choose a better copula for constructing the dependence needs more study. The interaction can be further divided into different categories (e.g., synergistic, unique, redundant), and different or smaller time scales can be analyzed. Also, more physical mechanisms behind the system can be integrated into the interaction characterization.

ACKNOWLEDGMENTS

This study was supported by the National Key Research and Development Program of China (2016YFC0401501, 2017YFC1502704), and the National Natural Science Fund of China (41571017, 51679118 and 91647203), and Jiangsu Province "333 Project" (BRA2018060).

REFERENCES

Goodwell, A.E. & Kumar, P. 2017. Temporal Information Partitioning Networks (TIPNets): A process network approach to infer ecohydrologic shifts, Water Resour. Res., 53, 5899–5919.

Ma, J. & Sun, Z. 2011. Mutual information is copula entropy. Tsinghua Science & Technology, 16(1), 51–54.

Nelsen, R.B. 1999. An introduction to copulas, Springer, New York.

Shannon, C.E. 1948. A mathematical theory of communication, The Bell System Technical Journal, 27, 379–423. ibid, 623–656.

Sklar, A. 1959. Fonctions de répartitionà n dimensions etleursmarges. Publ. Inst. Stat. Univ. Paris, 8, 229–231.

Risk Analysis Based on Data and Crisis Response Beyond Knowledge – Huang & Nivolianitou (eds)
© 2020 Taylor & Francis Group, London, ISBN 978-0-367-25146-8

Streamflow forecasting using long short-term memory network

Lingling Ni, Dong Wang* & Jianfeng Wu*
Key Laboratory of Surficial Geochemistry, Ministry of Education, Department of Hydrosciences, School of Earth Sciences and Engineering, Nanjing University, Nanjing, China

ABSTRACT: The effective prediction of streamflowis important for water resources planning and management. In this study, we explored the potential use of long short-term memory network (LSTM), and proposed an approach based ona network containing two-layer LSTM on top of a dense layer for multi-step ahead streamflow forecasting. Two streamflow datasets are used to evaluate the performance and applicability of the proposed approach. The prediction accuracy of LSTM is compared with that of multi-lay perceptron (MLP). The obtained results indicate that LSTM is applicable for time series prediction and is a superior alternative when longer time steps ahead prediction are expected.

Keywords: streamflow forecasting, long short-term memory network

1 INTRODUCTION

The reliable prediction of streamflow plays a significant role in reservoir management, risk evaluation, irrigation and flood prevention, and water planning and management (Liu et al., 2015). Considerable efforts have been made to develop models for streamflow and precipitation forecasting over the past few decades. Generally, these methods can be divided into two categories: process-based models and data-driven models (Zhang et al., 2015).

Process-based models usually have the advantage of assisting physically with understanding the hydrological process (Zhang et al., 2015). However, these models are subject to many simplification assumptions or excessive data requirements about the physics of the catchment (Mehr et al., 2013). Conversely, data-driven models mainly emphasize the direct role of historical observations. They have gained popularity for hydrological prediction due to their simplicity, rapid development time, fewer information requirements and ease of implementation in real time(Liu et al., 2015). Statistical models and machine learning methods are often employed to develop data-driven models.

Traditional statistical data-driven methods including multiple linear regression, autoregressive moving average (AMRA) and its variants, have been applied for streamflow forecasting since 1970s. Previous studies have shown that statistical models produce satisfactory predictions when the time series are linear or near-linear, but they perform badly in no stationary series because they poorly capture the nonlinear and non-stationary patterns hidden in the series (Zhang et al., 2016). However, hydrological forecasting is characterized by high complexity, dynamism and non-stationarity (Nourani et al., 2014). More recently, machine learning techniques have received considerable attention from hydrologists for their strong deep learning ability and suitability for modeling the complex and nonlinear process. A variety of machine learning models, such as artificial neural networks (ANN), support vector regression (SVR), genetic programming (GP), and adaptive neuro-fuzzy inference systems (ANFIS),

---

*Corresponding author: wangdong@nju.edu.cn, jfwu@nju.edu.cn

have shown their superior performance for forecasting nonlinear hydrological process (Nourani et al., 2014; Yaseen et al., 2015).

Recurrent neural network (RNN) is an advanced ANN specially designed to understanding temporal dynamics by involving feedback connections in the architecture to "remember" previous information (Elman, 1990). There have been advances in both RNNs architecture and ways of training them to be good at handling tasks that involve sequential inputs in various problem domains, such as natural language processing, machine translation, and time series modeling (LeCun et al., 2015; Shen, 2018). Kumar et al. (2004) conducted research using RNN for streamflow forecasting and found them to outperform traditional ANN for both single step ahead and multiple step ahead forecasting. Coulibaly and Baldwin (2005) investigated the effectiveness of RNN for no stationary hydrological time series forecasting, and showed that RNN can be a good alternative. RNN is considered very effective in modeling complex hydrological time series. However, learning with vanilla recurrent networks can be especially challenging due to the difficulty of learning long-range dependencies. The problem of vanishing and exploding gradients occur when back propagating errors across many time steps (Bengio et al., 1994; Hochreiter et al., 2001). There are many attempts to overcome this weakness of the vanilla RNN, and one of the solutions is long short-term memory (Chung et al., 2014).

The central idea behind LSTM architecture is a memory cell, which can maintain its state over time, and nonlinear gating units, which regulate the information flow into and out of the cell (Greff et al., 2017). LSTM networks have subsequently proven to be more effective than conventional RNNs. Stimulated by the success of LSTM on many domains related to sequential data, a few studies have explored the power of LSTM on hydrological problems and obtained promising results. Zhang et al. (2018) used LSTM to predict water table depth in agricultural areas and compared the performance with feed-forward neural network (FFNN). The results showed that LSTM outperformed FFNN. Zhang et al. (2018) built four different neural network models: multilayer perception (MLP), wavelet neural network (WNN), LSTM and gated recurrent unit (GRU, a variant of LSTM) for simulating and predicting the water level of combined sewage outflow structures. The comparison indicated that the LSTM and GRU presented superior capabilities for multi-step-ahead time series prediction. Kratzert et al. (2018) explored the potential of the LSTM to model the rainfall-runoff process of a large number of differently complex catchments for simulating runoff from meteorological observations. The results showed that LSTMs are able to predict runoff with accuracies comparable to the well-established benchmark model.

The objective of this study was to explore the potential of the LSTM in forecasting monthly hydrological time series and to develop a model containing a two-layer LSTM with a dense layer on top of advance the multi-step ahead prediction accuracy. Cuntan and Hankou monthly streamflow volume data are considered in this study and MLP was employed as the benchmark model.

2 LONG SHORT-TERM MEMORY (LSTM)

As LSTM is a special type of RNN, there is a need look briefly at RNN. Different from traditional ANN, RNN has a recurrent hidden unit, which forms a self-looped cycle to implicitly maintain information about the history of all the past elements of the sequence (see Figure 1) (Elman, 1990; LeCun et al., 2015; Lipton et al., 2015). At time t, hidden state h_t receives input from the current element x_t and also from previous hidden state h_{t-1}. In this way, an RNN can map an input sequence with elements x_t into an output sequence with elements y_t, with each y_t depending on all the previous $x_{t'}$ (for $t' \leq t$).

$$h_t = f(U_h h_{t-1} + W_i x_t + b_h) \tag{1}$$

$$y_t = f(W_o h_t + b_o) \tag{2}$$

where is f a nonlinear activation function, W_i, U_h, and W_o are the weights between layers, and b_h, and b_o are bias parameters.

The dynamics of the network across time steps can be unfolded it in time. RNN can be interpreted not as cyclic, but rather as a very deep network with one layer per time step and that shares the same weights across time steps (same parameters are used at each time step).

RNNs are very powerful dynamic systems but training then has proved to beproblematic because of the gradient vanishing/exploding. This problem can be solved by the structure of LSTM. The LSTM was initially proposed by Hochreiter and Schmidhuber (1997). Since then, several minor modifications to the original LSTM unit have been made. We follow the notation of Graves et al. (2013) (without peephole connections).

The LSTM replaces the ordinary recurrent hidden unit with a memory cell that acts like a gated leaky neuron: it has a connection toitself at the next step that has a weight of 1, but this self-connection is multiplicatively gated by another unit that learns to decide when to clear the content of the memory (LeCun et al., 2015). A LSTM unit can be seen in Figure 3.

The gates are a distinctive feature of the LSTM, which control the information flow within the LSTM cell. The first gate is the forget gate, to decide what information would be thrown away from the cell state, introduced by Gers et al, (2000). A 1 represents "completelykeep the previous cell state", while 0 "represents completelyget rid of it".

The next step is to decide what new information would be stored in the cell state. This contains two parts: an input gate, deciding which information is used to updated the cell state; and a tanh layer creating new candidate values that could be added to the state.

Then the cell state c_t is updated. The third gate, output gate controls the information of cell state that would be flow into the new hidden state. Finally, the new hidden state is calculated.

$$\text{Forget gate } f_t = \sigma(W_f x_t + U_f h_{t-1} + b_f) \tag{3}$$

$$\text{Input gate } i_t = \sigma(W_i x_t + U_i h_{t-1} + b_i) \tag{4}$$

$$\text{Potential cell state } \widetilde{c}_t = \tanh(W_{\widetilde{c}} x_t + U_{\widetilde{c}} h_{t-1} + b_{\widetilde{c}}) \tag{5}$$

$$\text{Cell } c_t = f_t \otimes c_{t-1} + i_t \otimes \widetilde{c}_t \tag{6}$$

$$\text{Output gate } o_t = \sigma(W_o x_t + U_o h_{t-1} + b_o) \tag{7}$$

$$\text{Hidden state } h_t = \tanh(c_t) \otimes o_t \tag{8}$$

where $\sigma(\cdot)$ represents the logistic sigmoid function, $\tanh(\cdot)$ is the hyperbolic tangent, $\otimes$ denotes element-wise multiplication, and (W_f, U_f, b_f), (W_i, U_i, b_i), $(W_{\tilde{c}}, U_{\tilde{c}}, b_{\tilde{c}})$, (W_o, U_o, b_o) define the set of learnable weights.

The following three performance measures are used to qualitatively evaluate the performance of the models developed: the root mean squared error (RMSE), Nash-Sutcliffe model efficiency coefficient (NSE), and mean absolute relative error (MARE),

$$\text{RMSE} = \sqrt{\frac{1}{n}\sum_{i=1}^{n}\left(y_{pred} - y_{obs}\right)^2} \tag{9}$$

$$\text{NSE} = 1 - \frac{\sum_{i=1}^{n}\left(y_{obs} - y_{pred}\right)^2}{\sum_{i=1}^{n}\left(y_{obs} - \overline{y_{obs}}\right)^2} \tag{10}$$

$$\text{MARE} = \frac{1}{n}\sum_{i=1}^{n}\left|\frac{y_{pred} - y_{obs}}{y_{obs}}\right| \tag{11}$$

where n is the number of observations, y_{pred} represents the predicted flow, y_{obs} is the observed flow and $\overline{y_{obs}}$ denotes the average of observed flow.

3   A LSTM MODEL FOR HYDROLOGICAL TIME SERIES FORECASTING

The procedure of the LSTM approach for hydrological time series forecasting is described as follows.

1) Let X, $t = 1, 2, \ldots, n$ denote the time series. Divide time series into training and testing sets, X_{train} and X_{test} respectively.
2) Decide the lag-time (q, the number of previous time steps to use as input variables) and the lead-time (p, the number of next time period we want to predict) and sliding windows method to create a combination of inputs and outputs. That is, $\left(X^{t+1}, X^{t+2}, \ldots, X^{t+p}\right) = f\left(X^t, X^{t-1}, \ldots, X^{t-q+1}\right)$.
3) Use training set to train the network (containing two-layer LSTM and a dense layer on top) for p-step ahead forecast.
4) Apply the trained network to predict p-step ahead values.

4   APPLICATION

Monthly streamflow volume data from Cuntan and Hankou stations in Yangtze River basin, China, were selected for the applications of the developed LSTM approach. The Yangtze River is the longer river in China and the third longest in the world, at 6280 km. Cuntan station is the inflow gauge point of the upper Yangtze River, and Hankou station is locatedat the middle Yangtze River. Both of the datasets span the period 1959–2008, consisting of 600 records. The last 10 years' records are used as testing data.

In this study, we explored the potential use of LSTM on 1-, 3-, 6-, 9-, and 12-steps ahead forecasting with previous streamflow data (3 month lag) to camper the performance of a three-layer MLP. All the hyper-parameters of neural network (NN) were set though trial-and-error procedures and a dropout method (Srivastava et al., 2014) was applied to prevent overfitting.

The RMSE, NSE, and MARE statistics of the LSTM and MLP in training and testing are given in Table 1. The best error measures are highlighted in red. Table 1 shows that the LSTM outperformed MLP in terms of all the performance criteria.

Table 1. Performance measures of LSTM and MLP for multi-step ahead forecasting in the training and testing periods at Cuntan and Hankou stations.

Time steps	Station			Cuntan			Hankou		
				RMSE	NSE	MARE	RMSE	NSE	MARE
1	LSTM	Training		89.23	0.84	0.18	156.48	0.78	0.21
		Testing		92.62	1.79	0.18	154.99	0.76	0.21
	MLP	Training		105.46	0.78	0.28	176.74	0.72	0.26
		Testing		99.97	0.76	0.28	169.95	0.71	0.26
3	LSTM	Training		95.78	0.82	0.21	170.7	0.74	0.26
		Testing		95.05	0.78	0.21	169.37	0.71	0.26
	MLP	Training		118.82	0.72	0.37	220.89	0.57	0.39
		Testing		112.47	0.69	0.37	212.89	0.55	0.39
6	LSTM	Training		101.1	0.80	0.21	175.49	0.73	0.27
		Testing		98.28	0.77	0.21	179.30	0.69	0.27
	MLP	Training		166.60	0.44	0.41	297.41	0.22	0.41
		Testing		158.86	0.40	0.41	287.05	0.21	0.41
12	LSTM	Training		103.25	0.78	0.22	177.66	0.72	0.28
		Testing		111.26	0.73	0.22	202.50	0.64	0.28
	MLP	Training		168.61	0.41	0.41	296.09	0.21	0.41
		Testing		170.28	0.37	0.41	305.48	0.18	0.41

RMSE, root mean squared error;NSE, Nash-Sutcliffe model efficiency coefficient; MARE, mean absolute relative error; LSTM, long short-term memory; MLP, multilayer perceptron.

As shown in Table 1, the model prediction accuracy reduced as the prediction steps increase. For the one-step ahead prediction, LSTM and MLP showed satisfying results, with low RMSE and MARE, and high NSE. When turning the prediction time steps to three, the performance of MLP deteriorated immediately. With longer time steps, MLP performance decreased continually, while the performance of LSTM did not decrease quickly with longer time steps.

5 CONCLUSIONS

Due to the nonlinear and nonstationary nature of streamflow, novel models are required to predict streamflow. Long short-term memory (LSTM) is a popular neural network (NN) suitable for sequential data. This paper explored the potential use of LSTM in streamflow forecasting, and proposed a model containing a two-layer LSTM with a dense layer on top to perform multi-step ahead prediction. The model was applied to predict monthly streamflow at Cuntan and Hankou station of Yangtze River, and compared with MLP. The obtained results indicate that LSTM is applicable forstreamflow forecasting, and significantly out performs MLP. LSTM is a superior alternative when longer time steps ahead prediction are expected.

ACKNOWLEDGMENTS

This study was supported by the National Key Research and Development Program of China (2017YFC1502704, 2016YFC0401501), and the National Natural Science Fund of China (41571017, 51679118 and 91647203), and Jiangsu Province "333 Project" (BRA2018060).

REFERENCES

Bengio, Y., Simard, P. and Frasconi, P., 1994. Learning long-term dependencies with gradient descent is difficult. IEEE transactions on neural networks, 5(2): 157–166.

Chung, J., Gulcehre, C., Cho, K. and Bengio, Y., 2014. Empirical evaluation of gated recurrent neural networks on sequence modeling. arXiv preprint arXiv:1412.3555.

Coulibaly, P. and Baldwin, C.K., 2005. Nonstationary hydrological time series forecasting using non-linear dynamic methods. Journal of Hydrology, 307(1–4): 164–174.

Elman, J.L., 1990. Finding structure in time. Cognitive science, 14(2): 179–211.

Greff, K., Srivastava, R.K., Koutník, J., Steunebrink, B.R. and Schmidhuber, J., 2017. LSTM: A search space odyssey. IEEE transactions on neural networks and learning systems, 28(10): 2222–2232.

Hochreiter, S., Bengio, Y., Frasconi, P. and Schmidhuber, J., 2001. Gradient flow in recurrent nets: the difficulty of learning long-term dependencies. A field guide to dynamical recurrent neural networks. IEEE Press.

Kratzert, F., Klotz, D., Brenner, C., Schulz, K. and Herrnegger, M., 2018. Rainfall-Runoff modelling using Long-Short-Term-Memory (LSTM) networks. Hydrol. Earth Syst. Sci. Discuss., https://doi.org/10.5194/hess-2018-247, in review.

Kumar, D.N., Raju, K.S. and Sathish, T., 2004. River flow forecasting using recurrent neural networks. Water resources management, 18(2): 143–161.

LeCun, Y., Bengio, Y. and Hinton, G., 2015. Deep learning. nature, 521(7553): 436.

Liu, Z., Zhou, P., Chen, X. and Guan, Y., 2015. A multivariate conditional model for streamflow prediction and spatial precipitation refinement. Journal of Geophysical Research: Atmospheres, 120(19).

Mehr, A.D., Kahya, E. and Olyaie, E., 2013. Streamflow prediction using linear genetic programming in comparison with a neuro-wavelet technique. Journal of Hydrology, 505: 240–249.

Nourani, V., Baghanam, A.H., Adamowski, J. and Kisi, O., 2014. Applications of hybrid wavelet–artificial intelligence models in hydrology: a review. Journal of Hydrology, 514: 358–377.

Shen, C., 2018. A transdisciplinary review of deep learning research and its relevance for water resources scientists. Water Resources Research, 54(11): 8558–8593.

Yaseen, Z.M., El-Shafie, A., Jaafar, O., Afan, H.A. and Sayl, K.N., 2015. Artificial intelligence based models for stream-flow forecasting: 2000–2015. Journal of Hydrology, 530: 829–844.

Zhang, D., Lindholm, G. and Ratnaweera, H., 2018. Use long short-term memory to enhance Internet of Things for combined sewer overflow monitoring. Journal of Hydrology, 556: 409–418.

Zhang, H., Singh, V.P., Wang, B. and Yu, Y., 2016. CEREF: A hybrid data-driven model for forecasting annual streamflow from a socio-hydrological system. Journal of Hydrology, 540: 246–256.

Zhang, J., Zhu, Y., Zhang, X., Ye, M. and Yang, J., 2018. Developing a Long Short-Term Memory (LSTM) based model for predicting water table depth in agricultural areas. Journal of hydrology, 561: 918–929.

Zhang, X., Peng, Y., Zhang, C. and Wang, B., 2015. Are hybrid models integrated with data preprocessing techniques suitable for monthly streamflow forecasting? Some experiment evidences. Journal of Hydrology, 530: 137–152.

Hydrometric network design based on Copula and entropy

Wenqi Wang, Dong Wang & Yuankun Wang
Department of Hydrosciences, School of Earth Sciences and Engineering, Nanjing University, Nanjing, China

ABSTRACT: Hydrometric data is important for water resources management and decision support. Hydrometric network provides direct informative data and therefore deserves careful consideration on its design and evaluation. Among most widely-used methods for hydrometric network design, information theory based methods have reached great attention and application. This study focuses on the hydrometric network (stream flow and precipitation) design. In addition, the estimation on basic measures from information theory have not been enough discussed. Herein, we compare binning estimation and copula entropy. It is found that while entropy measures exhibited similar results from different methods, the ranking results were different especially for stations in the middle of the ranking.

Keywords: hydrometric network design, entropy, Copula, mutual information, copula entropy.

1 INTRODUCTION

Gathering hydrologic data from hydrometric networks is the first step for water resources and management. Reliable and representative hydrometric data are fundamental and important for the effective management of water resources, which calls for careful evaluation and design of monitoring networks in hydrology and hydrometeorology. Information theory is widely used in the design of monitoring networks because it provides a quantitative measure of the information content within a hydrometric network (Markus et al., 2003; Li et al., 2012; Alfonso et al., 2014). The Copula theory offers a flexible way to construct a joint distribution independent from marginal distributions (Nelsen, 1999). The main advantage of this approach is that it is free from constraints of different marginal distributions, which is common in different hydrological variables. Due to this property, the application of copulas in bivariate and multivariable hydrological analysis has grown rapidly in the past decade.

Here we compared different hydrometric networks based on entropy and copula theory. The entropy theory is applied for maximizing information and minimizing redundancy of the monitoring network. Meanwhile, the copula theory is utilized for constructing dependence structure of precipitation and stream flow. Therefore, we can realize an integrated design for multivariable hydrometric networks. We apply the method to the design of a monitoring network in Wei River system, especially for precipitation and stream flow monitoring. The proposed method has great flexibility and can be applied for other hydrological variables and monitoring networks. To conclude, it provides a new perspective and direction for the design of multivariable hydrometric networks based on Copula and entropy.

2 BASIC MEASURES AND ESTIMATION METHODS

2.1 *Entropy measures*

The entropy theory (Shannon, 1948) based methods for designing hydrometric network apply several concepts, such as marginal entropy, joint entropy and mutual information. Definitions of these measures are given as follows.

$$H(X) = H(p_1, p_2, p_3, \ldots, p_N) = -\sum_{i=1}^{N} p_i \log p_i \tag{1}$$

where $p_i(0 \le p_i \le 1)$ is the probability of occurrence $x_i(i = 1, 2, \ldots, N)$ and $\sum p_i = 1$.
For bivariate case, the joint entropy between them can be defined as:

$$H(X, Y) = -\sum_{i=1}^{N} \sum_{j=1}^{M} p_{ij} \log p_{ij} \tag{2}$$

where the joint probability of x_i and y_j denoted as $p\,(x_i, y_j) = p_{ij}$, $i = 1,2, \ldots, N; j = 1, 2 \ldots, M$.
Mutual information describes the amount of information transmission or shared between two random variables, which is represented as:

$$I(X, Y) = \sum_{i=1}^{N} \sum_{j=1}^{M} p_{ij} \log\left(\frac{p_{ij}}{p_i p_j}\right) \tag{3}$$

2.2 *Mutual information estimation methods*

Binning estimator is widely used for estimating mutual information (MI). However, the choice of bin width (class interval) is not uniquely determined and estimation results are sensitive to bin width (Fahle et al., 2015; Wang et al., 2018). We determined the class interval following Mogheir et al. (2003) as:

$$NC = 1 + 1.33\ln(N) \tag{4}$$

where NC is number of classes and N is the number of observations for the variable.
For MI estimation using copula method, we generally adopt the following steps.

1. First normalize data to [0,1] as

$$\bar{X} = \frac{X - X_{min}}{X_{max} - X_{min}}, \quad \bar{Y} = \frac{Y - Y_{min}}{Y_{max} - Y_{min}} \tag{5}$$

2. Then compute Kendall τ of $(\bar{X}, \bar{Y})$
3. Estimate parameter of Copula function, one of the main methods for estimating Copula parameter is based on Kendall τ, since there exists an immediate relation between Copula parameter and Kendall τ as:

$$\tau = 4 \iint_{[0,1]^2} C(u, v) dC(u, v) - 1 \tag{6}$$

For example, for the Gaussian Copula we use in the study the distribution function is defined as:

$$C(u, v) = \int_{-\infty}^{\phi^{-1}(u)} \int_{-\infty}^{\phi^{-1}(v)} \frac{1}{2\pi\sqrt{1-\theta^2}} \exp\left(\frac{2\theta xy - x^2 - y^2}{2(1-\theta^2)}\right) dxdy \tag{7}$$

Then the parameter θ can be estimated by Kendall τ as:

$$\theta = \sin\frac{\pi}{2\tau} \tag{8}$$

4. Get copula density function $c(u, v)$ with the Copula parameter θ.
5. Estimate Copula entropy as:

$$H_c(u, v) = -\frac{1}{N}\sum_1^N \log c(u_i, v_i) \tag{9}$$

where $c(u_i, v_i), i = 1, \ldots, N$ are N samples from Copula density function.
6. Obtain the mutual information, which is equivalent to negative Copula entropy:

$$I(X, Y) = -H_c(u, v) = \frac{1}{N}\sum_{i=1}^N \log c(u_i, v_i) \tag{10}$$

3 STUDY AREA AND DATA

3.1 Study area

Wei River is the largest tributary of the Yellow River. The total length of the main stream is 818 km. There are many tributaries on both sides of the Wei River system, and most of the tributaries in the south bank originate from the northern foot of Qinling Mountains. Thus, the flow path of the river in the south bank is short and the slope is large, and thus runoff is large with low sediment concentration. Larger tributaries of Wei River are located on the north bank, mostly originating in the Loess Plateau and hilly area of northern Shaanxi. The river course is relatively wider with lower gradient, and the river flow is slower with relatively larger sediment concentration.

Totally 16 hydrometric stations constitute the hydrometric network. Daily precipitation (mm) and stream flow (m³/s) data from year 2008 to 2012 were used for the integrated hydrometric network design. The data was obtained from National Earth System Science Data Sharing Infrastructure, National Science & Technology Infrastructure of China (http://www.geodata.cn).

3.2 Data

Statistical properties of precipitation and stream flow data are provided in Table 2, which shows that the two variables (precipitation and stream flow) can have different statistical characteristics for different sites. Generally, the differences between individual stations can be more significant for stream flow compared with precipitation according to Table 1.

For direct and visual knowledge of the dependence between precipitation and stream flow, we first chose daily data (from 2008-01-01 to 2008-12-31) of station No.1 (Wushan station) as an example. Figure 2 presents the positive correlation between precipitation and stream flow,

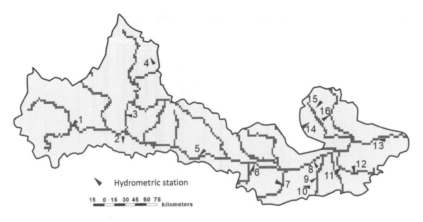

Figure 1.   Wei River basin and hydrometric network.

Table 1.   Statistical properties of daily precipitation and stream flow data in Wei River basin.

station index	precipitation			Stream flow			
	mean	std	maximum	mean	std	maximum	minimum
1	1.212	3.766	38.200	10.068	13.536	153.000	0.217
2	1.289	4.194	47.600	1.510	2.174	30.800	0.030
3	1.233	3.794	50.400	2.693	4.657	72.000	0.000
4	1.215	3.768	42.400	0.049	0.055	0.664	0.006
5	2.109	7.476	110.200	30.606	45.659	1313.000	0.266
6	1.819	6.235	92.200	66.069	141.760	1830.000	4.530
7	2.377	8.357	124.400	10.972	31.663	654.000	0.000
8	1.534	5.235	64.300	98.830	165.880	2890.000	9.950
9	1.843	6.014	66.000	7.490	19.196	431.000	1.060
10	2.437	8.370	115.600	3.747	9.024	195.000	0.267
11	1.692	5.659	72.200	14.537	32.872	542.000	0.185
12	2.300	7.536	73.600	5.878	17.317	286.000	0.472
13	1.443	5.087	59.400	175.505	258.794	4410.000	10.600
14	1.691	5.938	70.000	0.310	0.652	13.500	0.026
15	1.786	6.859	108.600	3.062	8.200	219.000	0.180
16	1.474	5.358	61.800	1.189	3.889	75.300	0.000

which is in line with the common sense that streamflow will naturally get larger when there is more rainfall recharge. Hence, the change in one variable may significantly indicate the information content of another variable.

4   RESULTS AND DISCUSSIONS

4.1   *Correlation of precipitation and stream flow*

To discuss the linear or nonlinear correlations between individual stations, we used the correlation coefficient for linear correlation comparison and MI estimated by the binning method for nonlinear correlation comparison. Figure 3 illustrates that linear and nonlinear correlations between stations showed differences, especially for stream flow stations. It can be seen from Figure 3a and 3c, that the correlation with other stations was relatively weaker for precipitation stations 1, 2, 3, 4 in the network. However, for stream flow stations the dependence could be more different for linear and nonlinear relationships.

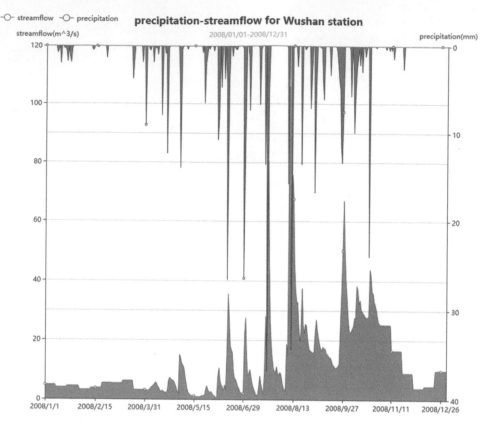

Figure 2. Dailyprecipitation-streamflow for Wushan station from 2008/01/01 to 2008/12/31.

For example, the correlation coefficient for station 4 with other stations was lower than for stations 6~11 but Bin MI for station 4 with other stations did not show significant differences from stations 6~11. In addition, stream flow correlation was more likely to be related to the spatial distribution of rivers. As is shown in Figure 3b and 3d, both linear and nonlinear correlations between stations 8 and 13 were relatively higher than between other stations (especially in Figure 3d). It is noteworthy that stations 8 and 13 are both located on the main stream of Wei River. In such a case, the streamflow at the midstream station 8 may have a direct influence on the downstream stations like station 13, so the streamflow data may be more correlated for these two stations. Similar results can be found for station pairs especially located on the mid-to-downstream, e.g., stations 6-13, 5-6, 6-8, etc. Going one step further, the pattern may be more clearly recognized from the MI indicator (Figure 3d). Prior analysis confirmed the spatial dependence pattern of hydrometric stations and the feasibility of MI indicator on capturing this pattern.

4.2 *Mutual information and rankings with different estimation methods*

We use Gaussian copula for mutual information estimation. The estimated entropy value was also compared with results from binning method (N = 11). From Figure 4, we can find that for most stations the relative mutual information was similar under the two estimation methods.

According to total redundancy, the ranking sequences were also provided for network design(see Table 2, the second row is by copula entropy and third row is by binning method). Based on redundancy minimization ranking criterion, other entropy values (conditional entropy and joint entropy) of the hydrometric network was computed with increasing number

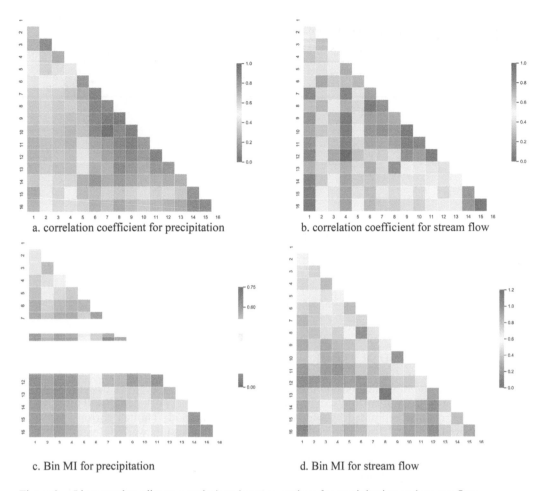

a. correlation coefficient for precipitation

b. correlation coefficient for stream flow

c. Bin MI for precipitation

d. Bin MI for stream flow

Figure 3.   Linear and nonlinear correlations between stations for precipitation and stream flow.

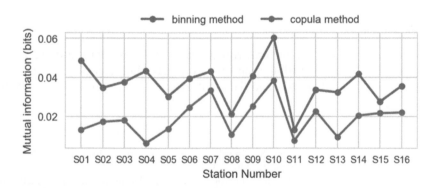

Figure 4.   Mutual information estimated by two methods.

of stations(see Figure 5, only binning method was presented). It can be concluded that the saturated entropy values for streamflow were earlier reached than for precipitation.

Table 2. Station rankings estimated by two methods.

Rank No.	1	2	3	4	5	6	7	8	9	10	11	12	13	14	15	16
Station No.	4	12	1	16	9	3	7	11	14	5	13	2	15	10	8	6
Station No.	4	11	1	16	12	5	3	9	13	14	2	7	10	15	6	8

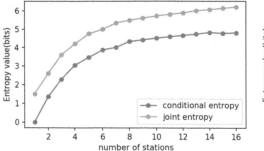 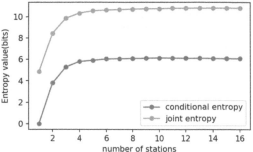

Figure 5. Entropy values with increasing number of stations.

5 CONCLUSION AND OUTLOOK

The impact of estimation methods on precipitation and streamflow network design was evaluated in this study. Copula entropy can be used in hydrometric network design and most results were similar to those estimated by widely used binning method. Though entropy measures exhibited similar results from different methods, the ranking results were different especially for stations in the middle of the ranking.

As accurate entropy estimation for higher dimensions becomes more difficult, we mainly adopted bivariate entropy measures (mutual information) in this study. The sensitivity and impact of entropy-based measures on hydrometric network design can be explored with more complex and effective multivariate entropy estimation methods in the future.

ACKNOWLEDGMENTS

This study was supported by the National Key Research and Development Program of China (2017YFC1502704, 2016YFC0401501), and the National Natural Science Fund of China (41571017, 51679118 and 91647203), and Jiangsu Province"333 Project" (BRA2018060).

REFERENCES

Alfonso, L., Ridolfi, E., Gaytan-Aguilar, S., Napolitano, F., & Russo, F. 2014. Ensemble Entropy for Monitoring Network Design, Entropy, 16, 1365–1375.
Li, C., Singh, V. P., & Mishra, A. K. 2012. Entropy theory-based criterion for hydrometric network evaluation and design: maximum information minimum redundancy. Water Resour. Res. 48, W05521.
Markus, M., Knapp, H.V., & Tasker, G.D. 2003. Entropy and generalized least square methods in assessment of the regional value of stream gages, J. Hydrol. 283(1-4).
Nelsen, R.B. 1999. An introduction to copulas, Springer, New York.
Shannon, C.E. 1948. A mathematical theory of communication, The Bell System Technical Journal, 27, 379–423.ibid, 623–656.
Fahle, M., Hohenbrink, T.L., Dietrich, O., & Lischeid G. 2015. Temporal variability of the optimal monitoring setup assessed using information theory, Water Resour. Res. 51, 7723–7743.

Wang, W.Q., Wang, D, Singh, V.P., Wang, Y.K., Wu, J.C., Wang, L.C., Zou, X.Q., Liu, J.F., Zou, Y, & He, R.M. 2018. Optimization of rainfall networks using information entropy and temporal variability analysis. J. Hydrol. 559:136–155.

Mogheir, Y., de Lima, J.L.M.P.,& Singh, V.P. 2003. Assessment of spatial structure of groundwater quality variables based on the entropy theory, Hydrol. Earth Syst. Sci. 7 (5), 707–721.

A Copula approach for multi-dimensional ecological risk assessment of polycyclic aromatic hydrocarbons in the surface sediments of Taihu Lake

Wenyue Liu & Dong Wang
School of Earth Sciences and Engineering, Nanjing University, Nanjing, China

ABSTRACT: The existent ecological risk assessment methods for polycyclic aromatic hydrocarbons (PAHs) treat each compound as directly independent of each other, without considering the possible correlations between them. To study the multi-dimensional joint distribution of PAHs with correlations, based on the principle of considering multivariate correlations, a Copula approach with adaptability of marginal distributions and a multivariate dependency model used for constructing the model of PAHs exposure concentrations based on Copula functions, then ecological risk assessments were carried out. The types of ecological risk sources of PAHs were analyzed and evaluated based on the results of the Copula model, enhancing the effectiveness and feasibility. Taking the surface sediments of Taihu Lake as examples and comparing with existing assessment methods, the Copula model shows that it takes into account the correlations between variables and makes the risk evaluation results more accurate compared to the risk quotient (RQ) method. And the results of model evaluation can directly reflect the types of risk sources and the degree of risk of each sample point. The results of model evaluation are more comprehensive, more effective and more feasible.

Keywords: polycyclic aromatic hydrocarbons (PAHs), Copula functions, multi-dimensional joint distribution, ecological risk assessment

1 INTRODUCTION

The ecological risk assessment of polycyclic aromatic hydrocarbons (PAHs) have been widely researched due to their properties: semi-volatile, persistent, carcinogenic, teratogenic and mutagenic with strong toxicity (Tang et al., 2005). In recent years, various evaluation methods for the ecological risks of toxic substances, especially PAHs, have emerged. The existent joint ecological risk assessment methods are mostly based on the simple addition of different pollutants or weighted by toxic equivalent factors (TEFs), which means that each PAH compound has been regarded as directly independent of each other, ignoring the correlations existing among them, and therefore the accuracy of the evaluation results obtained by the joint distribution has much room for improvement.

Copulas are functions that connect two or more one-dimensional marginal distributions on a multivariate joint distribution formed on [0, 1], which can be used to measure the dependence structure between variates characterized by corresponding margins and calculate the joint probability distribution (Liu et al., 2017). The Copula functions adaptively distribute the margins and calculate the multi-dimensional dependence model, which ensures the accuracy of the result and simplifies the calculation process. Thus, based on the Copulas, the joint distribution of PAHs is analyzed, considering their correlativity.

Taihu Lake is the third largest freshwater lake in China (Wang et al., 2018),and there are wide concerns about a risk assessment for persistent organic pollutants in the lake. Consequently, further ecological risk assessment of PAHs in surface sediments is necessary and can provide a more reliable reference for decision-makers to avoid the repercussions of PAHs pollution.

Concentrations of 16 priority control PAHs listed by USEPA (Wang et al., 2015) were studied in the surface sediments of Taihu Lake, considering (as far as possible) the consistency, representativeness and the number of samples in the same period, which were derived from studies published on https://link.springer.com/. In this article, the bivariate and multivariate correlation coefficient matrices were calculated to validate the correlations between variables, which is the prerequisite for applying Copula. Copula functions were applied to calculate the multi-dimensional exposure concentrations of various PAHs by generating the fittest joint probability distribution. Then, the synthesized concentrations were applied for ecological risk assessment of PAHs based on SSDs indices and risk source judgments.

2 METHODS

2.1 *Measure consistency*

Most PAHs in nature do not exist independently, and there are certain correlations between them. To measure the interdependency between variables, correlation coefficients that directly can indicate consistency must be calculated. The Spearman rank correlation coefficient is more suitable to characterize the correlations of the PAHs exposure concentrations, which are near to lognormal distributions, and the multivariate Kendall coefficient W is used to denote the multivariable consistency test.

2.2 *Copula functions and parameter estimates*

The d-dimensional Copula function C can be expressed as follows (Joe, 1997):

$$F(x_1, x_2, \ldots, x_d) = C(F_{X_1}(x_1), F_{X_2}(x_2), \ldots, F_{X_d}(x_d); \theta) \tag{1}$$

Where $F(\cdot)$ is a multivariate cumulative distribution function, F_{X_i} is the marginal distribution of X_i, and θ is the Copula parameter. From eq. (1), $F(\cdot)$ is determined by θ, and the semiparametric method (SP) is one of approaches to determine parameter θ of the Copula function; this calculation procedure is flexible and accurate (Xu et al., 2017). The semiparametric method is also used to select the best fitting Copula function from the Archimedean Copula family, which is an alternative Copula function to simulate the joint distribution of two-dimensional PAHs exposure concentrations, and the Copula with the best fitting result is selected to simulate the joint distribution of multi-dimensional distributions.

2.3 *Construction of multi-dimensional joint cumulative probability distribution of PAHs*

According to the single PAH exposure concentration sequence, obtain adaptive marginal distribution and structure [0, 1] mapping. After assorting variables legitimately, use 'fitcopula' or 'gofcopula' functions of the 'copula' package in R studio to estimate the parameter θ and fit bivariate and multivariate joint probability distributions of PAHs concentrations. Then, the comprehensive and reliable joint distribution models, fully considering the correlations between the variables of PAHs, could be established based on Copula functions.

2.4 *Ecological risk assessment*

PAHs are often produced in heat sources (natural or man-made fossil fuel combustion) or petroleum sources (distribution of petroleum and petroleum products) and can be identified based on the proportion of specific compounds and the number of benzene rings (Zhang et al., 2016). The ratios of PAHs exposure concentrations to the corresponding toxicity data reference values are called the level of risk quotients. The maximum permissible concentrations (MPCs) and negligible concentrations (NCs) of ecological risk indicators of SSDs can also be used as reference values, which can obtain the maximum permissible concentrations quotients (RQ_{MPCs}) and the NCs quotients (RQ_{NCs}), respectively, defined as follows (Cao et al., 2010):

$$RQ = C_{PAHs}/C_{QV} \qquad (2)$$

$$RQ_{NCs} = C_{PAHs}/C_{QV(NCs)} \qquad (3)$$

$$RQ_{MPCs} = C_{PAHs}/C_{QV(MPCs)} \qquad (4)$$

where, C_{PAHs} is the concentration of PAHs and C_{QV} means the corresponding quality value. Furthermore, the total MPCs quotients ($\sum RQ_{MPCs}$) and the total NCs quotients ($\sum RQ_{NCs}$) can be calculated and risks can be judged by the criteria.

Combined with SSDs indices (MPCs and NCs), the multi-dimensional joint probability distribution assessment model of PAHs exposure concentrations was established using Copula functions, to conduct comprehensive ecological risk assessment of exposure concentrations and compare with traditional evaluation results via the RQ method. Associated with the discrimination basis of source of PAHs to analyze the source of risk, the Copula method is more reasonable and accurate than others, which is conducive to the formulation of risk management plans. All calculations above were done using Excel and R Studio.

3   CASE STUDY

3.1   *Consistency analysis*

The experimental data of 28 groups of Taihu Lake surface sediment samples extracted in the literature (Lei et al., 2014) were taken as examples, and the concentrations of 16 PAHs was detected in all samples. Through statistical analysis of existing data, we calculated the correlation coefficients between two variables and the multivariate PAHs. Table 1 shows the bivariate PAHs Spearman rank correlation coefficients and the multivariate Kendall coefficients grouped according to the number of benzene rings. This indicates that all bivariate correlation coefficients are above 0.5 and some are even over 0.9, with the exception of Pyr, and all Kendall coefficients are more than 0.75, which signifies there are strong correlations between PAHs. For most PAHs, the compounds are not independent, and the correlations among them cannot be neglected, so it was appropriate to apply the Copula functions to construct the multi-dimensional joint distribution models of PAHs exposure concentrations in order to fully consider the correlations among them.

3.2   *Copulas selection*

Regarding Gumbel, Clayton and Frank Copula in the Archimedean Copula family as alternative functions, and taking Cramer-von Mises statistic (Sn) and AIC criterion (AIC) as fitting evaluation indicators (Xu et al., 2017), we selected the best Copula, which is the fittest function associating theoretical values and empirical results. The the smaller values of Sn and AIC, the better fitting the effect of Copula. According to all fitting results, the Gumbel Copula function was the best choice to calculate the multi-dimensional joint probability distribution of PAHs exposure concentrations.

3.3   *Construction of multi-dimensional probability distribution model*

The properties of PAHs are related to the number of benzene rings. In order to reduce the complexity of the high-dimensional Copula calculation and accelerate the computation, the multi-dimensional probability distribution of exposure concentrations of PAHs

Table 1. Correlation coefficients of PAHs.

PAHs	Nap	Ace	Acy	Fluo	Phe	Ant	Flua	Pyr	BaA	Chry	BbF	BkF	BaP	DBA	IP	BghiP
Rings	2	2	2	2	3	3	3	4	4	4	4	4	5	5	5	5
Nap	1															
Ace	0.738	1														
Acy	0.746	0.873	1													
Fluo	0.605	0.895	0.885	1												
Phe	0.522	0.781	0.851	0.927	1											
Ant	0.516	0.801	0.829	0.885	0.891	1										
Flua	0.589	0.640	0.795	0.755	0.850	0.720	1									
Pyr	0.575	0.267	0.511	0.201	0.284	0.317	0.505	1								
BaA	0.516	0.718	0.817	0.849	0.904	0.805	0.831	0.377	1							
Chr	0.615	0.605	0.821	0.722	0.834	0.664	0.955	0.573	0.864	1						
BbF	0.555	0.504	0.744	0.622	0.738	0.576	0.845	0.571	0.770	0.939	1					
BkF	0.576	0.587	0.799	0.696	0.798	0.660	0.861	0.545	0.820	0.948	0.988	1				
BaP	0.567	0.602	0.799	0.703	0.801	0.679	0.861	0.512	0.821	0.937	0.973	0.990	1			
DBA	0.566	0.668	0.810	0.777	0.817	0.777	0.730	0.399	0.849	0.823	0.880	0.922	0.922	1		
IP	0.592	0.540	0.752	0.634	0.713	0.599	0.776	0.539	0.730	0.882	0.967	0.966	0.962	0.924	1	
BghiP	0.585	0.506	0.722	0.596	0.679	0.567	0.748	0.517	0.701	0.856	0.948	0.943	0.940	0.911	0.990	1
W	0.825				0.881			0.790					0.950			

was constructed severally and divided according to the number of benzene rings. This is conducive to discerning the influence degrees of different PAHs sources on all sample points. By substituting 28 sets of data into the model construction step above, the multi-dimensional probability distribution models of PAHs exposure concentrations corresponding to different benzene ring numbers can be obtained. The SSDs indices are mapped into the Copula model, and the corresponding probability levels of each index can be obtained, which are compared with theoretical values to assess the ecological risk situations of PAHs in each sample.

3.4 *Ecological risk assessment of PAHs by RQ method*

According to the RQ method, the RQ_{MPCs} and RQ_{NCs} of a single type of PAH in each sample can be calculated, and then the integrated RQ can be obtained by summation, as showed in Figure 1, where the dotted line indicates $\sum RQ_{NCs}$ and corresponding standard value, and the full line shows the $\sum RQ_{MPCs}$ with their indices.

Figure 1 shows that each total negligible concentration quotient of all sample points is less than 800, which means that the ecological risk of all sample points is not in the high-risk class. The $\sum RQ_{MPCs}$ of sample points S5~S9, S13, S19 are all greater than 1, indicated a medium-risk level 2, where S7 is the largest, which means the S7 point has the highest ecological risk in all sites. The sampling points S5~S9 are all along the Meiliang Bay area of Taihu Lake, indicating that the risk of PAHs there is medium-risk level 2 and much higher than in other areas. The sample points S1~S4, S10~S12, S14~S18 and S20~S28 are all at low-risk levels. Applying the RQ method to assess holistic risk is seemingly reasonable, but these assessment results were derived from consumption of single risks, which neglects the correlations among PAHs.

3.5 *Multi-dimensional joint distribution of PAHs ecological risk assessment*

The integrated concentrations probability of PAHs in different benzene rings and the probability of each index of MPCs and NCs can be calculated according to the multi-dimensional probability distribution Copula model constructed above. The results are shown in Figure 2.

In Figure 2, the horizontal full line shows the NCs indices, and the dotted line showsthe MPCs standard values, with different colors corresponding to the number of different

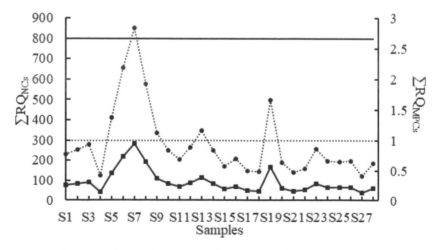

Figure 1.   Integrated risk assessment of PAHs by RQ method.

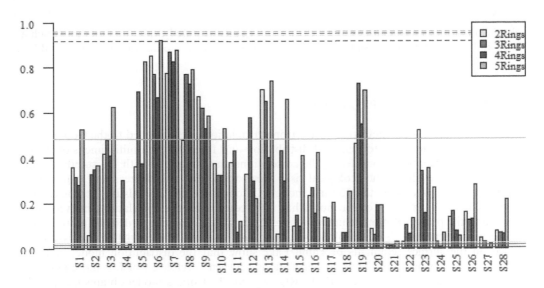

Figure 2.   Probability distributions and indices of PAHs exposed in different benzene rings.

benzene rings. The height of the bars expresses the multi-dimensional joint probability values of different ring number PAH exposure concentrations. Figure 2 reflects the probability values corresponding to the exposure concentrations of PAHs in each ring, and also shows the risk degrees, which can be directly compared with other samples, and can be more accurate. Figure 2 also shows that 2-ring and 5-ring PAHs in S6, and 3-ring and 5-ring PAHs in S7, have the largest multi-dimensional probabilities of exposure concentrations, which means that their ecological risks are the highest. The S4, S17~S18, S20~S22 and S24~S28 have lower ecological risks, but except for 5-ring, all other values are higher than NCs standards, indicating that the overall ecological risks of the above sample points are low but have potential risks. The risks of S5~S9, S13 and S19 are slightly higher than others. All in all, the frequencies of exposure concentrations of each ring at S6 and S7 are the highest, almost reaching to the standard line of MPCs, indicating that both have higher potential ecological risks.

Based on this Copula model, comprehensive risk analysis can be carried out on specific ring number results. When the total risk evaluations are obtained, the degrees of influences of each ring PAHs on the study area can be directly observed. Then, source analysis can be obtained from specific types of impact sources, which is conducive to the formulation and deployment of risk control measures. Figure 2 shows that for S5 and S8, the most important ecological risk comes from 3-ring PAHs, S6 and S9 are mainly derived from 2-ring PAHs, and S7 contains all PAHs rings.

Thus, the source of PAHs in the Meiliang Bay of Taihu Lake during the evaluation period is complex. Because low weight PAHs often are emitted from petroleum sources, vise are mainly originated in heat sources. The proportion of petroleum sources has occupied the predomination, while partial heat sources exist at the same time and both have high concentrations. This has a considerable negative influence on the water and surface sediments along the Meiliang Bay, with relatively high ecological risks. The 4-ring PAHs at S13 are slightly lower, indicating that S13 is marginally affected by heat sources of 4-ring PAHs, although its total ecological risk is higher here. Similarly, 2-ring PAHs, which are substantially generated by petroleum sources, have slightly lower effects on S19.

In summary, because the multi-dimensional probability distribution Copula model of PAHs exposure concentrations has considered the correlations among variables, even regarding it as a precondition, the simulation results obtained from distribution models can also accurately recognize the correlations within them. Consequently, the evaluation results obtained are more reasonable and accurate than other integrated evaluation methods such as RQ, and according to the simulation results the sources of ecological risks also could be analyzed.

4 CONCLUSION AND PROSPECT

The PAHs exposure concentrations with strong correlations were applied to the Copula functions, which are adaptive to marginal distribution and fitted to the multi-dimensional joint probability distributions considering the correlations and compared with the standard value of the SSDs to evaluate the ecological risk. Mean while, the risk degrees and main source types were analyzedusing the Copula model. Compared with the existing ecological risk assessment methods, the Copula model has fully considered the correlations between variables, and the principle of modeling is more realistic, which can be regarded asa more reliable reference to formulate risk control measures and deployments. The results of risk assessment based on the surface sediments of Taihu Lake show that the distribution model based on Copula functions has considered the correlations between variables, which is more appropriate for the situation. Compared with the RQ method, the Copula model obtains more evaluation results and more accuracy. More grouping methods (such as PAHs toxicity) can be applied in this model, and this requires further study.

ACKNOWLEDGMENTS

This study was supported by the National Key Research and Development Program of China (2016YFC0401501,2017YFC1502704), the National Natural Science Fund of China (51679118, 41571017, and 91647203), and the Jiangsu Province "333 Project" (BRA2018060).

REFERENCES

Cao, Z., Liu, J., Luan, Y., Li, Y., Ma, M., Xu, J. & Han, S. 2010. Distribution and ecosystem risk assessment of polycyclic aromatic hydrocarbons in the Luan River, China. *Ecotoxicology* 19: 827–837.

Joe, H. 1997. *Multivariate Models and Dependence Concepts.* New York: Chapman & Hall.

Lei, B., Kang, J., Wang, X., Yu, Y., Zhang, X., Wen, Y. & Wang, Y. 2014. The levels of PAHs and aryl hydrocarbon receptor effects in sediments of Taihu Lake, China. *Environ Sci Pollut Res Int* 21:6547–6557.

Liu, D., Wang, D., Singh, V. P., Wang, Y., Wu, J., Wang, L., Zou, X., Chen, Y. & Chen, X. 2017. Optimal moment determination in POME-copula based hydrometeorological dependence modelling. *Adv Water Resour* 105: 39–50.

Tang, L., Tang, X.Y., Zhu, Y.G., Zheng, M.H. & Miao, Q.L. 2005. Contamination of polycyclic aromatic hydrocarbons (PAHs) in urban soils in Beijing, China. *Environ Int* 31: 822–828.

Wang, D., Wang, Y., Singh, V. P., Zhu, J., Jiang, L., Zeng, D., Liu, D., Zeng, X., Wu, J., Wang, L. & Zeng, C. 2018. Ecological and health risk assessment of PAHs, OCPs, and PCBs in Taihu Lake basin. *Ecol Indic* 92: 171–180.

Wang, X., Chen, L., Wang, X., Lei, B., Sun, Y., Zhou, J. & Wu, M. 2015. Occurrence, sources and health risk assessment of polycyclic aromatic hydrocarbons in urban (Pudong) and suburban soils from Shanghai in China. *Chemosphere* 119: 1224–1232.

Xu, P., Wang, D., Singh, V.P., Wang, Y., Wu, J., Wang, L., Zou, X., Chen, Y., Chen, X., Liu, J., Zou, Y. & He, R. 2017. A two-phase copula entropy-based multiobjective optimization approach to hydrometeorological gauge network design. *JHydrol* 555: 228–241.

Zhang, D., Liu, J., Jiang, X., Cao, K., Yin, P. & Zhang, X. 2016. Distribution, sources and ecological risk assessment of PAHs in surface sediments from the Luan River Estuary, China. *Mar Pollut Bull* 102: 223–229.

Risk Analysis Based on Data and Crisis Response Beyond Knowledge – Huang & Nivolianitou (eds)
© 2020 Taylor & Francis Group, London, ISBN 978-0-367-25146-8

Return period analysis for extreme rainfall in the western region of the Taihu Lake Basin

Yuwei Tao, Yuankun Wang* & Dong Wang*

Key Laboratory of Surficial Geochemistry, MOE, Department of Hydrosciences, School of Earth Sciences and Engineering, State Key Laboratory of Pollution Control and Resource Reuse, Nanjing University, Nanjing, China

Dong Sheng

Hunan Water Resources and Hydropower Research Institute, Changsha, China

ABSTRACT: Climate change and rapid urbanization development in the western region of the Taihu Lake Basin, China, would change the precipitation and its variability, which are often related to water disasters. Under these circumstances, the need for accurate information about extreme rainfall events has been amplified. Based on the daily rainfall data of Danyang, Jintan, Liyang, and Yixing rainfall stations from 1961 to 2015, the annual maximum daily rainfall under different return periods was analyzed using the principle of maximum entropy (POME) method. The results show that annual maximum daily rainfall of more than 50 mm happened nearly every year at each station. When the return period was 100 years, the maximum rainfall at Danyang and Jintan was 269.51 mm and 261.68 mm, respectively, reaching the level for heavy rainstorms. We hope this study could provide a scientific reference for prevention and risk management of extreme climate in the western region of the Taihu Lake Basin.

Keywords: extreme rainfall, return period, western region of the Taihu Lake Basin

1 INTRODUCTION

In recent years, extreme climate and weather events frequently have occurred because of global warming derived from the positive concentration of greenhouse gases. Rainstorm, as one of the major meteorological disasters, has exerted significant influence on humans, property and the environment. Thus, extreme rainfall analysis has become a vital task in the fields of meteorology, hydrology and disaster.

Much effort have been devoted worldwide to the probability distributions of extreme rainfall (Zalina et al., 2002; Koutsoyiannis, 2004; Chu et al., 2009). For example, Zalina et al. (2002) reported that the generalized extreme value (GEV) showed very good descriptive and predictive abilities of the annual extreme rainfall series in Peninsular Malaysia. These studies play a crucial role in the prevention of flood damage caused by extreme rainfall events, as the occurrence of a heavy rainstorm once in 100 years would result in devastating disaster. The extreme rainfall data under specific return periods derived from probability distributions are considered to be invaluable tools for the construction of certain projects, such as drainage systems and dams.

Taihu Lake Basin (TLB), located in the Yangtze River Delta, is one of the most developed, densely populated and flourishing economic areas in China. The rapid urbanization process of TLB has a profound impact on the precipitation variability over recent decades. The

---

* Corresponding author: yuankunw@nju.edu.cn; wangdong@nju.edu.cn

western region of TLB hosts the main water supply of Taihu Lake and is also closely related to flood regulations for TLB. Previous studies have investigated the precipitation variability and its physical causes in this region (Yu et al., 2012; Lin et al., 2018). In the study, we applied the principle of maximum entropy (POME), which has shown better performance in deriving the probability distribution of variables to calculate the return period of extreme rainfall for forewarning extreme climates.

2 STUDY AREA AND DATA

Taihu Lake Basin (TLB) is located in the southern area of the Yangtze River Basin in China. It has a total area of 36 895 km^2, where plains account for 80% and the mountainous area in the west is around 20%.

The western region of TLB lies on the upper section, with a total area of 7896 km^2. It is subject to subtropical monsoon climate: dry and cold in winter but hot and humid in summer. The average annual rainfall is 1408 mm and the average evaporation from water surface is 884 mm. L-type hilly area makes up 32% of this region.

The daily rainfall data in the western region observed at Danyang, Jintan, Liyang and Yixing rainfall stations from 1961 to 2015 are used for the study. The location of rainfall stations are shown in Table 1.

3 METHOD: PRINCIPLE OF MAXIMUM ENTROPY (POME)

The principle of maximum entropy (POME) is used to derive the cumulative distribution function of annual maximum daily rainfall and further determine the 'return period' for extreme weather.

Entropy measures the uncertainty of a random variable or its variability. For a continuous random variable X with the probability density function (PDF) $f(x)$, the Shannon entropy H(X) can be expressed as follows (Shannon, 1948):

$$H(X) = -\int_a^b f(x) \log f(x) dx \tag{1}$$

The POME makes inferences on the probability distribution by maximizing entropy subject to given data and constraints (Jaynes, 1957). Thus, the entropy-based method can be written as:

$$\int_a^b f(x) dx = 1 \tag{2}$$

$$\int_a^b g_i(x) f(x) dx = E(g_i), (i = 1, 2, \ldots, m) \tag{3}$$

Table 1. Locations of rainfall stations in the western region of TLB.

Rainfall station	Latitude	Longitude	Height/m
Danyang	31.9833	119.6000	6.6
Jintan	31.7167	119.5500	6.9
Liyang	31.4333	119.4833	5.9
Yixing	31.3333	119.8167	16.4

where $g_i(x)$ is the selected or specified function with respect to the properties of interest (e.g., moments); $E(g_i)$ is the expectation of $g_i(x)$; and m is the number of constraints. By specifying $g_0(x) = 1$, Eq. (3) is the general form of constraints with Eq. (2) as a special case.

According to the POME, the probability distribution with the maximum entropy in Eq. (1), subject to the given constraints in Eq. (2) and (3), should be selected. In other words, for a given set of constraints, a unique distribution can be defined (Chen and Singh, 2018). Therefore, finding proper constraints is critical (Singh, 2013 and 2015) and leads to the mathematical optimization problem, which can be solved using the method of Lagrange multipliers. The Lagrange function L can be expressed as follows (Kapur, 1989):

$$L = -\int_a^b f(x)dx - (\lambda_0 - 1)[\int_a^b (f(x) - 1)dx)] - \sum_{i=1}^m \lambda_i[\int_a^b f(x)dx - E(g_i)] \tag{4}$$

where $\lambda = [\lambda_0, \lambda_1, ..., \lambda_m]$ are the Lagrange multipliers. By differentiating L with respect to f and setting the derivative to zero, the maximum entropy PDF can be derived as follows (Kapur and Kesavan, 1992):

$$f(x) = \exp(-\lambda_0 - \lambda_1 g_1(x) - \lambda_2 g_2(x) - \cdots - \lambda_m g_m(x)) \tag{5}$$

Using sample statistical moments as constraints has the advantage of reaching the universal PDF more accurately for the given data instead of assuming certain types of distribution based on a nonparametric approach (e.g., kernel density function). In general, $g_i(x)$ is set as x^i.

Accordingly, the cumulative distribution function (CDF) can be expressed as:

$$F(x) = \int_0^x f(x)dx \tag{6}$$

In this study, if the Lagrange multipliers with constraints of orders (denote m from 1 to 8) of moments were considered for more moments, this would lead to considerable computation-expense and less improvement in accuracy (Liu et al., 2017). All the derived POME-based PDFs were compared with the Akaike information criterion (AIC) to select the optimal probability distributions for annual maximum daily rainfall. The best fitted distribution is the one that has the minimum AIC. AIC can be expressed as:

$$AIC = -2log(\text{maximized likelihood for the model}) + 2(\text{numbers of fitted parameters}) \tag{7}$$

The Kolmogorov-Smirnov (K-S) statistic is another goodness of fit test measuring the distance between an empirical distribution function and a theoretical continuous distribution function, to check whether the selected optimal distribution is suitable to the respective empirical data.

In summary, the main objective is to obtain the return period of annual maximum rainfall using the POME method and a brief step-by-step description is given below:

Step 1. Obtain the series of annual maximum daily rainfall. Extract the annual maximum daily rainfall from daily rainfall for each year to form the initial data series X.

Step 2. Normalize X to the unified interval [0,1]:

$$X' = \frac{X - X_{\min}}{X_{\max} - X_{\min}} \tag{8}$$

Step 3. Derive the maximum entropy distribution given constraints of different orders m (from 1 to 8) of moments:

$$g_i(x) = x^i (i = 1, 2, 3, \ldots, m) \tag{9}$$

Step 4. Determine the optimal orders of moments by AIC.

Step 5. Obtain the optimal moment based cumulative distribution $F(x)$ using Eq. (6).

Step 6. Calculate the 'return period' $N(x)$ of annual maximum daily rainfall:

$$N(x) = 1/(1 - F(x)) \tag{10}$$

4 RESULTS AND DISCUSSION

The annual maximum daily data for the four stations are shown in Figure 1. From the 5-year moving average curves, no obvious trends are present in the maximum daily rainfall of all stations. The mean annual maximum rainfall at Jintan was 101.1 mm, higher than other stations. Annual maximum rainfall is usually more than 50 mm at each station per year. The largest maximum rainfall happened in 2003 at Danyang and 2015 at Jintan – both exceeded 250 mm.

According to the AIC, the optimal order of moments for each time series of annual maximum daily rainfall was determined as 4, 3, 3 and 3. The derived POME-based distributions, given the optimal moment of constraints, all passed the Kolmogorov-Smirnov (KS) test, indicating that theoretical cumulative probability distributions performed fairly well and fitted to their respective empirical data. In addition, the probability distributions of annual maximum at stations had lower AIC values compared to a commonly used P-III type PDF. It has been proved that the POME method is superior to P-III type distribution in capturing the shape of PDF.

Based on the optimal cumulative distribution derived from the POME method, the return period of annual maximum daily rainfall in the western region of TLB is determined using Eq. (10) (Figure 2).

Table 2 shows the estimated annual maximum daily rainfall under different return periods. It can be seen that when the return periods reached 5 years, the maximum rainfall of the four stations all exceeded 100 mm. Maximum rainfall under a 100-year return period at the four stations were 269.51 mm, 261.68 mm, 151.13 mm, and 174.85 mm. Noticeably, the maximum rainfall measurements at Danyang and Jintan were beyond 250 mm which is the level for heavy rainstorm. If the annual maximum daily rainfall was 100 mm, the estimated return period for the four stations were 3.5 years, 2.4 years, 3.30 years and 3.29 years, which were close to the observed data of 3.4 years, 2.4 years, 3.24 years and 2.89 years, respectively, indicating that the estimated return period is accurate and reasonable by using the POME method.

5 CONCLUSION

In this study, the annual maximum daily rainfall under the return period at Danyang, Jintan, Liyang and Yixing in the western region of TLB was investigated by using the POME method, which can capture the shape of the probability distribution. The results revealed that annual maximum daily rainfall of more than 50 mm nearly happened every year at each station. When the return period was 100 years, the maximum rainfall at Danyang and Jintan was 269.51 mm and 261.68 mm, respectively, reaching the level for heavy rainstorm. We hope the results of this study will act as a useful tool for designing the appropriate infrastructure for prevention of extreme climates in the western region of TLB.

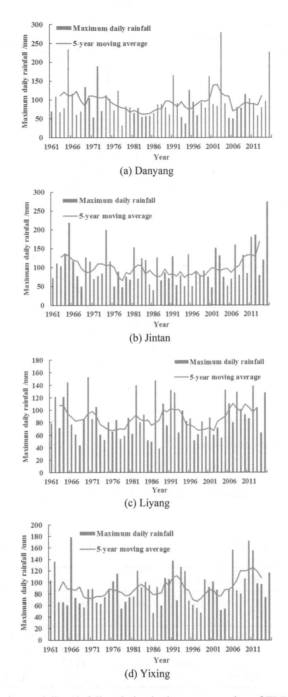

Figure 1. Annual maximum daily rainfall variation in the western region of TLB.

289

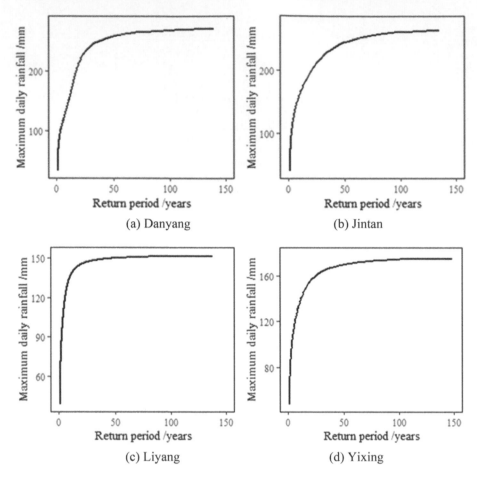

Figure 2.   Return period of annual maximum daily rainfall in the western region of TLB.

Table 2.   Annual maximum daily rainfall under return periods in the western region of Taihu Lake Basin.

	Return period (years)						
Maximum rainfall (mm)	5	10	20	30	50	100	150
Danyang	112.54	145.91	215.13	242.32	258.39	269.51	272.36
Jintan	131.25	164.15	199.40	222.90	245.23	261.68	266.08
Liyang	115.90	135.46	144.60	148.02	149.74	151.13	151.76
Yixing	113.00	136.03	156.42	164.32	170.34	174.85	176.17

ACKNOWLEDGMENTS

This study was supported by the National Key Research and Development Program of China (2016YFC0401501, 2017YFC1502704), and the National Natural Science Fund of China (41571017, 51679118 and 91647203), and the Jiangsu Province "333 Project" (BRA2018060). "Study on water ecology and water environmental protection in northern area of Dongting Lake"- Hunan Water Conservancy Science and Technology Plan" (2015-245-01).

REFERENCES

Chen, L., Singh, V.P. 2018. Entropy-based derivation of generalized distributions for hydrometeorological frequency analysis. *Journal of Hydrology* 557: 699–712.

Chu, P.S., Zhao, X., Ruan, Y., Grubbs, M. 2009. Extreme rainfall events in the Hawaiian islands. *Journal of Applied Meteorology and Climatology* 48(3): 502–516.

Jaynes, E.T., 1957. *Information Theory and Statistical Mechanics*. Physical Review 106.

Kapur, J.N., 1989. *Maximum-Entropy Models in Science and Engineering*. Biometrics.

Kapur, J.N., Kesavan, H.K. 1992. *Entropy Optimization Principles and their Applications*. Academic Press.

Koutsoyiannis, D., 2004. Statistics of extremes and estimation of extreme rainfall: II. Empirical investigation of long rainfall records. *International Association of Scientific Hydrology* 49(4): 591–610.

Lin, L., Liu, J.T., Gan, S.W., Sun, Q., Zhou, C.L. 2018. Study on spatio-temporal variation of precipitation in Western Region of Taihu Lake Basin based on TFPW-MK method. *Water resources and Power* 36(4): 1–5. (in Chinese).

Liu, D.F., Wang, D., Singh, V.P., Wang, Y.K., Wu, J.C., Wang, L.C., et al. 2017. Optimal moment determination in pome-copula based hydrometeorological dependence modelling. *Advances in Water Resources* 105: 28–50.

Shannon, C.E. 1948. *A mathematical theory of communication*. Bell Labs Tech.

Singh, V.P. 2013. Entropy theory and its application in environmental and water engineering. Wiley-Blackwell, Hoboken.

Singh, V.P. 2015. *Entropy theory in hydrologic science and engineering*. McGraw-Hill, New York.

Yu, W.J., Yan, Y.G., Zou, X.Q., 2012. Study on spatial and temporal characteristics of rainstorm in Taihu Lake Basin. *Journal of Natural Resources* 27(5): 766–777. (in Chinese).

Zalina, M.D., Desa, M.N., Nguyen, V.T.V., Hashim, M.K. 2002. Selecting a probability distribution for extreme rainfall series in Malaysia. *Water Science and Technology* 45: 63–68.

Entropy based multi-criteria evaluation for rainfall monitoring networks under seasonal effect

Heshu Li, Dong Wang* & Yuankun Wang

School of Earth Sciences and Engineering, State Key Laboratory of Pollution Control and Resource Reuse, Nanjing University, Nanjing, P.R. China

ABSTRACT: Entropy has been widely used for rainfall network evaluation or optimization, but limited researches focused on the seasonal effect. This paper proposed an entropy based multicriterion method for evaluation of rainfall monitoring networks. We applied the method to a rainfall monitoring network containing 45 stations in the southwestern Taihu Lake Basin of China. Two indexes, separately accounting for information content and redundancy, were integrated with ideal point method to form the evaluation criterion. Values of the objective function were then computed to identify the significant stations and areas. In particular, we employed the whole year, flood season and non-flood season series respectively for computation and generated corresponding results.

1   INTRODUCTION

A well-designed rainfall monitoring network can reflect the spatial-temporal variability of precipitation variables (Yeh et al. 2011). In the context of climate and land use changes, the requirement of optimal network design keeps increasing. Entropy directly defines information and quantifies uncertainty (Shannon 1948). Entropy based methods have been widely used in hydrometric network optimization. A fundamental consideration of network optimization is the stations selected should contain as much information as possible and share less transinformation. Many researches incorporated entropy in multiobjective methods for network optimization. Alfonso et al. used WMP approach for water level monitoring network design in polders (Alfonso et al. 2010). Li et al. (2012) presented a maximum information minimum redundancy (MIMR) criterion to design hydrometric networks. Alfonso et al. (2013) used ranking method to find more informative and less redundant monitoring network configurations. Samuel et al. (2013) proposed a combined regionalization and dual entropy-multiobjective optimization (CRDEMO) method for determining minimum network. Xu et al. (2015) used an entropy-based multi-criteria method to resample the rain gauging networks.

Many studies have noted the variability of the optimal results under the high variability of rainfall series and its effect on rainfall network design (Ridolfi et al. 2011; Wei et al. 2014; Fahle et al. 2015). These findings demonstrate the important first consideration of network objectives when determining the spatial and temporal sampling used to calculate entropy. However, research on seasonal effect is still limited, and more work is needed to provide robust guidance on this topic.

This study focused on the evaluation of rainfall networks. An entropy based multicriterion method was first presented to identify stations with more information content and less redundancy. The seasonal effect was explored by using three data sets to compute the index values. The method was applied to a rainfall monitoring network in the southwestern Taihu Lake basin of China.

---

*Corresponding author: wangdong@nju.edu.cn

2   STUDY AREA AND DATA

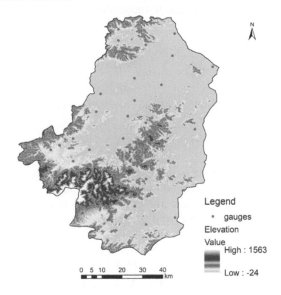

Figure 1.   Study area.

The study area is located in the southwest of the Taihu Lake Basin, China, covering an area of 5930.9 km² (Figure 1). The area is densely covered with crisscrossing river channels. Under a subtropical monsoon climate, it featured four distinct seasons with annual mean temperature measured as 16-18°C and annual precipitation measured as 1100-1150 mm. Daily rainfall data of 45 rainfall monitoring stations of 10 years (from 2007 to 2016) were used for evaluation.

3   METHODS

Suppose X is a random variable with observation values x_i $(i = 1,2,...,n)$ and probability density function $p(x_i)$, the entropy of X is defined as:

$$H(X) = -\sum_{i=1}^{n} p(x_i)\log_2 p(x_i)$$ (1)

The unit of entropy is bit. To measure the total uncertainty or information contained in two or more variables, joint entropy is defined. For the bivariate case, it is formulated as:

$$H(X, Y) = -\sum_{i=1}^{m} \sum_{j=1}^{n} p(x_i, y_j)\log_2 p(x_i, y_j)$$ (2)

where $p(x_i, y_j)$ is the joint probability density of X and Y. Similarly for the multivariate case, joint entropy (JE) is:

$$H(X_1 X_2, \ldots, X_N) = -\sum_{i=1}^{n1} \sum_{j=1}^{n2} \cdots \sum_{k=1}^{nN} p(x_{1,i}, x_{2,j}, \ldots, x_{N,k})\log_2 p(x_{1,i}, x_{2,j}, \ldots, x_{N,k})$$ (3)

where $p(x_{1,i}, x_{2,j}, \ldots, x_{N,k})$ is the joint probability density function.

Mutual information is defined to estimate the shared information between two variables X and Y, and it can be interpreted as the reduction in the uncertainty of X given the knowledge of Y:

$$I(X, Y) = \sum_{i=1}^{m} \sum_{j=1}^{n} p(x_i, y_j) \log_2 \frac{p(x_i, y_j)}{p(x_i)p(y_j)} \tag{4}$$

To investigate multivariate dependence, total correlation (TC) is defined as:

$$C(X_1 X_2, \ldots, X_N) = \sum_{i=1}^{N} H(X_i) - H(X_1 X_2, \ldots, X_N) \tag{5}$$

It measures the total amount of information shared by all the variables $X_1, X_2, \ldots, X_N$.

In this study, the rainfall series recorded by each station in the rainfall monitoring network is viewed as a random variable and there are 45 of them in total. For each station X_i, we use two indicators - the ratio of entropy RE_i and the ratio of correlation RC_i to measure its information capacity and redundancy. Specifically,

$$RE_i = \frac{H(X_i)}{H(X_1 X_2, \ldots, X_N)} \tag{6}$$

and

$$RC_i = \frac{I(X_i, X_1) + I(X_i, X_2) + \cdots + I(X_i, X_{i-1}) + I(X_i, X_{i+1}) + \cdots + I(X_i, X_N)}{C(X_1 X_2, \ldots, X_N)} \tag{7}$$

where N is the total number of stations. RE_i is the proportion of the information contained by station X_i in the total information of the network. Larger RE_i value indicates that more information is contained and thus the station is more important in the network. Analogously, RC_i measures proportion of the information shared between station Xi and the other stations takes up in the total correlation of the network. Larger RC_i value means more overlapped information, i.e., more redundancy is contained.

In order to realize multi-criterion evaluation, we integrate these two indexes with the ideal point method in multi-objective optimization. The objective function $G(X_i)$ is formulated as:

$$G(X_i) = \left[(RE_i - RE_{max})^2 + (RC_i - RC_{min})^2 \right]^{\frac{1}{2}} \tag{8}$$

where RE_{max}, RC_{min} are ideal solutions. For a specific station X_i, a smaller G value indicates that the two indexes are closer to the ideal point, meaning that it contains relatively more marginal information and less redundant information, and is more prominent in the network. On the contrary, a larger the G value indicates greater the distance from the ideal point and lower significance of a station.

The discretization procedure is conducted in preparation for computing the entropy related variables. The most frequently used methods are histogram discretization and mathematical floor function. Since the bin size is subjectively determined and has a significant effect on the computation results, the histogram method remains questionable. Thus the floor function is used here, which converts the continuous value x to a quantized value x_q– the nearest lowest integer multiple of a constant a:

$$x_q = a \left\lceil \frac{2x + a}{2a} \right\rceil \tag{9}$$

4 RESULTS

The RE, RC, and function G values were calculated for three data sets, i.e. the flood season (from April to September), the non-flood season (from October to March next year), and the complete recorded series concerning each whole year during 2007-2016. Table 1 shows the summary of the index values. JH and TC values of the whole-year series were between those of the flood season and non-flood season. Flood season featured larger values while non-flood season featured smaller values. This indicates that information content and redundancy were positively correlated with the precipitation. The same rule was followed by the RC index. The complete year-round series mostly got the minimum value of the statistics of RE, which indicates when the year-round series was separated to flood season and non-flood seasons, stations reached larger information content. Statistics of G values of the non-flood series always got maximums. Results suggest that the non-flood season had larger variation regarding information content and redundancy.

Figure 2 shows the RE values regarding the three series. Distribution of RE was highly correlated with the elevation of the area. In the southwestern mountain region, large-RE centers were observed in all three figures. Oppositely, the northern plain area saw small RE values. RE generally had consistent distribution for the whole year series and the flood season, while that of the non-flood season showed slight difference, which can be seen at the central part of the area, where the range of high RE region seemed to shrink. This indicates that non-flood season, which features less precipitation, could provide different knowledge of the significance of the stations and regions. Similar rule is also shown in Figure 3, the distribution of RC index. Series of the whole year and the flood season yielded generally consistent spatial variation of RC, while those of the non-flood season saw a more varied region of high RC values. It is seen that the significant center identified by RC, meaning more information redundancy, moved to a larger region at the central part of the area, which was a transition region between the mountain and plain. On the contrary, the northern and southern part were less redundant in information. RC index was much less correlated with elevation, meaning information redundancy was under the impact of more complicated factors and more unpredictable.

Distribution of the integrated objective G is shown in Figure 4. Similarly, series of whole year and flood season generated almost consistent results, and those of the non-flood season generated more varied distribution. Moreover, it is shown that distributions of G values of all the three scenarios largely resembled that of RC, rather than RE. The northern and southern areas obtained higher G values, thus could be regarded as more significant in the area considering both information content and redundancy, as the stations were measured as more "close" to the ideal state of performance of a station. By comparison, the central region was less significant. As measured by G, these parts contained more duplicated information that could lower the efficiency of the network.

Table 1.   Summary of index values.

Statistics		Whole year	Flood season	Non-flood season
JE		4.27	4.70	3.19
TC		45.84	52.49	38.59
RE	Min	0.23	0.24	0.26
	Max	0.30	0.32	0.33
	Range	0.07	0.07	0.07
	Mean	0.26	0.27	0.29
RC	Min	0.34	0.29	0.46
	Max	0.46	0.40	0.60
	Range	0.12	0.11	0.14
	Mean	0.41	0.35	0.54
G	Min	0.03	0.03	0.04
	Max	0.12	0.11	0.15
	Range	0.09	0.08	0.10
	Mean	0.08	0.08	0.10

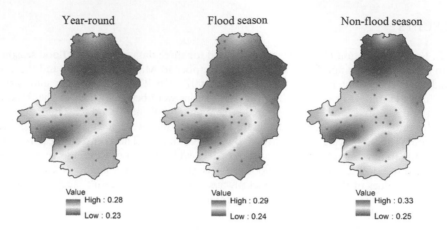

Figure 2. Distribution of ratio of entropy (RE) calculated for the series of the whole year, the flood season, and the non-flood season.

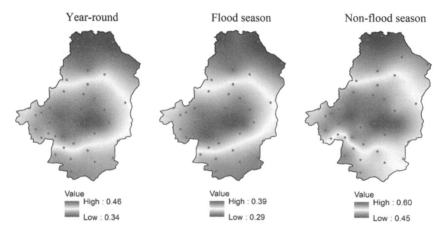

Figure 3. Distribution of ratio of correlation (RC) calculated for the series of the whole year, the flood season, and the non-flood season.

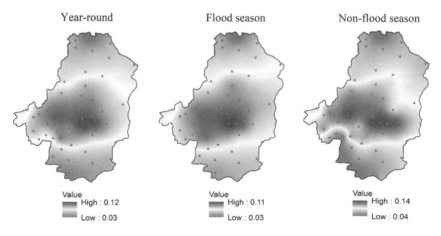

Figure 4. Distribution of objective function G calculated for the series of the whole year, the flood season, and the non-flood season.

5 CONCLUSIONS

This study proposed an entropy based multicriterion approach to identify stations and areas with more information content and less redundancy in the network. Rainfall series of the whole year, the flood season and the non-flood season were employed for computation to investigate seasonal effect. We found stations had larger variation in their performances regarding information content and redundancy during the non-flood season. Series of the whole year and the flood season yielded generally consistent spatial variation of index values, while those of the non-flood season exhibited larger variation. The southern and northern regions of the area featuring higher G values were regarded as more significant for the network.

ACKNOWLEDGMENTS

This study was supported by the National Key Research and Development Program of China (2016YFC0401501, 2017YFC1502704), and the National Natural Science Fund of China (41571017, 51679118 and 91647203), and Jiangsu Province"333 Project" (BRA2018060).

REFERENCES

Alfonso, L., Lobbrecht, A., & Price, R. 2010. Information theory-based approach for location of monitoring water level gauges in polders. *Water Resources Research* 46(3): 374–381.

Alfonso L., He L., Lobbrecht A., Price P. 2013. Information theory applied to evaluate the discharge monitoring network of the Magdalena River. *Journal of Hydroinformatics* 15(1): 211–228.

Fahle, M., Hohenbrink, T.L., Dietrich, O., & Lischeid, G. 2015. Temporal variability of the optimal monitoring setup assessed using information theory. *Water Resources Research* 51(9): 7723–7743.

Li, C., Singh, V.P., & Mishra, A.K. 2012. Entropy theory-based criterion for hydrometric network evaluation and design: maximum information minimum redundancy. *Water Resources Research* 48(48): 5521.

Ridolfi, E., Montesarchio, V., Russo, F., & Napolitano, F. 2011. An entropy approach for evaluating the maximum information content achievable by an urban rainfall network. *Natural Hazards and Earth System Science* 11(7): 2075–2083.

Samuel, J., Coulibaly, P., & Kollat, J. 2013. CRDEMO: combined regionalization and dual entropy-multi-objective optimization for hydrometric network design. *Water Resources Research* 49(12): 8070–8089.

Wei, C., Yeh, H.C., & Chen, Y.C. 2014. Spatiotemporal scaling effect on rainfall network design using entropy. *Entropy* 16(8): 4626–4647.

Xu, H., Xu, C. Lthun, N.R., Xu, Y., Zhou, B., & Chen, H. 2015. Entropy theory based multi-criteria resampling of rain gauge networks for hydrological modelling – a case study of humid area in southern china. *Journal of Hydrology* 525: 138–151.

Yeh, H.C., Chen, Y.C., Wei, C., & Chen, R.H. 2011. Entropy and kriging approach to rainfall network design. *Paddy & Water Environment* 9(3): 343–355.

Shannon, C.E., 1948. A mathematical theory of communication. *Bell System. Technical Journal* 27(3): 379–423.

Risk Analysis Based on Data and Crisis Response Beyond Knowledge – Huang & Nivolianitou (eds)
© 2020 Taylor & Francis Group, London, ISBN 978-0-367-25146-8

Developing a transinformation based multi-criteria optimization framework for rainfall monitoring networks

Heshu Li, Dong Wang* & Yuankun Wang

School of Earth Sciences and Engineering, State Key Laboratory of Pollution Control and Resource Reuse, Nanjing University, Nanjing, China

ABSTRACT: Rainfall monitoring networks provide fundamental information for water resources management and hydraulic engineering. In this study, we developed a multi-criteria approach for rainfall monitoring network optimization. The criteria comprise two transinformation based objectives accounting for overlapped information among selected stations and transferable information from selected to unselected stations, and a rescaled Nash-Sutcliffe Coefficient (NSC) objective measuring the total residual error. The approach was applied to a rainfall monitoring network in the Taihu Lake Basin of China. Pareto solutions under different scenarios were selected from massive candidate solutions for comparison.

Keywords: multi-criteria network optimization, transinformation, Pareto solutions

1 INTRODUCTION

Rainfall monitoring networks collect fundamental information for water resources management, reservoir operation and flood forecast and control. There has been considerable research related to the optimization of rainfall networks. In the cases of surface water (i.e., precipitation and streamflow) network design, researchers have developed approaches including statistical methods, entropy-based methods, expert recommendations and hybrid methods, as reviewed by Mishra and Coulibaly (2009) and Chacon-Hurtado (2017). Among the approaches, entropy based methods have been widely adopted as entropy terms can quantify the information content delivered in the network, and are flexible for establishment of multiple criteria. There a consensus that stations in the optimal network should have maximum total information and minimum redundant information (Keum and Coulibaly 2017; Keum 2018; Leach 2015; Leach 2016; Fahle 2015). Alfonso et al. (2010) combined total correlation with joint entropy to get optimal location of gauges. Li et al. (2012) proposed a maximum information minimum redundancy (MIMR) method, using joint entropy, total correlation and transinformation between selected and unselected stations as objective functions. Samuel et al. (2013) presented a combined regionalization and dual entropy-multiobjective optimization (CRDEMO) framework which maximizes joint entropy and minimizes total correlation of streamflow networks. Xu et al. (2015) proposed a multi-objective function to minimize transinformation of paired stations, reduce the percent bias of measure errors and increase the Nash-Sutcliffe Coefficient (NSC).

Mishra and Coulibaly (2010) proposed a transinformation (TI) index to evaluate Canadian watersheds, in which they used multiple regression method to aid the computation of transinformation between a single station and the network. TI index is especially useful in the identification of important stations and critical areas, which deserves further investigation.

---

*Corresponding author: wangdong@nju.edu.cn

In this study, we presented a multi-criteria optimization framework with focus on the transinformation among stations within a network. Multi-objective functions composed of two transinformation index (TI) based criteria and a statistical NSC were proposed. To transform the multiple rainfall data series of the station combination into a single one, we employed multiple linear regression (MLR) and inverse distance weight (IDW) for interpolation. Candidate network schemes under different scenarios were evaluated and Pareto solutions were obtained.

2 STUDY AREA AND DATA

The study area is a mountain-hilly-plain transition region located in the southwest of the Taihu Lake basin, covering an area of about 5930.9 km2 (Figure 1). The area is densely covered with crisscrossing natural and artificial river channels, the length of which averages 3.3 km per square kilometer. Under a typical subtropical monsoon climate, the area featured four distinct seasons with annual mean temperature measured 16-18°C and annual precipitation measured 1100-1150 mm. Daily rainfall data from 26 rainfall monitoring stations of 10 years (from 2006 to 2015) were used for network optimization.

3 ENTROPY THEORY

Given a random variable $X \in S$ with probability density function (PDF) $p(x)$, the entropy of X is defined as:

$$H(X) = -\sum_{i=1}^{n} p(x_i) \log_2 p(x_i) \tag{1}$$

where n is the sample number and $p(x_i)$ denotes the probability of an observation.

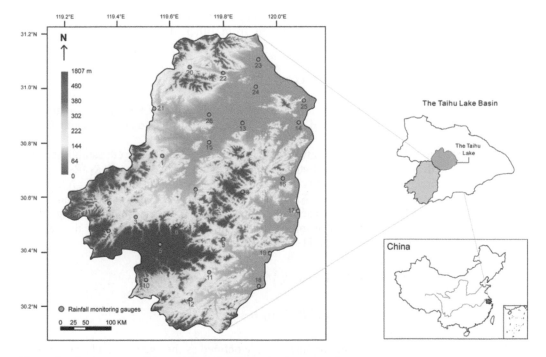

Figure 1. Study area and the existing rainfall monitoring network.

For d random variables $X_1 X_2, \ldots, X_d$, joint entropy is defined as:

$$H(X_1 X_2, \ldots, X_d) = -\sum_{i=1}^{n_1} \sum_{j=1}^{n_2} \cdots \sum_{k=1}^{n_d} p(x_{1,i}, x_{2,j}, \ldots, x_{d,k}) \log_2 p(x_{1,i}, x_{2,j}, \ldots, x_{d,k}) \quad (2)$$

where $p(x_{1,i}, x_{2,j}, \ldots, x_{d,k})$ denotes the joint probability of observations $(x_{1,i}, x_{2,j}, \ldots, x_{d,k})$.

There exists common information shared by the variables, which is measured by transinformation (also called mutual information), formulated as:

$$I(X, Y) = \sum_{i=1}^{m} \sum_{j=1}^{n} p(x_i, y_j) \log_2 \frac{p(x_i, y_j)}{p(x_i) p(y_j)} \quad (3)$$

In multivariate cases, total correlation is commonly used to measure the shared (overlapped) information. Total correlation is defined as the difference between the joint entropy and the sum of marginal entropy (McGill, 1954; Watanabe, 1960):

$$C(X_1 X_2, \ldots, X_d) = \sum_{i=1}^{d} H(X_i) - H(X_1 X_2, \ldots, X_d) \quad (4)$$

In network evaluation, joint entropy and total correlation are often combined to reach the maximum information content and meanwhile the minimum redundant information (MIMR):

$$\begin{cases} \max\{H(X_1, X_2, \ldots, X_d)\} \\ \min\{C(X_1, X_2, \ldots, X_d)\} \end{cases} \quad (5)$$

4 MULTI-CRITERIA NETWORK OPTIMIZATION FRAMEWORK

Suppose X denotes a rainfall series of a monitoring station, and $\hat{X}$ denotes a corresponding simulated series obtained through interpolation from other stations within the network. A transinformation index (TI) is defined as:

$$\text{TI} = I(X, \hat{X}) \quad (6)$$

Mishra and Coulibaly (2010) used multiple linear regression (MLR) for interpolation to obtain $\hat{X}$. In this study, we employed both MLR and the inverse distance weight (IDW) method.

To evaluate the effectiveness of a network, three objective functions were established. Let $(S_1, S_2, \ldots, S_N)$ denote the stations selected from the existing network, and $(R_1, R_2, \ldots, R_M)$ denote the remaining unselected stations. For each selected station $S_i (i = 1, 2, \ldots, N)$, we use all the other selected stations, i.e., $(S_1, S_2, \ldots, S_{i-1}, S_{i+1}, \ldots, S_N)$ for interpolation and obtain the simulated $\widehat{S}_i (i = 1, 2, \ldots, N)$. For each unselected station R_j, $\widehat{R}_j$ is simulated using all the selected stations $(S_1, S_2, \ldots, S_N)$. The information that can be delivered from the selected stations to S_i or R_j, denoted as TI_{S_i} or TI_{R_j}, are formulated as:

$$\text{TI}_{S_i} = I(S_i, \widehat{S}_i)(i = 1, 2, \ldots, N) \quad (7)$$

$$\text{TI}_{R_j} = I(R_j, \widehat{R}_j)(j = 1, 2, \ldots, M) \quad (8)$$

Accordingly, two TI based objective functions are defined as the average of TI_{S_i} and TI_{R_j}:

$$\mathrm{ATI}_S = \frac{1}{N}\sum_{i=1}^{N}\mathrm{TI}_{S_i} = \frac{1}{N}\sum_{i=1}^{N}I\left(S_i, \widehat{S_i}\right) \qquad (9)$$

$$\mathrm{ATI}_R = \frac{1}{M}\sum_{j=1}^{M}\mathrm{TI}_{R_j} = \frac{1}{M}\sum_{j=1}^{M}I\left(R_j, \widehat{R_j}\right) \qquad (10)$$

ATI_S measures the average overlapped information of a certain network scheme. ATI_R measures the average transferable information from the selected to the unselected stations. Networks featuring small ATI_S and large ATI_R values are regarded as efficient for cutting redundancy and avoiding information loss due to the reduction of stations.

Moreover, the Nash-Sutcliffe-Coefficient (Nach and Sutcliffe, 1970) is incorporated as a measure of the relative magnitude of the residual variances, which is formulated as:

$$NSC = 1 - \frac{\sum_{t=1}^{n}(s_t - p_t)^2}{\sum_{t=1}^{n}(p_t - \bar{p})^2} \qquad (11)$$

where n denotes the sample size. s_t is the sampled areal mean rainfall calculated by the selected stations. p_t are the "true" areal mean rainfall from all existing stations. $\bar{p}$ is the average p_t.

We take its negative logarithm to rescale the value:

$$LNSC = -\log(1 - NSC) \qquad (12)$$

In conclusion, the multi-criteria objective function is formulated as:

$$\begin{cases} min\{\mathrm{ATI}_S\} \\ max\{\mathrm{ATI}_R\} \\ max\{LNSC\} \end{cases} \qquad (13)$$

where each criterion accounts for an optimization target, i.e., the least redundant information, the most transferable information, and the least residual error of the system.

5 CASE STUDY: OPTIMIZATION OF A RAINFALL MONITORING NETWORK IN THE SOUTHWEST TAIHU LAKE BASIN

Pareto solutions are frequently used in multi-criteria decision making for their well balance among the objectives. In this study, for each scenario with certain number of stations and interpolation method, we generated 5000 random schemes as candidate solutions, which covers a broad range of combinations of stations with various landform type and distances between them. The Pareto solutions are presented in Figure 2. All the three objective functions values have increasing trends as the station number increases, which is true for both the MLR and IDW scenarios. This implies that when more stations are incorporated into the network, there will be more information transferred among the selected stations and between the selected and unselected stations. Meanwhile, magnitude of the total residual errors of the rainfall series measured by LNSC is reduced. It is also noticed that variation of LNSC corresponding to MLR method is larger than that of IDW, making the solutions more distinct from each other. Also, in both plots there are several "outliers" with relatively higher ATI_S and lower ATI_R values, which are against the objectives. These solutions are featured with high LNSC values, indicating that the incorporation of the third objective has a direct effect on decision.

To investigate if the common solutions have advantages over those produced with one method, we calculated the average ATI_S and ATI_R values of all the Pareto solutions derived from the two methods. Results are shown in Table 1. MLR always produces higher average

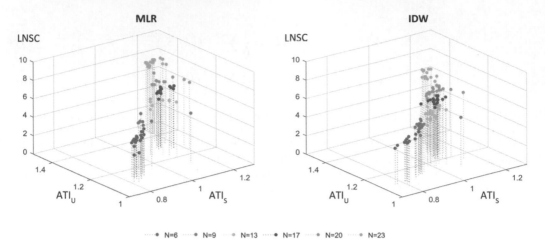

Figure 2.   Objective function values of the Pareto solutions derived from two interpolation methods.

Table 1.   Objective function values of the MLR-derived, IDW-derived and common Pareto solutions.

Number of stations selected (N)	Derivation of Pareto solutions	Number of Pareto solutions	Objective function values and interpolation methods used					
			Average ATI_S		Average ATI_R			
			MLR	IDW	MLR	IDW	LNSC
6	MLR	14	**0.85**	0.84	**1.13**	1.08	3.88
	IDW	6	0.91	0.84	**1.13**	**1.11**	4.30
	Common	7	0.87	**0.81**	**1.13**	1.10	**4.40**
9	MLR	5	**0.95**	0.94	**1.20**	1.14	4.68
	IDW	13	1.00	**0.93**	1.18	**1.15**	4.89
	Common	2	1.04	1.02	1.18	1.14	**5.32**
13	MLR	3	**1.06**	1.03	**1.25**	1.17	5.47
	IDW	14	1.08	**1.01**	1.23	**1.19**	5.18
	Common	5	1.08	1.04	1.24	1.18	**5.89**
17	MLR	5	**1.13**	1.08	1.29	1.21	6.09
	IDW	9	1.15	**1.07**	1.27	**1.24**	5.99
	Common	8	1.14	**1.07**	**1.30**	1.22	**6.32**
20	MLR	6	**1.16**	1.11	**1.34**	1.22	6.44
	IDW	14	1.18	**1.10**	1.31	**1.27**	6.57
	Common	6	1.18	1.11	1.32	1.24	**6.90**
23	MLR	13	**1.20**	1.13	**1.43**	1.32	7.53
	IDW	7	1.21	**1.12**	1.38	1.31	7.02
	Common	10	1.21	1.13	1.42	**1.34**	**7.87**

ATI_S and ATI_R values, which proves the above assumption. For example, for the 14 MLR-derived Pareto solutions with six stations selected, average ATI_S and ATI_R are measured as 0.85 and 1.13, both exceeding those IDW-derived ones measured as 0.84 and 1.08. Comparison within each scenario show that the optimal (best) average ATI_S and ATI_R values of Pareto solutions can be provided by any of the three resources. However, the best (maximum) LNSC values are always given by the common Pareto solutions. The results indicate that the

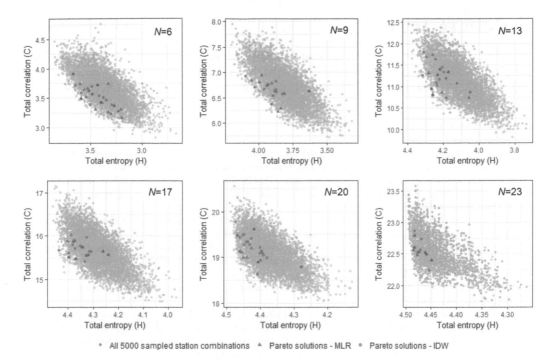

Figure 3. Total entropy (H) and total correlation (C) values of candidate network schemes under the six scenarios.

common solutions can significantly improve *LNSC*, while have no obvious advantage concerning the other two objectives.

To assess the effectiveness of the multi-criteria optimization approach, we evaluated the Pareto solutions obtained using MIMR criterion. Figure 3 exhibits the joint entropy (H) and total correlation (C) values computed for the schemes, where Pareto solutions produced by the MLR and IDW are highlighted. Most of themfeatured large H and small C, which means satisfying the aim of more total information and less redundant information. Several "outliers" with low H and high C are observed. This can be explained as the effect of the *LNSC* objective which strengthens cutting residual errors. The comparison illustrates that multi-criteria incorporated *NSC* can well balance the requirements concerning both information content and statistic errors.

6 CONCLUSIONS

This study proposed a multi-criteria approach containing three objective functions, i.e., ATIS, ATIR and LNSC for rainfall monitoring network optimization. The approach was applied to perform rainfall network optimization (reduction) in a mountain-hilly-plain transition area in the Taihu Lake Basin, China. We obtained Pareto solutions under different scenarios. The Pareto solutions were generally consistent with MIMR objectives.

The approach shares similar idea with previously used MIMR or CRDEMO methods, while focuses more on the information transferred to individual stations instead of the information content or redundancy of the whole network. TI based criteria can help analyzing specific status of each single station, and copula entropy based estimation of transinformation surpasses data discretization method in avoiding uncertainty. Moreover, the incorporation of LNSC supplemented the entropy based criteria.

ACKNOWLEDGMENTS

This study was supported by the National Key Research and Development Program of China (2017YFC1502704, 2016YFC0401501), and the National Natural Science Fund of China (41571017, 51679118 and 91647203), and Jiangsu Province"333 Project" (BRA2018060).

REFERENCES

Alfonso, L., Lobbrecht, A., & Price, R. (2010b). Optimization of water level monitoring network in polder systems using information theory. *Water Resources Research*, 46(12), 595–612.

Chacon-Hurtado, J.C., Alfonso, L., & Solomatine, D. (2017). Rainfall and streamflow sensor network design: a review of applications, classification, and a proposed framework. *Hydrology and Earth System Sciences Discussions*, 1–33.

Fahle, M., Hohenbrink, T.L., Dietrich, O., & Lischeid, G. (2015). Temporal variability of the optimal monitoring setup assessed using information theory. *Water Resources Research*, 51(9), 7723–7743.

Keum, J., & Coulibaly, P.. (2017). Information theory-based decision support system for integrated design of multivariable hydrometric networks. *Water Resources Research* , 53(7), 1–21.

Keum, J., Coulibaly, P., Razavi, T., Tapsoba, D., Gobena, A., & Weber, F., et al. (2018). Application of snodas and hydrologic models to enhance entropy-based snow monitoring network design. *Journal of Hydrology*, *561*, 688–701.

Leach, J.M., Kornelsen, K.C., Samuel, J., & Coulibaly, P. (2015). Hydrometric network design using streamflow signatures and indicators of hydrologic alteration. *Journal of Hydrology*, 529, 1350–1359.

Leach, J.M., Coulibaly, P., & Guo, Y. (2016). Entropy based groundwater monitoring network design considering spatial distribution of annual recharge. *Advances in Water Resources*, S0309170816302263.

Li, C., Singh, V.P., & Mishra, A.K. (2012). Entropy theory-based criterion for hydrometric network evaluation and design: maximum information minimum redundancy. *Water Resources Research*, 48 (48), 5521.

Mishra, A.K., & Coulibaly, P. (2009). Developments in hydrometric network design: a review. *Reviews of Geophysics*, 47(2), 2415–2440.

Mishra, A.K., & Coulibaly, P. (2010). Hydrometric network evaluation for Canadian watersheds. *Journal of Hydrology*, 380(3), 420–437.

Samuel, J., Coulibaly, P., & Kollat, J. (2013). CRDEMO: combined regionalization and dual entropy-multiobjective optimization for hydrometric network design. *Water Resources Research*, 49(12), 8070–8089.

Xu, K., Yang, D., Xu, X., & Lei, H. (2015). Copula based drought frequency analysis considering the spatio-temporal variability in southwest china. *Journal of Hydrology*, 527, 630–640.

Risk Analysis Based on Data and Crisis Response Beyond Knowledge – Huang & Nivolianitou (eds)
© 2020 Taylor & Francis Group, London, ISBN 978-0-367-25146-8

A natural environment risk assessment for ship navigation on the 21st-century Maritime Silk Road

Xin Li* & Ren Zhang
College of Meteorology and Oceanography, National University of Defense Technology, Nanjing, China

Yang Lu
The Army Engineering University of PLA, Nanjing, China

Ju Wang, Mei Hong & Lizhi Yang
College of Meteorology and Oceanography, National University of Defense Technology, Nanjing, China

ABSTRACT: This study built a risk index system and assessment model based on risk assessment theory, for the complex natural environment threats faced by ship navigation on the 21st-century Maritime Silk Road. Then, taking into account the risk of high winds, the study predicted a disaster risk assessment test based on previous meteorological and hydrological data, terrestrial elevation data and relevant shipping report information. The results show that, in winter, the risk of high winds for ship navigation is relatively higher in most areas around the South China Sea and especially around the Taiwan Strait and the Bashi Strait, while the risk is generally below medium level in the Indian Ocean. In summer, the risk of high winds is relatively higher in most areas around the Arabian Sea, and especially in the northeast of Somalia, while it is usually below medium level in other areas.

Keywords: Maritime Silk Road, risk index, natural environment, South China Sea, Taiwan Strait, Bashi Strait, Indian Ocean, Arabian Sea, Somalia

1 INTRODUCTION

The ancient Maritime Silk Road has been an important bridge for both economic and cultural exchanges between the East and the West since it first opened in the Qin and Han Dynasties. In October 2013, Chinese President Xi Jinping proposed a strategic idea of building a 21st-century Maritime Silk Road to last well into the future. At present, the 21st-century Maritime Silk Road maintains routes in mainly three directions. The northern route goes through Korea, Japan and the Bering Strait to the Far East of Russia and then on to the Arctic Ocean. The southern route goes through Indonesia to Australia. The western route goes southward from China's coastal ports, through the South China Sea, the Malacca, Lombok and Sunda Straits, and along the Northern Indian Ocean, to the Persian Gulf, the Red Sea, the Gulf of Aden, and other sea areas. Under the background that China has become the second fastest growing economy in the world, and especially in the global political and economic world, the opening and expansion of the 21st-century Maritime Silk Road will undoubtedly enhance the strategic security of China. In particular, the western route holds extremely important strategic value. On the one hand, solely relying on ASEAN and its member countries, the western route can radiate to the surrounding areas and the other South Asian regions, and can extend as far as the Middle East, East Africa, North

---

* Corresponding author: lixin_atocean@sina.cn

Africa and Europe. It covers the main trade routes between China and Southeast Asia, South Asia, the Middle East, Africa and Europe, and plays an important role in China's future economic development. The western route is also China's main import channel for petroleum and other mineral materials, consisting of over 80% of the freight of China's maritime energy imports. If this channel is blocked or cut off, China's industry, agriculture and transportation industry could no longer operate normally, and any military activities would be threatened.

However, while recognizing the strategic significance of this route, we cannot ignore the risks. First, the natural environment of the coast along the Maritime Silk Road is severely polluted. Tropical cyclone, high winds, high waves, thunderstorms and other natural disasters occur frequently. The complex water channel of the strait also contains many reefs and shoals. All will pose potential threats to ships navigating the area; second, the cultural and political situation of the Maritime Silk Road is very complex. There are ongoing fierce disputes over the sea areas, and foreign powers are also intervening. Armed conflict may break out at any moment; Piracy is rampant. Religious extremism and terrorism are growing; Political forces in some coastal countries are complex, with many uncertainties in the current political situation and policies. This paper focuses on the analysis of natural environment risks of the Maritime Silk Road.

2 DATA AND INFORMATION

The geographic data in this paper mainly includes the terrestrial elevation data, marine topographic data, and administrative division data. The terrestrial elevation terrain applies SRTM30 data that was jointly developed by NIMA (National Imagery and Mapping Agency) and NASA (National Aeronautics and Space Administration) with an accuracy of 30%. The ocean terrain is obtained by extracting the marine part of the Terrain Base data set, jointly founded by the National Geophysical Data Center of the United States of America and the World Solid Geophysical Data Center with are solution of 5'*5'. After re-sampling, splicing and revising the SRTM30 and Terrain Base data, they were integrated into the geographic information platform to enable us to obtain the base topographic and geomorphologic map of the investigated area, and on this basis a sketch map of the main routes along the Maritime Silk Road was drawn up (Figure 1). According to the map, the geographical environment of the related sea areas along the Maritime Silk Road is relatively complex. Among them, the complex topographic features and narrow water channels of the straits have adverse effects on the navigation safety of vessels.

The marine hydrological and meteorological data supplied in this paper, mainly includes ICOADS (Integrated Ocean-Atmosphere Data Set), near-real-time ship reporting data, CMA-ST tropical cyclone path set and tropical storm data of the Indian Ocean, which was compiled by the New Delhi Storm Center of the Indian Meteorological Agency. Through the statistical analysis of this data (not shown), the main risk factors affecting the Maritime Silk Road were verified as tropical cyclones, thunderstorms, high winds, high waves and low visibility.

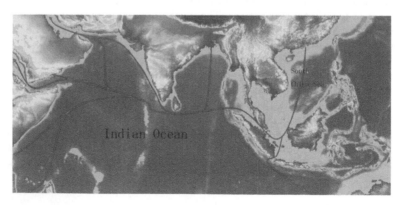

Figure 1.   The topographic features of the main routes and related areas along the Maritime Silk Road.

3 THE THEORY AND METHOD OF RISK ASSESSMENT

3.1 *The natural environment risk of the Maritime Silk Road*

There is no uniform definition of risks. The United Nations Disaster Relief Coordinator (UNDHA) defined all natural disaster risks as follows (United Nations, 1991, 1992): "Risk is the expected loss of life, property and economic activities caused by a natural disaster in a given region and period of time". It describes the nature and connotation of risk comprehensively and accurately. This definition has been recognized by many Chinese scholars, foreign scholars and international organizations.

Based on the concept of disaster risk, this paper defines the risk as "the possibility of disastrous events on or around the Maritime Silk Road and the severity of the consequences it may cause". We must ensure that these disastrous events generally refer to all the factors that may cause danger or harm. Among them, the concept of "disaster" is basically equivalent to risk, such as disaster-causing factor, also known as a risk-causing factor.

According to the above definition, referring to the analysis methods of natural disaster risks (Shi, 1991; Huang, 2005; Zhang et al., 2006; Xie, 2007; Ge et al., 2008), the natural environment disaster and risk factors of the coastal area of the Maritime Silk Road can be summarized:

(1) Disaster-causing factor and risk: This means the factors that can cause disasters or destructive effects. The greater the abnormality these factors may cause, the higher the possibility of damage and increased damage to human social and economic activities accordingly. The human social and economic system may have to bear the higher disaster risk from the risk source. In disaster research, this nature of risk source is usually called risk, which is mainly determined by the scale (intensity) and frequency (probability) of any catastrophic activities.

(2) Disaster-bearing bodies and their vulnerability: In risk analysis, the object bearing the disaster is called disaster-bearing body. Hazardous possibility does not necessarily mean factual disaster, because disaster is relative to the behavior body – humans and their social and economic activities. Only when a risk source is likely to endanger a certain social and economic object, a disaster-bearing body, will the risk-bearing body bear the disaster risk relative to the risk source and risk carrier. Therefore, it is generally referred to as vulnerability or fragility (Xu et al., 2006).

(3) Disaster prevention and mitigation ability: This mainly refers to the abilities of human society to monitor, preempt and forecast future disasters, the protective measures for disaster-bearing bodies, and the emergency response capabilities to deal with these disasters. With the improvement and progression of science and technology, and emergency response abilities, disaster prevention and mitigation ability are also gradually improving.

The above-mentioned three basic elements are the foundation for the natural environment risk assessment of the Maritime Silk Road.

3.2 *Assessment method*

At present, there are various methods of risk assessment used in China and overseas. In summary, they roughly include qualitative risk assessment methods, quantitative risk assessment methods, and comprehensive risk assessment methods that combine equalitative and quantitative methods. Typical qualitative risk assessments methods include factor analysis methods, logical analysis method, historical comparison methods and the Delphi method. Typical quantitative risk analysis methods include probability and statistics methods, risk index methods, analytic hierarchy processes, fuzzy synthetic evaluation methods, artificial neural network evaluation methods, grey relational degree methods and evaluation methods, solely based on information diffusion. This paper applies the comprehensive risk assessment method combining qualitative and quantitative methods, including risk index method, analytic hierarchy process, fuzzy synthetic evaluation methods, grey relational degree methods, and probability and statistics methods. Because of the limit on length, no more details are mentioned. The core technologies in the evaluation mainly include the index quantification method, weight calculation method and risk level classification methods.

4 TECHNICAL PROCESS AND INDEX SYSTEM

4.1 Assessment idea and technical process

According to the analysis of the assessment methods and the content in the previous section, based on many documents and group discussions, the research idea was further clarified, as shown in Figure 2. Therefore, the main steps of natural environment risk assessment and risk management for ship navigation on the Maritime Silk Road are as follows (Figure 2):

(1) Collect and sort out the data for research and to integrate them into the GIS platform.
(2) Risk identification.
(3) Establish an index system and assessment model.
(4) Risk assessment tests based on the collected data.
(5) Design and implement emergency response measures and make a residual risk analysis.
(6) Risk control.
(7) Supervise and evaluate the response results.

Among them, the key technologies are to establish an index system and assessment model.

4.2 Index system

According to the above-mentioned analysis, the natural environment factors harming the ship navigation on the Maritime Silk Road include storms, high waves, poor visibility, tropical cyclones and thunderstorms, etc. For channel, ship and other disaster-bearing bodies, the density of water transport, nature of the ships, material value, average wind resistance, water depth of the channel and the number of reefs and shoals, all constitute the vulnerability factors. The

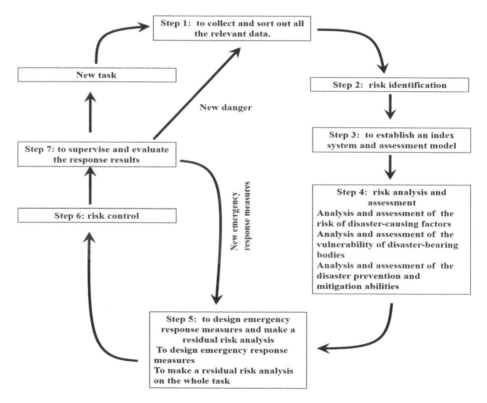

Figure 2.   The technological process of natural environment risk assessment and risk management of the Maritime Silk Road.

disaster prevention and mitigation ability mainly depend on the forecasting and an early warning system, regional disaster prevention facilities and the perfection and proficiency of all emergency plans. On this basis, an index system was built for natural environment risk assessment for ship navigation on the Maritime Silk Road, as shown in Table 1, taking high winds as an example.

5 ASSESSMENT MODEL AND TEST RESULTS

According to risk theory, risk is a comprehensive measure of the possibility and severity of hazard accidents (United Nations, 1991, 1992). The index to measure the risk is R(Risk), which results from the interaction of accident threat T (Threat), vulnerability of disaster-bearing body V (Vulnerability) and disaster prevention and mitigation ability D (Defense). Considering that the above three factors are parallel, we should adopt the multiplication model when calculating the degree of risk:

$$R = T \times V \times (1 - D) \tag{1}$$

Combining the above-mentioned index system and considering the fusion principle of secondary indexes, we should set the mathematical expression for risk assessment of a single type natural disaster as follows:

$$R_i = 100 \left(\prod_{j=1}^{J_i} T_{ij}^{\omega_{ij}} \right)^{\omega_T} \times \left(\sum_{k=1}^{K_i} \omega_{ik} V_{ik} \right)^{\omega_V} \times \left(1 - \sum_{l=1}^{L_i} \omega_{il} D_{il} \right)^{\omega_D} \tag{2}$$

In the equation, R_i is the risk index of disaster i; T_{ij} is the sub-level index j of the risk of this disaster; ω_{ij} is the corresponding weight; V_{ik} is the sub-level index k of the vulnerability of this disaster; ω_{ik} is the corresponding weight; D_{il} is the sub-level index l of the ship's disaster prevention and mitigation ability against this disaster, ω_{il} is the corresponding weight. J_i, K_i and L_i are the numbers of the sub-level indexes of the risk, vulnerability and disaster prevention and mitigation ability of disaster i respectively.

Table 1. The index system of natural environment risk assessment for ship navigation on the Maritime Silk Road.

Level 1 Index	Level 2 Index	Level 3 Index	Description
Disaster risk of high winds A1	Risk B1	Frequency of high winds C1	The higher, the more dangerous
		Intensity of high winds C2	The higher, the more dangerous
	Fragility B2	Density of ships C3	The higher, the more fragile
		Tonnage of ships C4	The smaller, the more fragile
		Distance from haven C5	The bigger, the more fragile
	Disaster prevention and mitigation ability B3	High winds monitoring technology C6	The higher, the greater
		High winds prediction level C7	The higher, the greater
		Formulation of defense plans against storms C8	The more completed, the greater
		Implementation of defense plans against storms C9	The more proficient, the greater

In this paper, the risks of various disasters are determined by the intensity and frequency, so $J_i = 2$. ω_T, ω_V and ω_D are weights of risk, vulnerability and disaster prevention and mitigation ability, respectively. This paper applies 0.5, 0.3 and 0.2 respectively to these, because usually people pay more attention to the risk of disasters. It should be emphasized that the values of all indicators in the above equation are standardized values. The weight of each factor was obtained by using an expert scoring method and analytic hierarchy process.

Take the risk of high winds as an example: Comparing the two factors of risk, the frequency of high winds is more important than intensity, so the weight of the two factors are 0.6 and 0.4, respectively. Comparing the three factors of vulnerability, the density of a ship is more important than its tonnage and distance from haven, so the weights of the three factors are 0.5, 0.25 and 0.25 respectively. Comparing the four factors of disaster prevention and mitigation ability, the high winds prediction level is slightly more important than the other three factors, so its weight is 0.3, and the other three factors are 0.2. Therefore,

$$R_1 = 100 \times \left(C_1^{0.6} \times C_2^{0.4}\right)^{0.5} \times \left(0.5C_3 + 0.2C_4 + 0.25C_5\right)^{0.3}$$
$$\times \left(1 - 0.2C_6 - 0.3C_7 - 0.2C_8 - 0.2C_9\right)^{0.2} \tag{3}$$

The comprehensive natural environment risk of multiple disasters can be expressed as below:

$$R = \sum_{i=1}^{M} \omega_i R_i \tag{4}$$

In the equation, R is the comprehensive natural environment risk index, ω_i is the corresponding weight of disaster and M is the number of disasters. Among them, the corresponding weights of various disasters are based on the historical disaster database and obtained by a 1–9 scale method and analytic hierarchy process (AHP) (Li et al., 2012).

According to the basic geographic information, data, historical hydrological and meteorological data and shipping data collected in this paper, we can apply Equation (2) and Equation(4) for the risk assessment tests of single disasters and multiple disasters, respectively, and then obtain the risk values at each grid point. On this basis, we can grade all risk values in the investigated area (grading standard as Table 2) and draw the corresponding risk grading map by using GIS technology. Taking the risk of high winds as an example, we substitute the standardized C1–C9 values at each grid point into Equation(3) to obtain the risk value of high winds at each grid point, and then we obtain the risk grading map of high winds in the main sea areas of the Maritime Silk Road (Figure 3) by grading according to Table 2. According to Figure 3, the risk of high winds in the South China Sea is obviously higher than that in other sea areas during winter, and generally reaches a level2 risk. Among them, the Taiwan Strait, the Bashi Strait and the south eastern coast of Vietnam all reach a level1 risk, mainly because the above-mentioned sea areas are in the high value area of winter storm and is busy with shipping transport. The natural environment risks of the Indian Ocean in winter is obviously smaller than that of the South China Sea, generally below level3, mainly because the storms are weaker in the Indian Ocean during winter. In summer, the higher risks are concentrated in the northern–southwestern areas of the Arabian Sea, above level2, with the north eastern sea area of Somalia reaching level1. This is not only related to the high frequency of high winds during the summer time, but also the busy shipping lanes. In addition, some areas in the northeast and southwest of the South China Sea, and some areas in the center and west of the Bay of Bengal and the areas along the Indian Ocean route in the south of the Bay of Bengal, are also at a higher risk.

Table 2. Grading of the natural environment risk levels on the Maritime Silk Road.

Evaluation Score	Grade	Characteristic description
≥80	Level1	Quite highrisk, quitehigh probability of disaster impact or damage, and quiteserious loss
60–80	Level2	High risk, a high probability of disaster impact or damage, and serious loss
40–60	Level3	Moderate risk, not a high probability of disaster impact or damage, and generally serious loss
20–40	Level4	Low risk, a low probability of disaster impact or damage, and little loss
≤20	Level5	Quite lowrisk, quite a low probability of disaster impact or damage, and quitelittle loss

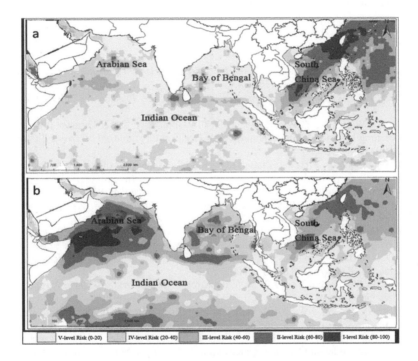

Figure 3. Risk grading map of high winds on the Maritime Silk Road. a. January b. July.

6 SUMMARY AND DISCUSSION

Focusing on the strategic position of the Maritime Silk Road, this study firstly analyzed the natural geographical environment and climatic characteristics of the main coastal along the Maritime Silk Road based on the elevation data, historical hydrological and meteorological data and shipping transport data. On this basis, aiming at the natural disasters that may affect ship navigation on the Maritime Silk Road, a risk index system and assessment model was built for the natural environment of the Maritime Silk Road and a single disaster risk assessment test was designed using historical data based on the risk assessment and risk management theory.

However, there are still problems in the current assessment. First, the index system is not perfect and the vulnerability evaluation index of the disaster-bearing body is far too simple.

The next step is to strengthen the details of the vulnerability index and consider multiple disaster-bearing bodies, and especially the important basis of the Maritime Silk Road port construction. Second, the assessment methods and models need to be further improved upon, mainly including the fusion method of indexes and the determination method of weight between the indexes. We can try to replace the traditional AHP method with new methods such as a cloud model, Bayesian network, fuzzy logic reasoning, fuzzy integral (Yuan et al., 2013) and projection pursuit. Third, we should make concerted efforts to build greater risk analysis and an early warning system, based on the geographic information platform, and builds a practical digital navigation and emergency response service platform for the Maritime Silk Road, based on historical data and combining forecast and early warning products.

ACKNOWLEDGMENTS

This project was supported by the National Natural Science Foundation of China (41605051).

REFERENCES

Ge, Q.S., Zou, M. and Zheng, J.Y. et al., 2008. *Integrated Assessment of Natural Disaster Risks in China* (in Chinese). Beijing: Science Press, 136–234.

Huang,C.F.2005. *Risk Assessment of Natural Disaster –Theory and Practice* (in Chinese). Beijing: Science Press, 3–19.

Li, X., Hong, M., Wang, B., Zhang, R., Ge, S.S. and Qian, L.X. 2012. Disaster assessment and risk zoning concerningthe South China Sea and Indian Ocean safety. *Journal of Tropical Oceanography* (in Chinese), 31(6): 121–127.

Liu, J.F., Yu, M.G., Zhang, X.H. and Zhao, H.Q. 1998.Characteristics of wind wave field and optimum shipping line analysis in North Indian Ocean. *Tropic Oceanology* (in Chinese), 17(1): 17–25.

Shi, P.J. 1991.Theory and practice on disaster system research. *Journal of Natural Disasters* 11: 37–42.

United Nations Department of Humanitarian Affairs. 1992. *Internationally Agreed Glossary of Basic Terms Related to Disaster Management*, DNA/93/36, Geneva.

United Nations Department of Humanitarian Affairs. 1991. *Mitigating Natural Disasters: Phenomena Effects and Options—a Manual for Policy Makers and Planners*. New York: United Nations, 1–164.

Xie, M.L. 2007. Risk factors analysis and evaluation clue of meteorological disasters. *Meteorology and Disaster Reduction Research* (in Chinese), 30(2): 57–59.

Xu, S.Y., Wang, J., Shi, C. and Yan, J.P. 2006. Research of the natural disaster risk on coastal cities (in Chinese). *Acta Geographica Sinica*, 61(2), 127–138.

Yuan, L.L., Liu, P. and Si, J.W. 2013. Application of fuzzy integral in the marine standard system evaluation. *Marine Science Bulletin* (in Chinese) 30(2): 57–59.

Zhang, J.Q., Okada, N. and Tatano, H. 2006. Integrated natural disaster risk management: comprehensive and integrated model and Chinese strategy choice. *Journal of Natural Disasters* 15(7): 29–37.

Risk Analysis Based on Data and Crisis Response Beyond Knowledge – Huang & Nivolianitou (eds)
© 2020 Taylor & Francis Group, London, ISBN 978-0-367-25146-8

Correlation study of safety and emergency management information systems based on social network analysis

Yi Zhou*, Xuewei Ji, Aizhi Wu, Fucai Yu & Yonghua Han
Beijing Academy of Safety Science and Technology, Beijing, China

Liping Fang & Yanyan Zhang
Beijing Towery Computer Software System Technical Limited Company, Beijing, China

ABSTRACT: With the purpose of using the big data accumulated in the information systems of Beijing Emergency Management Bureau (BEMB) to help the urban safety and emergency management, a social network analysis (SNA) model was used to study the data correlations between nine information systems in BEMB. The results showed that there were 59 related data items with a correlation coefficient bigger than 0.1. The association network map designed with the Ucinet software reflected that the system node of administrative law enforcement had the biggest degree centrality and was the most active participant in the whole network. The data node hidden hazard numbers had the biggest betweenness centrality and controlled the data flow of the overall network. It can be concluded that the SNA method quantified and visualized the correlations between the different information systems, which provides an important reference for promoting government information integration and optimization.

Keywords: safety and emergency management, government information integration, social network analysis

1 INTRODUCTION

At present, the governmental information constructions in the field of safety and emergency management of Beijing have achieved a great deal, which promotes the efficiency and refinement of management affairs. At the same time, the big data accumulated by the information systems brings new perspectives and methods for urban safety management. The State Council of China and the Beijing Municipal Government (BMG, 2016) also put forward clear requirements for big data platform construction and data applications in this area. The prime minister has taken a series of actions to speed up the integration of governmental information systems, with the aim of breaking the "fragmented and chimneys like systems, isolated information island"[1] situation. For this, the Beijing Emergency Management Bureau (BEMB) has been a forerunner in safety and emergency big data platform construction and data access, aggregation and integration. Now, using big data to optimize the information systems and to help the management decision-making are the most practically valuable things and need to be the focus for the future.

Some of the existing studies on governmental information systems integration focus on the management method, for example the ability to construct collaborative innovation

---

*Corresponding author: zhouyi_0502@sina.com
1. This quote is from Chinese Premier Keqiang Li, which describes the information systems are isolated like chimneys.

communities in the integration and sharing projects of government information systems (An et al. 2019), or the multi agent information coordination mechanisms in safety production supervision (Zhang et al. 2017). Other scholars focus on the information platform construction technologies, for example, the data sharing technology between heterogeneous hydrological information systems based on a web service (Qin et al. 2018). Zou (2015) studies the E-government information resources sharing technology based on cloud computing. Regarding the data correlations in different systems, studies were conducted on an intergovernmental diffusion network of interconnected risks in disastrous weather (Li P. et al. 2018), and on the semantic organization of E-government information resources based on linked data (Lv Y.Z. et al. 2012). In fact, there must be a sharing or correlation relationship between governmental information systems. Quantifying and visualizing this relationship is significant for the integration of information systems and risk management of the city.

2 SOCIAL NETWORK ANALYSIS MODEL FOR INFORMATION SYSTEMS

2.1 *Problem to be solved*

Taking the BEMB as an example, there are 21 information systems, 83 information resource tables and 4,373 data resource items. During the big data platform construction processes, it was found that some data items were repeatedly collected and uploaded in different systems, and they were in different completeness states and update frequencies. There is a correlation relationship in large amounts of these data items. With the method of data mining to dig the deep relationship and quickly find the key data, promoting the coordination between different systems will be helpful to the risk evaluation and management of our city.

If social actors are regarded as nodes, then social networks are the collection of these nodes and the connections between them. Social network analysis (SNA) is a method to study the network relationships between social structures. In contrast with the previous assumption that individual behaviors are independent, it focuses more on the relationship between social entities (John S. 2000). Many statistical methods can be used to reveal the overall structure of the network and the hidden relations.

SNA methods are used in many areas, especially in the fields of information science (Cao et al. 2018), knowledge management (Liu et al. 2019) and interpersonal relationships (Wang 2010). For the information dissemination and sharing, the processes in a small network or group were analyzed in most of the existing studies (Di 2016; Yi et al.). In this article, we considered all the safety and emergency management information systems and analyzed the correlations between business data, so as to provide guidance for the integration and optimization of information resources and for the safety and emergency management businesses.

2.2 *Data sources and processing method*

In this paper, nine information systems about the safety work management affairs in the BEMB were selected: administrative license system (S1); administrative law enforcement system (S2); full-time safety supervisor inspection system (S3); hidden hazard management system (S4); accidents management system (S5); major hazards management system (S6); safety work standardization management system (S7); work safety liability insurance system (S8); and system for enterprises data (S9). Based on enterprise ID and enterprise name, the correlated data source items in two different systems were listed, and the correlation coefficients were calculated with the data of all these systems in 2018. Three methods of Pearson correlation coefficient, spearman correlation coefficient and Kendall correlation coefficient were selected according to the data types. Part of the correlation relationships are shown in Figure 1, and the correlation coefficient kept two decimal places.

Table 1. Part of the correlation relationships between some data source items.

No.	Information systems	Data items	Type of correlation coefficient	Value of correlation coefficient
1	S1	Inspection times for law enforcement officials	Pearson	0.22
	S3	Inspection times for full-time safety supervisors		
2	S1	Document numbers for administrative law enforcement inspection	Pearson	0.77
	S3	Inspection times for full-time safety supervisors		
3	S1	Question numbers listed in the administrative law enforcement inspection plan	Pearson	− 0.20
	S3	Number of unqualified items found by safety supervisors		
4	S1	Total money penalized in administrative penalty general cases	Spearman	− 0.65
	S5	Indirect cause of accident		
5	S3	Inspection times for full-time safety supervisors	Pearson	0.40
	S4	Hidden hazard inspection times		
6	S8	Safety liability insurance company	Kendall	− 0.18
	S9	Enterprise type		

2.3 The association matrix

The correlation relationships between the nine information systems were systematically sorted, and 59 related items were found with a correlation coefficient bigger than 0.1. With the correlation coefficients of every two data items, an incidence matrix with 50 rows and 50 columns was formed, as shown in Figure 2. The data in the table are the absolute value of the correlation coefficient.

3 RESULTS AND DISCUSSION

3.1 The visual social network map

With the help of SNA software Ucinet and Netdraw, a network map shown in Figure 1 was conducted for the nine information systems, according to the association matrix in Figure 2. This undirected network diagram visualizes the relationships between the data items and gathered them together. Each graph in the diagram represents a node, and there are nine system nodes and 50 data item nodes in total. The size of the node represents its centrality in the entire network, that is, the association degree with other nodes. It can be seen from Figure 1 that systems of administrative law enforcement and enterprise have more correlations with other systems. As can be seen from the large numbers of nodes and links in Figure 1, there are relations between systems and data items with varying degrees. The length of the linking line here has no practical significance, while its thickness represents the absolute value of the corresponding correlation coefficient.

3.2 The association mechanism for the systems based on centrality theory

The position and importance analysis of actors in the social network is an important aspect of SNA research. Bavelas proposed the concept of centrality, and Leavitt and Freeman et al. (Lin, 2009) developed the centrality theory, which provide theoretical support to quantify the

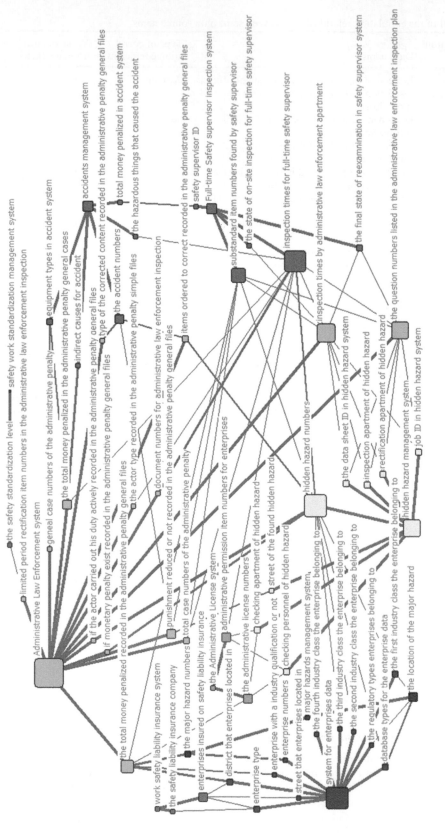

Figure 1. The association network map for the nine information systems.

Table 2. The incidence matrix formed by the 59 related data items.

	A	B	C	D	E	......	F	
A	0	0	0.14	0.22	0		0	A: Inspection times by administrative law enforcement apartment
B	0	0	0.14	0	0		0	B: The question numbers listed in the administrative law enforcement inspection plan
C	0.14	0.14	0	0	0		0	C: Inspection apartment of hidden hazard
D	0.22	0	0	0	0.40		0	D: Inspection times for full-time safety supervisors
E	0	0	0	0.40	0		0.71	E: The hidden hazard numbers
......								F: If monetary penalty exists, recorded in the adminis-
F	0	0	0	0	0.71		0	trative penalty general files

actor's position in the network. There are three main methods to quantify the centrality: degree centrality, betweenness centrality and closeness centrality.

Degree centrality describes the centrality of a node in the network. The bigger the degree centrality value, the closer the node is to the center of the network and the greater the power it has. Conversely, the smaller the value is, the more edge position it locates. Calculated by the Ucinet software, the system node of the administrative law enforcement system has the biggest degree centrality, with the absolute degree centrality of 13 and relative degree centrality of 22.8. That means, among the 59 actors group composed by nine system nodes and 50 associated data item nodes, this system node has a direct connection with 13 other nodes. Second, the node of enterprise system has a higher degree centrality, with absolute degree centrality at 11 and relative degree centrality at 19.3. This result matches the phenomena of nodes' size in Figure 1, so the node size corresponds to the node degree centrality value.

Betweenness centrality refers to the interval degree between one node and other nodes in the network, describing its value as a mediator. A bigger betweenness centrality means that more nodes need to pass through this node on the way of contacting other individuals. If one node is on the shortest path of connecting the other two nodes, it has a big betweenness centrality. If the betweenness centrality is 0, it means that the node cannot control any node and is at the edge of the network. According to the calculation results of Ucinet software, the top two nodes are the number of hidden hazards and the inspection times for full-time safety supervisors. The relative betweenness centrality values are 0.752 and 0.627, respectively. It shows that there are more short paths passing through these two nodes, and they can largely control the data flow of the whole network.

Closeness centrality examines the centricity of node by the distances to other nodes. The smaller the value, the shorter the distance between the node and other nodes, and the higher the centrality. With the results of Ucinet, the smallest three nodes are the system nodes of administrative law enforcement, the full-time safety supervisor and the data time of inspection times for full-time safety supervisor. The closeness centralities are 1.724, 1.724 and 1.754, respectively, with little difference.

4 CONCLUSIONS

In view of the potential applications of big data in government decision-making, the association rules of the safety and emergency information systems were explored based on the SNA method, which was an attempt at data mining of the safety and emergency management affairs. The conclusions are as follows.

Sharing and correlation relationships widely existed between the data in the independent information systems for safety and emergency management affairs. Correlation coefficient calculation, depending on the large amount of basic data, quantified the relationship between different systems.

The association network map for the systems built according to SNA method greatly visualized the correlation relationship of safety and emergency management information systems.

The results showed that the systems for administrative law enforcement and full-time safety supervisors had the highest participation in the whole network.

According to the SNA centrality theory, the system node administrative law enforcement has the biggest degree centrality, which means it is at the core of the whole network. The data node hidden hazard numbers has the biggest betweenness centrality, which indicates that it can control the data flow of the overall network to a large extent.

ACKNOWLEDGMENTS

This project was supported by Beijing Municipal Science & Technology Commission (Z181100009018003, Z181100009018010) and National Key R&D Program of China (Grant No. 2018YFC0809900).

REFERENCES

An X.M., Guo M.J., Wei W. 2019. Research on the ability construction of collaborative innovation community in the integration and sharing project of government information system. *Information studies: Theory & Application* 1: 1–10.

Beijing Municipal Government. 2016. *Beijing big data and cloud computing development action plan (2016–2020)*. Available at: http://www.beijing.gov.cn/zhengce/wenjian/192/33/50/438650/233281/index.html.

Cao X., Li X.S., Chen Y.X. 2018. Comparative analysis of domestic and international collaborative research in library and information fields based on knowledge graph. *Journal of Preventive Medicine Information* 34(8): 1110–1115.

Di Y.M. 2016. *The research of construction team safety information sharing efficiency based on social network analysis*. Nanjing: Southeast University.

John Scott. 2000. *Social Network analysis: A handbook*. London, Sage Publications, p. 83.

Li P., Li W.H. 2018. A study on intergovernmental diffusion network of interconnected risks in disastrous weather. *Chinses Public Administration* 2: 114–119.

Lin J.R. 2009. *Social network analysis: Theory, analysis and application*. Beijing Normal University Publishing Group, pp. 107–126.

Lv Y.Z. 2012. Semantic organization of E-government information resources based on linked data. *Library and Information Service* 56(21): 143–147.

Liu C.Y., He L.N. 2019. Research on hot topics of China's think thanks in 2014–2018 based on keyword co-occurrence and social network analysis. *Information Research* 3: 118–125.

Qin H., Zhang J., Gao J. et al. 2018. A data sharing technology between heterogeneous hydrological information systems. *Studies and Discussions* 11(28): 40–48.

Wang F.F. 2010. *The communication structure and information transmission study of the SNS virtual community*. Dalian: Dalian University of Technology.

Yi L.L., Wang Y.K., Huang M.Y. 2019. Government information sharing mechanism: A research based on social network analysis taking the information circulation of public service in D city of Guangdong Province as an example. *Journal of Intelligence* 3: 1–10.

Zhang Y.W. 2017. Multiagent information coordination mechanism in safety production supervision-taking De-Qing county Zhejiang province as research sample. *Nanchang: Nanchang University*.

Zou Q.L. 2015. E-government information resources sharing research based on cloud computing. *Fuzhou: Fujian Normal University*.

Research on risk perception of tourism utilization of intangible cultural heritage—tourist perspectives

Juan Wang, Fei Wen* & Dunli Fang
School of Tourism, Huangshan University, Huangshan, China

ABSTRACT: Broad engagement by stakeholders is an important way to manage risks. Under the background of rapid urbanization in China, with the changes in society and living environments, the protection and inheritance of intangible cultural heritage has become a social issue of wide concern. Tourism survival is one of the most effective modes of intangible cultural heritage protection and the tourist is the vital stakeholder in intangible cultural heritage tourism. Tourists' perception of destination risk is a significant factor affecting their decision-making processes. Based on the tourist perspective, taking Huangshan City as an example, this article adopts the negative importance-performance analysis (IPA) method to quantitatively analyze the perceived risks in the process of utilizing intangible cultural heritage tourism. Targeted measures can then be taken to resolve the negative feelings of tourists. Through rational tourism utilization of intangible cultural heritage, we hope to retain traditional memories and revive traditional cultures.

Keywords: risk perception, tourism utilization, intangible cultural heritage, tourists

1 INTRODUCTION

Intangible cultural heritage (ICH) is the living inheritance of human wisdom and civilization, and the individuality and vitality of tourist destinations. Under the background of rapid urbanization in China, with the changes in society and living environments, the protection and inheritance of ICH has become a social issue of great concern. Tourism survival is one of the most effective modes of conservation and inheritance (Wang & Tian, 2010). However, in the process of ICH tourism survival, tourists feel some risks, such as lacking cultural cognition and cultural identity, over-commercialization, lacking authenticity, low levels of tourism production, and so on.

Tourists' perception of destination risk is an important factor that influences their decision-making process (Sönmez & Graefe, 1998). To some extent, the sensitivity and fragility of ICH enhances the perceived risk by tourists, and this study analyzes the influencing factors that influence this perception. As a result, some countermeasures have been taken to dissolve the negative emotions of tourists, such as worry, anxiety and fear. In order to better retain the traditional memories and rejuvenate traditional culture, we should make use of ICH in a reasonable and healthy way for tourism utilization.

2 RISK SOURCE IDENTIFICATION

The "likelihood of adverse events" is the heart of risk (Huang, 2005). By judging the likelihood of adverse events, risk assessment provides a basis for avoiding, mitigating and managing risks (Wilson & Crouch, 1987).

---

*Corresponding author: 7788915@qq.com

From the perspective of epistemology, Huang (2001) divides risk into four basic types: real risk, statistical risk, predictive risk and perceived risk. Perceived risk is the part of the risk that people can feel and originated from nuclear safety risks in the 1960s (Sjöberg, 2000). In recent years, the study of tourism perceived risk has deepened, from social security risk (Chew & Jahari, 2014) to traffic risk (Fuchs & Reichel, 2011), cultural risk (Qi & Gibson, 2009), environmental risk (Cheng, Zhou, Wei, & Wu., 2015) and other specific research areas of perceived risk. Li et al. (2014) focused on the risk assessment of tourism development, starting from the emic risk, the etic risk to the environmental risk, and took Tianjin as an example for empirical analysis. However, it is rare to study tourism risk from the perspective of tourists' perception.

In order to obtain the literature related to ICH tourism, searches were performed in English language databases with "intangible heritage tourism" as the keyword. Then, annals of tourism research and tourism management were searched for the literature about cultural tourism, heritage tourism and ethnic tourism, and 69 articles in English were obtained by artificial screening according to the actual content. Searching for "intangible cultural heritage tourism" on the China National Knowledge Infrastructure obtained 105 Chinese documents. By combing through the relevant research literature, the following four risk sources were determined: over-commercialization, lacking authenticity, participation, and cultural identity.

3 RESEARCH DESIGN

3.1 Research procedures

This study follows six steps:

(1) The four sources of tourists' perceived risk in tourism utilization were identified from the related literature.
(2) Through brainstorming and visitor interviews by the research team, the perceived elements corresponding to these four sources of risk were listed.
(3) The research team designed the evaluation index system, performed the pre-investigation with ICH tourists, adjusted the evaluation index system according to the visitor feedback, and deleted semantic repetitions or elements that easily cause ambiguity. This resulted in 18 indexes to be included in the evaluation using a Likert scale of 1–5.
(4) In June 2018, a questionnaire survey was conducted in Tunxi Old Street and Liyang Water Lane in the national historical and cultural district of Huangshan City.
(5) Based on the investigation data, the research team explored the tourists' perceived risk and differences in the use of ICH.
(6) Decision-making analysis was carried on, and the risk response strategies were proposed.

400 questionnaires were distributed and 336 were recovered, with an effective recovery rate of 84%. The Cronbach's coefficient of negative strength and perception evaluation were 0.854 and 0.889, respectively. The reliability was good.

3.2 Research method

The modified negative importance-performance analysis method is used for quantitative analysis. This method is derived from the importance-performance analysis (IPA). According to the level of importance and performance, the IPA method can establish a four-quadrant model and take the corresponding evaluation score of the study object into account to determine the development strategy of the future. However, Cheng et al. (2015) proposed the "negative IPA" matrix, which can be used for the negative evaluation of the research object, and applied it to the study of inbound tourists' perception of environmental risk in China. We modified the model to make it appropriate for the name and interpretation of the abscissa and the ordinate.

The modified model consists of a negative perception evaluation axis (x-axis) and a negative strength evaluation axis (y-axis). In quadrant I, the risk occurrence probability and

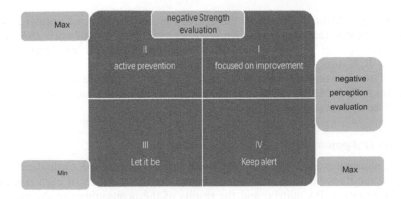

Figure 1. Negative importance-performance analysis.

negative influence are both high, and is "focused on improvement". In quadrant II the risk occurrence probability is high but there is weak negative influence, so it is necessary to take "active prevention". In quadrant III, the risk occurrence probability and negative influence are both low, and the measure is "let it be". In quadrant IV the risk occurrence probability is low but there is high negative influence, so the evaluation is "keep alert" (Figure 1).

4 RESULTS AND ANALYSIS

4.1 Evaluation of negative risk perception and analysis of strength differences

The tourists' perceived negative perception evaluation and the negative strength evaluation of 18 risk factors were calculated according to the average value of four risk types, and the results are shown in Figure 2.

The important risk factors that tourists consider to affect the quality of experience are basically the same as those felt in the process of culture tourism, mainly manifested in over-commercialization and lack of authenticity. The average value of both is between 3.51 and 3.6. The risks of lack of participation and cultural identity are not as high: the average value of both is between 3.33 and 3.44.

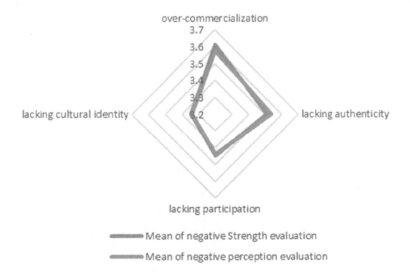

Figure 2. Mean grouping factor comparison grid on ICH risk perceptions.

There is no significant difference in the evaluation and strength of negative risk perception, which indicates that the Social Amplification of Risk Framework (SARF) used to explain the perceptions of nuclear pollution risk, terrorism risk and haze risk are not applicable to ICH tourism utilization. This may be, on the one hand, because the risk perception of the use of ICH tourism by tourists is not as intense and fearful as the public's perception of the above-mentioned risks, and on the other hand because many tourists may not be truly aware of the risks.

4.2 *Negative IPA positioning and coping strategy analysis*

The negative strength evaluation mean (3.48) and negative perception evaluation mean (3.47), were taken as the basis for four-quadrant division. The mean values of each risk factor were placed in the negative IPA matrix, and the results of the positioning are shown in Figure 3. The numbers correspond to the risk factor numbers in Table 1.

The first quadrant covers 1–9, the greatest risks currently felt in tourism consumption, indicating that tourists consider these risks to occur frequently and have a serious negative impact. Among these, the problem of excessive commercialization is the most prominent. There is no doubt that survival and then development is the fundamental principle of tourism development, but if the pursuit of economic benefits is excessive, it will give tourists the impression of profit and increase the perceived risk of tourists. The coping strategy is "focus on improvement".

First, the tourism management department should rationally divide the layout of commercial forms, with centralized distributions of business types: streets for restaurants and bed and breakfasts, historical streets, calligraphy and painting streets, traditional folk areas, and other distinct themes. Second, the type of ICH commodities should be strictly regulated, the quantity of each type should be controlled, the characteristics should be highlighted and homogeneous competition should be avoided. Third, market supervision management and penalties should be strengthened to guide handicraft makers to operate in good faith, severely crack down on illegal acts of selling counterfeit goods, and trade associations should be set up to strengthen self-discipline and jointly resist unreasonable pricing.

At above, three countermeasures have been proposed to prevent the over-commercialization from the historical block management angle. In order to solve this problem fundamentally, creative thinking must be used.

(1) Develop an innovate mode, adopting the "four in one" mode to integrate government guidance, the market operations, ingenious designs and brands, developing more medium- and low-grade cultural tourism derivatives to attract young people's attention and purchasing power.

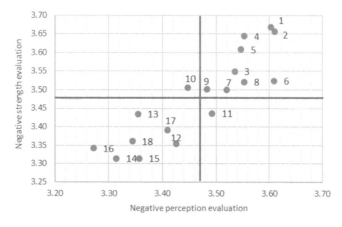

Figure 3.   IPA grid of tourists ICH tourism risk negative perceptions.

Table 1. The sorting of negative strength and perception evaluations of ICH risk perceptions.

Coding	Risk elements	Negative strength evaluation			Negative perception evaluation		
		Order	Mean	SD	Order	Mean	SD
1	The commercial atmosphere is so heavy it is repugnant	1	3.67	0.945	2	3.61	0.912
2	Handicraft ICH is "commercialized"	2	3.66	0.871	6	3.61	0.927
3	Acting ICH "stage-only"	5	3.55	0.921	8	3.55	1.044
4	Fear of being tricked into spending money	3	3.64	0.943	1	3.60	0.906
5	Cannot distinguish the authentic from the fake	4	3.61	0.902	4	3.55	0.963
6	The price is too high for consumption	6	3.52	1.000	5	3.55	0.924
7	Cannot experience the "original"	10	3.50	0.921	9	3.48	0.882
8	The culture is deviated from its native environment	7	3.52	0.934	3	3.53	1.022
9	ICH tourism products homogeneity	9	3.50	0.935	11	3.49	0.879
10	The development level is low	8	3.51	0.903	7	3.52	0.935
11	Lack of participatory cultural tourism activities	11	3.44	0.941	10	3.45	0.887
12	The ICH tourism activity lacks the power of shock value	15	3.35	0.943	13	3.35	0.948
13	Cannot learn cultural knowledge about Huizhou	12	3.44	1.061	12	3.43	0.961
14	Never heard of the ICH of Huizhou	17	3.31	0.930	14	3.31	0.933
15	Cannot appreciate the ICH of Huizhou	18	3.31	1.002	16	3.27	0.990
16	Do not agree with the cultural aesthetics of the local people	16	3.34	0.957	18	3.34	0.976
17	Huizhou culture is too heavy	13	3.39	0.940	17	3.41	0.961
18	Huizhou culture is too profound	14	3.36	0.942	15	3.36	1.003

(2) The operation methods should be innovated to adapt the mode of compound operation: reducing the commercial environment, increasing the leisure atmosphere and the tourism explication system of ICH, encouraging the tourists to study the cultural knowledge of Huizhou, enhancing the ability to identification non-heritage handicrafts, and the appreciation ability of the performance ICH.

(3) Innovating marketing methods should mix "online and offline" marketing, using experiential marketing, event marketing, brand marketing, new media marketing and other means comprehensively to expand sales, rather than relying on physical stores for sales.

As sales increase, businesses will come to realize that tourists in physical stores in historic blocks are more important than actual purchases because they focus on increasing the number of potential consumers online, and online sales are crucial to the survival of handicrafts in the digital age.

In the second quadrant, there is only one risk factor, the ICH tourism development is low, which suggests that this risk factor does not make a large impact on tourism activities while tourists often feel the ICH tourism development is low-grade. Taking the city of Huangshan as an example, the best-selling item for wealthy tourists was the Huangshan small-baked cake, and in the old streets of Tunxi the sauce was the best-selling product. There are also a lot of small commodity markets that are wholesale, the same as the tourist goods of other tourist

323

destinations. In fact, the emblem culture has a rich connotation: the bamboo carvings in the "Tiangong Fong" shop, the bookmark, the bamboo woven handbag, where the artistic integrity and the practical functions are successfully combined. That is just one of the four carvings of Huizhou, which include brick, wood, stone, and bamboo. According to the negative IPA positioning, the coping strategy is "active prevention". Taking handicrafts as an example, in the era of quality tourism and in the development of high-end and middle-end products, it is necessary to enhance the "translation" ability of culture: on the one hand, to seek inspiration from historical culture, on the other hand, to endow tourism commodities with practical functions and modern aesthetics. It not only respects history, but also has aesthetics for life.

The third quadrant includes the lack of cultural identity, not enough shock value, unable to learn about Huizhou cultural knowledge. It shows the probability is low that tourists generally believe these risks occur in the non-exhaustive tourism utilization, and the negative impact is small. At present tourists' negative perception of these risks is not obvious so the corresponding strategy is "no priority." The perception of tourists is related to their cognition. At present, the main concern is that the economic interests are not damaged, and the attention to the social and cultural interests is not key. Non-legacies are the representatives of Chinese traditional culture. With the development of cultural construction, most tourists have a good traditional cultural foundation and have universal recognition of the modern inheritance of Chinese traditional culture. For the means of presentation and learning, tourists are also willing to accept flexible and diverse ways. In the future, people's cultural identity, pride and confidence should be increased through experience, interaction, participation and other forms of tourism utilization.

The fourth quadrant also has only one risk element: lack of cultural tourism activities that can be involved. This shows that this risk is not often felt, but once tourists do feel it, there will be a great impact on tourists' tourism quality. The strategy is to be vigilant. Interactive participation can enhance the emotional exchange between tourists and destinations, satisfy tourists' curiosity and desire for learning, deepen tourists' impressions of the area, and also solve the problems of over-commercialization and development at lower levels in the process of tourism utilization of ICH. Activities should be designed based on the specific contexts, such as a Huizhou ink-making experience where tourists can participate in the production process.

5 CONCLUSIONS AND DEFICIENCIES

Tourists' perception of destination risk is a significant factor affecting their tourism decision-making process. Based on the tourist perspective and taking Huangshan City as an example, this article adopts the negative IPA method to quantitatively analyze the perceived risks in the process of using intangible cultural heritage tourism. Targeted measures can then be taken to resolve the negative feelings of tourists. Through rational tourism utilization of ICH, we hope to retain traditional memories and revive traditional cultures.

This study takes Huangshan City as an example; the questionnaire survey was taken in the historical commercial street area, so the over-commercialization risk is at the top of the tourist's perceptions. A follow-up study can extend this to the traditional village tourism area to enhance the universality of the research result. The second deficiency is that the questionnaire was only distributed to the Chinese domestic tourists, who have a certain understanding of the Huizhou culture, which may be the reason that the cultural identity is high. A follow-up could be conducted with international tourists to enrich the research conclusions.

ACKNOWLEDGMENTS

This project was supported by the Project of Cultivating Tourism Youth Professionals of Anhui Province (AHLYZJ201615), Key Projects of Humanities and Social Sciences Research in Colleges and Universities in Anhui Province(SK2017A0371), the Humanities and Social Sciences Research Project of Anhui Provincial Department of Education (SK2013B490), the

First-class (brand) major construction project in Anhui Province (2018ylzy081), the Research project of Huangshan university (2015xsk003), and the Project of Cultivating Tourism Youth Professionals of Anhui Province (AHLYZJ201415).

REFERENCES

Cheng, D., Zhou, Y, Wei, X. and Wu, J. 2015. A study on the environmental risk perceptions of inbound tourists for China using negative IPA assessment. *Tourism Tribune* 30(1): 54–62.

Chew, E.Y. and Jahari, S.A. 2014. Destination image as a mediator between perceived risks and revisit intention: A case of post-disaster Japan. *Tourism Management* 40: 382–393.

Huang, C.F. 2001. *Natural Disaster Risk Analysis*. Beijing: Beijing Normal University Press.

Huang, C.F. 2005. *Theory and Practice of Natural Disaster Risk Assessment*. Beijing: Science Press.

Qi, C.X., Gibson, H.J. and Zhang J.J. 2009. Perceptions of risk and travel intentions: the case of China and the Beijing Olympic Games. *Journal of Sport & Tourism* 14(1): 43–67.

Sjöberg. 2000. Factors in risk perception. *Risk Analysis* 20 (1): 1–11.

Sönmez, S.F, and Graefe, A. R. 1998. Influence of terrorism risk on foreign tourism decisions. *Annals of Tourism Research* 25 (1): 112–144.

Wang, D. and Tian, Y. 2010). Tourism survival: a choice of pattern about the inheritance of non-material culture heritage. *Journal of Beijing Second Foreign Studies College* 1: 16–21.

Wilson, R. and Crouch, E.A.C. 1987. Risk assessment and comparisons: an introduction. *Science* 236 (4799): 267–270.

Li, Y., Wang, Q. and Li, Z. 2014. Evaluation on the risk of tourism development of intangible cultural heritage: a case study of Tianjin city. *Area Research and Development* 33(5): 88–93.

Risk Analysis Based on Data and Crisis Response Beyond Knowledge – Huang & Nivolianitou (eds)
© *2020 Taylor & Francis Group, London, ISBN 978-0-367-25146-8*

Farmers' risk perception and its impact on climate change adaptation behavior

Jianjun Jin*, Rui He, Xinyu Wan, Foyuan Kuang, Jing Ning, Yuhai Wang & Xia Hu
Faculty of Geographical Sciences, Beijing Normal University, Beijing, China

ABSTRACT: This study measures farmers' risk perception on climate change and explores the effects of farmers' risk perception on their climate change adaptation behavior. Farmers' risk perception was elicited by using a questionnaire survey in Dazu district of China. This study divided risk perception into five dimensions, climate change risk probability perception, severity perception, efficacy perception, self-efficacy perception, costs perception. Our results show that most farmers agreed that climate change has severely affected their lives. Measures taken cannot mitigate the impact of climate change. The ability to cope with climate change is relatively low, and the cost of taking measures is relatively high. In addition, our estimation results indicate that farmers' risk perception have significant effects on their adaption behavior. Specifically, risk probability perception, severity perception, and self-efficacy perception have positive significant effects. The efficacy perception and cost perception have negative significant effects. In addition, farmers' socio-economic factors (education, household income, household size and years in farming) have significant effects on their climate change adaptation behavior. The findings of this study provide implications for policy makers on the promotion of farmers' adaption behavior of climate change in China and other developing world.

Keywords: risk perception, farmer, climate change adaptation, China

1 INTRODUCTION

Climate change has seriously disturbed the natural conditions of the world's climate, and has had a series of negative impacts on the natural and human social systems (Woods et al., 2017). How to mitigate and adapt to the adverse effects of climate change has attracted extensive attention. Agriculture is one of the most serious sectors that were affected by climate change (Schmid et al., 2016). Agricultural production is extremely vulnerable to the impact of climate change (FAO, 2016). In order to maintain productivity in the face of climate change, farmers need to respond with effective adaptation strategies (Chase et al., 2015; Ashraf et al., 2014; Bryan et al., 2013).

Previous studies have shown that public perception of risk largely guides their behavior. Perception is one of the key components of the adaptation process (Slovic, 1987; Tam, 2013; Feng et al., 2019). Lu and Chen (2010) found that in addition to the basic characteristics of farmers such as age and education level, farmers' awareness of climate change has a significant impact on whether they take active adaptive behavior. Li et al. (2015) found that the response measures of farmers in different regions to ecological degradation are different, and farmers' perception of ecological degradation has a significant impact on the response strategies in the lower reaches of Shiyang River, China. Wang et al. (2017) found that farmers

---

* Corresponding author: jjjin@bnu.edu.cn

have different perceptions of climate change such as precipitation and temperature. Qi et al. (2017) found that the public perception of climate change was uncertain and there were obvious regional differences. Moreover, adaptation strategies are context specific and changer from area to area and even within particular societies (Alam et al., 2017). This paper focuses on local level knowledge and perception of adaptation in China.

Chongqing is one of the four municipalities directly under the Central Government of China, and it is also an important food production base in China. It is important to understand farmers in Chongqing because of their significant role in the national and global food system, and their contribution as well as vulnerability to global climate change. However, there are few studies on farmers' climate change risk perception and climate change adaptation decision-making in Chongqing. Based on the existing research, this paper takes Dazu District of Chongqing as a typical research area, uses questionnaire survey method to measure farmers' climate change risk perception, and to explore the impact of farmers' risk perception on their climate change adaptation strategies.

This paper is structured as follows: Section 2 describes the study area, the survey design and data collection. Section 3 presents the empirical results and discussion. The last section provides a summary and some policy implications.

2 MATERIALS AND METHODS

2.1 Study area

Dazu District of Chongqing City is located in Huazhong Double-single-cropping rice cultivation area (68% of the total rice area in China), which belongs to the subtropical warm and humid monsoon climate. The annual average temperature is 17.0-18.8 C, the average precipitation is 1000-1400 mm, the annual average sunshine time is 700-1500 h, the average relative humidity is about 80%, the frost-free period is 200-350 days, and the rice growing season is 210-260 days. The study area is a national commodity grain base county and an advanced county in grain production. It has a strong representative in agricultural production in China. Drought, flood, high temperature, cold wave, storm, hail and other disasters occur in varying degrees every year, which has a great impact on agricultural production and farmers' livelihoods in the study area. In July 2016, the region suffered a severe rainstorm, which was the largest daily rainfall in July in the local meteorological records.

2.2 Survey design

The questionnaire used in this study was based on a series of focus group discussions among some experts on climate and social science, local government officials and farmers to collect more detailed information on the current climate change situation, and to obtain opinions on the design of the questionnaire. A pre-survey on farmers in the study area was conducted to avoid any ambiguity in the questions and to ensure that all the questions in the survey were understandable. Based on the feedback from the pilot pre-survey, some clarifications and modifications were made.

In order to elicit farmers' perception on climate change adaptation, this paper draws on the Social Cognitive Model of Individual Active Adaptation to Climate Change (MPPACC) proposed by Grothmann et al. (2005). It decomposes farmers' perception of climate change adaptation into climate change risk perception and adaptation perception. Risk perception is decomposed into severity perception and possibility perception. Adaptation perception is decomposed into adaptive efficacy perception (referring to the adaptive action taken to protect themselves or others). Effectiveness or anticipated out come from threats, self-efficacy perception (referring to the perceived ability of individuals to implement or execute adaptive actions), and adaptive cost perception (referring to the anticipated cost of implementing adaptive actions). In the actual questionnaire design, the research divides climate change risk perception into climate change risk event perception, risk source perception and impact perception based on the principle of operability of indicators. Risk event recognition refers to farmers' understanding of climate change events and their understanding of the causes. Climate change risk sources refer to the types of climate change from which farmers perceive risks, such as drought and floods. Cognition of impacts is mainly farmers' overall understanding of the impacts of climate disasters on their lives. This paper also investigates the extent of crop losses caused by climate change and the possibility of farmers' predicting crop losses in the coming growth period. Respondents were asked: What are the top five risks you have faced in agricultural production in the past 10 years? In view of the "degree and possibility of loss" and on the premise of reducing crop yields of farmers, this survey designed the following two questions: (1) How many losses have been caused to your crops by climate change in the past year on average? (2) According to your existing knowledge and experience, what is the probability of crop loss in this upcoming growth period?

The questionnaire also sought information on farmers' response strategies towards climate change. Through sorting out and analyzing data of the pre-survey, the research group found that the adaptation measures commonly adopted by local farmers focused on planting new seed varieties, diversified planting, adjusting fertilization behavior, improving irrigation methods/frequency, and adjusting pesticide behavior.

The last section of the questionnaire was to collect socioeconomic information of the respondents and their households.

2.3 *Sample and data collection*

Data were collected using face-to-face interviews between July and August 2018. In this survey, the analysis unit was the household and the household head was the key informant. The simple random sampling method and the multiple stage stratified random sampling method was combined to select the respondent. Firstly, based on the town size, population and geographical location, two townships (streets/communities) were randomly selected. Then six villages were randomly selected from the two townships. In each selected village, 30-50 household heads were randomly chosen from a complete list of households in the selected villages. Finally, 475 farmers were interviewed, which provided 462 valid questionnaires after excluding missing or inconsistent responses.

3 RESULTS AND DISCUSSION

3.1 *Socioeconomic characteristics of respondents*

Of the total valid respondents, forty-nine percent were male and the rest fifty-one percent were female. The average age of the respondents surveyed was about 59 years old. The average level of education was close to primary school. Some respondents (27.19%) did not even receive any education and could not read. Approximately 45.18% of the respondents were primary school educated. Only 17.1% of the respondents received junior middle school education. The average household size was 5.51 persons/household, and the average household owned farmland area is 0.22ha. Among the households surveyed, the average monthly household income was 4097CNY (about 611USD).

3.2 *Farmers' risk perception on climate change and adaptation*

Approximately 80.26% of the respondents surveyed had heard of climate change, 58.33% believed that industrial exhaust was the cause of climate change, 55.70% believed that air pollution also had an impact, 54.39% believed that automobile exhaust was the cause, 40.35% believed that people's unfriendly behavior to the environment contributed to climate change, 38.16% believed that too many factories, 34.21% of the respondents thought urban development caused climate change, 29.39% of the respondents thought that social development has an impact on climate change, 22.37% thought that climate change is related to national construction and development, and 11.41% of the respondents were not sure about the causes of climate change. It can be seen that most farmers know about climate change, but not all farmers have heard about it. As for the causes, industrial emissions, air pollution, automobile emissions and people's unfriendly behavior to the environment were considered to be the top four causes of climate change.

For the risk source, different farmers perceived differently. For the question: What are the top five risks you have faced in agricultural production over the past 10 years? the results showed that the increase of pests and weeds, drought, irregular rainfall, extreme high temperature/temperature rise, and soil quality deterioration were the top five risks facing farmers in agricultural production, and flood ranked sixth. In view of the first risk, 55.26% of respondents thought drought had a greater impact, followed by the increase of diseases, insects and weeds, and irregular rainfall.

The survey results show that 91.63% of the households surveyed had suffered drought in the past decade, 74.89% thought that pests and weeds had increased, 52.42% had suffered irregular rainfall, 51.98% had suffered extreme high temperature/temperature rise, 35.24% had suffered floods, 34.36% believed that soil quality had deteriorated, 19.82% believed that snow fell, and 14.98% had suffered extreme low levels. The top six climatic disasters suffered by local farmers were drought, increased pests and weeds, irregular rainfall, extreme high temperature/temperature rise, flood, soil and poor quality.

About 75.33% of the respondents believed that the climate disasters they had suffered in the past 10 years had a negative impact on them. Among those who were badly affected by climate disasters, 88.89% thought that their crop yield was reduced, 57.89% thought their household income was reduced, 38.60% thought their production cost of planting was increased, 36.26% thought that the family expenditure was increased, 32.75% thought that their property was damaged, 32.16% thought that the living pressure was increased, 22.81% thought that their health was affected and 19.88% of respondents thought that their crop production period had changed. It can be seen that farmers were generally concerned about the impact of climate disasters on their agricultural output, household income and agricultural production costs.

In view of the degree and possibility of loss, the choice of farmers ranged from 3% to 50%, and the average result was 30.68%. This shows that climate change has reduced the average output of farmers by one third, which has brought huge losses to their agricultural production. In view of the problem of "what is the probability of crop loss in the coming growth period", the choice of farmers was between 0-60%, and the average filling-in result was 23.20%.

The MPPACC model was used to further analyze farmers' perception. The results show that 68.86% of the respondents agreed or strongly agreed that climate change has seriously affected their lives, 23.25% of farmers choose neutrality and 7.89% of farmers disagree. It can be seen that most farmers believed that climate change has seriously affected their lives. About 62.28% of the respondents agreed or strongly agreed that the possibility of further changes in the future was great, 33.33% of the respondents chose general, indicating that the respondents were pessimistic about future climate change, which was consistent with the previous studies. Roughly 49.56% of the respondents believed that measures could not mitigate the impact of climate change, and only 7.90% of the respondents believed that measures could mitigate the impact of climate change. Approximately 41.67% of respondents thought that their ability to cope with climate change was very low, and only 7.90% of respondents were optimistic about their ability to cope with climate change. 75.88% of respondents thought that the cost of taking countermeasures was relatively high, and 4.82% thought that it was low. Therefore, it can be seen that respondents pay more attention to the cost of adopting climate change adaptation measures.

3.3 Farmers' climate change adaptation measures

The survey results show that most respondents have taken various measures to adapt to climate change. Only 14 respondents have not taken any decision. The results show that the most popular adaptation strategy in the study area was planting new varieties. Approximately 72.81% of farmers have planted new seed varieties, which may be due to the efforts of seed suppliers or the mutual reference of farmers' planting behavior. About 65% of the households adjusted their fertilization behavior; 42.54% of the households chose improved irrigation methods/frequencies. About 38.16% of farmers choose diversified planting.

This paper also analyzed the initial motivation of farmers to take measures. The results show that 69.74% of the respondents said they were acting on their own, 17.54% said they were learning from their neighbors, 6.14% said they were encouraged and supported by the government, 4.82% said they were promoting agricultural machinery stations, and 1.75% said they learned from television or networks. It can be seen that local farmers adopted adaptation measures mainly by themselves, followed by the demonstration role of the surrounding people.

3.4 *The link between farmers' perception and adaptation behavior*

This paper utilized the binary probit regression model to explore the link between farmers' perception and their choices of climate change adaptation strategies. The risk perception variable is the social perception model (MPPACC) of Grothmann et al. (2005), which involves climate change risk perception and adaptation perception. Risk perception is decomposed into severity perception and possibility perception variables, and adaptive perception is decomposed into adaptive efficacy perception, self-efficacy perception and adaptive cost perception variables. This paper also included some socioeconomic variables as control variables, including age, education, farming years and so on.

The results show that the perception of possibility and self-efficacy have positive effects on farmers' choices of climate change adaptation strategies, while the perception of adaptive effectiveness and cost perception has negative effects on farmers' willingness to adopt adaptive decision-making. For controlling variables, farmers with longer farming years are more willing to adopt climate change adaptation behavior. This may be because farmers with longer farming years have more agricultural experience and more dependence and attention on agricultural production, so they are more willing to adopt behavior to cope with climate change. The larger the family size, the more willing farmers are to take action to cope with climate change, because the larger the population, the heavier the burden of households, so they are more willing to adopt adaptive behavior. The more educated farmers are, the more willing they are to adopt climate change adaptation behavior. The more educated farmers are, the higher their knowledge structure is. This is consistent with common sense. Farmers with higher incomes are more reluctant to take action to cope with climate change. This may be because farmers with higher incomes are less dependent on agriculture and pay less attention to agricultural production. In addition, the happier the farmers are, the more willing they are to adopt climate change adaptation behavior.

4 CONCLUSION

This study provided insights into how local farmers believed the risk of climate change and their choices of climate change adaptation in Dazu District, Chongqing City, China. Our results show that most farmers agreed that climate change has severely affected their lives. Measures taken cannot mitigate the impact of climate change. The ability to cope with climate change is relatively low, and the cost of taking measures is relatively high. In addition, our estimation results indicate that farmers' risk perception have significant effects on their adaption behavior. Specifically, risk probability perception, severity perception, and self-efficacy perception have positive significant effects. The efficacy perception and cost perception have negative significant effects. In addition, farmers' socio-economic factors (education, household income, household size and farming years) have significant effects on their climate change adaptation behavior. The findings of this study provide implications for policy makers on the promotion of farmers' adaption behavior of climate change in China and other developing world.

ACKNOWLEDGMENTS

This project was supported by the National Natural Science Foundation of China (41671170 and 41771192).

REFERENCES

Alam G.M.M., Alam K, Mushtaq S., 2017. Climate change perceptions and local adaptation strategies of hazard-prone rural households in Bangladesh. *Climate Risk Management*, 17: 52–63.

Ashraf M. Jayant K.R., Muhammad S., 2014. Determinants of farmers' choice of coping and adaptation measures to the drought hazard in northwest Balochistan, Pakistan. *Natural Hazards*, 73: 1451–1473.

Bryan E., Claudia R., Barrack O., Carla R., Silvia S. Mario H. Adapting agriculture to climate change in Kenya: Household strategies and determinants. *Journal of Environmental Management*, 2013, 114: 26–35.

Chase S., Joost V. Thomas T., Ariella H., David M., Abrar C. Exploring farmer preference shaping in international agricultural climate change adaptation regimes. *Environmental Science & Policy*, 2015, 54: 463–474.

FAO, 2016. The State of Food and Agriculture: Climate Change, Agriculture and Food Security. Available at: http://www.fao.org/3/a-i6132e.pdf.

Feng Ling, Chen Da, Gao Shan, et al. 2019. Responding to Global Warming: Mitigation Policies and Actions of Stakeholders in China's Tourism Industry. *Journal of Resources and Ecology*, 10(1): 94–103.

Grothmann T., Patt A. Adaptive capacity and human cognition: The process of individual adaptation to climate change. *Global Environmental Change*, 2005, 15: 199–213.

IPCC (Intergovernmental Panel on Climate Change). 2007. A Report of Working Group One of the Intergovernmental Panel on Climate Change-Summary for Policy Makers.

Li Wenmei, Zhao Xueyan, Guo Fang, Zhang Liqiong, 2015. Farmers' perception and adaptive behavior to environmental degradation in the lower reaches of Shiyang River. *Chinese Journal of Eco-Agriculture*, 23(11), 1481–1490. (in Chinese).

Lu Y., Chen S., 2010. An analysis on farmers' perception and adaptive behavior. *Chinese Rural Economy*, 7: 75–86 (in Chinese).

Qi Xinhua, Yang Ying, Jin Xingxing, Liu Guanqiu, Li Damou, Pan Danlin, Qi Xi, 2017. Rural household perceptions and adaptations to climate change based on an investigation and comparison of two middle and eastern villages in China. *Acta Ecologica Sinica*, 37(01), 286–293. (in Chinese).

Schmid Julia C., Andrea Knierim, Ulrike Knuth. Policy-induced innovations networks on climate change adaptation -An expost analysis of collaboration success and its influencing factors. *Environmental Science & Policy*, 2016, 56: 67–79.

Slovic P., 1987. Cognition of risk. *Science*, 236: 280-285.

Tam Jordan, Mc Daniels, Timothy L. Understanding individual risk cognitions and preferences for climate change adaptations in biological conservation. *Environmental Science & Policy*, 2013, 27: 114–123.

Wang Chengchao, Yang Yusheng, Pang Wen, Hong Jing, Xie Jianbin, 2017. A Review on Farmers' Perceptions and Adaptation of Climate Change and Variability. *Scientia Geographica Sinica*, 37(06), 938–943. (in Chinese).

Woods B., Nielsen H., Pedersen A., Kristofersson D., 2017. Farmer Perceptions of Climate Change and Responses in Danish Agriculture. *Land Use Policy*, 65: 109–120.

Risk Analysis Based on Data and Crisis Response Beyond Knowledge – Huang & Nivolianitou (eds)
© 2020 Taylor & Francis Group, London, ISBN 978-0-367-25146-8

Risk assessment of the transform from old industrial plant area into the creative industrial parks

Qian Wu & Mengqi Sun
School of Civil Engineering, Xi'an University of Architecture and Technology, Xi'an, China

Li Wang
School of Management, Xi'an University of Architecture and Technology, Xi'an, China

ABSTRACT: With the vigorous promotion of old industrial building recycling projects in China, creative industrial parks have become the mainstream of many transformation modes. The safety of the construction structures and environmental policy are the potential risks that affecting the operation and maintenance management of creative industrial parks. In this paper, through spot investigation, principal component analysis is used to construct the risk index system. According to the stakeholders' assessment of risk factors, CRITIC model is used to calculate the influence degree of the risk of building structure safety, management, economic and environmental policy. It is concluded that building structure safety is the main factor affecting the risk of creative industrial park during its operation and maintenance period. Suggestions are put forward for each risk level to provide reference for risk assessment in operation and maintenance period.

Keywords: creative industry park, stakeholders, operation and maintenance management, risk assessment

1 INTRODUCTION

At present, the research on the regeneration of the old industrial plant area for the creative industrial parks project focuses on the decision-making, design, construction and completion acceptance stages. However, as the regenerative creative industrial parks project is gradually put into use, a unified creative industrial park operation and maintenance management model has not yet been formed, and the operation and management experience of the similar park is lacking, thus the risk of operation and maintenance management has emerged. Literature research finds that the risks affecting operation and maintenance management mainly include: In the actual operation process, the technical means is single inefficient, and there is no regular and effective structural safety inspection (Wu and He et al. 2017); stakeholders do not pay attention to the management concept of operation and maintenance period; lack of relevant management standards and norms (Wang 2018); capital cost investment in operation and maintenance stage is too large .

The degree of influence of stakeholders on operation and maintenance management is an important basis for risk assessment during operation and maintenance. Therefore, based on the perspective of stakeholders, this paper conducts a risk assessment of the operation and maintenance management of the old industrial plant regeneration project as a creative industrial park, analyzes the main factors affecting the risk during the operation and maintenance period, and proposes targeted recommendations.

2 SELECTION OF RISK EVALUATION INDEX

According to the different causes of risk, and considering the actual situation of the operation and maintenance period of creative industrial park, the risk level of this paper is divided into four major risks: building structure safety, management, economy, environment and policy. The selection of secondary indicators should be comprehensive, practical and pertinent. It can not only take into account the commonness of stakeholders, but also reflect the particularity of creative industry park. On the basis of the existing literature, 24 risk factors affecting the operation and maintenance of creative industry park are collected and sorted out by consulting the relevant scientific research personnel of universities and combining with the actual situation. Separately, 24 risk factors were named variables 1, 2, 3,... 24. The specific risk factors are shown in Table 1.

This paper uses the method of survey scales to design the questionnaire, considering the stakeholders involved in the operation and maintenance stage comprehensively. The main objects of the questionnaire are the managers and technicians of building developers, related property organizations and residential businesses. Due to the limited number of building development professionals in the park, in order to ensure the validity of the questionnaires, 40 professionals were selected according to certain criteria to issue questionnaires. All 40 questionnaires were collected and valid.

3 OPTIMIZATION OF RISK ASSESSMENT INDICATORS

The optimization process is mainly to better determine the comprehensive and accurate risk factors for the regeneration of the old industrial plant area into the creative industry park. The reliability and validity analysis of the data analysis of the questionnaires of the above survey subjects was carried out by SPSS software.

Table 1. Risk factors for the regeneration of the old industrial plant area into the operation and maintenance period of the creative industrial park.

Level 1 indicator	No.	Level 2 indicator
Safety structure of	1	Lack of effective detection mechanism for structural instability
building structure level	2	Damaged reinforcement of beams and columns
	3	Durability of building structure
	4	Re-laying of building infrastructure
	5	Development and use of new products and technologies
Management level	6	Training for contract management personnel
	7	Effective performance and settlement management of contracts
	8	Dispute resolution
	9	Related management configuration
	10	Risk prevention awareness of relevant managers
	11	Managerial competence
	12	Property management agency management
	13	Security personnel information feedback channel
Economic level	14	Operating funds cannot be put in place in time.
	15	Maintenance cost overrun
	16	Promotion fee
	17	Facility maintenance and maintenance costs
	18	Safety facility renewal fee
Environmental and	19	garbage and sewage treatment
policy level	20	Surrounding resource utilization
	21	Adverse effects of the use environment
	22	Inappropriate use of partitions
	23	National policy guidance
	24	Park policy orientation

3.1 Using SPSS for reliability analysis and validity analysis

Reliability Analysis is to ensure the stability of the data provided by professionals in this creative industry park and the reliability of the results obtained. The reliability analysis was performed using the Cronbach coefficient of the SPSS software (Gao and Shao 2017).

According to the range of values of the above table coefficients, for indicators with coefficients above 0.7, the measures taken are to accept and retain these indicators. Conversely, if the coefficient is less than 0.7 and below, then the indicator will be deleted or changed (Xie 2016).

Validity Analysis is also an important part of the evaluation of the indicator. Through the questionnaire for the operation period of the creative industry park, it is judged whether the results of the risks of each level studied in this paper are effective. The questionnaire used in this paper is based on the particularity of the regeneration of the old industrial plant area into a creative industrial park, and fully reviews and absorbs the research of the predecessors, and further improves the expression of the items in the questionnaire in combination with the status quo. Therefore, the content validity analysis of the questionnaire is in line with the requirements. In this study, the structural validity analysis was mainly tested, and the exploratory factor analysis (EFA) test was used to prove the structural validity of the scale (Gao and Shao 2017).

3.2 Index optimization results

According to the results generated by the questionnaire statistics, the SPSS software is used to calculate the coefficient of the risk indicator under each risk level, and the index is retained or deleted according to the range of the value of the coefficient. The reliability analysis is carried out one by one for each dimension to obtain the measurement results. After analysis, the Cronbach's Alpha is 0.972. The Alpha of Cronbach based on standardized projects is 0.971. The coefficients of the four levels studied in this paper are all greater than the standard of 0.7. It indicates that the indicators have good internal consistency reliability, and the CITCs are all greater than 0.5, indicating that these index variables are in line with the research requirements, and deleting these indicators will not cause an increase in the coefficient values, and also indicate that these indicators have good reliability (Tang and Shao et al 2017).

On the premise that the reliability meets the requirements, an exploratory factor analysis is conducted on the questionnaire. The exploratory factor analysis using SPSS22.0 is performed on the scales of KMO and Bartlett's spherical test. The value of KMO is 0.880, which is greater than 0.7, and Bartlett's spherical test value is significant (Sig.<0.001), indicating that the questionnaire data meets the premise requirements of factor analysis. Therefore, the analysis is further carried out. The principal component analysis method is used to extract the factor, and the common factor is extracted with the eigenvalue greater than 1 as the factor. When the factor is rotated, the maximum variance method is used for factor analysis. The cumulative interpretation variance is 79.797%, both of which are greater than 50%. The load is greater than 0.5. Through the above analysis, it shows that the risk indicators of the creative industry parkin the operation and maintenance period have good structural validity.

Table 2. Reliability statistics.

Cronbach's Alpha	Alpha of Cronbach based on standardized projects	Number of items
0.972	0.971	24

Table 3. KMO and Bartlett Certification.

Kaiser-Meyer-Olkin measures sampling suitability.	0.880
Bartlett's spherical check is about chi square	770.377
Df	190
Significant	.000

4 RISK FACTOR ASSESSMENT

After the above indicator selection and optimization process, the results of 24 indicator optimizations are all retained. Combined with the selected indicators, the CRITIC model is used to calculate the weight of each evaluation index and evaluate the risk.

4.1 CRITIC weighting model construction

Suppose that the number of risk levels to be evaluated is m, the number of evaluation indicators is n, and b_{ij} represents the score of the jth evaluation index of the i-th level, and the evaluation score matrix B of the risk factors to be evaluated is:

$$B = \begin{bmatrix} b_{11} & \cdots & b_{1n} \\ \vdots & \ddots & \vdots \\ b_{n1} & \cdots & b_{mn} \end{bmatrix} \tag{1}$$

The steps of constructing the CRITIC weighting model are as follows:
1) Dimensionless processing. In order to eliminate the impact of data dimension, the score data needs to be dimensionless. If it is a positive indicator, it is treated by formula (2); if it is a negative indicator, it is treated by formula (3). The matrix E after the dimensionless processing is obtained (Cao and Liu et al. 2018).

$$e_{ij} = \left(b_{ij} - b_j^{\min} \right) / \left(b_i^{\max} - b_j^{\min} \right) \tag{2}$$

$$e_{ij} = \left(b_j^{\max} - b_{ij} \right) / \left(b_j^{\max} - b_j^{\min} \right) \tag{3}$$

where e_{ij} is the dimensionless processed data; $b_j^{\max}$ and $b_j^{\min}$ respectively represent the maximum value and the minimum value of the j th evaluation index.

2) Calculate the standard deviation j of the variability representative value within the evaluation index.

3) Calculate the correlation coefficient r_{ij} between each index, and calculate the conflicting representative value R_{ij} of the evaluation index according to formula (4).

$$R_j = \sum_{i=1}^{n} \left(1 - r_{ij} \right) \tag{4}$$

4) Calculate H_j according to formula (5). Let H_j denote the amount of information contained in the j-th evaluation index, and the larger the amount of information included in the j evaluation indicators, the greater the relative importance of the index (Fu and Dou 2018).

$$H_j = \sigma_j \sum_{i=1}^{n} \left(1 - r_{ij} \right) = \sigma_j R_j \tag{5}$$

(5) Calculate the weight according to formula (6).

$$\omega_j = \frac{H_j}{\sum_{i=1}^{n} H_j} \tag{6}$$

4.2 Weight analysis

According to the data obtained from the questionnaire, the score is scored in the form of a 5-level scale, as shown in Table 4 .

Table 4. Score table.

Level of influence	Very small	Smaller	General	Larger	Very large
Score	1	2	3	4	5

(1) Score according to four levels (C_1), (C_2), (C_3), and (C_4). The score table is converted into an evaluation score matrix B. Since the indicators in this paper are all positive indicators, matrix E is obtained by non-quantization processing according to formula (1).

$$E(C_1) = , E(C_2) = , E(C_3) = , E(C_4) =$$

$E(C_1)$	$E(C_2)$	$E(C_3)$	$E(C_4)$
0.8000	0.8438	0.8000	0.8333
0.4500	0.7188	0.7500	0.4583
0.6500	0.8438	0.4500	0.6667
0.5000	0.6563	0.5000	0.7917
0.7500	0.7500	0.8500	0.7500
0.2500	0.4063	0.4000	0.4583
0.6500	0.8438	0.7000	0.6667
0.8000	0.7188	0.7000	0.7083
0.7500	0.9375	0.6500	0.7500
0.8500	0.7813	0.8500	0.7083
0.6500	0.6563	0.6500	0.6667
0.0000	0.0000	0.0000	0.0000
0.6000	0.7500	0.7000	0.6250
0.7500	0.7813	0.6000	0.5417
0.7500	0.7500	0.7500	0.9583
0.7000	0.8750	0.6500	0.8333
1.0000	0.6250	0.7000	0.7917
0.5500	0.5938	0.6000	0.6250
0.6500	0.7500	0.5000	0.6667
0.5000	0.5313	0.5500	0.4583
0.9000	0.2500	0.3000	0.2083
0.9500	0.8750	0.7500	0.7917
0.5000	0.6250	0.5000	0.4167
0.6500	0.3125	0.3500	0.2083
0.7500	0.9375	0.6500	0.7917
0.8500	1.0000	0.8500	0.7500
0.0000	0.0000	0.4000	0.2500
0.9000	0.9063	0.7500	0.9167
0.9000	0.8750	0.8500	0.8333
0.3000	0.4375	0.5000	0.3750
0.4000	0.0313	0.1500	0.3750
0.6500	0.9375	0.9000	0.8333
0.2000	0.2500	0.3500	0.1250
0.9000	1.0000	0.7000	0.9583
0.3000	0.4688	0.5500	0.4583
0.9500	0.8438	0.6000	0.9583
0.6500	0.6250	0.6000	0.5833
0.7000	0.6250	0.8500	0.8750
0.5000	0.7188	0.6500	0.7083
0.5000	0.7813	0.7000	0.5417

(2) Calculate the standard deviation j and the correlation coefficient r_{ij} between the indicators, and calculate R_{ij} according to the formula (4).

$$j = (0.2475 \qquad 0.2671 \qquad 0.1979 \qquad 0.2409),$$

$$R_{ij} = (0.9171 \qquad 0.6119 \qquad 0.7602 \qquad 0.6339)$$

(3) Calculate H_j and weight ω_j according to formulas (5) and (6) respectively. The calculation results are as follows:

$$H_j = (0.2270 \qquad 0.1635 \qquad 0.1505 \qquad 0.1527)$$
$$\omega_j = (0.3273 \qquad 0.2356 \qquad 0.2169 \qquad 0.2202)$$

4.3 Analysis of results

In summary, the calculation and analysis shows that in the risk factors of the operation and maintenance period of a creative industrial park in Xi'an, the standard deviation of risk at the management level is larger than that of other levels, indicating that the difference of the indicators under the management level is large, and the data sample reflects the amount of information. According to the correlation coefficient, the main risk influencing factors are from the safety level of the building structure, followed by the management level and the environmental policy level, and finally the economic level.

5 CONCLUSIONS AND SUGGESTIONS

Based on the data obtained from this case and the above analysis results, the following suggestions are put forward for the risk of the operation and maintenance period of the Creative industry park regeneration project.

(1) Forming a dynamic and long-term safety detection mechanism for building structures

In the operation and maintenance period, the relevant management departments should strengthen the daily structure verification work, carry out security precautions from air defense, physical and technical defense, regularly inspect and appraise the structure in the creative industry park, and timely repair and reinforcement of the structure, and replace the structure without bearing capacity (Wu and Wang et al 2014); at the same time, we can adopt the technology of implanting detection chips in the structure. The damage condition and degree of structure, bearing capacity of foundation are tested, and feedback information is provided from time to time. The dynamic and long-term structural safety detection mechanism established in this way is more conducive to the operation and maintenance of creative industrial parks.

(2) Improving the relevant operation and maintenance management mechanism

Regeneration of old industrial plants into the operation and maintenance stage of creative industrial park is facing another problem: there is no unified operation and maintenance management mechanism among developers, property and merchants, and there are great differences in management. Therefore, to a certain extent, the establishment of a perfect and feasible operation and maintenance management mechanism has become the key factor for the success of the regeneration project of creative industrial park, and also the key to its revitalization.

(3) Crowd funding

Although the relevant policies and regulations of our country are not perfect, we can draw lessons from the policies already issued at home and abroad. Industrial park actively attract foreign investment by introducing businessmen to finance projects. At the same time, we should develop the space for public creation, strive for the support of the government, create the space for independent innovation and entrepreneurship, and standardize the later operation and management of creative industry park.

(4) Establishment of operational and maintenance period standards for old industrial regeneration projects

In China, the standards for the operation and maintenance period of old industrial regeneration have not yet been issued, which leads to no clear guidance. Therefore, it is urgent to prepare the standards and review the standards. It is suggested that the old industry research teams of relevant universities should compile the standards and organize experts from various industries to review the standards so as to apply them to the operation and maintenance of the project as soon as possible.

REFERENCES

Wu, Q., He, X.D. and Jia C.Y. 2017. Analysis of influencing factors of safety assessment of old industrial building structure. *Industrial Safety and Environmental Protection*, 43(05): 56–58.

Wang, Y.L. 2018.Research on operation and maintenance cost management of industrial park based on functional transformation of old industrial buildings. *Xi'an University of Architecture & Technology*.

Gao, F., Shao, L. 2017. The Chinese Eastern Railway industrial heritage value evaluation index optimization research based on SPSS analysis. *Urbanism and Architecture*, (10): 121–123.

Xie, J.Q. 2016. Analysis of industrial modernization factors in Chongqing districts and counties based on SPSS. *Economic Forum*, (12): 17–22.

Tang, Y.X., Shao, L., Wang, X.2017. Evaluation of Harbin urban landscape characteristics based on data analysis. *Journal of Northeast Forestry University*, 45(01): 65–70.

Cao, Z.C., Liu, Y.S., Li, M.Y., Su, D. 2018. Evaluation of prefabricated building green degree based on CRITIC and TOPSIS method. *Building Energy Efficiency*, 46 (09): 37–40+58.

Fu, B., Dou, X.C. 2018. Comprehensive benefit evaluation of modern agricultural science and technology park based on CRITIC method. *Journal of Shandong Agricultural University* (*Social Science Edition*), 20(03): 52–56+74+151.

Wu, Q., Wang, C., Zheng, D.Z., Ma, H.F. 2014. Research on the risk factors of reuse of old industrial buildings in Xi'an based on principal component analysis. *Industrial Construction*, 44(10): 61–63+90.

Risk Analysis Based on Data and Crisis Response Beyond Knowledge – Huang & Nivolianitou (eds)
© 2020 Taylor & Francis Group, London, ISBN 978-0-367-25146-8

How to define and manage stakeholders in sea use licensing in China

Shufen Liu & Zengkun Wang*
National Ocean Technology Center, Tianjin, China
Future TV Co., Ltd., Tianjin, China

ABSTRACT: This paper, taking the administrative management of sea use licensing in China as an example, defines the stakeholders herein, including sea use applicant(s) and other stakeholders under management, as well as other administrative departments. It analyzes the interest appeal, social responsibility and obligations of different stakeholders. In combination with the current management system of administrative licensing of sea use, interest coordination approaches are investigated, including the interest coordination of those regulated, the opinions of management departments, and the intervention of government departments.

Keywords: sea use licensing, stakeholder, interest appeal, social responsibility, obligation, China

1 INTEREST APPEAL DEVELOPMENT OF STAKEHOLDER THEORY

The concept of stakeholders originated from corporate governance in the 1930s, and the original intention was that companies should serve stakeholders. In the 1960s, scholars from Stanford research institute gave a clear definition of stakeholders: for enterprises, there were some interest groups without whose support the enterprises could not survive. In the 1970s, stakeholder theory was gradually recognized and accepted by scholars and enterprises, and its influence expanded gradually. The application of stakeholder theory was no longer limited within corporate governance, but extended to engineering investment, public decision-making, risk assessment, pollution prevention and other fields.

Chinese scholars began research on stakeholders in the 1990s, also associated with enterprise theory and company governance. The most representative research was by Yang Ruilong on "shared ownership" and stakeholders "co-governance mechanism" (Riulong 1997, 1998, 2000, 2001), and by Li Weian on the "boundary of corporate governance" (Weian 1998, 2001). In 2002, the China Securities Regulatory Commission (CSRC) published a Code of Corporate Governance for Listed Companies, which absorbed the concept and theory of stakeholders in "Chapter VI – Stakeholders".

In addition, the study of stakeholder theory has been applied and developed in other fields: by Yang Zhihong in the marine environment pollution damage compensation mechanism; by Hao Liang in the arable-land heavy-metal pollution control policy and the empirical research of shareholder mutual feedback mechanism; by Jin Yanrong in the reform of the administrative examination and approval system; by Feng Xin in agricultural water research; by Yang Jianguo in the food safety research; and by Wang Feng in the study of the risk assessment model for social stability; as well as administrative management.

---

*Corresponding author: wangzengkun@126.com

2 DEFINITION OF STAKEHOLDERS

Scholars define corporate stakeholders in broad and narrow senses. The broad concept provides enterprise managers with a comprehensive stakeholder analysis framework, emphasizing the relations between stakeholders and corporate and, of course, including extensive shareholders such as creditors, employees, suppliers, customers, and even the community, the environment, the media and others whose activities have direct or indirect influence on enterprise. In a narrow sense, stakeholders refer to those who have direct influence on the enterprise.

Unlike land, ocean water is mobile, and marine pollution may spread with ocean currents. Therefore, in the marine development and utilization activities, in a broad sense the stakeholders should include the users of all the sea areas where the ocean current passes, and even the public in the related coastal areas. However, this study is aimed at sea use licensing, which is a type of government management behavior. If the research scope is expanded without restriction and the scope of stakeholders is expanded thereafter, it will greatly increase the cost of risk analysis and prevention for managers and sea area use applicants.

Reasonable definition of stakeholders in sea use licensing can effectively improve the work efficiency of government departments and reduce the costs of risk analysis and prevention for sea use applicants. As a result, this article studies the stakeholders in the sea use licensing, and the scope of the study is narrow in extent. Therefore, this article defines the stakeholders in the sea use licensing in a narrow sense, that is, only those directly related to the sea use licensing management are stakeholders.

3 STAKEHOLDERS IN SEA USE LICENSING AND THEIR INTEREST APPEAL

3.1 *Concept of sea area use licensing*

In China, the Law on Administration of the Use of Sea Areas stipulates that the "sea areas in China's territorial sea belong to the state, and the State Council exercises the ownership of the sea areas on behalf of the state"; "no units or individuals may appropriate, trade or transfer sea areas by other illegally means"; and "the right to the use of sea areas must be obtained in accordance with law in sea area utilization". According to these regulations, in China those sea area users must apply to government department for the use of sea areas, and only after the application is approved can they obtain the right to the use of sea areas, with different terms for different purposes of sea area utilization, such as construction projects.

The administration of the sea use licensing discussed herein starts from the materials prepared by the applicant and ends with the issuance of the sea area use permits.

3.2 *Stakeholders involved in the management of sea area use licensing*

The utilization of sea area will affect its marine environment, as well as the marine environment and utilization of surrounding waters. Therefore, the application materials required include the sea use verification report, the marine environmental impact assessment report, the agreement(s) with licensed user(s) of surrounding waters that may be affected, as well as the approval(s) of relevant management department(s) of public facilities like navigation channels (if affected). Therefore, stakeholders involved in the management of sea area use licensing include the sea area use licensing management department, the sea area use applicant, licensed user(s) of surrounding sea areas, and the management department(s) of surrounding public facilities.

3.3 *Interest appeal of stakeholders in sea area use licensing*

Regarding the licensing authority, the appeal of management is to safeguard the interests of the country on sea resources (we can think of this as safeguarding the public interest). In

China, the environmental impact assessment management and the sea area use licensing management are exercised, respectively, by the ecological environment management department and the natural resources management department. They represent and maintain the public interest which, in terms of the use of sea areas, can be expressed as the "orderly, moderate, green and efficient" utilization of sea areas.

Regarding the sea use applicant, the appeals mainly include the clear and simple application procedure, the short term licensing processing, the clear and accessible application materials, and the low cost of preparing the application materials.

Regarding the licensed user(s) of surrounding sea areas, the hope is that the impact shall be as small as possible and shall be reasonably compensated.

Regarding the relevant public facilities management department(s), the intention is to minimize the impact on public facilities. If affected, relevant measures should be taken to offset the impact of sea area use on public facilities.

4   RELATIONSHIP AMONG STAKEHOLDERS OF SEA AREA USE LICENSING

Stakeholders of sea area use licensing can be divided into two types: management departments and non-management departments. The management departments herein include the natural resources management department, the ecological environment management department and the surrounding public facilities management department. The non-administrative departments include the sea use applicant and the licensed user of the surrounding sea areas. Their relationship is shown as in Figure 1.

Among the management departments, the Ministry of Natural Resources and the natural resources management department of local government are the acceptors and verifiers of the application for permission to use the sea areas, as the department in charge of sea use licensing. The Ministry of Ecological Environment and the local ecological environment management departments examine and verify the marine environmental impact assessment reports. Both the two Ministries are administrative departments of the Chinese government, of the same tier, but with different duties.

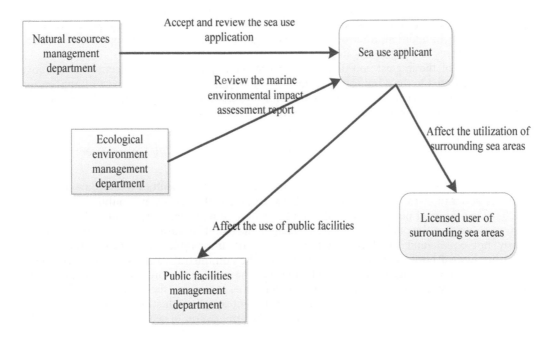

Figure 1.   Relationship among stakeholders of sea area use licensing.

The public facilities management department has administration duties, so we classify it into the management department, but it does not assume any management functions in sea area use licensing. Like the licensed user of the surrounding sea area, the public interest it represents may be affected by the sea utilization under application.

The applicant for the use of sea areas is subject to the management behavior imposed in the permission for the use of sea areas, and to coordination with the licensed user(s) of surrounding sea areas and the public facilities management department(s).

The licensed users of surrounding sea areas, however, do not belong to the governed group in this sea area use licensing procedure, they are under negotiation relationships with the applicant, but they do not have direct contact with the administrative department herein.

Thus far, this article has performed an analysis of interest appeal, social responsibility and obligation of different stakeholders herein, in combination with the current management system of administrative licensing for sea area utilization. In addition, it will further analyze the ways to coordinate the interests in administrative licensing of sea use, including the interest coordination between the regulated groups, the solicitation of opinions among management departments, and the intervention by government department(s).

5 RECONCILABLE INTERESTS AND COORDINATION CHANNELS

In the administration of permission for the use of sea areas in China, the reconcilable interests include those between the applicant and the licensed user of adjacent sea areas, as well as those between the applicant and the public facilities management department.

5.1 *Coordination between the applicant and the licensed user of adjacent sea areas*

The interest conflicts between the applicant and the licensed user of adjacent sea areas shall be properly coordinated through consultation. The reason for this is that outside the relationship of impacting and impacted, the applicant has an obligation to compensate the licensed user of surrounding sea areas if they are affected, but the method and amount of compensation must be negotiated. In addition, after obtaining the right to use sea areas, the applicant is obliged to take measures to mitigate any impact.

The result of the interest coordination between the applicant and the licensed user above shall be submitted to the administrative department of natural resources as part of application materials. If the applicant and the licensed user above cannot reach agreement, the application for the use of the sea area shall not be licensed.

5.2 *Interests coordination between the applicant and the public facilities management department*

The conflict of interests between the applicant and the public facilities management department is actually of the nature of conflict between personal interests and public interests. As the representative of public interest, the public facilities management department has the obligation to protect the public interest from being damaged. The status of the two sides is obviously not equal, and equal consultations are nearly impossible. As a result, the interest coordination of both sides needs intervention by the administrative department in charge of sea use licensing management. It is suggested that the natural resources management department, in the process of performing management duties, conducts such coordination of interests between two sides above through inter-department opinion soliciting. If this channel can't solve the problem of the loss of public interest, the application of sea area use should not be licensed.

REFERENCES

Li Yang, Wang Hui. 2004. Dynamic Development and Enlightenment of Stakeholder Theory. *Modern Finance and Economics*, 7, 32–35.

Yang Zhihong, Li Shuangjian, Zhou Yipu. 2014. Compensation Mechanism for Marine Environmental Pollution Damage from the Perspective of Stakeholders. *Marine Economy*, 6, 13–21.

Hao Liang, Li Yingming, Zhang Conglin, Zhao Zilin. 2018. Empirical Study on the Policy of Farmland Heavy Metal Pollution Control and the Mutual Feedback Mechanism Among Stakeholders – Based on Research on a Pilot Area. *China Environmental Management*, 4, 21–27.

Jin Yanrong. 2016. A Brief Analysis of the Reform of Administrative Review and Approval System From the Perspective of Stakeholders. *Administration and Law*, 26–31.

Feng Xin, Jiang Wenlai. 2018. Progress and Prospect of Agricultural Water Stakeholder Research in China. *China Agricultural Resources and Zoning*, February 2018.

Yang Jianguo, Lin Lin. 2008. Logical Reflection and Remodeling of Food Safety Governance – From the Perspective of Stakeholders. *Journal of Fujian Agriculture and Forestry University* (Philosophy and Social Science Edition), 21(5), 56–62.

Wang Feng, Hu Xiangming. 2012. Research on Risk Assessment Model of Social Stability of Major Projects – Perspective of Stakeholders. New Vision, 4, 58–62.

Risk Analysis Based on Data and Crisis Response Beyond Knowledge – Huang & Nivolianitou (eds)
© 2020 Taylor & Francis Group, London, ISBN 978-0-367-25146-8

Lack of coordination in communications between scientific society and authorities during earthquakes: Case study of L'Aquila, Italy

Ploutarchos Kerpelis*
Earthquake Planning and Protection Organization (E.P.P.O.), Athens, Greece

Aspasia Karamanou
Region of Attica, Athens, Greece

ABSTRACT: In 2012, seven members of the High-Risk Prevention Commission in Italy, were sentenced to six years in prison for failing to meet their obligations to adequately analyze seismic risk and provide clear, correct and complete information that might have saved many lives. This article attempts to systematically examine the L'Aquila trial (court decisions and bibliographic documents) to provide reliable lessons and safe conclusions on the psychological, social and legal aspects of disasters and how the uncertainty of natural hazards can be recognized, addressed and communicated to the public. It also outlines a detailed portrait of how scientists, bureaucrats and the media interact in such uncertainties.

Keywords: risk communication, risk perception, scientific uncertainty, risk governance

1 INTRODUCTION

In the spring of 2009, a devastating earthquake occurred at the Italian city of L'Aquila and the surrounding areas. In addition to the tragedy of human and material loss, the destruction led, first, to six scientists and the Deputy Head of Civil Protection being accused of manslaughter through negligence. A few days before the earthquake, the population were told there was no danger and so were not properly prepared, causing global interest and various reactions. This is a complex and controversial issue that has been interpreted in many different ways. Many publications, annotations and conclusions were produced around the issues that affect how scientific advice is provided and how decisions are made in situations of high uncertainty.

1.1 The earthquake of L'Aquila

On April 6, 2009, at 3:32 local time, an earthquake occurred in central Italy (Magnitude Mw = 6.3 or Ml = 5.9, focal depth of about 8 km). The focus was at Tornimparte, a small village located 7 km northeast of L'Aquila, (the capital of the Abruzzo region). The earthquake caused significant damage, although it was classified as a moderate-scale earthquake. Many aftershocks occurred. The majority of damage occurred in the medieval city of L'Aquila and the surrounding villages. This is a densely populated urban area, 309 people died, about 1600 were injured and 66,000 were homeless, in the post-seismic period. According to official reports, 20,000 buildings had serious damage or collapsed (nearly 70% of total buildings). The temporary estimate of losses was over 10 billion euros (Global Risk Myiamoto, 2009; Grimaz & Maiolo, 2010; Kaplan et al., 2010).

---

*Corresponding author: pkerpelis@oasp.gr

1.2 The seismicity of the area and the social reactions before the earthquake

The area of L'Aquila has been identified as a high-risk area according to the new Italian Seismic Risk Map, which represents the fundamental tool of preventing and mitigating seismic hazard. In 2004, the region was classified as Zone I, rather than Zone II (Cocco et al., 2015). Historically, great earthquakes had occurred in the area, such as the earthquake of 1703 that killed more than 3000 people (Boschi et al., 2000).

The vulnerability risk for public buildings in the city was high and this was already known to local authorities, as it was the subject of a special study published by the National Civil Protection Agency in 1999 called the *Barberi report* and the Abruzzo region was one of the areas described. The study was published immediately to the scientific community, and then was forwarded to decision-makers (civil protection officers, municipalities, etc.) and to the public (Cocco et al., 2015).

From October 2018, a seismic cluster occurred in the city, including a series of small seismic vibrations, with progressively increasing intensity and frequency. A small amount of damage occurred to buildings and schools were closed many times for precautionary purposes. The population was upset and scared.

During pre-seismic period, Giampaolo Giuliani, a technician working at the National Laboratory of Astrophysics at Gran Sasso as an amateur seismologist, made public appearances and numerous warnings. He claimed he could predict great earthquakes, on the basis of gas measurements (called radon), contrary to the opinion of the scientific community. On 27/3/2009, Giuliani warned the Mayor of L'Aquila about the possibility of an earthquake within the next 24 hours, causing mass terror. A few days later, in a second public prediction, he warned the mayor of the neighboring city of Sulmona (50 km from L'Aquila) about an imminent earthquake in the next 6–24 hours. The public was upset and there was a widespread state of anxiety (approaching, in some cases, even panic). This situation was causing public order problems. People in cars equipped with speakers (who were self-identified as civil Protection officials) spread rumors of an imminent earthquake. The local press gave plenty of space to all the assumptions that had been made about the possible development of a seismic cluster.

The unverifiable prediction for an imminent earthquake in the city led to allegations of "causing fear" and false alarms. The Head of the National Department of Civil Protection (NDCP), Guido Bertolaso, reported Giuliani to the authorities and banned his public statements, forcing him to remove his findings from the internet.

On 30 March 2009, an earthquake (magnitude 4.1) occurred in the L'Aquila area, exacerbating the already tense state of social reactions and fear. The Mayor of L'Aquila asked the Italian government to declare a state of emergency in the area. The Head of NDCP asked for the convergence of the National Committee for the Prediction and Prevention of Major Hazards (NCPPMH), which was a scientific advisory body for all civil protection actions (Law 225/1992). The members were summoned to L'Aquila on March 31, 2009. The committee was chaired by Professor Franco Barberi (Deputy Head of NDPC) and consisted of six of the best geo-scientists in Italy.

The Commission meeting was held "so as to provide the citizens of the Abruzzo region with all the information available from the scientific community about seismic activity in the past few weeks" (Political Protection Press Release). The purpose of the meeting, was as follows: i) make an objective assessment of seismic events' progress, in relation to what can be predicted and, ii) discuss and provide information on widespread alarm in the population (Meeting draft report, 2009).

2 THE TRIAL AND THE VERDICT

On October 22, 2012, seven members of the committee were found guilty of multiple manslaughter of 29 people by negligence and the multiple unjustified injuries of four people. After a 781-page verdict and about 31 hearings, they were sentenced to six years in prison and fines amounting to several million euros (Tribunale di L'Aquila, 2012). The trial lasted 13 months, attracting the attention of the international community.

The members of the committee were attributed with "negligence, recklessness and clumsiness". They did not fulfill their institutional obligations to provide clear, correct and complete information that could have saved many people. The risk assessment was carried out with insufficient attention, in a superficial, approximate and ineffective manner in relation to their assigned duties, in violation of Law about Prevention and Prediction (Law 225/1992).

Further, responsibilities were attributed by the court for misapplication of the legal framework on information and communication by public authorities (DPCM 23582/2006 and Law 150/2000). The members of the committee were convicted of violating the precautionary principle. In EU environmental policy, the precautionary principle stipulates that when an activity creates threats to the environment or human health, precautionary measures should be taken even if the cause–effect relationship has not been scientifically confirmed (EUR–Lex, nd).

The committee members assessed the risk and provided over-reassuring information to the public. This reassuring message of the committee was favored as the chosen communication policy, but as a result the residents changed their previously established precautions that would save their lives. The citizens did not move away from their homes during the precursor vibrations on the day of the great earthquake, because of the committee members' announcements, and as a result they were killed by collapsing buildings (causality).

3   THE RISK ASSESSMENT AND THE INFORMATION PROVIDED

The risk indicators were individually analyzed by the committee without a holistic approach (statistics, historical data, etc.). The information provided was incomplete, inaccurate and contradictory about the nature, causes, risks and future developments of seismic activity. This information (public statements, television and radio interviews, etc.) shaped the verdict of the court:

- Unspecified arguments such as "It is unlikely that an earthquake such as that of 1703 will occur soon, although this cannot be totally excluded" (Boschi).
- Information that causes confusion through the question of whether or not the prediction is achievable e.g., "Any provision has no scientific basis" and "It cannot be said that a sequence of small-scale seismic vibrations is a precursor to a powerful earthquake" (Barberi).
- Serious inconsistencies, such as a failure to predict seismic risk forecasts and estimates for an unlikely increase in the size of earthquakes, e.g. "Swarm earthquakes tend to be of the same size and are very unlikely to grow in size" (Barberi) and "Nobody knows if next week (the earthquakes) are larger in size since it is known that earthquakes cannot be predicted" (testimony).
- Swarm earthquake activity was normal and not deviating from the usual measurements, because it was a seismic area (testimony).
- Excessively reassuring but incorrect and inaccurate information, e.g. "I told the Mayor of Sulmon: There is no (seismic) risk. The scientific community continues to confirm that this (swarm earthquakes) is a favorable situation because there is a continuous release of energy. There are mostly intense events, but not very intense" (De Bernardinis).

The committee did not associate pre-earthquakes with the possibility of a major earthquake, (indicating that seismic activity is common). The scientific community does not accept the theory of energy dissipation when small earthquakes occur (Cocco et al., 2015).

The court accepted that current knowledge does not allow the prediction of earthquakes (time, size and location) and that effective prevention and mitigation of seismic risk is the application of newer seismic rules to old buildings. Therefore, the committee's judgments were based on poor risk assessments and incorrect information.

The incorrect communication policy chosen to inform the population shaped the guilty verdict for the committee members. More specifically: a) the purpose of the committee's convergence (from the outset) was the complacency of the public opinion rather than the information (telephone conversation on March 31, 2009); b) the committee voluntarily assumed the responsibility for public information and communication with the public; c) other people were allowed to attend the meetings (indefinitely and an unknown number of committee members); and d) members of the committee and other participants of the meeting were allowed to participate in the press conference (to maximize the communication impact of the meeting). A press conference was given by Region Abruzzo, civil protection official and president of the committee. The behavior of committee members was not scientific but regulatory.

The committee did not address its advice to the Civil Protection Department but directly to the public (incorrect communication strategy). Providing information to the public is the responsibility of the National Civil Protection Service, directly or through its local departments at regional, provincial and municipal level. The dissemination of information reproduced to the public became "simplified", "directly" and "without any filter". The "power" and persuasion of the scientific community had enhanced the effectiveness of reassurance through transmitted messages.

5 RISK COMMUNICATION AND DECISION-MAKING

In L'Aquila, the population received the message there would be "no destruction", but destruction did happen. The contradictory position was: "there is no way to predict earthquakes, but we predict non-earthquake." The committee members claimed that swarm earthquakes was completely normal, and it was a neutral event or even a positive event because energy was released.

The basic purpose of risk communication is more than just the transmission of a message. Several theoretical models have been developed describing how the risk message or information is being developed, how risk perceptions are shaped and how decisions are made, providing the foundation for coordinating and effectively communicating risk management (Reynolds & Seeger, 2005; CDC, 2012).

People tend to identify physical dangers through various subjective perceptions and beliefs with a multidimensional way of thinking (Risk Perception). Faulty personal perceptions of individuals based on subjective risk assessments often act as an important vulnerability factor. The perception of risks is recognized as essential and crucial to a successful disaster reduction strategy (Slovic et al., 1974,1981; Kahneman et al., 1982; Finucane et al., 2000).

The cornerstone of risk communication is the "strengthening of trust" (Wray et al., 2006). It is a cognitive shortening of the logical decision process, considering that the decisions taken and the instructions given are accurate, although there is extreme mistrust.

People at risk consider their homes as "safe", nature as a "positive moral force", and society as "responsible protector". Trust is linked to acceptable decisions and the creation of a favorable climate to reduce social vulnerability (Smart & Vertinsky, 1977; Rosenthal & Hart, 1991; Pearson & Clair, 1998).

Risk communication recommendations often use scientific expertise to guide public reactions. Science and decision-making are considered separate sectors with very different forms of legalization (Merton, 1973; Fischoff, 1995, 2013; McNie, 2007). Others claim that there is a co-progress between science and society (Jasanoff, 2004). For each of these considerations, relevant arguments are developed: positive and negative responsibility, scientific integrity and morality, independence, etc. (Elliott, 2006; Douglas, 2009).

Decision-makers and others prefer simple answers and clear suggestions as to what to do, which may not be available or perhaps will change over time (Cash et al., 2006; McNie, 2007; Sarewitz & Pielke, 2007; Ravetz et al., 2013). Scientists have difficulty in predicting some physical (Clark, 1980). It is necessary for science to use new technologies, which are not always available. An unverifiable prediction strikes the credibility of the whole scientific community.

Uncertainty may also derive from the "contrast between different assumptions" as expressed by scientists (Denis, 1995) or by the appearance of "self-appointed experts" who will offer unwanted advice with the tendency to exaggerate the size of the risk (Thornburgh, 1987). Incorrectly, the media has been given indirect power, which results in psychological rather than rational risk assessment and of course errors in the decision-making process (Denis, 1995). People have greater confidence in scientists and their explanations, compared to politicians whose perception is perceived as biased (Rosenthal & Hart, 1991). The view of scientists quite often conflicts with political and economic interests.

Frequent errors in making decisions under uncertainty fuels the assumption of public authorities that the dissemination of emergency information should be limited to avoid excessive reactions and public panic, called "the information targeting paradox" (Otway & Wynne 1989). The distinction between fears and panic of the population is difficult for public authorities. Public authorities may think that people will adopt life-threatening behaviors (Quarantelli & Dynes 1972; Drabek, 1986). At the same time, false negative or positive warnings can have enormous negative consequences and authorities are usually afraid of both situations (Woo, 2013).

6 THE IMPACT OF SOCIAL THEORIES ON THE TRIAL

In L'Aquila's trial, the opinion of the committee members that human choices are influenced by free will during an emergency was not accepted by the court. Antonello Ciccozzi supported that scientific communication has a strong persuasion and Roberto Gialdini argued that we are unaware of how we handle information provided by the authorities. Most scientific information is considered valid (Rigas, 1991).

According to the verdict, an understanding of risk is needed, which in the case of L'Aquila did not occur due to the incomplete evaluation of all risk indicators by the members of the committee. Inhabitants who stayed in their homes during the pre-seismic vibrations had a false perception of risk. People in situations of danger or something unknown act instinctively on the basis of the social representations and act under the instructions of the competent authorities. Social and cultural forms acquired as a result of education and social rules are capable of shaping its individual choices. The mechanisms that existed were collective removal of "earthquake fear" and collective adherence to the suggestions and assessments of the participants in the meeting.

7 GLOBAL REACTIONS: SCIENCE IN TRIAL

There has been global rage and reactions by the scientific community to the accusation (more than 30 associations and 5,000 scientists signed an open protest; it was featured by media outlets such as the BBC, New York Times, etc.). Seven Italian scientists were facing unfair criminal charges for failing to predict the earthquake, although the prediction of future earthquakes remains technically impossible. Science was on trial mainly because of the inability of lawyers to understand concepts such as uncertainty and probability. But the verdict explicitly stated that the committee members were not convicted because they did not predict the earthquake.

Statements of solidarity indicated that the scientific presentation was clear and precise, while the supposed reassuring message to the public did not arise at the expert meeting or in the official reports. Furthermore, there were fears that Italian scientists would be more reluctant to take on counsel roles (due to the consequences of the sentence) (American Geographical Union, 2010; Cressey, 2010; Leshner, 2010; Nosengo, 2010; Aspinall, 2011; Hall 2011; Kluger, 2012; Davis, 2012; Marzocchi, 2012; Prats 2012; Sturloni, 2012; Amato et al., 2013; Cocco et al., 2015; Benessia & De Marchi, 2017).

The L'Aquila case was interpreted in various ways and perspectives. The failure to implement international disaster risk reduction policies was highlighted (Imperiale & Vanclay 2018). It was argued that the reaction of the population and the lack of preparedness and awareness of seismic risk was due to insufficient initiatives in previous years to prevent and mitigate local authorities to increase society's resilience to seismic danger and educate the population (Cocco et al., 2015; Imperiale & Vanclay, 2018). Seismologist John Mutter argued that the committee should be censored (or imprisoned) because they did not remind people of the importance of seismic exercises (Hall, 2011).

Concepts were discussed such as acceptable risk, a causal link between the experts' statements and the behavior of inhabitants and, more generally, the question of scientific evidence (Zalin and Butti 2013, Tallacchini 2013, Masera 2013). The media were accused of vague ambiguous and inconsistent messages during the pre-seismic period.

The proceedings of the committee meeting (March 31, 2009) had not been published. In practice, communication with the public was done by civil protection officials through the media (Cocco et al., 2015). Instead of presenting the information at a press conference, what was broadcast was an interview of the Deputy Head of Civil Protection (given just before the committee meeting) saying that there was no risk and the scientific community had confirmed this (Corte di Appello del L'Aquila, 2014; Imperiale & Vanclay, 2018). It has been stated that social and anthropological sciences contribute to better disaster management (Benadusi, 2016), and the convictions concerned the failure of risk communication in order for citizens to receive appropriate information and make informed and safe choices (Prats, 2012; Ropeik, 2012; Sturloni, 2012; Woodman, 2013; Scolobig, et al,. 2014).

In the case of L'Aquila, the scientists acted simultaneously as counselors and decision-makers. Overlapping roles affected the entire communication process, leading scientists to become jointly responsible for deciding how to communicate risk to the population. In Italy, the mayor is the ultimate authority responsible for issuing a warning after consultation with provincial and regional authorities (Article 225/1992). Professor Aspinall argued that the committee was trapped in deciding possible actions: "evacuation or nothing" (Cartlidge, 2012).

L'Auqila's case was a prime example of the paradox of risk communication and confusion of roles. National and local authorities were worried that people would overestimate anything except a reassuring message. As a result, the risk management problem was shaped from the point of view of public control rather than public security (De Marchi, 2013). The convergence of the Commission proved to be a "political trick" to calm public feelings (Imperiale & Vanclay, 2018), and involved the indirect transfer of the moral and legal responsibility of failure to science (Benessia & De Marchi, 2017). There have been mixed or overlapping roles between scientists and decision-makers related to the overlapping of competences and the need for emergency managers to reassure the public and control the over-reaction of residents (Scolobig et al., 2014; DeVasto, 2015).The separation of roles and the different degrees of legitimacy was also discussed (Benessia & De Marchi, 2017), and this confusion of roles is supported by Bochi (2013), who completely separates the advisory work of scientists from the responsibility of other authorities.

The psychological, social and institutional conditions caused an unprecedented sequence of psychological and legal consequences for the population, scientists and civil servants involved in the assessment of risk situations and communications (Oreskes, 2015). The chain of information provided to citizens was so unusual that it caused the absurd "prediction of a non-earthquake" (Alexander, 2010, 2014).There was uncertainty of the risk, its detection and communication during emergency situations (Benessia & De Marchi, 2017), and the logical fallacies and psychological effects that arose from a pre-seismic swarm earthquakes (Ciccozi, 2014; DeVasto et al., 2015).

8 THE FINAL VERDICT

The first instance judgment was appealed. After a lengthy procedure, the Court of Appeal, announced on November 10, 2014 that it was overturning the conviction (Corte di Appello del L'Aquila, 2014). The six scientists were not convicted due to the absence of evidence.The Deputy Chief of Civil Protection and member of the High-Risk Committee was not relieved of liability but his sentence was reduced to 2 years and his financial responsibility was limited to legal costs (without any other compensation for loss of life or injury). The Supreme Court judgment issued on November 20, 2015 confirmed the decision of the Court of Appeal (Corte di Cassazione, 2016).

Many issues have been redefined as follows. a) The convergence of the High-Risk Committee was not formal because it was not convened by the normal procedure. It was also closed to the public. There is no evidence that scientists were aware of the political protection press release or its intention to make public statements. b) The scientists only provided advice on the protection policy. c) No regulation determines what constitutes a proper risk assessment. The scientists provided a scientifically correct analysis of the hazards associated with swarm earthquakes. Vulnerability and exposure assessments are not relevant to the purposes of the meeting. e) Only civil protection officials are responsible for the information provided to the public, and only they should have considered the influence that the institutional communication would have on the public.

9 CONCLUSION AND DISCUSSION

L'Aquila was a city of high-risk and vulnerability. Despite facing a threat, the public received from the public authorities the message that there was no danger. There was confidence in the authorities but they did not do what they had the duty to do and there were many errors with unfavorable results. The problem was not that they did not predict the earthquake, but that they did not take preventive and preparedness measures and their design proved inadequate. Some of the factors that led to failure in crisis management were psychological risk assessment, cognitive failures, heuristic mechanisms, limited rationality and logical mistakes in decision-making, dangerous improvizations and ignorance of the disaster research literature, etc. Emergency management always requires the examination of a vast array of interrelated issues: political, legal, economic, organizational, social, psychological, and scientific science (Dynes, 1970).

The case of L'Aquila was particularly tragic and complicated, as the risk that could not be predicted turned into a real disaster. How could the information available reassure the population under uncertain situations? Effective warnings should inform people about possible protection measures that need to be taken, and about the potential risk, benefits and cost of their decisions, thus allowing them to make correct and responsible choices (Fischhoff, 2013).

The broad publicity facilitated a series of discussions, providing lessons-learned, advice and guidelines. However, there is a repeat of the same mistakes and errors in each disaster. It is now required through established and clear procedures to enhance the quality of organizational learning, and appropriate scientific findings should be adopted by promoting solutions that prevent repeating the same mistakes.

REFERENCES

Alexander, D. 2010. *The L'Aquila earthquake of 6 April 2009 and Italian Government policy on disaster response. Journal of Natural Resources Policy Research*, 2(4): 325–342.

Amato A., Cocco M., Cutrera G., Galadini F., Margheriti L., Nostro C. and Pantosti D. (2013). *The L'Aquila trial. Geophysical Research* Abstract, Vol. 15 EGU2013-12140.

American Geographical Union. 2010. AGU Statement: Investigation of Scientists and Officials in L'Aquila, Italy, is Unfounded. Eos, *Transactions of the American Geophysical Union* 91 (28): 48.

Aspinall, W. 2011. Check your legal position before advising others. *Nature*, 477: 251.

Benadusi, M. 2016. The Earth will Tremble? Expert knowledge confronted after the 2009 L'Aquila Earthquake. *Archivio Antropologico Mediterraneo*.

Benessia, A., and De Marchi Bruna 2017. When the earth shakes … and science with it. The management and communication of uncertainty in the L'Aquila earthquake. *Futures*.

Boschi, E., Guidoboni E., Ferrari G., Mariotti D., Valensise G., and Gasperini, P. 2000. Catalogue of Strong Italian Earthquakes from 461 B.C. to 1997. *Annals of Geophysics* 43, 609–868.

Boschi, E. 2013. L'Aquila's aftershocks shake scientists. *Science* 341 (6153). p.1452

Cartlidge, E. 2012. Aftershocks in the Courtroom. *Science* 338 (6104) (December 10): 184–188. doi:10.1126/science.338.6104.184.

Cash, D., Borck, J, and Patt, A. 2006. Countering the 'loading dock' approach to linking science and decision making: a comparative analysis of ENSO forecasting systems. Science, *Technology, and Human Values* 31: 465–494.

Centers for Disease Control and Prevention (CDC). 2013. Ed. *Crisis Emergency Risk Communication*. US Department of Health and Human Services.

Ciccozzi, A. 2014. Il terremotodel L'Aquila e il processoalla Commission eGrandiRischi: note antropologiche. Antropologia applicata. *Pensa Lecce* pp.123–176.

Clark, W.C. 1980. Witches, foods and wonder drugs: historical perspectives on risk management. in Schwing, R.C. and Alerts, W.A. (Eds), *Societal Risk Assessment: How Safe is Safe Enough*, Plenum Press, New York, NY, 1980, pp. 287–313.

Cocco, M., Cultrera G., Amato A., Braun T., Cerase A., Margheriti L., Bonaccorso A., Demartin M., De Martini P.M., Galadini F., Meletti C., Nostro C., Pacor F., Pantosti, D. Pondrelli S., Quareni F. and Todesco M. 2015. *The L'Aquila trial*. Geological Society: London. Special Publications, vol. 419, issue 1, pp. 43–55.

Corte di Cassazione 2016."Sentenzasulricorsoproposto dal ProcuratoreGeneralepresso la Corte di Appellodell'Aquila et al.", *Corte Suprema di Cassazione*, IV SezionePenale, Roma. Available at: www.giuris prudenzapenale.com/2016/04/01/terremoto-laquila-la-sentenza-della-cassazionesulla/.

Corte di Appellodell'Aquila 2014, "Sentenzanelprocessopenale a carico di Barberi Franco, De Bernadinis Bernardo, Boschi Enzo, Selvaggi Giulio, Calvi Gian Michele, Eva Claudio, Dolce Mauro", *Corte di Appello*dell'Aquila, L'Aquila, available at: https://processoaquila.files.wordpress.com/2015/03/sen tenza-appello-aq-1-1001.pdf.

Davis, G.H. 2012. Italian Court Action Likely to Harm Efforts to Mitigate Earthquake Losses. *Geological Society of America*.

Denis, H. 1995."Scientists and disaster management", *Disaster Prevention and Management*, 4:14–19.

De Marchi, B. 2013. Risk governance and the integration of scientific and local knowledge. In Fra Paleo, U. (ed.) *Risk Governance. The articulation of hazard, politics and ecology*. Berlin: Springer.

DeVasto, S. Graham S.S. and Louise Zamparutti L. 2015. *Journal of Business & Technical Communication*. https://doi.org/10.1177/1050651915620364.

Drabek, T.E. 1986. *Human system responses to disaster: an inventory of sociological findings*. N.Y.: Springer-Verlaag.

Dynes, R.R. 1970. *Organized Behavior in Disaster*. D.C. Heath: Lexington, MA.

Elliott K.C. 2006. An Ethics of Expertise Based on Informed Consent. *Science and Engineering Ethics* 12 (4): 637–661.

EUR-Lex (nd). The precautionary principle. Available at: https://eur-lex.europa.eu/legal-content/EL/TXT/?uri=LEGISSUM%3Al32042.

Finucane, M.L., Alhakami, A., Slovic, P., and Johnson, S.M. 2000. The affect heuristic in judgments of risks and benefits. *Journal of Behavioral Decision Making* 13: 1–17.

Fischoff, B. 1995. Risk perception and communication unplugged: twenty years of process. *Risk Analysis* 15:137–145.

Fischoff, B. 2013. The sciences of science communication. *Proceedings of the National Academy of Sciences*. 110: 14033–14039.

Global Risk Myiamoto. 2009. L'Aquila Italy M6.3 earthquake. *Earthquake Field Investigation Report*, 6 April. Available at: http://miyamotointernational.com/wp-content/uploads/Italy-EQ-Report.pdf.

Grandori, G. and Guagenti, E. 2009. Prevedereiterremoti: la lezionedell'Abruzzo. *IngegneriaSismica*, XXVI (3): 56–61.

Grimaz, S. and Maiolo, A. 2010. The impact of 6th April 2009 L'Aquila earthquake (Italy) on the industrial facilities and lifelines. Consideration in terms of NaTech risk. *Chemical Engineering Transactions* 19, 279–284, doi:10.3303/CET1019046.

Hall, S.S. 2011. Scientists on trial: At fault? *Nature*, 477: 264–269.

Jasanoff, S. (ed). 2004. States of knowledge. The co-production of science and the social order. London/New York: Routledge.

Imperiale, A.J., and Vanclay, F. 2018. Reflections on the L'Aquila trial and the social dimensions of disaster risk. *Disaster Prevention and Management: An International Journal*.

Kahneman, D., Slovic, P. and Tversky, A. 1982. Judgment under uncertainty: *Heuristics and biases*. Cambridge University Press. New York.

Kaplan, H, Bilgin, H., Yilmaz, S., Binici, H, and Oztas, A., 2010. Structural Damages of L'Aquila (Italy) earthquake, *Natural Hazards and Earth System Sciences*, 10, 499–507.

Kluger, J. 2012. Scientific Illiteracy: Why The Italian Earthquake Verdict Is Even Worse Than It Seems. *Time Magazine.*

Leshner, A.I. 2010. Letter to the president of the Italian Republic. *American Association for the Advancement of Science (AAAS).* Retrieved from: https://www.aaas.org/sites/default/files/s3fs-public/10_06_29earthquakelettertopresidentnapolitano.pdf.

Oreskes, N. 2015. How earth science has become a social science. *Historical Social Research* 40(2). pp.246–270.

Marzocchi, W. 2012. Putting science on trial. *Physics World* December 2012: 17–18.

Mc Nie, E. 2007. Reconciling the supply of scientific information with user demands: an analysis of the problem and review of the literature. *Environmental Science and Policy* 20: 17–38.

Meeting draft report, 2009. Draft of the minutes report of the 31 March 2009 *HRC meeting.* Retrieved from: https://processoaquila.files.wordpress.com/2012/11/3b-bozza-cgr-310309eng.pdf.

Merton, R.K. 1973 [1942]. The Normative Structure of Science. In: R.K. Merton, *The Sociology of Science: Theoretical and Empirical Investigations* (pp 267–278). Chicago: University of Chicago Press.

Nosego, N. 2010. Italy puts seismologists on the dock. *Nature* 465:992.

Prats, J. 2012. The L'Aquila earthquake: Science or risk on trial? *Significance* 9:13–16. doi:10.1111/j.1740-9713.2012.00615.x.

Rigas, A. 1991. From the attitudes to social representations: a review of the recent bibliography. *Social Research Survey*, 80(80): 156–165. doi:http://dx.doi.org/10.12681/grsr.630.

Ropeik, D. 2012. L' Aquila verdict: A judgement not against science, but against o failure of science communication. *Scientific American.*

Rosenthal, U. and 't Hart, P. 1991. Experts and Decision Makers in Crisis Situations. *Knowledge: Creation, Diffusion, Utilization* 12(4): 350–372.

Reynolds B., & Seeger M.W. (2005). Crisis and Emergency Risk Communication as an Integrative Model. *Journal of Health Communication.* 10(1). 43–55.

Otway, H. and Wynne, B. 1989. Risk Communication: Paradigm and Paradox. *Risk Analysis* 9(2) 143.

Quarantelli, E.L. and Dynes, R.R. 1972. When disaster strikes (It isn't much like what you've heard or read about). *Psychology Today*: 67–70.

Ravetz, J., Funtowicz, S., & Economics, I. (2013). Post-Normal Science. In: Simon Meisch, Johannes Lundershausen, Leonie Bossert, Marcus Rockoff (Hrsg.) *Ethics of Science in the Research for Sustainable Development*, pp. 99–112. Glashütte, Germany: Nomos.

Sarewitz, D. and Pielke, R. 2007. The neglected heart of science policy: reconciling supply of and demand for science. *Environmental Science and Policy* 10: 5–16.

Scolobig, A., Mechler, R., Komendantova, N., Liu, W., Schroeter, D. and Patt, A. 2014. The co-production of scientific advice and decision making under uncertainty: Lessons from the 2009 L'Aquila earthquake. Italy. *Planet@Risk* 2(2): 71–76.

Slovic, P., Kunreuther, H. and White, G.F. 1974. Decision processes, rationality and adjustment to natural hazards. edited by G. F. White, *Natural hazards, local, national, and global*, New York, Oxford University Press. pp. 187–205.

Slovic, P., Fischhoff, B. and Lichtenstein, S. 1981. Facts and Fears: Societal Perception of Risk. in Advances in *Consumer Research* Volume 08. eds. Kent B. Monroe, Ann Abor: Association for Consumer Research, pp: 497–502.

Smart, C. and Vertinsky, I. 1977. Designs for crisis decision units. *Administrative Science Quarterly* 22(4): 640–657.

Sturloni, G. 2012. A lesson from L'Aquila: The risks of science (mis)communication. *Jcom* 11(4).

Thornburgh, R. 1987. The Three Mile Island experience: ten lessons in emergency management. *Industrial Crisis Quarterly* 1(1): 5–14.

Tribunale di L'Aquila 2012. SezionePenale, Dott. M. Billi. Sentenza n. 380/2012. Retrieved from: https://processoaquila.files.wordpress.com/2013/01/sentenza-grandi-rischi-completa-1.pdf.

Woo, G. 2013. *Calculating catastrophe.* LondonImperial College: Press.

Woodman, A. 2013. Facilitating informed choice: The communication obligations of scientists to the public and the 2009 L'Aquila Earthquake. Unpublished doctoral dissertation. Cambridge, U.K.: Trinity College, University.

Wray, R., Rivers, J., Whitworth, A., Jupka, K., Clements, B. 2006.Public Perceptions About Trust in Emergency Risk Communication: Qualitative Research Findings. *International Journal of Mass Emergencies and Disasters.* 24(1): 45–75.

Risk Analysis Based on Data and Crisis Response Beyond Knowledge – Huang & Nivolianitou (eds)
© 2020 Taylor & Francis Group, London, ISBN 978-0-367-25146-8

Swarm of UAVs as an emergency response technology

Ilias Gkotsis, Agavi-Christina Kousouraki & George Eftychidis
Center for Security Studies (KEMEA), Athens, Greece

Panayiotis Kolios & Maria Terzi
KIOS Research Center, Nicosia, Cyprus

ABSTRACT: The SWIFTERS project capitalizes on unmanned aerial vehicle (UAV) swarms to study, design and test, cooperation strategies that support the allocation of evacuation operations to UAVs in an intelligent way, with the ultimate goal of improving response efficiency and reducing evacuation times. Primarily, SWIFTERS aim to shape a clear picture of individual and common needs to be addressed, regarding the deployment of UAV platforms for emergency evacuation. An updated list of possible roles that UAVs in a swarm may have during emergency evacuation is presented and analyzed; including several UAV uses during past disaster events. For this, both the roles identified in the first part combined and the literature review were considered. Another significant part of the project is the list of indicators that will be used further to assess the gains of introducing UAV swarms in the evacuation process and disaster response in general. These key quality indicators (KQIs) will also be used to identify the scenarios that can benefit more, in which the SWIFTERS solution will be tested and evaluated.

Keywords: unmanned aerial vehicle, swarms, evacuation, disaster management and response, civil protection

1 INTRODUCTION

Unmanned aerial vehicle(UAV) swarms in emergency response have several possible applications, with the most frequent to be search and rescue. A swarm can cover a region of interest quickly and requires a limited number of resources (e.g. one operator) (Munoz-Morera et. al., 2015). Another application is exploration, as swarms of simple and small vehicles cancan high-risk buildings and sites rapidly, whereas large vehicles cannot. The challenge is how to control multiple vehicles that cooperate automatically to finish a task. New challenges imposed by intelligent swarms have attracted much research in the last decade.

Undoubtedly, applying unmanned aerial systems (UAS) to civil protection activities can be beneficial (EENA, 2015). Broadly speaking, these services can be classified based on two main roles: a helper and an informer/observer. A helper UAS can help first responders by shipping, quickly and effectively, important dispensable equipment payloads, such as life vests, defibrillators, first-aid kits, thermal blankets, water and food or any combination of these items, depending on the requirements of the situation. An observer UAS can maximize situational awareness and aid civil protection personnel in deciding or planning their actions.

Additionally, a fully autonomous UAS can locate and track an object of interest and provide information and data to the control station when necessary. It can also provide first responders a bird's eye view, helping them prioritize their search and rescue efforts. During evacuation procedures, situational awareness (SA) is a key necessity that enables prioritization of actions and allows responders to focus on tasks that need immediate attention. Time is a crucial parameter and promptly available and highly deployable equipment is directly linked with mission performance. In that respect, UAS platforms increase operational effectiveness and decrease operational costs and response time.

Growth in the demand for UAS platforms (the FAA expects sales of UAS for commercial purposes to grow by 2020), combined with their applicability in different operations, has increased the availability of different categories and types of UAS platforms. This has resulted in a choice deluge that complicates end-user decisions in selecting and operating the best UAS platform.

The first stage of the SWIFTERS project defines the complete picture of operational needs of civil protection related authorities and other relevant stakeholders regarding UAS platforms. First, the existing regulations and the most common and important UAS roles during emergency response, are presented. These roles have been further elaborated with information deriving from a qualitative analysis of an end-user requirements questionnaire. After conducting research on the needs, the SWIFTERS project presented the main key quality indicators (KQIs) for evacuation procedures with swarms of drones, either in simulation or in real-time crisis management.

These will enable decision makers to (a) assess the consequences of incidents and crisis scenarios, (b) compare possible impacts resulting from alternative actions, (c) support strategic decisions on capabilities and related investments, (d) optimize the deployment of the UAVs dedicated to evacuation, and (e) improve action plans for preparedness and response phases. They are valid in the scope of the whole scenario – either cumulative or pertinent to the final world state at the end of the crisis.

2 NEEDS ASSESSMENT

In the framework of the SWIFTERS project, in order to define the complete picture of end-user needs, a questionnaire regarding the use of UAS and swarms during civil protection missions was shaped and distributed to both project end users and external stakeholders. The questionnaire was answered by members of the Civil Defense, Police and Fire Services from Cyprus, Italy and Greece.

According to end users' responses, the most important type of disaster was related to fire, followed by flood and landslides. One-third of the respondents also selected drought management and sea and mountain rescue (as missions) that can occur irrelevant of or without any hazard.

The most significant types of missions were search and rescue and real-time information and situation monitoring. Of almost similar importance to the end users was humanitarian aid-supply delivery, surveillance and reconnaissance. One-third of the participants suggested other missions, such as immigrant management, security and public safety.

During emergency evacuation procedures, the end users stated that locating a person and operating during the night were the most important task they perform, with 18% and 15% of the responses, respectively, then locating a group was the next most important task.

The most essential payloads for end users were optical cameras, videos and data links, which were highly evaluated by all. GPS and IR/thermal camera were the next essential payloads. One of the end users also defined a high-power light source as an essential type of payload.

3 ROLES OF DRONES IN SWARMS

The use of more than one drone in the evacuation operations after a disaster could be helpful, faster and less expensive. Having a swarm of drones means that each drone will have a specific role, so for the evacuation stage, seven major roles have been identified, which are presented and analyzed below.

3.1 *Real-time information*

One of the first information needs is the confirmation of the impact and extent of damage resulting from the disaster. For incidents such as earthquakes, tornadoes, hurricanes, and flash flooding, the need to expeditiously locate survivors is critical and warrants immediate access to resources that can assist in locating and guiding emergency responders to initiate rescue efforts. Once life saving measures have commenced, the magnitude of the disaster must be determined to guide the formation of the evacuation team experts to assist in determining incident priorities and initial mission assignments.

3.2 Reconnaissance and mapping (RAM)

The need for regular mapping of disaster-prone areas cannot be overstated. One or several drone flights are needed to cover a designated mapping area. Flood maps help coordinate disaster response efforts after events, 3D topographical mapping can help identify areas prone to mudslide, and high-resolution visual imaging can support first responders to flag critical infrastructure that needs secured immediately after a disaster. Through RAM, civil protection personnel, conscripts and volunteers can gather information, survey large areas and make informed decisions on a concrete situation and consider risks associated to respective missions.

Drones offer numerous advantages in these cases, as they are more readily deployable than satellites or manned aircraft and can create higher resolution maps. Researchers at the University of New Mexico and San Diego State University, for example, are developing a drone-based remote sensing suite that will detect very fine damage to transportation infrastructure after disasters such as hurricanes, earthquakes, or floods. Some small drones can even be deployed while natural disasters are still taking place.

3.3 Primary humanitarian aid and supply delivery

Severe weather events, earthquakes, sabotage and other natural and man-made disasters can destroy water lines, roadways, bridges, oil and gas pipelines, power plants, transmission lines, and other infrastructure. Disruptions in the normal operation of critical infrastructures can have significant impacts on people safety and can make evacuation efforts more difficult.

In many of these cases, UAVs can deliver or release (after adjusting custom made equipment, such as buckets, claws, etc.) urgent medical and other supplies to those in need, including first-aid kits, bottles of water, and tools (Klausen, Fossen & Johansen, 2015; Klausen et.al., 2015), mobile phones, beacon markers, etc.

3.4 Search and Rescue (SAR)

Deployment of drones in SAR aims in particular at increasing observational awareness of ground teams to improve safety, and aiding in locating missing people. SAR operations are highly codified and planned to ensure optimal impact despite the complex environments of emergency responses. They are equally as complex as all other operations in the disaster response. Under proper and correct control, a swarm of UAVs can help with actions such as detection, recognition, giving information about the position and condition of people involved, and indicating the path for the rescuers. The small portable UAV systems have the benefit of the scanning buildings or debris for survivors very quickly and this could be helpful for the rescue teams to rapidly cover massive amounts of affected areas. To be useful, the UAVs and trained staff must be available within 24–48 hours of a sudden-onset disaster. In addition, these technical approaches must be integrated into the protocols for rescue teams, agreements for importing disaster response equipment, the International Search and Rescue Advisory Group (INSARAG) guidelines and other areas of policy.

3.5 Restoring telecommunication

Utility infrastructure often fails or becomes unusable during emergencies and when disasters strike. A reliable communication infrastructure is crucial for an efficient and successful emergency response system. An indicative example includes Hurricane Sandy (October 2012), during which high storm surge levels caused extensive flooding that triggered power outages and complete telecommunications breakdown. Hence, offering an alternative utility infrastructure through UAS aerial platforms has been of great interest to civil protection personnel and relevant stakeholders at strategic, tactical and operational level (Barbatei, Skavhaug & Johansen, 2015). Identified complementary utility solutions include primarily telecommunication services (i.e., unmanned vehicles acting as a relay for information exchange), for illuminating dark areas during the night, as notice boards to display inform and replaying of acoustic messages. Airborne drones could serve as temporary airborne warning and control systems-(AWACS) platforms, sending Wi-Fi and cell phone coverage across an area stricken by downed power lines and damaged cell towers.

3.6 *Surveillance*

This role refers to the process of using UAS platforms and their respective payloads (e.g., cameras, radiometric sensors, etc.) to detect and assess vulnerabilities, hazards, and natural/man-made/technological emergencies.

UAVs equipped for real-time telemetry of video imagery, sensor support data, and GPS/INS navigation, can provide autonomous surveillance of defined perimeters, video tracking and active following of targets of interest, and real-time cueing to other imaging UAVs.

In a disaster response for evacuation, incident commanders must function to effectively coordinate personnel and resources, often with delayed and inaccurate information on the hazards and evolving conditions at the scene. Coordinated autonomous UAVs address these problems by rapidly providing simultaneous views of the scene at different resolutions using fewer cameras. They can traverse into sites that are unreachable or too dangerous for first responders and locate missing people, and send recognition and the geo-reference of their position in real time with sufficient accuracy. The UAVs can also be redirected to locations of interest selected by the user. Video and GPS coordinates can be sent through a network to other responders.

3.7 *Logistics and package delivery*

Information collection from drones may be just the first step. Smaller UAV systems, such as those that can carry up to a few kilograms, are unlikely to deliver useful quantities of relief goods efficiently.

Drones provide an alternative logistical support after a disaster, as they can fly above destroyed infrastructure. Currently, rescue workers may devote substantial time to establishing logistical support for other responders, ferrying equipment and essential supplies like food and water. Drones can also deliver tools, building materials, survival kit, and any other materials needed to help the evacuation efforts. Additionally, real-time maps created by drone-based sensors can be used by logistics companies to determine the most efficient route for the ground forces to find and evacuate victims.

4 KEY QUALITY INDICATORS

During disaster response operations such as evacuations, KQIs are means of encapsulating a complex reality in a single measurable construct that offers a systematic approach to discuss and quantitatively evaluate the efficiency of disaster management, including the use of UAVs in swarms during this type of procedure (Engelbach et al., 2014). The most appropriate KQIs will be used to assess the gains of introducing UAV swarms in the evacuation process. These quality indicators will demonstrate the advantages of using swarms over single drone operations (McCune and Madey, 2014). They will also be used to identify which scenarios can benefit more from UAV swarms, rather than a single UAV, depending on evacuation time, response time, human costs, capacity and resource planning, cost effectiveness, situation indicators, operational efficiency, speed – flow –density and evacuation route.

A swarm of UAVs can cover the area in parallel, reducing the overall time of the mission. Typically, due to a limited number and short flight time of the UAVs, it will not be possible to cover the whole area at once. This may cause extra overheads to the operators, who will need to frequently replace and recharge batteries.

To optimize the total distance covered by the different UAVs, and therefore, the exploration time and the time needed for the batteries to be changed, we used the well-studied vehicle routing problem (VRP). The problem can be formulated as follows. Given a set of locations, determine a set of routes, each performed by a single vehicle that starts and ends in a depot, while the operational constraints are satisfied, and the overall cost is minimized. In our case, the set of locations are those where we want to obtain information. The set of routes are those that will be covered by the UAVs and they will start and finish in our base station depot. The operational constraint is the maximum time that can be employed on a route, which is the flight time duration. The VRP is an NP-hard problem, so the complexity increases with the number of locations.

The VRP can be formulated as an integer linear programming. Let $G = (V, A)$ be a complete graph with $V = \{0, \ldots, n\}$ denoting vertices of the graph and A the arc set. The vertex 0 corresponds to the initial point where all vehicles start, also called depot. The vertices with index $i = 1, \ldots, n$ correspond to the locations that need to be visited. Figure 1 shows a sample graph, with the blue points denoting the vertices, gray lines representing all possible connection between vertices, and the red lines showing the solution to the problem.

The cost associated with each arc $(i, j) \in A$ is represented as $c_{i,j}$ and denotes the cost of going from node i to node j. In the considered problem, the cost is the time to go from one point to another. The UAV velocity is supposed to be constant. Then, by assuming a known Euclidean distance between points, it becomes possible to calculate the required time. The time needed to collect the information in node i is then added to the cost. The resulting optimization problem is then shown in Equations (1)–(8).

The binary variable x_{ij} is equal to 1 if arc $(i, j) \in A$ belongs to the optimal solutions and 0 otherwise. Figure 4 shows that when x_{ij} is equal to 1, the arc is drawn with red, and is gray otherwise. Equations (2) and (3) establish that only one arc enters and leaves each vertex. Equations (4) and (5) constrain that K vehicles will leave and enter the depot. Equation (6) describes the vehicle distance constraint and the connectivity of the solution. No route can be created that is not connected to the depot. Function r(S) is the minimum number of vehicles

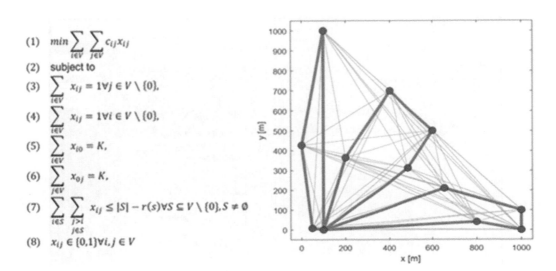

$$(1) \quad min \sum_{i \in V} \sum_{j \in V} c_{ij} x_{ij}$$

$(2) \quad \text{subject to}$

$$(3) \quad \sum_{i \in V} x_{ij} = 1 \forall j \in V \setminus \{0\},$$

$$(4) \quad \sum_{i \in V} x_{ij} = 1 \forall i \in V \setminus \{0\},$$

$$(5) \quad \sum_{i \in V} x_{i0} = K,$$

$$(6) \quad \sum_{j \in V} x_{0j} = K,$$

$$(7) \quad \sum_{i \in S} \sum_{\substack{j > i \\ j \in S}} x_{ij} \leq |S| - r(s) \forall S \subseteq V \setminus \{0\}, S \neq \emptyset$$

$(8) \quad x_{ij} \in \{0,1\} \forall i, j \in V$

Figure 1. Routes generated using the vehicle routing problem.

357

needed to serve a subset of customers S. It is this equation where the battery duration is introduced. To obtain the value of this function, a multi-agent Traveling Salesman Problem (m-TSP) is solved with heuristics. Note that the number of constrains defined in Equation (7) increases exponentially with the number of customers.

Based on this formulation, the algorithm was solved using exact methods. However, the time needed to solve the problem increase exponentially with the number of locations [6], [7]. In an emergency situation every second counts, so having a solution in the minimum time is critical. Therefore, we decided to use heuristics. These methods are faster than exact algorithms, but the algorithm will run in two phases. First, it created a solution using one of the heuristic methods mentioned above. Second, the obtained solution was improved using met heuristic methods. The selection of the used algorithm depends on the geometry of the problem. We tried different combinations to find the optimal result. It is important to know that met heuristics methods run indefinitely so it is necessary to specify a running time. When the routes were created, a feasibility algorithm was executed to check if there were collisions. We detected that most of the collisions occurred on the way from the depot to the first waypoint of the route and to the last waypoint to the depot. Therefore, we omitted these collisions and use a different height to execute these path legs.

6 SWIFTERS PLATFORM

The SWIFTERS platform was designed to provide a modular, extendable, scalable, stable and reliable infrastructure to enable end users' functional and non-functional requirements. It is focused on enabling the connection; communication and control of a UAV swarm and on hosting multiple algorithms such as UAV swarm routing and computer vision. The system architecture presented in Figure 2 consists of five main layers with different components. Each layer is briefly presented below and discussed in an individual section.

The robot operating system (ROS) is a framework consisting of a collection of tools, libraries, and conventions widely used in the implementation of robotic platforms. In SWIFTERS, ROS is used as a communication framework between the different modules of the system such as the UAVs and the graphical user interfaces.

The computation graph is the peer-to-peer network of ROS processes that are processing data together. The basic computation graph concepts of ROS are nodes, master, parameter server, messages and topics, all of which provide data to the graph in different ways.

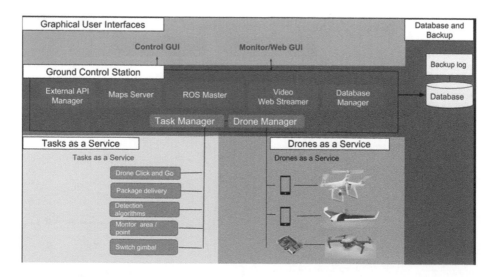

Figure 2. The SWIFTERS system architecture.

7 CONCLUSION AND DISCUSSION

This paper provides an update regarding the results of the SWIFTERS project. These results include the following: (a) the presentation of main existing regulations related to using UAVs, possible restrictions and future challenges; (b) an updated list of roles that UAVs in a swarm need to have during emergency evacuation; (c) an analysis of end users' overall needs regarding the use of a swarm of UAVs in emergency response on the evacuation stage; (d) KQIs related to UAVs and evacuation; and (e) an explanation of the algorithm designed for searching and area mapping evacuation operation. The presented project results support the development of a software package, which will be released under open-source license, featuring emergency operation planning capabilities enabled by UAV swarms, such as task allocation to individual UAVs of the swarm (e.g. monitoring emergency event progress, identifying stranded survivors, marking the evacuation path etc.), and path planning for each UAV. In doing this, the SWIFTERS project will support the allocation of evacuation operations to UAVs in an intelligent way with the ultimate goal of improving response efficiency and reducing evacuation times.

ACKNOWLEDGMENTS

The work has received funding from the European Union Civil Protection Call for proposals 2017 for prevention and preparedness projects in the field of civil protection and marine pollution under grant agreement – 783299 – SWIFTERS.

REFERENCES

Barbatei, R.G., Skavhaug, A. and Johansen. T.A. 2015. Acquisition and relaying of data from a floating wireless sensor node using an Unmanned Aerial Vehicle. International Conference on Unmanned Aircraft Systems (ICUAS).

European Emergency Number Association (EENA). 2015. RPAS and the Emergency Services, Operations Document 2015.

Engelbach W., et al. 2014. Indicators to compare simulated crisis management strategies. 5th International Disaster and Risk Conference IDRC 2014. Integrative Risk Management – The role of science, technology & practice, 24–28 August 2014. Davos: Switzerland.

Klausen, K., Fossen, T.I., and Johansen, T.A. 2015. Nonlinear control of a multirotor UAV with suspended load. 2015 International Conference on Unmanned Aircraft Systems (ICUAS).

Klausen, K., Fossen, T.I., Johansen, T.A. and Aguiar, A.P. 2015. Cooperative path-following for multirotor UAVs with a suspended payload. 2015 IEEE Conference on Control Applications (CCA 2015).

McCune R.R. & Madey G.R. 2014. Control of artificial swarms with DDDAS. *Procedia Computer Science* 29:1171–1181, Nevada: USA.

Munoz-Morera J., et. al. 2015. Assembly planning for the construction ofstructures with multiple {UAS} equipped with robotic arms. 2015 International Conference on Unmanned Aircraft Systems (ICUAS). 9–12 June 2015. Denver: USA.

Risk Analysis Based on Data and Crisis Response Beyond Knowledge – Huang & Nivolianitou (eds)
© 2020 Taylor & Francis Group, London, ISBN 978-0-367-25146-8

Development of agricultural risk management system in Japan and adapting it for China

Yueqin Wang & Sijian Zhao*

Agricultural Information Institute of Chinese Academy of Agricultural Sciences, Key Laboratory of Digital Agricultural Early-warning Technology, MOA, Beijing, China

ABSTRACT: Both Japan and China are facing serious agricultural risks due to frequent natural disasters. Japan is highly experienced in all aspects of agricultural risk management. There are many similarities between China and Japan in the agricultural production environment and model. This paper systematically studies the Agricultural Mutual Relief System, Government support policies and Income Insurance System in Japan's Agricultural Disaster Compensation System. It will provide a useful reference for improving China's agricultural insurance system and contribute to the improvement of agricultural risk management.

Keywords: Japan, agricultural risk management, Agricultural Disaster Compensation System, agricultural mutual relief, income insurance

1 INTRODUCTION

Agricultural risks refer to the possibility that a farmer's final benefit is lower than the expected return due to external factors beyond their control. Agricultural risks mainly come from the geographical environment, climate, biological threats and external uncertainties such as agricultural input and output. Agriculture, especially the planting industry, is directly and closely dependent on nature and is always a high-risk activity. Frequent changes in climatic conditions are the main source of agricultural production risks. Various natural disasters like drought, floods, freezing, earthquakes, hail, storms, etc., can lead to serious yield losses. Due to the inherent vulnerability and longer production cycle, agricultural risks have the characteristics of strong seasonality, wide range, frequent occurrence and tremendous loss. Once agricultural disasters occur, agricultural production will drop and farmers' incomes will decrease. It would even affect food security and social stability. This highlights the importance of agricultural risk management.

Agriculture is the foundation of a national economy. China is a country that is regularly exposed to serious natural disasters. According to the National Bureau of Statistics of China, direct economic losses caused by natural disasters reached 301.87 billion yuan in 2017. The disasters affected 144 million people and 18.48 million hectares of crops were damaged. Both Japan and China are located in the Asian monsoon region with many natural disasters, especially meteorological events. Japan is similar to China in that many farmers are involved in small-scale production which makes them vulnerable and too weak to resist risks. Once disasters occur, it is difficult for farmers to restore reproduction on their own strength. Japan has an established agricultural risk management system with the Agricultural Disaster Compensation System, whose Agricultural Mutual Relief System provides a reliable risk guarante efor agriculture. In order to meet the diverse demands of farmers, Japan has set up an income

---

*Corresponding author: Dr. Zhao Sijian, engaged in agricultural risk management and agricultural insurance research, Address: No.12 Zhongguancun South St., Haidian District Beijing P.R. China, Email: zhaosijian@caas.cn

insurance scheme beginning in 2019. At present, Japan's agricultural risk management system includes both the original Agricultural Disaster Compensation System and the start-up Income Insurance System. It is of great significance to study the history and establishment of Japan's agricultural risk management system in order to adapt them for use in China.

Based on analyzing the agriculture and agricultural risks in Japan, this paper systematically studies the development process of Japan's Agricultural Disaster Compensation System. This section mainly demonstrates the characteristics of agricultural risks and why agricultural risk management is necessary. The second section is about the agriculture and agricultural risks in the study area and reveals the urgency of agricultural risk management in Japan. The third section analyzes the development process of Japan's Agricultural Disaster Compensation System from three aspects: agricultural mutual relief, organizational operations and government support policies. The fourth section mainly uses data to illustrate the implementation of Japan's agricultural mutual relief. The fifth section introduces income insurance, which is currently a hot issue in Japan. The final section is the adaptation of Japan's agricultural risk management system to China.

2 AGRICULTURE AND AGRICULTURAL RISKS IN JAPAN

The proportion of arable land in Japan is relatively small. Only 30% of all land is suitable for agricultural production. Unfortunately, the cultivated land area has been declining in recent years. The main grain crop in Japan is rice, which includes paddy field rice and upland rice. Figure 1 shows how the planted area of rice in Japan has declined almost yearly from 1.7 million hectares in 2001 to 1.5 million hectares in 2016. The growth rate for most years has been negative. Japan's food self-sufficiency rate has been declining for many years, and has fallen below 40% in recent years. It has had to rely on imports to solve the problem of food shortage (He et al., 2014). In addition to the decline in planted area and self-sufficiency rate, the agricultural population in Japan is also declining. In 2016, the number of agricultural workers in Japan fell below 2 million for the first time, which is 60% less than in 1990 (Jia, 2018). The average age of agricultural workers is 66.6 years old and the population is aging, so Japan is introducing diversified policies to encourage young people to participate in agricultural work.

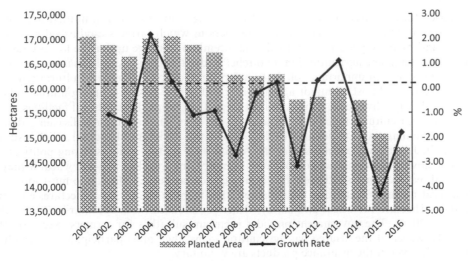

Figure 1. The planted area and growth rate of rice in Japan from 2001 to 2016.

Source: Ministry of Agriculture, Forestry and Fisheries of Japan (MAFF)

Table 1. Economic losses caused by major natural disasters in Japan from 2008 to 2014 (Million Dollars).

| Year | Disasters | | | | | Total |
	Typhoon	Rainstorm	Heavy Snow	Earthquake	Others	
2008	21	97		311	258	688
2009	337	201			102	640
2010	20	402	87		132	641
2011				11,456		11,456
2012	216	813	136		122	1,287
2013	389	519	1,452		297	2,657
2014	104	457	0	21		582

Note: The values in the table are the total losses of crops, agricultural land and agricultural facilities.
Source: MAFF, Futoshi Okada 《Sustainable Growth in Crop Natural Disaster Insurance: Experiences of Japan》

Japan is located in the Asian monsoon region and volcanic belt around the Pacific Rim. Its status as a cluster of islands and its exposure to the monsoon climate makes it vulnerable to frequent natural disasters such as earthquakes, typhoons, volcanos, tsunamis, floods and low temperatures. Table 1 shows the economic losses caused by natural disasters from 2008–2014 ranging from 582 million dollars to 11.46 billion dollars. The Great East Japan Earthquake in 2011 caused unprecedented economic losses of up to 11.46 billion dollars, of which 93.3% were damage to agricultural land and facilities, and 1.6% were crop losses (Futoshi, 2016). According to MAFF statistics, the proportion of individual farmers in all agricultural business entities in 2017 was 95% (Mu & Zhao, 2019). This demonstrates that Japanese agriculture is still largely comprised of small-scale operations under individual farmers. Natural disasters with high frequency and large losses lead to uncertainties for individual farmers' income in Japan. It is urgent for farmers to improve their capacity for agricultural risks management with the help of the government.

3 AGRICULTURAL DISASTER COMPENSATION SYSTEM

3.1 *Mutual relief insurance products*

The Agricultural Disaster Compensation System began in 1947. It is a mutual relief insurance based on mutual assistance between farmers in which farmers established a common reserve fund by paying a certain mutual relief premium to make up for the losses caused by natural disasters. The agricultural mutual relief under the Agricultural Disaster Compensation System plays an important role in transferring and diversifying agricultural risks. The insurance covers almost all major agricultural products in Japan including mutual relief of agricultural crops, livestock, fruits and fruit trees, field crops, horticultural facilities, greenhouses and agricultural machinery. It also covers risks such as natural disasters, diseases and insect damage and accidents. Table 2 shows the agricultural mutual relief insurance products in Japan. A characteristic of the Agricultural mutual relief is strong political support. The government imposes compulsory insurance on crops like rice and wheat if they are planted above a certain sized area. For example, rice farmers in Hokkaido must join in the agricultural mutual relief when their planted area is more than 0.3 ~ 1 hectare, and more than 0.4 ~ 1hectare for wheat and barley. Other products can join voluntarily (Long, 2006; Fumihiro, 2014). The government provides premium subsidies for farmers except on their houses. However, if the insured has agricultural loans, they must participate in agricultural mutual relief even if the insurance products are voluntary.

Table 2. The agricultural mutual relief insurance products in Japan.

	Products	Insured risks	Underwriting	Premium Subsidy
Agricultural crops	Rice: paddy field rice and upland rice Wheat and barley	Losses caused by disasters, diseases and insect damage	Compulsory when exceeds a certain area	Yes
Livestock	Cattle: dairy cattle and beef cattle Horses Pigs: breeding pigs and fattening pigs	Dead and disused, diseased and injured	Optional	Yes
Fruits and fruit trees	Fruit-Crop and fruit tree: designated citrus, apple, grape, etc.	Yield and quality losses caused by disasters and diseases and insect damage for fruit-crop, death for fruit tree	Optional	Yes
Field crops	Potatoes, soybeans, sugarcane, sugar beet, tea, sweet corns, sericulture, etc.	Losses caused by disasters, diseases and insect damage, fire	Optional	Yes
Horticultural facilities	Specific facilities and subsidiary facilities, agricultural products in the insured facilities	Losses caused by disasters	Optional	Yes
Agricultural property	Farmer's house and agricultural machinery, etc.	Property losses caused by disasters and fire for farmer's house, live stock housing and agricultural machinery	Optional	No

Source:MAFF

3.2 *Organizational operation*

The Agricultural Disaster Compensation System is a risk sharing mechanism with inter-dependent farmers, which is to divert the uncertainty risk among different entities. The Agricultural Disaster Compensation System has a long history and has gradually formed a unique three-level organizational structure of "MAFF-Prefectural Federation of AMRAs-Agricultural Mutual Relief Associations (AMRAs)" and a risk protection system of "mutual relief – insurance – reinsurance", which provides a comprehensive guarantee (Deng & Zheng, 2015). The Agricultural Mutual Relief System is shown in Figure 2: (1) The first level is AMRAs, which provide services for insured farmers. Farmers pay AMRAs mutual relief premiums and become members. When crops are damaged, AMRAs pay farmers for mutual relief indemnity. This is a mutual relationship between farmers and AMRAs; (2) The middle level is prefectural federation of AMRAs, which is formed by the combination of AMRAs and plays a transitional role in the whole system. AMRAs pay the prefectural federation of AMRAs insurance premiums. When disasters or damages occur, AMRAs can receive insurance indemnity from the prefectural federation of AMRAs. This is an insurance relationship between AMRAs and the prefectural federation of AMRAs; (3) The third level is the MAFF agricultural mutual aid reinsurance special account. The prefectural federation of AMRAs pays MAFF reinsurance premiums, and the government sets up a special reinsurance account to provide reinsurance for the prefectural federation of AMRAs. The government is in a reinsurance relationship with the prefectural federation of AMRAs. The government provides mutual insurance premium subsidies for farmers and provides an administration and operating subsidy (A&O subsidy) for AMRAs and prefectural federations of AMRAs. In recent years, due to the aging population, it has been merging AMRAs with prefectural federations of AMRAs which has resulted in a reduction in the number of AMRAs. For example, from 2013 to 2017, the number of prefectural federations of AMRAs decreased from 38 to 17

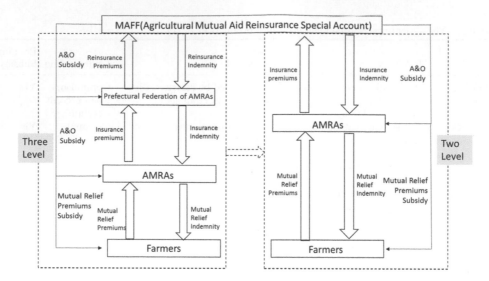

Figure 2. The agricultural mutual relief system in Japan.

Source: MAFF, Yueying Mu and Peiru Zhao, Japan's Agricultural Mutual Relief System and Implementation of Agricultural Income Insurance.

and the number of AMRAs decreased from 241 to 141. As a result, the two-level structure has emerged as the right side of Figure 2. At present, both three-level and two-level structures exist at the same time in Japan. Under the two-level system, AMRAs directly get reinsurance from the government, and directly receive compensation after disasters, which omits inter-mediate links and is conducive to an orderly and efficient operation (Mu &Zhao, 2019).

3.3 Government support policies

In order to improve the food self-sufficiency rate, the Japan government has implemented pro-tection policies for agriculture and greatly supports the Agricultural Disaster Compensation System. This paper mainly introduces the government support policies from four aspects: legislation, financial subsidy, reinsurance and risk management.

(1) Legislation. With the development of agricultural industrialization, the government of Japan has gradually recognized the importance of agricultural risk dispersion and post-disaster management, and is continuously exploring and improving agricultural risk manage-ment through legislation. The development of the Agricultural Disaster Compensation System legislation is shown in Figure 3. The Agricultural Disaster Compensation Law in 1947 marks the formal establishment of the Agricultural Disaster Compensation System, which protects farmers' crops and livestock losses caused by extreme weather events, pests and diseases (Fumihiro, 2014). In order to adapt to the development of agricultural mutual relief, Japan revised the laws on premium rates, subsidies, and participation methods in 1957, 1985, and 2003, and expanded coverage and insurance products in 1949, 1951, 1971, and 1979. Through a continuous revision of legislations, Japan provides a strong institutional guarantee for the sustainable development of agricultural mutual relief insurance. In 2014, the government first proposed agricultural income insurance, and added it to the law in 2017. At the same time, the Agricultural Disaster Compensation Law was changed to the Agricultural Insurance Law (Jiang & Fei, 2018; Mu, 2018).

(2) Financial Subsidy. Japan has frequent natural disasters and there are high agricultural losses. Therefore, the agricultural mutual relief insurance premium rate is about ten or even hundreds times higher than other property and accidental injury insurance. If farmers were to participate as individuals, the insurance participation rate would be greatly reduced (Fumi-hiro, 2014). Hence the Japan government provides premium subsidies for farmers. Japan's

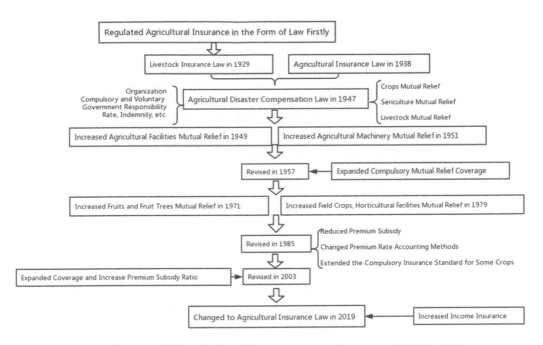

Figure 3. The development of the agricultural disaster compensation system legislation in Japan.

Table 3. Japan's agricultural mutual relief premiums and subsidies in 2016.

Item		Mutual Relief Premiums (thousand yen)			Subsidy Ratio (%)
		Total	Government Share	Farmers Share	
Agricultural crops	Paddy field rice	8,744,239	4,371,777	4,372,463	50.00
	Upland rice	1,829	915	915	50.03
	Wheat and barley	11,159,961	5,951,510	5,208,451	53.33
	Total	**19,906,029**	**10,324,202**	**9,581,829**	**51.86**
Livestock	Dairy cattle	37,578,618	18,345,147	19,233,471	48.82
	Beef cattle	20,387,368	9,297,273	11,090,095	45.60
	Horses	682,681	283,943	398,738	41.59
	Breeding pigs	315,160	121,584	193,576	38.58
	Fattening pigs	1,670,387	668,070	1,002,317	39.99
	Total	**60,634,214**	**28,716,017**	**31,918,197**	**47.36**
Fruits and fruit trees	Fruit-crop	4,262,335	2,131,156	2,131,179	50.00
	Fruit tree	96,645	48,322	48,322	50.00
	Total	**4,358,980**	**2,179,478**	**2,179,501**	**50.00**
Field crops		10,859,387	5,972,503	4,886,884	55.00
Horticultural facilities		6,287,617	3,082,089	3,205,528	49.02

Source: MAFF

agricultural mutual relief premiums and subsidies in 2016 is shown in Table 3. The subsidy ratio is calculated as follows:

$$Subsidy\ Ratio(\%) = \frac{Government\ Shared\ Premium}{Total\ Premium} \times 100 \qquad (1)$$

In general, farmers only need to pay about 50% of the total premium, and the rest is borne by the government. The mutual relief premiums for livestock are highest, followed by agricultural crops and field crops. The subsidy ratio of agricultural crops and field crops is more than 50%, and the premium paid by the government is higher than the farmers' share. However, for livestock and horticultural facilities, the subsidy ratio is less than 50% and the burden on farmers is slightly higher. The subsidy ratio for fruits and fruit trees is 50%, and the premium is shared by farmers and the government equally. The government also provides an A&O subsidy for operating organizations like AMRAs and prefectural federations of AMRAs (as shown in Figure 2), which includes operating expenses, claims adjusting, etc., which can ensure the normal operation of these organizations.

(3) **Reinsurance.** Agricultural risks have the characteristics of large risk, high frequency, tremendous loss and regional effect. This means that insurers are very likely to face huge compensation claims. Once a catastrophe occurs, they will face severe business risks. As shown in Figure 2, the government provides a multi-level reinsurance protection for the mutual relief organizations. In the three-level organizational structure, the prefectural federation of AMRAs reinsures for AMRAS, and the government provides reinsurance for the prefectural federation of AMRAs. Under the two-level structure, the government directly provides reinsurance for AMRAs. If there are large disasters, such as large-scale infectious diseases and natural disasters, the government will undertake the whole compensation (Sun & Wang et al., 2007). The reinsurance policy provides a good bottom-up guarantee for post-disaster recovery and stable farmers' lives (Jiang & Fei, 2018).

(4) **Strengthen Loss Control.** Farmers are faced with various risks in addition to natural disasters in agricultural production activities. In order to resolve and reduce these risks, AMRAs have carried out some measures such as crop growth prediction, pest control, soil diagnosis, and animal epidemic prevention. MAFF is responsible for the guidance and supervision of all insurance business, and provides technical support such as statistical data, premium rate determination, and insurance products research (Sun& Wang et al., 2007).

4 IMPLEMENTATION OF AGRICULTURAL MUTUAL RELIEF

Referring to Table 4, we analyze the implementation of Japan's agricultural mutual relief from 2012 to 2016.The insured rate is as follows:

$$Insured\ Rate(\%) = \frac{Insured\ acreage\ or\ heads}{Planted\ area\ or\ heads} \times 100 \qquad (2)$$

In terms of the insured rate, there has been a slow increase in the coverage of most crops and livestock from 2012–2016. The compulsory insurance requirement for rice, wheat and barley

Table 4. Amount of mutual relief, insured rate of staple crops and livestock in 2012–2016.

		2012	2013	2014	2015	2016
Rice	Amount of mutual relief (billion yen)	1085.4	1097.2	1080.4	1015.7	957.3
	Insured rate (%)	94	94	95	97	98
Wheat and barley	Amount of mutual relief (billion yen)	139.8	118.5	114.1	110.1	124.8
	Insured rate (%)	95	95	95	95	98
Soybeans	Amount of mutual relief (billion yen)	42.3	42.9	45.7	47.6	54.0
	Insured rate (%)	76	77	78	79	80
Beef cattle	Amount of mutual relief (billion yen)	340.3	325.2	347.3	367.4	411.8
	Insured rate (%)	85	85	86	88	89
Fattening pigs	Amount of mutual relief (billion yen)	14.5	14.8	14.6	20.3	19.9
	Insured rate (%)	21	22	21	-	25

Note: Rice includes paddy field rice and upland rice.
Source: MAFF

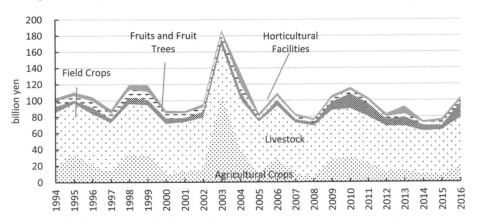

Figure 4.   Mutual relief indemnity for farmers in 1994–2016.

Note: Mutual relief indemnity of agricultural crops is sum of paddy field rice, upland rice, wheat, and barley. The mutual relief indemnity of fruits is the sum of fruit-crops and fruit trees. Source: MAFF.

has led to a more than 90% insured rate. Field crops like soybeans have been developing rapidly and reached an 80% insured rate in 2016. The insured rate of different livestock varies greatly with beef cattle being very high at close to 90%but fattening pigs is only at about 20%. In terms of risk protection, the amount of mutual relief for rice, wheat and barley slightly decreased, but soybeans and livestock have increased significantly. As shown in Figure 4, the compensation varies annually with natural disasters. In 2003, it was affected by strong typhoons and cold air, so the compensation was the highest. From a historical perspective, livestock has received the largest compensation because it faces the highest risk, accounting for more than half of the whole payment. The next highest is agricultural crops, which accounts for about 20%. The indemnity of field crops, fruits and fruit trees, horticultural facilities is relatively small. In 2016, the indemnity of livestock reached 58.23 billion yen, crops 20.54 billion yen, field crops 18.32 billion yen, fruits 3.51 billion yen and horticultural facilities 3.19 billion yen.

5   INCOME INSURANCE SYSTEM

The Agricultural Disaster Compensation System only compensates losses caused by natural disasters and does not cover risks caused by price fluctuations. Research shows that the main reason for the low participation in fattening pigs mutual relief is that its price fluctuates greatly. In Japan, the number of people engaged in agriculture is declining because of the aging population. The number of large-scale business entities is increasing, but the agricultural mutual relief covers only single products, which makes it difficult to meet the diverse demands of farmers. Therefore, Japan has introduced agricultural income insurance in order to protect farmers' low-income risks. This system can not only compensate for losses caused by natural events, but also compensates for losses caused by fluctuations in prices. It is conducive to improving the enthusiasm of agricultural operators and establishing a more comprehensive agricultural insurance system. It was introduced in small-scale pilot schemes and added to the legislation in 2017. By 2019, it was officially implemented.

Farmers can voluntarily participate in income insurance or the current agricultural mutual relief, but they are not allowed to participate in multiple insurance plans at the same time (Keiko, 2018). Only individuals and organizations that have "Blue Return" in the past five years and can properly manage agricultural production are eligible to participate in income insurance. "Blue Return" means that agricultural producers keep their accounts in accordance with double-entry bookkeeping and offer income and income tax details to the authorities

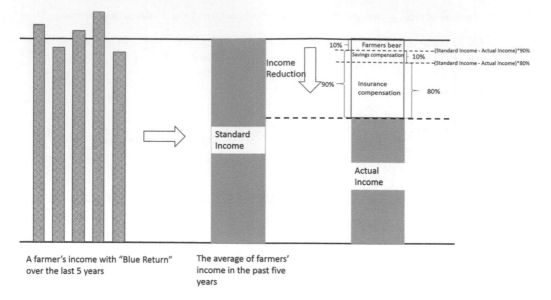

A farmer's income with "Blue Return" over the last 5 years

The average of farmers' income in the past five years

Figure 5.   Income insurance compensation in Japan.

Source: Keiko Fujibayashi, MAFF 2018 Budget Proposal Shifts Spending to Income Insurance; Yueying Mu and Peiru Zhao, Japan's Agricultural Mutual Relief System and Implementation of Agricultural Income Insurance.

(Zhang & Xia, 2013). The "Blue Return" can help accurately measure income for income insurance. As shown in Figure 5, the standard income is the average of farmers' incomes in the past five years. If the actual income during the insurance period is less than 90% of the standard income, farmers can receive indemnity. That is to say, 10% of income reduction will be borne by the farmers themselves. The other 90% will be compensated by savings and insurance compensation. The savings compensation can retain this compensation for the next premium that farmers need to pay. The insurance compensation is up to 80% of the income reduction, and the remaining 10% can be saved. Different protection levels have different insurance rates. The government provides subsidies for 50% of insurance compensation and 75% of savings compensation. To ensure that farmers can receive indemnity even in extreme cases, the government provides reinsurance for income insurance (Fumihiro, 2017).

6   LESSONS FOR CHINA

Agricultural insurance is indispensable in China's agricultural risk management system. Since the China government started to provide premium subsidies in 2007, China's agricultural insurance has developed rapidly and is now the world's second largest agricultural insurance market after the United States. Although China's agricultural insurance has clearly developed, it is undeniable that there are still many problems in the agricultural insurance system. The development of agricultural risk management and agricultural mutual relief insurance in Japan can be used to guide their development in China:

(1) **Compulsory and voluntary insurance.** In order to guarantee grain food supply, Japan enforces compulsory insurance for rice and wheat and voluntary insurance for livestock, fruits and field crops. China should learn from Japan's experience and carry out compulsory crop insurancedue to changes in the population and cropland. In addition, China can institute price insurance and income insurance for aquaculture and featured agricultural products in order to meet the diversified demands of new agricultural business entities. They can participate in the insurance schemes voluntarily.

(2) **Establish rural agricultural insurance organizations.** As direct management organizations in rural areas, AMRAs manage farmers' agricultural insurance, collect premiums, promote insurance products and pay indemnity. They also communicate with the higher authorities to get reinsurance and improve protection for farmers. China's lack of rural agricultural insurance organizations is a big shortcoming. At present, only agricultural insurance companies operate in rural areas, and there are no nonprofit organizations active. In order to solve the problems of scattered and small-scale production, China can learn from Japan's agricultural mutual relief insurance system and establish cooperatives like AMRAs in the countryside They can be directly responsible for agricultural insurance operations and build a bridge between the government and farmers.

(3) **Revise and improve legislation constantly.** The worldwide experience of agricultural insurance shows that a sound legal system guarantees the development of agricultural insurance. In order to meet the needs of agricultural development, the Japan government has revised the Agricultural Disaster Compensation Law constantly to cover the needs for operating organizations, insurance products, premium rate, etc. A specialized legislation of China's agricultural insurance i.e. the Agricultural Insurance Regulations was implemented in 2013. Better legal regulations are yet to be enacted and refined with the development of agricultural insurance.

(4) **Diversify government support policies.** The Japan government's support policies for agricultural insurance mainly includes premium subsidies, reinsurance, A&O subsidy, and loss control. China's current support policies are relatively simpler. Although the premium subsidy ratio has reached 80%, the layered A&O subsidy and reinsurance system have yet to be established.

ACKNOWLEDGEMENTS

This study was funded by grants provided by the Natural Science Foundation of China (41471426). The MOE Project "The Assessment of Agricultural Production Risk" is based on a mixed data(Grant No. 17JJD910002) from the Key Research Institute of Humanities and Social Sciences at Universities.

REFERENCES

Fumihiro, K. 2014. Agricultural Disaster Compensation System in Japan. Available at: http://ap.fftc. agnet.org/ap_db.php?id=249 (accessed April 18, 2014).

Fumihiro, K. 2017. Introduction to the Income Insurance System of Japan. Available at:http://ap.fftc. agnet.org/ap_db.php?id=827 (accessed December 18, 2017).

Futoshi, O. 2016. Sustainable Growth in Crop Natural Disaster Insurance: Experience of Japan, FFTC-RDA International Seminar on Implementing and Improving Crop Natural Disaster Insurance Program. Jeonju, Korea.

Jia, Y.Y.2018. The Japanese Experience of Small-scale Farmers' Agricultural Insurance and Its Enlightenment to China. *World Agriculture* 11: 190–195.

Jiang, SH.ZH and Fei, Q. 2018. An Analysis of Japanese Mutual-aid Agricultural Insurance System. *Contemporary Economy of Japan* 3704: 23–34.

Keiko, F. 2018. MAFF 2018 Budget Proposal Shifts Spending to Income Insurance. Available at:https:// www.fas.usda.gov/data/japan-maff-2018-budget-proposal-shifts-spending-income-insurance.

Long, W.J. 2006. The experience of Japan's agricultural insurance. *China Insurance* 09: 42–46.

Mu, Y.Y and Zhao, P.R. 2019. Japan's Agricultural Mutual Relief System and Implementation of Agricultural Income Insurance. *World Agriculture* 03: 4–11.

Mu, Y.Y. 2018. Japan's agricultural mutual relief system and income insurance. Beijing: Agricultural Information Institute of CAAS.

Sun, W.L, Wang R.B, et al. 2007. Japan's practice of developing policy-based agricultural insurance and its reference to China. *Issues in Agricultural Economy* 11: 104–109.

Zhang, Z.Y., Xia, Z.L. 2013. The Design and Enlightenment of Japanese Blue Return System. *Taxation Research* 4: 3–7.

Risk Analysis Based on Data and Crisis Response Beyond Knowledge – Huang & Nivolianitou (eds)
© 2020 Taylor & Francis Group, London, ISBN 978-0-367-25146-8

Stakeholders' engagement in emergency management—an overview from the Danish offshore wind sector

Dewan Ahsan
Department of Sociology, Environmental and Business Economic, Danish Centre for Risk and Safety Managment, University of Southern Denmark, Esbjerg, Denmark

Soren Pedersen
Faculty Unit of Offshore Wind Farm Operation and Maintenance, Fredericia Maskinmesterskole, Esbjerg, Denmark

ABSTRACT: Offshore wind sector is a promising source of renewable energy. Denmark is the world first country who initiated to produce energy from offshore wind source. As offshore wind industry is a fast-growing sector, emergency related to occupational health and safety and asset damage is one of the challenging issues for this sector. Therefore, this article focuses to provide an overview on emergency management procedure in offshore wind in Denmark and to identify the role of various key stakeholders in emergency management process.

Keywords: emergency management, offshore wind energy, Denmark.

1 INTRODUCTION

In the energy sector, sustainability is becoming a key factor as conventional energy producing sector is considered as one of major challenges for the protection of environment. To overcome this challenge, global electricity production from renewable sources is gradually increasing. In this aspect, offshore wind energy is playing a significant role. In Denmark, presently 12 Offshore Wind Farms (OWF) are operating and four more OWFs are coming in future (Table 1).

Though there has been a huge technological improvement in wind turbines, the wind energy sector, especially offshore wind industry, is not yet an absolutely safe technology and human beings are needed to operate and maintain the offshore wind farms. The offshore technicians are exposed to not only extreme weather conditions, but they also need to work under confined work conditions. So, despite maintaining high levels of safety protocol, offshore workers are still exposed to several risks (Garcia and Bruschi, 2014).

Managing and cooperating with stakeholders to achieve the company's strategic goal (for its long-term sustainability) is becoming an important issue in any business. Like any other business, emergency management in offshore wind industry is involved with many stakeholders. So, stakeholder management is directly relevant to the organization's sustainability (Freeman, 1984; Ahsan and Pedersen, 2018). Then the obvious question is, who are the key stakeholders for an offshore wind industry to work with (manage) regarding emergencies?

Therefore, the objectives of this empirical research are to i) provide an insight in the overall emergency management system in the Danish OWF and ii) to identify the key stakeholders involved in the emergency management plan for the Danish OWF.

Table 1. Present and future Offshore Wind Farms in Denmark.

Offshore wind farm	Year in operation	Turbines (Numbers)	Total capacity (MW)
Vindeby(Decommissioned)	1991	11	5
Tunø Knob	1995	10	5
Middelgrunden	2000	20	40
Horns Rev I	2002	80	160
Rønland	2003	8	17
Nysted	2003	72	166
Samsø	2003	10	23
Frederikshavn	2003	3	8
Horns Rev II	2010	91	209
Avedøre Holme	2009/10	3	11
Sprogø	2009	7	21
Rødsand II	2010	90	207
Anholt	2013	111	400
Horns Rev III	2020	49	400
Vesterhav Syd	Yet to decide	20	170
Vesterhav Nord	Yet to decide	21	180
Kriegers Flak	Yet to decide	72	600

2 THEORIES AND METHODOLOGIES

2.1 Emergency management cycle

Emergency Management (EM) consists of four phases; Mitigation, Preparedness, Response and Recovery (Haddow et al, 2013). Briefly, the objective of the preparedness phase is to ensure that response and rescue plans are efficiently capable to work in case an emergency happen. While in response phase, rapid actions are taken to control the emergency. On the other hand, the aim of the recovery phase is to bring the post emergency situation back to normal state as before. Whereas, in mitigation phase necessary steps are taken to ensure an effective preparedness to reduce the human and economic losses.

2.2 Theory of stakeholder analysis

The stakeholder theories are generally divided into two categories i.e. normative and instrumental. Principally, normative theories focus on very broad aspects and are based on the notions of norms, values and ethics (Friedman and Miles, 2006). Whereas, the instrumental theory is seen as the connection between how an organization manages its stakeholders, and how they achieve the organizational goals (Clarkson, 1995; Friedman and Miles, 2002). Instrumental theory helps an organization's management to identify the core stakeholders, on whom a firm largely depends to run its business, and who have more influence to achieve the firm's corporate goal. This research followed the concept of instrumental stakeholder theory to identify the key stakeholders who are involved in OWF emergency management.

2.3 Data collection procedure

The primary data was collected through interviewing various stakeholders like government agencies, OWF operators, contracting & service companies and offshore technicians working in the OWF industry. For this study 18 interviews have been performed using the "snowball" sampling technique. The secondary data were collected from the literary survey, information from governmental and company websites and company brochures.

3 RESULTS AND DISCUSSION

3.1 *Emergency management procedure in Offshore Wind Farms*

All the Danish OWFs must have their own Emergency Response Plan (ERP) to handle the different emergencies which can arise in the OWF. An ERP must comply with the specific legal requirements set by the four regulatory agencies: i) Danish Workforce Agency (DWA), ii) Danish Transport, Construction and Housing Agency (DTCHA), iii) Danish Environmental Protection Agency (DEPA) and iv) Danish Maritime Agency (DMA). The DWA is the agency concerned regarding the workplace safety on the offshore wind turbine, but DWA has no specific regulation for OWF ERP. Therefor all the OWF operators follow their own Emergency Management Plans (ERP).

Still today, there is no standardized emergency response plan (ERP) for OWFs in Denmark. All the operators have made their own specific ERP for their OWFs. This research reveals that to handle a similar emergency, different operators use different ERPs, which creates complexity for the offshore technicians. It would be easier and beneficial for the offshore technicians, if they could have the opportunity to work with a standardized ERP, which could be applicable for all OWF operators.

Generally, the operator emergency management plan is divided into three lines; $1^{st} - 2^{nd} - 3^{rd}$. The 1st line is the OWF ERP which must be operational without external assistance. The offshore technician teams of 2 to 3 technicians comprise the 1st line offshore rescue setup, and they will handle the incident on the wind turbine until additional help is required, which is illustrated in Figure 1.

Once the need for help increases the OWF ERP is activated (1st line) with the emergency call to the emergency coordination center which can either be the CTV available offshore or a 24 hours' surveillance office. The emergency coordination takes over after the emergency call and informs the entire OWF (e.g. offshore technicians, CTVs, etc.). When needed emergency coordination can contact the Joint Rescue Coordination Center (JRCC) and/or Radio Medical for additional help. If there is a serious injury (e.g. life threatening) of a technician or serious damage to the asset (e.g. fire, blow out etc.) and there is a need for additional support to handle the emergency, then the 1st line activates the company 2nd and 3rd lines depending on the seriousness of injury or asset damage and to inform the press and/or the public.

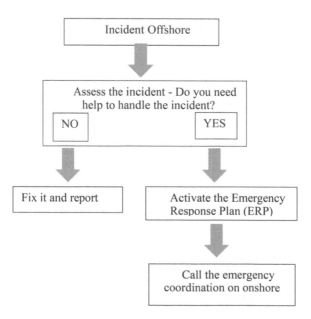

Figure 1.   The tasks needed to be performed by the offshore technicians in an emergency.

This article has identified nine key stakeholder groups; Operator, Operator's own technicians; Turbine supplier, Governmental agencies, Subcontractor's manpower, Grid owner, Vessel supplier, Industry association and Helicopter rescue service who are directly involved in the OWF emergency management. The specific roles of these key stakeholders are presented in Table 2.

It is identified that all the regulatory agencies are involved only in mitigation phase as they have the responsibility to amend existing legislations and/produce, propose new regulations to mitigate/avoid the emergency situations based on the changing environment, new demands and enhancement of knowledge. While JRCC, CTV and Service Offshore Vessel are intensively involved with emergency preparedness and response phases. On the other hand, as the key investors, operators and grid owner are primarily responsible to handle the emergencies in OWF. Therefore, they need to take part in all four emergency management cycles; mitigation, preparedness, response and recovery. One interesting finding is that all governmental agencies are reactive when emergencies happen in the OWF. The Operator's Emergency Response Plan (ERP) requires contact to the JRCC, and the JRCC has actually no obligation to inform the DWA as the

Table 2. Key stakeholders' involvement in offshore wind farms' emergency management (EM).

Organization/Company	Role in EM	How do they work	Involvement in EM cycle
Danish Transport, Construction and Housing Authority (DTCHA)	Legislation for helicopter CTV	Approve rescue setup for helicopter rescue provider	Mitigation
Danish Workforce Agency (DWA)	Legislation for workplace safety on offshore wind turbines	Perform offshore accident investigation if chosen	Mitigation
Danish Maritime Agency (DMA)	Legislation for All Maritime Vessels (SOV, CTV)	Approve vessels for offshore wind farms	Mitigation
Danish Environmental Protection Agency (DEPA)	Legislation for the environment of offshore wind farms	Perform investigation of environmental incidents in offshore wind farms if chosen	Mitigation
Danish Energy Agency (DEA)	Approval of offshore wind farms	Requirement of 1st line emergency response plans for construction and O&M	Mitigation
Operator own emergency management system	1st line emergency management	Offshore Wind Farm Emergency Management Unit	Mitigation Preparedness Response Recovery
Joint Rescue Coordination Centre (JRCC)	Public rescue service for offshore wind farms	Offshore Wind Farm Rescue Helicopter Service	Preparedness Response
Crew Transfer Vessel (CTV) Service Offshore Vessel (SOV)	Helicopter, Crew boat, SOV	Offshore Wind Farm Rescue transport service	Preparedness Response
Grid Owner	1st line emergency management	Offshore Substation Emergency Management Unit	Mitigation Preparedness Response Recovery

onshore 112 (public emergency assistance) must inform the DWA if a workforce accident occurs in OWF, so the DWA can take necessary action.

4 CONCLUSION

Nine key stakeholders have been identified by this study who are actively involvement in managing emergency in the Danish OWF. However, this research noticed that one of the big challenges in the offshore wind energy industry is the lack of cooperation and knowledge sharing among the key stakeholders. However, for the enhancement of the OWF's emergency management plan, all stakeholders need to work in an integrated way, especially strong cooperation and strategic alliances between operator-operator and operators-grid owners are vital to achieve the goal zero accident.

REFERENCES

Ahsan, D., Pedersen. S., 2018. The influence of stakeholder groups in operation and maintenance services of offshore wind farms: Lesson from Denmark. Renewable Energy 125: 819–828.

Clarkson, M.B.E., 1995. A Stakeholder Framework for Analyzing and Evaluating Corporate Social Performance. Academy of Management Journal 20: 92–118.

Freeman, R.E., 1984. Strategic management: A stakeholder approach. Boston: Pitman.

Friedman, A.L., Miles, S., 2006. Stakeholders: Theory and Practice. Oxford University Press.

Friedman, A.L., Miles, S., 2002. Developing Stakeholder Theory. Journal of Management Studies 39: 1–21.

Garcia, A.D., Bruschi, D., 2014. A risk assessment tool for improving safety standards and emergency management in Italian onshore wind farms. Sustainable Energy Technologies and Assessments 18: 48–58.

Haddow, G.D., Bullock, J.A., Coppola, P.D., 2013. Emergency Management, Butterworth-Heinemann. pp. 61–66. In: Hayes, J., 2014. The Theory and Practice of change manageme.

Risk Analysis Based on Data and Crisis Response Beyond Knowledge – Huang & Nivolianitou (eds)
© 2020 Taylor & Francis Group, London, ISBN 978-0-367-25146-8

An empirical study on risk perception of hazardous chemicals

Tiezhong Liu, Jinyang Dong & Yan Chen
School of Management and Economics, Beijing Institute of Technology, Beijing, China

Hubo Zhang
China Electronics Standardization Institute, Beijing, China

ABSTRACT: This study investigated the factors that affect risk perception and protective action of hazardous chemicals and the path of influence in China. The research collected 1,700 valid questionnaires from six cities in China. Through literature analysis and exploratory factor analysis, the four parts that affect risk perception and protective action are identified. The structural equation modeling method was used to determine the logical relationship between the various elements.

Keywords: risk communication, risk perception, hazardous chemicals, protective action

1 INTRODUCTION

It has become a huge challenge to raise public awareness of hazardous chemicals in China. For individual behavior, which could primarily be driven by perception and not by facts, risk perception is generally viewed as a vital factor to predict individual behavioral under risk situations (Kettle and Dow, 2016). Correct risk perception can reduce individuals' blind fear of risk, increase the trust and cooperation between individuals, and then encourage them to make correct risk responses (Liu et al., 2014). There is much literature related to risk perception of natural disasters, but risk perception of chemical hazards is quite different, as it is a type of technical disaster. Generally, individuals would consider the harm caused by the technical disasters of hazardous chemicals as unacceptable (Zhu, 2016). However, although most research focuses on risk perception of natural disasters such as floods and landslides, a small number of studies focus on risk perception of hazardous chemicals (Birkholz et al., 2014; Bodoque, 2016).

This article discusses the factors that affect the public's risk perception of hazardous chemicals, as well as the impact path. Data on risk perception of hazardous chemicals were collected from six different cities in China, and were analyzed with following objectives: (1) to determine the key factors that affect individual risk perception and the corresponding impact path; (2) to establish the relationship between risk perception and protective action of hazardous chemicals; (3) to propose measures and recommendations that will improve individual risk perception of hazardous chemicals.

2 THEORETICAL BACKGROUND

2.1 *Risk perception*

Risk perception belongs in the field of psychology, refers to the individual's perception of various objective risks existing in the outside world, and emphasizes the influence of experience acquired by the individual on intuition and subjective perception (Lee 2015). Slovic (1987) believes that risk perception is an intuitive attitude and the feeling of an individual when faced with a specific risk. Due to psychological factors such as fear, there is often a large gap between subjective risk perception and objective risk, that is, between risk perception and actual risk. At the same time, insufficient understanding of risk will lead to the public perception of risk bias. In a word, the concept of "risk" means different things to different people.

2.2 *Psychological factors affecting risk perception*

There are many paradigms for studying risk perceptions, such as heuristics, geography, and political science, but most studies use the psychometric paradigm approach. There are a series of mental measurement models to explore the relationship between different psychological factors and risk perception. Of course, many scholars have studied risk perception from other psychological aspects, such as trust, responsibility, empathy, controllability and so on. Among them, trust is an important issue in risk perception research. Relevant research shows that risk perception has an important relationship with public psychology. Therefore, we will use trust and responsibility as psychological factors that affect risk perception of hazardous chemicals.

2.3 *Ability factors affecting risk perception*

In addition to psychological factors, risk-related knowledge, skills and other abilities also affect an individual's risk perception. Some scholars have suggested that the difference between the experts' and the public's understanding of the same event is largely due to inaccurate understanding of the information by the public (Roder 2016). At the same time, if the public can master the risk-response skills more effectively, their risk perception will be reduced correspondingly (Miceli 2008). Some scholars call this "controllability." Therefore, we will use knowledge and skills as ability factors that affect the risk perception of hazardous chemicals.

2.4 *Behavior intentions*

Behavior intentions mean ideas and intentions to respond to a risk situation. Promoting the public to take a reasonable risk-response behavior is an important topic of risk research. Many risk perception studies have found that risk perception and behavior intentions have a close relationship, and risk perception can better predict public behavior. Awareness of risk and awareness of self-responsibility are important factors that motivate the public to take protective action. At the same time, protection motivation theory views self-preservation behavior as motivated by four factors: the perceived severity of a threat, the perceived probability of the occurrence, the perceived effectiveness of any recommended response, and the perceived ability to implement a response. Therefore, we still believe that the main factors affecting the public behavior intentions include their risk perception of hazardous chemicals, psychological factors, and ability factors.

Figure 1 shows the basic risk perception model of hazardous chemicals, created by reorganizing the logical relationships among the various elements.

Assumptions for the paths in the basic model are discussed in the following sections.

3 BASIC HYPOTHESIS

3.1 *Psychological factors*

Many studies have found that psychological factors can significantly affect the public's risk perceptions, and the most important factor may be trust. This is divided into the trust in the subject and the trust in the object. The former is trust in the government, individuals, etc., and the latter include the message itself, emergency measures and so on. Liu et al. conducted a sample survey to investigate the public's trust in different sources of food safety information and found that the level of trust directly affects the public perception of food safety.

In addition to trust, responsibility also has a major impact on the public's perception of risk. Using the factor analysis method, Liobikiene and Juknys (2016) analyzed the factors influencing Lithuanian public environmental risk perception and behavior and found that

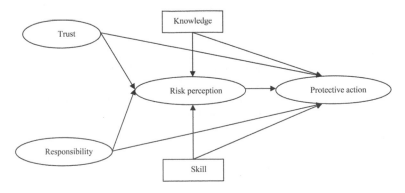

Figure 1. Hypothesized risk perception structural equation model.

those who are more responsible are more sensitive to environmental risks and are more willing to take action. At the same time, there is reason to believe that in China, where the concept of family and friends is stronger, responsibility is an important factor that influences risk perception and protective action. Accordingly, the following hypotheses are proposed:

H1a. The level of trust is positively related to risk perception.

H1b. The level of trust is positively related to protective action.

H2a. A stronger sense of responsibility leads to a stronger risk perception.

H2b. A stronger sense of responsibility leads to more aggressive protective action.

3.2 *Ability factors*

Although psychological factors have a significant impact on risk perception, individual ability factors cannot be ignored. Ability is the objective capability of the public to deal with related risks. It is usually related to education level, related experience, etc. Botzen et al. (2009) quantified hazard knowledge by asking respondents about the causes of a flood. They found that individuals with a little knowledge of the causes of floods have lower perceptions of flood risk. In particular, the amount of knowledge people have about technology is related to their risk perception regarding technology.

In addition to knowledge, the public emergency response skills can affect their risk perception and protective action. At the same time, an individual's disaster preparedness is also closely linked to their skill level. Based on previous studies, the following hypotheses have been formulated:

H3a. People who have more knowledge about hazardous chemicals have a greater risk perception.

H3b. People who have more knowledge about hazardous chemicals are more willing to take protective action.

H4a. People who have more skills with hazardous chemicals have a greater risk perception.

H4b. People who have more skills to handle the risks associated with hazardous chemicals, are more willing to take protective action.

3.3 *Risk perception*

A growing number of studies have examined the factors that drive private flood mitigation behavior, and in particular risk perceptions. Related research has shown that there is a close relationship between risk perception and behavior. People's risk perception of technology is also an important factor that can help policymakers understand why the public boycott technologies that are assessed by experts to have high levels of safety. Therefore, the final hypothesis is as follows:

H5. The higher the level of people's risk perception of hazardous chemicals, the more likely they are to take protective action.

3.4 *Protective action*

People can adapt to floods by taking various adaptive measures, such as raising one's home above the highest flood level, using sand bags or taking out flood insurance. Most studies believe that risk perception can predict the public behavior in the face of risk, but this prediction deviation sometimes appears relatively large. This shows that risk perception and behavior are not in complete correspondence. Among these, a very important reason is that different individuals have different abilities to deal with the risks. Therefore, the public ultimately take hazardous chemicals protective action that depends on the psychology, ability and risk perception, and other factors.

4 RESEARCH METHOD

4.1 *Questionnaire design*

The questionnaire was based on risk perception, hazardous chemicals and other related literature, and we also conducted an expert review. In addition to the demographic section, the initial questionnaire comprised four parts with 24 items: psychological factors, ability factors, risk perception of hazardous chemicals, and protective action of hazardous chemicals. The questionnaire results were measured using a five-point Likert scale. After confirming the initial questionnaire, a small scale trial with 50 samples was conducted to adjust some of the contents to ensure that the contents of the questionnaire can be understood by the public.

4.2 *Construct measurement*

4.2.1 *Psychological factors*

The psychological factors that affect the public's risk perception are complex, such as trust, responsibility, empathy, fear and so on. The public trust is particularly important in technology risk. Many studies examine the important impact of trust on risk perception. Past research (mainly on flood) provides a strong foundation for the hypothesis that greater trust in others will be associated with lower risk perceptions and greater acceptance of the protective action. In the meantime, the causes of many cases of anti-PX projects in China are the public's lack of trust in the government and enterprises. Therefore, we use as a measure of trust the public's assessment of the professionalism of hazardous chemicals handled by governments, enterprises, media and research institutes and the reliability of published information.

The public perception of the dangers of hazardous chemicals is also affected by the sense of responsibility. China is a society that places great emphasis on individual relationships. The most important of these social relations include family relations, kinship and friendships. Responsibility has a significant impact on human perception and behavior (He and Liang 2012). So, we used the sense of responsibility toward family, kin and friends as a measure of responsibility.

4.2.2 *Ability factors*

Knowledge and skills are important factors that affect risk perception. Therefore, we use the amount of public participation in hazardous chemicals seminars as a measure of knowledge and the public participation in hazardous chemicals training as a measure of skills.

4.2.3 *Risk perception*

Risk perception is regularly measured according to the following scales: individuals' expectations about the likelihood of personal, physical and social impacts caused by hazard, people's perceptions of the likelihood of hazard, outrage factors, and institutional trust. In practice, the possibilities and consequences are often used in theory to describe the risk. "Perceived risk," as defined here, refers to the combined measurement of "perceived probability" and "perceived consequences" of a certain event or activity. Therefore, this study used the possibility and consequences of the risk of hazardous chemicals as a measure of risk perception.

4.2.4 *Protective action*

Lindell and Perry (2012) suggest classifying adaptive measures according to the phases of the hazard life cycle: (1) mitigation, (2) preparedness, and (3) recovery. The three phases each occur at a different time, relative to the actual flood event. Corresponding to the protective actions taken over the three cycles above, in the research we used "develop a family emergency plan, evacuate, buy insurance" as a measure of protective action.

4.3 *Data collection and analysis methods*

This research selected a total of six Chinese cities as study samples: Chongqing, Changzhou, Lanzhou, Tianjin, Zibo and Zhangzhou. All of these cities have been impacted by hazardous chemicals in the past five years. The public in these cities are more aware of the dangers of hazardous chemicals than other urban population. Finally, we collected 1,700 valid questionnaires. This study used a structural equation model as the main research method. The structural equation modeling (SEM) of the hypothesized constructs used Analysis of Moment Structure (AMOS), Version 20.

5 RESULTS

5.1 *Reliability analysis and validity analysis*

We analyzed the reliability and validity of the questionnaire. First, using software SPSS20.0 test analysis, we found that the questionnaire *Cronbach's* α coefficient was 0.814, showing the reliability of the questionnaire met the requirements. Second, as the content of the questionnaire has been discussed in many rounds and evaluated by experts in the field, it was considered that the content validity of the questionnaire met the requirements. Therefore, only the KMO test was used to examine the validity of the questionnaire. Using software SPSS20.0, test analysis found that the KMO value of the questionnaire was 0.912, with a significant p-value of 0.00, indicating the validity of the questionnaire met the requirements.

5.2 *Structural equation model (SEM)*

Structural equation modeling (SEM) explores the relationship between exogenous variables and endogenous variables. The SEM model must be fitted with four fit index categories, as shown in detail in Table 1. In this process, nine covariance relations were added for the SEM model to meet all of the criteria. The final output results are shown in Table 2.

6 DISCUSSION AND CONCLUSION

(1) Result of Hypothesis 1
First, H1a is not supported, because there is no positive correlation between trust and risk perception of hazardous chemicals. However, there is a negative correlation between trust and risk perception. The results show that the public's trust in government, enterprises, etc.

Table 1. Final model fitted criteria.

Fit index	Evaluation criteria	Test result
Chi-square test X2/df	<3.00[54]	2.828
Absolute fit RMSEA	<0.10[55]	0.033
Incremental fit CFI	>0.90[56]	0.976
Parsimonious fit PNFI	>0.50[57]	0.737

Table 2. Summary of tests of hypotheses.

Hypothesis	Estimate	S.E.	C.R.	p	Hypothesis test result
H1a. Trust→Risk perception	-0.096	0.037	-2.571	0.010	Direct influence
H1b. Trust→Protective action	0.412	0.042	9.824	0.000	Direct influence
H2a. Responsibility→Risk perception	0.354	0.085	4.172	0.000	Direct influence
H2b. Responsibility→Protective action	0.521	0.080	6.492	0.000	Direct influence
H3a. Knowledge→Risk perception	0.117	0.022	5.265	0.000	Direct influence
H3b. Knowledge→Protective action	-0.024	0.019	-1.245	0.213	Indirect effect
H4a. Skill→Risk perception	0.053	0.016	3.367	0.000	Direct influence
H4b. Skill→Protective action	-0.022	0.015	-1.426	0.154	Indirect effect
H5. Risk perception→Protective action	-0.001	0.037	-0.024	0.981	Indirect effect

will reduce risk perception of hazardous chemicals to a certain extent. We think trust is used as a shortcut to reduce the necessity of making rational judgments based on knowledge, by selecting trustworthy experts whose opinion can be considered as accurate. So, if the public trust them more, the public's perception of hazardous chemicals will be lower.

Second, H1b is supported, as trust has a significant positive impact on the public's opinion of hazardous chemicals protective action. This conclusion supports the view of Eiser et al. that the public's trust in the government will enable them to respond positively to the government's proposal for protective action and will therefore promote them to take protective action.

(2) Result of Hypothesis 2

First, H2a is supported, which shows that the responsibility of protecting family and friends has a significant positive impact on an individual's risk perception of hazardous chemicals.

Second, H2b is supported, which shows that the responsibility of protecting family and friends has a significant positive impact on the public's opinion of protective actions for hazardous chemicals.

(3) Result of Hypothesis 3

First, H3a is supported, which shows that knowledge has a significant positive impact on individual risk perception of hazardous chemicals. It can be deduced that risk perception will be increased if there is more knowledge of hazardous chemicals. This is probably because most of the public knowledge of hazardous chemicals is related to bad effects of chemicals and there is little knowledge on how to reduce the risk by themselves.

Second, H3b is not supported, because there is no positive correlation between knowledge and protective actions. If the public only know about the nature of hazardous chemicals and few know how to reduce the risks effectively, then they have a sense of powerlessness in dealing with the risks of hazardous chemicals and are unlikely to take effective protective action.

(4) Result of Hypothesis 4.

First, H4a is supported, as skills have a significant positive impact on individual risk perception of hazardous chemicals. In China, most of the public skills to deal with the hazards of hazardous chemicals are obtained through the training of government or unit organizations, where the public are shown the significant impact of the hazardous chemicals. Therefore, the public's risk perception of hazardous chemicals will be strengthened as skills improve.

Second, H4b is not supported, because there is no positive correlation between skills and protective actions. This contradicts our psychological feelings. Compared with the display of excessive hazards in the training process, government and other agencies simply teach the corresponding emergency protective measures. As with the reasons why knowledge influences protective action, this may lead the public to master some basic skills, but they may feel personally incapacitated when a real hazardous chemical accident occurs.

(5) Result of Hypothesis 5.

H5 is not supported, because there is no positive correlation between risk perception and protective actions. In this study, we used the probability and consequences to measure risk perception, which are usually used to represent the size of the risk. However, there is a criteria gap between public assessment and the judgment of risk. Therefore, the risks in the expert's opinions do not necessarily represent the public's risks. As a result, risk perception in the expert's sense cannot predict related behaviors. This is consistent with the public performance in a variety of hazardous chemical incidents in China. The government assesses individual risk perception prior to the implementation of a hazardous chemical project, but many times the final assessment is not the same as the actual response from the public. Therefore, we do not think risk perception of hazardous chemicals can predict the public behavior accurately.

This paper analyzes the impact of psychological and ability factors on risk perception and protection of hazardous chemicals, and the specific path of influence between psychological factors, ability factors, and risk perception and protection actions. It can be concludes that psychological factors (responsibility) and ability (knowledge, skills) factors have positive effects on risk perception; psychological factors (trust) have negative effects on risk perception; psychological factors have a positive effect on protective actions; and there is no significant correlation between ability factors and protective actions. There is no significant correlation between "expert" risk perception and protection actions. The above findings show that there is a clear difference between the risk perception of hazardous chemicals and that of natural disasters.

ACKNOWLEDGMENTS

This paper is sponsored by National Social Science Foundation of China (No. 16BGL175) and National Key R & D Projects of China (No. 2017YFF0209604-2).

REFERENCES

Birkholz, S., Muro, M., Jeffrey, P., and Smith, H.M. 2014. Rethinking the relationship between flood risk perception and flood management. *Science of The Total Environment* 478, 12–20.

Bodoque, J.M., Amerigo, M., Garcia, J.A., Cortes, B., Ballesteros-Canovas, J.A., and Olcina, J. 2016. Improvement of resilience of urban areas by integrating social perception in flash-flood risk management. *Journal of Hydrology* 541, 665–676.

Botzen, W.J.W., Aerts, J.C.J.H., and Van den Bergh, J.C.J.M. 2009. Dependence of flood risk perceptions on socioeconomic and objective risk factors. *Water Resources Research* 45(10), 10440.

He, X.L., and Liang, H., 2012. The theoretical discussion and policy recommendations to promote the reform of the family doctor system. *Chinese Journal of Health Policy* 6, 3–8.

Kettle, N.P., and Dow, K. 2016. The role of perceived risk, uncertainty, and trust on coastal climate change adaptation planning. *Environment and Behavior* 48, 579–606.

Lee, T.M., Markowitz, E.M., Howe, P.D., Ko, C.Y., and Leiserowitz, A.A. 2015. Predictors of the public climate change awareness and risk perception around the world. *Nature Climate Change* 5, 1014–1020.

Lindell M.K., and Perry R.W., 2012. The Protective Action Decision Model: Theoretical Modifications and Additional Evidence. *Risk Analysis* 32(4), 616–632.

Liobikiene, G., and Juknys, R., 2016. The role of values, environmental risk perception, awareness of consequences, and willingness to assume responsibility for environmentally-friendly behaviour: the Lithuanian case. *Journal of Cleaner Production* 112, 3413–3422.

Liu, R.D., Pieniak, Z., and Verbeke, W. 2014. Food-related hazards in China: consumers' perceptions of risk and trust in information sources. *Food Control* 46, 291–298.

Miceli, R., Sotgiu, I., and Settanni, M. 2008. Disaster preparedness and perception of flood risk: A study in an alpine valley in Italy. *Journal of Environmental Psychology* 28(2), 164–173.

Roder, G., Ruljigaljig, T., Lin, C.W., and Tarolli, P. 2016. Natural hazards knowledge and risk perception of Wujie indigenous community in Taiwan, *Natural Hazards* 81(1), 641–662.

Slovic P., 1987. Perception of Risk. *Science* 236, 280–285.

Zhu, W.W., Wei, J.C., and Zhao, D.T., 2016. Anti-nuclear behavioral intentions: The role of perceived knowledge, information processing, and risk perception. *Energy Policy* 88, 168–177.

Risk Analysis Based on Data and Crisis Response Beyond Knowledge – Huang & Nivolianitou (eds)
© 2020 Taylor & Francis Group, London, ISBN 978-0-367-25146-8

Research on the evaluation of hotel operation risk in mountain scenic resorts based on AHP and fuzzy comprehensive evaluation

Yurong He

Faculty of hospitality, Huangshan University, Huangshan, China

ABSTRACT: Hotel operation risk assessment is an important part of security management. The hotel is faced with more uncertain risks in the process of operation, because of the special natural geographical conditions and market conditions of mountainous scenic resorts. According to the characteristics of multi-level and multi-factor influence of operational risk, this paper establishes the index system of risk evaluation. Based on the conventional analytic hierarchy process (AHP), the weight of each index factor is determined and the operation risk assessment model of mountain scenic resorts hotel is established. This paper uses this model to evaluate the operational risk of 6 hotels in Huangshan scenic resorts. This method is used to test the survey data of 6 hotels in Huangshan scenic resorts. The results show that the calculation process is scientific and reasonable, and the final evaluation result is scientific.

Keywords: operation risk, AHP, fuzzy comprehensive evaluation, mountain type scenic resorts, hotels

1 INTRODUCTION

Hotel operations risk assessment is an important part of hotel security management. The judgment result of the influence degree of risk factors is directly related to the decision-making of operational risk management of mountain scenic resorts hotels. Because of the special natural geographical conditions and market conditions of the mountain scenic resorts, hotels in scenic resorts face more uncertain risks in the process of operation. The impact of various risk factors needs to be considered comprehensively in the process of risk assessment.

The core of AHP is to quantify the decision maker's experience judgment. It is an evaluation and decision-making method that combines qualitative analysis with quantitative analysis. It adopts multilevel processing of multiple factors to determine the weight of each factor, so as to enhance the accuracy of decision-making basis. It is more practical in the case of complex target structure and lack of statistical data (a; a).The relationship between risk factor assessment indicators is very complex, and the risk also has the characteristics of transmission. It is difficult to accurately estimate various risk factors based on historical data or data, and most of them are subjective estimates of relevant professionals. This kind of subjective estimation has fuzziness and uncertainty. Therefore, it is advisable to combine the analytic hierarchy process (AHP) and fuzzy comprehensive evaluation method to evaluate the operational risk.

2 LITERATURE REVIEW

Research on hotel operating risks, foreign scholars mainly focus on quantitative analysis and study the influencing factors of operating performance, while domestic researches are mainly qualitative. Research is the main aspects of the hotel group consolidation and merger and acquisition risk, financial risk, cultural risk, liability risk, human resources risk, system risk and legal risk, the property type hotel investment risk, the cost risk to the economy hotel investment, brand control risk and speculative risk research and the detailed risk assessment and preventive measures (a)□a proposed that the evaluation methods of hotel operation risk, includes risk degree evaluation method, risk matrix analysis method and fuzzy comprehensive evaluation method. He also proposed the method of probability statistics to predict future risk losses by collecting market and financial risk data in the past hotel operation (a). In China, the systematic research on hotel operation risk analysis and risk prevention is relatively deficient.

Based on the above analysis, the study of hotel operation risk in foreign countries is earlier and deeper. Domestic research is more superficial. Although the relevant risk assessment indicators have been established, there are few studies on the ranking degree of each risk assessment index. Hotel operators cannot take targeted management measures according to the degree of impact of risk factors on the operation.

3 ESTABLISHMENT OF EVALUTION MODE BASED ON AHP AND FUZZY COMPREHENSIVE EVALUTION

3.1 Establishing operational risk evaluation index system

Risk index system is the premise and foundation of risk assessment. On the basis of combing the hotel operational risk indicators in the existing research and discussing with professional managers and hotel management experts, the antecedent and consequence relationships among the evaluation factors in the comprehensive evaluation system were determined, and the author constructed the operational risk evaluation index system of hotels in mountain scenic resorts. The system includes 7 first-level indicators, Namely, "natural disaster risk", "market risk", "financial risk", "human resource risk", "policy and legal risk", "brand reputation risk", "real asset risk" and 25 secondary indicators, as shown in Table 3.

3.2 Establishing the hierarchical structure model of evaluation indicators

According to the evaluation criteria of "level 1" and "level 2" of operational risk of hotels in mountain scenic resorts, the hierarchical structure model is concluded by summarizing the elements of operational risk of mountain scenic area hotels. The first layer is the target layer A, namely the overall risk level of mountain scenic area hotels. The second layer is the criterion layer B, which is composed of seven level 1 evaluation indicators, expressed as $\{U_1, U_2, U_3, \ldots, U_m\}$. Let the factors at the first level U_i contain n sub factors. The factor set at the second level is expressed as $U_i = \{U_{i1}, U_{i2}, U_{i3}, \ldots, U_{in}\}$, where U_{ij} represents the sub-factors of the second level of j class under the risk factors of the first level of i class. The number of levels 2 under each level 1 risk factor is not equal. So different i correspond to different n. The corresponding values of the seven first-level evaluation indexes of the second layer are given on the scale of 1-9. They are 9, 7, 5, 3 and 1. Using 8, 6, 4, 2, 1\2, 1\4, 1\6, and 1\8 indicates the score of the index grade between two adjacent grades, as shown in Table 1.

Table 1. Scale of relative importance of indicators.

Assignment	Illustrate
9	U_i is more important than U_j
7	U_i is much more important than U_j
5	U_i is obviously more important than U_j
3	U_i is slightly more important than U_j
1	U_i and U_j are equally important

3.3 Determine the weight of risk evaluation index system

The weight set of risk index system reflects the position and importance of each evaluation index in the comprehensive evaluation (a). Because of the mountain type scenic resorts the influence degree of the hotel operational risk evaluation is vague and difficult to accurate quantification, the author uses the Delphi method to assign values to the degree of relationship U_{ij} between various factors and indicators and then uses AHP algorithm to realize quantitative evaluation index. The specific steps are as follows:

3.3.1 Construct judgment matrix A

(1) Matrix is used to represent the pleadings of the first-order indicator set

$$
A = \begin{bmatrix}
1 & \frac{1}{2} & \frac{1}{3} & \frac{1}{4} & \frac{1}{4} & \frac{1}{5} & \frac{1}{4} \\
2 & 1 & \frac{1}{2} & \frac{1}{2} & \frac{1}{2} & \frac{1}{3} & \frac{1}{3} \\
3 & 2 & 1 & \frac{1}{2} & 2 & \frac{1}{2} & \frac{1}{4} \\
4 & 2 & 2 & 1 & 3 & \frac{1}{2} & \frac{1}{2} \\
4 & 2 & \frac{1}{2} & \frac{1}{3} & 1 & \frac{1}{2} & \frac{1}{3} \\
5 & 3 & 2 & 2 & 3 & 1 & \frac{1}{3} \\
4 & 3 & 4 & 2 & 4 & 2 & 1
\end{bmatrix}
\tag{1}
$$

(2) Finding the weight set

Normalize A by column:

$$
\tilde{A} = \begin{bmatrix}
0.043 & 0.030 & 0.032 & 0.038 & 0.040 & 0.044 & 0.084 \\
0.087 & 0.060 & 0.048 & 0.038 & 0.066 & 0.073 & 0.110 \\
0.130 & 0.121 & 0.097 & 0.076 & 0.099 & 0.110 & 0.084 \\
0.174 & 0.121 & 0.194 & 0.152 & 0.099 & 0.110 & 0.167 \\
0.174 & 0.121 & 0.048 & 0.050 & 0.099 & 0.110 & 0.110 \\
0.218 & 0.181 & 0.194 & 0.304 & 0.199 & 0.221 & 0.110 \\
0.174 & 0.181 & 0.387 & 0.608 & 0.400 & 0.442 & 0.334
\end{bmatrix}
\tag{2}
$$

For $\tilde{A}$ input row sum:

$$
\tilde{W} = [\, 0.303,\ 0.52,\ 0.717, 1.017, 0.712, 1.49,\ 2.526 \,]^{T}
\tag{3}
$$

To $\tilde{W}$ normalized processing, full weight set, with the following weight vector said:

$$
W = [\, 0.042,\ 0.071, 0.098, 0.140, 0.098,\ 0.205,\ 0.347 \,]
\tag{4}
$$

(3) Consistency test for the judgment matrix of first-order index weight

To calculate the maximum eigenvalue of the weight vector of the first-order index:

$$
\lambda_{\max} = \frac{1}{n} \sum_{i=1}^{n} \frac{(AW)_i}{W_i} \quad (i = 1, 2 \dots 7)
\tag{5}
$$

Table 2. Average random consistency RI calibration values.

The dimension	1	2	3	4	5	6	7	8	9	10	11
RI	0	0	0.52	0.89	1.12	1.26	1.36	1.41	1.46	1.49	1.52

$$While\ AW = [0.303,\ 0.52,\ 0.717,\ 1.017,\ 0.712,\ 1.49,\ 2.526]^{T} \tag{6}$$

Substituting (6) into equation (5) to get the following result:

$$\lambda_{max} = \frac{1}{7}\left(\frac{0.303}{0.042} + \frac{0.52}{0.071} + \frac{0.717}{0.098} + \frac{1.017}{0.140} + \frac{0.712}{0.098} + \frac{1.49}{0.205} + \frac{2.526}{0.347}\right) = 7.27 \tag{7}$$

In order to prevent the results from being inaccurate due to the large deviation of subjective judgment, CR should be used for consistency test, when CR<0.1, the hierarchical ordering has satisfactory consistency. Meaning passes the consistency test, Otherwise, the judgment matrix needs to be adjusted. The formula is:

$CR = \frac{CI}{RI}$, $CI = \frac{\lambda_{max}-n}{n-1}$, where RI is the values and shown in Table 2:

$$Consistency\ index\ CI = \frac{\lambda_{max} - n}{n - 1} = \frac{7.27 - 7}{7 - 1} = 0.045 \tag{8}$$

While n = 7, RI = 1.32

$$CR = \frac{CI}{RI} = \frac{0.045}{1.32} = 0.034 < 0.1 \tag{9}$$

Similarly, the weight set of second-level indicators can be obtained:

$$W_{21} = [0.181, 0.41, 0.41]$$
$$W_{22} = [0.281, 0.368, 0.113, 0.237]$$
$$W_{23} = [0.201, 0.151, 0.367, 0.281]$$
$$W_{24} = [0.272, 0.460, 0.087, 0.197]$$
$$W_{25} = [0.160, 0.537, 0.303]$$
$$W_{26} = [0.456, 0.175, 0.89, 0.289]$$
$$W_{27} = [0.6197, 0.284, 0.097]$$

The maximum characteristic root of the judgment matrix of the weight set of second-level indicators is calculated, and the consistency test is carried out:

$$\lambda1_{max} = 3, CR = 0 < 0.1$$
$$\lambda2_{max} = 4.002, CR = 0.001 < 0.1$$
$$\lambda3_{max} = 4.003, CR = 0.001 < 0.1$$
$$\lambda4_{max} = 4.029, CR = 0.097 < 0.1$$
$$\lambda5_{max} = 3.023, CR = 0.021 < 0.1$$
$$\lambda6_{max} = 4, CR = 0 < 0.1$$
$$\lambda7_{max} = 3, CR = 0 < 0.1$$

According to the calculation results, the judgment matrix and weight of the first-level index and second-level index are reasonable, as shown in Table 3.

Table 3. Operation risk evaluation index weight system of mountain resort hotels.

Level indicators	Weight coefficient	The secondary indicators	Weight coefficient
Natural disaster risk	0.042	Geological hazard risk	0.181
		Meteorological hazard risk	0.410
		Biohazard risk	0.410
Market risk	0.071	Market positioning risk	0.281
		product risk	0.368
		Distribution channel risk	0.113
		Risk of tourism demand discrepancy in peak and low season	0.237
Financial risk	0.098	Hotel investment error risk	0.201
		Stock market, fund and financing risk	0.151
		Financial corruption risk	0.367
		Capital recovery risk	0.281
Human resource risk	0.140	Risk of insufficient human resources	0.272
		Risk of Lack of professionals	0.460
		Employee turnover risk	0.087
		Low moral quality of employees	0.179
Policy and legal risks	0.098	National policy changes	0.016
		compensation for customer's property loss and personal injury	0.537
		Compensation for employee injuries	0.303
Brand reputation risk	0.205	Poor service carries reputational risks	0.456
		Hotel or staff responsible for accidents	0.175
		Poor service by online bookmakers carries reputational risks	0.089
		network bad review brings reputation risk	0.289
Physical asset risk	0.347	Risk of loss due to fire	0.620
		Risk of loss due to equipment failure	0.284
		Risk of theft	0.097

3.4 The fuzzy judgment of hotel operation risk in mountain scenic resorts

The basic principle of fuzzy evaluation is to describe the fuzzy boundary of each factor by membership degree on the basis of determining the grade and weight of evaluation factors, using the principle of fuzzy relation synthesis, constructing fuzzy evaluation matrix, and finally determining the grade of evaluation object through multi-layer compound operation (a).

3.4.1 Construct the evaluation set

The risk evaluation set is composed of the evaluation results made by experts for different evaluation index factors. Supposing the risk assessment set is composed of s kinds of decisions, expressed as T={T1, T2, T3, …… Ts}. According to the impact of risk factors on the operation of mountain scenic resorts hotels and the actual situation, the evaluation set is divided into 5 levels , that s = 5. T = {T1, T2, T3, T4, T5} = {lower risk, low risk, general risk, high risk, higher risk}. Each of these equal fractions occupies a fifth of the length and the corresponding risk interval is: [0, 0.2], [0.2, 0.4], [0.4, 0.6], [0.6, 0.8], [0.8, 1.0].

3.4.2 Establish fuzzy evaluation matrix R

When the research domain is a finite set, the relation R can be expressed by a matrix. When R is a fuzzy relation, R is called a fuzzy relation matrix:

$$R_i = \begin{bmatrix} r_{11} & \cdots & r_{1m} \\ \vdots & \ddots & \vdots \\ r_{n1} & \cdots & r_{mn} \end{bmatrix} \tag{10}$$

($i = 1,2,3,4,5,6,7$; $n = 7$, $m = 7$), the r_{ij} represents the membership degree of u_i to the comment level T_j.

3.4.3 *Fuzzy comprehensive evaluation algorithm*

The M (., +) model of fuzzy mathematics is adopted to calculate layer by layer from the low level to the high level, and the single factor evaluation matrix of the first-level index is established, $B_1 = A_1.R_1$, $B_2 = A_2.R_2 \ldots B_7 = A_7.R_7$, among them $A_1 \sim A_7$ is the weight set of secondary indicators.

Fuzzy evaluation matrix of the final result is:

$$B = A^T.R \tag{11}$$

The B represents the fuzzy comprehensive evaluation result, A represents the risk factor set, and R represents the evaluation set. According to the maximum subordination principle, the result of risk grade evaluation is obtained.

4 EMPIRICAL ANALYSIS OF OPERATIONAL RISKS OF MOUNTAIN RESORT HOTELS

4.1 *Operation risk calculation of hotels in Huangshan scenic resorts*

According to the method introduced above, further empirical analysis is made on the operational risk in Huangshan scenic resorts hotel. Based on the specific content of the comprehensive evaluation index system of operational risk of mountain scenic resorts hotels in Table 3, 30 hotel industry experts and three hundred hotel employees from six hotels in Huangshan scenic resorts rated the impact of the 25 secondary indicators on hotel operations on a five-point scale. Summarize the scoring data with EXCEL. According to the scoring results, a judgment table of operation risk degree of Huangshan scenic resorts hotel is formed, as shown in Table 4.

According to the fuzzy comprehensive evaluation model and the fuzzy evaluation algorithm, the evaluation results of the second-level operational risk index of hotels in Huangshan scenic resorts are obtained, that is:

B1 = (0.300, 0.259, 0.141, 0.100, 0);
B2 = (0.124, 0131, 0.191, 0.300, 0.287);
B3 = (0.171, 0.152, 0.281, 0.084, 0.298);
B4 = (0, 0.099, 0.172, 0.313, 0.413);
B5 = (0.323, 0.285, 0.227, 0.315, 0.092);
B6 = (0.119, 0.241, 0.211, 0.351, 0.334);
B7 = (0.034, 0120, 0.145, 0.271, 0.418);

According to the above results, formula 11 is used to calculate the final evaluation result B, which can be obtained as follows:

B = (0.119, 0.168, 0.215, 0.274, 0.328)

The above results show that operational risk membership of hotel is respectively 0.119, 0.168, 0.215, 0.274 and 0.328. As can be seen from the above results, the maximum value is 0.328, and the corresponding risk level is the lower risk.

Table 4. Operation risk judgment table of hotel in Huangshan scenic resorts.

Risk categories	Degree of risk (proportion)					Synthetic weight
	Lower risk	Low risk	Medium risk	High risk	Higher risk	
Geological hazard risk	0.3	0.3	0.1	0.1	0	0.007
Meteorological hazard risk	0.2	0.3	0.2	0.1	0	0.017
Biohazard risk	0.4	0.2	0.1	0.1	0	0.017
Market positioning risk	0.15	0.15	0.3	0.3	0	0.02
product risk	0.1	0.1	0.1	0.3	0.4	0.026
Distribution channel risk	0.4	0.25	0.2	0.15	0	0.008
Risk of tourism demand discrepancy in peak and low season	0	0.1	0.2	0.3	0.4	0.017
Hotel investment error risk	0.4	0.3	0.25	0.05	0	0.02
Stock market, fund and financing risk	0.6	0.3	0.1	0	0	0.015
Financial corruption risk	0	0.05	0.1	0.4	0.45	0.036
Capital recovery risk	0	0.1	0.2	0.3	0.4	0.028
Risk of insufficient human resources	0	0.1	0.1	0.35	0.45	0.038
risk of Lack of professionals	0	0.1	0.2	0.3	0.4	0.06
Employee turnover risk	0	0.1	0.2	0.3	0.4	0.012
Low moral quality of employees	0	0.1	0.2	0.3	0.35	0.025
National policy changes	0.35	0.3	0.2	0.1	0.05	0.053
compensation for customer's property loss and personal injury	0.3	0.3	0.25	0.15	0.1	0.03
Compensation for employee injuries	0.35	0.25	0.2	0.1	0.1	0.03
Poor service carries reputational risks	0.1	0.1	0.1	0.3	0.4	0.093
Hotel or staff responsible for accidents	0.4	0.2	0.1	0.1	0	0.036
Poor service by online bookmakers carries reputational risks	0	0.1	0.15	0.45	03	0.018
network bad review brings reputation risk	0	0.05	0.15	0.4	0.4	0.059
Risk of loss due to fire	0.05	0.1	0.15	0.3	0.4	0.215
Risk of loss due to equipment failure	0.05	0.1	0.15	0.3	0.6	0.099
Risk of theft	0.6	0.3	0.1	0	0	0.034

4.2 Result analysis

The study uses the fuzzy comprehensive evaluation method to evaluate the operational risk impact of six hotels in Huangshan scenic resorts, and concludes that the operational risk level belongs to the lower risk level. The first-level indicators that have the greatest impact on the overall risk are "risk of loss of physical assets", "Brand reputation risk"and "Human resource risk" in order. Among the secondary indicators, "risk of loss caused by fire" has the greatest impact on the overall risk, followed by "reputation risk caused by poor service ", "risk of lack of professionals", "risk of national policy change" and" network bad review brings reputation risk ". This is consistent with the actual operation situation of the six hotels. Therefor the feasibility and practicability of the evaluation method are verified. Based on this result, in determining the risk control measures for the operation of hotels, the risk prevention measures should be taken from the aspects of the internal management of the mountain scenic area hotels, the introduction and training of professional and technical personnel and the brand reputation management.

The results also show that the "natural disaster factors" will bring certain operational risks to the hotel, and the index weight is 0.042, indicating that the industry believes that the natural disaster risk has little impact on the hotel operation, which is different from the risk experts' general risk perception of mountain scenic resorts. There are several reasons as follows: First is that those hotels belong to Huangshan tourism co., LTD. The operator does not need to bear the risk of natural disasters. Second is the hotel is basically out of operation when natural disasters occur, so there is no risk of loss. Third, different groups have different degrees of recognition of natural disaster risk. In the future, the awareness of natural disaster risk and prevention skills training of employees should be strengthened. The prevention of natural disaster risk factors should be a part that cannot be ignored in the hotel operation risk management system.

5 CONCLUSIONS

Based on the joint discussion by hotel industry experts and sufficient survey data obtained through questionnaires, this paper establishes the evaluation index system of mountain scenic resorts hotel operation risk according to the operation characteristics of hotels in mountain scenic resorts. AHP method is used to calculate index weight which overcomes the problem of subjective arbitrariness of index determination, then, fuzzy mathematics theory is used to establish the operational risk assessment model of hotels in mountain scenic resorts. This method is used to test the survey data of 6 hotels in Huangshan scenic resorts. The results show that the calculation process is scientific and reasonable, and the final evaluation result is scientific.

Because different roles have different perceptions of risks, and the sample size is limited, the operational risk assessment system constructed still has some deficiencies. In the future, the actual situation of hotel operation can be combined to further screen and modify the risk assessment indicators.

ACKNOWLEDGEMENTS

This work is supported by Anhui Quality Engineering Project (2015jxtd037) and Anhui Quality Engineering Project (2017ppzy36).

REFERENCES

Xu, S.B. 1998. *The Principle of Analytic Hierarchy Process (AHP) is a Practical Decision-Making Method.* Tianjing: Tianjin university press.

Wu, W.A., Zhang, Z.W. and Wang, Q.H. 2017. Based on fuzzy comprehensive evalua tion and AHP information security risk assessment model. *Journal of Chongqin University of Technology* (Natural Science),(7): 156–160.

He, Y. R. 2017. *Research on Operational Risk Factors of Hotels and Building System of Risk Management in Mountain Scenic Resorts*, CR2017, 11–17.

Luo, X.Y. 2011. Comprehensive evaluation of the hotel operational risk. *Journal of Hubei University of Economics* (Humanities and Social Science), 8(7): 57–59.

Luo, X. Y. 2013. Hotel operation risk probability estimation. *Journal of Chifeng University* (Natural Science Edition), 29(11): 80–81.

Li, Y.F, Yhui, J.A and Zhu, P.F. 2018. Investment risk evaluation for tourism endowment estate based on fuzzy AHP. *Journal of Dalian Minzu University*, 20(2): 165–169.

Li, L and Chen X.J. 2008. Study on the risk of expansion hotel and avoiding strategy. *Hotel Modernization*, (9): 34–36.

Diao Z.G. 2011. Risk control and management of hotel industry in China. *Humanities & Social Sciences Journal of Hainan University*, 29(3): 61–66.

Risk Analysis Based on Data and Crisis Response Beyond Knowledge – Huang & Nivolianitou (eds)
© 2020 Taylor & Francis Group, London, ISBN 978-0-367-25146-8

Effects of major hurricanes in Atlantic Canada from 2003 to 2018

Luana Souza Almeida, Floris Goerlandt* & Ronald Pelot
Maritime Risk and Safety Group, Department of Industrial Engineering, Dalhousie University, Halifax, Canada

ABSTRACT: Atlantic Canada is a geographical area that is susceptible to be reached by hurricanes. In the past 15 years, it was hit by several hurricanes, among which Igor (2010) and Juan (2003) caused major impacts. These hurricanes caused different types of damage such as toppled trees, power outages, and the destruction of bridges and transportation infrastructure. While there is generic knowledge about the effects of hurricanes, there is a lack of systematic analysis and understanding of the particular effects of hurricanes Igor and Juan. This article contributes to building this understanding, using selected case studies and through application of a cause-and-effect analysis. The analysis aims to identify factors associated with hurricanes that should be considered when managing risk in the immediate response phase. The focus is on the disruptive effects on the supply chain of delivering goods to affected communities, and on the actions taken by response organizations, but other issues are addressed as well. The results are relevant for understanding the issues which risk management strategies should address, and can also serve as a basis for decision support models.

Keywords: cause and effect analysis, risk management, hurricanes, Atlantic Canada

1 INTRODUCTION

The study of past natural disasters plays a key role in the risk management field, because knowledge developed from historical data can help during the decision-making processes related to future emergency response. Of particular concern are the impacts associated with hurricanes. According to Klotzbach et al. (2018), hurricane Irma made landfall in 2017 in central and north America and was responsible for 44 direct fatalities as well as economic damage of around $50 billion. In the same year, hurricane Harvey reached the United States and caused an estimated damage of $125 billion (Klotzbach et al., 2018).

The Atlantic region of Canada is also susceptible to hurricanes. However, according to Environment Canada (2013a), when hurricanes reach Atlantic Canada, they are in a different stage of hurricane denoted as "post-tropical" cyclones. In 2003, hurricane Juan hit Halifax (Nova Scotia) and many houses had no power for hours, house damage was widespread, streets were blocked, and rail tracks were damaged (Environment Canada, 2014a). Another infamous hurricane that hit Atlantic Canada was hurricane Earl in 2010 which caused a large-scale power failure due to falling trees. Weeks afterwards, hurricane Igor hit Newfoundland and several communities were isolated for nearly 11 days due to the destruction of bridges. (Masson, 2014)

The potential impacts caused by a hurricane are expected accordingly to its classification on the Saffir-Simpson scale (National Hurricane Center). Most reports related to hurricanes such as Lawrence et al. (2005) and Beven and Blake (2015) address only general impacts such as the total damage and fatalities. However, there is little systematic understanding of the impacts caused by hurricanes in Atlantic Canada, linking specific hazards such as rain and wind, with their impacts.

---

*Corresponding author: floris.goerlandt@dal.ca

This paper aims to support this understanding by presenting a cause-and-effect analysis of the major hurricanes that made landfall in Atlantic Canada in the past 15 years. The analysis is based on four case studies and historical data for each category of hurricane. The impacts are organized into a fishbone diagram (Ishikawa, 1990). The purpose of the analysis is to map the specific impacts of hurricanes of different severity categories. Such knowledge is useful to develop risk management strategies, and can also serve as a basis for developing models for emergency preparedness and response.

The present paper is divided into six parts. Section 2 presents the literature review, section 3 introduces the methodology adopted, section 4 presents the results, section 5 presents the discussion, and section 6 concludes.

2 LITERATURE REVIEW

There are relatively few studies that assess the impacts of past hurricanes in Atlantic Canada. Farquhar (2004) presents an interesting case study about hurricane Juan. In this study, the events related to the hurricane are documented in chronological order from the day that the hurricane was announced by meteorologists up to when the hurricane made landfall in Nova Scotia.

Taylor et al. (2008) presents the case study of the impacts caused by hurricane Noel in 2007 on the Atlantic coast. In this study, the impacts on the geology of eight beaches is studied, because storms like Noel can increase the intensity of the landward recession of the coast related to the rise of the sea level. Blake et al. (2011) introduce a report of the deadliest and costliest hurricanes, however the information is presented in terms of fatalities and costs and it is oriented only to United States. Environment Canada (2013c) presents a description of notable hurricanes in Atlantic Canada, however the main consideration is fatalities. Environment Canada (2013b) and Environment Canada (2014a) report on the impacts caused by hurricanes Igor, Earl, and Juan. The meteorological perspective and the impacts are reported in general terms. A different point of view of the historical hurricane-related data is presented by Environment Canada (2013a) where the impacts are described according to different types of threats such as wind and rain. Masson (2014) presents an overview of hurricane Igor that hit Newfoundland. In this study, the meteorological perspectives of the hurricane are introduced, followed by a description of the impacts. The participation of stakeholders during the emergency response is also briefly described. Other sources of information related to the hurricanes in Atlantic Canada are the annual reports issued by hurricane centers and the local newspapers. In relation to annual reports, Lawrence et al. (2005) presents an overview of all the hurricanes that crossed the Atlantic region in 2003, however the description of impacts is not structured, and it is in general terms.

3 METHODOLOGY

According to the International Electrotechnical Commission (IEC, 2009) a cause and effect analysis organizes the possible contributory factors into major classes in order to identify possible causes of undesired events. This is especially important in the assessment of hurricanes since it structures the available data. This paper focuses on four past hurricanes in the Atlantic Region: Juan, Noel, Earl and Igor. Environment Canada (2013c) considers hurricanes Juan and Igor as the two most notable hurricanes in the past 15 years. Furthermore, according to Beven and Blake (2015), Newfoundland was hit by hurricane Earl weeks before hurricane Igor, which explains why the consequences of hurricane Earl are also considered in the analysis. Finally, hurricane Noel is also included in this study because of its impact on the Atlantic coast. In 2007 hurricane Noel created high waves and a powerful storm surge across the province of Nova Scotia. One scope limitation of this paper is that it assesses only the impacts of hurricanes in terms of physical damage but does not consider the meteorology variations such as height of waves, pressure, and so on. This section is divided as follows: first, the fishbone diagram is explained, then the classification classes are introduced.

3.1 *Cause and effect diagram guidelines*

The basic steps to construct the cause-and-effect diagram follow the guidelines provided by IEC (2009). The cause and effect diagram, proposed by Ishikawa (1990), is composed of a central line that leads to the major undesired event, and categories of events are linked to this major line. Initially, the effect or undesired event to be analyzed is established. Next, the main categories of causes are defined. These classes are selected according to the context of each type of event. Then, possible causes are inserted as branches of pre-determined classes. In this paper, the categories are presented in the next section.

3.2 *Undesired event and major categories*

In the context of natural disasters, the undesired event is the destruction caused by the natural hazards. For the present purposes, the events in focus are the destructive impacts of hurricane categories of different severity levels, caused by different natural processes associated with the hurricane occurrence. The total destruction caused by a hurricane is a sum of several types of processes such as uprooted trees due to powerful winds and flooding caused by excessive rain. In this paper, the major categories are classified according to Environment Canada (2013a). There are four possible threats: wind, rain, storm surge, and ocean waves.

According to Environment Canada (2013a), wind is a threat that has its characteristics changed when a hurricane moves from a stage of tropical to post-tropical cyclone. There are three main differences: in the post-tropical cyclones case, the maximum wind speed drops off, moves out from the storm center, and the area of stronger winds increases. According to the Saffir-Simpson Scale (National Hurricane Center), in a hurricane category 1, the wind gusts from 119 to 153 km/h, and in a category 2, the speed of the wind is between 154 and 177 km/h.

The second threat considered in this paper is rain. According to Environment Canada (2013a), this category is critical in terms of potential of damage because it happens much more often. When a tropical cyclone changes to post-tropical cyclone, the intensity of the rainfall acquires a characteristic of thunderstorm. Chen (2011) states that the precipitation of a post-tropical cyclone can be substantial and cause severe flooding. It is also stated that it is difficult to predict the precipitation of a post-tropical cyclone.

The third threat is related to the impacts caused by a storm surge. According to Zheng (2009) storm surges occurs when high winds push the water from the ocean toward the shore raising the normal sea level because of the low pressure at the center of the hurricane. According to Environment Canada (2013a), storm surges happen both in tropical cyclones and post-tropical cyclones, and it is a big concern for coastal areas. The National Oceanic and Atmospheric Administration (2014) states that storm surges can increase the water level by 30 feet, and it can affect rivers and lakes by increasing their water level too. Storm surges are not exclusive to oceans because, according to Danard et al. (2003), high waves can also occur in a closed small water body such as lakes and rivers.

Ocean waves constitute the fourth threat category, which is similar to storm surges. On the one hand, a storm surge is the high wave on the coast, and on the other hand, ocean waves are the development of high waves in the sea (Environment Canada, 2013a). Between 1900 and 1950, ocean waves were responsible for around 75% of the fatalities in Canada during a hurricane, however this scenario has changed in the past few years because of the raised awareness of sailors (Environment Canada, 2013a).

4 RESULTS

The first set of analyses examined the impact of a hurricane category 2, in particular hurricane Juan. The second set of analyses included the impacts of hurricanes Noel, Earl and Igor that were registered by Environment Canada (2013a) and Environment Canada (2014a) as category 1.

The first analysis, relating to hurricanes category 2, is based on Farquhar (2004), Environment Canada (2014a) and the Nova Scotia Management Office (2003). Figure 1 presents the fishbone diagram for this category, with sources shown in Table 1. Based on the data reported by Farquhar (2004), there was no damage related to rain and ocean waves. In fact, Environment Canada (2013a) mentions that the rainfall during hurricane Juan was low. From Figure 1, it can also be observed that most impacts were related to the wind and the consequences associated with falling trees.

The second analysis focuses on category 1 hurricanes. This analysis includes the impacts of hurricanes Earl, and Igor. As shown in Figure 2, the high volume of rain washed away bridges and important roads such as Trans-Canada highway. Consequently, communities were cut off (Masson, 2014) and around 100 communities (see Figure 3) were impacted (CBC, 2010).

As mentioned by Masson (2014) and CBC (2010), military forces had to step in and help reconstruct the bridges that were washed out in order to deliver resources such as food and water to the isolated communities. According to Environment Canada (2013b) the major impacts of hurricane Noel are associated with storm surges, while the impacts of hurricane Igor are connected to flooding.

Table 1. Sources of impacts for hurricanes category 1 and 2.

ID	Source	Hurricane
[A]	Nova Scotia Management Office (2003)	Juan
[B]	Farquhar (2004)	Juan
[C]	Environment Canada (2014a)	Juan
[D]	Environment Canada (2013b)	Earl and Igor
[E]	Masson (2014)	Igor
[F]	Environment Canada (2014b)	Noel
[G]	CBC (2017)	Igor
[H]	CBC (2010)	Igor
[I]	Taylor et al. (2008)	Noel

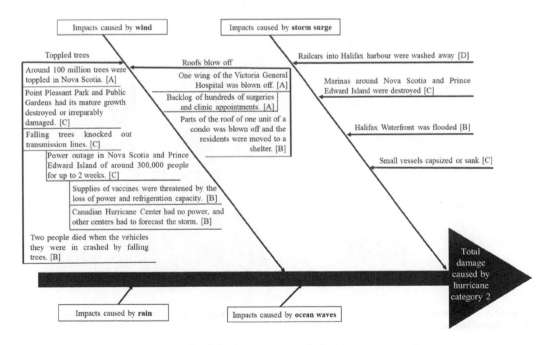

Figure 1. Cause and effect analysis of the damage caused by hurricane category 2. (sources are available in Table 1).

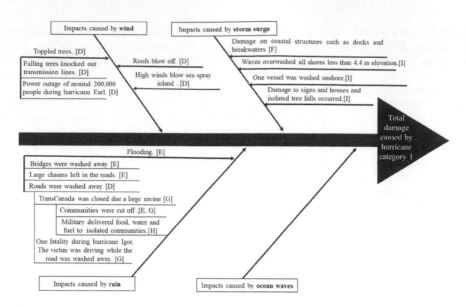

Figure 2.   Cause and effect analysis of the damage caused by hurricane category 1.
(sources are available at Table 1).

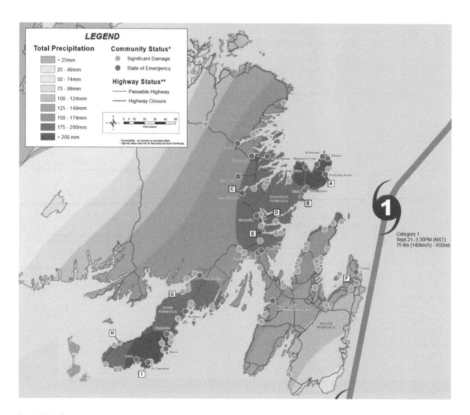

Figure 3.   Hurricane Igor: community and highway status during the emergency response phase
(Newfoundland and Labrador Statistics Agency).

It is notable that in all studies events, toppled trees and flooding blocked the roads. Consequently, a map with the roads that were blocked in each hurricane can provide a valuable information in terms of emergency response, for instance for planning possible routes which can be used for delivery of supplies. Figure 3 presents this type of assessment based only on the roads that were damaged during hurricane Igor. Another important point is to analyze post-hurricane operability of essential logistics infrastructure such as ports and airports.

Based on the results presented in Section 4, this section will discuss further improvements of the research and the application of results to the supply chain perspective.

First, the current study found that the association of the impacts with the categories can be harder depending on the characteristic of the disaster. In the case of hurricane Juan, it was easy to associate the toppled trees only with the wind because the reports were clear about the low rainfall. However, if both rain and wind are severe, it is difficult to separate the impacts of each threat. Further work can be performed in terms of cumulative risk assessment, studying the consequence caused by the union of two hazards, such as wind and rain.

Another important finding is the low impact caused by ocean waves. This result reflects the one presented by Environment Canada (2013a). It is important to state that even though there was no impact associated with ocean waves in this study, it does not mean that it is not a threat. According to Environment Canada (2013a), the highest and fastest-building ocean waves were found in Canadian waters during hurricanes, but mariners are being effectively warned to protect themselves during natural disasters. This result also reinforces the fact that the level of preparedness influences the total impact of hurricanes.

Another finding is that even though Figure 2 has more listed impacts than Figure 1, it is important to remember that three hurricanes are studied under category 1, and only one in category 2. Also, even though some impacts are the same for both categories (i.e. toppled trees), the consequences of hurricane category 2 are more severe than for category 1. For example, a comparison between the quantity of people without power after hurricane Juan is greater than during hurricane Earl as shown in Figure 1 and Figure 2.

A further study with more focus on impacts is therefore suggested. The study cases used in this paper reported general impacts of the hurricanes, however detailed information can add even more value to this structured analysis of the disasters.

5 CONCLUSION

The aim of the present research was to propose an impact assessment of different categories of hurricanes that made landfall in Atlantic Canada since 2003. The impacts were structured according to the fishbone diagram proposed by Ishikawa (1990).

Four hurricanes were studied: Juan, Noel, Earl and Igor. The cause and effect analysis showed that toppled trees and flooding are the major causes that block roads during the emergency response phase. Storm surges can cause severe damage in the coastal areas, especially in ports and marinas. In terms of supply chain response, an assessment map based on blocked roads during past hurricanes is suggested in order to predict different possible paths to flow emergency resources such as water and food.

The results presented in this paper can be used to address possible areas to be studied in depth during the planning of an emergency response. For example, a detailed assessment of the blocked roads caused by toppled trees can help to improve the preparedness for future hurricanes. Finally, the study reveals that the level of preparedness of society can minimize the impacts of the hurricanes, such as what happened with avoiding impacts caused by ocean waves.

ACKNOWLEDGMENTS

This paper is a part of a project denoted by Shipping Resilience: Strategic Planning for Coastal Community Resilience to Marine Transportation Risk (SIREN). This project is financially supported by Marine Environmental Observation Prediction and Response (MEOPAR), and the Province of British Columbia.

REFERENCES

Beven, J., & Blake, E. (2015). Atlantic Hurricane Season of 2010*. *Monthly Weather Review*, 143 (9),3329–3353.

Blake, E. S., Landsea, C. L., Gibney, E. J. (2011). The deadliest, costliest, and most intense United Stated tropical cyclone from 1851to 2010 (and other frequently requested hurricane facts). Retrieved March 23, 2019, from https://www.nhc.noaa.gov/pdf/nws-nhc-6.pdf

CBC (2010). Military begins Igor relief effort. Retrieved from https://www.cbc.ca/news/canada/newfound land-labrador/military-begins-igor-relief-effort-1.913754

CBC. (2017). N. L. slammed by Hurricane Igor. Retrieved from https://www.cbc.ca/news/thenational/ n-l-slammed-by-hurricane-igor-1.1780145

Chen, G. (2011). A Comparison of precipitation distribution of two land falling tropical cyclones during the extratropical transition. *Advances in Atmospheric Sciences*, 28(6),1390–1404.

Danard, M., Munro, A., & Murty, T. (2003). Storm Surge Hazard in Canada. *Natural Hazards*, 28 (2),407–434.

Environment Canada. (2013a). Learn about hurricanes: Hazards and impacts. Retrieved March 26, 2019, from https://www.canada.ca/en/environment-climate-change/services/hurricane-forecasts-facts/learn/ hazards-impacts.html

Environment Canada. (2013b). 2010 Tropical Cyclone Season Summary. Retrieved March 26, 2019, from http://www.ec.gc.ca/ouragans-hurricanes/default.asp?lang=En&n=2A6E3A33-1

Environment Canada. (2013c). Notable Canadian Tropical Cyclones. Retrieved March 28, 2019, from http://www.ec.gc.ca/ouragans-hurricanes/default.asp?lang=En&n=CC8A7AA0-1

Environment Canada. (2014). Canadian Tropical Cyclone Season Summary for 2003. Retrieved March 26, 2019, from http://www.ec.gc.ca/ouragans-hurricanes/default.asp?lang=en&n=DCA5B0C3-1

Environment Canada. (2014). Canadian Tropical Cyclone Season Summary for 2007. Retrieved from https://www.ec.gc.ca/ouragans-hurricanes/default.asp?lang=en&n=99F43FA9-1

Farquhar, L. (2004). *Nova Scotia's Vulnerability to Hurricanes: A Case Study of Hurricane Juan, September 29th, 2003*.

International Electrotechnical Commission (2009) *Risk management – Risk assessment techniques* (Standard No. 31010). Retrieved from https://www.iso.org/obp/ui/#iso:std:iec:31010:ed-1:v1:en

Ishikawa, K. (1990). *Introduction to quality control*. Tokyo: 3A Corporation.

Klotzbach, P., Schreck, C., Collins, J., Bell, M., Blake, E., & Roache, D. (2018). The Extremely Active 2017 North Atlantic Hurricane Season. *Monthly Weather Review*, 146(10),3425–3443.

Lawrence, Avila, Beven, Franklin, Pasch, Stewart, & Lawrence, M. (2005). Atlantic Hurricane Season of 2003. *Monthly Weather Review*, 133(6),1744–1773.

Masson, A. (2014). The extratropical transition of Hurricane Igor and the impacts on Newfoundland. *Natural Hazards*, 72(2),617–632.

National Oceanic and Atmospheric Administration. (2014). Storm Surge. Retrieved March 27, 2019, from https://oceanservice.noaa.gov/podcast/may14/mw125-stormsurge.html

National Hurricane Center. (n.d.). Saffir-Simpson Hurricane Wind Scale. Retrieved from https://www. nhc.noaa.gov/aboutsshws.php

Newfoundland and Labrador Statistics Agency. (n.d.). Hurricane Igor. Retrieved March 28, 2019, from https://www.stats.gov.nl.ca/Maps/PDFs/HurricaneIgor.pdf

Nova Scotia Emergency Management Office. (2003). A Report on the Emergency Response to Hurricane Juan. Retrieved March 26, 2019.

Taylor, R. B., Frobel, D., Forbes, D. L., & Mercer, D. (2008). Impacts of Post-tropical Storm Noel (November, 2007) on the Atlantic Coastline of Nova Scotia. Retrieved from http://publications.gc.ca/ site/archivee-archived.html?url=http://publications.gc.ca/collections/collection_2016/rncan-nrcan/ M183-2-5802-eng.pdf

Zhang, J. (2009). A Vulnerability Assessment of Storm Surge in Guangdong Province, China. *Human and Ecological Risk Assessment: An International Journal*, 15(4),671–688.

Study on system testing approach-based risk evaluation of subsea tree

Xiaobing Yuan
College of Mechanical and Electronic Engineering, China University of Petroleum, Qingdao, China
Risk Management Research Department, Shenzhen City Public Safety Technology Institute, Shenzhen, China

Baoping Cai & Guoming Chen
College of Mechanical and Electronic Engineering, China University of Petroleum, Qingdao, China

ABSTRACT: Subsea Tree is one of the most important well control devices in offshore oil and gas production activities. If it fails, inestimable loss can be caused for the staff on the production facility, the ocean environment and other consequences. This paper presents a risk evaluation and identifies the major risk of the subsea tree, and develops a system testing approach to evaluate and assess the reliability of the subsea tree. This study performs a system testing approach to evaluate and assess the reliability of the subsea tree and obtains innovative achievements with important practical values.

Keywords: subsea tree, risk evaluation, testing approach

1 INTRODUCTION

In the early stages of the safety assessment of equipment, the method used is mainly quantitative risk assessment (QRA) and failure modes and effects analysis (FMEA). This paper focuses on risk evaluation and identifies the major risk of the subsea tree, and develops a system testing approach to evaluate and assess the reliability of the subsea tree. However, completing a full qualification test of a subsea tree and its various components involves not only testing the unit itself, but also a review of the field system design basis, general operational (functional) requirements, and considerations of materials and design calculations. The best way to ensure operational reliability is to test each component (FAT) to assure and verify that it operates within its full design capacity. In addition, testing of the completely assembled unit (SIT) ensures that the subsea tree unit meets requirements and will function as expected throughout its design life.

2 RISK ANALYSIS METHODOLOGY

Figure 1 shows a model that can be divided into three layers. The three-tier structure divides the whole network node into cause, incident and accident, which is the central concept for constructing this model. Cause refers to the behavior that may trigger an abnormal event, which may or may not be wrong. Incident refers to those who can cause accidents or abnormal events, and sometimes an incident does not intuitively show any serious consequence. Accident refers to those with serious consequences, threatening personal safety and/or resulting in economic losses. When there is a cause, the likelihood of an incident will increase. An incident is something people do not want to see, and once the event has not been ignored by people, and cannot be self-regulation, then an accident may occur.

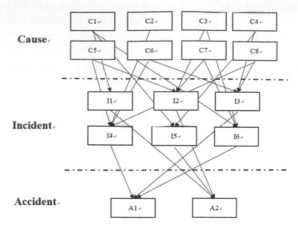

Figure 1. Typical three-tier risk analysis.

The graphs of the network nodes represent the variables, and the arrows with directions indicate the dependencies and causal relationships between the nodes. This quantitative relationship is established by the conditional probability table. In general, the conditional probability table is determined by historical data and expert judgment. At the same time, once the corresponding nodes change, the nodes associated with them also change accordingly, and the real-time status of each node can be changed by the influence of each other.

3 RISK ANALYSIS AND EVALUATION

After determining the events that may occur, the next step is to divide them into the three-tier structure of the cause, the incident and the accident.

(1) Cause: in the subsea tree system, the cause of the accident can be divided into three categories. The first cause is the failure of the components, such as corrosion and aging of the components. During the work of the tree, the device may also be faulty. When these events occur, the likelihood of an accident will increase significantly. Therefore, when these events are found, effective means should be taken to remedy the incident to prevent a more serious accident. The second cause is human error, such as mis-use in the operation or maintenance of the subsea tree. Generally, human error can be further divided into organizational errors, group errors and individual errors. The third cause is the cause of nature, for example, a too high pressure can cause leakage of flammable substances, or other harsh conditions such as earthquakes, waves, ocean currents, and so on.

(2) Incident: Some incidents that occur on the subsea tree can lead to more serious consequences, such as equipment failure and abnormalities. The operation of the subsea tree should ensure that the various subsystems are stable, and if the equipment is abnormal it may cause some danger, resulting in equipment damage and system failure, or even a fire or explosion.

(3) Accident: Although accident do not occur often, the loss they bring can be huge. After these accidents, there are not only immeasurable economic losses, resulting in a huge waste of energy, but also mass pollution of the environment areas.

4 SYSTEM TESTING APPROACH

4.1 Design basis document

API Specification 17D is the primary approach that defines the design, operation and testing of subsea production systems. Once the exact field parameters for subsea tree are identified, a specific field design basis document (DBD) must be developed, which describes the specific requirements of the field. A functional specification can then be developed by the operator that will define the general requirements to the subsea tree manufacturer. This functional specification will also provide the basis for the design verification and validation of the testing program.

The development of a DBD for the subsea tree is a crucial step in ensuring that the system will fulfill its required function in a safe and reliable manner. It is in this DBD that the requirements of the overall system are set forth. It defines the overall field operating and environmental conditions, loads, and the equipment functionality requirements.

4.2 *Components testing*

The subsea tree also contains other flow control and isolation components such as, annulus access valves, wing valves, cross over valves, flow isolation valves, etc. All of these components have to be designed to the requirements of the DBD and all have to be design verified and validation tested based on the functional specification, to insure system operation.

All of the subsea tree components must be designed for the specific loading and environmental conditions they will be subjected to in the field, as defined by the subsea tree functional specification and the field DBD. In addition, the components in the subsea tree must be selected so that they can work in concert and be coupled with a control system capable of long-term operation in the harsh subsea environment. Each individual component is tested by the manufacturer as part of the factory acceptance test (FAT) before it is incorporated into the subsea tree. The design of the subsea tree and its various components is based on the following:

- *Design basis document*: Defines the field requirements and operating parameters including internal and external pressure, temperature, production flow rate and corrosivity (by operator).
- *Operational functionality*: Ensures that the system will fulfill the required functions listed in the DBD and that interferences do not exist between components.
- *Constructability*: Ensures that the components and the subsea tree system, as a whole, can be manufactured and assembled. Consideration is also needed to allow for component replacement and system upgrade without removing the tree.
- *Stress integrity analysis:* Ensures that the components are sized properly and there is adequate strength and operational safety.
- *Material assessment*: Ensures that the metallic and polymeric (e.g. seals) components are appropriate for the external environment and internal production flow and resist corrosion for the intended life of the field.
- *FMEA (failure modes and effects analysis)*: Identifies all potential failure modes, determines weak links in the system, and ensures system redundancies are included where needed.
- *Testing*: Ensures there is physical validation under anticipated operational conditions for materials integrity and general equipment functionality.

4.3 *Factory Acceptance Test (FAT)*

FAT is performed on all component/assemblies to be delivered. This is to demonstrate compliance with specifications and sufficient strength for safe operation. The tests are performed in accordance with developed FAT procedures written by their respective suppliers. Test definition sheets for FAT procedures are issued for company review and approval in due time prior to commencing any testing. The FAT of an assembly is generally built using the FAT of the different components; hence the assembly FAT does not necessarily repeat all component FATs. FAT is, as a general rule, performed at the supplier's premises. FAT shall confirm that the equipment is fit for purpose and meets all of the design requirements. It shall ensure that critical internal and external interfaces have been validated. FATs will always be carried out prior to further use. Prior to the start of any testing, a detailed test procedure for each test shall be developed, which should contain, as a minimum:

- Step-by-step definition of the test activities
- Equipment schematics for all components
- Acceptance criteria for each individual test
- Test equipment calibration record sheet
- Punch list and witness record sheet
- Safety review and work authorization sign-off by site safety official
- Final acceptance sign-off of FAT sheet
- Safety response, spill containment and emergency contact notification information

As part of the test, the subsea control module (SCM) shall be interfaced and all hydraulic/electric line communications verified with the subsea tree, including using the SCM primary and backup communications systems.

Figure 2 shows a typical hydrostatic bore test schematic to be used during a FAT. The schematic clearly defines what portion of the subsea tree bore shall be considered part of the test. It also defines valve locations, pressure supply points, and pressure transducer locations for the test. This schematic clearly defines the extent of the proposed test limitation. Similar schematics should be developed for each test to identify test areas and define pressure containment components (valves, sensors, etc.). A typical procedure record sheet shall be utilized for all tested components. The sheet identifies the component and defines the specifications of the component and, in general, the governing codes, regulations and other procedural requirements for the test.

4.4 Extended FAT (EFAT)

Extended FATs or Sub-System Tests are normally performed on all systems in a series. This testing will be performed prior to the equipment being included in the SIT. EFAT will be

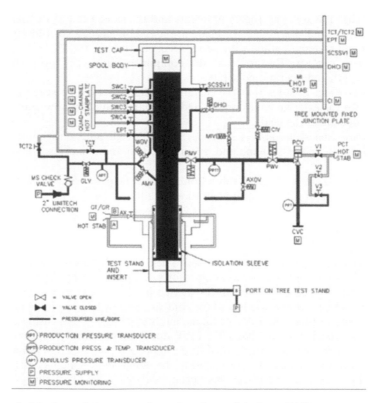

Figure 2. A typical hydrostatic bore test schematic to be used during a FAT.

performed on a complete system or sub-system. The EFAT shall demonstrate the operability of the sub-system, confirm internal and external interfaces, and verify that functional and performance requirements are met.

Each component of the test shall be verified and an Inspection witness records sheet completed. The completed witness records sheet shall be retained in the project documentation system.

4.5 *System Integration Test (SIT)*

The system integration test (SIT) is typically conducted when the entire subsea tree system has been fabricated and assembled. All equipment and Assembly FATs and EFATs should have been successfully completed prior to start of the SIT. The SIT shall replicate field installation and operation procedures as closely as possible and shall capture all possible offshore installation scenarios to minimize the potential for offshore non-productive time.

The purpose of the SIT is to prove and verify the installation/operation procedures and to simulate the sequence of offshore operations including installation and operation of the system. The SIT is land-based and performed on a complete subsea tree system including the subsea tree, tubing hangar, subsea controls, test jumper kits, etc. along with any associated tooling. It should consist of a test at the system level to demonstrate correct interface and operation.

Figure 3 shows a typical test configuration layout of a subsea production tree and a water injection tree in preparation for SIT. All associated components including the production manifold, production SDU, satellite SDU, EFL/HFL, UTH are shown. Test service pumps, MCS, HPU, HYD and ancillary power services are all identified. The layout of the components will require adequate space at the testing location for easy access and allow operational observation.

ROV accessibility and function checks at the following locations, using tooling where appropriate, should be performed during the SIT: well hubs, flowline hubs, HFL/MQC plates, EFL receptacles and valve overrides. Function checks at each multi-quick connect plate

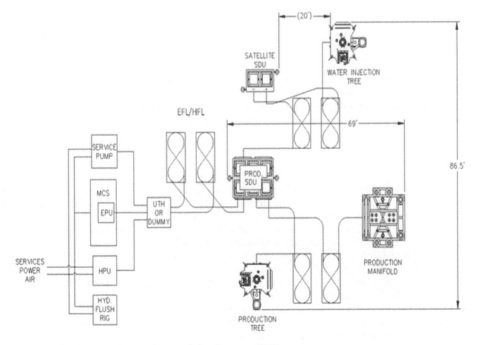

Figure 3.   Typical production and water injection tree SIT layout.

locations, using a subsea tree control module or test device shall be also considered as part of the SIT.

The SIT procedure may be written such that the equipment is left configured and pressure tested in the appropriate condition, at the end of the SIT, thus avoiding further work. All activates of the SIT shall be carefully planned and documented. The company witness should be clearly indicated and recorded. Records should be retained in the project document filing system.

A schedule shall be provided for the various testing completed in a typical SIT. These durations will be different for each component tested and will depend on the system complexity. In addition, the time durations must revised should problems be encountered during the testing effort. It is recommended that SIT be performed according to procedures for installation of the equipment and using key support personnel.

Once the subsea tree and its ancillary equipment have been tested and validated to the functional specification, the entire system will then be ready for offshore installation.

5 SUMMARY AND OTHER CONSIDERATIONS

The testing process for a subsea tree is a complex and time-consuming task. It should be noted that many of the tests included in the codes, standards, regulations, and recommended practices are function/performance tests. Function/performance tests are not sufficient to qualify a design unless the individual components and subassemblies were also tested. It is critical that:

(1) Documentation for subsea tree subassembly designs requiring API Certification should be provided by the equipment provider. A subsea tree assembly cannot be tested to API standards and then rated for subsea use unless all of its components have been verified to a design basis and validated by testing.

(2) Before testing a subsea tree assembly, appropriate quality analysis (QA) documents must be provided by the equipment provider, demonstrating that each subassembly was successfully tested prior to assembly on the subsea tree.

(3) Results from the subsea tree FAT, EFAT, and SIT are essential for an integrity management plan to manage the operating integrity of the subsea tree during its field life.

(4) The qualification of a piece of equipment includes establishing an approach for determining the failure point for the equipment. In general, this will require both analysis and testing. Failure points established through analysis need to be verified through testing. This could entail conducting tests that can be used to verify analysis techniques. If failure points are to be established through testing, prototypes for failure testing will need to be constructed. Construction and testing of these prototypes should be included in the project schedule in order to form realistic expectations for completion of an operational subsea tree.

REFERENCES

1. A. Bobbio, L. Portinale, M. Minichino, and E. Ciancamerla. 2001. Improving the analysis of dependable systems by mapping fault trees into Bayesian networks. *Reliability Engineering & System Safety* 71(3): 249–260.
2. IEC. 2010. *Electric/Electronic/Programmable Electronic safety-related systems, parts 1–7. Technical report*, International Electrotechnical Commission. Issue, 2010.
3. B. Cai, L. Huang and M. Xie. 2017. Bayesian Networks in Fault Diagnosis. *IEEE Transactions on Industrial Informatics* 13(5): 2227–2240.
4. J. Yang and S.J. Cao. 2008. The research of Situation and Trend for Deep Water Oil Drilling Technology. *Oil Drilling Process* 2: 10–13.
5. Y. Dou, Z. Guan. 2006. Offshore Drilling Development Overview. *Offshore Oil* 2: 64–67.
6. ASME. 2005. *Boiler and Pressure Vessel Code, Section VIII*, Divisions 1, 2, and 3 - Qualification. The American Society of Mechanical Engineers.

7. API. 2013. *Specification Q1, Specification for Quality Management System Requirements for Manufacturing Organizations for the Petroleum and Natural Gas Industry*. Ninth Edition. The American Petroleum Institute.

8. API. 2013. *Specification 6A, Specification for Wellhead and Christmas Tree Equipment*. 20th Edition, Addendum 3. The American Petroleum Institute.

9. API. 2014. *Standard 6X, Design Calculations for Pressure Containing Equipment*. First Edition. The American Petroleum Institute.

10. ISO. 2015. *Recommended Practice for Wellhead Surface Safety Valves and Underwater Safety Valves for Offshore Service* (BS EN ISO 10423).

11. API. 2007. *Recommended Practice for Installation, Maintenance and Repair of Surface Safety Valves and Underwater Safety Valves Offshore*. Fifth Edition. The American Petroleum Institute.

12. API. 2012. *Design and Operation of Subsea Production Systems-Subsea Wellhead and Tree Equipment*. 2nd Edition. The American Petroleum Institute.

13. API. 2004. *Recommended Practice for Remotely Operated Vehicle (ROV) Interfaces on Subsea Production Systems*, Second Edition. The American Petroleum Institute.

14. API. 2012. *Specification 20E, Alloy and Carbon Steel Bolting for Use in the Petroleum and Natural Gas Industries*, First Edition. The American Petroleum Institute.

15. API. 2004. *Recommended Practice 75, Recommended Practice for Development of a Safety and Environmental Management Program for Offshore Operations and Facilities*, Third Edition. The American Petroleum Institute.

16. B. Cai, Y. Liu, Q. Fan, Y. Zhang, S. Yu, Z. Liu, and X. Dong. 2013. Performance evaluation of subsea BOP control systems using dynamic Bayesian networks with imperfect repair and preventive maintenance. *Engineering Applications of Artificial Intelligence* 26(10): 2661–2672.

17. B. Cai, Y. Liu, Y. Ma, Z. Liu, Y. Zhou, and J. Sun. 2015. Real-time reliability evaluation methodology based on dynamic Bayesian networks: A case study of a subsea pipe ram BOP system. *ISA Transactions* 58:595–604.

18. Q. Feng, X. Bi, X. Zhao, Y. Chen, and B. Sun. 2017. Heuristic hybrid game approach for fleet condition-based maintenance planning. *Reliability Engineering & System Safety* 157: 166–176.

19. Q. Feng, W. Bi, Y. Chen, Y. Ren, and D. Yang. 2017. Cooperative game approach based on agent learning for fleet maintenance oriented to mission reliability. *Computers & Industrial Engineering* 112: 221–230.

20. T. Huang, J. Yan, M. Jiang, W. Peng, and H. Huang. 2016. Reliability analysis of electrical system of computer numerical control machine tool based on Bayesian networks. *Journal of Shanghai Jiaotong University* (Science) 21(5): 635–640.

21. J. Mi, Y. Li, Y. Yang, W. Peng, and H. Huang. 2016. Reliability assessment of complex electromechanical systems under epistemic uncertainty. *Reliability Engineering & System Safety* 152: 1–15.

Risk Analysis Based on Data and Crisis Response Beyond Knowledge – Huang & Nivolianitou (eds)
© 2020 Taylor & Francis Group, London, ISBN 978-0-367-25146-8

Risk of photovoltaic systems on building fire safety in China

Yangyang Mu, Jin Li, Changxing Ren, Wang Zhang & Xin Zhang
Tianjin Fire Research Institute of MEM, Tianjin, China
National Center for Fire Engineering Technology, Tianjin, China

ABSTRACT: The use of photovoltaic (PV) systems mounted on buildings is growing considerably in China. These operate at high voltages with the potential to cause or promote fires. However, assessment research is lacking on the fire risk of PV systems to buildings. In this article, we present three main sections about PV buildings: fire behavior testing, the fire protection of buildings, and fire suppression measurements. Mainly we focus on the development to current standards of the fire classification rating testing of PV modules and basic fire-protection requirements as well as extinguishment. We also discuss the potential safety hazards and associated problems. Finally, we propose steps to mitigate the fire risk of PV systems installed on buildings.

Keywords: building, photovoltaic system, fire risk, fire classification rating

1 INTRODUCTION

Developments in renewable energy are increasing considerably in many countries, due to the risk of the global warming and the depletion of fossil fuels. Photovoltaic (PV) is the third renewable energy source in terms of global capacity. According to the International Energy Agency (IEA) forecast, total global solar PV capacity could exceed 1500 gigawatts (GW) by 2030 and will grow to 4000 GW by 2050. The PV industry has become a strategic industry in China and achieved a yearly growth rate of around 40 GW in the last two years with the support of the government (Yang & Zhao, 2018). Statistically, the Chinese domestic PV capacity exceeded 174 GW in the last year and total capacity will reach 201 GW by 2019, which is the fastest growing market in the world. Figure1 shows the change of total PV capacity in China from 2010 to 2019.

Generally, PV systems are mounted on the rooftops of existing buildings. However, installation involves several safety problems, especially fire risk (Murata et al., 2003). In fact, the diffusion of PV systems has led to several accidental combustion cases around the world. As a result, it is worthwhile identifying the risk of PV systems to buildings in China and propose plans that to be implemented in further research.

2 PV SYSTEM FIRES

There were 298 reported interventions of fire departments related to PV systems in Italy (2011) and 390 related fires in Germany in 2012 (Manzini et al., 2015b). Meanwhile, PV system fires in public have caused billions of dollars economic losses in China over the past few years.

Table 1 shows typical domestic PV system fires in the last two years. Based on the analysis of previous fire cases related to PV systems, the potential causes mainly include the following.

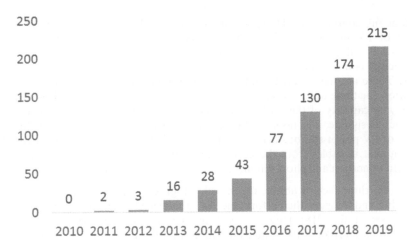

Figure 1. The change of total PV capacity in China from 2010 to 2019 (GW).

Table 1. Description of typical domestic PV system fires happened in 2017 and 2018.

Date	Location	Survey	Accident cause
May, 2018	Beijing, China	The rooftop PV power station of a primary school caught fire, caused by an electrical fault.	Electrical fault
April 3rd, 2018	Gansu, China	The 550 V high-voltage transformer box of a country PV power plant caught fire.	Electrical fault
February, 2018	Hebei, China	A PV power plant under construction, whose capacity would be 10 MW, ignited and the modules amounting to 3 MW were damaged. This was caused by fire from the construction site and led to direct economic losses exceeded five million.	Fire from the construction site
January, 2018	Henan, China	A fire happened in a 100 MW PV power plant under construction in the south of Henan province. Approximately 1.2 MW of modules were damaged. The fire ignition source was flying flame and the fire lasted for more than seven hours.	Flying flame caused by human error
December, 2017	Zhejiang, China	A PV power plant of 20 MW capacity was ignited by fire and almost 1000 panels, as well as some cables and combiner boxes, were destroyed. The direct economic losses exceeded 1 million Yuan.	Fire left over from local people
December 5th, 2017	Shannxi, China	A fire in a PV power plant in Dingbian county, Shannxi province. The transformer and voltage regulator were destroyed and around 120 electrical storages were damaged to varying degrees.	Electrical fault
May, 2017	Shanxi, China	The inverter of a rooftop PV power station was ignited, and the fire was propagated to the whole system.	Direct current arcing

(1) Arcing. Most of PV fires are electrical fires and the main cause of electrical fires is arcing. According to statistics on the fires at PV power stations, those caused by direct current arcing account for more than 40% (Nehme et al., 2017). The equipment at greater risk of fire includes combiner boxes, inverters, transformers and distribution cabinets, etc. Reasons for this include loose connections, poor contacts and poor-quality construction in the PV system, which may induce arcs and ignite the back sheet of PV panels.

(2) Spontaneous combustion. There are many factors that may cause spontaneous combustion of PV modules, such as the hot spot effect, poor production quality, aging and thermal radiation, etc. The hot spot effect is one of the most common causes of spontaneous combustion.

(3) Cables. There are many cables and electrical equipment in a PV system, and the power generation for these depends on solar radiation. The load of the electrical equipment and the carrying capacity of the cables also fluctuates with the change of solar radiation, and they reach design peak at noon and decrease to zero at night. It has been shown that the fire risk of PV power stations is higher than that of thermal power plants, caused by accumulating heat in cables from the current fluctuation. In addition, the overheating under long-term operation and short-circuiting of cables can also lead to fire accidents.

Aimed at the three factors above, much technological effort has been made to reduce the harm of arcing and improve the performance of cables (Reil et al., 2011; Falvo & Capparella, 2015; Kristensen et al., 2018). However, with the development of research, we conclude that some of the potential risks of PV systems to buildings' fire safety must be addressed.

3 POTENTIAL RISKS OF PV SYSTEMS TO BUILDINGS

3.1 *Behavior testing of modules and back sheets*

The PV cells family used for commercial applications can be divided into three types: single crystal silicon PV cells, polycrystalline silicone PV cells, and thin-film PV cells. From the past fire accidents, we can see that all these cells were involved in the combustion experience.

Current PV modules pose the following potential safety risks.

(1) The fire safety regulation for backing materials should be improved. As required by the FM Global Insurance Company, the back panels for the PV modules in a rooftop PV power station should be of fireproof grade A or B (Manzini et al., 2015a). However, in China there are currently no standards for the fireproof performance of backing materials for PV modules. As a result, low quality fortifications leave potential fire hazards.

(2) The flame-retardant properties of backing materials are varied. Insulation back sheets made of polymer composite materials, typically a sandwich structure featuring fluorine film + polyethylene terephthalate (PET) + fluorine film, are adopted in the majority of crystalline silicon PV cells, and the fluorine film is the key to determining the flame-retardant properties of the back sheet. However, domestic enterprises add low fluorine content to the materials because of cost and the filming process, resulting in poor flame resistance, which is a weak fireproofing link in the whole module.

(3) The fire risk of using packaging adhesive for crystalline silicon PV modules is overlooked. Ethylene-vinyl acetate copolymer (EVA) is a commonly-used packaging material in the packaging process of crystalline silicon PV cells, but is a typical combustible with a melting temperature of 123°C and a combustion grade of B2 (Yang et al., 2015; Nair & Kulkarni, 2018). According to previous studies, EVA is the most likely component to be ignited in the PV cell, whose total released combustion heat accounts for 60% of the PV cell, releasing a mass of flue gas and posing a high risk of fire.

(4) The present study on the fire spread properties of PV modules is deficient. Rooftop PV power stations and PV curtain walls are new uses of PVs. However, there is presently no fireproof design and regulation for PV buildings in China. What's worse, there is a lack of studies on the effects of PV modules on building fire protection. In particular, there is still a gap in the knowledge on the fire spread of PV modules in practical applications, and current studies on the combustion performance of PV cell panels are mainly concentrated on crystalline silicon PV cell panels, with a lack of combustion characteristic parameters of thin-film PV modules.(Spataru et al., 2013; Tommasini et al., 2014; Despinasse & Krueger, 2015) Therefore, it is of great significance to study the fire risk of PV power stations.

3.2 *Technical testing standards of PV modules*

Great importance has been attached at home and abroad to the standardized construction of the PV industry. The following two aspects can be observed from analyzing the similarities and differences in domestic and foreign PV industries. First, a well-developed PV standard system has been established overseas (Manzini et al., 2015a) and the IEC standards and ANSI UL standards are the two main PV test protocols globally. Of these, the IEC standards IEC 61215 and IEC 61730 are applicable to the fire risk testing of PV modules, proposing safety evaluation methods for PV modules, such as hot spot endurance tests, bypass diode thermal performance tests, temperature resistance tests, and fireproof performance tests, etc. In addition, UL 1703 and UL 790 specify the fireproof test methods for the distributed roof-mounted PV systems used for buildings or connected to buildings, and classify the fire ratings of the PV system into Grade A, Grade B and Grade C (Dhere et al., 2012). In comparison, UL standards put more emphasis on fire safety tests for PV modules, while IEC standards focus on product performance. Table 2 lists the main technical standards of PV systems from the IEC and UL.

Table 2.   Technical standards of PV system from IEC and UL.

Edition	Title	Details
IEC 61215-1-2016	Terrestrial photovoltaic (PV) modules – Design qualification and type approval – Part 1: Test requirements	IEC 61215-2016, which includes five related standards, lays down special requirements for testing the electrical and thermal char-
IEC 61215-2-2016	Terrestrial photovoltaic (PV) modules – Design qualification and type approval – Part 2: Test procedures	acteristics of the module as well as the ability to withstand prolonged exposure in climates. This standard is intended to apply to all terrestrial flat-plate module materials, such as crystalline silicon module types as well as thin-film modules.
IEC 61730-1:2016	Photovoltaic (PV) module safety quali-fication – Part 1: Requirements for construction	IEC 61730-1:2016 lays down the fundamen-tal construction requirements for photovol-taic (PV) modules in order to provide safe
IEC 61730-2:2016	Photovoltaic (PV) module safety quali-fication – Part 2: Requirements for testing	electrical and mechanical operation. Spe-cific topics are provided to assess the pre-vention of electrical shock, fire hazards, and personal injury due to mechanical and environmental stresses. IEC 61730-2 defines the requirements of testing to verify the safety of PV modules.
UL 1703:2002 (2018 approved)	Standard for Flat-Plate Photovoltaic Modules and Panels	UL 1703:2002 provides detailed measure-ments to determine the fire performance characterization and system fire class rating of modules and panels, which is vital to the fire protection of buildings. This standard applies to all flat-plate photovoltaic mod-ules and panels installed on or integral with buildings or to be freestanding.
UL 790: 2004 (2018 approved)	Standard for Standard Test Methods for Fire Tests of Roof Coverings	UL 790: 2004 describes the requirements of the relative fire characteristics for both materials and components as roof coverings exposed to simulated fire sources from out-side a building on which the coverings are installed. The intermittent-flame, burning-brand, and flying-brand test and the spread-of-flame test are requirements for PV module safety qualification according to the UL790 standard.

Second, the standardization of domestic PV systems should be accelerated. There is a need for research into standards of PV modules. Implemented standards mainly focus on technical and construction regulations, lacking safety test standards for PV modules. Furthermore, some existing standards have not been revised for a long time – even since the 1990s. China only issued GB/T 20047.1-2006 "Photovoltaic (PV) module safety qualification—Part 1: Requirements for construction", which is identical to IEC 61730:2005, in 2006 to stipulate the requirements for construction of PV modules. Nevertheless, test standards relating to the fire-proof performance of PV modules has not yet been issued, let alone the application standard of PV technology in buildings. In reality, the reaction to fireproof performance of PV modules has been mainly tested through drawing from other building materials and products, such as GB 8624-2012 "Classification for burning behavior of building materials and products".

3.3 *Codes for fire protection of distributed PV power stations*

Domestic distributed PV power stations, in general, are added on the basis of existing buildings, which may greatly increase the fire risk of buildings. Unfortunately, no corresponding safety measures have been proposed for the structure, fire resistance grade, fire separation distance and corresponding fire-fighting facilities of the buildings. This is specifically reflected as follows.

(1) There is no clear regulation on building fire protection. Current fire-protection standards used in most PV power plants are applicable merely to large-scale PV power plants, while the fire-protection design of the distributed PV power stations built on civil and industrial buildings is mainly implemented through referring to the GB 50016-2018 "Code for fire-protection design of buildings". Nonetheless, active and passive fire precautions in the buildings of distributed PV power plants are not matched with the actual fire risks, while fire protections of the buildings and power generation systems are not covered by the existing standards. Consequently, new fire-protection problems may be raised. For instance, the fire resistance grade of PV buildings and fire separations with surrounding buildings are mainly defined in accordance with the property of the main building by referring to GB 50016-2018, while the effect of PV power stations on the fire risk of buildings is ignored.

(2) The combustion performance of external thermal insulation of buildings should be strengthened. As can be seen from past fires, roof-mounted insulation materials would be ignited in the case of fire in PV power stations. Also, the outbreak of fire in the PV systems on the external walls would result in fire spreading upwards rapidly, or even spreading inside the buildings through doors, windows, and holes on the external walls, and cause more damage. Hence, the combustion performance of the external thermal insulation in the distributed PV power generation system buildings should have higher requirements.

(3) The fire risk of PV glass curtain walls is a pressing research topic. A PV curtain wall is a novel structure. According to the survey, it is found that the fireproof design criteria for the construction of PV curtain walls are in line with the fortification requirements for the glass curtain walls in GB 50016-2018. There are no fireproof design requirements and standards specially prepared for PV curtain walls. In the meantime, the fire risk of PV curtain walls will be far higher than the non-combustible glass curtain walls due to the particularity of the PV curtain wall. In addition, the fire spreading properties of tPV curtain walls is still unknown, which undoubtedly increased the fire risk of such buildings.

4   CONCLUSION AND DISCUSSION

The application of photovoltaic systems face many risks in China, such as fire behavior of materials, testing standards of modules and the fire protection of PV systems. Reports have shown that the backing sheets of PV panels are flammable and it is necessary to identify the features of PV modules and backing sheets regarding this fire risk. There is also a deficiency of studies on the fire spread properties of PV modules. Thus, research would be of great significance.

It is evident that foreign fire prevention organizations have a series of technical testing standards for PV modules, while Chinese organizations have not proposed test standards related to the fireproof performance. The standardization of PV systems should be developed in harmony with their application in China.

The current codes do not cover distributed PV power stations mounted on buildings and the technical requirements for fire resistance grade, fire separation distance and extinguisher facilities of the buildings still remain unclear. There is also urgent research needed on the fire risk of PV glass curtain walls.

ACKNOWLEDGMENTS

This project was supported by the Basic Scientific Research Foundation of Tianjin Fire Research Institute (2018SJ05).

REFERENCES

Despinasse M-C, Krueger S. 2015. First developments of a new test to evaluate the fire behavior of photovoltaic modules on roofs. *Fire Safety Journal* 71, 49–57.

Dhere NG, Backstrom R, Dini D, Wohlgemuth JH. 2012. Firefighter safety and photovoltaic installations research project. In: Proceedings of SPIE – The International Society for Optical Engineering.

Falvo MC, Capparella S. 2015. Safety issues in PV systems: Design choices for a secure fault detection and for preventing fire risk. *Case Studies in Fire Safety* 3, 1–16.

Kristensen JS, Merci B, Jomaas G. 2018. Fire-induced reradiation underneath photovoltaic arrays on flat roofs. *Fire and Materials* 42, 316–323.

Manzini G, Gramazio P, Guastella S, Liciotti C, Baffoni GL. 2015a. The Fire Risk in Photovoltaic Installations - Test Protocols For Fire Behavior of PV Modules. *Energy Procedia* 82, 752–758.

Manzini G, Gramazio P, Guastella S, Liciotti C, Baffoni GL. 2015b. The Fire Risk in Photovoltaic Installations – Checking the PV Modules Safety in Case of Fire. *Energy Procedia* 81, 665–672.

Murata K, Yagiura T, Takeda K, Tanaka M, Kiyama S. 2003. New type of photovoltaic module integrated with roofing material (highly fire-resistant PV tile). *Solar Energy Materials & Solar Cells* 75, 647–653.

Nair SS, Kulkarni AM. 2018. Experimental Study on the Flammability of Photovoltaic Module Backsheets. *International Journal of Scientific & Engineering Research* 9, 98–102.

Nehme B, Msirdi NK, Namaane A, Akiki T. 2017. Analysis and Characterization of Faults in PV Panels. *Energy Procedia* 111, 1020–1029.

Reil F, Vaassen W, Sepanski A, et al. 2011. Determination of fire safety risks at PV systems and development of risk minimization measures. *Proceedings of the 26th European Photovoltaic Solar Energy Conference and Exhibition, 2011*. Hamburg, Germany.

Spataru S, Sera D, Blaabjerg F, Mathe L, Kerekes T. 2013. Firefighter safety for PV systems: Overview of future requirements and protection systems. *Proceedings of the Energy Conversion Congress & Exposition, 2013*.

Tommasini R, Pons E, Palamara F, Turturici C, Colella P. 2014. Risk of electrocution during fire suppression activities involving photovoltaic systems. *Fire Safety Journal* 67, 35–41.

Yang FF, Zhao XG, 2018. Policies and Economic Efficiency of China's Distributed Photovoltaic and Energy Storage Industry. *Energy* 154.

Yang HY, Zhou XD, Yang LZ, Zhang TL. 2015. Experimental Studies on the Flammability and Fire Hazards of Photovoltaic Modules. *Materials (Basel)* 8, 4210–4225.

Analysis of the dangerous consequences of storing pentane in polyurethane foam enterprises

Yue Wang, Jin Li, Dong Lv, Nan Jiang & Yan Zhang
National Center of Fire Engineering Technology, Tianjin, China

ABSTRACT: In the production of polyurethane foam, using pentane instead of HCFC for foaming, which reduces the destruction of the ozone layer, on the other hand increases the risk in the process of polyurethane foam production and raw material storage because of low flash point (-35°C) of pentane. In this paper, the dangerous consequences of storing pentane and foaming materials containing pentane in polyurethane foam manufacturing enterprises are analyzed. The scenes of explosion after leakage are simulated in four kinds of situation, which are pentane stored in a completely closed space, the pressure relief area of 9m^2 and 18m^2 and completely open space. The flame propagation range, explosion overpressure, pressure wave propagation range and temperature change are calculated. Finally, it can be found that the maximum overpressure is 861.1kPa in the case of completely closed space. When in a fully open space, the maximum overpressure is 2.03kPa. When the pressure relief area is 9m^2 and 18m^2, the maximum overpressure are 16.50kPa and 24.06kPa respectively. Considering the factors of blockage outside pressure relief opening and indoor ventilation, pressure relief area of 9m^2 is adopted as the most reasonable way of pressure relief.

Keywords: pentane, leakage, flame propagation, explosion overpressure, pressure wave, temperature change

1 INTRODUCTION

Polyurethane foam(Cao, Liao et al. 2019) is an important insulating and waterproof material. Due to its good physical, electrical and chemical resistance and other properties, polyurethane foam is widely used in insulation layer of refrigerators and freezers box, and insulation materials of cold storage and refrigerated vehicles and pipe insulation materials. Polyurethane foam is formed by reaction of polyether with isocyanate (Ma and Spencer 2018) and foaming agent. HCFC-141b (difluoromono-chloroethane) (Le et al. 2012), as a transitional technology to replace CFC foaming agent, is generally recognized and used by the polyurethane foam enterprises around 1994. However, HCFC compounds still contain chlorine elements, and ozone depletion potential (hereinafter referred to as ODP value (Awad and Oboh-Ikuenobe 2019) is not zero, which still has a destructive effect on the ozone layer. With the deepening of the implementation process of the Montreal Protocol on Substances that Deplete the Ozone Layer, HCFC compounds are gradually phased out, and the ultimate goal is to completely adopt the zero-ODP foaming agent. To achieve this goal, the replacement technology of hydrocarbon foaming agent represented by pentane (Ferkl et al. 2017) is developing continuously, and more and more enterprises try to use this technology to replace the original foaming agent. Pentane is a polyurethane foaming agent with zero ODP value and little greenhouse effect. As an alternative to HCFC, many enterprises have modified or built new production lines to meet the foaming process of polyurethane foam (Demirel and Tuna 2019). Pentane is flammable and explosive, which vapor could form explosive mixture with air. The explosive mixture could occur combustion in case of open fire, high heat, or contacting with strong oxidants.

In the process of storing pentane and the raw materials containing pentane in the polyurethane foam enterprises, if pentane leaks and be ignited, it will bring serious dangerous consequences (Gao et al. 2018). Therefore, this paper mainly carries out a quantitative and qualitative analysis of the relevant risk consequences after the leakage of the raw materials involved in the storage of pentane and the production of pentane, so as to improve the safety production capacity and the stability of the equipment (Kim et al. 2013).

2 MAJOR HAZARD ANALYSIS

According to the physical and chemical properties of pentane in the Hazardous Chemical Materials Safety Technical Book, it can be found that the flash point of pentane is around -35°C, and the flash point of the premixed pentane portfolio polyether material is also around 0°C. According to the classification of the dangerous liquid, these two kinds of materials belong to Class A dangerous liquid (da Cunha et al. 2018). In general, in order to facilitate production, these two kinds of materials will be placed in the intermediate warehouse of the production plant and stored in plastic tons of barrels. In the daily production process, if the tons of barrels are impacted by external objects, the tons of barrels containing the premixed pentane portfolio polyether material will leak out and form explosive gas mixture with the air (Zhang and Yan 2019). The intermediate warehouse is a relatively closed space, in which the ignition of mixed gas will lead to a serious explosion accident, which will bring serious damage to buildings, production equipment and operators (Zhang et al. 2019).

3 GEOMETRIC MODEL AND SIMULATION SCENES

3.1 Geometric model

According to the actual investigation of the plant production of polyurethane foam manufacturing enterprises, the production plant is simplified as the following model, as shown in Figure 1. The geometric size of the production plant is 40m*30m*5m. There are twelve windows on the wall around the production plant, including three windows on the short side and 9 windows on the top of the production plant. The production plant is equipped with three production lines, which is 25 m in length, and has four foaming machines for producing polyurethane foam. The intermediate warehouse for storing the premixed pentane portfolio polyether material is the part in red color in Figure 1, with the size of 10m*8m*3m. The intermediate warehouse is located near the workshop with doors and windows nearby.

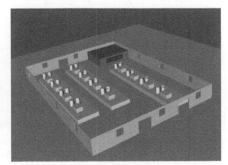

External Structure   Internal Structure

Figure 1.   The geometric model of production plant.

| Completely Closed | Pressure Relief Area of 9 m² | Pressure Relief Area of 18 m² | Completely Open |

Figure 2.   Leakage explosion scenes during storage.

3.2   *Explosion scenes*

According to the investigation of different manufacturing enterprises, the intermediate ware-house can be divided into three types: completely closed without any pressure relief opening, partially closed with pressure relief port and completely open.

This paper chooses four explosion scenes of intermediate warehouse, among these three types, which are the completely closed space without any pressure relief opening, the pressure relief area of 9m², the pressure relief area of 18m² and the completely open space, as shown in Figure 2. In the explosion scenes, it is assumed that gaseous pentane mixed with air, in which the concentration of pentane is within the explosion limit, fills the entire intermediate warehouse after the leakage of the premixed pentane portfolio polyether materials. It is also assumed that the mixed explosive gas is ignited at the center of the intermediate warehouse. The international quantitative risk calculation software FLACS is used for the quantitative calculation of the simulation scenes, and parameters such as explosion pressure and flame temperature are calcu-lated to quantitatively evaluate the dangerous consequences of the explosion.

4   CALCULATION RESULTS AND DANGEROUS CONSEQUENCE ANALYSIS

FLACS, a commercial simulation software, is used to conduct quantitative simulation calcula-tion for four explosion scenes. The change of maximum explosion pressure with time in the four explosion scenes is shown in Figure 3. The specific conditions of the maximum explosion

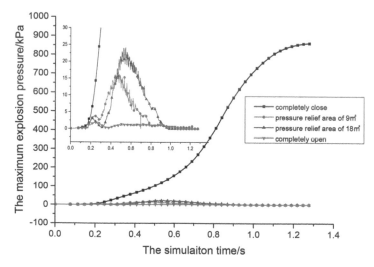

Figure 3.   The change of maximum explosion pressure with time in the four explosion scenes.

412

Table 1. The specific conditions of maximum explosion pressure and maximum flame temperature.

Explosion Scenes	Maximum Explosion Pressure /kPa	Maximum Flame Temperature /°C
Completely Closed Space	861.1	2322.6
Pressure Relief Area of 9m^2	16.50	2025.85
Pressure Relief Area of 18m^2	24.06	2035.05
Completely Open Space	2.03	2008.95

pressure and maximum flame temperature are shown in Table 1. For a completely closed space, flame and explosion wave did not spread out, and the scope of influence is only in the intermediate warehouse, while flame and explosion shock wave spread out in other scenes, as shown in Figure 4 and 5.

4.1 *Dangerous consequence analysis of completely closed space*

When the premixed pentane portfolio polyether materials leak and explode in the completely closed space, the maximum explosion pressure can reach 861.1kPa and the highest temperature of flame can reach 2322.6°C. Under the condition of this maximum explosion pressure, the large steel frame structure of the building will be destroyed; the brick wall will collapse; seismic reinforced concrete damage; small house collapse. If there is an explosion exploded in

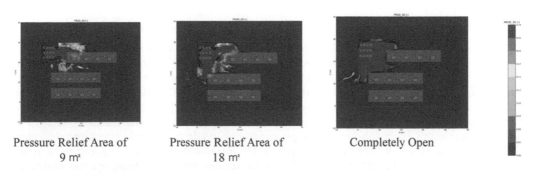

Pressure Relief Area of 9 m²	Pressure Relief Area of 18 m²	Completely Open

Figure 4. Maximum spread of flame.

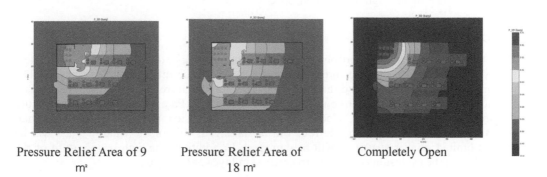

Pressure Relief Area of 9 m²	Pressure Relief Area of 18 m²	Completely Open

Figure 5. Overpressure cloud picture of explosion.

the completely closed space, the whole room will collapse and part or all of the shell of the plastic tons of barrels in the intermediate warehouse will be shattered and cause further combustion or explosion accident. The plastic tons of barrels contain premixed pentane composite polyether room. For personnel, all personnel will die under this huge explosion pressure condition. In addition to explosion damage, the high temperature of the flames during the explosion can also burn entire skin layer of the body.

4.2 *Dangerous consequence analysis of the space with pressure relief area of 9 m²*

When the premixed pentane portfolio polyether materials leak and explode in the space with pressure relief area of 9m², the explosion shock wave spreads from the center of the intermediate warehouse. The window on one side of the intermediate warehouse is used as a pressure relief opening for pressure relief. The pressure of explosion wave rises fastest along the pressure relief opening and the flame also spurt out from the pressure relief opening and spreads around. The maximum explosion pressure and the maximum flame temperature appear in the intermediate warehouse. The maximum explosion pressure can reach 16.50kPa and the highest temperature of flame can reach 2025.85°C. Under the condition of this maximum explosion pressure, 50% of the brick will be damaged; the structure of the plant will be seriously damaged; the concrete wall will collapse; the warehouse cannot continue to be used. The explosion wave also could cause the following consequences: all the glass of the windows and doors of the compressed surface will be broken; window frame will be damaged; asbestos plate will be crushed; steel plate or aluminum plate will be wrinkled and tightening failures. In the event of an explosion, the intermediate warehouse will collapse and the glasses will be shattered and send shards flying into the air, which maybe injure or kill people nearby. For personnel, the explosion wave can cause minor damage to the body. In addition, the high temperature of the flames during the explosion can also burn entire skin layer of the body.

4.3 *Dangerous consequence analysis of the space with pressure relief area of 18 m²*

When the premixed pentane portfolio polyether materials leak and explode in the space with pressure relief area of 18 m², the explosion shock wave spreads from the center of the intermediate warehouse. The doors and windows in the horizontal and vertical directions shown in Figure 4 are used as pressure relief opening for pressure relief. The flame during the explosion will also spurt out from the two pressure relief ports in the horizontal and vertical directions and spread around. The highest flame temperature appears in the intermediate warehouse, and the maximum explosion pressure appears at 8m outside the vertical pressure relief port. The maximum explosion pressure can reach 24.06kPa and the highest temperature of flame can reach 2035.05°C. The main reason why the maximum explosion pressure appears outside the pressure relief opening is that there is a relatively long and narrow space outside the pressure relief opening, which has a blocking effect on the blast wave. The explosion wave will rise further after passing through the blocking space.

4.4 *Dangerous consequence analysis of completely open space*

When the premixed pentane portfolio polyether materials leak and explode in the completely open space, the explosion wave spreads from the center of the intermediate warehouse. The maximum explosion pressure and the maximum flame temperature appear in the intermediate warehouse. The maximum explosion pressure can reach 2.03kPa and the highest temperature of flame can reach 2008.95°C. Under the maximum explosion pressure, noise more than 143dB will be produced; part of the roof will be damaged; 10% of the window glass will be broken; limited smaller structures will be damaged. In addition, the high temperature of the flames during the explosion can also burn entire skin layer of the body.

5 CONCLUSION

According to calculation results and dangerous consequence analysis of four kinds of leakage and explosion scenes, it can be found that:

(1) When the intermediate warehouse of the premixed pentane portfolio polyether material is completely closed, the maximum explosion pressure will reach 861.1kPa if the premixed pentane portfolio polyether material inside the warehouse leaks and explosion occurs. This maximum explosion pressure, which is enough to destroy the entire intermediate warehouse, is also devastating to the surrounding buildings, production equipment and the body of personnel.

(2) The explosion pressure can be significantly reduced by setting a pressure relief opening in the intermediate warehouse of the polyurethane foam production plant. Generally, the maximum explosion pressure will appear in the intermediate warehouse. When there is a relatively long and narrow space outside the pressure relief opening, the explosion pressure will significantly increase when explosion wave passes through this long and narrow space. The overpressure of this process will cause great damage to the building, production equipment and the body of personnel.

(3) When the intermediate warehouse of the premixed pentane portfolio polyether materials is a completely open space, the maximum explosion pressure is about 25% of the same condition in the space with pressure relief area of $18m^2$. Although the completely open storage method can effectively reduce the damage to human body, buildings and production equipment caused by explosion overpressure, it is not conducive to daily ventilation and the discharge of trace leakage of pentane and other combustible vapors in plastic tons of barrels of the premixed pentane portfolio polyether material. It will significantly increase the probability of ignition after the leakage of the premixed pentane portfolio polyether material.

(4) In addition to the damage caused by the explosion shock wave, the maximum flame temperature will reach over 2000°C in four kinds of explosion scenes, which also could cause damage to the body of personnel. The high temperature of flame will burn entire skin layer of the body. The high temperature flame will also bake the leaking tons of barrels, further accelerating the leakage and leading to a secondary fire and explosion accident. The high temperature flame will also ignite the surrounding combustibles, including the finished polyurethane foam production.

(5) Considering the factors of blockage outside pressure relief opening and indoor ventilation, pressure relief area of $9m^2$ is adopted as the most reasonable way of pressure relief.

ACKNOWLEDGMENTS

This project was supported by the Science and Technology Program of Tianjin, China (17YFZCSF00970).

REFERENCES

Awad, W. K. and F. E. Oboh-Ikuenobe (2019). "Paleogene-early Neogene paleoenvironmental reconstruction based on palynological analysis of ODP Hole 959A, West Africa." Marine Micropaleontology 148: 29–45.

Cao, Z.-J., W. Liao, S.-X. Wang, H.-B. Zhao and Y.-Z. Wang (2019). "Polyurethane foams with functionalized graphene towards high fire-resistance, low smoke release, superior thermal insulation." Chemical Engineering Journal 361: 1245–1254.

da Cunha, S., V. Gerbaud, N. Shcherbakova and H.-J. Liaw (2018). "Classification for ternary flash point mixtures diagrams regarding miscible flammable compounds." Fluid Phase Equilibria 466: 110–123.

Demirel, S. and B. E. Tuna (2019). "Evaluation of the cyclic fatigue performance of polyurethane foam in different density and category." Polymer Testing.

Ferkl, P., M. Toulec, E. Laurini, S. Pricl, M. Fermeglia, S. Auffarth, B. Eling, V. Settels and J. Kosek (2017). "Multi-scale modelling of heat transfer in polyurethane foams." Chemical Engineering Science 172: 323–334.

Gao, Z. M., Y. Gao, W. K. Chow, Y. Wan and C. L. Chow (2018). "Experimental scale model study on explosion of clean refrigerant leaked in an underground plant room." Tunnelling and Underground Space Technology 78: 35–46.

Kim, E., J. Park, J. H. Cho and I. Moon (2013). "Simulation of hydrogen leak and explosion for the safety design of hydrogen fueling station in Korea." International Journal of Hydrogen Energy 38(3): 1737–1743.

Le Bris, K., J. McDowell and K. Strong (2012). "Measurements of the infrared absorption cross-sections of HCFC-141b (CH3CFCl2)." Journal of Quantitative Spectroscopy and Radiative Transfer 113(15): 1913–1919.

Ma, P. and J. T. Spencer (2018). "Cyclodimerization of isocyanates promoted by one large vertex metallaborane." Polyhedron 149: 148–152.

Zhang, L., H. Ma, Z. Shen, L. Wang, R. Liu and J. Pan (2019). "Influence of pressure and temperature on explosion characteristics of n-hexane/air mixtures." Experimental Thermal and Fluid Science 102: 52–60.

Zhang, Q. and Q. Yan (2019). "Experimental study on explosion process of flour deposits/air mixture in horizontal pipelines." Powder Technology 346: 273–282.

Risk Analysis Based on Data and Crisis Response Beyond Knowledge – Huang & Nivolianitou (eds)
© 2020 Taylor & Francis Group, London, ISBN 978-0-367-25146-8

A fuzzy extension to the risk situation awareness provision indicator (RiskSOAP)

Apostolos Zeleskidis & Ioannis M. Dokas
Civil Engineering Department, Polytechnioupoli, Xanthis, Greece

ABSTRACT: This study presents an extension of the Risk Situation Awareness Provision (RiskSOAP) indicator to allow the incorporation of fuzzy safety requirements. RiskSOAP is a comparison-based indicator that expresses to what extent a system, based on its design and the state of its components during operation, understands the presence of threats and vulnerabilities. The proposed extension utilizes fuzzy linguistic variables to translate real world system data to design vectors with values of [0, 1], diverging from the crisp values used by the original RiskSOAP methodology. Euclidean distance is then used to replace the dissimilarity between the ideal and the real design vectors. The proposed fuzzy RiskSOAP indicator was applied to a case study of a typical railway crossing and its results were compared against the results of the crisp RiskSOAP indicator applied on the same system.

Keywords: safety, vulnerabilities, RiskSOAP, fuzzy, linguistic variables, Euclidean distance

1 INTRODUCTION

With the motive of developing new tools and methods for designing and developing systems that are aware of their vulnerabilities and react to prevent accidents and losses, Chatzimicailidou and Dokas (2016) introduced the RiskSOAP methodology and its respective indicator to measure the capability of a complex socio-technical system to provide its agents with Situational Awareness (SA) about the presence of threats and vulnerabilities and enables analysts to assess distributed SA. The RiskSOAP methodology uniquely combines three methods: (1) the STPA hazard analysis (Leveson, 2013), (2) the EWaSAP early warning sign identification approach (Dokas, Feehan & Imran, 2013), and (3) a dissimilarity measure for calculating the distance between binary sets.

However, a limitation of the RiskSOAP indicator is that it neglects that its variables, which indicate if a safety constraint at the time of the analysis is satisfied or not, may have a truth value that ranges between '0' and '1' (i.e. not satisfied and fully satisfied) (Chatzimicailidou & Dokas, 2016).

This paper introduces a way to address this problem. Specifically, all possible fuzzy safety requirements are categorized in three distinct membership function types: discrete fuzzy, continuous fuzzy and strictly crisp. The values of these fuzzy safety requirements, that belong in the [0,1] set, change over time during the system operation phase according to their assigned fuzzy membership function. Then, the Euclidean distance is utilized to calculate the dissimilarity measure between the comparable non-binary sets.

The approach is illustrated with the case study of a railway crossing. This crossing consists of barriers, a buzzer alert and a light alert. Cars can cross and are alerted of incoming trains by those means. It was found that 43% of the railway crossing safety requirements can be expressed as fuzzy safety requirements. As a result when the Fuzzy RiskSOAP approach was applied its indicator value diverged by 0.7% to 13% from the crisp Risk-SOAP value.

2 METHODS

2.1 *The RiskSOAP indicator*

The RiskSOAP indicator provides a measure of the inherent capability of a system to shape and maintain its situational awareness of threats and vulnerabilities. This is achieved by measuring the dissimilarity of at least two binary vectors. The vectors have $n \times 1$ dimensions where each row represents a different safety requirement defined by a hazard analysis applied to that system. Thus, n equals the total number of system safety specifications.

The first vector depicts the ideal system state (i.e. the system fulfills all safety specifications). This vector is called the "ideal" vector "i" and all its elements take the value 1. The value 1 denotes that the safety specification is fully satisfied. The other vectors may depict the actual state of the system at different time periods. These vectors are called "real " or "r". Their elements take values of either 0 or 1 according to if the specification in question is satisfied or not.

The dissimilarity between each "r" vector against the "i" vector is measured by the Rogers-Tanimoto dissimilarity measure using the formula that is depicted in Table 1. The terms "S00", "S01", "S10", and "S11" in Table 1 denote the total number of the corresponding (0,0), (0,1), (1,0), and (1,1) pairs of binary integers in Figure 1, of the two compared vectors. Figure 1 conveys that in order for vectors to be compared, they have to have the same number of rows (Chatzimichailidou, 2015), meaning one cannot compare the vectors of two different systems.

Its dissimilarity can range from 0 to 1. The value 1 expresses that the two vectors (i.e. ideal vs real) are dissimilar in full. This means that all safety specifications are met in the ideal state but in the real state all safety specifications have been violated. As mentioned by (Chatzimichailidou, 2015) the analyst can apply different dissimilarity measures to calculate values for the RiskSOAP indicator, such as crisp Euclidean distance. The RiskSOAP methodology is graphically depicted in Figure 2.

Table 1. The Rogers-Tanimoto dissimilarity measure formula.

Rogers-Tanimoto	$\dfrac{2S10 + 2S10}{S11 + S00 + 2S10 + 2S01}$

Source: Chatzimichailidou, 2016

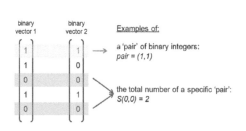

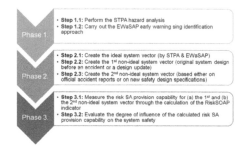

Figure 1. Graphical representation of the pairs for the dissimilarity measure.

Source: Chatzimichailidou, 2015

Figure 2. The RiskSOAP methodology.

Source: Chatzimichailidou, 2015

2.2 STAMP, STPA and EWaSAP

STPA is a hazard analysis technique that encapsulates the principles of the STAMP model of Leveson's (2011). Because STPA is a top-down approach to system safety, it can be used to generate safety requirements and constraints of existing systems or systems early in the development phase (Leveson, 2011). STPA is a rigorous method through which the analyst identifies inadequate control actions and examines scenarios or paths to accidents instead of calculating probabilities of failures and events or estimating severity of outcomes (Leveson, 2011). STPA also identifies causal factors not fully handled by traditional hazard analysis methods, such as software errors, component interactions, decision-making flaws, inadequate coordination and conflicts among multiple controllers, poor management and regulatory decision-making (Leveson, 2011). Safety is thus treated as a dynamic control problem, rather than a component reliability problem (Journal of Safety Studies ISSN 2377–3219 2016, Vol. 2, No. 2)

EWaSAP is an add-on to STPA (Dokas et al., 2013) and its aim is to provide a structured method for the identification of early warning signs to accidents. Furthermore, EWaSAP introduces an additional type of control action i.e. the awareness action. An awareness control action allows a controller to provide warning messages and alerts to other controllers inside or outside the system boundary, whenever data indicating the presence of threats or vulnerabilities is perceived and comprehended. STPA and EWaSAP could be performed as one process (Dokas, Feehan, & Imran, 2013).

2.3 Fuzzy logic

In mathematical terms, fuzzy logic utilizes the real numbers that exist between 0 and 1 to express that a proposition can be true to a degree. Fuzzy sets are comprised of a domain set (the complete set of possible values) and a membership function. A membership function is a function that maps the domain set to membership values (i.e. degrees of truth).Examples of fuzzy sets can be seen in Figure 4 and an example of classical crisp sets in depicted in Figure 3.

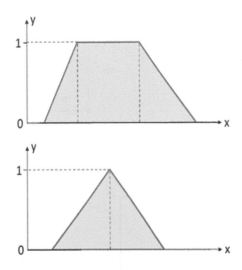

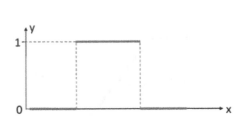

Figure 3.   Example of crisp sets.
Source: Botzoris and Papadopoulos, 2015

Figure 4.   Examples of fuzzy sets (trapezoidal Membership function above, triangular Membership function below.
Source: Botzoris and Papadopoulos, 2015.

3   THE FUZZY RISKSOAP METHODOLOGY AND PROCESS

The Fuzzy RiskSOAP methodology begins by applying the STPA and EWaSAP analyses to the system of interest as per the crisp RiskSOAP. Based on the number of STPA and EWaSAP safety requirements the "i" vector with dimension n×1 is formed so that n is equal to the number of the safety requirements. Then, the values in the "i" vector are all set to 1 as in the crisp RiskSOAP approach. The analyst then must set the values for the "real" vector. Those values will result from the membership functions assigned to each safety requirement produced by STPA and EWaSAP. There are at least three general categories of membership function types. Therefore, the analyst should first assign the proper membership function type to each safety requirement based on the "nature" of the requirement.

The three types of membership functions are shown below.

1) **Discrete fuzzy safety requirement** (discrete values of membership function)

Examples where an analyst should assign a discrete fuzzy membership function to a requirement would be any type of requirement where multiple parts (more than one) of the system have to adhere to the same criteria. See Figure 5.

2) **Strictly crisp safety requirement** (crisp values of membership function)

Examples where an analyst should assign a strictly crisp membership function to a safety requirement is where a specific element of the system must exist or not. Considering the instance of an airplane and its safety requirement "There must be an autopilot installed in the aircraft" the autopilot can either exist or not, meaning the requirement can only take the values 0 or 1. See Figure 6.

3) **Continuous fuzzy safety requirement** (trapezoidal, triangular membership function e.tc)

Examples where an analyst should assign a trapezoidal fuzzy membership function to safety requirements include delays or other concepts of time, environmental conditions, performance of humans and other system elements. For these types of safety requirements there is typically a consensus on which values represent the optimum or the ideal and the worst or the non-ideal state. There are however unclear boundaries on the values that represent the intermediate states. See Figure 7.

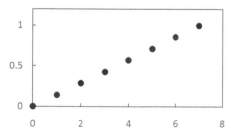

Figure 5.   Example of discrete fuzzy membership function.

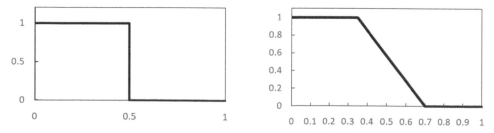

Figure 6.   Example of strictly crisp membership function.

Figure 7.   Example of continuous fuzzy trapezoidal membership function.

Based on the assigned membership functions on each safety requirement, the values of the safety requirements for the "r" vector at any time can be produced. These values can range between 0 and 1 expressing the degree of satisfaction of each safety specification at time t. With the values of the "i" and the "r" system vectors available, the analyst can apply the following formula that is based on the Euclidean distance between the vectors to calculate their dissimilarity. The distance between the "i" and "r" vectors is defined with the equation:

$$d(\vec{i} - \vec{r}) = \sqrt{\frac{\sum_{j=1}^{n} (1 - x_j)^2}{n}} \tag{1}$$

Where $\vec{i}$ denotes the ideal vector, $\vec{r}$ the real vector, x_j the degree of confidence of the safety specification j in the real vector and n the number of safety specifications.

4 CASE STUDY

A scenario of operation of a typical railway crossing system is utilized to demonstrate the fuzzy RiskSOAP indicator. The railway crossing in this case study is composed of an automated controller, train tracks, the train itself, train sensors, barriers blocking passage through the crossing as well as numerous alerts for drivers for when a train passes.

The STPA and EWaSAP analysis which was conducted on the system (Kirizakis, 2018) identified 87 safety specifications. Thirty-eight out of the 87 specifications were categorized as not strictly crisp (44%). Twelve out of the 38 not strictly crisp specifications were categorized as discrete fuzzy (32%) and the remaining 26 were categorized as continuous fuzzy (68%). Samples of crisp and fuzzy safety specifications are shown in Table 2.

The scenario unfolds as follows. Two trains pass through the crossing within a smalltime difference. At that time the controller has medium to high delays at enforcing the control actions (i.e. close the barrier and turn on the buzzer and light alert systems) and the lights and buzzer alerts are malfunctioning. The first train passes and causes damage to the crossing surface. Then a car comes through the crossing and gets stuck on the tracks because of the damage at the crossing surface. While the second train is about to pass the car manages to leave the crossing. Afterwards, the crossing surface is fixed by the maintenance group together with the lights and the buzzer alert.

To demonstrate the fuzzy RiskSOAP approach and compare its results against its crisp approach this scenario is divided into seven time steps, as shown in Table 3.

Table 2. Indicative examples of STPA and EWaSAP safety specifications.

#		SAFETY SPECIFICATION	MF TYPE
1	C9	Open barriers or Close light driver alert or Close sound driver alert should not be given when there is an object stopped on the crossing	Discrete fuzzy n = 3
2	C10	The Railway crossing surface should not be damaged when a train is passing the crossing	Continuous fuzzy (Trapezoidal)
3	C15	The Controller should Open the barriers when a train has just passed the crossing	Strictly Crisp
4	C21	The Controller should not open the barriers when the light and sound driver alert systems are cutoff and a train is passing the crossing	Discrete fuzzy n = 3
5	C44	The Controller should not give for a long time span the activation of the sound driver alert system if a train is not anticipated to pass the crossing	Continuous fuzzy (Trapezoidal)
6	C54	The Controller should activate the sound driver alert system when an object and a train are passing the crossing	Strictly Crisp
7	C72	The higher hierarchical levels should adequately manage the system by either updating the software or maintenance	Strictly Crisp

Table 3. Scenario of operation events sequence.

T	DESCRIPTION OF EVENT
0	A train is approaching the crossing while the crossing has a malfunctioning light and sound alert
1	First train passes and breaks the crossing surface
2	Car comes through and gets stuck on the tracks because of the broken crossing surface
3	Second train is coming close to the crossing
4	Car comes through and gets stuck on the tracks because of the broken crossing surface
5	Car manages to leave the crossing before the train passes
6	The train passes while the crossing surface is damaged
7	Maintenance crew arrives and fixes surface and alert malfunctions

5 RESULTS

Because the Rogers-Tanimoto dissimilarity measure and Euclidean distance are based on different assumptions (i.e. Rogers-Tanimoto measures the 0s and 1s in each vector, whereas Euclidean distance measures the distance between n dimensional vectors), the direct comparison of the arithmetic results may be misleading. This is shown in Figure 8 where Rogers-Tanimoto, Crisp Euclidean and Fuzzy Euclidean distances are plotted and Rogers-Tanimoto values are lower than both Euclidean distances but the behaviors over time are almost identical. This indicates that the Euclidean distance can also be used to produce the values of the RiskSOAP capability.

To assess the added value of applying the fuzzy logic in RiskSOAP, the crisp Euclidean and the fuzzy Euclidean distances were applied to the "r" vectors of the operational scenario as Figure 8 shows and the results were compared. The vector on the left depicts a sample of the ideal "i" vector, the next vector depicts a sample of the real "r" vector created by the crisp RiskSOAP and the right vector depicts a sample of the same safety requirements as the last vector but created with the fuzzy RiskSOAP (chapter 3).

To quantify the difference made by interpreting safety requirements with fuzzy logic, the average distance of the two functions was 4% and the maximum distance between them 13%.

6 CONCLUSIONS AND FUTURE WORK

In this paper a fuzzy logic extension of the RiskSOAP indicator was proposed. The fuzzy approach utilizes fuzzy linguistic variables to translate real world system data to design vectors with values in [0, 1], diverging from the crisp values used by the original RiskSOAP methodology. Then the Euclidean distance is used to calculate the dissimilarity of the ideal and the real design vectors.

Based on the results of the case study it is concluded that the Euclidean distance and the Rogers-Tanimoto dissimilarity measure behave almost identically in providing a value to the RiskSOAP capability. This means that the Euclidean distance can be reliably used for the RiskSOAP indicator.

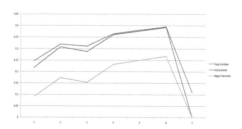

Figure 8. Graph of the RiskSOAP indicator calculated with 3 different approaches for the scenario of operation on the case study.

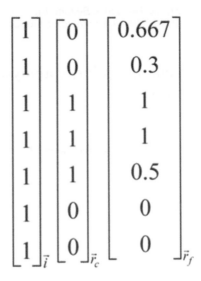

Figure 9. A sample of the ideal and real vectors depicting the violation or not of the safety requirements in Table 2 with crisp and fuzzy values for t = 4.

Moreover, by utilizing fuzzy logic to create the values of the vectors a better degree of accuracy can be achieved in measuring the capability based on which a system can be self-aware of the presence of flaws, threats and vulnerabilities. A noticeable example of this is shown at the final time step t = 7 in the case study. The maintenance crew fixes the malfunctions of the system, but the delays of the controller on enforcing control actions are built in the system. So, while the crisp approach shows that the system has no distance from the ideal the fuzzy indicates a small differentiation between them.

With the proposed fuzzy approach, the gap of the original RiskSOAP which neglects that its variables, which indicate if a safety constraint at the time of the analysis is satisfied or not, may have a truth value that ranges between '0' and '1' (i.e. not satisfied and fully satisfied), can be addressed.

REFERENCES

Botzoris, G.N. & Papadopoulos, B.K. 2015. *FuzzySets: Engineering Design and Management Applications*. Thessaloniki: sofiapress (In Greek).

Chatzimichailidou, M. M. 2015. *RiskSOAP: A Methodology for Measuring Systems' Capability of Being Self-Aware of Their Threats and Vulnerabilities*, PhD. Thesis, Xanthi: Democritus University of Thrace.

Chatzimichailidou, M.M. & Dokas, I.M. 2016. Introducing RiskSOAP tocommunicate the distributed situation awareness of a system about safetyissues: An application to a robotic system. *Ergonomics* 59: 1–14.

Chatzimichailidou, M.M. & Dokas, I.M. 2016. RiskSOAP: Introducing and applying a methodology of risk self-awareness in road tunnel safety. *Accident Anal. Prevention* 90:118–127.

Dokas, I.M., Feehan, J. & Imran, S. 2013 .EWaSAP: An early warning sign identification approach based on a systemic hazard analysis. *Safety Sci.* 58: 11–26.

Kyrizakis K. 2018. *Safety Specifications of Railway Crossing using STPA*, BSc Thesis, Xanthi: Democritus University of Thrace (In Greek).

Leveson, N.G. 2011. *Engineering a safer world: Systems thinking applied to safety*. Cambridge: MITPress.

Salmon, P.M., Stanton, N.A., Walker, G.H., Baber, C., Jenkins, D.P., McMaster, R., Young, M.S. 2008. What really is going on? Review of situation awareness models for individuals and teams. *Theoretical Issues in Ergonomics Science* 9(4): 297–323.

Risk Analysis Based on Data and Crisis Response Beyond Knowledge – Huang & Nivolianitou (eds)
© 2020 Taylor & Francis Group, London, ISBN 978-0-367-25146-8

Risk-based fault assessment and priority analysis for safety management in industrial systems

Guozheng Song*

Department of Mechanical and Industrial Engineering, Norwegian University of Science and Technology, Trondheim, Norway

Faisal Khan

Centre for Risk, Integrity and Safety Engineering (C-RISE), Faculty of Engineering and Applied Science, Memorial University of Newfoundland, St. John's, Canada

Nicola Paltrinieri

Department of Mechanical and Industrial Engineering, Norwegian University of Science and Technology, Trondheim, Norway

ABSTRACT: In the industry 4.0 era, large numbers of sensors are employed in industrial plants for safe production. However, the sensors may generate numerous alarms which cannot be effectively interpreted and managed by employees. Thus, industrial plants need a tool to help employees respond to alarms in a proper and timely way. Some works have attempted to isolate faults of an alarm based on occurrence likelihood of the faults. However, if employees only focus on the most likely fault, the real fault with severe consequences can be ignored, which easily causes accidents. The current work proposes a risk-based assessment system of faults to overcome the drawback. Moreover, some research has prioritized alarms to facilitate safety management. However, alarms cannot convey information of faults in a straightforward way, and thus employees have to interpret alarms based on their knowledge and experience before responding. The current work identifies and prioritizes faults, and also pre-stores checking and fixing measures of the faults in the management system; thus, it can effectively guide practical safety management in a straightforward way. The framework of fault management and proposed methods are explained in this paper.

Keywords: alarm management, risk-based fault assessment, fault priority analysis, data-driven method, Bayesian network

1   INTRODUCTION

Alarms and corresponding responses of operators constitute a critical protection layer, preventing the evolution of deviation into accidents (Goel et al., 2017; Stauffer and Clarke, 2016). In this safety layer, various sensors are employed to monitor abnormal situations and the detected information is displayed on the Human-Machine Interface (HMI), assisting operators to estimate the safety and production states (Goel et al., 2017; Urban and Landryová, 2016). Operators interpret detected information based on their experience and knowledge and then take actions to return abnormal situations to normal states (Goel et al., 2017). Essentially, operators estimate the states and make real-time decisions and they are key players in the safe operation of a plant (Goel et al., 2017). Thus, proper responses of

---

*Corresponding author: guozheng.song@ntnu.no

operators are crucial for safe production. Poor alarm management, such as judgmental error and ineffective responses to alarms, has become one of the leading causes in industrial incidents (Goel et al., 2017; Stauffer et al., 2010). Practical cases have shown how ineffective operator response contributes to accidents. For example, a methyl chloride leak occurred at the DuPont plant in Belle, WV, USA in 2010 (CSB, 2011). This leak was caused by a ruptured disc. The burst sensor on the ruptured disc had alarmed the fault, but operators considered it as a false alarm and ignored it. As a result, approximately 2,000 pounds of methyl chloride were released to the atmosphere over five days. Poor alarm recognition and management has been identified as a contributing cause of this accident (CSB, 2011). In another case (National Transportation Safety Board, 2012), an oil pipeline ruptured in Marshall, Michigan, USA on July 25, 2010. Multiple alarms were immediately generated when the pipeline ruptured downstream. However, operators believed that the alarms were caused by a combination of column separation and erratic pressures generated due to shutdown instead of a rupture. Thus, this spill was not discovered for over 17 hours. The spilled amount was estimated to be 843,444 gallons of crude oil.

The existing alarm management system only provides alarm information (Urban and Landryová, 2016). Operators may have an incorrect interpretation of the alarm and take inappropriate responses (CSB, 2011; National Transportation Safety Board, 2012). Thus, the research to guide effective response to detected alarms is important for practical management. Recognizing this need, some works have studied fault isolation to help employees diagnose the fault given detected alarms (Cai et al., 2017a). Wang et al. (Wang et al., 2019) proposed a simplified BN-based method for the fault diagnosis of a diesel engine fuel injection system. Cai et al. (Cai et al., 2016; Cai et al., 2015; Cai et al., 2017b) presented BN, object-oriented BN and DBN to conduct fault isolation of subsea systems. Liu et al. (Liu et al., 2015a) introduced a development approach of BN for fault diagnosis which includes an operational procedure layer. The method was applied to diagnose the faults in the procedure of closing a subsea blowout preventer. Liu et al. (Liu et al., 2015b) conducted fault diagnosis of a solar assisted heat pump system using incomplete data and expert knowledge. These works have used the following diagnosis criteria:

(1) If the difference between posterior probability and prior probability of a fault node is equal to or larger than a certain threshold (e.g., 60%), then the fault is isolated (Cai et al., 2016; Cai et al., 2017b; Wang et al., 2019).
(2) If the difference between posterior probability and prior probability of a fault node is a - threshold percent higher than the second largest one, then the fault is isolated (Wang et al., 2019).
(3) If a fault has the largest posterior probability and the probability is larger than a certain threshold, then the fault is isolated (Cai et al., 2015).
(4) If the difference between the largest fault posterior probability and the second biggest one is larger than a certain threshold, then the fault with the largest posterior probability is isolated (Cai et al., 2015).
(5) If a fault has the largest posterior probability, then the fault is the most suspected one (Liu et al., 2015b).

Essentially, these works attempted to estimate the likelihood of potential faults of being the real one to cause the alarm. The diagnosed fault is the most likely fault (sometimes most likely multiple faults) to cause the alarms. These works consider the likelihood of faults as the assessment criterion. In the above-mentioned practical case (CSB, 2011), the burst sensor had many unreliable records, while disc rupture rarely occurred. This is the reason why operators treated the alarm as a false alarm. If this case is analyzed using previous methods with the assessment criterion of fault likelihood, there is a large chance that false alarm is the diagnosed fault, having the same result as the estimation of operators. Then operators focus on the diagnosed result (false alarm) and the real fault (disc burst) is still missed. This makes us reconsider whether the existing assessment system of faults, only considering the likelihood of potential faults, is adequate for safety management. Actually, even though the most likely fault is diagnosed, there is no guarantee that it is the real fault. This means that the assessment result itself has uncertainty. Once other potential faults with severe consequences are the real faults rather than the diagnosed one, a severe accident can easily occur. Thus, the existing fault assessment system is inadequate to guide effective alarm management. Besides above works of fault isolation, various model-based quantitative methods were reviewed in (Venkatasubramanian et al., 2003). These methods also attempted to infer the occurrence likelihood of various faults and specify the actually existing faults. Above-mentioned limitation still exists for these reviewed works.

The number of process sensors and alarms has greatly increased in industrial plants over past decades (Goel et al., 2017). This requires operators to manage alarms in a quick way. To cope with this challenge, previous works have attempted to prioritize the alarms. Fink et al. (Fink et al., 2004) called for alarm priority in nuclear industry, based on required immediacy of operator actions and the impact of the condition on safe plant operation. It also explained alarm display strategies according to priority level. Basu et al. (Basu et al., 2013) proposed a method to rank the alarms of buses according to severity. Foong et al. (Foong et al., 2009) developed an alarm prioritization system with fuzzy logic for an oil refinery. Bijoch et al. (Bijoch et al., 1991) expanded the functions of the NSP Intelligent Alarm Processor, enabling it to prioritize alarms. These works facilitate alarm management by highlighting critical alarms. However, multiple potential faults are often behind an alarm (Cai et al., 2015). Such works cannot directly guide safety management, since even if critical alarms have been recognized, operators need to interpret the alarms and check (exclude or confirm) potential faults based on their experience and knowledge. Furthermore, because of the diagnosis time consumed by other potential faults, the faults with severe consequences may not receive timely attention and some of them can evolve into a catastrophic accident in a short time. Thus, the research to assess and prioritize potential faults from risk perspective is essential for practical management.

After identifying the gaps in alarm management, we proposed a risk-based fault assessment and priority analysis approach. The proposed work aims to solve the problems in the operator management stages shown in the dash line of Figure 1. The work has a new assessment system, including the likelihood of a fault of being the real one, time for the fault to evolve into the unacceptable likelihood of an accident and the consequences of the accident. The proposed work also extends the priority analysis from alarm level to fault level, which provides a straightforward guidance for practical management. In the assessment system, when a fault has higher likelihood of being the real fault and takes a shorter time to evolve to the unacceptable likelihood of an accident with more severe accident consequences, it has a higher priority. To the authors' knowledge, no work has presented a quantitative framework for assessment and priority analysis of faults using the proposed assessment system. Moreover, the measures to check (i.e., exclude or confirm) and fix the potential faults are included in the management framework. The proposed framework can avoid missing real critical faults, and the priority analysis of faults and the inclusion of measures can directly guide effective safety management.

This work is organized as follows. Section 2 presents the conceptual framework of risk-based fault management, while methodology is explained in Section 3. Section 4 provides result and conclusions are presented in Section 5.

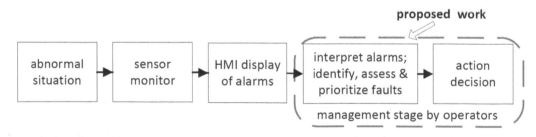

Figure 1.   Different stages of alarm management.

2 CONCEPTUAL FRAMEWORK

The conceptual framework of the proposed work is shown in Figure 2. When alarms are detected, potential faults triggering the alarms are identified. Then these identified faults are assessed according to three criteria: likelihood of being the real fault, time for the fault to evolve to an unacceptable probability of an accident and consequences of the accident. The management priority of faults is decided based on the three parameters which will be explained in Section 3 in detail. The fault with a higher likelihood of being the real one, a shorter time evolving into an accident with an unacceptable probability and more severe potential consequences, has higher priority. Employees need to manage such faults first. To facilitate safety management for operators, the measures for fault checking and fixing are also included in the framework. Employees can exclude (or confirm) the potential faults and fix the real one following the guidance of these measures. Figure 2 also explains how employees can use the results of the proposed work to manage faults. Employees check the prior fault first. If it is the real one, they fix the fault with the guidance of recommended reparation measures. In contrast, if it is not the real one, employees need to check the prior one within the remaining faults.

The new factors of the framework are that it establishes a risk-based system to assess and prioritize faults, and it integrates the measures of fault checking and fixing into the alarm management framework. This new assessment system avoids ignorance of the critical faults which can evolve into accidents in a short time. The priority analysis of faults enables employees to cope with critical faults first, avoiding management delay. Moreover, the included measures give employees a direct guide for the proper actions to check and fix faults. The works included in this framework have great practical meaning to guide employees' safety management.

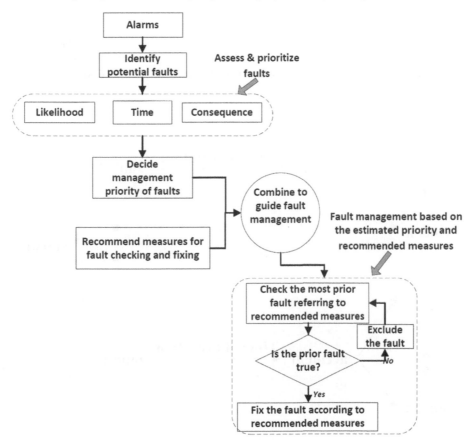

Figure 2.   Conceptual framework of risk-based fault management.

3 METHODOLOGY

The fault assessment and priority analysis can be achieved through the methods shown in Figure 3. For a detected alarm, three steps are included to assess and prioritize the potential faults, and the work in each step is explained as follows.

Step 1: Identify the potential faults of an alarm and analyze the likelihood (P) of the potential faults of being the real faults. For a detected alarm, the potential faults are identified with a fault tree model (FT) as shown in the left graph of Figure 4. With an organized form, the FT can be easily developed and understood, and it is expert in cause identification of an undesired event (e.g., an alarm). However, it has limitations for quantitative calculation (Khazad et al., 2011; Song et al., 2016). Thus, after fault identification, the FT is converted to BN (as shown in the middle graph of Figure 4) (Khazad et al., 2011). With the function of backward inference, BN is an excellent tool for fault diagnosis (Cai et al., 2014). Variables of BN are represented using nodes, and dependent variables are linked by arcs. The quantitative relationships between nodes are represented using conditional probability tables (CPTs) (Song et al., 2016). Assume a node y has n parent nodes $x_1, x_2 \ldots, x_n$, and each node has two states, true (T) and false (F); if we observe $y = T$ and infer the state of x_1, the backward inference principle of BN is shown in Eq. (1).

$$\begin{cases} P(x_1 = T | y = T) = \frac{P(x_1 = T, y = T)}{P(y = T)} = \frac{\sum_{x2,x3,\cdots,xn \in \{T,F\}} P(x_1 = T, x_2, x_3, \cdots, x_n, y = T)}{\sum_{x2,x3,\cdots,xn \in \{T,F\}} P(x_1, x_2, \cdots, x_n, y = T)} \\ P(x_1, x_2, \cdots, x_n, y = T) = P(y | \cap_{i=1}^{n} x_i) * \prod_{i=1}^{n} P(x_i | \cap_{j=1}^{i-1} x_j) \end{cases} \quad (1)$$

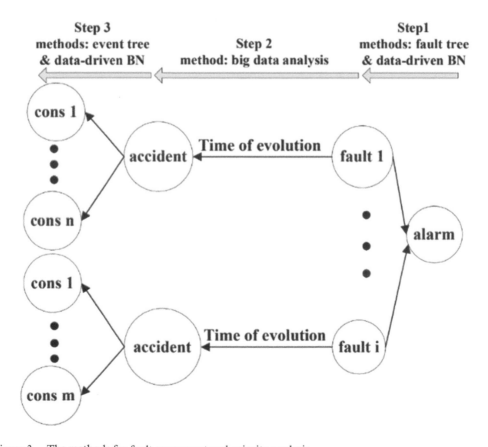

Figure 3. The methods for fault assessment and priority analysis.

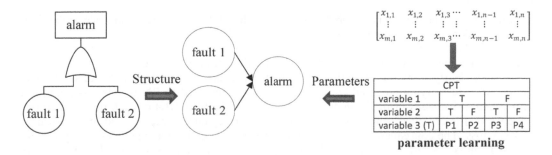

CPT				
variable 1	T		F	
variable 2	T	F	T	F
variable 3 (T)	P1	P2	P3	P4

parameter learning

Figure 4.   Schematic graph of BN development for fault likelihood calculation.

In the simple BN of Figure 4, the posterior probability calculation of fault 1 (L_1), given the occurrence of an alarm (A), is shown in Eq. (2).

$$P(L_1|A) = \frac{P(L_1)}{P(A)} = \frac{P(L_1, L_2, A) + P(L_1, L_2', A)}{P(L_1, L_2, A) + P(L_1, L_2', A) + P(L_1', L_2, A) + P(L_1', L_2', A)} \qquad (2)$$

According to rule in Eq. (1), Eq. (2) can be converted to Eq. (3).

$$P(L_1|A) = \frac{P(L_1) * P(L_2|L_1) * P(A|L_1, L_2) * P(L_1) * P(L_2'|L_1) * P(A|L_1, L_2')}{\left\{ \begin{array}{l} P(L_1)*P(L_2|L_1)*P(A|L_1,L_2)*P(L_1)*P(L_2'|L_1)*P(A|L_1,L_2')+ \\ P(L_1^2)*P(L_2|L_1')*P(A|L_1',L_2)*P(L_1')*P(L_2'|L_1')*P(A|L_1',L_2') \end{array} \right\}} \qquad (3)$$

where A is alarm; L_1 and L_2 are the occurrence of fault 1 and fault 2, while L_1' and L_2' are nonoccurrence of fault 1 and fault 2. The prior probabilities in Eq. (3) have been defined and the conditional probabilities can be obtained from the conditional probability tables (CPTs) of BN.

The structure of BN is cstablished based on FT, and the parameters are obtained through parameter learning, as shown in Figure 4. The parameter learning of BN from training data facilitates the practical application of BN (Liao and Ji, 2009). The parameter learning refers to the estimation of CPTs according to observed data (Ji et al., 2015). CPTs are important components of BN and they represent quantitative relationships among variables (Song et al., 2016). There are various learning approaches, such as the Expectation Maximization algorithm, Gibbs sampling and the Gaussian approximation method (Ji et al., 2015; Liao and Ji, 2009). The parameter learning can be achieved using Genie software (BayesFusion LLC, 2019).

Step 2: Analyze the time parameter from a fault to an accident. This step analyzes the time length from occurrence of a fault to the time point when the fault evolves into an accident with an unacceptable likelihood. If the fault occurs but is not managed in time, the occurrence probability of the accident increases over time. In this work, we set the threshold of occurrence

likelihood of an accident as 0.01. This means that if the occurrence probability of an accident caused by the fault is equal to 0.01 at a time point, operators must fix the fault before this moment. The length of time (T_t) between the occurrence of a fault (T_0) and the moment when the occurrence likelihood of the accident caused by the fault equals 0.01 (T_1) is defined as the second assessment parameter of the fault. It is the red segment (refer to web version for color) in Figure 5, and its mathematical expression is Eq. (4). This time length can be obtained directly through historical data analysis.

$$\begin{cases} F(t = T_0) = 1 \\ P(t = T_1) = 0.01 \\ T_t = T_1 - T_0 \end{cases} \qquad (4)$$

where $F(t)$ is the occurrence probability of the fault, while $P(t)$ is the occurrence probability of the accident evolved from the fault.

Step 3: Analyze the consequence of the accident caused by the fault. The potential consequences are identified using an event tree (ET) (See the left graph of Figure 6).

ET can clearly show the escalation process of consequences based on states of safety barriers. After identifying potential consequences, ET is converted to BN to relax its limitation for quantitative assessment (Song et al., 2016). The probability (P_{ci}) of each potential consequence (C_i) is obtained using BN. Each potential consequence is represented in the form of money loss. The consequence (C) of the accident is the weighted average loss of the potential consequences (see Eq. (5)), which is the third parameter of fault assessment. The parameters (CPTs) of the BN for consequence assessment are also learned from data.

$$C = \sum_{i=1}^{n}(P_{ci} \times C_i) \qquad (5)$$

where n represents the number of potential consequences and i is the i-th potential consequence.

Note that the parameters of Eq. (5) come from BN instead of ET. Since BN considers the dependency of variables and has a flexible logic (Khakzad et al., 2011; Song et al., 2016), the result of BN can be different from that of ET.

After obtaining the three assessment parameters, the assessment index (R) of a fault is calculated as equation (6) which is adapted from (Chang et al., 2011). The bigger index a fault has, the higher priority it has.

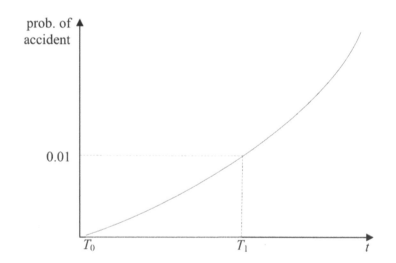

Figure 5. Time parameter for fault assessment.

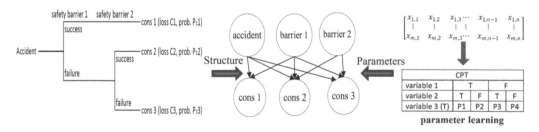

Figure 6. Schematic diagram of BN development for consequence analysis.

$$R = P \times C \times 100^{1-T_t/60} \tag{6}$$

where P is the likelihood of a potential fault of being the real fault; T_t is the time length between the occurrence of a fault and the moment when the accident caused by the fault has an unacceptable occurrence likelihood; C refers to the weighted average loss for the potential consequences of the accident.

The checking and fixing measures of potential faults are integrated into the alarm management framework. Employees can refer to these measures for their responses. The measures are proposed using historical records and the experience of a group of experts. These experts are process operators, safety managers and manufacture engineers who design the targeted systems. For each potential fault, experts review previous measures for the same fault from historical records and propose more measures. These measures are assessed using an analytic hierarchy process (AHP) (Sipahi and Timor, 2010) by the experts. The selected measures are pre-stored in the fault management system. When the prior fault is decided, corresponding measures for the fault are referred.

4 RESULT

The work aims to guide employees proper responses to alarms. A risk-based management framework is presented to achieve the goal. The framework not only analyzes the likelihood of faults of being the real ones, but also considers faults' severity through analysis of evolution time and consequences. After fault assessment with the new assessment system, the priority of faults can be decided according to a quantitative index. In practice, accidents have been caused by inappropriate operator responses due to ignorance or a delayed response to critical faults. The proposed work of fault assessment and priority analysis can inform employees of potential problems and which problem needs prior management. Thus, the current work can help avoid the catastrophic accidents caused by ineffective operator responses. Moreover, the checking and fixing measures are assessed by experts and the selected measures are included in the framework to guide practical management.

Methods have been proposed to achieve assessment and priority analysis of faults. FT is used to identify the potential faults of an alarm, while ET is applied to analyze the escalation process of consequences of an accident caused by the fault. Since FT and ET have limitations for quantitative calculation, the probabilities of faults and the consequences of the accident are calculated using BN. The structures of BNs are converted from FT and ET, while their parameters are learned from data. The data-driven BN can reduce the uncertainty caused by subject input. Another fault assessment parameter (evolution time from occurrence of a fault to an unacceptable probability of an accident) is calculated from data directly. The three parameters are used to calculate an assessment index of faults which decides fault priority. The checking and fixing measures for faults are selected by experts using AHP. These measures are pre-stored in the fault management system.

5 CONCLUSIONS

This work proposed a conceptual framework for fault management in industrial processes. A risk-based assessment system of faults is presented. This system can avoid ignorance of the critical faults. Moreover, fault priority is analyzed to guarantee the management of critical faults in time. The checking and fixing measures are pre-stored in the management framework; thus, the proposed framework can directly guide fault management. Furthermore, potential methods to achieve functions of the framework are explained. FT and ET are used to identify the potential faults and consequences of accidents evolved from the faults. They are then converted to BN for the probability calculation of faults of being the real faults, and for quantitative assessment of the weighted average losses of the accident caused by the fault. The time interval from fault occurrence to the moment when the probability of the accident becomes unacceptable is analyzed from the historical data. The assessment index of faults is calculated based on the three parameters. The measure assessment is achieved by experts using AHP. This work can help employees understand alarms and deal with critical faults in time.

In future work, a case study will be conducted to demonstrate the merits of the proposed management framework.

REFERENCES

Basu, C., Das, K., Hazra, J., 2013. Enhancing Wide-Area Monitoring and Control with Intelligent Alarm Handling, 4th IEEE PES Innovative Smart Grid Technologies Europe (ISGT Europe), Copenhagen.

BayesFusion LLC, 2019. GeNIe User's Manual. Available at https://support.bayesfusion.com/docs/. Accessed on 28.05.2019.

Bijoch, R.W., Harris, S.H., Vollunann, T.L., Bann, J.J., Wollenberg, B.F., 1991. DEVELOPMENT AND IMPLEMENTATION OF THE NSP INTELLIGENT ALARM PROCESSOR. IEEE Transactions on Power Systems 6, 806-812.

Cai, B., Huang, L., Xie, M., 2017a. Bayesian Networks in Fault Diagnosis. IEEE Transactions on Industrial Informatics 13, 2227-2240.

Cai, B., Liu, H., Xie, M., 2016. A real-time fault diagnosis methodology of complex systems using object-oriented Bayesian networks. Mechanical Systems and Signal Processing 80, 31-44.

Cai, B., Liu, Y., Fan, Q., Zhang, Y., Liu, Z., Yu, S., Ji, R., 2014. Multi-source information fusion based fault diagnosis of ground-source heat pump using Bayesian network. Applied Energy 114, 1-9.

Cai, B., Liu, Y., Ma, Y., Liu, Z., Zhou, Y., Sun, J., 2015. Real-time reliability evaluation methodology based on dynamic Bayesian networks: A case study of a subsea pipe ram BOP system. ISA Trans 58, 595-604.

Cai, B., Liu, Y., Xie, M., 2017b. A Dynamic-Bayesian-Network-Based Fault Diagnosis Methodology Considering Transient and Intermittent Faults. IEEE Transactions on Automation Science and Engineering 14, 276-285.

Chang, Y., Khan, F., Ahmed, S., 2011. A risk-based approach to design warning system for processing facilities. Process Safety and Environmental Protection 89, 310-316.

CSB, 2011. INVESTIGATION REPORT: METHYL CHLORIDE RELEASE JANUARY 22, 2010.

Fink, R., Hill, D., O'Hara, J., 2004. Human Factors Guidance for Control Room and Digital Human-System Interface Design and Modification: Guidelines for Planning, Specification, Design, Licensing, Implementation, Training, Operation and Maintenance. USDOE Office of Nuclear Energy, Science and Technology, Niger, U.S.

Foong, O.M., Sulaiman, S.B., Rambli, D.R.B.A., Abdullah, N.S.B., 2009. ALAP: Alarm Prioritization System For Oil Refinery, Proceedings of the World Congress on Engineering and Computer Science San Francisco, USA.

Goel, P., Datta, A., Mannan, M.S., 2017. Industrial alarm systems: Challenges and opportunities. Journal of Loss Prevention in the Process Industries 50, 23-36.

Ji, Z., Xia, Q., Meng, G., 2015. A Review of Parameter Learning Methods in Bayesian Network, Advanced Intelligent Computing Theories and Applications, pp. 3-12.

Khakzad, N., Khan, F., Amyotte, P., 2011. Safety analysis in process facilities: Comparison of fault tree and Bayesian network approaches. Reliability Engineering & System Safety 96, 925-932.

Liao, W., Ji, Q., 2009. Learning Bayesian network parameters under incomplete data with domain knowledge. Pattern Recognition 42, 3046-3056.

Liu, Z., Liu, Y., Cai, B., Zheng, C., 2015a. An approach for developing diagnostic Bayesian network based on operation procedures. Expert Systems with Applications 42, 1917-1926.

Liu, Z., Liu, Y., Zhang, D., Cai, B., Zheng, C., 2015b. Fault diagnosis for a solar assisted heat pump system under incomplete data and expert knowledge. Energy 87, 41-48.

National Transportation Safety Board, 2012. Enbridge Incorporated Hazardous Liquid Pipeline Rupture and Release, Marshall,Michigan, July 25, 2010.

Sipahi, S., Timor, M., 2010. The analytic hierarchy process and analytic network process: an overview of applications. Management Decision 48, 775-808.

Song, G., Khan, F., Wang, H., Leighton, S., Yuan, Z., Liu, H., 2016. Dynamic occupational risk model for offshore operations in harsh environments. Reliability Engineering & System Safety 150, 58-64.

Stauffer, T., Clarke, P., 2016. Using alarms as a layer of protection. Process Safety Progress 35, 76-83.

Stauffer, T., Sands, N.P., Dunn, D.G., 2010. ALARM MANAGEMENT AND ISA-18 – A JOURNEY, NOT A DESTINATION, Texas A&M Instrumentation Symposium, Houston, TX, USA.

Urban, P., Landryová, L., 2016. Identification and Evaluation of Alarm Logs From the Alarm Management System, 17th International Carpathian Control Conference (ICCC). IEEE, Tatranska Lomnica, Slovakia.

Venkatasubramanian, V., Rengaswamy, R., Yin, K., Kavuri, S.N., 2003. A review of process fault detection and diagnosis Part I: Quantitative model-based methods. Computers and Chemical Engineering 27, 293-311.

Wang, J., Wang, Z., Stetsyuk, V., Ma, X., Gu, F., Li, W., 2019. Exploiting Bayesian networks for fault isolation: A diagnostic case study of diesel fuel injection system. ISA Trans 86, 276-286.

Risk Analysis Based on Data and Crisis Response Beyond Knowledge – Huang & Nivolianitou (eds)
© 2020 Taylor & Francis Group, London, ISBN 978-0-367-25146-8

Pool fire of liquefied natural gas

Mattia Carboni, Valerio Cozzani, Tommaso Iannaccone, Gianmaria Pio & Ernesto Salzano*
Department of Civil, Chemical, Environmental, and Materials Engineering, University of Bologna, Bologna, Italy

ABSTRACT: The increased relevance of liquefied natural gas (LNG) as a fuel for transportation systems has raised safety concerns. Pool fires of significant amount of LNG have been identified as the main scenario in case of accidental release of cryogenic liquid. Besides, large-scale experiments are discouraged due to the complexity, hazard and costs. Consequently, several numerical studies have been performed to this topic. Considering the nature of the investigated scenario, large eddy simulation (LES) models should be preferred. However, several simplifications, regarding natural gas composition, its combustion chemistry or turbulence sub-models, are often adopted. To this aim, a comprehensive numerical study on the effect of these assumptions on LNG pool fire characterization was conducted in terms of either temperature distribution or surface emitter power. The accuracy of the obtained data was evaluated by comparing numerical results with small-scale data from literature.

Keywords: integrated risk, multi-hazards, joint probability distribution, disaster function, information diffusion technique

1 INTRODUCTION

The increasing demand for cleaner and more convenient energy sources has favored the exploration and exploitation of remote reservoirs. In this light, the utilization of cryogenic systems for long term transportation has been largely incentivized. Among the others, natural gas is considered as a promising greener, low-carbon, fuel alternative for transportation, including navigation, heavy-duty trucks and aviation (Perpignan et al., 2018), especially when stored in the liquefied form as liquefied natural gas (LNG) (Zhiyi and Xunmin, 2019). Furthermore, low temperature combustion has the potential to meet the stringent regulations concerning the automotive sector (Agarwal et al., 2017). Despite of these advantages, the lack of knowledge on the phenomenological aspects occurring at low temperature has raised several concerns regarding the safety aspects and has inspired European projects as SUPER-LNG (2019), which involves several harbors in the Mediterranean sea. In the last decades, several data have been collected to characterize accidental scenario due to large release of LNG (Zhang et al., 2018), e.g. Burro and Coyote test series (Puttock et al., 2012). The effect of mitigation systems (e.g. water mist, water curtain and expanding foams) has been investigated, as well (Rana et al., 2008; Qi et al., 2011; Liu et al., 2012; Kim et al., 2013). On the other hand, the complexity of the investigated scenarios and its hazardous nature have led to a dearth of laboratory-scale experimental data. Hence, in lieu of measurements, numerical approach is preferred (Gavelli et al., 2008). However, many hypotheses are commonly posed to simplify the investigated problem and make computational costs affordable. Among the others, LNG is commonly assumed as pure methane for the sake of numerical evaluation. However, it should be noted that in real cases the amount of ethane and/or propane cannot be neglected (BSI, ISO 13686:2013). Other substances (e.g. larger

*Corresponding author: ernesto.salzano@unibo.it

Table 1. Composition examples of real LNG mixtures.

Mixture	CH$_4$ [% v/v]	C$_2$H$_6$ [% v/v]	C$_3$H$_8$ [% v/v]	Others [% v/v]
Mix 1	100	0	0	0
Mix 2	70	18	8	4

alkanes, nitrogen or carbon dioxide) can be found in smaller tenor. Typical LNG compositions were shown in Table 1. Methane, ethane and propane are distinguished from the other species composing the mixture. On the contrary, all the other species (mainly heavier alkanes, carbon dioxide and nitrogen) were considered as a single, inert compound and referred as "Others".

It is worth mentioning that Mix 1 was included as a benchmark composition to be compared against mixtures representative of real conditions, whereas Mix 2 represents the heavier composition allowed by ISO 13686:2013 (BSI, ISO 13686: 2013). Hence, any mixtures referred as natural gas must be included in the given ranges. As cited above, however, computational fluid dynamics studies commonly assumes the adoption of pure methane as representative of natural gas (Betteridge, 2018). Regardless the analyzed composition, literature agrees that pool fire, flash fire, vapor cloud explosion, fireball or gas dispersion represent the main scenarios in the case of the accidental release of pressurized LNG on land. Rapid phase transition (RPT) should be however included for release on water, exclusively (Aneziris et al., 2014). The occurrence of each scenario can be determined upon event tree quantification, based on well-established procedure (Haag and Ale, 2005) (Figure 1).

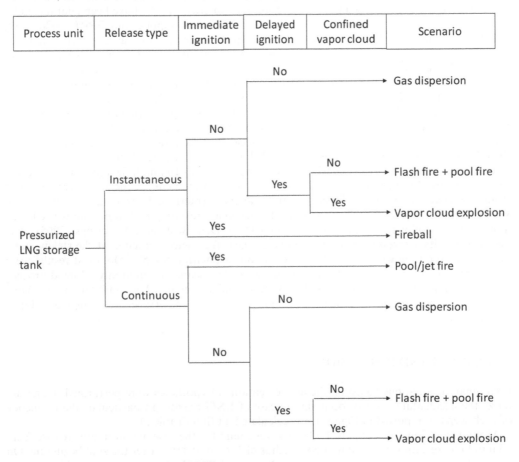

Figure 1. Event tree for pressurized LNG storage tanks.

435

Within the outcomes of the LNG event trees, particular attention should be paid to pool fire scenario, since pseudo steady-state can be achieved, thus resulting in conditions potentially leading to second cascading events (i.e. domino effect) (Jujuly et al., 2015). For the sake of the pool fire modelling, the accurate evaluation of the evaporation rate represents an essential step (Liu et al., 2012). Nevertheless, few experimental data are available to evaluate and valuate the existing evaporation rate models, developed for non-cryogenic fuels (Nguyen et al., 2017). Preliminarily, pool fire can be assumed to have a cylindrical shape, tilted by wind, and characterized by surface emitter power (SEP) (Raj, 2007). Several parametric correlations for the estimation of fire dimensions, i.e. diameter and height, as a function of pool diameter exist (Babrauskas, 1983). However, empirical-based parameters are still missing for cryogenic liquids.

For these reasons, this study was devoted to the analysis of the effects of LNG pool dimension on small- and medium- scales accidental fire scenarios triggered after the release of LNG on concrete ground.

2 METHODOLOGY

The open-source computational fluid dynamics (CFD) model named fire dynamic simulator (FDS), which is based on large eddy simulation approach (LES), has been implemented. Additional information on the code can be found in the technical reference guide of the abovementioned software (McGrattan et al., 2004). Preliminary analyses have suggested the adoption of horizontal axes of 3 pool diameter and vertical axis of 10 pool diameter, with a maximum of 60 m. Moreover, the validity and robustness of the numerical results have been guaranteed by proper cell sizes (δ) to flame size (D) ratio (NUREG-1824-suppl.1 and EPRI-3002002182, 2016) (i.e., from 16 to 40). Structured, parallelepiped meshes with an average cell volume of about $1.6 \times 10^{-3} m^3$ have been then considered for all simulations. Concrete, having density equal to 2100 kg/m³ and specific of 0.88 kJ/kg/K was defined to estimate the release on land. Storage conditions and size of commercial LNG truck were considered for preliminary analyses (Cryogas M&T Poland S.A. and IVECO). More specifically, initial liquid pressure of 9 bar, tank volume of 0.525 m³ with a filling level of 90 %v/v (representing the fill up value commonly adopted for this kind of configurations), ambient temperature of 288.15 K and Pasquill stability class 2F were assumed. Definition of accidental scenarios was carried out following the release categorization as proposed in purple book (Haag and Ale, 2005).

Two release categories were considered: small release from 10 mm hole and a large leak due to the catastrophic rupture of LNG tank. Loss of containment frequencies were retrieved from technical literature (or similar research papers) (American Petroleum Institute, 2008). Well-established source models were applied to calculate resulting pool dimensions needed as input for the following numerical modelling (Van Den Bosh and Weterings, 2005). The obtained results, expressed in terms of radiative heat flux where valuated at the steady state, intended as the time when the heat flux variations are within the 5%. Obtained results were considered for the estimation of stand-off distances by means of commonly adopted procedure and threshold values. More specifically, threshold values of 12.50, 7.00 and 5.00 kW/m² were considered, representing structural damage, fatality and irreversible damage (injury), respectively.

3 RESULTS AND DISCUSSION

Preliminary evaluations based on the abovementioned hypotheses were performed to characterize the accidental scenarios occurring in case of LNG release. In particular, the frequency of each scenario reported in Figure 1 was calculated, at first (Table 2).

Moreover, a flow rate equal to 1.138 kg/s was found for the case of continuous release from 10 mm hole, resulting in maximum pool radius of 1.93 m at 150 s after the spill beginning. On the contrary, maximum pool radius of 3.23 m was observed in case of catastrophic rupture. Starting from these results, further analyses were devoted to the pool fire characterization,

Table 2. Event tree quantification in terms of scenario frequency [event/y].

Continuous release				Catastrophic rupture			
Pool fire/ jet fire	Vapor cloud explosion	Flash fire + Pool fire	Gas dispersion	Fireball	Vapor cloud explosion	Flash fire + Pool fire	Gas dispersion
$1.00 \cdot 10^{-5}$	$6.00 \cdot 10^{-8}$	$9.00 \cdot 10^{-8}$	$9.85 \cdot 10^{-6}$	$5.00 \cdot 10^{-7}$	$8.00 \cdot 10^{-9}$	$1.20 \cdot 10^{-8}$	$4.80 \cdot 10^{-7}$

being the most frequent accidental scenario in case of release of LNG. The initial pool diameter considered for the implementation in CFD code was variated within the range 3.00 – 7.00 m to include in this analysis different filling level, as well. The well-known pool fire stages of ignition, propagation, steady state and decay were observed. However, it should be noted that the transient time to achieve the steady state condition is weakly affected by the pool diameter and it was found smaller than 7.0 s. Figure 2 shows the temperature distribution with respect to position at symmetric plane and steady state for pool diameter of 3.0 m. This result should be intended as representative of the investigated scenarios.

In the following, further insights on the steady state characterization were given. Pool fire were characterized by evaluating the overall radiative heat flux as a function of horizontal distance at given height of 1.5 m for all the investigated conditions. Similar trends where observed in terms of heat flux with respect to distance from the pool center for different pool radius. Indeed, radiative heat flux shows non-monotonic trend for distances lower that pool radius and decreases at increasing distances outside the pool. More specifically, in the former condition, a maximum value can be observed. For this reason, the quantity ϑ, as defined in the Eq. 1), was introduced for the characterization of the radiative heat flux emitted in case of pool fire. This parameter is intended as the position where the maximum heat flux can be observed:

$$\vartheta = x - (\alpha r) \tag{1}$$

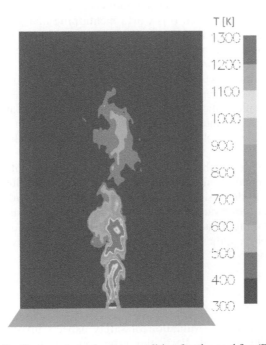

Figure 2. Temperature distribution at steady state condition for the pool fire (D = 3.0 m).

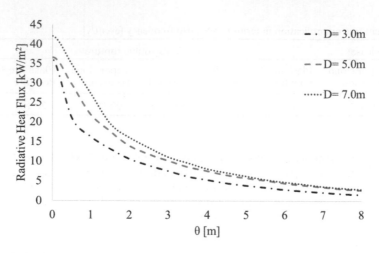

Figure 3. Radiative heat flux with respect to θ for different pool diameter.

Table 3. Safety distance (x) for different threshold values for the heat radiation produced by LNG LNG pool fire with given diameter d. Results in brackets are expressed in dimensionless form (x/r).

Threshold value	Safety distance from pool center (m)		
kW/m²	d = 3.0 m	d = 5.0 m	d = 7.0 m
5.0 (injury)	6.6 (4.4)	7.9 (3.2)	7.7 (2.2)
7.5 (fatality)	5.1 (3.4)	6.0 (2.4)	6.3 (1.8)
12.0 (structural damage)	3.6 (2.4)	4.5 (1.8)	4.8 (1.4)

where x is the horizontal distance from the pool centre, r is the pool radius and α assumes constant value for a given condition. It is worth mentioning that α variates within the range 0.6 – 1.0, meaning that maximum values are always in proximity of the pool border. In order to evaluate the effect of size on pool fire characterization, radiative heat fluxes as a function of ϑ for different pool diameter were reported (Figure 3).

Results indicate that the effect of diameter decrease at higher values, suggesting the existence of asymptotic parameters representative of the scale effect. Threshold values for personal injuries, in the absence of specific individual protection, as reported in Table 3, were adopted for the estimation of safety distances. Results were normalized with respect to pool diameter to make them comparable, as well. Hence, either absolute or normalized values were reported in Table 3.

Quite obviously, the greater the pool diameter, the higher the safety distance is expected. However, it should be highlighted that no clear trends can be individuated for larger diameter. On the other hand, the dimensionless value allows for the individuation of almost linear trend. This is particularly relevant for the development of empirical correlations aiming the characterization of pool fire. In accordance with the abovementioned observations, regardless the threshold values selected, the effect of diameter is reduced at elevated values, i.e. at diameter included in the range 5.0 m – 7.0 m.

4 CONCLUSION

This work was devoted to the evaluation of safety aspects involving the utilization and transportation of LNG on land. Particular attention was paid to small- and medium-scale tanks, with particular reference to the LNG truck configuration, where pressurized tanks of 525 l with maximum liquid level of 90 %v/v were adopted. Either small leakage (i.e. continuous release due to the presence of hole with 10 mm of diameter) or large leakage (i.e. catastrophic rupture) from LNG tanks at different filling levels were investigated. Diameters lower than 4.00 m were found for the former case, whereas a maximum diameter of 7.00 m was indicated for the latter. The nature of the final events potentially occurring and their frequency, as calculated in this work, have suggested further investigation on pool fire scenario. For these reasons, computational fluid dynamic has been applied to characterize the LNG pool fire resulting from the investigated conditions. The effect of time on pool fire development was characterized. It was found that the steady state condition was reached at the time of about 7.0 s for all the investigated pool sizes.

The effect of pool diameter on the emitted radiation at steady state condition has been evaluated, as well. Results show that the maximum heat flux value increases with the pool diameter as expected. Moreover, this value was reached in the proximity of the pool perimeter, at distances variating within the range $0.6r - 1.0r$.

Threshold values for structural damage (i.e. potentially leading to domino effect), fatality or irreversible damages were considered. It was found that the effect of pool size on safety distances cannot be expressed in terms of distance, whereas it should be conveniently indicated by normalizing the distance with respect to the pool size.

ACKNOWLEDGMENTS

Authors gratefully acknowledge the Adriatic-Ionian Programme INTERREG V-B Transnational 2014-2020. Project #118 – SUPER-LNG "SUstainability PERformance of LNG-based maritime mobility" for financial support.

REFERENCES

SUPER-LNG, 2019. Adriatic-Ionian Programme INTERREG V-B Transnational 2014-2020. Project #118: SUPER-LNG - "SUstainability PERformance of LNG-based maritime mobility,".

Agarwal, A.K., Singh, A.P., Maurya, R.K., 2017. Evolution, challenges and path forward for low temperature combustion engines. Prog. Energy Combust. Sci. 61, 1–56. https://doi.org/10.1016/j.pecs.2017.02.001.

American Petroleum Institute, 2008. Risk-Based Inspection Technology, API RP 581, second. ed, API Recommended Practice 581. API, Washington, D.C.

Aneziris, O.N., Papazoglou, I.A., Konstantinidou, M., Nivolianitou, Z., 2014. Integrated risk assessment for LNG terminals. J. Loss Prev. Process Ind. 28, 23–35. https://doi.org/10.1016/j.jlp.2013.07.014.

Babrauskas, V., 1983. Estimating large pool fire burning rates. Fire Technol. 19, 251–261. https://doi.org/10.1007/BF02380810.

Betteridge, S., 2018. Modelling large LNG pool fires on water. J. Loss Prev. Process Ind. 56, 46–56. https://doi.org/10.1016/j.jlp.2018.08.008.

BSI, B.S.I., 2013. BSI Standards Publication Natural gas — Quality designation (ISO 13686 : 2013).

Cryogas M&T Poland S.A., IVECO, n.d. Iveco Stralis Natural Power: Report on Testing of IVECO LNG Vehicles in Poland.

Gavelli, F., Bullister, E., Kytomaa, H., 2008. Application of CFD (Fluent) to LNG spills into geometrically complex environments. J. Hazard. Mater. 159, 158–168. https://doi.org/10.1016/j.jhazmat.2008.02.037.

Haag, U. de, Ale, 2005. Purple Book - Guidelines for Quantitative risk assessment. Purple B.

Kim, B.K., Ng, D., Mentzer, R.A., Sam Mannan, M., 2013. Key parametric analysis on designing an effective forced mitigation system for LNG spill emergency. J. Loss Prev. Process Ind. 26, 1670–1678. https://doi.org/10.1016/j.jlp.2013.01.007.

Liu, Y., Gao, X., Olewski, T., Vechot, L., Sam Mannan, M., O 'connor, M.K., 2012. Modelling the vaporization of cryogenic liquid spilled on the ground considering different boiling phenomena. Icheme Symp. Ser. No. 158.

Masum Jujuly, M., Rahman, A., Ahmed, S., Khan, F., 2015. LNG pool fire simulation for domino effect analysis. Reliab. Eng. Syst. Saf. 143, 19–29. https://doi.org/10.1016/j.ress.2015.02.010.

McGrattan, K., Hostikka, S., McDermott, R., Floyd, J., Weinschenk, C., Overholt, K., 2004. FDS User's Guide. https://doi.org/10.6028/NIST.SP.1019.

Nguyen, L.D., Kim, M., Choi, B., 2017. An experimental investigation of the evaporation of cryogenic-liquid-pool spreading on concrete ground. Appl. Therm. Eng. 123, 196–204. https://doi.org/10.1016/j.applthermaleng.2017.05.094.

NUREG-1824-suppl.1, EPRI-3002002182, 2016. Verification and validation of selected fire models for nuclear power plant applications. Supplement 1.

Perpignan, A.A.V., Gangoli Rao, A., Roekaerts, D.J.E.M., 2018. Flameless combustion and its potential towards gas turbines. Prog. Energy Combust. Sci. 69, 28–62. https://doi.org/10.1016/j.pecs.2018.06.002.

Puttock, J.S., Colenbrander, G.W., Blackmore, D.R., 2012. Maplin Sands Experiments 1980: Dispersion Results from Continuous Releases of Refrigerated Liquid Propane, in: Heavy Gas and Risk Assessment — II. https://doi.org/10.1007/978-94-009-7151-6_9.

Qi, R., Raj, P.K., Mannan, M.S., 2011. Underwater LNG release test findings: Experimental data and model results. J. Loss Prev. Process Ind. 24, 440–448. https://doi.org/10.1016/j.jlp.2011.03.001.

Raj, P.K., 2007. Large hydrocarbon fuel pool fires: Physical characteristics and thermal emission variations with height. J. Hazard. Mater. 140, 280–292. https://doi.org/10.1016/j.jhazmat.2006.08.057.

Rana, M.A., Cormier, B.R., Suardin, J.A., Zhang, Y., Sam Mannan, M., 2008. Experimental Study of Effective Water Spray Curtain Application in Dispersing Liquefied Natural Gas Vapor Clouds. Aiche J. 27, 345–353. https://doi.org/10.1002/prs.

Van Den Bosh, C.J.H., Weterings, R.A.P., 2005. Methods for the Calculation of Physical Effects (Yellow Book).

Zhang, B., Laboureur, D.M., Liu, Y., Gopalaswami, N., Mannan, M.S., 2018. Experimental Study of a Liquefied Natural Gas Pool Fire on Land in the Field. Ind. Eng. Chem. Res. 57, 14297–14306. https://doi.org/10.1021/acs.iecr.8b02087.

Zhiyi, Y., Xunmin, O., 2019. Life Cycle Analysis on Liquefied Natural Gas and Compressed Natural Gas in Heavy-duty Trucks with Methane Leakage Emphasized. Energy Procedia 158, 3652–3657. https://doi.org/10.1016/j.egypro.2019.01.896.

Risk Analysis Based on Data and Crisis Response Beyond Knowledge – Huang & Nivolianitou (eds)
© 2020 Taylor & Francis Group, London, ISBN 978-0-367-25146-8

Failure analysis of gas pipelines under surface load

Yan Li*
Beijing Academy of Safety Science and Technology, Beijing, China

Jian Shuai
China University of Petroleum-Beijing, Beijing, China

ABSTRACT: According to the action mechanism of external loads, a test device which can be used to measure the deformation of pipe subjected to external loads is designed. The test device is free from underground installation and without excavation measurement. To illustrate the rationality of the test device, compare the test results with theoretical calculations and the result agrees well. Based on the test data, the deformation characteristics and failure modes of the pipe with defects under external loads are analyzed. The factors affecting the bearing capacity of the pipe with defects are discussed, and the failure mode of the pipe with defects under external loads is stiffness failure determined by deformation.

Keywords: failure analysis, defective pipe, external loads, deformation feature.

1 INTRODUCTION

In municipal engineering construction, pipelines often need to be buried under roads for engineering purposes, which subjects the pipes to multiple loads. For the city gas pipeline, the external load formed from vehicles and deposits above the road will act on the pipe. When thin-walled steel pipe is subjected to vertical load oval deformation, which may lead to difficult situations for cleaning the pipe work. In severe cases, the pipe will lose carrying capacity due to insufficient rigidity. Statistics have indicated that metal loss is one of the most common factors threaten pipeline safe operation. Metal loss results in a decrease of the pipe wall thickness and expansion over time. As a result, it may lead to perforating, rupturing, and eventually threatening the pipeline integrity. This study mainly focused on the remaining strength of defective pipes under internal pressure and formed systematic assessment guidelines. There are numerous simulations and experimental research aimed at improving the conservatism of the assessment method. However, compared with long-distance pipeline, the pressure in a gas pipeline is lower. The external load makes a significant impact on safety operation of a gas pipeline while the influence of vertical load on stiffness of the pipe with metal loss remains unresolved. The mechanical behavior of defective pipes under external load is a pressing research point.

Research on pipe deformation behavior under vertical load began with the Iowa formula developed by Spangler in 1947, which has been widely used in pipe design standards in many countries (ALA, 2001; ASTM, 2011). In the derivation of the formula, Spangler made an ideal hypothesis about the load distribution and pipe deformation shape. In addition, the soil properties are obtained from test data using a semi-empirical method (Spangler, 1964). Influenced by early research condition, the results of Iowa formula show more conservative under some operation environment (Khatri et al. 2015). With the advance of computer and electronic apparatus technology, researchers have investigated the interactions between pipe and soil using laboratory

---

*Corresponding author: yanli-2014@hotmail.com

tests. Yoo found large differences in the actual load distribution of the Spangler formula under different soils and burial conditions (Yoo et al. 1999). Masada and Tian changed the load distribution of the Spangler formula and developed a modified calculation method (Masada 2000; Tian et al. 2015). Later, Masada focused on the effect of soil on the deformation of thermoplastic pipes in a backfill process and introduced a calculation method of peaking the deflection of the pipe in the back fill process (Masada et al. 2007). Similarly, Howard proved that both load and soil compaction have an effect on pipe deformation and that the deformation after initial backfill and that subjected to long-term load are different (Howard et al. 2011). Considering this view, Howard introduced a time lag factor and load lag factor based on the deflection lag factor in the Spangler formula, and proposed a maximum defection calculation method for pipes during the backfill process. Jin considering the fractural feature in the stability study of glass-reinforced thermosetting plastic pipes and established a safety assessment method and thickness optimal equation by using the load distribution of Spangler theory (Jin et al. 2013).

The external load can affect the structural design and failure mode of buried pipe, particularly for defective pipe, the influence of metal loss on stiffness will interact with external load, ultimately lead to decrease the carrying capacity of pipe. Therefore, the stiffness analysis is important for guarantee the safety operation of buried gas pipeline. Analytical methods, experimental methods, and numerical modeling methods can be used to study the mechanical behavior of defective pipes and a systematic prediction method of remaining strength. Nevertheless, there is still insufficient research on the deformation behavior of defective pipes. Without considering the effect of defects, the early theory only covered vertical deflection calculation of pipes without defects. In addition, the present experimental research mainly focuses on the interaction between pipe and soil. This paper tested pipe subjected to external load, a testing device has been designed and a series of tests have been done. The radial deformation and strain variation have been measured and the effects of defects on the deformation of pipes have been analyzed.

2 EXPERIMENTAL PROCEDURE

The testing device is composed of four components: strain testing systems, the pipe specimen, a fixed device and a screw jack, as shown in Figure 1. The pipe specimen was placed in the two semicircle mountings in the middle of the fixed device and with rubber to distribute the load. The width of the rubber is equal to one-sixth of the pipe circumference and the thickness of rubber is 80mm. The load is exposed from the bottom up of the pipe through the rubber and mounting, that is, the upper semicircle mounting is fixed on the fixed device and the screw jack is placed under the lower semicircle mounting. When the handle of the screw jack is turned, the lower semicircle mounting moves up gradually, driven by the sleeve, and therefore achieves the purpose

Figure 1.   Testing device.

Figure 2.   Groove type defect.

of the reverse loading. The load exposed on the pipe can be read from the dynamometer connected with the screw jack.

Five pipe specimens were tested: one pipe without defects and four pipes with defects of different sizes. The dimensions of all the pipe specimens is 508×15mm. The material grade is Q235b and the specific minimum yield strength and tensile strength are 235MPa and 370MPa, respectively. The defect is an manufactured outer surface groove, as shown in Figure 2, along the circumferential direction of pipe; the depth of the defect remains the same across the width of the defect. The defect sizes are listed in Table 1. The experiments included cross-sectional deformation measurements by Vernier caliper and strain tests by a resistance strain measuring system DH3816Nin combination with a BFH120-3AA strain gauge. According to the one-quarter bridge connection method, strain gauges are arranged along the pipe circumferential direction and on the same material plate for temperature compensation.

Figure 3 illustrates the location of strain gauges in the pipe without and with defects, respectively. The red represents the strain gauges located on the internal surface and the blue represent

Table 1.   Parameters of pipe specimens.

NO.	Diameter	Thickness	Width/°	Depth/%T
T15c0d0			0	0
T15c30d20			30	20
T15c30d50	508	15	30	50
T15c60d20			60	20
T15c60d50			60	50

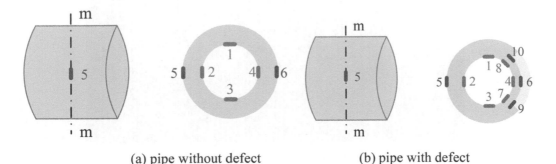

(a) pipe without defect   (b) pipe with defect

Figure 3.   Strain measurement location.

Figure 4.   Test pipe with strain gauge.

the strain gaugeslocatedon the outer surface. The locations are the same for measuring points 1–6 in the pipe without and with defects. Measuring points 7–10 are located in the defect edge of the inner or outer surface of pipe, and the test pipe with strain gauges are shown in Figure 4.

In order to study the deformation behavior of the pipe, a deformation measurement is taken after certain pressures. Figure 5 shows the non-deformed and deformed shape of the pipe. The pipe becomes oval under the external load: horizontally elongated and vertically shortened. For quantification purposes, using the rate of diameter variation in the horizontal and vertical direction, that is, using the ratio of diameter difference before and after deformation. The curves of the rate of diameter variation change with an external load on the pipe without defects are shown in Figure 6, and there is larger vertical deformation than horizontal. As external load increases, pipe deformation goes through the following three stages: linear increasing, nonlinear deformation and deforming destroying. In the initial linear increasing phase, the pipe deformation increases linearly with load. As deformation reaches a certain range, the pipe enters a

(a) Non-deformedshape (b) Deformed shape

Figure 5.   Shape of pipe in different stages.

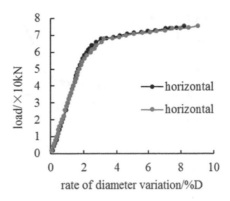

Figure 6. The rate of diameter variation of the pipe without defects changing with external load.

nonlinear deformation stage, where relatively a small load can cause a large deformation and the slope of the curve reduces gradually. With the load continuing to increase, the curve becomes almost a straight line and the pipe will lose carrying capacity due to instability.

Figure 7 shows the curves of the rate of diameter variation change with the external load on four pipes without defects. Similar to the pipe without defects, pipe deformation also goes through three stages.

The strain variety curves of the T15c0d0 pipe without a defect are shown in Figure 8. The filled and hollow circles represent measuring points on the internal and outer surface of pipe, respectively. The black, red, blue and green represent measuring points on the top, left, bottom and right of the pipe, respectively. A similar trend in strain variation can be observed

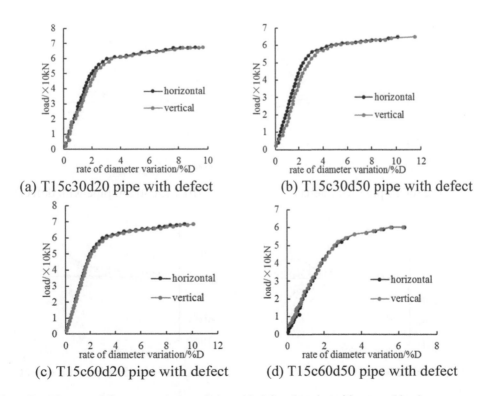

Figure 7. The rate of diameter variation of pipe with defect changing with external load.

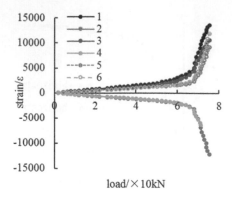

Figure 8.   Strain distribution of the pipe without a defect.

in comparisons of diameter variation: strain increasing linearly in the early stage and then entering a nonlinear stage, the load begins increasing slowly and a small load can cause strain to increase greatly. After that, the strain increases rapidly. Under the external load, the strain distributes as bilateral symmetry. The strain at measuring point measuring points 2 and 4 is negative, and the corresponding areas are subjected to compressive load. While of the strain at measuring point measuring points 5 and 6 is positive, the corresponding areas are subjected to tensile load. The bending state of inner and outer surface are opposite, but the strain on the internal surface is larger due to the compression causes by external load acting on the cross section combined with a bending moment.

The strain variation curves of the pipes with defects are shown in Figure 9. The markers are the same as the pipe without a defect; solid lines 7 and 8 represent the measuring point

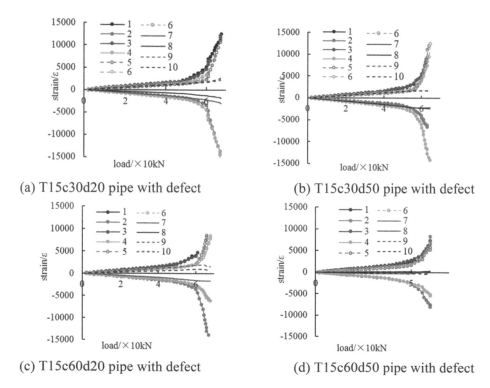

(a) T15c30d20 pipe with defect

(b) T15c30d50 pipe with defect

(c) T15c60d20 pipe with defect

(d) T15c60d50 pipe with defect

Figure 9.   Strain distribution of pipe with defect.

446

located in the defect edge on the inner surface of the pipe and the dotted lines 9 and 10 represent the measuring point located in the defect edge on the outer surface of the pipe. The strain variation of pipes with defects are similar to the pipes without defects. In the pipe T15c30d20, T15c30d50 and T15c60d20, the bending state of defect edge are the same as horizontal measuring points. The magnitude and distribution of strain are almost the same as the linear stage of the pipe without a defect, and the nonlinear stage does not appear. In pipe T15c60d50, the strain of the measuring point on the defect edge is apparently smaller and even close to 0.

3   TEST VERIFICATION

In order to verify the experiment model, we compared the test result of the pipe without defect with the calculations by the Iowa formula. Considering the Iowa formula does not apply to the nonlinear deformation stage, the comparison takes the elastic data in the first stage. The Iowa formula is shown as equation 1.

$$\Delta X_s = \frac{ZK_aW_cD^3}{8EI + 0.061E'D^3} \tag{1}$$

where, ΔX_s = diameter deformation of pipe, mm; I = inertia moment of pipe cross section, $I = t^3/3$; t = thickness of pipe, mm; D = diameter of pipe, mm; Z = time lag factor, 1.5 for lack of data; K_a = bedding factor of soil; E = modulus of pipe; E' = modulus of soil reaction. W_c = vertical load on top of pipe, consisting of soil load W_1 and vehicle load W_2 as equation (2) and (3) show, respectively, N/mm.

$$W_1 = \gamma DH \tag{2}$$

$$W_2 = \frac{\mu_d Q_{vk} D}{(a + 1.4H)(b + 1.4H)} \tag{3}$$

The comparison result is shown in Figure 10. The load using the Iowa formula is obtained from the test that represents specific load conditions by formula (2) and (3) conversion. The calculations are consistent with the experiment result that proved the test device is appropriate to simulate a pipe subjected to external load.

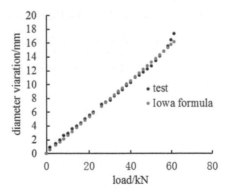

Figure 10.   Result comparison between experiment and Iowa method.

4 CONCLUSION

In order to study the deformation behavior and strain distribution of pipes subjected to external load, a test device is designed and a series of tests are performed. Through analysis, some conclusions are drawn as follows:

(1) Using the test device illustrated in this paper to simulate the pipe subjected to external load caused by vehicles and ground deposits is reasonable and the results can reflect the key characteristic of strain variation and deformation behavior. The experiment results achieve high conformity with calculations obtained by the Iowa formula, which verifies the correctness of the test device. In addition, the device also provides a convenient way to replace the field test and improve the economic efficiency and practicability.
(2) According to the experiment data, the deformation behavior of the pipe subjected to external load can be summarized: the deformation of pipe undertakes linear and nonlinear variation, and failure is due to a loss of stability.

ACKNOWLEDGMENTS

This work was supported by National Key R&D Program of China (Grant No. 2018YFC0809900) and Beijing science and technology Plan program (Z181100009018003, Z181100009018010).

REFERENCES

ALA. 2001. *Guidelines for the design of buried steel pipe*. ALLIANCE A L. American: American Society of Civil Engineers.
ASTM. 2011. *Standard Test Method for Determination of External Loading Characteristics of Plastic Pipe by Parallel-Plate Loading*. ASTM. United States: American Society for Testing and Materials.
Khatri, D.K., Han, J., Corey, R., et al. 2015. Laboratory evaluation of installation of a steel-reinforced high-density polyethylene pipe in soil. *Tunnelling and Underground Space Technology* 49: 199–207.
Howard, A., Walsh, T., Dean, S.W. 2011. Deflection Lag, Load Lag, and Time Lag of Buried Flexible Pipe. *Journal of Astm International* 8: 102–108.
Jin, N.J., Hwang, H.G., Yeon, J.H. 2013. Structural analysis and optimum design of GRP pipes based on properties of materials. *Construction & Building Material* 38: 316–326.
Masada, T. 2000. Modified Iowa Formula for Vertical Deflection of Buried Flexible Pipe. *Journal of Transportation Engineering* 126: 440–446.
Masada, T., Sargand, S.M. 2007. Peaking Deflections of Flexible Pipe during Initial Backfilling Process. *Journal of Transportation Engineering* 133: 105–111.
Spangler, M.G. 1964. Pipeline Crossings Under Railroads and Highways. *American Water Works Association* 56: 1029–1046.
Tian, Y., Liu, H., Jiang, X., et al. 2015. Analysis of stress and deformation of a positive buried pipe using the improved Spangler model. *Soils & Foundations* 55: 485–492.
Yoo, C.S., Chung, S.W., Lee, K.M., et al. 1999. Interaction between Flexible Buried Pipe and Surface Load. *Journal of KGS* 15: 83–97.

Risk Analysis Based on Data and Crisis Response Beyond Knowledge – Huang & Nivolianitou (eds)
© *2020 Taylor & Francis Group, London, ISBN 978-0-367-25146-8*

The development trends in catastrophes across the world

Xiaojun Pan*
Academy of Disaster Reduction and Emergency Management, Faculty of Geographical Science, Beijing Normal University, Beijing, China

Chengyi Pu
School of Insurance, Central University of Finance and Economics, Beijing, China

Shilong Peng
Foreign Languages School of Sichuan Tourism University, Chengdu, China

ABSTRACT: There are three trends in the development of catastrophes worldwide. First, the frequency of catastrophe grows higher and has increased by five times from 1970 to 2013, and there are now almost 180 catastrophes per year, on average. Second, the economic loss of a single catastrophe and the total loss caused by all catastrophes are constantly rising. North America, Europe and Asia are the regions that suffer the most loss, as total economic loss there accounts for over 90% of the world's total loss. Third, the casualties have become increasingly serious. A catastrophe that causes a death toll of over 100,000 used to occur once very ten years but now occurs once every four or even two years.

Keywords: catastrophe, economic loss, casualty, development trend, the world

1   INTRODUCTION

Since the turn of the century, the greenhouse effect has grown constantly worse and many disasters have taken place across the globe: major natural disasters such as earthquake, flood, drought, and typhoon have occurred more frequently, and some have occurred simultaneously (Sigma, 1999–2017). The number of those disasters has been increasing (the black line in Figure 1) and the loss of life and property is greater. The deterioration of the ecological environment and vulnerability of the biological environment have caused major economic loss that increases year by year (the blue line in Figure 1), and humans are facing unprecedented challenges of existence and development.

With vast land and diversity of the natural environment, China suffers different disasters all year round and is one of world's particularly big catastrophes-stricken areas. The number of particularly big catastrophes in China accounts for around 16% of the world's total (1991–2017 *China Statistical Yearbook*). From 1991 to 2017, those disasters took place frequently, some of which occurred simultaneously, and the number has risen greatly (the black line in Figure 2). The direct economic loss is huge and is increasing year by year (the blue line in Figure 2). Flood and earthquake are the most destructive disasters.

The research on the trends in catastrophes mainly focuses on catastrophology and environmentology. Many scholars state that with a greater greenhouse effect and increasingly active crustal movement (Shi, 1996), the interval of two successive meteorological disasters or geological disasters is shortening (Yan, 2011; Yang et al., 2012), various catastrophes occur

*Corresponding author: Xiaojun Pan, E-mail:1002343190@qq.com

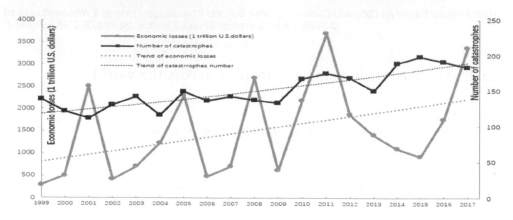

Figure 1. Trends in global catastrophe and economic losses (1 trillion US dollars), 1999–2017.
Sources: (1) Sigma, 1999–2017. (2)https://www.ncdc.noaa.gov/billions/

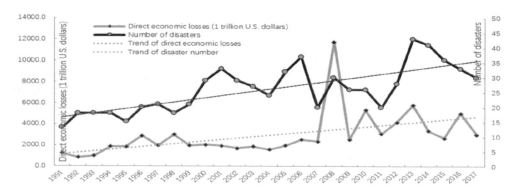

Figure 2. Trends in the number of catastrophes and economic losses (1 trillion US dollars) in China, 1991–2017.

Source: http://www.emdat.be/database

withincreasing frequency,and due to increased concentration of the population and property, casualties and economic loss caused by catastrophes willcontinue to rise.

The research on catastrophe risk measurement, based on mathematical operations, mainly predicts the possible damages caused by catastrophe, and some advancements have been made by other countries. With developing computer technology, the mathematical modeling for catastrophe risk measurement has been simplified. Karen (1986) used the Monte Carlo Modelto simulate catastrophe andgenerated random results to evaluatecatastrophe loss, arguing that the probability theorycould not predict earthquake loss with the data in hand, and the theorycouldorganize related statistics only after catastrophe. Hsieh (2004) used the theory of generalized extreme value distribution to verify catastrophe loss in the US in the 1990s and demonstrated the validity of the theory for predicting catastrophe loss. Li (2010) tried to estimate catastrophe loss with Bayesian Network and a Spatial Analysis Method. After estimating earthquakes and hurricanes inthe US and analyzing those statistics, Braun (2011) found that the catastrophe loss followed the process of Burr Double Random Poisson.

In summary, we found that many other scholars have studiedthe same issue. However, their studies are relatively simple, only analyzing onearea, such as insurance business variety, operation institution, or operating model. In those papers, there is alack of research on the overall impact of trends in catastrophes on the world insurance industry andon the whole picture of this industry.

Therefore, we attempted to comprehensively analyze the trends in catastrophes across the world. This thesis is an integration of past literature and innovation and has great theoretical significance.

In this paper, we introduce the concept and classification of catastrophe, analyze three characteristics of catastrophe compared with general disasters, and then discuss three factors that affect the probability of catastrophe occurrence and economic loss. Finally, assisted by charts and illustration, we analyze trends in catastrophes from the perspectives of China and international society.

2 CHARACTERISTICS AND TREND OF GLOBAL CATASTROPHE

The word "catastrophe" originated from the ancient Greek word "καταστροφή", which was originally translated into "meteor", and later "tragic ending". In this paper, we define catastrophe as a natural disaster with losses of over 15 million US dollars and over 200 casualties. The data has been deducted from the impact of inflation.

With the changes in economic and social development and from different angles, the concept of catastrophe is constantly evolving. Governments, international organizations and insurance institutions pay close attention to catastrophe and have given their own definitions (Table 1).

There are many types of catastrophe classification standards. The most representative is set in terms of cause and frequency of catastrophein Table 2.

Table 1. Definition of catastrophe by organizations across the world.

Name of Agency and Organization	Definition
United Nations (UN)	Social dysfunction brought to human, materials and the environment, which is also beyond the tolerance of their own resources.
United Nations International Decade for Disaster Reduction (UDNIDDR, 1994)	The disaster causes property loss of over 1% of the national income; the affected population exceeds1% of the national population; the death toll exceeds100.
Belgian Center for Epidemiology and Disaster Research (CRED, 2013)	After a natural disaster occurred in a certain area, the disaster is a catastrophe if one of the following caseswas identified: over 10 residents were reported dead; over 100 people were affected by the disaster; the state (provincial) government declared a state of emergency; the affected area sought international assistance.
US Insurance Service(ISO, 2009)	Events that affect many policyholders and insurers. Incidents that cause direct insurance loss on assets of more than $25 million and affect many insulants and insurers.
Munich Re (2009)	If a natural disaster affects a certain area beyond the latter's self-rescue capacity and one of these conditions is recognized, the disaster is a catastrophe: the affected area has to rely on inter-regional or international assistance; thousands of people are killed; hundreds of thousands of people are homeless; the overall economic loss is huge; considerable insurance loss.
Swiss Re (2013)	After a disaster, if one of these conditions was identified, the disaster is a catastrophe: a shipping disaster with an insurance loss of over $19.3 million; an aviation disaster with an insurance loss of over $38.6 million; and other disasters with an insurance loss of over $48 million; The total economic loss is more than $96 million; the combined death and missing population are greater than 20; the injured population is greater than 50; and the homeless population is greater than 2000.

Source: 1. http://www.emdat.be/criteria-and-definition.2. Swiss Re, Natural Disasters and Man-made Disasters in 2013: Large-scale Losses are caused by Floods and Hail; "Haiyan" Attacked the Philippines, Sigma, 2014, Vol.1: 2.

Table 2. Catastrophe classification.

Criteria	Name	Definition	Characteristics	Example
Cause	Natural catastrophe	Catastrophic events cause by natural forces	The amount ofloss depends on the intensity of natural forces and frangibility of affected objects.	Earthquake, typhoon, flood, etc.
	Man-made catastrophe	Catastrophic events caused by human activities	Only a small area or a certain large target is affected; only a small number of the affected objectsare covered by insurance.	Major fire, traffic accident, terrorist activities, etc.
Frequency	Normal catastrophe	More than once a year	Although the occurrence ofsuch catastrophe is predictable to some extent, the exact number and intensity of occurrence cannot be determined, so the actual loss often is larger than expected, causing abnormal financial fluctuation of insurance company of the same year.	Seasonal storm, heavy rain, etc. in the coastal areas of Guangdong, China.
	Abnormal catastrophe	Little probability; lack ofcharacteris-ticsof catastrophe	The catastrophe will not occur for a long time, but if it occurs, the loss will be very large, and may even exceed the insurance company's actuarial expectation of that year, which may lead to the bankruptcy of some small and medium-sized insurance companies.	Hurricane Andrew, Indonesian Tsunami, Wenchuan. Earthquake, etc.

Source: Zhigang Zhou: Risk Insurability Theory and National Management of Catastrophe Risk, Fudan University, 2005:3.

2.1 Characteristics of global trends in catastrophes

2.1.1 Natural factors increasing catastrophe risk

According to the latest release of the World Meteorological Organization, the global average temperature in 2018 was 0.38°C higher than the 1981–2010 average, which is about 1.0°C higher than the pre-industrial level (average of 1850–1900)and was the fourth warmest year-since complete meteorological observation records began. The past five years (2014–2018) are the warmest five years since complete meteorological observations began.

In 2018, the average temperature in the Asian region was the fifth highest since1901, andChina also had an unusually warm year. From 1951 to 2018, China's annual average temperature increased by 0.24°C every 10 years, and therateof temperature rise was significantlyhigher than the global average.

According to the National Oceanic and Atmospheric Administration (NOAA), in the past 30 years every decade has beensignificantly warmer than the previous one, and sets a new record of the world's highest temperature. Studies have shown that due to global warming, natural disaster-inducing factors are more active, and the frequency and scope of natural disasters and loss they cause will further increase. Due to the warming of the climate and the increase innatural variation, the risk of catastrophes such as strong typhoon, heavy rain, flood and major drought has increased. Climate change may cause uneven distributions ofprecipitation, the global pattern of "flood in the south and drought in the north" will remain, and climate change may also lead to the expansion ofarid areas and the rise of sea levels.

At the same time, theglobal geological plate movement is more frequent. The Asia-Europe plate, the Pacific plate and the American plate are moving at speeds of several centimeters to tens ofcentimeters per year and are squeezing each other or stretching themselves. As a result ofthe movement, the crust is broken or displaced, therebyincreasing the probability of a major earthquake.

2.1.2 *Human factors increasing catastrophe risk*

Humans survive and develop by acquiring, utilizing and destroying natural resources. Humans have unreasonably exploited water, forestsand mineral resources, which has aggravated soil erosion and desertification, and increased the frequency and intensity of crustal movement, which may incur large-scale mudslides, floods, earthquakes and other natural disasters. Therefore, environmentaldestructionbyhumansresults in natural disasters. As early as 2007, the assessment report by the United Nations Intergovernmental Panel on Climate Change (IPCC) stated that anthropogenic emission of greenhouse gases is a major cause of climate warming and has a profound impact on climate change and disaster occurrence. The growing population and the overly rapid economic development have constantly deteriorated humans' living environments, directly or indirectly causing disasters, and exacerbating the disaster and increasing the probability.

2.1.3 *Population and property intensification increasing catastrophe loss*

Due to such factors as the economy, cultural and personal reasons, populations and property have densified in recent years. Against thatbackground, once a catastrophe occurs it can lead to major property damage and casualties. If the affected area is isolated, there are few residents, or the property value is low, the damage caused is relatively slight. If the affected area has a large population, and large and concentrated wealth, disasters can cause catastrophic loss. For example, coastal areas boast an advanced economy, a large population and concentrated properties. However, these areas are frequently stricken by natural disasters, such as hurricane and typhoon, and the loss is often enormous.

2.2 *Global trends in catastrophes*

2.2.1 *Increasing frequency of catastrophe*

Since the 1970s, with the rapid development of the global economy, the areas of human activities have expanded largely. The environment has been constantly contaminated, and problems such as soil erosion, the greenhouse effect, and the hole in the ozone layer have become increasingly serious. The frequency of natural disasters has been rising, and social and economic development and humane living conditions have been threatened. Figure 3 shows that during the period from 1970 to 2013, the number of natural catastrophes in the world fluctuated constantly but also increased constantly. Especially in the late 1980s, the number of catastrophes increased greatly, and as of 1993 the number of catastrophes has quadrupled. In

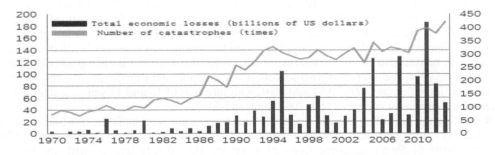

Figure 3.   Number of global natural catastrophes and their economic losses from 1970 to 2013.
Source: Database of Sigma Explorer

the next 17 years, the number remained at 135 per yearonaverage. By 2010, the number of catastrophes had rocketed to 172, never going down, and then continued to increase for several years. In 2013, there were 187 natural catastrophes globally, and the world was entering a period of high-frequency catastrophes.

2.2.2 *More economic loss and regular regional distribution of catastrophe*

With the rapid increase in the frequency and number of catastrophes, the economic loss has also risen sharply. Figure 3 shows the economic loss caused by global catastrophes since 1970. In 1994, the annual global economic loss caused by catastrophe exceeded, for the first time, $100 billion, and generally showed a growing trend after 1994. The average annual economic loss of natural catastrophe from 1994 to the present day has reached $140 billion, six times the average annual economic loss from 1970 to 1993. Among these years, 2011 is worthy of particular focus, as the economic loss exceeded $410 billion, which is the largest economic loss caused by natural catastrophe in availablerecords. The lossesthat year accounted for 0.31% of global GDP, and most was incurred by Japan in the"3.11 Earthquake" and its subsequent massive tsunami, which set a record for earthquake magnitude in Japan. In February 2011, the earthquake in Christchurch, New Zealand caused an economic loss of about $15 billion, which increased the global economic losscaused by earthquake to more than $230 billion, breaking the previous record. In addition tothose most devastating earthquakes in world history, Thailand also experienced the most severeflood in recent decades in 2011. The flood caused a huge loss to the Thai manufacturing industry which, therefore,was disconnected from the international supply chain.

The economic loss caused by a single natural catastrophe also showed a growing trend. Table 3 shows the top 30 natural catastrophes in global economic loss between 1970 and 2014. In terms of the number of catastrophes with economic loss exceeding $15 billion, there were zero in the 1970s, three from 1981 to 1990, five from 1991 to 2000, and nine from 2001 to 2010. During the period from 2011 to 2013, there werefive natural catastrophes causing such great economic loss. This shows that a single catastrophe is increasingly destructive, and the economic loss is getting becoming more serious.

In addition, due to the frequency of natural catastrophe, the intensity of disasters and the property density in the affected areas, the economic loss shows regular regional distribution. In Figure 4, North America, Europe, and Asia are the three regions most affected by natural catastrophe, and economic loss in North America and Asia has increased significantly in recent years. For example, in 2013 the economic losses in Asia, North America and Europe were $62 billion, $32 billion and $33 billion, respectively,accounting for 44.29% and 22.86% and 23.57% of total global economic loss, whilethe loss in other continentsaccounted for less than 10%.

2.2.3 *More serious casualties caused by catastrophe*

In addition toeconomic loss, more important is that the lives of residents in the affected areas are also greatly threatened by catastrophes. Between 1970 and 2013, there were 10 years in which the death toll from catastrophes exceeded 50,000 across the world. The highest was in 1970, when the number of deaths reached 371,000. The total death toll in the 44 years was 2,441,769, with an average of 55,495 people losing their lives each year due to natural catastrophes. Figure 5 shows that, since the end of the 1970s, the toll due to disasters has generally increased. Because of the density of population and the intensity of disasters in affected areas, the interval betweentwo successive catastrophes causing heavy causalities has gradually shortened. Before 2004, the interval of two successive catastrophes causing deaths of over 100,000 waslonger than 10 years (1976, 1991 and 2004, with intervals of 15 and 13 years), and now the interval is fouror even two years (e.g. 2004, 2008 and 2010).

Regarding the casualties caused by a single catastrophe, earthquake is the most destructive and becomes more destructive as the population density increases. According to statistics in the first issue of Sigma magazine from 2014, among the 40 catastrophes with the largest number of victims in the world there were 21 earthquakes, accounting for more than half of the total catastrophes. The number of hurricanes and floods were seven and five, respectively.

Table 3.   Top 30 natural catastrophes by global economic loss between 1970 and 2014.

Ranking	Time	Country	Catastrophe	Economic loss*	Death Toll
1	Aug 29, 2005	US	Hurricane Katrina	1250	1833
2	Jan 17, 1995	Japan	Earthquake	1000	5297
3	May 12, 2008	China	Earthquake	850	87476
4	Oct 28, 2012	US	Hurricane Sandy	500	54
5	Aug 5, 2011	Thailand	Flood	400	813
6	Feb 27, 2010	Chile	Earthquake	300	562
7	Sep 12, 2008	US	Hurricane Ike	300	82
8	Jul 1, 1998	China	Flood	300	3656
9	Jan 17, 1994	US	Earthquake	300	60
10	Feb 23, 2004	Japan	Earthquake	280	40
11	Aug 24, 1992	US	Hurricane Andrew	265	44
12	Jan 10, 2008	China	Blizzard	211	129
13	Jun 20, 2012	US	Drought	200	0
14	Aug 17, 1999	Turkey	Earthquake	200	17127
15	Nov 23, 1980	Italy	Earthquake	200	4689
16	May 29, 2010	China	Flood	180	1691
17	Sep 15, 2004	US	Hurricane Ivan	180	52
18	Sep 23, 2005	US	Hurricane Rita	160	10
19	Aug 13, 2004	US	Hurricane Charlie	160	10
20	May 20, 2012	Italy	Earthquake	158	7
21	Feb 22, 2011	New Zealand	Earthquake	150	181
22	Aug 1, 1995	South Korea	Flood	150	68
23	Oct 24, 2005	US	Hurricane Wilma	143	4
24	Sep 21, 1999	Taiwan, China	Earthquake	141	2264
25	May 20, 2011	US	Hurricane	140	176
26	Dec 7, 1988	Soviet Union	Earthquake	140	25000
27	Jan 12, 1994	China	Drought	137	0
28	Jun 30, 1996	China	Flood	126	2775
29	Jul 16, 2007	Japan	Earthquake	125	9
30	Jun 24, 1993	US	Flood	120	48

Source: Statistics from EM-DAT of Belgium CRED.
* Unit: hundred million US dollars (2013 price standard).

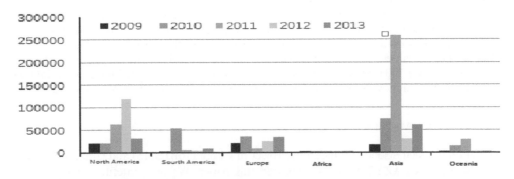

Figure 4.   Economic losses caused by natural catastrophes on continents in 2009–2013 ($1 million).
Source: Sigma data collation of Swiss reinsurance from 2010 to 2014.

In terms of geographical factor, Asia is the most affected region, which is almost certainly related to its high population density. Taking earthquake as an example, according to the statistics of 2009 Geological Natural Catastrophes by Munich Reinsurance Company (Table 4), seven of the world's ten worst earthquakes since 1990 have occurred in Asia.

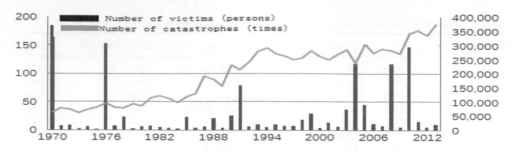

Figure 5.   Number of global natural catastrophes and casualties from 1970 to 2013.
Source: based on the Sigma Explorer database.

Table 4.   Earthquakes with the highest death toll since 1900.

Country	Year	Catastrophe	Death toll
China	1976	Earthquake	242000
China	1920	Earthquake, Landslide	235000
Indonesia, Sri Lanka, Thailand, India	2004	Earthquake, Tsunami	220000
Haitian	2010	Earthquake	200000
Japan	1923	Earthquake	142800
Pakistan, India, Afghan	2005	Earthquake	88000
Italy	1908	Earthquake, Tsunami	85925
China	2008	Earthquake	84000
China	1932	Earthquake	77000
Peru	1970	Earthquake, Landslide	67000

Source: Munich Re Topic GEO Natural Catastrophes 2009 Analysis, Assessment, Position

3   CONCLUSION

First, the frequency of catastrophe is getting higher, and has increased by five times from 1970 to 2013, with almost 180 catastrophes in each year on average.

Second, both the economic loss of a single catastrophe and the total loss caused by all catastrophesare constantly rising. North America, Europe and Asia suffer the most loss, as the total economic loss in these regions accounts for over 90% of the world's total loss.

Third, the casualties have become increasingly serious. Previously, acatastrophe with a death toll of over 100,000 occurred once very ten years (in 1976, 1991 and 2004, at intervals 15 and 13 years), but now occurs once every fouror even two years (in 2004, 2008 and 2010).

ACKNOWLEDGMENTS

The National Social Science Foundation Project of China (16BJL115) and China Earthquake Administration (CEAZY2019JZ13) supported this paper. We are grateful to reviewers and editors for their insightful comments on this article.

REFERENCES

Braun, A. 2011. Pricing catastrophe swaps: A contingent claims approach. Insurance: Mathematics and Economics (3): 520–536.
Hsieh, P.H. 2004. A data-analytic method forecasting next record catastrophe loss. The Journal of Risk and Insurance (71): 309–322.

Karen, M.C. 1986. A Formal Approach to Catastrophe Risk Assessment and Management. Casualty Actuarial Society, Arlington, USA.

Li, L.F., Wang, J.F. and Leung H.T. 2005. Using spatial analysis and Bayesian network to model the vulnerability and make insurance pricing of catastrophic risk. International Journal of Geographical Information Science (5): 1759–1784.

National Bureau of Statistics of China, 2018, 2017 China Statistical Yearbook, China Statistics Press. (Sigma, 1999-2017).

Swiss Re. Institute. 2018. Natural disasters and man-made disasters in 2017: a record year of losses. Sigma (1): 2–27.

Shi, Y.F. 1996. Development trend of natural disasters in china under the influence of global warming. Journal of Natural Disasters 5(2): 102–116.

Yan, J.P., Bai, J., Su K.H. et al. 2011. Symmetry and some major natural disaster trends research. Geography Research 30 (7): 1159–1167.

Yang, J.F., Zhang, C.G. and Feng, Y.Y. 2012. Trends in long-term changes in geological disasters at home and abroad. Geological Hazards and Environmental Protection 23(1): 7–12.

Risk Analysis Based on Data and Crisis Response Beyond Knowledge – Huang & Nivolianitou (eds)
© *2020 Taylor & Francis Group, London, ISBN 978-0-367-25146-8*

Improvement of risk communication in the Internet of Intelligences

Rundong Wang & Chongfu Huang*

Key Laboratory of Environmental Change and Natural Disaster, Ministry of Education, Beijing Normal University, Beijing, China

State Key Laboratory of Earth Surface Processes and Resources Ecology, Beijing Normal University, Beijing, China

Academy of Disaster Reduction and Emergency Management, Faculty of Geographical Science, Beijing Normal University, Beijing, China

ABSTRACT: Risk communication is important for risk analysts who collect information and experiences from stakeholders who participate in the Internet of Intelligences. This article reviews some risk communication cases from China and in the United States. From this, is possible to conclude that the convergence communication approach has better communication effects than the crisis communication approach, and the public participation and satisfaction ratio are high, so the convergence communication approach and Internet of Intelligences can be combined.

Keywords: Internet of Intelligences, risk communication, convergence communication, crisis communication, public participation

1 INTRODUCTION

In the Internet era, it has become possible to have an online platform for risk analysis and management, by efficiently using data, models, and sharing intelligences. The Internet of Intelligences (IOI) (Huang, 2015), consisting of a set of agents, a network, and a model, provides a new alternative way to construct such platform. In fact, an IOI is also a risk communication platform and can only effectively eliminate the barriers of data sharing by adopting a good communication method.

Risk communication is an exchange of risk information. Thus, a series of complexities of communication problems results from the complexities of the alternatives, objectives, and uncertainties inherent in risks and their regulation (Caron et al., 1995).

Members of the public often like to use simple terminology to describe risk issues, such as safe or not safe, normative or nonstandard. Risk regulators can't use such simplifications: they must examine and analyze all possible actions to address the risks. When analyzing a pharmaceutical factory's safety management, for example, regulators must consider multiple options, from unpopular "inaction" decisions to regulations that may lead to a complete closure of the factory. Of course, communicating about a series of alternatives is much more difficult than a simple "yes-no" decision.

In addition to the complexity of alternatives, risk regulators must address multiple conflicting goals. When regulating emissions from a pharmaceutical plant, health risk will be a key issue, but other issues to be considered include potential environmental damage, socioeconomic impact, equipment costs, and political impact. The multiplicity of these goals and the need to balance them are difficult to communicate, especially for the public who are used to asking for "both" when faced with difficult trade-offs.

*Corresponding author: hchongfu@bnu.edu.cn

Risk issues itself are also complex, as solving these issues requires interdisciplinary and extensive knowledge. The knowledge of chemists and environmentalist who needs to assess the danger of emissions in pharmaceutical plants, for example, should combine with knowledge about the possible impact of emissions on political, economic and industrial facilities. No one can be expected to be an expert in all of these disciplines. Risk issues require an interdisciplinary team to solve them. The team's leaders usually have some in-depth knowledge in selected areas, but only experience in other areas. Therefore, risk analysts have some difficulties in communicating with the public due to the breadth and depth of knowledge.

Huang defined risk as a future scenario related to some kind of adverse events (Huang and Ruan, 2008). We can conclude from this definition that the fourth complexity inherent in risk problems is the uncertainty over the source of the risk, the field of influence, and the object of action. However, analysts communicate risk decisions to the public with uncertainty, which often leads to mistrust: even governments and policy makers cannot tolerate this uncertainty.

Therefore, the complexity of risk issues determines the necessity for risk communication.

2 CASE ANALYSIS

Risk communication has many aspects. With the rapid development of new media and the full application of modern media technology, the impact of the field of risk communication is almost subversive. Social media has become an important channel for citizens to participate in risk communication. In Europe and the United States, Twitter has become an information dissemination platform for government and public communication, and the Chinese government social media is mainly based on Microblog. Therefore, this case study defines the government social media in China and the United States as Microblog and Twitter, respectively.

2.1 Case selection

Based on factors such as the time of case and data acquisition, this study selected as examples the typhoon Megi in Fujian in 2016 and the flood caused by the hurricane in Louisiana in 2016. Typhoon Megi was generated on September 23, 2016 and dissipated on September 29, 2016 (Wikipedia). On August 8, 2016, officers of the Louisiana State Department began to warn about a hurricane, and from August 11 to August 19, Louisiana experienced an unprecedented flood. Because the research cases are natural disaster events, the main management of these events is the professional department of the local government dealing with emergencies—police, the fire department, and the medical department. The performance of these departments' social media accounts represents their risk communication abilities. The study selected the following accounts: the Fuzhou Government Official Microblog @Fuzhou release, @Fuzhou police, @Fuzhou fire, @Louisiana police, @Louisiana fire and @Louisiana EMS.

2.2 Risk communication approaches

The crisis communication approach is that risk communicators should do everything they can to mobilize stakeholders to take appropriate action (Regina et al., 2016). For example, in the event of a flood outbreak, it is necessary to release information to urge the stakeholders to evacuate to a higher place and to avoid obstructing the work of rescue workers. A senior risk communicator stated that, "The only thing you need to tell the stakeholders is that they need to leave, and everything else has nothing to do with them." For this purpose, risk communicators pass on information such as the probability of a risk and other alternative plans are meaningless. This approach insists that the organization knows what is best for the stakeholders.

In the convergence communication approach (Rogers et al., 1981), communication (including risk communication) is an iterative, long-term process and the values of risk communication organizations and stakeholders (culture/experience and social context) all contribute to this process. When the organization publishes formation, the stakeholders will process the information and give feedback within their own understanding (such as "We don't believe you!" "What is this?" and "What do you want me to do?"). The organization then processes the feedback and responds by posting supplemental or revised information. Through this continuous cycle of information exchange, the two parties will slowly reach a consensus. The inspiration for risk communicators is that the stakeholders must be involved in the process of risk communication, and this process must be a two-way dialogue, not a monologue from the organization. Continuous feedback and explanations are important for effective risk communication.

China favors a crisis communication approach and the US tends toward a convergence communication approach.

2.3 *Communication performance indicators*

Xie et al. (2005) proposed analysis indicators of the communication effect: online comments, interactions and shares, because the behavioral level (the actual actions of netizens after browsing Microblog) cannot be reflected in the content analysis. The study uses the public shared mean and commented mean to measure their actions. In order to more clearly describe the attitude of netizens, the public's emotional comments are divided into positive comments and negative comments. Interaction can be measured by "news from the public", "public questioning rate" and "official response rate". An explanation of these indicators is shown in Table 1.

2.4 *Communication content*

This study uses the media framework theory to divide communication into fact framework, action framework, and accountability framework (Zhong, 2017), as shown in Table 2.

2.5 *Data analysis*

By searching the government official Microblog for keyword "Megi" and examining the relevant news about the hurricane in Louisiana in 2016 on Twitter, we can learn that the amount of information released by the two countries in risk communication is similar. Fuzhou government social media released 201 microblogs, and Louisiana social media posted 206 tweets.

As can be seen from Table 3 and Table 4, in terms of framework, China was more likely to follow the Fact framework (46%), and the United States was more likely to follow the Action framework (47%). In terms of content, both countries were more concerned about disaster information, response measures, and inspiring deeds. China was more focused on disaster information, and rumor control, and the United States focused on attention to suggest, invitation, and accountability.

From the performance of different content in public use behavior, the highest average shares were rumor control for China and disaster information for the United States. In

Table 1. Explanation of these indicators.

Indicators	Explanation
Shares	Number of Microblog shares
Positive comments	Number of positive comments
Negative comments	Number of negative comments
News from the public	News obtained from the public
Public questioning	There are public questions in the Microblog comment area
Official response	Official response to public questions

Source: Xie et al. (2005)

Table 2. Media framework theory.

Framework	Content	Description
Fact framework	Disaster information	The actual situation in the disaster area, such as the number of victims and injured, and weather conditions
	Public information	Opening or closing of public facilities, telephone number for evacuation sites
	Rumor control	Clarify rumors
	Suggest	Suggesting public action
	Leadership ability	The action or achievement of the leader in response
	Response measures	Disaster reduction measures
Action framework	Institutional cooperation	Collaboration among institutions
	Invitation	Invite the public to participate in information gathering or rescue activities
	Comfort	Reassure public
Accountability framework	Accountability	Investigate the responsibility of government officials
	Inspiring deeds	Deeds of disaster relief heroes

Source: Zhong (2017)

Table 3. Public use behavior in Fuzhou.

Content	N	Proportion	Forwarded mean	Positive comments mean	Negative comments mean	Positive comments ratio
Disaster information	71	0.35	117.27	20.69	35.46	0.37
Public information	18	0.09	122.11	2.72	7.33	0.27
Rumor control	4	0.02	331.50	16.00	130.50	0.11
Suggestions	1	0.00	220.00	13.00	29.00	0.31
Leadership ability	10	0.05	32.30	4.20	10.40	0.29
Response measures	48	0.24	163.10	41.10	28.06	0.59
Institutional cooperation	15	0.07	44.40	8.80	10.33	0.46
Invitations	2	0.01	171.00	46.50	34.00	0.58
Comfort	2	0.01	49.00	8.50	3.00	0.74
Accountability	6	0.03	309.67	30.33	65.83	0.32
Inspiring deeds	24	0.12	290.92	25.92	3.75	0.87

Source: @Fuzhou release, @Fuzhou police and @Fuzhou fire

general, the public shared mean in the United States was much higher than in China. The Chinese public received a positive emotional impact from response measures, invitations, comfort and inspiring deeds information, while other content attracts more negative comments than positive comments. The United States is significantly different from China at this point, as positive comments on all content are much higher than negative comments, at above 82%.

Finally, in China there were 0 news items from the public, 58 Microblog questions and four answers from officials, and in America there were 26 news items from the public, 46 tweeted questions and five answers from officials. Through calculations, we can find the interactive performance, as shown in Table 5. From the public questioning rate, it is clear that the public in both

Table 4. Public use behavior in Louisiana.

Content	N	Proportion	Forwarded mean	Positive comments mean	Negative comments mean	Positive comments ratio
Disaster information	27	0.13	9267.19	245.70	1.89	0.99
Public information	26	0.13	428.42	3.35	0.58	0.85
Rumor control	0	0.00	0.00	0.00	0.00	0.00
Suggestions	12	0.06	1115.75	9.08	0.58	0.94
Leadership ability	13	0.06	560.00	18.38	0.38	0.98
Response measures	46	0.23	318.89	4.52	0.35	0.93
Institutional cooperation	13	0.06	740.77	12.62	0.23	0.98
Invitations	19	0.09	1779.00	5.11	1.11	0.82
Comfort	4	0.02	1756.75	279.50	0.00	1.00
Accountability	21	0.10	108.10	1.76	0.10	0.95
Inspiring deeds	22	0.11	114.59	8.14	0.23	0.97

Source: Xie et al. (2015)

Table 5. Interactive performance.

	News from the public	Public questioning rate	Official response	Official response rate
China	0	0.29	0.02	0.07
America	0.11	0.22	0.02	0.11

countries wanted to participate in risk communication, but the US government social media preferred to share public information, and the official response rate was higher than Chinese.

3 CONCLUSION AND DISCUSSION

Combined with the complexity of risk issues and the necessity of risk communication, we found that the shared tweets mean in the United States is much higher than the equivalent in China, and the American public has a higher rate of positive comments for the government in risk communication. This had much to do with the different risk communication approaches adopted by China and the United States.

This example gave great confidence to the construction of the IOI. Risk analysts can collect the public's experiences and lessons on accidents or incidents through the IOI, and let the public participate in risk communication. They can also release some popular science articles and disaster mitigation measures on the IOI to improve the public's risk communication abilities.

REFERENCES

Bodoque. D. Amerigo. G. and Olcina. 2019. Enhancing Flash Flood Risk Perception and Awareness of Mitigation Actions through Risk Communication: A Pre-post Survey Design. *Journal of Hydrology* 568: 769–779.

Caron C. Kandice L. Salomone and Billie J.H. 1995. Improving risk communication in government: research priorities. *Risk Analysis* 15(02): 127–135.

Gao X., Zhang S.Z., Yang G.L. and Duo Y.Q. 2011. Overview on research status of risk communication. *Journal of Safety Science and Technology* 7(05): 148–152.

Hua Z.Y. 2017. Risk communication: concept, evolution and principles. *Journal of Dialectics of Nature* 39(3): 97–103.

Huang C.F., Liu A.L. and Wang. Y. 2010. A discussion on basic definition of disaster risk. *Journal of Natural Disasters* 19(06): 8–16.

Huang, C.F. 2015. Internet of intelligences can be a platform for risk analysis and management. *Human and Ecological Risk Assessment* 21(5): 1395–1409.

Liu, L.Y. and Fan, Y.D. 2012. *Research on Major Natural Disaster Risk Communication*. Beijing: Surveying and Mapping Press. (in Chinese).

Regina E.H. and Andrea H.M. 2016. *Risk Communication: A Handbook for Communicating Environmental, Safety and Health Risks*. Beijing: Communication University of China Press. (in Chinese).

Rogers E.M. and Kincaid D.L. 1981. Communication networks: toward a new paradigm for research. Journal of Communication 32(4): 188–191.

Shang Z.H. 2017. Current situation and progress of natural disaster risk communication research. *Safety and Environmental Engineering* 24(6): 30–36.

Wikipedia. 2016. Typhoon Megi. https://en.wikipedia.org/wiki/Typhoon_Megi_(2016).

Xie Q.H., Tang S.K. and Chu J.X. 2015. Research on the influencing factors of crisis communication in American government Microblog. *China Radio* 8: 89–93.

Xie X.F., Wang H., Ren J. and Yu Q.Y. 2005. Analysis of Folk-Centered Risk Communication during SARS Crisis. *Applied Psychology* 2: 104–109.

Zhong J.Y. 2017. Microblog research on incidents in media framework theory. http://media.people.com.cn/n1/2017/0112/c409679-29019073.html.

Risk Analysis Based on Data and Crisis Response Beyond Knowledge – Huang & Nivolianitou (eds)
© 2020 Taylor & Francis Group, London, ISBN 978-0-367-25146-8

An approach to comply the geospatial information diffusion filling in gap units in Internet of intelligences for emergency rescue

Yiran Yin & Chongfu Huang

Key Laboratory of Environmental Change and Natural Disaster, Ministry of Education, Beijing Normal University, Beijing, China

State Key Laboratory of Earth Surface Processes and Resource Ecology, Beijing Normal University, Beijing, China

Faculty of Geographical Science, Academy of Disaster Reduction and Emergency Management, Beijing Normal University, Beijing, China

ABSTRACT: The disaster data plays a fundament role for the government to implement disaster relief and reconstruction. However, after natural disasters, especially major natural disasters, the disaster data on some administrative units cannot be obtained in time, which seriously restricts the accuracy and timeliness of disaster relief. To this end, this paper proposes to build an Internet of intelligences (IOI) system for emergency rescue, which uses information diffusion on geographical space to achieve a preliminary assessment of the disaster data on the gap units. The platform is build by using PHP programming language. The advanced system is supposed to have several functional modules including multi-terminal information collection, disaster online assessment, and disaster progress display. This paper summarizes the existing IOI technologies and focuses on the technical path of the disaster online evaluation, in order to achieve this goal, a database structure of IOI for emergency rescue is well designed.

Keywords: Internet of intelligences, information diffusion, disaster assessment, PHP

1 INTRODUCTION

The timeliness and accuracy of natural disaster relief decisions are related to people's life safety and property loss. However, disaster information is often incomplete in a period of time following a disaster, which poses a severe challenge to disaster relief work. For example, within 2 hours after the earthquake, it is difficult for the rescue team to reach the earthquake site and report information. This time period is usually referred as the "black box period".

The collection and rapid assessment of natural disaster information is a worldwide scientific problem and has become one of the focal issues of current disaster prevention and mitigation research. To this end, researchers worldwide have developed some online or offline platform to support disaster relief decision-making. PAGER developed by USGS, in general, can publish preliminary assessment results within 30 minutes after the earthquake. HAZUS developed by FEMA can perform multiple natural disasters assessment including earthquake, flood and hurricane. EMA-DLA designed by EMA can predict indirect economic loss as well as direct economic loss. Nevertheless, the assessment results of these systems don't include practical disaster data.

After the occurrence of natural disasters, we are not faced with a complete blind spot of disaster data, but partially known. Here we use the term gap unit to represent the geographic unit where the disaster data is still unknown. As time goes by, disaster information will be gradually gathered and updated, and the number of gap units will become less and less. The disaster damage is related to local socio-economic attributes and natural geographical attributes, through proper mathematical method, we can abstract a math model among those indicators. In fact, on account of the similarity among adjacent units, their math models mentioned before can also have a certain correlation. Therefore, we may find a mathematical method to estimate the data on the gap unit using known data and in this way the rapid disaster assessment on gap units can be quickly realized.

2 INTERNET OF INTELLIGENCES FOR EMERGENCY RESCUE

2.1 *Concept of Internet of intelligences*

The original idea of Internet of intelligences (IOI) was generated by Huang when he studied how to conduct risk analysis online services (Huang, 2011). The strict definition is given by Huang in 2015, that is, let *A* be a set of agents, *N* be a network used by *A*, and *M* is a model that processes the information provided by *A*, then the ternary body <*A, N, M*> is called an IOI, where an intelligent agent refers to an individual which has the ability to observe, interpret, reason, and solve problems. Compared with other platforms, IOI system emphasizes the diversi- fication of data sources. In theory, all individuals or organizations with terminals that can log on to an IOI system and submit data, which has natural geographical attributes and is geographically well dispersed. The key to the problem is how to process the information collected in this way and then obtain the results that meet the reliability requirements.

2.2 *Review of IOI development*

Since the concept of IOI has been proposed, ten platforms have been developed, and four of them have been applied in the field of natural disasters including the IOI for macroscopic anomaly analysis of earthquakes, the IOI for Wenzhou typhoon disaster risk analysis, the IOI for marine environmental risk analysis and the IOI for risk analysis of Yonghe Irrigation District, Sichuan, China. Among them, the IOI for macroscopic anomaly analysis of earthquakes discusses how to structure the flexible information provided by the earthquake victims to obtain the risk assessment results (Wang et al, 2012). The IOI for Wenzhou typhoon disaster risk analysis is the first IOI platform to be put into practical use (Guo et al, 2014). It gathers multiple participants come from municipal governments, meteorological bureaus, research institutes, aquaculture companies, insurance companies to get both practical and theoretical advises and produces a analysis report by cross-process all the information. The IOI for Yonghe Irrigation District risk analysis introduces WebGIS technology for the first time (Sun, 2017; Sun, 2018), and developed an IOI platform with self-made geographic map, which laid the foundation for the application of IOI in geography and natural disaster related fields. The existing platforms have already discussed some data processing methods for flexible information such as unstructured texts. However, the methods above are limited in quantitative assessment.

2.3 *The idea of IOI for emergency rescue*

In order to realize the rapid assessment of the disaster data on the gap unit, we propose to build an IOI for emergency rescue, where the agents refers to the masses and the disaster administrators who have the latest disaster information, and the information processing model refers to the mathematical method applied to evaluate disaster damage on gap units.

In order to realize the core target of online disaster evaluation, the workflow is designed as demonstrated in Figure 1. IOI for emergency rescue consists of three main modules, that is, multi-terminal information acquisition module, disaster rapid evaluation module and disaster progress display module. After the disaster occurs, the disaster administrators update the disaster database as a user of the platform, while all of the disaster information stored in the database is open to the public through a disaster progress map. In this process, there must be some units whose disaster data is acquired relatively late. The disaster administrators can lock the target unit to perform the information processing model to calculate the unknown values.

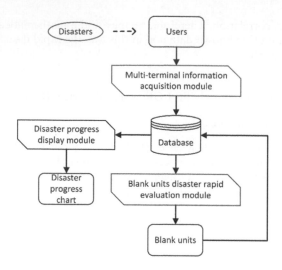

Figure 1. Workflow of IOI for emergency rescue.

3 INFORMATION PROCESSING MODEL OF IOI FOR EMERGENCY RESCUE

3.1 *Demand analysis*

It is well known that extrapolating unknown date from the known is a common idea for dealing with incomplete data. As a kind of spatial data with geographical attributes, the disaster data should be estimated using spatial data method.

Spatial interpolation methods such as Kriging method and inverse distance weight method are wildly used in geography to predict the data in the absence of observations, equivalently transforming discretely distributed observation points into continuous data surfaces (Haining, 2009).

Spatial interpolation methods are usually applied to the processing of physical quantities such as temperature and precipitation, but it is not suitable for dealing with disaster data. This is because depending on the complexity of the disaster system, disaster data often does not have continuity in space. Geographically weighted regression model is a local regression model that is commonly used to study the quantitative relationship between two or more variables with spatial distribution characteristics (Fortheringham et al, 1997; Brunsdon et al, 2010). Once this quantitative relationship is determined, it can be used as a predictive model. Compared with the global regression model, the explanatory variable parameter of the geographically weighted regression is a function of the geographical location, so it can effectively reveal the spatial heterogeneity of the variable relationship. However, geospatial weighted regression is a linear regression method, and the assumption of linear relationship is insufficient to describe the complexity of the disaster system, so it cannot be used for the information processing module in IOI for emergency rescue. In summary, we hope that there is a method to demonstrate both complex quantitative relationships and spatial heterogeneity.

3.2 *Geospatial information diffusion*

Huang proposed the concept of geospatial information diffusion at the 8th Annual Conference of the Risk Analysis Council of China Association for Disaster Prevention (Huang, 2018), and thenfurther defined the concepts involved in this method and detailed calculation formulas (Huang, 2019). Geospatial information diffusion can be regarded as a method to fill the lacking space data. It uses background data and known observations to estimate the data on the gap units. The background data here refers to the index value that affects the observation value of a unit, including two parts of socioeconomic attributes and natural geographical attributes.

Table 1. Observations and background data on research area.

Unit	Background Data				Observation
g_1	z_{11}	z_{12}	...	z_{1t}	w_1
g_2	z_{21}	z_{22}	...	z_{2t}	w_2
...		...			...
g_{n-q}	z_{n-q1}	z_{n-q2}	...	z_{n-qt}	w_{n-q}
g_{n-q+1}	$z_{n-q+1,1}$	$z_{n-q+1,2}$	... $z_{n-q+1,t}$		Unknown
...		...			...
g_n	z_{n1}	z_{n2}	...	z_{nt}	Unknown

In order to better illustrate how geospatial information diffusion method are applied to naturaldisaster data estimation, a disaster scenario is assumed. As shown in Table 1, it is assumed that there are n administrative units in the research area, and the indicator w_i is used to measure the disaster damage, and the background data set Z is composed of z_i total of t indicator values. Assuming that $n-q$ of the w_i values on the units are known at this time, the q unknown w_i values can be calculated by the following formula

$$w_i = f(Z, w_1, w_2, \ldots, w_{n-q+1}) \tag{1}$$

where $i = n-q, n-q + 1,\ldots,n$. Due to limited space, this paper only briefly describes the primary principles of geospatial information diffusion, and doesn't introduce the specific calculation process.

The geospatial information diffusion method is an extension of the information diffusion method in the probability space. The former is based on the fuzzy set theory as the latter, and the process of constructing the information matrix is equivalent to constructing a spatial function "polygon", so the non-linear complex relationship between variables can be explained. Therefore, the geospatial information diffusion method satisfies the two basic requirements of the disaster rapid evaluation model required in IOI for emergency rescue and we try to apply this new theory to the IOI platform.

4 TECHNICAL PATH

4.1 Data base structure design

In the geospatial information diffusion model, there are two parts as input, that is, the background data of all the research units and the known disaster indicator observations, and the output of the model is the estimated value of disaster indicator. Input and output above constitute the main body of the IOI platform database. In addition to background data and disaster data, the database also needs to manage information about users and administrators. Cooperating with evaluation model, the database structure is carefully designed as shown in Figure 2.

In database structure, the database consists of five main subjects: administrator information, registered user information, background data, reported disaster data, and predicted disaster, corresponding to five database tables respectively. Among them, the administrator refers to the platform maintenance technicians, we set up a special account for them and authorize advanced management rights. The registered user refers to persons who process an account of the platform and thus are allowed to insert, update, delete and read the data in database, and in the meanwhile to carry out the calculation of lacking disaster indicators. The background data table and the two disaster data tables are recorded with the mark of their unit names. The former is used to record the natural geographical attributes and socioeconomic attributes of each unit, and the latter is used to record the disaster data of each unit. Considering the actual application scenario of the database tables, each table has different fields and corresponding

467

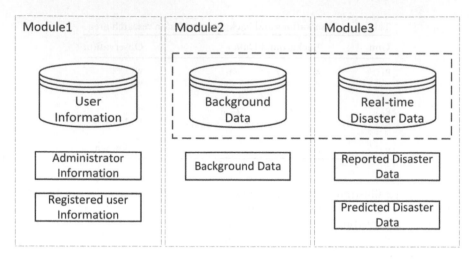

Figure 2. Database structure.

attributes. Taking two disaster data tables as an example, Table 2 shows the structure of them, including field name, data type and length. In addition to recording the disaster data submitted by the users and calculated online, the operator's identity and operation time are also recorded for verification. Out of explanation, Table 2 shows three quantitative indicators: affected population, casualties, and direct economic losses. When the platform is actually applied to a certain area, indicators need to be adjusted as needed.

4.2 *Development tools*

After years of technical accumulation, the IOI platform has formed a relatively mature basic technical framework, that is, using WAMP Server to build a network environment, using MySQL to build a database, and using PHP programming language to build the platform. Framework is a part of the design of an application in a given problem area, a reusable design of an entire or partial system, and concludes a set of methods of interaction between abstract artifacts and component instances. The Yii framework is a high-performance PHP language framework, rich in functionality and free, especially suitable for rapid Web development, and has been widely used in the field of PHP development. Therefore, we choose Yii framework to develop IOI for emergency rescue.

In the previous section, we reviewed the existing IOI platform. From a technical point of view, most existing platforms only contain simple data operation logic, and few use complicated calculating methods as information processing models. It is exactly the first technical key point in building IOI for emergency rescue. IOI for Yonghe Irrigation District risk analysis firstly used the open source GIS software QGIS to create geographic layer and embed it into the platform by means of Leaflet code, thus realizes the interactive display of

Table 2. Structure of disaster data tables.

Field Name	Datatype (Len)	Annotation
user_name	varchar (50)	Name of registered user
unit_name	varchar (50)	Name of unit
submit_time	date	Submission time
affect_pop	float (6)	Affected population
casualty_pop	float (6)	Casualty population
direct_loss	float (6)	Direct economic loss
. . .	. . .	. . .

vulnerability map of the research area (Sun, 2017; Sun, 2018). There was a period of time when map module was built based on API technology. Although API technology can hardly meet the customized requirements of specific map features, displaying the data in the database on the map layer using API is much easier than using Leaflet. To meet the two needs above at the same time is the second key technical point of building IOI for emergency rescue.

4.3 *Technical path of online calculating*

To realize disaster rapid evaluation on gap units using Yii framework, we must deeply understand the working principle of Model-View-Controller (MVC) mode.

The design concept of MVC mode is to separate the input, processing, and output of an application in the form of Model, View, and Controller. Among them, the View is the interface that the user sees and interacts with; the Model is the main part of the MVC mode, which contains the operation data and logic, and is responsible for accessing and updating the database data; the Controller is responsible for receiving the request from the browser and deciding which View or Model to call. In the MVC mode, the technical path of online lacking disaster data evaluation is shown in Figure 3. When the user issues a calculation request, the Controller receives the request and invokes the disaster evaluation model. And eventually, the calculation results will be displayed in disaster progress map of View.

5 CONCLUSION AND DISCUSSION

5.1 *Conclusion*

The geospatial information diffusion method provides a new idea for filling spatial missing data. It can not only describe the non-linear relationship between multivariates with geographical attributes, but also reflect the spatial heterogeneity of this relationship. Therefore, this paper proposes to try to apply this method in IOI fore mergency rescue, so as to achieve rapid assessment of the disaster damage and provide decision-making support for the disaster emergency management. The gap units disaster evaluating model is the core of the IOI for emergency rescue. To achieve

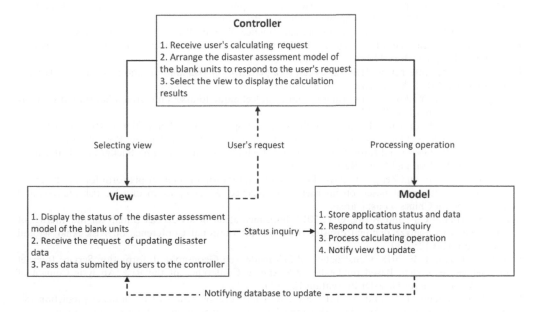

Figure 3. MVC design mode of online disaster evaluation module.

this function, it needs to establish a good interaction operation with the database and program the high dimensional matrix operation, which is one of the key technologies. Customizing the specific map elements and realizing the superposition of database data on the map is the second key technical point for building the IOI for emergency rescue.

5.2 *Discussion*

The key to guarantee the reliability of the evaluation results is to select appropriate indicators to quantify the disaster damage and to select appropriate influencing factors as independent variables to construct the mathematical relationship on each unit. Firstly, the characteristics of target disaster need to betaken into account when selecting quantitative indicators of disaster situation; secondly, the researcher's perspective of measuring the disaster situation will also affect the selection of indicators, such as focusing on the physical characteristics of the hazard factors or the casualties and losses caused by the disaster. Finally, different research regions have different characteristics of the pregnancy environment, and the disaster evaluation model should be adapted to local conditions and have regional characteristics. In fact, the unit here is not only limited to the administrative unit, but can also be the unit composed of multiple administrative units or one part of separated from an administrative unit.

The reliability of the data itself also affects the results of disaster evaluation. It can be improved by constraining the information source. For example, giving access to disaster administrators who have been professionally trained as registered users, while for non-registered users, they can only browse disaster data without modifying the database. Of course, problem of data reliability requires deeper research, even multi-disciplinary research. The best solution is to build a set of filter conditions based on certain scientific methods, and then verify the data before using it.

REFERENCES

Fotheringham, A.S., Charlton, M. and Brunsdon, C. 1997. Geographic information systems: A new(ish) technology for statistical analysis. New techniques and technologies for statistics II 141–147.

Brunsdon, C., Fotheringham, A.S. and Charlton, M.E. 2010. Geographically Weighted Regression: A Method for Exploring Spatial Nonstationarity. Geographical Analysis 28(4):281–298.

Guo, J., Ai F.L., Jiang, W.G. and Huang, C.F. 2014. Flexible geographic information in intelligence of internet for risk analysis of natural disaster and it's application. 8th Annual Conference of the Risk Analysis Council of China Association for Disaster Prevention 199–205.

Haining. R. 2009. Spatial data analysis theory and practice. Wuhan: Wuhan University Press.

Huang, C.F. 2011. Internet of intelligences in risk analysis for online services. Journal of Risk Analysis and Crisis Response 1(2): 110–117.

Huang, C.F. 2015. Internet of intelligences can be a platform for risk analysis and management. Human and Ecological Risk Assessment 21(5):1395–1409.

Huang, C., Huang, Y. 2018. An information diffusion technique to assess integrated hazard risks. Environmental Research 161:104–113.

Huang C.F. 2018. Information Diffusion on Geographical Space and its Application in Risk Analysis. 8th Annual Conference of the Risk Analysis Council of China Association for Disaster Prevention 1-7.

Huang C.F. 2019. Geospatial Information Diffusion Technology Supporting by Background Data. Journal of Risk Analysis and Crisis Response 9(1): 2–10.

Su, W., Huang, C.F. and Zeng F.L. 2014. Preliminary exploration of internet platform for marine environment risk assessment based on internet of intelligence. 2th symposium disaster risk analysis and management in Chiness coastal areas.

Sun, Y.N., Huang C.F. and Wang, Q.Y. 2017. Implementation of a web based map for the internet of intelligences to analyze regional disaster risks. 6th International Conference on Risk Analysis and Crisis Response 125–130.

Sun, Y.N., Huang C.F. 2018. Construction of Self-made web Geographic Information Service Internet of intelligences system Based on Leaflet. 8th Annual Conference of the Risk Analysis Council of China Association for Disaster Prevention 125–130.

Wang, W.D., Huang, C.F. and Ai, F.L. 2012. A tentative exploration of earthquake prediction. 5th Annual Conference of the Risk Analysis Council of China Association for Disaster Prevention.

The toxic effect of two types of nanoparticles on *Pseudourostyla cristata*

Xiaohuan Zhao
School of Life Science, East China Normal University, Shanghai, China

Zhiwei Gong
School of Physics and Materials Science, East China Normal University, Shanghai, China

Yang Lv, Xilei Gao, Xinpeng Fan & Bing Ni*
School of Life Science, East China Normal University, Shanghai, China

ABSTRACT: AgNPs and TiO$_2$NPs are the most common nanomaterials in industrial production and our daily life. Here, we used the single-cell protozoan ciliate *Pseudourostyla cristata* as the test organism to study AgNPs and TiO$_2$NPs against their acute toxic effect on *P. cristata* by using the probability unit method, AO-PI double fluorescent staining, Fourier transform infrared spectroscopy (ATR-FTIR), transmission electron microscopy (TEM), and enzyme activity assay. The results showed that the 24h-LC$_{50}$ of AgNPs and TiO$_2$NPs against *P. cristata* was 1.21 × 10^{-5} mg/L and 138.75 mg/L, respectively. The 24h-EC$_{50}$ were 3.55 × 10^{-6} mg/L and 88.85 mg/L. This indicates that the acute toxicity of AgNPs on *P. cristata* is stronger than TiO$_2$ NPs. AO-PI double fluorescent staining and TEM showed that TiO$_2$NPs, but not AgNPs, caused damage to the cell membrane. However, the high-oxidation of AgNPs and TiO$_2$NPs on the constituents of the cell membrane component (amide I, -COO-, v_a (PO^{2-}), δ (C-OH) of the polysaccharide, etc.) was both detected by ATR-FTIR. Besides, AgNPs and TiO$_2$NPs initially increased the activity of the cell superoxide dismutase (SOD) and then reduced it while the activity of the catalase (CAT) continued to increase. The above results indicate that the acute toxicity of AgNPs to *P. cristata* is much stronger than that of TiO$_2$NPs, but TiO$_2$NPs is more resistant to the structure and composition of cell membranes, suggesting that other mechanisms of toxicity are present. *P. cristata* is sensitive to the toxicity of AgNPs and TiO$_2$NPs, so it can be used as a model organism to predict the environmental risk of nanomaterials in the early stage.

Keywords: AgNPs, TiO$_2$NPs, *Pseudourostyla cristata*, acute toxicity, toxic mechanism

1 INTRODUCTION

With the rapid development of nanotechnology, various kinds of nanomaterials are increasingly used in many fields and their release into the water environment has an impact on aquatic organisms, food chains, and health of animals in contact with water (Krzyżewska et al., 2016). Nano-silver and nano-titanium dioxide particles are commonly found in human daily activities. Nano-silver is widely used in household articles, fungicides and other products (Mohanty et al., 2012). Nano-titanium dioxide has high thermal stability, insolubility, and photo catalytic properties, so it is often used in the treatment of sunscreens, paints, paper, wastewater decontamination and other products

---

*Corresponding author: bni@bio.ecnu.edu.cn

(Robichaud et al., 2009). Many studies have shown that nano-silver and nano-titanium dioxide particles have certain effects on aquatic organisms. Previous research on these species have mainly focused on some aspects such as biological growth, development, reproduction, ecotoxicity and cytotoxicity such as changes in cell ultrastructure and activity of the antioxidant enzyme, etc. Bang et al. (2011) found that TiO_2NPs of various particle sizes reduced the number and reproductive rate of *Daphnia magna*. Heinlaan et al. (2008) only measured the LC_{50} and EC_{50} of TiO_2NPs against *Vibrio fischeri* and *Daphnia magna*, but did not explore its mechanism of toxicity. Simona S et al. (2016) evaluated the toxicity of AgNPs and TiO_2NPs to *Marine Phytoplankton*, and found that the two nanomaterials have an inhibitory effect on algae population growth. Sarkar et al. (2014) observed from an ultrastructural point of view that AgNPs caused damage to the ultrastructure of the posterior ovary of zebrafish (*Danio rerio*). A study of cell biochemical indicators by Peng et al. (2010) explored the oxidation caused by TiO_2NPs with ATR-FTIR. Shu et al. (2018) found that AgNPs caused a decrease in microbial population and CAT activity. A large number of experimental data show that AgNPs and TiO_2NPs have acute toxic effects on model organisms of tetrahymena, aphids, algae and fish. They can cause some significant changes in physiological and biochemical indicators, but few researchers have studied the toxicity effects and mechanisms of these two materials on environmentally sensitive protozoa (Griffitt et al., 2008). Therefore, *Pseudourostyla cristata*, a single-cell protozoan which is more sensitive and widely used in the environment, was used as an experimental animal to study the acute toxic effects of nano-silver and nano-titanium dioxide particles on its cells. We explored the mechanism of action of these two nanomaterials on cells primarily to provide a basis for ecological risk assessment.

2 MATERIALS AND METHODS

2.1 *Culture of P. cristata and preparation of nanoparticle suspension*

The *P. cristata* was collected from a small lake in the suburbs of Shanghai. In an artificial constant temperature incubator (temperature: 26°, humidity: 76% RH), the *P. cristata* was fed with fermented liquid of wheat to conduct a pure culture. Every three days, a self-made glass pipette is used to absorb impurities and my coderm, and finally cells in the vegetative phase are obtained.

Nanomaterials were purchased from Aladdin (the particle size of AgNPs and TiO_2NPs is 30 nm and 5-10 nm respectively). We weighed out 3 g of nanoparticles and suspended them in ultrapure water to prepare a stock solution and used ultrasonic wave (power 60 w) to prevent any aggregation of the NPs.

2.2 *Determination of 24h-LC_{50} and 24h-EC_{50}*

In the preliminary experiment, six experimental concentration gradients (including the control group) were set. A nano-suspension of each concentration was mixed with 30 cells (with a small amount of medium) and injected into a concave dish which was then placed in a wet box. The number of statistical cells was observed after being cultured for 24 hours at 26°C and 76% RH. We set five parallel groups, counted the numbers and took the average of these values as the test result.

2.3 *AO-PI double fluorescent staining*

AO (acridine orange) and PI (propidium iodide) stock solutions (Beyo time) were diluted to 1% respectively and mixed in a 1:1 ratio. The cell liquid was then mixed with the mixed dye solution in a 1:19 ratio and observed under the fluorescence microscope (ZEISS Imager Z.2). The resulting mix is then checked.

2.4 ATR-FTIR

The blank control group and the treated group were separately collected, centrifuged (2000 rpm, 5 min), frozen for 12 h, vacuum freeze-dried, mounted, and detected by the Fourier infrared spectrometer.

2.5 Determination of enzyme activity

Refer to the SOD and CAT test kit instructions (purchased from the Nanjing Jiancheng Bioengineering Institute)

2.6 Statistical analysis

Using Origin 8 software, the logarithm value of the nanoparticle mass concentration (C, mg/L) or lgC was used as the independent variable X, and the cell death rate (DR, %) or cell reproduction rate (RR, %) was used as the dependent variable Y. Graphpad software was used to produce Figures 1 and 7.

3 RESULTS

3.1 24-hour acute toxicity of two nanomaterials to P. cristata

3.1.1 24h-LC_{50} determination

The *P. cristata* cells were treated with AgNPs and TiO$_2$NPs respectively, and the change in the number of cells in 24 hours is shown in Figure 1. The control group could multiply normally while the number of cells in the treatment groups decreased significantly, but the response time of cell death caused by the two nanoparticles were different. In the AgNPs treatment, the number of cells began to decrease 6 hours later while in the TiO$_2$NPs treatment the reduction of cell numbers started after 12 hours. The number of cells in each group after 24 hours of treatment was counted and the mortality rate was then calculated. Using the probabilistic unit method, the 24 h acute toxicity regression curve of two nanomaterials against *P. cristata* can be obtained as shown in Figure 2. The mortality of *P. cristata* is linear with the logarithm of the concentration of AgNPs and TiO$_2$NPs. The linear relationship equation of the 24h acute toxicity of AgNPs on *P. cristata* was Y = 119.82366X + 998.80714 (R^2 = 0.95364). Further calculation showed that the 24h-LC_{50} value of the AgNPs against *P. cristata* was 1.21 × 10^{-5} mg/L, and the 24h-LC_0 value was 4.62 × 10^{-6} mg/L while the safe concentration was 1.21 × 10^{-6} mg/L. The linear relationship equation of the 24h acute toxicity of TiO$_2$NPs on *P. cristata* was Y = 929.54676X - 1941.30306 (R^2 = 0.96428). The 24h-LC_{50} value was 138.75 mg/L, 24h-LC_0 value was 122.59 mg/L and the safe concentration was 13.88 mg/L.

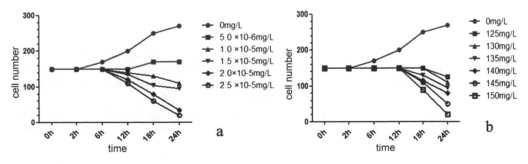

Figure 1.   Line diagram of the variation in cell numbers of *P. cristata* treated with different concentrations of AgNPs (a) and TiO$_2$NPs (b).

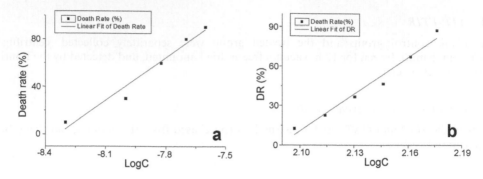

Figure 2. The 24h acute toxicological relationship curve between AgNPs (a) and TiO$_2$NPs (b) on *P. cristata*. (Determination of 24h-L$_{50}$ value).

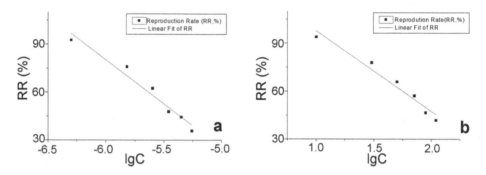

Figure 3. The 24h acute toxicological relationship curve between AgNPs (a) and TiO$_2$NPs (b) on *P. cristata*. (Determination of 24h-EC$_{50}$ value).

3.1.2 *24h-EC$_{50}$ determination*

We set the concentration gradient based on pre-experimental data and the toxic effects of two nanomaterials on the reproduction rate of *P. cristata* were studied. As shown in Figure 3, each logarithm of the concentration of AgNPs and TiO$_2$NPs has a linear relationship with the reproduction rate of *P. cristata* within a certain concentration gradient. For AgNPs, the linear relationship equation was Y = -54.84004X - 248.84083 (R^2 = 0.95266) and the 24h-EC$_{50}$ value was 3.55 × 10^{-6}mg/L. The linear relationship equation between TiO$_2$NPs and the 24h chronic toxicity of *P. cristata* was Y = -50.28599X + 147.990093 (R^2 = 0.95214) and the 24h-EC$_{50}$ value was 88.85 mg/L.

3.2 *Destruction of P. cristata cell membranes by two kinds of nanomaterials*

3.2.1 *Ultrastructural observation*

Compared with the control group (Figures 4a & 4b), the membranes of most cells in the AgNPs treatment group at a concentration of 24h-EC$_{50}$ were well preserved without damage (Figure 4d). But the membranes of a small number of cells were damaged to some extent and some mitochondria were slightly deformed (Figure 4c).

After treatment with TiO$_2$NPs at a concentration of 24h-EC$_{50}$, the cell membranes were broken to varying degrees. Most of the cell membranes were slightly broken, as shown in Figure 4e. A small number of cell membranes was severely broken, but no mitochondria with deformation or structural damage were observed (Figures 4e & 4f).

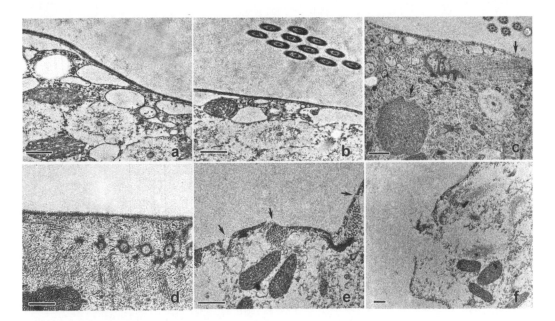

Figure 4.   Ultrastructural observation of *P. cristata* cells observed by transmission electron microscopy. a/b: The cell membrane structures of the blank control group were intact; c/d: 24h-EC_{50} concentration of AgNPs had no obvious damage to the ultrastructure of the cell membranes; e/f: TiO_2NPs at a concentration of 24h-EC_{50} damaged the cell membranes. bar = 0.5μm.

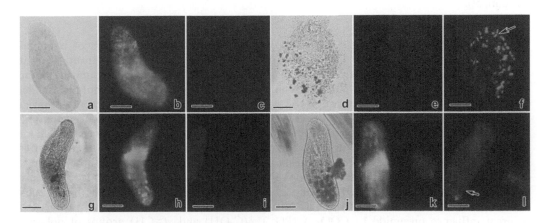

Figure 5.   AO-PI double fluorescent staining results.

Figures a, d, g, and j were the blank control group, the damaged and the 24h-EC_{50} concentration of the two nanoparticles treated with the *P. cristata* cells after 24 hours under the bright field; the b, e, h, and k images were, in turn, the cells under the four conditions dyed by the AO; the c, f, i, and l were cells under four conditions dyed with PI; the g, h, and i were AgNPs particle treatment groups, the j, k, l for TiO_2NPs particle treatment; the h, k, and b images consistently showed green fluorescence; the cells in the f and l diagrams showed red fluorescence, and the arrows in Figure f showed the nucleus dyed by PI. bar = 50μm.

3.2.2 *AO-PI double fluorescent staining*

The results showed that the blank control group emitted green fluorescence only under the excitation light of 488 nm, which is the result of staining of the AO dye through the cell membrane of living cells (Figure 5b). Ruptured cell membranes exhibited red fluorescence at the

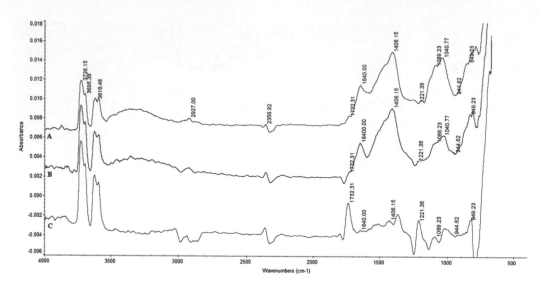

Figure 6.   Fourier infrared spectroscopy of *P. cristata.*

excitation light at 560 nm (Figure 5f, the nucleus indicated by the arrow), which is the result of the binding of PI dye to nuclear DNA. The cells treated with AgNPs at a concentration of 24h-EC$_{50}$ still showed green fluorescence but did not have red fluorescence (Figures 5h & 5i); TiO$_2$NPs at 24h-EC$_{50}$ concentration showed weak red fluorescence of *P. cristata* cells (Figure 5l).

3.2.3 *ATR-FTIR*

The infrared spectrum results of *P. cristata* after treatment with two kinds of nanoma-terials are shown in Figure 6. The significant peaks of the blank control group were at 3726 cm^{-1}, 2927 cm^{-1}, 1640 cm^{-1}, 1406 cm^{-1} and 1040 cm^{-1}, which corresponded to the protein primary amide group N-H stretching vibration, v_a(-CH$_2$-), phospholipid amide, a -COO-group, and a polysaccharide. The v_a(PO^{2-}), v_s(PO^{2-}) and γ(=CH) groups showed small peaks at 1221 cm^{-1}, 1089 cm^{-1} and 849 cm^{-1}, and all peaks and their corresponding groups are summarized in Table 1. Figure 1 shows that the 24h-EC$_{50}$ concentration of these two nanomaterials had little effect on the protein amide group at 3726 cm^{-1}. After treatment with TiO$_2$NPs at a concentration of 24h-EC$_{50}$, a band appeared at 1732 cm^{-1}, corresponding to the RHC=O group. AgNPs at a concentration of 24h-EC$_{50}$ only reduced the peak of the δ(C-OH) group of the polysaccharide and did not have much influence on the remaining groups.TiO$_2$NPs particles with 24h-EC$_{50}$ concentration had a great influence on amide I, -COO-, v_a(PO^{2-}), δ(C-OH) and γ(=CH) groups of polysac-charides, and even some functional groups disappeared. Both nanomaterials increased the peak intensity of the v_a(PO^{2-}) and γ(=C-H) groups, and they also caused the peak shape at v_a(PO^{2-}) to change. Further, the peaks of δ (C-OH) of the -COO- group and the polysaccharide were all lowered, and the peak of δ(C-OH) of the polysaccharide of the TiO$_2$NPs particle-treated group nearly disappeared.

3.3 *Determination of SOD and CAT enzyme activity*

The SOD and CAT activities of *P. cristata* were measured by the two nanomaterials at an EC$_{50}$ concentration and measured within one breeding cycle (24 h). Compared with the blank control group, the activity of SOD enzymes increased initially and then decreased with time. The activity of the CAT enzyme gradually increased (Figure 7). AgNPs caused the change of SOD and CAT enzyme activities to be more significant.

476

Table 1. Summary of the ATR-FTIR peaks of *P. cristata* in the spectra.

Wavelength (cm^{-1})	Functional groups	Wavelength (cm^{-1})	Functional groups
849	$\gamma(=C-H)$	1640	Amide I
944	H-C=C-H	1732	RHC=O
1040	δ(C-OH) of carbohydrates	2356	CO_2
1089	$v_s(PO_2^{-})$	2927	$v_a(-CH_2-)$
1221	$v_a(PO_2^{-})$	3726	N-H stretching vibration of protein primary amide group
1406	-COO^{-}		

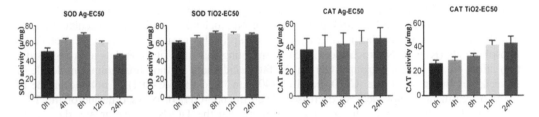

Figure 7. Effect of two kinds of nanomaterials on SOD and CAT enzyme activities of *P. cristata* in 24h (0 h (control group), 4 h, 8 h, 12 h, and 24 h).

4 DISCUSSION

4.1 *Comparison of cytotoxicity of AgNPs and TiO₂NPs against P. cristata*

From the above experiment, we know that the acute toxicity of AgNPs to *P. cristata* is stronger than TiO$_2$NPs. However, the oxidative damage of TiO$_2$NPs to the cell membrane is stronger than that of the AgNPs particles, which may be due to the interaction of the silver ions (Ag$^+$) released by the AgNPs in the water environment and the phosphorus groups in the DNA, resulting in the interruption of DNA replication (Li et al., 2014). AgNPs also has a strong lethal effect on bacteria, fungi and eukaryotic single-celled organisms (Marambio-Jones & Hoek, 2010), so the lethal effect of AgNPs on cells is stronger than that of TiO$_2$NPs. The lower concentration of AgNPs has little effect on the cell membrane structures, while TiO$_2$NPs mainly weaken the vitality of the cells by attacking the cell membranes. The lower concentration of TiO$_2$NPs will not cause damage to the internal structure of the cells (Swetha et al., 2010), but it may cause great damage to the structure and composition of the cell membrane.

4.2 *Membrane damage and oxidative stress to P. cristata by two nanomaterials*

The results of AO-PI double fluorescent staining showed that compared with the only green fluorescence of the sample treated with AgNPs at 24h-EC$_{50}$ concentration, the TiO$_2$NPs at 24h-EC$_{50}$ concentration destroyed the cell membrane structures, which in turn caused the PI dye to enter the cell and causing the nucleus to show a red fluorescence. Studies have shown that AO can act as an indicator of intracellular acidic structure, and then bind to lysosomes that induce apoptosis to make early apoptotic cells show an orange-yellow fluorescence (Shatilovich et al., 2015). However, there was no orange-yellow fluorescence in this experiment, indicating that the two nanomaterials at 24h-EC$_{50}$ concentration did not cause early apoptosis of cells. Only the TiO$_2$NPs caused different degrees of damage to the cell membrane, while the

AgNPs did not cause damage to the cell membrane. Therefore, it is speculated that the TiO_2 NPs mainly damage the cells by destroying the cell membrane structure.

Using the ATR-FTIR technique to analyze the chemical composition of the cell membrane, the following conclusions were obtained. As we know, the membrane consists of phospholipids and protein, and phospholipids are composed of a hydrophilic circular head and two hydrophobic fatty tails, which are mainly composed of the phospholipid amide I band, the -COO- group and the δ(C-OH) group of the polysaccharide. They are located on the surface of the cell membrane and exposed to the environment so the peaks in the control group are most obvious (Kiwi & Nadtochenko, 2005). The 24h-EC_{50} concentrations of the two nanomaterials had little effect on the protein amide group at 3726 cm^{-1}, and it is possible that these two materials will not change the protein components in the cell membrane (Bai et al., 2015). Both can increase the peak intensities of the $v_a(PO^{2-})$ and $\gamma(=CH)$ groups, and the two nanomaterials also cause change in the peak shape at $v_a(PO^{2-})$, which may be related to the formation of hydrogen (Selle et al., 1999). In the infrared spectrum of the cells treated with TiO_2NPs, a band appeared at 1732 cm^{-1}, corresponding to the RHC=O group, suggesting that the TiO_2NPs stimulated the cell membrane to produce the RHC=O group (Kiwi & Nadtochenko, 2005). It is speculated that it may be formed by the oxidation of a hydroxyl group (-OH) group in the fatty acid chain of phospholipids (Kiwi & Nadtochenko, 2005), manifesting that the TiO_2NPs has a strong oxidation effect on the hydroxy group (-OH) in the fatty acid chain of the phospholipid molecule of the cell membrane. Also, the -COO- band of 1406 cm^{-1} and the δ(C-OH) of polysaccharide of the two treatment groups were all decreased. Even the δ(C-OH) of polysaccharide of the TiO_2 NPs particle treatment group disappeared, indicating that the TiO_2NPs have greater damage to the polysaccharide on the cell membrane surface. The above results further verified that TiO_2 NPs has a strong oxidation effect on the cell membranes of *P. cristata*. At the same time, it was also proved that although the damage caused by the AgNPs to the cell membranes was not observed under the fluorescence microscope or transmission electron microscope, the AgNPs also caused a small oxidative damage to the functional groups constituting the cell membrane.

Studies have shown that when organisms such as *Euplotes crassus* (Kim et al., 2014), *Tetrahymena thermophila* (Ferro et al., 2015), and *Tigriopus japonicus* (Kim et al., 2011) are subjected to a mild environmental stress, they will response to it strongly and the activity of SOD enzymes will increase first This is the same trend showed by the results of this study. The two nanomaterials of EC_{50} cause an oxidative stress reaction of the organism and induce the production of SOD enzymes to protect the organism (Wang et al., 2017). As the stress time increases, the SOD enzyme activity gradually returns to normal levels. This may be due to the organism gradually adapting to the environment and becoming detached from the oxidative stress state (Üner et al., 2005). The activity of the CAT enzyme also continuously increased in this study. It is speculated that AgNPs and TiO_2NPs cause an increase in glutathione (GSH) (Vlahogianni et al., 2007). It may also be a self-protective response formed by cells against the production of hydrogen peroxide (H_2O_2) (Fridovich, 1989).

4.3 *P. cristata is used to evaluate the toxic effects of nanomaterials*

In previous studies, Kennedy et al. (2010) studied the toxicity of AgNPs to aquatic organisms and found that the 48h-LC_{50} value of AgNPs to *Daphnia magna* was 1.8-97µg/L. Clément et al. (2013) discovered that the 24h-L_{50} value of TiO_2NPs to *Brachionus plicatilis* was greater than 20g/L while Zhu et al. (2008) measured the 48h-LC_{50} value of TiO_2NPs to *Danio rerio* embryos to be 610mg/L. Griffitt et al. (2008) measured that the 48h-L_{50} value of AgNPs for zebrafish larvae was 7.20 mg/L. The LC_{50} and safe concentration values of AgNPs and TiO_2NPs measured in this study were lower than those of *Daphnia magna* and zebrafish. In addition, the two nanomaterials used in this experiment and the previous experiments have lower EC_{50} values for these model organisms (Wang et al., 2014). Therefore, based on the results of the 24h-LC_{50}, safe concentration and 24h-EC_{50} measured in this experiment, the *Pseudourostyla cristata* test organism can be used as a pattern creature for early prediction of the environmental risk of nanomaterials.

5 CONCLUSIONS

The probabilistic unit method was used to evaluate the acute and chronic toxic effects of AgNPs and TiO$_2$NPs on *P. cristata* in this study. Fluorescence microscopy and Fourier transform infrared spectroscopy were first combined with transmission electron microscopy and enzyme activity determination was used to explore the mechanism of the toxicity of two nanomaterials initially. Studies have shown that AgNPs is more toxic to *P. cristata* than TiO$_2$NPs. In addition, *P. cristata* is more sensitive to these two nanomaterials compared with other aquatic organisms, which suggest that the protozoan ciliate *Pseudourostyla cristata* can be used as a model organism for early prediction of environmental risks of nanomaterials. From the result of AO-PI double fluorescent staining, ATR-FTIR, and TEM, TiO$_2$NPs is more destructive to the cell membrane. These three experimental techniques mutually validated and complemented each other, and jointly verified the effects of two kinds of nanoparticles on the cells.

ACKNOWLEDGEMENT

This work was supported by the National Natural Science Foundation of China (No. 31672249, 31572223).

REFERENCES

Bai, Y.K., Yu, L.W., Zhang, L., Fu,J., Cao, W.L. 2015. Research on application of fourier transform infrared spectrometry in the diagnosis of lymph node metastasis in gastric cancer. *Guang pu xue yu guang pu fen xi Guang pu* 35(3), 599–602.

Bang, S.H., Le, T., Lee, S.K., Kim, P., Kim, J.S., Min, J. 2011. Toxicity Assessment of Titanium (IV) Oxide Nanoparticles Using Daphnia magna (Water Flea). *Environmental Health and Toxicology* 26, e2011002.

Clément, L., Hurel, C., Marmier, N. 2013. Toxicity of TiO$_2$ nanoparticles to cladocerans, algae, rotifers and plants – Effects of size and crystalline structure. *Chemosphere* 90(3), 1083–1090.

Ferro, D., Bakiu, R., De Pittà, C., Boldrin, F., Cattalini, F., Pucciarelli, S., Miceli, C., Santovito, G. 2015. Cu,Zn Superoxide Dismutases from Tetrahymena thermophila: Molecular Evolution and Gene Expression of the First Line of Antioxidant Defenses. *Protist* 166(1), 131–145.

Fridovich, I. 1989. Superoxide dismutases. An adaptation to a paramagnetic gas. *J. Biol. Chem* 264(14), 7761–7764.

Heinlaan, M., Ivask, A., Blinova, I., Dubourguier, H.C., Kahru, A. 2008. Toxicity of nanosized and bulk ZnO, CuO and Ti$_2$ to bacteria Vibrio fischeri and crustaceans Daphnia magna and Thamnocephalus platyurus. *Chemosphere* 71(7), 1308–1316.

Griffitt, R.J., Luo, J., Gao, J., Bonzongo, J.C., Barber, D.S. 2008. Effects of particle composition and species on toxicity of metallic nanomaterials in aquatic organisms. *Environmental Toxicology & Chemistry* 27(9), 1972–1978.

Kennedy, A.J., Hull, M.S., Bednar, A.J., Goss, J.D., Gunter, J.C., Bouldin, J.L., Vikesland, P.J., Steevens, J.A. 2010. Fractionating nanosilver: importance for determining toxicity to aquatic test organisms. *Environ Sci Technol* 44(24), 9571–9577.

Kim, B.M., Rhee, J.S., Park, G.S., Lee, J., Lee, Y.M., Lee, J.S. 2011. Cu/Zn- and Mn-superoxide dismutase (SOD) from the copepod Tigriopus japonicus: Molecular cloning and expression in response to environmental pollutants. *Chemosphere* 84(10), 1467–1475.

Kim, S. H., Kim, S. J., Lee, J. S., Lee, Y. M. 2014. Acute effects of heavy metals on the expression of glutathione-related antioxidant genes in the marine ciliate Euplotes crassus. *Mar Pollut Bull* 85(2), 455–462.

Kiwi, J., Nadtochenko, V. 2005. Evidence for the mechanism of photocatalytic degradation of the bacterial wall membrane at the Ti$_2$ interface by ATR-FTIR and laser kinetic spectroscopy. *Langmuir* 21(10), 4631–4641.

Krzyżewska, I., Kyzioł-Komosińska, J., Rosik-Dulewska, C., Czupioł, J., Antoszczyszyn-Szpicka, P. 2016. Inorganic nanomaterials in the aquatic environment: behavior, toxicity, and interaction with environmental elements. *Arch Environ Prot* 42.

Li, L., Wu, H., Peijnenburg, W.J.G.M., van Gestel, C.A.M. 2014. Both released silver ions and particulate Ag contribute to the toxicity of AgNPs to earthworm Eisenia fetida. *Nanotoxicology* 9, 792-801.

Marambio-Jones, C., Hoek, E.M.V. 2010. A review of the antibacterial effects of silver nanomaterials and potential implications for human health and the environment. *J Nanopart Res* 12, 1531-1551.

Mohanty, S., Mishra, S., Jena, P., Jacob, B., Sarkar, B., Sonawane, A. 2012. An investigation on the antibacterial, cytotoxic, and antibiofilm efficacy of starch-stabilized silver nanoparticles. *Nanomedicine: Nanotechnology, Biology and Medicine* 8(6), 916-924.

Peng, L., Wenli, D., Qisui, W., Xi, L. 2010. The envelope damage of Tetrahymena in the presence of Ti_2 combined with UV light. *Photochemistry & Photobiology* 86(3), 633-638.

Robichaud, C.O., Uyar, A.E., Darby, M.R., Zucker, L.G., Wiesner, M.R. 2009. Estimates of Upper Bounds and Trends in Nano-TiO_2 Production as a Basis for Exposure Assessment. *Environ Sci Technol* 43(12), 4227-4233.

Sarkar, B., Netam, S. P., Mahanty, A., Saha, A., Bosu, R., Krishnani, K. K. 2014. Toxicity evaluation of chemically and plant derived silver nanoparticles on zebrafish (Danio rerio). *Proceedings of the National Academy of Sciences, India Section B: Biological Sciences* 84(4), 885-892.

Selle, C., Pohle, W., Fritzsche, H. 1999. FTIR spectroscopic features of lyotropically induced phase transitions in phospholipid model membranes. *J Mol Struct* s 480–481:401-405.

Shatilovich, A., Stoupin, D., Rivkina, E. 2015. Ciliates from ancient permafrost: Assessment of cold resistance of the resting cysts. *Eur J Protistol* 51(3), 230–240.

Shu, K.H., Zhang, L., Ling-Li, W.U., You-Bin, S.I., Liu, Q.X. 2018. Effects of silver nanoparticles on microbial communities and enzyme activity in four soils. *Journal of Agro-Environment Science* 173(4): 554-558.

Simona S. 2016. Toxic effect of different metal bearing nanoparticles (ZnO NPs, TiO_2 NPs, SiO_2 NPs, Ag NPs) toward marine phytoplankton. Università degli studi di Napoli Federico II.

Swetha, S., Santhosh, S.M., Balakrishna, R.G. 2010. Synthesis and Comparative Study of Nano-TiO_2 Over Degussa P-25 in Disinfection of Water. *Photochemistry & Photobiology* 86(3), 628–632.

Üner, N., Oruç, E., Sevgiler, Y. 2005. Oxidative stress-related and ATPase effects of etoxazole in different tissues of Oreochromis niloticus. *Environ Toxicol Phar* 20(1), 99–106.

Vlahogianni, T., Dassenakis, M., Scoullos, M.J., Valavanidis, A. 2007. Integrated use of biomarkers (superoxide dismutase, catalase and lipid peroxidation) in mussels Mytilus galloprovincialis for assessing heavy metals' pollution in coastal areas from the Saronikos Gulf of Greece. *Mar Pollut Bull* 54(9), 1361–1371.

Wang, C., Pan, X., Fan, Y., Chen, Y., Mu, W. 2017. The oxidative stress response of oxytetracycline in the ciliate *Pseudocohnilembus persalinus*. *Environ Toxicol Phar* 56, 35–42.

Wang, J., Wang, W.X. 2014. Significance of physicochemical and uptake kinetics in controlling the toxicity of metallic nanomaterials to aquatic organisms. *Journal of Zhejiang Universityence A* 15(8):573–592.

Zhu, X., Zhu, L., Chen, Y., Tian, S. 2008. Acute toxicities of six manufactured nanomaterial suspensions to Daphnia magna. *J Nanopart Res* 11(1), 67–75.

Risk Analysis Based on Data and Crisis Response Beyond Knowledge – Huang & Nivolianitou (eds)
© *2020 Taylor & Francis Group, London, ISBN 978-0-367-25146-8*

Risk analysis of large-scale gasoline spill fire based on experiment and numerical simulation

Yan Zhang
Tianjin Fire Research Institute of Ministry of Emergency Management, Tianjin, China

Changxing Ren & Jin Li
Tianjin Fire Research Institute of Ministry of Emergency Management, Tianjin, China
National Center for Fire Engineering Technology, Tianjin, China

Jie Wang
National Center for Fire Engineering Technology, Tianjin, China

Tao Xue
Tianjin Fire Research Institute of Ministry of Emergency Management, Tianjin, China

ABSTRACT: Fire hazards from the leaking of petroleum are a frequent occurrence in the petrochemical industry and cause great harm to industrial and residential settings. A spill fire experimental platform was designed and constructed aimed at reducing the fire hazards from leaks during the storage and transportation of liquid fuel. With both experimentation and computational fluid dynamics (CFD) calculation methods, the spill fire property of gasoline on a 3° slope was studied. The mechanism of the fire development properties, thermal radiation intensity and the temperature field, etc., were analyzed and obtained. Results indicate that for the temperature testing point with a height of less than 1 m, the experimental results showed a agreement with the CFD calculation. For the points above 1 m, certain differences exist between the results of the two methods, because the wind velocity setting in the CFD calculation cannot totally agree with the field data. In addition, the thermal radiation field was obtained using the CFD calculation. Generally speaking, the error between the experimental results and CFD calculation remained in an acceptable range.

Keywords: Risk Analysis, Storage and Transportation, Spill Fire, Burning characteristics

1 INTRODUCTION

Fire accidents occur frequently, induced by leakage accidents in the process of oil storage and transportation. In case of a major leakage accident, a massive amount of material might flow along the ground under gravity, which could lead to a spill fire. It is a characteristic of spill fire that the burning area expands rapidly, which makes it difficult to control and extinguish. For many years, liquid fuel fire has been studied by scholars with fixed area pans constructed from steel, but there are different combustion characteristics between a pool fire with a fixed area and a spill fire, such as burning rate and burning area. In recent years, research on spill fires has focused on the characteristics of combustion and spreading. For instance, the Federal Aviation Administration carried out a full-scale fire experiment of jet fuel leakage on a slope to study spill fires on warships or aircraft carriers. Gottuk et al. [1] and Putorti [2] both demonstrated that the burning rates achieved in fuel spill scenarios, with liquid depths to 1 mm, are approximately one-fifth of that measured in steady-state burning scenarios. Mealy and Benfer [3] studied spill fires, focusing on spill dynamics and fuel burning dynamics. Spill and spill fire dynamics testing was performed to characterize the depth of liquid present atop a substrate when spilled in an unconfined scenario and to identify the factors affecting the depth and burning dynamics of the liquid, such as liquid type, substrate, spill volume, ignition delay, and substrate temperature [4,5]. Rujun [6] designed an experiment comprising a horizontal tank with a volume of $3m^3$ surrounded by spill fire to study the hazard on the tank. Quanyi [7] built a flowing liquid fuel fire burning test platform with an adjustable slope and carried out an n-heptane fire test to study the effect of gradients on combustion characteristics.

In this study, a spill fire experimental platform was designed and constructed, and using experiment and CFD calculation methods, the spill fire property of gasoline on a 3° slope was examined. The mechanisms of the fire development properties, thermal radiation intensity, and the temperature field etc. were analyzed and obtained.

2   EXPERIMENTAL SET UP AND PROCEDURE

The spill fire experimental platform, as shown in Figure 1a, was mainly composed of a fuel supply system, an oil trench with slope of 3°, a collecting plate, temperature sensors and a radiation heat flow meter. The length, width and depth of the oil trench, constructed from steel, were 3 m, 1 m, and 0.2 m, respectively. The collecting plate (depth 0.5 m), was set up at the end of the oil trench to collect the liquid waste generated during the experiment. The fuel supply system consisted of a barrel, outlet pipeline and valves, and the flow of liquid fuel was adjusted by controlling the valves. The temperature sensors were arranged above the oil trench at heights of 0.5 m, 1.0 m, 1.5 m and 2.0 m (shown in Figure 1b). Gasoline was chosen as the liquid fuel in the test, with

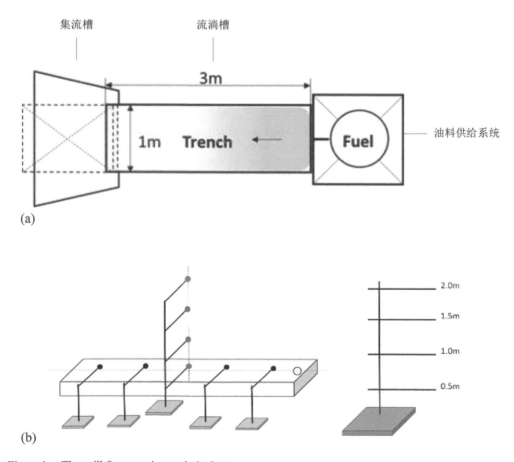

Figure 1.   The spill fire experimental platform.

density 0.78 g/cm³ and flash point –15°C. During the test, the liquid fuel was issued from the barrel and flowed into the oil trench through the outlet pipeline. The liquid fuel in the trench was ignited immediately by an igniter when the oil flowed into the trench. As the leaked liquid fuel continued to increase, the fire gradually expanded to the whole trench. The data acquisition and control system gathered experimental data, including the temperature and heat radiation intensity of the monitoring points. The test was recorded by a digital video camera about 8 m from the test platform.

3 EXPERIMENTAL RESULTS

When the gasoline in the oil trench was ignited by the igniter, the liquid fuel flowed and burned at the same time, which is a salient feature of spill fire. The combustion area of gasoline increased gradually, and the height of the flame elevated rapidly, with the result that the thermal radiation and the temperature of the monitoring points increased gradually. The above phenomenon showed that the spill fire was in the development stage. As time went on, the spill fire gradually tended to the stable combustion stage, where the flame height and combustion area fluctuated in a fixed interval and the flame height was so large that it was beyond the scope of the recording equipment. In this stage, the combustion rate of the gasoline was almost equal to the liquid fuel supply rate. As the fuel was gradually consumed it entered the attenuation stage because the liquid fuel supply rate was not enough to meet the combustion rate of the gasoline. In the attenuation stage, the flame height and combustion area were gradually reduced. The spill fire development process was shown in Figure 2, consisting of the ignition stage, the rapid development stage, the stable combustion stage and the attenuation stage.

In this experiment, four thermocouples were set horizontally above the oil trench, which were numbered 1#, 2#, 3#, 4# from left to right. In addition, the other four thermocouples were set vertically and numbered 5#, 6#, 7#, 8# from top to bottom, as shown in Figure 1b. During the test, the eight thermocouples recorded the temperatures of different locations in the process of the spill fire. To analyze the variation of the temperature of monitoring points with time in the process of a full-size spill fire, the data recorded by four thermocouples in the vertical direction were selected and the plot was shown in Figure 3. As shown in the plot, after the gasoline flowing from the barrel in the fuel supply system was ignited, the temperature of the 8# thermocouple rose rapidly and soon reached the highest value. As the fuel was gradually consumed, the spill fire was in the attenuation stage and the temperature of the

Figure 2.   The development of spill fire.

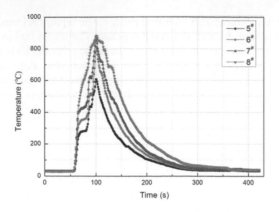

Figure 3.   The temperature of monitoring points in the vertical direction.

monitoring points dropped gradually. In addition, the temperature of the 8# thermocouple in the low position, was higher than the temperature of the 8# thermocouple in the high position. For the 8# thermocouple, the highest temperature was 912°C. For the 5# thermocouple, the highest temperature was 608°C. In the test, the temperature recorded by the eight thermocouples was significantly affected by the environment, such as wind. The temperature of the monitoring points might be higher than the temperature recorded by the thermocouples.

4   NUMERICAL SIMULATION OF THE SPILL FIRE

A full-size test would have significant problems, such as high costs, long test cycles and high security and environmental risks. Besides, experimental research needs a certain number of tests to improve the persuasiveness and reliability, which was conflicting with the existing problems of full-scale experiment [9–11]. The numerical simulation calculation has many advantages such as lower test costs, safety and reliability, and can achieve a certain degree of reliability using the existing technology, which has been adopted by many scholars. For those reasons, scholars increasingly combine experimental research with numerical simulation calculations for scientific research. Kameleon FireEx is currently the leading international fire calculation software in gas diffusion, torch and fire simulation and has gradually become an industrial standard in the field of combustion and fire [12]. The software has a combustion diffusion module, which can calculate liquid spill fire. To simulate the spill fire, the governing equations for this problem included a mass conservation equation, a momentum conservation equation, and an equation of conservation of energy. The combustion process was described by Eddy Dissipation Model.

The flame diagram of the development process of gasoline spill fire obtained by numerical calculations is shown in Figure 4. It could be seen that the simulated calculation spill fire also experienced the ignition stage, the rapid development stage, the stable combustion stage and the attenuation stage. The flame diagram was roughly the same as the recording of the full-size test. In addition, the wind velocity setting in the CFD calculation cannot totally agree with the field data because the wind speed and direction of the test site are always in a changing state, which resulted in some differences in the flame morphology and smoke spread. In general, the whole combustion development process obtained from the simulation calculation was basically consistent with the full-size test results.

To verify the reliability of the numerical calculation, the temperature recording by two thermocouples, including 1# and 2# above the oil trench, were selected. The temperature of two points located in the 1# and 2# of test were shown in Figure 5, as well as the numerical calculation result. As shown in Figure 5, the temperature recording by the 1# and 2# thermocouple and the simulation results were in good agreement. The highest simulated temperature calculated was 932.0°C and the highest temperature at 1# obtained in the test was 873.0°C, with an

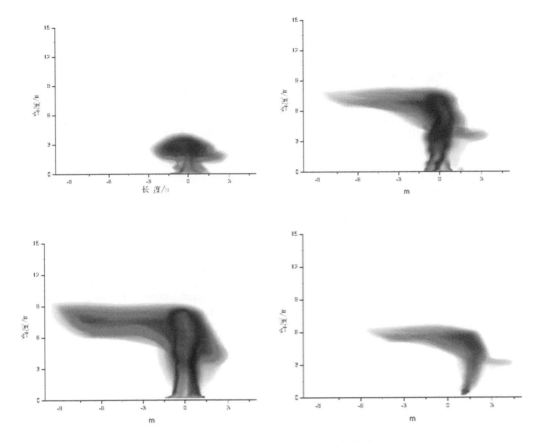

Figure 4.   Flow fire development process obtained by numerical calculation.

error of 6.7%; The highest simulated temperature calculated was 982.0°C and the highest temperature at 1# obtained in the test was 898.0°C, with an error of 9.3%. The boundary conditions of simulation calculation could not fully describe the test scenarios, such as wind speed and direction and flowing surface roughness, which could lead to the error in the two methods. This was also due to the changing of the wind speed and direction during the test, which resulted in the fluctuating flame. The fluctuating flame could affect the temperature field, which may have lead to the misfit after 150 seconds. In general, the error between the experimental results and CFD numerical calculation is within the acceptable range.

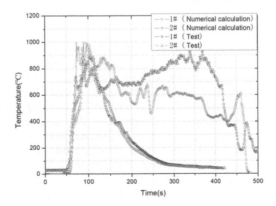

Figure 5.   The temperature of 1# and 2# obtained by the two methods.

5 RISK ANALYSIS OF SPILL FIRE

The change of burning area with time is the biggest difference between flowing fire and pool fire, which bring great risks to the people and equipment around the spill fire. After the ignition, the burning area of the gasoline spill fire increased with approximately linear characteristics [7]. As the burning area continued to increase, the thickness of gasoline began to decrease and the consumption by the combustion increased, which resulted in a gradual decrease in the spread rate of the spill fire. At this time, the burning area of gasoline flowing fire reached the maximum, which filled the whole oil trench. Then, the combustion area was gradually reduced, which was visually reflected as the retrogression of the flame.

Thermal radiation is the other fatal threat of spill fire. Monitoring points of thermal radiation intensity were set at heights of 0.5 m, 1 m, 1.5 m and 2 m to record the changes of thermal radiation in the process of spill fire after ignition. The thermal radiation at the four locations during the spill fire was measured, as shown in Figure 6. It could be seen that in the initial stage and development stage of spill fire, the thermal radiation intensity of the four locations increased rapidly until reaching the maximum value. As the consumption and leakage of the flowing fire fuel reached a dynamic balance, the fire gradually entered a stable combustion stage, and the fluctuation of thermal radiation intensity also decreased until it reached a stable range. In the steady combustion stage of spill fire, the thermal radiation intensity of 1m was stable at 46.5 kw/m^2, and that of 2 m was stable at 54.4 kw/m^2. The thermal radiation intensity at 3 m away from the trench reached at peak of 36.4 kw/m^2, where the fatality rate within 1 min is 100%.

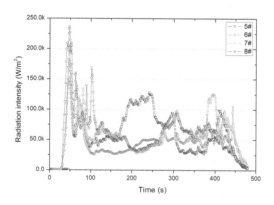

Figure 6. The radiation intensity of 5# to 8# obtained by the numerical calculation.

6 CONCLUSION AND DISCUSSION

The development characteristics of gasoline spill fire were studied by building a test platform and by using numerical calculation. The reliability and accuracy of the calculation software was verified by comparing and analyzing the test results of the two methods. In general, the error between the experimental results and CFD numerical calculation is within the acceptable range. The study followed the characteristic parameters of spill fire in the trench, such as flow area and temperature field, radiation field, as well as the development process of flow fire spread. The spill fire development process consisted of the ignition stage, the rapid development stage, the stable combustion stage and the attenuation stage. After the ignition, the burning area of the gasoline spill fire increased with approximately linear characteristics at first and then the increase rate of the burning area decreased gradually. The burning area of gasoline spill fire reached the maximum, which filled the whole oil trench, and the combustion area was gradually reduced. The thermal radiation intensity at 3 m away from the trench reached a peak of 36.4 kw/m^2, where the fatality rate within 1 min is 100%.

ACKNOWLEDGMENTS

This project was supported by Tianjin Science and Technology Planning Project (17YFZCSF00970).

REFERENCES

1. Gottuk D, Scheffey J, Williams F, Gott J, Tabet R. 2001. *Optical Fire Detection for Military Aircraft Hangars: Final Report on OFD Performance to Fuel Spill Fires and Optical Stresses*. Naval Research Laboratory, NRL/MR/6180000-8457R.
2. Putorti A. 2001. *Flammable and Combustible Liquid Spill Burn Patterns*. National Institute of Justice, NIJ-604-00.
3. Mealy CL, Benfer ME, Gottuk DT. 2011. *Fire Dynamics and Forensic Analysis of Liquid Fuel Fires*. Bureau of Justice Statistics.
4. Mealy CL, Benfer ME, Gottuk DT. 2014. Liquid fuel spill fire dynamics. *Fire Technol* 50(2): 419–436.
5. Benfer ME. 2010. *Spill and Burning Behavior of Flammable Liquid*. University of Maryland, College Park.
6. Wang RJ. 2006. Study on the effects of the spilling oil fire on the nearby tank. *Journal of Safety Science and Technology* 4: 1–55.
7. Liu QY, He YH, Zhang H. 2018. Effects of different substrate slope on burning characteristics of n-heptane spill fire. *Fire Science* 27(02): 85–91.
8. Li Y, Hong H, Jian S, et al. 2018. Experimental study of continuously released liquid fuel spill fires on land and water in a channel. *Journal of Loss Prevention in the Process Industries* 52: 21–28.
9. Qiao Y, West HH, Mannan MS, et al. 2006. Assessment of the effects of release variables on the consequences of LNG spillage onto water using FERC models. *Journal of Hazardous Materials* 130(1–2): 155–162.
10. Dang WY. 2017. Research on the full-surface pool fire of large LNG storage tank. *Fire Science and Technology* 36(05): 606–609.
11. Zhang Y, Bai Y, Wang CL, et al. 2015. Temperature field numerical simulation of large floating roof tank based on SIMPLE algorithm. *Technical Supervision of Petroleum Industry* 31(09): 39–43.
12. Zhao JL, Tang Q, Huang H, Su BN, Li YT, Fu M. 2015. Quantitative risk assessment of external floating roof tank areas based on numerical simulation. *Journal of Tsinghua University* (Sci & Technol) 55(10): 1143–1149.

Risk Analysis Based on Data and Crisis Response Beyond Knowledge – Huang & Nivolianitou (eds)
© *2020 Taylor & Francis Group, London, ISBN 978-0-367-25146-8*

Analysis of evolution law for aviation emergency rescue processes after earthquakes

Xiuyan Zhang

Institutes of Science and Development, University of Chinese Academy of Sciences, Beijing, China
China-France Joint Research Center of Applied Mathematics for Air Traffic Management, Civil Aviation University of China, Tianjin, China

Hong Chi

Institutes of Science and Development, University of Chinese Academy of Sciences, Beijing, China

Xiaobing Hu

China-France Joint Research Center of Applied Mathematics for Air Traffic Management, Civil Aviation University of China, Tianjin, China

ABSTRACT: Understanding the evolution law for aviation emergency rescues after earthquakes is very important for the rational dispatch of rescue aircraft and improving rescue efficiency. Based on the mechanisms of earthquake evolution, this paper discusses evolution law for aviation emergency rescues after earthquakes. The proposed methodology consists of three steps: (1) analysis of the common evolution mechanisms of emergency events to determine the evolution mechanisms of earthquakes; (2) macro evolution law analysis to describe the stages of aviation emergency rescue processes; (3) micro evolution law analysis to clarify the mission priorities of each stage of the aviation emergency rescue process. A numerical example is presented based on the large-scale earthquake that occurred on May 12, 2008 in Sichuan to demonstrate the evolution law.

Keywords: Aviation Emergency Rescue, Event Mechanism, Earthquake, Evolution Law

1 INTRODUCTION

Aviation emergency rescue has become an important rescue method after earthquakes. The advantages of aviation emergency rescue are particularly obvious in areas where ground rescue is difficult after earthquakes. However, due to incomplete information after the earthquake, it is extremely difficult to effectively dispatch rescue resources. The root cause of this is that there is insufficient research on the evolution law of emergencies such as earthquakes, so the correct implementation process of aviation emergency rescue cannot be determined according to the evolution law of events. In order to make the most of the precious and limited resources after earthquakes and avoid wasting resources due to command coordination, it is necessary to study evolution law of the aviation emergency rescue process.

The study of the mechanism of emergencies must start with the inherent law of emergencies and regard emergencies as a dynamic and living process of development and change [1,2]. Only by analyzing the mechanism and essence of emergencies and deconstructing them can we keep unchanged and respond to all changes [3]. Chen proposed that the general mechanism of emergency management can be divided into four levels: principled mechanism, theory mechanism, procedure mechanism and operational mechanism [4]. After mastering the mechanism of events, it is possible to react quickly and effectively after disasters occur and take the most reasonable and timely response measures to achieve the ultimate goal of dealing with emergencies and reducing losses [5][6]. However, at present, emergency management in China focuses more on the guiding principles and there are few studies on the theory, procedure and operational mechanisms of emergency management [7].

Guo took the Wenchuan earthquake as an example and proposed a grid management model to study the emergency rescue model after the disaster [1]. From the perspective of the government, Xue believed that the process of emergency response is essentially an interactive process of relevant information, in which time composition, information sources, decision-making methods and execution methods are closely related [8]. Yao pointed out that managers of emergencies must adjust their management activities dynamically according to the results of phased disposal and the development trends [9]. Some scholars apply scenario data after a disaster to a decision analysis model [10][11] and others have proposed optimization algorithms to optimize the post-disaster relief resource scheduling model and minimize the response time and improve resource utilization [12][13]. Others draw lessons from HEMS (Helicopter Emergency Medical Service) scheduling model, based on the experience of different rescue incidents, using the large data analysis method, and ultimately propose the optimal scheduling scheme [14].

At present, there is much research conducted on the evolution mechanism of emergencies in China and abroad. There is a general understanding of the evolution and development mechanism of earthquakes, but there is a lack of research on the mechanism of earthquake emergency rescue, especially the mechanism of aviation emergency rescue after earthquakes. Therefore, according to the evolutionary law of emergencies, this article will propose the evolutionary law of aviation emergency rescue after earthquakes from macro and micro perspectives, which will provide guidance for emergency rescue work.

The remainder of the paper is organized as follows. Section 2 presents the solution methods; the common evolution mechanism of earthquakes is analyzed in Section 3; Section 4 analyzes the macro and micro evolution law for aviation emergency rescue process; and Section 5 concludes this paper.

2 SOLUTION METHODS

Evolution law of aviation emergency rescue focuses more on procedure and operational mechanisms. Only when considering earthquake mechanisms and combining this with the characteristics of aviation emergency rescue can we determine the evolution law of aviation emergency rescue after earthquakes.

The solution proposed here is shown in Figure 1. First, the common evolution mechanism of earthquakes is analyzed; then, according to the macro law, the different stages of

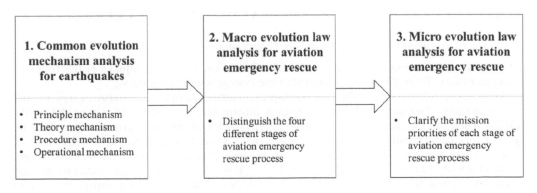

Figure 1.   Solution methods.

emergency rescue are divided into stages; and finally, the key tasks of each stage are proposed through micro-analysis.

3 COMMON EVOLUTION MECHANISM ANALYSIS FOR EARTHQUAKES

3.1 *Analysis of common evolution mechanism of emergency events*

The mechanism originally refers to the internal structure and working principle of the machine. It has been widely used in natural and social phenomena, referring to the law of internal organization, operation and changes. The mechanism is the foundation for any system.

Through the analysis of the mechanism of emergency events, we can find the source of events, and identify the objective law and the original driving force to promote the development of events. The evolution mechanism of emergency events can be divided into four levels according to the degree of detail: principle mechanism, theory mechanism, procedure mechanism and operational mechanism.

The principle mechanism is a simple description of the characteristics of emergency events, which can be described by suddenness, vagueness, contingency and inevitability. Theory mechanism includes the occurrence mechanism and evolution mechanism of events, which can be divided into four categories: transformation, spread, derivation and coupling. Procedure mechanism refers to the gradual development of emergencies along the optimal path, which consumes the least energy and maximizes disasters and can be divided into four basic forms: chain, radiation, migration and convergence, which correspond to transformation, spread, derivation and coupling, respectively. Operational mechanism is based on procedure mechanism.

3.2 *Analysis of earthquake evolution mechanism*

From the perspective of energy release and transmission, the evolution of earthquakes can go through four stages: energy accumulation, energy release, path selection and recession.

The energy released by earthquakes comes from the heat energy inside the earth, which promotes the convection of material in the mantle and forms plate movement, which is the direct cause of earthquakes. The energy accumulated by earthquakes is mainly released by the medium in the form of earthquake waves. The more medium involved in vibration, the greater the energy released by earthquakes and the greater the magnitude of earthquakes. The impact of earthquakes on disaster-stricken areas is limited by many factors, such as topography, weather, the characteristics of population distribution in disaster-stricken areas, and the effect of emergency rescue. An earthquake's evolutionary path cannot be exactly the same every time it occurs. As the effects of earthquakes and other secondary disasters caused by earthquakes fade away, the evolution of earthquakes enters the stage of recession.

4 EVOLUTION LAW ANALYSIS FOR AVIATION EMERGENCY RESCUE

4.1 *Macro evolution law analysis for aviation emergency rescue*

In the energy accumulation stage, because the formation of the earthquake is the result of plate tectonic collision, it is not considered in the emergency rescue process.

The energy release stage, when an earthquake occurs, releases enormous energy in just a few minutes. Grass-roots departments must inform senior departments about the earthquake. Then, senior departments can make judgments based on the information provided by grass-roots departments on which emergency plans should be initiated, and what level of rescue forces need to be mobilized to rescue to disaster areas. In the path selection stage, searching and rescuing buried survivors is the key task. All the emergency rescue forces and materials should be deployed with the safety of human life as the first priority. Heavy and light machinery and manual rescue should be used together to rescue the most survivors in a limited time. In the period of mitigation and recession of emergencies, usually about 10 days after the earthquake, the corresponding emergency rescue work has shifted to the post-disaster disposal stage. The main task of the rescue forces is to ensure that the people in the disaster-stricken areas have food, water, clothing, shelter, education and sickness to be treated.

4.2 *Micro evolution law analysis for aviation emergency rescue*

The path selection stage of emergencies is the key node of evolution, which determines the final outcome of events. Therefore, this article focuses on the microevolution mechanism of emergency rescue work in the path selection stage.

4.2.1 *Analysis of micro evolution mechanism of chain path aviation emergency rescue*

The chain path evolution of emergencies indicates that the occurrence of primary event A directly leads to the occurrence of secondary event B. The two events have a sequential relationship in time and a causal relationship in logic, and the relationship is one-way. We usually focus more on the occurrence of the primary event A and the transformation path of event A to event B. Event B does not require much attention.

The typical result of earthquake chain path evolution is landslide, which will block road traffic in the affected area and block rivers to form barrier lakes. At this time, emergency rescue should prevent the transformation of primary events to secondary events through chain path and cut off the process of path transmission. The main task of aviation emergency rescue is to put in large-scale machinery, which needs heavy helicopters with large takeoff weight and heavy load to complete the task.

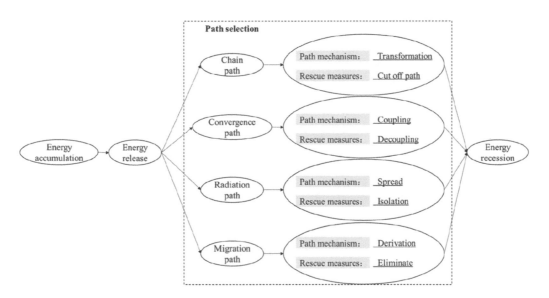

Figure 2. Micro evolution law analysis for aviation emergency rescue.

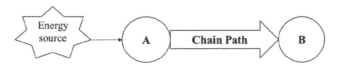

Figure 3. Chain path.

4.2.2 *Analysis of micro evolution mechanism of convergence path aviation emergency rescue*

(1) Cyclic coupling

The condition of cyclic coupling is that the initial power source comes from the outside, and the secondary events at all levels are correlated in turn and diffuse in one direction. In the case of earthquake, mountain collapse can be regarded as the primary event A, road block can be regarded as secondary event B, bad weather can be regarded as primary event A1, vehicle detention can be regarded as secondary event C. An earthquake led to mountain collapse, which blocked the highway. At this time, there happened to be bad weather, which led to static vehicles.

(2) Co-generation coupling

The condition of co-generation coupling is that the diffusion power generated by the primary event is synchronized, and there is a coupling relationship among the events at the same level. Landslides can be regarded as event A, rainstorms as event B, and debris flows as event C. In the case of earthquakes, landslides occur, which loosens the soil on the mountain. At this time, the disaster-stricken area encounters a once-in-a-century rainstorm.

(3) Suppression coupling

Suppression coupling refers to the coupling mode in which primary events delay and inhibit secondary events caused by primary events at the same level in the process of the diffusion of emergencies. Event A denotes a superior event, which promotes the next level event C, and Event B denotes another superior event, which inhibits the next level event C. The condition for the occurrence of hair suppression coupling is that the initial diffusion of the primary and suppressed events is synchronized, and the primary and suppressed events are unidirectional diffusion for the sustained events, and there is coupling between them.

(4) Promotion coupling

Promotion coupling refers to the coupling mode of emergencies in the process of diffusion, under the joint action of primary events and secondary events caused by them, leading to another secondary event. The occurrence condition of promotion coupling is that the endurance event is reversible, and the primary and promotion events are unidirectional diffusion to the endurance event.

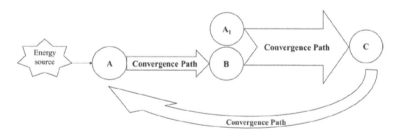

Figure 4.   Cyclic coupling.

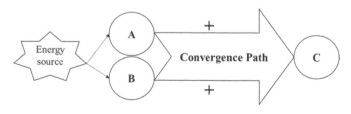

Figure 5.   Co-generation coupling.

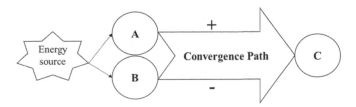

Figure 6.   Suppression coupling.

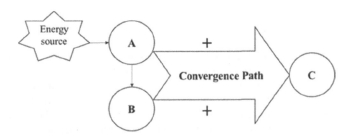

Figure 7.   Promotion coupling.

Emergency rescue work focuses on "decoupling" by requisitioning strategy, which refers to changing the state characteristics of the factor itself, including changing the structure, energy, speed and other attributes, so that the factor cannot play a coupling effect. The main task of aviation emergency rescue is intelligence reconnaissance in the disaster area. The most suitable aircraft type is an unmanned reconnaissance aircraft, which is light, flexible and easy to take off and land.

4.2.3 *Analysis of micro evolution mechanism of convergence path aviation emergency rescue*
The spread of emergencies means that a primary event triggers multiple secondary events of the same type, and the superposition of multiple events produces more serious consequences. For multiple aftershocks, the effect of aftershocks spreading depends not only on the resistance of the terrain to earthquake energy, the degree of excitation and the energy itself, but also on the energy that has not been released by previous earthquakes. For this type of spread mode, we can adopt an "isolation" strategy to reduce casualties and property losses. Isolation strategy works to isolate the original events from the follow-up events and reduce the impact on the follow-up events. The main task of aviation emergency rescue is to transfer the affected people. The most suitable type of helicopter is a multi-functional helicopter that can quickly transport personnel and materials.

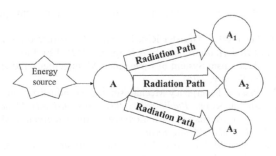

Figure 8.   Radiation path.

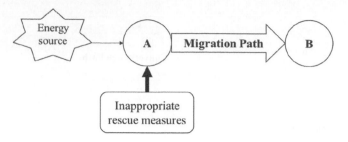

Figure 9.   Migration path.

4.2.4 *Analysis of micro evolution mechanism of migration path aviation emergency rescue*

The derivation mechanism of emergencies is that people take inappropriate measures that lead to another emergencies. A typical example of derivative evolution after earthquakes is the disposal of barrier lakes. In order to eliminate the potential crisis of a barrier lake that may erupt at any time, the disaster relief headquarters decide to carry out directional blasting. This decision may cause the towns downstream of the barrier lake to be flooded, resulting in people going missing or being trapped. For the derivative events of earthquakes, we adopt the "elimination" strategy to deal with them. "Elimination" refers to the adoption of other strategies to eliminate the hazards of measures taken. At this time, the main task of aviation emergency rescue is to rescue the trapped people. The most suitable type of helicopter is a light, flexible and stable hovering light helicopter.

5   CONCLUSIONS

In this paper, the macroevolution mechanism of emergencies is analyzed. From the perspective of the energy release of emergencies, the possible energy release stages of earthquake evolution are considered. The characteristics of aviation emergency rescue are combined with the evolution of aviation rescue strategies with different microevolution mechanisms, and the universal law of aviation emergency rescue evolution after earthquake is obtained.

Earthquake emergency rescue is a complex system engineering that requires people to find more common characteristics of earthquake evolution based on the specific analysis of each event's own characteristics.

ACKNOWLEDGMENT

Thanks to the support of the National key research and development plan 2016YFC0802601.

REFERENCES

[1] Guomiao Miao. 2011. Research on grid management of unconventional emergencies. Wuhan University of Technology.
[2] Shuzhen Li, Alateng Tuya. 2015. Disaster Risk Research Literature on Statistics Analysis in China Journal Net. *Journal of Risk Analysis and Crisis Response*, 5(2): 129–140.
[3] Chen An. 2009. The Mechanism System of Emergency Management and Emergency Management. *Science, Technology and Society*, 24(5).
[4] Zhou Dan. 2014. Analysis of Emergency Mechanism and Standardization of Public Safety: A Case Study of MERS Epidemic in Korea. *Public Safety Business Continuity Management*, 3.
[5] Chen An. 2011. The Origin, Mechanism, Characteristics and Emergency Management Principles of Emergency Events. *The World of Management*, 2.
[6] Fusheng Yu. 2013. Decision-Making Model in the Environment of Complex Structure Data. *Journal of Risk Analysis and Crisis Response*, 3(2): 103–109.

[7] Chi Fei.2015. Research and Analysis on the Current Situation of Emergency Management in China. *The World of Management*, 4.

[8] Xue Kexun. 2004. Research on Government Emergency Response Mechanism. *Administration in China*, 2.

[9] Yao Jie. 2005. Dynamic Game Analysis in Emergency Management. *Management Review*, 17(3).

[10] Zhao Ping. 2017. Multi objective optimization of railway emergency rescue resource allocation and decision. *International Journal of System Assurance Engineering and Management*, 2.

[11] F. Fiedrich. 2006. An HLA-based multi agent system for optimized resource allocation after strong earthquakes. *Winter Simulation Conference*, Washington D.C., 486–492.

[12] C. Kessler. 2015. Helicopter emergency medical service: motivation for focused research, *CEAS Aeronaut*, 6.

[13] Shuo Yan. 2013. Emergent disaster rescue methods and prevention management. *Disaster Prevention and Management*, 22(3).

[14] Georgette Eaton. 2018. HEMS dispatch: A systematic review. *Trauma*, 20(I).

Risk Analysis Based on Data and Crisis Response Beyond Knowledge – Huang & Nivolianitou (eds)
© 2020 Taylor & Francis Group, London, ISBN 978-0-367-25146-8

New approach of forming the confinement barrier by phenolic foam for "stopping" the liquid chemical spills and flowing fires

Nan Jiang & Changxing Ren
National Center for Fire Engineering Technology, Tianjin, China
Tianjin Fire Research Institute of MEM, Tianjin, China

Yinhe Sun
Tianjin Institute of Metrological Supervision and Testing, Tianjin, China

Jin Li, Dong Lv, Yan Zhang & Yangyang Mu
National Center for Fire Engineering Technology, Tianjin, China
Tianjin Fire Research Institute of MEM, Tianjin, China

ABSTRACT: Liquid chemical spills that occur on land could lead to serious safety issues. A kind of confinement material consisted of phenolic foam with good fire resistance and chemical resistance was put forward. A barrier (65 cm-wide, 26 cm-high) was constructed by phenolic foam. According to the water-resistance and oil-resistance experiments, we found that the barrier could block the spill movement completely. Mixed liquid of gasoline and ethylene glycol was chosen as the experimental fuel. The barrier remained its form under fire (maximum temperature > 850°C) for 30 min. The temperature inside the barrier was below 60°C, which showed the good heat insulation. Under fire condition for 30 min, part of the barrier became carbonization but not be burned. The flame border on the top of the barrier reached 41 cm away from the fire side of the barrier. In the whole process, the experimental fuel was stopped in one side of the barrier. In addition, some key factors to assess this confinement technology were put forwarded. Hence, as the confinement barrier, the use of phenolic foam provided a rapid and efficient approach for stopping the liquid chemical spills and flowing fires.

Keywords: Confinement Barrier, Liquid Chemical Spills, Flowing Fires, Phenolic Foam

1 INTRODUCTION

The increasing development of chemical industry has greatly stimulated the demand of chemical storage and logistics. Meanwhile, as increasing quantity of chemical storage and logistics, the risk of flammable and explosive chemicals increases quickly. Current method for control the risk of flammable and explosive chemicals in China mainly depends on the unified management by gathering the enterprises into chemical industry park, which leads to the highly centralized dangerous source. If liquid chemical spills on land, it could lead to serious safety issue. For example, liquid spills into urban areas could exhibit rapid horizontal movement and cause the fire or explosion.

Current disposal methods of liquid chemical leakage mainly included the control of leakage source, the control of leakage source and the collection of leakage. In the recommended practice of NFPA 471, these methods are also recommended for disposing the leakage of liquid chemical: 1) the control of leakage source mainly depends on the sealing technology; 2) the treatment of leakage mainly depends on containment, cover, dilution and adsorption. For the leakage of liquid chemical, it is better to use the method of foam coverage for lowering vapor volatilization. Generally, this method needs controlling liquid chemical in a certain area before covering. Therefore, the confinement technology for controlling liquid hazardous chemical is important for this process.

Liquid spill behavior on land is visible to be predicted. Thus, confinement technology is accomplished primarily by capturing the liquid chemical spill in some kinds of depression or excavation or some type of physical barrier. Among these techniques, spill barrier techniques show more flexibility. Using barriers could stop the liquid spill movement or divert the spill. Confinement barrier could be placed in the path of the liquid spill to stop the movement and hold it for further disposal. There are many types of barriers for chemical liquid spill confinement, such as natural barriers (depressions, ditches and basins, et al.), readily available materials (sand, soil, and gravel, et al.) and some other commercially materials (mats, plugs, sorbents, gels and foams, et al.). Some manufacturers produced small portable sprayer systems with a quick-setting polyurethane foam. In about one minute, the foam forms a light confinement barrier on concrete and other hard surfaces, which could erect a confining dike ahead of the spill.

In this paper, we put forward a kind of confinement material consisted of phenolic foam. We chose phenolic foam as the confinement material for blocking chemical liquid spill on land, which mainly attributed to the following reasons: (1) water-resistance; (2) oil-resistance; (3) good fire resistance; (4) good chemical resistance. Therefore, the effects of experimental conditions (such as water, oil and fire) on phenolic foam were investigated.

2   MATERIALS AND EXPERIMENTS

2.1   *Materials*

Phenolic foam, gasoline and ethylene glycol.

2.2   *Water-resistance experiment*

The water-resistance experiments were done in the steel sealed box with two sides open (top side and front side), which shown in Figure 1. In this process, phenolic foam was formed in the hatch area and block the front side. After 30 mins, about 140L water was pour into the box. the height of liquid level was 0.2 m. The change of the height of liquid level was monitored by the camera in the whole process.

2.3   *Oil-resistance, fire resistance and chemical resistance experiment*

Oil-resistance, fire resistance and chemical resistance experiment were done in the same steel sealed test platform (shown in Figure 2). The size of the test platform is 3 m × 1 m × 0.2 m (Length × width × depth). The thickness of steel plate is 3 mm. The size of phenolic foam formed was 0.65 m × 1 m × 0.26 m (Length × width × height). The burning area in the test platform is 1.4 m-long (the other size is same as the test platform). There were 6 K-type thermocouples around the test platform. The front view is shown in Figure 3. 1# and 2# thermocouples were inside the confinement barrier of phenolic foam, which close to the bottom of

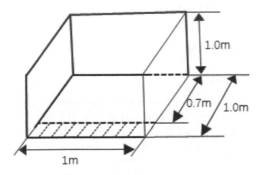

Figure 1.   Test platform for water-resistance experiment (front view).

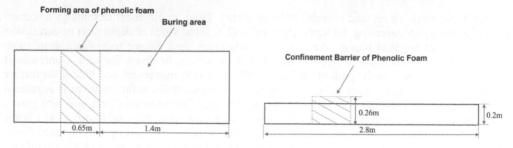

Figure 2.   Schematic diagram of Test platform (left: top view; right: front view).

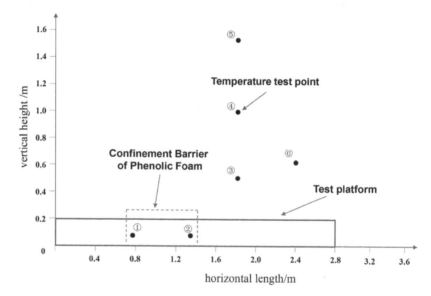

Figure 3.   The location of temperature test point. (front view, •was the temperature test point).

the test platform and 0.05 m from two edges. 6# thermocouple was 0.6 m above the bottom. 3#, 4# and 5# thermocouples were in the same vertical plane, 0.5 m, 1.0 m and 1.5 m above the bottom respectively.

The typical procedure was as follows: phenolic foam expands and hardens in the forming area; 50 L ethylene glycol and 5 L gasoline were poured into the test plat-form; when the liquid level had stabilized, the fuel was burned. The whole burning process burning continuous time about 30 mins.

3   RESULTS AND DISCUSSION

3.1   *Analysis of water-resistance*

As shown in Figure 4, confinement barrier of phenolic foam expands and forms in the preset framework. The height of the confinement barrier was 0.7 m. 1h after the addition of 140 L water (liquid level, 0.2 m), the liquid level decreased 0.015 m which showed good water-resistance. However, because of the expand force of phenolic foam, the steel plate deformed at the top of the test platform. If increasing the liquid level to 0.5 m, water leaked from the gap between confinement barrier and steel plate. Therefore, the gap between confinement barrier and steel plate (existing wall) need to be sealed, lest liquid spill.

Figure 4. Water-resistance experiment (front view).

3.2 *Analysis of temperature outside the confinement barrier*

Figure 5 showed the change of temperature with the increasing time outside the confinement barrier. As shown in the figure, at the beginning, the temperature rose rapidly. The highest temperature in the burning process was above 850°C (3# thermocouple). The highest temperature of 4# and 6# thermocouples were about 650°C. The highest temperature of 5# thermocouple was close to 500°C. After reaching the highest temperature, it decreased to a relatively stable stage. Except for 3# thermocouple, the temperature of other thermocouples decreased to 150°C. The temperature of 3# thermocouple decreased to 500°C for about 800 s, then deceased to 300°C. Because 3# temperature test point was closer to the confinement barrier, its temperature showed the fire resistance of the materials.

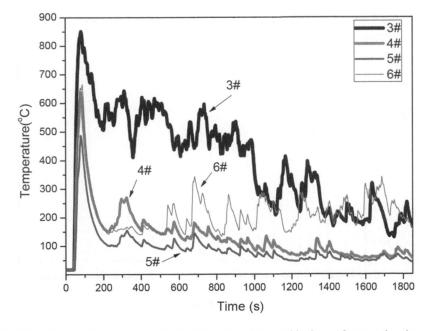

Figure 5. The change of temperature with the increasing time outside the confinement barrier.

The temperature of 6# thermocouple decreased to 150°C, then increased to 250°C. That's because the increasing temperature led to the change of steel plate's shape at the bottom of test platform, and further led to the bulge of steel plate. The remaining fuel flows from the middle to the two sides. Therefore, because more fuel existed in the two sides, the temperature was higher than that in the middle.

3.3 Analysis of temperature inside the confinement barrier

Because the foaming process of phenolic foam is an exothermic process, the temperature inside materials is higher than room temperature. The inside temperature of phenolic foam after fully formed is about 50°C. Figure 6. showed the change of temperature with the increasing time inside the confinement barrier. When the fuel burned, 1 # thermocouple remained 50°C in the whole process. Therefore, at this experiment condition for 30 mins, confinement barrier had the ability for blocking the flowing fire.

At the beginning, the temperature of 2 # thermocouple was about 60°C. When burning for 200s, the highest temperature rose to 700°C. Then, the development of temperature had three stages: 1) kept 650°C between 200-500 s; 2) temperature decreased between 500-600 s; 3) kept 320°C between 600–950 s. That's because the char forming of phenolic foam, which led the removement of 2 # thermocouple from the confinement barrier. 2 # thermocouple was burned by the fire directly, and led to the raise of temperature. Meanwhile, 2 # thermocouple was closer to the fire than 3 # thermocouple, which led to higher temperature than that of 3 # thermocouple. The results of 2 # and 3 # thermocouple, showed that this confinement barrier could block the flowing fire but needed a certain thickness.

3.4 Analysis of flame border of the confinement barrier

As shown in Figure 7, flame border moved toward the confinement barrier with increasing time. In this process, the material decomposed and happened charring. Because char particles blew by the fire wind, the flame border moved move along the Carbonized part of confinement barrier. We chose 3 mins as the gap for recording the movement of flame border on the confinement

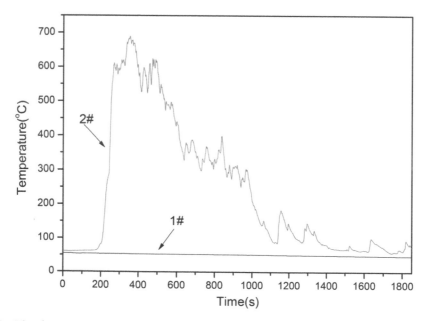

Figure 6. The change of temperature with the increasing time inside the confinement barrier.

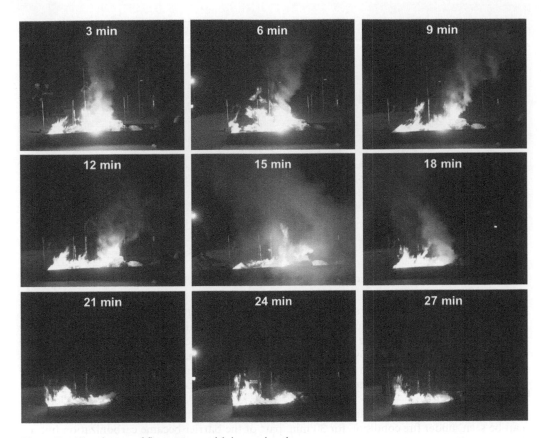

Figure 7.   The change of flame status with increasing time.

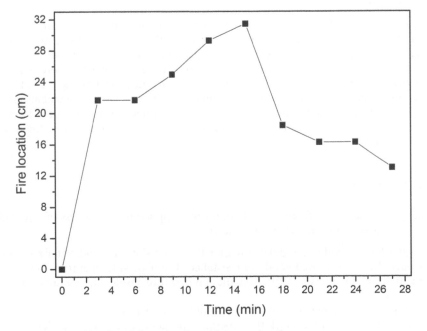

Figure 8.   The change of border of flame border with increasing time.

Figure 9.   The situation of confinement barrier and test platform when fire was extincted.

barrier (Figure 8). As it can be seen, flame border moved forward in the first 15 mins. When the burning time was 3 mins, the flame border on the top of the barrier reached 21 cm away from the fire side of the barrier. Then the movement speed slower and reached 41 cm. After 15 mins, flame border stayed still and then moved backward. That's because the lack of fuel. When fire extinct, flame border reached 35 cm. In the whole process, the experimental fuel was stopped in one side of the barrier. Therefore, 65 cm-thickness of confinement barrier had good fire resistance and chemical resistance.

    showed the situation of confinement barrier and test platform when fire was extincted. As it can be seen, under fire condition for 30 min, part of the barrier became carbonization but not be burned. The rest of confinement barrier could be moved easily.

3.5   Key factors of confinement technology

Although the use of confinement barrier formed by phenolic foam as a confinement technology is not complicated, some important factors must still be considered to assess this method to confine the liquid chemical spill. Not simply supply a phenolic foam as confinement barrier. Different fuel and site conditions need different operating parameter. Therefore, we chose some factors as the key factors to assess this confinement technology, such as water-resistance, oil-resistance, fire resistance and chemical resistance. The confinement barrier must be formed in a short period of time for effective spill confinement, meaning the foaming time is the first factor to be considered. Because of the carbonization of phenolic foam, a certain thickness is needed.

4   CONCLUSIONS

In summary, an efficient confinement technology was applied in the process of liquid chemical spill. The main conclusions are drawn as follows:

(1)   The water level only decreased 0.015m after 1h, which showed good water resistance.
(2)   This phenolic foam showed good fire resistance. The highest temperature outside the confinement barrier is above 850°C, but the temperature inside the confinement barrier is below 60°C.
(3)   In the whole burning experiment, confinement barrier was not burned.
(4)   Confinement barrier could block the experimental fuel in one side of barrier.

Based on the results above, some key factors were put forwarded, including water-resistance, oil-resistance, fire resistance, chemical resistance, foaming time and thickness of confinement barrier.

ACKNOWLEDGMENTS

This work was supported by the Tianjin Key Technology R&D Program of China (17YFZCSF00970).

REFERENCES

NFPA 471: 2002. Recommended Practice for Responding to Hazardous Materials Incidents.
Lees, Frank. 2012. Lees' Loss prevention in the process industries (Third Edition). Butterworth-Heinemann. 1–104.
GA/T 970: 2011. Guide for disposal of hazardous chemical leakage accident.
Fingas M. 2016. Oil spill science and technology. Gulf professional publishing. 303–426.
Li P, Cai Q, Lin W, et al. 2016. Offshore oil spill response practices and emerging challenges. *Marine Pollution Bulletin*. 110(1): 6–27.
Nyankson E, Rodene D, Gupta RB. 2016. Advancements in crude oil spill remediation research after the Deepwater Horizon oil spill. *Water, Air, & Soil Pollutio*. 227(1): 29.

Risk Analysis Based on Data and Crisis Response Beyond Knowledge – Huang & Nivolianitou (eds)
© 2020 Taylor & Francis Group, London, ISBN 978-0-367-25146-8

Research on the characteristics of road transportation of hazardous chemicals in China

Ronghua Zhao*

Safety Research Center of City Operation, Beijing Academy of Safety Science and Technology, Beijing, China

Ran Yu

GESIS - Leibniz Institute for the Social Sciences, Cologne, Germany

Jinlong Zhao

School of Emergency Management and Safety Engineering, China University of Mining & Technology, Beijing, China

Xuewei Ji

Safety Research Center of City Operation, Beijing Academy of Safety Science and Technology, Beijing, China

Chengwu Li

School of Emergency Management and Safety Engineering, China University of Mining & Technology, Beijing, China

ABSTRACT: In this paper, we analyzed statistics about 142 accidents with hazardous materials that happened during the road transportation process in the past 20 years. The characteristics and laws of these accidents were summarized from the time of occurrence, the cause of the accidents, the type of Hazmat/road, and the physical appearance. The results showed that the accidents had unique time-phase characteristics, were especially likely to occur in March and December, and were particularly frequent at 9 am. More than half of the accidents were related to flammable liquids, and most led to leakage. Human factors are the most common causes of accidents.

Keywords: hazardous material (Hazmat), road transportation, accidents

1 INTRODUCTION

At present, the transportation of hazardous materials (Hazmat) in China mainly relies on road transport. Accidents frequently occurred in the transportation process, causing catastrophic damage to life, property and the environment. Hazmat transportation accidents, which are often caused by traffic accidents, malfunctions of vehicles or equipment, usually lead to leakage, fire, deflagration or explosion. On March 29, 2005, a vehicle containing 40ts of liquid chlorine collided with a truck during the transportation process, resulting in a large leakage of liquid chlorine, causing 29 deaths, more than 400people poisoned, and tens of thousands of people evacuated. On March 1, 2014, two vehicles loaded with methanol were rear-end, causing 31 deaths and 42 cars burned. More Hazmat transportation accidents resulted in severe social harm. Aseries of realistic

---

*Corresponding author: rhzhao@foxmail.com

problems were exposed, such as Hazmat safety supervision, emergency response capabilities, risk assessment, emergency rescue, etc.

Research on Hazmat road transportation in previous studies has focused on the statistical analysis of accidents, risk assessment, transportation route selection, transportation supervision, etc. Fabiano, et al. (2002) investigated 3222 accidents involving Hazmat in Italy from 1926 to 1997 and found that 41% of the accidents were transportation accidents, which indicated that the transportation accidents accounted for a large proportion statistically and should be taken seriously. Ohtani et al. (2005) collected and summarized data on Hazmat accidents in Japan and analyzed the main causes of road transport accidents involving Hazmat. Roberto, et al. (2000) proposed an LPG road and rail transportation risk analysis method based on personal risk and social risk. The method used an FN curve to analyze the personal and social risks brought by LPG road and rail transportation for individuals and departments along the way.

With the increasing number of Hazmat road transportation accidents, some countries have introduced laws and regulations (Seveso III) requiring quantitative risk assessment of Hazmat transport devices. Two quantitative risk assessment methods are popular in Europe. The first is the worst accident scenario method. In the risk assessment process, only the worst possible accident scenario is considered to determine whether it is acceptable. If the worst case is acceptable, the solution is acceptable. The second is the probability-based risk assessment method. In the risk calculation process, careful consideration is given to the probability of occurrence of accidents and the consequences of accidents. Personal and social risks are used as indicators to determine whether the plan is acceptable according to the judgment criteria. Probability-based risk assessment method is gradually becoming widely used, which mainly relates to frequency of accidents, accident development process and risk calculation, to determine whether the transport device meets the requirements in conjunction with the risk criteria acceptable.

The frequency of Hazmat transportation accidents occurrence is usually based on statistical analysis of accident cases. Commonly used data usually comes from the CONCAWE database, or the Oreda database, etc. Also, many scholars refer to the Purple Books and Yellow Books, which are officially published by the European Union, to accurately determine the probability of accidents in different transport devices. Scholars such as Hui Lv, Laijun Zhao, Chunlin Xin et al. (Lv and Peng, 2017; Zhao, et al., 2009; Xin and Wang, 2012; Yang, et al., 2010) have conducted statistical analyses of the road traffic accident data of Hazmat in China, and summarized the different characteristics. However, there is still a lack of specialized Hazmat road transportation accident database for analysis in China.

For the study of the accident development process, the commonly used accident models include Hazmat leakage model, fire model and explosion shock wave model to analyze the diffusion process of toxic and harmful gases, thermal radiation and the distribution of explosion shock waves. Liu et al. (2011) used a boiling liquid extended steam explosion model to analyze the detailed consequences of the propane tanker explosion accident on the highway and determined the scope of the explosion. Sun et al. (2003) conducted a multi-aspect analysis of Hazmat road transportation accidents and proposed feasible suggestions. Wang et al. (2017) conducted a systematic study on the influencing factors of Hazmat in the road transportation process and analyzed the causes of accidents from several aspects, such as management reasons, human errors, equipment and facilities defects, road conditions and environmental impacts. Wu and Zhao et al. (Wu, et al., 2015; Zhao, 2016; Wang, et al., 2005; Leonelli, et al., 2000) analyzed the main influencing factors of Hazmat road transportation risk. On this basis, the risk evaluation index system of Hazmat road transportation was established and evaluated to find the weakness of current transport enterprises. Scholars such as Xiongjun Yuan, Chaogang Tang, Laijun Zhao, etc. (Yuan, et al., 2014; Tang and Yu, 2018) have studied the mechanism of Hazmat road transportation accidents.

Risk calculation mainly determines the probability of casualties based on the distribution of dangerous and harmful factors, combined with the personnel vulnerability model and the location of personnel. In addition to traditional model research, large eddy simulation, CFD simulation, Fluent simulation, etc., have also been used in recent years.

To reveal accident features and avoid damage efficiently, a study of 142 typical accidents from 1991 to 2019 in China was carried out, the characteristics of accidents were summarized and the correlations between the risks and the influencing factors were studied. This paper aims to provide an updated survey on the situation in this field in China by analyzing road accidents, including their causes, consequences, severity, and frequency. Based on the analysis of these data, preventive measures applied to reduce this frequency are also recommended.

2 METHODOLOGY

2.1 Accident information source

There are no specific databases about Hazmat road transportation accidents in China. The accidents involved in this article were found on the Chemical Accident Information Network of the Emergency Management Department, and the China Chemical Safety Association website. However, most of accidents only collected preliminary information (e.g., time, brief incidents, and source). In addition, detailed information was gathered from papers and the Internet. Nevertheless, some accidents without sufficient information were excluded. Finally, 142 accidents were screened out as research subjects and stored in the database. For each accident, these pieces of information were placed in the fields: source (name of the paper,web site, or other sources used); time, location, type of road, immediate causes of the accident (sequence of events that led to the accident); physical appearance (e.g., leakage, explosive, toxic, flammable, etc.); classification of substances in accordance with the Chinese classification and code of dangerous goods (GB 6944-2012);type of accident (time sequence of phenomena after the accident), and type of pollution eventually caused, and severity.

2.2 Method

Statistical analysis was performed using descriptive analysis. Descriptive analysis is the determination of conditional probability. In this method, the probability of an event is defined as the number of ways an event can occur, divided by the number of all possible results of observations.

3 RESULTS AND DISCUSSION

3.1 Distribution of the accidents over time

Accidents may be characterized by year, month, day, and time. Different types of accidents also have unique time-phase characteristics. This study first analyzed the trend of all accidents over a month, as shown in Figure 1. Then the phase characteristics of hour were analyzed, as shown in Figure 2.

As seen from Figure 1, the accident frequencies were higher from March to May, July to August, and October to December than in other months. This may be related to the characteristics of the seasons, which indicates that the temperature and the weather possibly affect accidents. The hot weather affects the physical and mental health of the workers, causing physical strength to drop and emotional impetuousness. High temperatures can also cause spontaneous combustion of flammable Hazmat, which can result in accidents. The road may be slippery when it rains or snow, leading to traffic accidents. In addition, extreme weather conditions can increase the instability of vehicles. The incidence of accidents was higher in March and December, which suggests that accidents are more likely to happen before and after the Spring Festival, when workers are preparing for vacation or are starting work again.

Figure 2 shows the accidents had significant time-phase differences. The transportation accidents of Hazmat were frequent at hours 0, 3, 9, 13, 16, 20, and 21, and there was significant rise at 9. This may be related to fatigued driving, poor traffic conditions, etc. The traffic conditions were often congested on the way to and from work, so simultaneous traffic accidents are easily caused. Moreover, workers were in a state of exhaustion after a long-time driving, and their safety awareness was reduced, which may lead to traffic accidents.

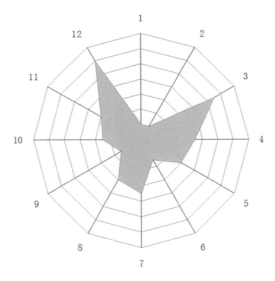

Figure 1.　Distribution of accidents by month.

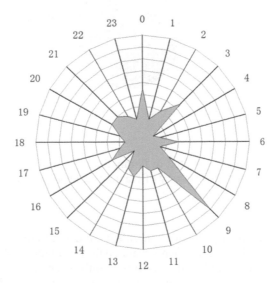

Figure 2.　Distribution of accidents byhour.

3.2　Types of Hazmat

The types of Hazmat involved in the accidents, based on the classification given by the List of Dangerous Goods (GB 12268-2012), were identified for 142 cases. The distribution among the different categories is shown in Figure 3. To determine the type of Hazmat associated with a particular feature of a substance (e.g., toxicity, corrosion, flammability, etc.), a statistical analysis of incidents that involved at least one Hazmat was carried out. As more than one feature of Hazmat considered may be present in any accident (e.g., 'flammable' and 'toxic' simultaneously), the sum of all these incidents is more than the number of cases. As seen from Figure 3, the most common accidents were related to flammable liquids, corrosive substances, and toxic substances. More than half of the accidents were related to flammable liquids.

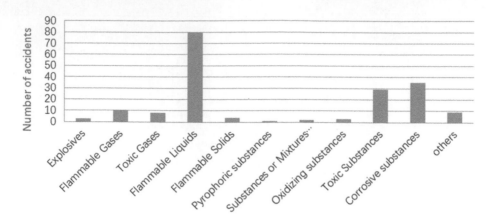

Figure 3. Accidents associated with the types of Hazmat.

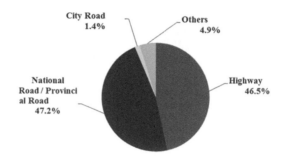

Figure 4. Accidents by the type of road.

Figure 3 also shows that accidents related to flammable substances, corrosive substances, and toxic substances accounted for the majority of Hazmat transportation accidents in China.

3.3 *Types of road*

There is also a certain correlation between accidents and road conditions. Statistics on the types of roads involved in the 142 accidents were analyzed. Figure 4 shows the type of road where the accidents occurred and that accidents mainly occurred on highways, provincial roads and national roads, with very few accidents on city roads and other roads. China mainly relies on highways for road transportation of Hazmat, where the accidents occurred frequently.

3.4 *Causes of accidents*

The road transportation system for hazardous chemicals generally consists of five elements: Hazmat, transport vehicles, related personnel, roads, and the environment. If the five elements form a harmonious unity, the system will operate safely; otherwise it will lead to an accident. Related to these five factors, four categories of possible causes were considered: human error; vehicle, packaging and equipment defects; road and environmental conditions; and management factors. According to the analysis shown in Figure 5, 63.4% of road accidents were caused by human errors (which often show an impact or collision between vehicles). Hence, human factors are the most common causes of accidents. The same result was also confirmed in the literature. The second common causes consisted of various types of vehicle, packaging

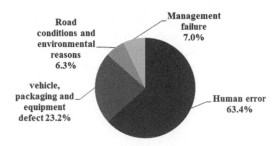

Figure 5.   Causes of accidents.

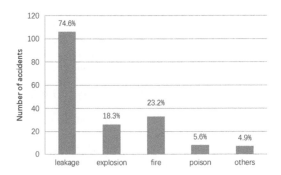

Figure 6.   Physical appearance of the accidents.

and equipment defects (23.2%). Road and environmental conditions and management factors cause almost equivalent numbers of accidents.

Based on the previous analysis, 63.4% of accidents were initiated by human errors (e.g. fatigued driving) and most Hazmat accidents that occurred on general roads were caused by traffic accidents. Therefore, improving the maintenance of trucks and equipment, road conditions, and the training of relevant staff must be done to minimize road accidents.

3.5   Physical appearance of the accidents

The physical appearance of 142 accidents was classified into five incidence types: leakage, fire, explosion, poison, and others. Each accident may fall under one or more of these types (Figure 6). For example, an accident might consist of leakage that then causes poisoning, or a release might cause an explosion followed by a fire. As seen in Figure 6, most accidents lead to leakage (74.6%), followed by fires (23.2%) and explosions (18.3%). As a matter of fact, most accidents do start with leakage. The percentage of leakage cases without further events (fire, explosion, poison) is highest at 51.4%, followed by leakage-fire sequences (9.9%), leakage-explosion sequences (6.3%), leakage-poison sequences (4.2%), explosion-fire sequences (4.2%), and leakage-fire-explosion sequences (3.5%).

4   CONCLUSION AND DISCUSSION

Based on the work above, these conclusions can be obtained. (1) Statistical accidents have unique time-phase characteristics. The accident frequencies were higher from March to May, July to August, and October to December, and particularly higher in March and December, which may be related to the characteristics of the seasons. The Hazmat transportation accidents were more frequent at hours 0, 3, 9, 13, 16, 20, 21 and there was significant rise at 9 am.

The reasons may for this be related to fatigued driving, poor traffic conditions, etc. (2) Most of the Hazmat transportation accidents were related to flammableliquids, corrosive substances, and toxic substances. More than half of the accidents were related to flammable liquids. (3) Accidents mainly occurred on highways, provincial roads and national roads. (4) Human factors are the most common causes of accidents. (5) Most accidents lead to Hazmat leakage.

ACKNOWLEDGMENTS

This project was supported by Beijing Natural Science Foundation (9194033), and the National Key R&D Program of China (Grant No. 2018YFF0301005).

REFERENCES

Fabiano, B., Curro, F., Palazzi, E. et al. 2002. A framework for risk assessment and decision-making strategies in dangerous good transportation. *Hazardous Materials* 93:1–15.

Ohtani, H., Kobayashi, M. 2005. Statistical analysis of dangerous goods accidents in Japan. *Safety Science* 43(5/6): 287–297.

Roberto, B., Cinzia, F. and Barbara M. 2000. Risk analysis of LPG transport by road and rail. *Loss Prevention in the Process Industries* 13:27–31.

Lv, H. and Peng, M. 2017. The regularity of road transport accidents of dangerous chemicals in Beijing. *Safety* 38(11):15–18.

Zhao, L.J., Wu, P. and Xu, K. 2009. Statistic analysis and countermeasures on dangerous chemical accidents in China. *China Safety Science Journal* 19(7):165–170.

Xin, C.L., and Wang, J.L. 2012. Review on historical analysis of accident in the transportation of hazardous materials. *China Safety Science Journal* 22(7):89–94.

Yang, J., Li, F.Y., Zhou, J.B. et al. 2010. A survey on hazardous materials accidents during road transport in China from 2000 to 2008. *Journal of Hazardous Materials* 184:647–653.

Liu, M. 2011. *Theory and Method of Accident Risk Analysis.* Beijing: Peking University Press.

Sun, M., Wu, Z.Z., Zhang, H.Y. 2003. Cause analysis of accidents in transporting dangerous chemicals on highway and their preventive measures. *China Safety Science Journal* 13(8):22–24.

Wang, X.L. 2017. Analysis on influencing factors of hazardous chemical materials road transport accidents and safety measures. *Journal of Highway and Transportation Research and Development* 34(10):115–121.

Wu, J.Z., Fan, W.J. 2015. Risk evaluation system of dangerous goods transport. *Journal of Highway and Transportation Research and Development* 32(12):6–11.

Zhao, Y.C. 2016. Study on the construction of evaluation index system of road transportation of dangerous goods. *Logistics Sci-Tech* 39(5):83–86.

Wang, Y.H, Tong, S.J., et al. 2005. Risk Analysis on Road Transport System of Dangerous Chemicals. *China Safety Science Journal* 15(2):8.

Leonelli, P., Bonvicini, S., Spadoni, G. 2000. Hazardous materials transportation: a risk-analysis-based routing methodology. *Journal of Hazardous Materials* 71(1/3): 283–300.

Yuan, X.J., Bi, H.P., et al. 2014. Study on Evolution Mechanism for Hazardous Chemical Leakage Accident. *Industrial Safety and Environmental Protection* 40(2):21–24.

Tang, C.G., Yu, J. 2018. Construction on causal mechanism analysis model of hazardous chemicals disaster accidents. *Guangzhou Chemical Industry* 46(8):140–143.

Classification and code of dangerous goods (GB 6944-2012), Implementation Date: November 1, 2012.

List of dangerous goods (GB 12268-2012), Implementation Date: November 1, 2012.

Review of risk assessment guidelines for LNG ship bunkering

Olga Aneziris, Ioanna Koromila & Zoe Nivolianitou
Institute of Nuclear & Radiological Sciences and Technology, Energy & Safety, National Center of
Scientific Research "Demokritos", Athens, Greece

ABSTRACT: The purpose of this paper is to analyze in detail the existing guidelines issued for assessing the risks appearing during the bunkering process of LNG-fueled ships. Due to the potential of major accidents in the use of LNG affecting both human life and the environment, it is necessary to carry out risk assessment studies both for the storage installation and for the various operations that take place during the bunkering of ships. In this context, several studies have been developed focusing on the establishment of risk assessment guidelines for the safe use of LNG in port bunkering processes. The Seveso Directive 2012/18/EU is among the key legal documents for assessing risk and setting the requirements for safety studies. In addition, there are some International and European standards that propose risk assessment methodologies applied in the field under consideration, while obviously the classification societies and other industrial associations have dealt with the development of relevant guidelines. All the above guidelines are being reported, analyzed and reviewed in this paper.

Keywords: guidelines, LNG ship, risk assessment, Seveso Directive, bunkering

1 INTRODUCTION

The use of liquefied natural gas (LNG) as an alternative marine fuel has increased since 2000 when the first LNG-fueled ship, the passenger ship MV Glutra, was put into operation. This trend is due to the International Maritime Organization's (IMO) demand for implementing strict measures to reduce gas emissions from ships, especially sulfur and nitrogen oxides. This requirement has been established in the framework of the international convention for prevention of pollution from ships, the well-known MARPOL 73/78 (IMO, 1997) demanding the repletion of the conventional fuels by other more environmentally friendly energy solutions; LNG constitutes such a solution.

In order for ships to be fueled by LNG, special facilities should be installed and operated in ports supplying ships with this fuel. LNG is a hazardous material and if it is released in the environment major accident may occur, affecting human life and the environment. Indeed, the very low temperature of LNG constitutes a significant hazard that can affect both materials (i.e. tank walls and ship structure) and people when it comes into contact with it, causing cracks and frostbite, respectively. LNG is flammable and maybe ignited if it coexists with oxygen. An LNG fire can mainly be developed as follows: LNG is released in the environment and forms a pool of boiling LNG, LNG evaporates and produces a vapor cloud. LNG may be ignited either in the pool, creating a pool fire, or in the vapour form creating jet fires, flash fires or explosions.

For addressing such issues, it is necessary to carry out risk assessment studies for both the installation and for the various operations taking place during the bunkering of ships. The present paper aims at providing and discussing the existing framework for risk assessment for storage, handling and bunkering of LNG at ports.

The structure of the current paper is as follows: Section 2 describes the general framework for the use of LNG in maritime operations. Section 3 introduces the legislation applicable to LNG safety in ports and also provides a comprehensive overview of the existing guidelines on risk assessment, while section 4 presents risk assessment approaches of significant interest. Finally, section 5 presents a discussion of the work being performed and section 6 contains the conclusions of this research.

2 LNG BUNKERING AT PORTS

An LNG bunkering process could be divided into three sides: the onshore, the offshore and the ship-shore interface. On the onshore side, namely the port side, the storage of LNG is included. Storage may be either permanent when LNG is provided by a terminal, or temporary when the LNG is stored in portable tanks on land or on trucks. The temporary means of storage are used in popular ports where relatively small quantities of LNG are needed and are far from a terminal. On the offshore side, there exist the receiving ship powered by LNG as well as the buffer, ship which may supply the receiving ship. Last but not least is the ship-shore side which includes the piping facilities. Figure 1 illustrates the bunkering process.

Based on these processes, the bunkering of LNG powered ships could be carried out in any of the following three well-known ways: *ship-to-ship* through the bunkering ship; *truck-to-ship* through a land-based truck; and *tank-to-ship* either through a portable storage tank or direct from the LNG terminal. *Truck-to-ship* seems to be the most common way of bunkering since it operates in the majority of LNG ports; among others this method is applicable to the ports of Amsterdam, Rotterdam, Antwerp, Zeebrugge, Barcelona, Marseille, Singapore, Yokohama, and Vancouver. On the other hand, fewer ports use the *tank-to-ship* and *ship-to-ship* bunkering, such the ports of Stockholm, Rotterdam and Zeebrugge. Significant is that a combination of methods can be used in a port; for instance, the Zeebrugee port provides ships the opportunity to bunker LNG in any of the aforementioned ways.

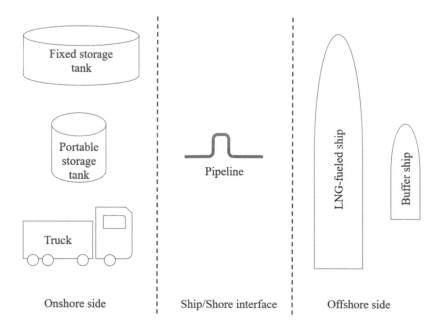

Figure 1.   LNG bunkering process.

3 RISK ASSESSMENT AT PORTS

3.1 *Requirements for risk assessment*

LNG bunkering ports comprise a hazardous installation which may cause major accidents affecting the human life, the environment, the installation structural integrity and the economy, as well. In order to control and prevent such accidents, risk assessment studies should be conducted. In this context, the European Commission has introduced the Directive 2012/18/EC, also known as "Seveso III", establishing mandatory requirements for performing risk assessment on facilities using hazardous substances, including LNG ports (EC, 2012). Seveso III does refer to all lower-tier port installations (i.e. fixed or portable storage tanks) capable of storing at least *50 tons* of LNG, as well as to upper-tier establishments storing over *200 tons* imposing even stricter requirements.

On the offshore side, ships using LNG as fuel should comply with the detailed prescriptive requirements contained in the International Code of Safety for Ship Using Gases or Other Low-flashpoint Fuels (IGF). The IGF code requires a risk assessment to be carried out for ensuring that risks arising from the use of such a fuel affecting persons onboard, the environment, the structural strength or the integrity of the ship are addressed (IMO, 2015).

3.2 *Standardization bodies guidelines*

For enhancing the performance of risk assessment, as required by Seveso Directive and IGF Code, several guidelines have been produced. The International Standardization Organization (ISO) has issued the technical specification ISO/TS 16901 on guidance for risk assessment in the design of onshore LNG installations including the ship/shore interface (ISO, 2015a). This technical specification actually suggests a common approach and guidance for assessing the major hazards emerged to LNG port facilities using risk-based methods to enable a safe design and operation, disregarding the environmental and navigational risks. These guidelines include examples of tolerable levels of risk focusing on the specific needs scenarios and practices within the LNG industry. The assessment of risk to personnel, material, reputation and environment during LNG bunkering operation sat port is guided by the ISO 18683 (ISO, 2015b). Indeed, there are guidelines for developing both qualitative and quantitative risk assessments (QRA), including calculation of safety and security zones, determination of risk matrix, development of risk acceptance criteria and conducting safety report. Moreover, ISO 20519 (ISO, 2017) introduces risk assessment for bunkering, approaches for estimating safety zones and LNG bunker checklists. Guidelines are provided for the risk assessment before a bunkering operation is conducted in order to ascertain whether the situation, regarding the installation, the condition of the LNG-fueled ship, the expected sea state, the marine traffic etc., is acceptable for LNG bunkering. Such an assessment should be performed again when the considered conditions change.

The European Committee for Standardization (CEN) has established standards requiring a risk assessment to be carried out for LNG installations designed and operated within European territory. EN 1473 defines the guidelines for the design, construction and operation of onshore LNG installations storing more than 200 tons LNG and requires the estimation of acceptable levels of risk for life and property outside and inside the boundaries of the establishment (CEN, 2016). For installations with a small storage capacity, between *5 tons* and *200 tons*, the EN 13645 has been issued including risk assessment requirements and examples (CEN, 2002).

The American National Fire Protection Association (NFPA) has introduced the standard for storage and handling of LNG (NFPA 59A, 2019) providing guidance for siting, design, construction, maintenance and operation of facilities that produce, store and handle LNG including a special chapter for small scale LNG facilities and for QRA of LNG plant siting together with an equipment failure rate database.

3.3 *Associations guidelines*

Besides the main standardization organizations, risk assessment guidelines on the safe LNG bunkering and storage have issued by various associations providing significant support to all parties involved. The International Association of Classification Societies (IACS) have issued Rec. 142 guidelines for, among others, providing recommendations on the responsibilities, procedures and equipment required for LNG bunkering operations and setting harmonized minimum basic recommendations for bunkering risk assessment, equipment and operations (IACS, 2017). The bunkering operations risk assessment is undertaken in accordance with ISO/TS 18683. Furthermore, IACS has issued Rec. 146 recommendations on the assessment of risk according to the requirements of the IGF Code on eliminating any adverse effect concerns the ship side (IACS, 2016).

Classification societies have, individually, published several guidelines such as those published by DNV-GL concerning the "Development and operation of liquefied natural gas" (DNV-GL, 2015). These guidelines present safety systems of bunkering facilities, risk assessment methods both qualitative and quantitative, for LNG bunkering facilities and Safety management system requirements. Guidance on requirements for Simultaneous operations during LNG bunkering operations are also addressed in the risk assessment guidelines. ABS has issued the LNG Bunkering Technical and Operational Advisory (ABS, 2017) providing guidance on technical and operational challenges of LNG bunkering for both the bunker vessels and the receiving vessel. The latter provides guidance for safety and risk assessment and includes initiating events which can lead to major accidents.

In addition, the specialized society for gas as a marine fuel (SGMF) has composed a guidance to all parties involved in the bunkering of LNG-fueled ships (SGMF, 2017). It aims to ensure that LNG-fueled ships are refueled with the highest levels of safety, integrity and reliability recognizing that there are potential differences in culture and understanding between suppliers and users that do not exist in the wider LNG transportation industry. Risk assessment approach, identification of hazard areas, reference to safety and security zones and responsibilities, of involved stakeholders, for LNG bunkering operations are discussed. Furthermore, all technical requirements and procedures taking place during pre-bunkering and bunkering phases are mentioned. The society of international gas tanker and terminal operators (SIGTTO) has published guidelines for the design of LNG installations and operations in port areas including, risk assessment (SIGTTO, 2003).

Finally, the European maritime safety agency (EMSA) has recently issued a very useful guidance on LNG bunkering to port authorities and administrations (EMSA, 2018). This is developed in the frame of the implementation of Directive 2014/94/EU respecting to LNG as a marine fuel and it is suggested as complementary to other reference documents. Detailed information is included on the risk assessment performance involving all sides, namely the onshore and offshore side as well as the bunkering interface.

4 RISK ASSESSMENT APPLICATIONS

Quantitative risk assessment methods focusing on the various hazards involved in LNG bunkering operations have been developed by many researchers. Jeong et al. (2017 & 2018) introduce a quantitative method for determining the safe exclusion zone for LNG bunkering procedures. According to this method, all bunkering methods are taken into account along with the bunkering capacity, population and tolerable risk criterion. The following steps have been considered as part of the Integrated Quantitative Risk Assessment (IQRA) method: a) four cases of bunkering capacity are identified, b) the scenario analysis identifying all undesirable scenarios is conducted by employing event tree analysis, c) the frequency of occurrence of each undesirable event is estimated, d) the consequences are calculated tasking into account the liquid release rate, the spreading and evaporation of LNG (i.e. pool fire, flash fire, or explosion) and the evaluation of the effects on human health. Results are plotted using F-N curves and a tolerable risk assessment is performed. Jeong et al.'s work is completed by demonstrating the methodology through a simple application. Among the significant findings is that the total annual time required for bunkering consist a critical factor in frequency analysis, as well as the human presence should be strictly limited within the safety exclusion zone.

Stokes et al. (2013) acknowledged the existence of new risks in the application of new technologies and focused on the need for novel ships using LNG as a fuel. According to this study, the human element consists the most important factor affecting the safety of such emerging designs and operating requirements. Indeed, there is a discussion on the approach of the competence management systems provided to assist the crew and the other personnel involved in all LNG bunkering processes, including potential hazards and corresponding human errors have been made in similar sectors (e.g. the aviation). Fan et al. (2013) introduced the risk of LNG leakage occurring when bunkering operations on pontoons placed in the Yangtze River in China. The undesirable consequences, namely heavy gas dispersion, flash vapor fire, jet fire and pool fire, were discussed. Further consideration was given to acceptable thermal radiation flux limit criteria for the people, pontoon's steel superstructure and storage tank. Parihar et al. (2011) developed a methodology for assessing the consequences of LNG release at deepwater port facilities. The consequences model includes determining the accidental release scenario (e.g. valve failure), the bouncing breach scenarios, the leakage magnitude as well as the resulting thermal hazard and the affected area.

Sun et al. (2017) performed a numerical analysis of hazardous consequence during *ship-to-ship* LNG bunkering using computational fluid dynamics (CFD) models. Different types of hazard, such as vapor dispersion and fire radiation were investigated to analyze the material effectiveness in both the LNG bunker and the cargo vessel. As can be seen, dispersion behavior could be different due to different LNG discharge positions. It is concluded that the complicated accident scenario could be effectively investigated by using dynamic simulations, and it could also provide a reliable assessment on possible mitigation strategies. Finally, Park et al. (2018) examined the LNG leakage dispersion characteristics calculating the safety zone in a *ship-to-ship* LNG bunkering process. In particular, CFD simulations were carried out to estimate specific scenarios where a hypothetical LNG bunker ship refueled two typical large ocean vessels. It has been demonstrated that wind speed and direction, ship geometry, loading conditions and gas leakage rate and duration affect the extent of the safety zones.

5 DISCUSSION

Risk assessment performance at ports storing and operating LNG as marine fuel is demanded for ensuring safety. Table 1 presents the aforementioned guidelines in accordance with the application area. Three application areas, based on safety requirements, are recognized: a) port design and operation, b) LNG-fueled ship bunkering, and c) bunkering operations. As indicated, each guideline focuses on a specific area, with the exception of the guide published by EMSA which constitutes a comprehensive guidance on assessing safety in LNG storage and bunkering.

Based on the requirements of risk assessment, researchers have focused on the use of quantitative methods for calculating hazards, consequences and risks applied to LNG operations at ports. Indeed, consequences models calculating LNG releases as well as safe exclusion zones have been developed. Finally, the most recent studies support and utilize the use of CFD models for assessing consequences in the field of LNG ship bunkering.

Table 1. Guidelines according to the applicable area.

Application area	Guidelines
Port design and operation	ISO 16901, EN 1473, EN 13645, SIGTTO, EMSA
LNG-fueled ship bunkering	ISO 20519, IACS Rec. 142, EMSA
Bunkering operations	ISO 18683, IACS Rec. 146, DNV-GL, ABS, SGMF, EMSA

Table 2. Developed methodology in the selected literature.

Document	Hazard	Method
Jeong et al. (2017)	Pool fire, jet fire, flash fire, explosion	IQRA for estimating six "Safety zones" (*5-200m*)
Jeong et al. (2018)	Pool fire, jet fire, flash fire, explosion	IQRA for estimating seven "Safety zones" (*5-200m*) considering also the population
Fan et al. (2013)	Pool fire	Fire models for estimating "Hazard zones"
Parihar et al. (2011)	Pool fire, distance to LFL	CFD for estimating "Hazard zones"
Sun et al. (2017)	Pool fire	CFD for estimating consequences
Park et al. (2018)	Jet fire, flash fire	CFD for qualitatively (small, medium, large) estimating "Safety zones" for the worst case scenarios in terms
ISO 20519 *(2017)*	Flash fire, distance to LFL	Deterministic approach (*5-250m*) and also Risk-based approach
SGMF (2017)	Pool fire, jet fire, flash fire, explosion, distance to LFL	BASIL model, estimation of distance to LFL

Table 2 presents the methodology developed for either calculating the consequences derived from an LNG release or proposing the acceptable limits of the safety and hazard zones. Many researchers have focused on pool fires, while ISO 20519 presents the major steps to estimate hazard and safety zones.

6 CONCLUSIONS

This paper deals with the need to perform risk assessment studies for ensuring safety of LNG storage and operations at ports as well as of ship bunkering operations. All the relevant guidelines developed in accordance with the requirements of the Seveso Directive and the IGF code are presented. In addition, resent scientific literature on quantitative risk assessment methods are cited and discussed.

AKNOWLEDGEMENTS

The authors gratefully acknowledge the financial support of Adriatic-Ionian Programme INTERREG V-B Transnational 2014-2020. Project #118: SUPER-LNG "SUstainability PERformance of LNG-based maritime mobility".

REFERENCES

ABS. 2017. *LNG Bunkering: Technical and Operational Advisory.*
CEN. 2002. EN 13645: *Installations and equipment for liquefied natural gas. Design of onshore installations with a storage capacity between 5 t and 200 t.*
CEN. 2016. EN 1473: *Installation and equipment for liquefied natural gas. Design of onshore installations.*
EC. 2012. *Directive 2012/18/EU of the European Parliament and of the Council of 4 July 2012 on the control of major-accident hazards involving dangerous substances, amending and subsequently repealing Council Directive 96/82/EC.*
DNV-GL. 2015. *DNVGL-RP-G105: Development and operation of liquefied natural gas bunkering facilities* (Recommended practice).
EMSA. 2018. *Guidance on LNG Bunkering to Port Authorities and Administrations.*
Fan, H., Zhang, H. & Xu, J. 2013. Assessment of the hazard distance of pool fire for LNG bunkering pontoon. *Ship Building of China* 54 (4):186–195.
IACS. 2016. IACS *Rec 146: Risk assessment as required by the IGF Code.*

IACS. 2017. *IACS Rec 142: LNG Bunkering Guidelines.* 2nd Edition.

IMO. 1997. *Resolution MEPC. 75(40) Amendments to the Annex of the Protocol of 1978 relating to the International convention for the prevention of pollution from ships, 1973.* London: IMO.

IMO. 2015. *Resolution MSC. 391(95) Adoption of the International code of safety for ships using gases or other low-flashpoint fuels (IGF code).* London: IMO.

ISO. 2015a. *ISO/TS 16901: Guidance on performing risk assessment in the design of onshore LNG installations including the ship/shore interface.* Geneva: ISO.

ISO. 2015b. *ISO/TS 18683: Guidelines for systems and installations for supply of LNG as fuel to ships.* Geneva: ISO.

ISO. 2017. *ISO 20519: Ships and marine technology – Specification for bunkering of liquefied natural gas fuelled vessels.* Geneva: ISO.

Jeong, B., Lee, B.S., Zhou, P. & Ha, S.M. 2017. Evaluation of safety exclusion zone for LNG bunkering station on LNG-fuelled ships. *Journal of Marine Engineering and Technology* 16 (3):121–144.

Jeong, B., Lee, B.S., Zhou, P. & Ha, S.M. 2018. Determination of safety exclusion zone for LNG bunkering at fuel-supplying point. *Ocean Engineering* 152:113–129.

NFPA 59A. 2019. *Standard for the production, storage, and handling of liquefied natural gas (LNG).*

Parihar, A., Vergara, C. & Clutter, J.K. 2011. Methodology for consequence analysis of LNG releases at deepwater port facilities. *Safety Science* 49 (5):686–694.

Park, S., Jeong, B., Yoon, J.Y. & Paik, J.K. 2018. A study on factors affecting the safety zone in ship-to-ship LNG bunkering. *Ships and Offshore Structures* 13:312–321.

SGMF. 2017. *Gas as a marine fuel, safety guidelines,bunkering.* London: SGMF.

SIGTTO. 2003. *LNG operations in port areas - essential best practices for the industry.*

Stokes, J., Moon, G., Bend, R., Owen, D., Wingate, K. & Waryas, E. 2013. Understanding the human element in LNG bunkering. *ASME/USCG 2013 3rd Workshop on Marine Technology and Standards, MTS 2013,* pp. 105–111.

Sun, B., Guo, K. & Pareek, V.K. 2017. Hazardous consequence dynamic simulation of LNG spill on water for ship-to-ship bunkering. *Process Safety and Environmental Protection* 107:402–413.

Risk Analysis Based on Data and Crisis Response Beyond Knowledge – Huang & Nivolianitou (eds)
© 2020 Taylor & Francis Group, London, ISBN 978-0-367-25146-8

Developing a risk-based method for predicting the severity of fire accidents in road tunnels

Panagiotis Ntzeremes*, Konstantinos Kirytopoulos & Vrassidas Leopoulos
Sector of Industrial Management and Operations Research, School of Mechanical Engineering, National Technical University of Athens, Athens, Greece

ABSTRACT: Fire accidents are considered serious events for tunnel safety since they can cause heavy losses as well as serious damage to the tunnel infrastructure and facilities. Although risk assessment prepares the tunnel system to deal with potential fire accidents, an adequate response by tunnel operators depends also on the information about a particular fire event. This paper proposes a novel risk-based method that supports tunnel operators in assessing the criticality of potential fire events. The structure of the proposed method is as follows. Initially, the factors that determine the criticality of a fire event are featured and the system parameters that affect these factors are identified. Subsequently, the method performs multiple simulations by changing the examined parameters randomly and thus the relation between the factors and the parameters arises. The outcome facilitates tunnel operators to predict promptly the severity of potential fire events and make better-informed decisions.

Keywords: Road tunnel, fire risk, safety, stochastic modeling

1 INTRODUCTION

Undoubtedly, fire safety is a matter of great importance when designing a road tunnel system (PIARC, 2017). The knowledge already gained from previous accidents justifies this point (Barbato et al., 2014). Reports have indicated that fire accidents result in heavy losses of life, personal injuries and destruction of tunnel infrastructure and facilities (Voeltzel & Dix, 2004). In order to come up with better fire safety strategies, tunnel managers focus primarily on the general design of the tunnel system, which relates to the geometrical configuration, for example the avoidance of large slopes or the use of fireproof materials. Apart from the design of the system, different technical safety systems such as mechanical ventilation can contribute to increase the level of safety (PIARC, 1999).

Nevertheless, the biggest challenge is to evaluate the performance of the whole tunnel system in case of an accident involving fire. Therefore, risk assessment has been incorporated in tunnel safety systems over the last fifteen years (Ntzeremes & Kirytopoulos, 2019). Through the systemic approach of risk assessment, the whole tunnel system is examined and the interrelations among its subsystems emerge. As a result, potential deficiencies are identified. The ultimate goal of risk assessment is to assess the tunnel system proactively and, if needed, to enable tunnel managers to select additional measures (Ntzeremes & Kirytopoulos, 2019).

Although the risk management systems in place prepare the tunnel system to confront potential fire accidents, an adequate response by tunnel operators depends also on the information about the particular fire event (Capote et al., 2013). The aim of this paper is to provide a method to aid toward exploiting real-time information during a fire. The proposed

---

*Corresponding author: ntzery@mail.ntua.gr, pntzeremes@gmail.com

method synthesizes the results from fire scenarios that are examined during risk assessment by combining them with the real-time data from the specific fire accident. Thus, the tunnel operator obtains better and faster information about the potential severity of the specific accident.

2 FIRE INCIDENTS AND RISK ASSESSMENT

Fire accidents in the past resulted in devastating effects for both users and tunnel infrastructure and facilities (Beard & Carvel, 2012). These tragedies have come to pinpoint that if a fire accident occurs, the associated consequences can be much higher than elsewhere due to the particularities of the tunnel environment.

Because of their enclosed environment, tunnels are less exposed to extreme weather conditions, e.g. snowfall, that usually constitute a serious source of risk. Furthermore, they usually lack junctions, advertisement signages and pedestrian movement, which are either causal or aggravating factors in case of accidents (Nævestad & Meyer, 2014). However, studies have concluded that the first and foremost reason for the lower accident rate is that users tend to drive more carefully when passing through tunnels (Nævestad & Meyer, 2014; Kirytopoulos et al., 2017).

Despite the aforementioned benefits, the enclosed environment of tunnels is responsible for the high catastrophic impact, especially when fire events occur. The research on tunnel fires have provided analysts with some valuable information regarding their specific characteristics (Ingason et al., 2015). Generally, fires in tunnels differ from open to environment fires in two aspects i.e. the heat feedback from the surroundings and the availability of oxygen for the combustion. Thus, fire evolution is characterized by higher heat release rates. Experiments have shown that the same fire can have a maximum heat release rate of 1.5 to 4 times higher in tunnels than in open air (Beard & Carvel, 2012; Ingason et al., 2015). As a result, the tunnel environment demonstrates a higher air temperature. These temperatures expose users to a large amount of radiant and convective heat, which threaten their lives. Additionally, tunnel fires generate a large amount of irritant smoke that is extremely dangerous when inhaled by the users. The developed aerodynamic disturbances can also push the irritant smoke upstream of the fire location. Hence, users trapped near the fire location are surrounded by toxic smoke while their evacuation is hindered (Ntzeremes & Kirytopoulos, 2018). In addition, rescue teams face difficulties in quickly approaching the accident location and users have to deal with the fire consequences by themselves. Nevertheless, conducted studies have showcased a lack of knowledge in users in behaving appropriately during fire accidents i.e. awareness of using tunnel facilities or delay in moving toward the nearest emergency exit, etc. (Kirytopoulos et al., 2017).

Since the number of tunnels continues to rise, which increasesthe amount of goods and passenger numbers passing through tunnels, fire safety is high in the priority considerations of both policymakers and tunnel managers (Ntzeremes & Kirytopoulos, 2019). As such, risk assessment has been recognized as a valuable tool in order to enhance the fire safety of road tunnels. The reason for its usefulness is related to the peculiarities of tunnel systems. Safety analysts and tunnel managers must be aware of the performance of the whole system during an incident in order to make the best decisions regarding the safety level of a tunnel system. This need is becoming more important since tunnels constitute complex systems. Nowadays, tunnels are the most sophisticated elements of modern road networks. A quick overview of tunnel facilities together with the associated interdependencies highlight this aspect. The existence of complex surveillance systems for monitoring the traffic flow in order to identify potential incidents or the use of advanced technological systems for mitigating fire consequences such as mechanical ventilation or a fixed firefighting system, are some indicative examples (Ntzeremes & Kirytopoulos, 2019).

The complexityof tunnel systems is further increased due to the socio-technical attributes of these infrastructures. Relevant studies on this issue have indicated that ensuring an acceptable level of safety depends not only on the accurate operation of the technical equipment but also on designing a properorganizational structure as well asexamining in detail the human involvement. The hazards posed by the last two factors can exacerbate the fire consequences despite the prompt operation of the technical installations (Kazaras et al., 2012).

Finally, road tunnels are normally part of motorways and urban highways. As a consequence, potential critical events like fires, apart from the serious threat to the elements of the tunnel system itself (namely the users, the facilities, the infrastructure, the vehicles and the traffic), can also interrupt the proper operation of the wider road network for an extended time. Thus, road tunnels are regarded as critical infrastructures (Ntzeremes & Kirytopoulos, 2019).

Prior to the presentation of the risk assessment framework, it is important to define clearly the term risk. In general, there is no agreed definition of risk. The glossary of the Society for Risk Analysis provides a sum of six qualitative definitions and six metrics (Aven et al., 2018). Regarding the road tunnel area, risk is based on the identified scenarios. Therefore, risk follows the definition of Kaplan and Garrick (1981) in which itis defined as a set of scenarios s_i, each of which has a scenario-occurrence probability p_i and a scenario-occurrence consequence c_i. The consequence factor c_i refers to the losses and injuries among users. The total risk can be viewed as a multidimensional quantity that includes the pairings of all the possible scenarios n through the following equation:

$$\text{Risk} = [(p_1 * c_1), \ldots, (p_n * c_n)] \tag{1}$$

The risk assessment framework for road tunnels consists of three discrete phases following a top to bottom sequence. The first phase is the risk analysis that includes the system definition, the hazard identification and the risk estimation. Subsequently, the risk evaluation phase follows in which the estimated risks are evaluated based on the established risk criteria. If the evaluation finds the risk out-of-tolerance, the risk treatment phase is responsible for the selection of additional measures. Finally, the tunnel system is re-assessed in order to estimate the impact of additional measures on the tunnel level of safety.

Adopting the aforementioned principles, risk assessmentis aimed at estimating the fire risksfollowing a scenario analysis. In general, the structure of the scenario analysis consists of two distinct factors, which are: the scenario development and the associated consequence analysis (Ayyub & McCuen, 2011). However, risk assessment methods adopt different approaches depending on the type of the transported goods that are involved in the fire scenario. According to the agreement for transport of dangerous goods by road (UNECE, 2015), goods are separated between dangerous and non-dangerous goods. Fire risk derives from accidents involving dangerous goodsand directly impact a significant numberof users. The extremely high heat release rate or the possibility of explosion indicate an increased fatality rate. Therefore, this type of risk is recognized as a societal risk. To address theserisks, risk assessment estimates the level of safety of the tunnel, which is expressed through frequency/number of fatalities curves and/or expected value, If the risk is unacceptable, the safety analyst'sconcernis primarily to employ additional measures in order to raise barriers that can prevent the spark of the fire and secondly mitigate the consequences.

In contrast, fire accidents without the involvement of dangerous goods result in fires that can provide an opportunity to the trappedusers to rescue themselves if they behave appropriately and if the response of the tunnel system is adequate (Ntzeremes & Kirytopoulos, 2018). Therefore, this type of risk is recognized as individual risk. To address these risks, risk assessment methods estimate the level of safety of the tunnel, which is expressed in terms of losses associated with the examined scenario. If the risk is unacceptable, the safety analyst selects additional measures for protecting users and mitigating the losses among them despite the spark of the fire. Another reason that leads analysts to focus on protective measures is the regulative requirements of the countries. The lackof accident databases and the need for establishing a common safety strategyhave led the countries to impose in their regulations specific fire scenarios in which tunnel systems should be examined.

This approach unavoidably directs risk assessment to estimate the risk level of a tunnel by taking into account only the consequence part.

3 PROPOSED METHOD

3.1 *Predicting the severity of potential fire events*

In this section a novel risk-based method is proposed that supports tunnel operators in assessing the criticality of fire events. Following a scenario-based approach, the aim of the proposed method is to provide tunnel operators with the required data about the potential severity of a fire event. To do so, the method acts in conjunction with the existing tunnel facilities in order to receive real-time data. As a result, the control room of the tunnel can be informed promptlyabout the severity of the potential fire event and thus can make better-informed and real-time decisions.

It should be stated that the proposed method focuses primarily on predicting the severity of fires without the involvement of dangerous goods becausewhen dangerous goods are not present, the decisions of the tunnel operator can contribute even more to the successful self-evacuation process of the users being trapped near the fire location. Ultimately, the tunnel operator is the one that notifies the users, manages the operation of the mechanical ventilation and informs the authorities about the criticality of the emergency situation, until the emergency services take control.

3.2 *Process application*

The structure of the method, depicted in Figure 1, consists of two layers and follows a top to bottom process sequence. Each of these layers includes a certain number of discrete steps. The

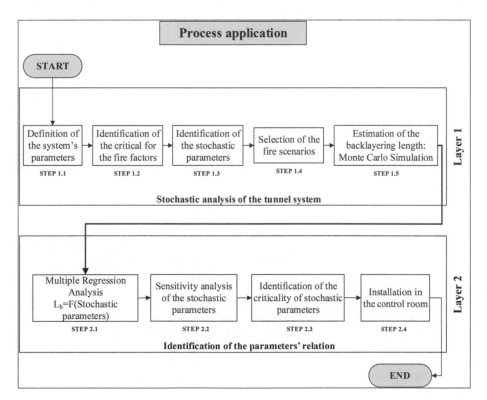

Figure 1. Process application of the proposed method.

final step of the process refers to the potential of its installation as a supplementary system in the control room of the tunnel.

3.2.1 *Stochastic analysis of the tunnel system*

The first layer of the process (**Layer 1**) deals with the stochastic analysis of the tunnel system. Unavoidably, each tunnel system has its unique particularities, which could be its type (rural or urban) associated with the traffic condition, its structural characteristics, etc. (Ntzeremes & Kirytopoulos, 2018). Thus, the first step (**Step 1.1**) of this layer aims at examining the system in order to define the system's parameters.

Having defined the parameters of the tunnel system, the factors that influence the fire behavior should be identified (**Step 1.2**). The term factor is employed in order to indicate not the parameters but the groups of parameters together with their interrelations. Considering that the tunnel system is in a state of readiness to address any potential accidents, the technical installationsof the system are supposed to respond without failure regardlessof the accident magnitude. For the same reason, the tunnel personnel should react adequately. Hence, the factors that affect the evolution of fire are included in the traffic and the environmental conditions of the tunnel.

Step 1.3 deals with the identification of those parameters included in the aforementioned factors and which influence the fire behavior. Regarding the traffic conditions, there are two parameters that are crucial, the annual average daily traffic expressed in vehicles per hour and the percentage of heavy good vehicles in the traffic volume. The significance of these parameters accounts for their influence on the piston effect. Piston effect refers to the force-dairflow in the tunnel and practically is the only deterring force towardback layering till the mechanical ventilation is fully activated (Ingason et al., 2015). Its importance is particularly significant in case of unfavorable differencesin pressure between the tunnel portals. The magnitude of the piston effect plays a crucial role inwhether the backlayering develops and, if so, in its length. As far as the environmental conditions are concerned, the difference in pressure between tunnel portals and the ambient temperature are identified. The first one due to its influence on the backlayering while the second one affects the air temperature of the tunnel air flow. Since all these parameters are subjected to changes, they should be treated stochastically.

The fire behavior depends on the selected fire scenario. The fire scenario determines the vehicles that take part in the accident and thus thematerials and loads to be burned. Various standardized fire scenarios are imposed by the countries' regulations. As a result, there is an extended variety of standardized heat release rates, e.g. a vehicle fire usually has 8 to 10 MW maximum heat release rate while fires from lorries or heavy good vehicles may go from 50 to 150 MW. Since the method follows the scenario-based approach, the set of fire scenarios should be defined (**Step 1.4**). In order to select these scenarios, the unique characteristics of the risk assessment methods as well as the country's normative provisions should be taken into account.

The last step of this layer focuses on the estimation of the criticality of the backlayering (**Step 1.5**). In this method, the criticality of the backlayering is established based on two aspects i.e. its length and in its duration. Based on the Monte Carlo simulation, the proposed method performs, as depicted in Figure 2, multiple simulations employing a one-dimensional computational fluid dynamics software by changing the examined parameters (refer to **Step 1.3**) randomly (their density functions are known) and for each fire scenario separately (refer to **Step 1.4**).

The approach followed in order to estimate the criticality of the backlayering is presented in more detail in Figure 3. Based on the unique characteristics of the examined tunnel system, the analyst has to definesome checkpoints, which are illustrated as black dots in Figure 3. Subsequently, in each iteration the derived data from the computational fluid dynamics software are checked to see whether they are into the backlayering region. At the end of the process, a table for each fire scenario is derived that includes the information regarding the backlayering.

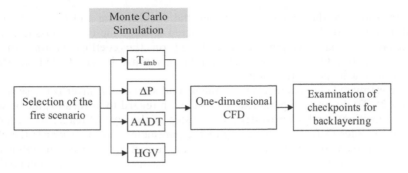

Figure 2.   The process for examining the criticality of backlayering.

Equation 2 shows the utility of these tables. Let L_b^k be the sample of backlayering observations on the *k-th* fire scenario. Additionally, n and j represent the checkpoints, like the four black dots in Figure 3, and the iterations of the Monte Carlo Simulation, respectively. Then, by examining the columns of each sample, the percentage of the scenarios out of the total number of simulations that develop backlayering is estimated.

$$L_b^k = \begin{bmatrix} L_{b11} & \cdots & L_{b1j} \\ \vdots & \ddots & \vdots \\ L_{bn1} & \cdots & L_{bnj} \end{bmatrix}, \qquad L_{bnj} = \begin{cases} 1, & \textit{backlayring detected} \\ 0, & \textit{no backlayering detected} \end{cases} \tag{2}$$

By examining the rows, the severity of backlayering is evaluated. For instance, in Figure 3 the analyst selects the four depicted checkpoints ($n = 4$) that correspond to the rows of the first repetition. If all points are in the backlayering region then the severity of fire consequences is considered significant. However, if for the checkpoints (x_i, t_1) and (x_i, t_i) no backlayering is detected, then the length is considered lower and the severity decreases. In addition, if only the point (x_i, t_1) is found in the region, then the mechanical ventilation is considered to address backlayering and the impact is related with the selection of t_i.

3.2.2 *Identification of the parameters' relation*
Having estimated the tables of the backlayering for all the selected fire scenarios, the second layer of the process aims at identifying the relation among the parameters and the backlayering (**Layer 2**). To do so, a multiple regression analysis is conducted in

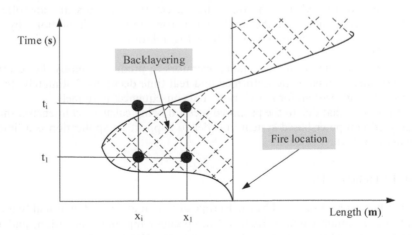

Figure 3.   Identification of points of interest.

Step 2.1. In doing so, the correlation between the backlayering and the stochastic parameters is discovered. If the correlation is considered strong, the characteristic curve of the binary variable of backlayeringis created based on the coefficients that the multiple regression estimates. Otherwise, further simulations must be conducted in Step 1.5 in order to achieve a higher correlation.

Using the estimated characteristic curve, the second step of this layer is the sensitivity analysis (**Step 2.2**). To do so, each parameter changes with regard to its mean value between -50% and +50%. Sensitivity analysis provides the next step of the layer (**Step 2.3**) with the required information in order to find the criticality of the parameters, in other words the effect their variation has on backlayering fluctuation. Thus, the tunnel operator can be informedin real-time if the system conditions would aggravate the consequences in case of potential fire accidents. The real-time parameters refer to both the traffic conditions (real-time vehicles per hour and percentage of heavy good vehicles) and the environmental conditions (real-time difference of pressure at tunnel portals and ambient temperature).

Having the required formulas and data, the last step of this layer (**Step 2.4**) indicates the potential to design andinstall a supplementary computer-based system in the control room of the tunnel.

4 CONCLUSIONS

Road tunnels constitute a key element of the road transportation system. Therefore, they must ensure an acceptable level of safety for the public. Fire accidents particularly are the foremost critical events since they can cause heavy losses, serious damages to the tunnel infrastructure and facilities, and interrupt network's operation for a lengthy duration.

Therefore, risk assessment has been employed by tunnel managers and safety analysts in order to enhance the preparedness of road tunnel systems against fire accidents. Through risk assessment, the overall performance of the tunnel system in case of fire is examined and additional safety measures are implemented, if needed. Although risk management systems prepare the tunnel system to deal with potential fire accidents, an adequate response by tunnel operators depends also on the information about the particular fire event. However, existing methods do not deal with this problem in a rigorous way and there are no research outputs related to fine-tuning such decisions.

This paper proposes a novel risk-based method that supports tunnel operators by considering the current state of the system to assume what the criticality of an accident would be, if it happens. The structure of the method is as follows. Initially, the factors that determine the criticality of a fire event are featured. Subsequently, the system's parameters are investigated and the ones that affect these factors are identified. Based on the Monte Carlo approach, the method performs multiple simulations by changing the examined parameters randomly and thus the relation between the factors and the parameters arises.

As a result, the outcome facilitates tunnel operators to predict promptly the severity of the fire event and thus to make better-informed and real-time decisions. Ultimately, the method can work as a warning system for a fire's potential consequences in tunnels.

It is worth to note that due to the proposed method being calculation intensive, the absence of an integrated computer-based system in order to support its application is a limitation as well as an area for further research.

ACKNOWLEDGMENTS

This research has been co-financed by the European Union and Greek national funds through the Operational Program Competitiveness, Entrepreneurship and Innovation, under the call RESEARCH – CREATE – INNOVATE (project code:T1EDK-02374).

REFERENCES

Aven, T., Ben-Haim, Y., Andersen, B.H., Cox, T. et al. 2018. Society for Risk Analysis: Glossary, Herndon, VA: SRA.

Ayyub, B. & McCuen, R., 2011. *Probability, Statistics, and Reliability for Engineers and Scientists.* 3rd ed. New York: CRC Press.

Barbato, L., Cascetta, F., Musto, M. & Rotondo, G., 2014. Fire safety investigation for road tunnel ventilation systems – An overview. *Tunnelling and Underground Space Technology* 43: 253–265.

Beard, A. & Carvel, R., 2012. *Road Tunnel Fire Safety.* 2nd ed. London: Thomas Telford.

Capote, J., Alvear, D., Abreu, A., Cuesta, A. & Alonso, V., 2013. A real-time stochastic evacuation model for road tunnels. *Safety Science* 52: 73–80.

Ingason, H., Li, Y.Z. & Lönnermark, A., 2015. *Tunnel Fire Dynamics.* 1st ed. New York: Springer.

Kazaras, K., Kyritopoulos, K. & Rentizelas, A., 2012. Introducing the STAMP method in road tunnel safety assessment. *Safety Science* 50: 1806–1817.

Kirytopoulos, K., Kazaras, K., Papapavlou, P, Ntzeremes, P. & Tatsiopoulos, I., 2017. Exploring driving habits and safety critical behavioural intentions among road tunnel users: A questionnaire survey in Greece. *Tunnelling and Underground Space Technology* 63: 244–251.

Nævestad, T. O. & Meyer, S., 2014. A survey of vehicle fires in Norwegian road tunnels 2008–2011. *Tunnelling and Underground Space Technology* 41: 104–112.

Ntzeremes, P. & Kirytopoulos, K., 2018. Applying a stochastic-based approach for developing a quantitative risk assessment method on the fire safety of underground road tunnels. *Tunnelling and Underground Space Technology* 81: 619–631.

Ntzeremes, P. & Kirytopoulos, K., 2019. Evaluating the role of risk assessment for road tunnel fire safety: A comparative review within the EU. *Journal of Traffic and Transportation Engineering,* 6 (3): 282–296.

PIARC, 1999. *Fire and Smoke control in Road Tunnels,* Paris: World Road Association.

PIARC, 2017. *Design Fire Characterisics for Road Tunnels,* Paris: World Road Association.

UNECE, 2015. *Agreement of Dangerous goods in Roads (ADR),* Geneva: United Nations Economic Commission for Europe.

Voeltzel, A. & Dix, A., 2004. A comparative analysis of the Mont Blanc, Tauern and St. Gotthard tunnel fires. *Routes/Roads magazine, PIARC Virtual Library,* pp. 18–34.

Risk Analysis Based on Data and Crisis Response Beyond Knowledge – Huang & Nivolianitou (eds)
© 2020 Taylor & Francis Group, London, ISBN 978-0-367-25146-8

Analysis and prevention countermeasures of safety accidents in catering industry based on situational factors

Yonghua Han*, Xuewei Ji, Fucai Yu, Yi Zhou & Yan Liu
Beijing Academy of Safety Science and Technology, Beijing, China

Chi Zhang
Anrun International Insurance Brokers co. LTD, Beijing, China

Tao Chen
Institute for Public Safety Research, Tsinghua University, Beijing, China

ABSTRACT: In order to understand the accidents in the production process of catering workers, the description data of the process of work in the catering industry report of liability insurance of work safety are collected. A scenario of work safety accidents is constructed. Based on the model, scenario elements in each record data are extracted. Injured persons, types of injuries, related devices, and related media are classified. The occurrence frequency of injured persons, injured parts, injury forms, related devices, related media and the accident site are statistically analyzed. In addition, the network correlation analysis is carried out for the risk scenario elements of customers and employees respectively. The results show that customers are mainly in the stairwell, restaurant dining area and toilet falls, bruises and other bodily injuries out of control injuries. And employees are mainly injured by cutting, scratching, pressing and other equipment in the kitchen operating space. Finally, the possible causes and scenarios of different injury types were analyzed, and corresponding accident prevention suggestions were proposed.

Keywords: catering industry, elements of a scenario, industrial injury, measures.

1 INTRODUCTION

The catering industry is a food production and operation industry that provides a variety of drinks, food, consumption places and facilities to consumers through instant processing, commercial sales and service labor (National Bureau of Statistics of China,2017). The catering industry is closely related to everyone. According to the National Bureau of Statistics of China (National Bureau of Statistics of China,2018), China's catering industry revenue in 2017 exceeded 3.9 trillion RMB, and the number of employees exceeded 2 million. At present, research on the safety of the catering industry focuses on food safety (Han et al., 2018; Liu et al., 2018; Zhang et al. 2017) and health research (Xu et al., 2016; Han et al., 2017), in addition to fire (Wei et al., 2009; Yao, 2018), explosion (Luo and Fu, 2018), poisoning and asphyxia (Song et al. 2009). There is less concern about the safety of general accidents of personal injury during production.

By literature research, it is found that the research on work safety of the catering industry mainly carries out research from two ways. One is to study the behavior of employees in the form of follow-up surveys or questionnaires. For example, D. Gleeson (Gleeson, 2001) studied the

---

*Corresponding author: hyh19891102@163.com

incidence, nature, and causes of injuries and illnesses in humans based on a 10-month survey of 315 students at a food school in Ireland. Sun-Jung Shin et al (Shin et al., 2017) conducted a questionnaire survey of chefs at Daegu School in South Korea, and analyzed the characteristics of the investigators, the frequency of accidents, and safety-related behaviors based on 307 questionnaires. Young-Woong Song (Song et al., 2018) based on the survey data of 321 people, quantitative analysis of kitchen risk factors identification and prevention strategies. The other is to construct a framework for evaluation analysis from a macro level, such as Chen Qiang and Wan Ki Chow (Chen and Chow, 2007) to discuss occupational health issues in China's catering industry. The first research method is highly targeted, and the research content is more concrete. It can conduct research on the production safety of a certain type of specific place and group. Due to the difficulty of tracking research, the workload is very large, and the number of samples obtained is small. The versatility of the study conclusions is not obvious. The second type of research belongs to the management system or operation method based on the analysis. The reliability of the research results depends on the depth of the researchers' perception of the industry, and the research conclusion lacks the support of necessary data.

Since 2015, B City has piloted the promotion of liability insurance of work safety (Beijing emergency management bureau, 2015), and implemented it in the whole industry in July 2017. As of September 2018, there have been more than 56,000 insurance policy data, and more than 20,000 insurance policies in the catering industry. According to the statistics of insurance companies, there were more than 3,100 claims records, and the catering industry recorded the most, with 1,218 records. The claim records described the simple process of accidents. The data covers various sizes and types of catering in a city. The data is authentic and reliable, and has a wide range of versatility. Based on the text description of the accident, this paper conducts information mining, constructs the scenario model of work safety accidents, and proposes prevention measures for production safety accidents in the catering industry.

2 SCENARIO FACTOR ANALYSIS METHOD OF RISK

We regard the occurrence of production safety accidents as a process. When someone enters a certain space, certain behaviors may occur, which may involve certain related things, and eventually cause physical injury or loss of property. In this process, the state of space may have some influence on persons and things, and there may be some interaction between persons. Based on this point of view, we regard person as the research subject, and the uninjured person enters a certain situation and becomes the injured person. The accident scenario model is shown in Figure 1. According to Figure 1, the relevant scene elements of the accident are

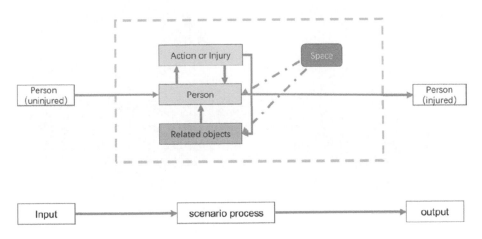

Figure 1.   Work safety accident scenario model.

extracted for statistical analysis. The statistical elements include: type of injured person, injured body part, type of injury, related objects, and space. Considering the lack of information in the claims record, when a scene element is extracted, if a certain element is missing from a record, it is recorded as a null value.

3   STATISTICS ON THE FACTORS OF WORK SAFETY ACCIDENTS IN CATERING INDUSTRY

The factors of the 1218 insurance records of the catering industry were extracted and the frequency of insurance was counted. The category of injured people can be classified into two categories, employees and customers. According to statistics, the ratio of the two risks is about 4:1, as shown in Figure 2a. The injured part is related to the protective measures taken. The statistical results of the injured part are shown in Figure 2b. The hand injury accounts for more than half of all injured parts. According to the text description, there are 1067 records that can determine the type of injury, and up to 39 forms of injury. The form of injury described by the text is related to the form of injury and has certain behavioral characteristics, which are used to characterize behavior in the text (such as cut, with the directivity of the action.) The damage is divided into 4 categories according to the energy or materials form that causes the damage. According to the energy or material form of the injury, the injury is divided into 4 categories. First, the loss of the person's loss of control over the balance of the body (referred to as the loss of control) may be caused by changes in the potential or kinetic energy of the person due to problems in the space. Second, the injury caused by the operation of certain equipment or equipment by a certain equipment (referred to as equipment injury). Third, the direct cause of injury is a certain medium, such as hot water, hot oil, etc. Fourth, other quantities are less likely to be attributed to the above three types of injuries based on available information. The statistics of the four types of injuries are shown in Table 1. There are four types of risk mentioned in the record, namely kitchen (224), restaurant dining area (26), stairs and elevator (41), toilet (14) (the number in brackets indicates the number of times recorded in the risk record). There are 32 ground slips mentioned in the state description of the space. For the risk related objects can be divided into two categories, one for the corresponding media damage, and one for the corresponding device injury. The related media are divided into three categories, and the statistical results are shown in Table 2. The kitchen equipment is divided into six categories, and the statistical results are shown in Table 3.

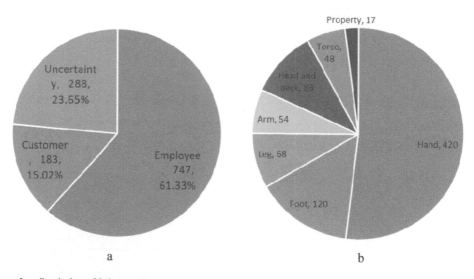

Figure 2.   Statistics of injury types and injury sites of personnel in the catering industry.

Table 1. Classified statistics of injury forms of safety accidents in catering industry.

Type of Injury	Quantity	Text Descriptions
Out of Control Injury	439	Fall (360), Bruise (24), Bruise (23), Sprain (19), Bruise (12), Contusion (1)
Instrument Injury	376	Cut (105), Scratch (76), Crush (46), Cut (38), Bruise (38), Puncture (16), Bruise (15), Pinch (13), Whip (10), Scratch (4), Bruise (4), Crush (3), Strain (3), Saw Injury (2), Bruise (1), Wound (1), Cut (1)
Media Injury	220	Scalds (172), Fire Burns (17), Poisoning (17), Streaks (4), Infections (3), Explosions (2), Electric Shocks (2), Injuries (2), Earthquakes (1)
Other Injury	32	Material Damage (20), Fainting (4), Scratch (3), Disease (2), Sudden Death (1), Disappearance (1), Sting (1)

Table 2. Statistics of media related to safety accidents and injuries in catering industry.

Medium Form	Quantity	Specific Description
Liquid	78	Hot Water (30), Hot Soup (15), Steam (2), Oil (27), Soy Milk (2), Alkaline Water (1), Disinfectant (1)
Gas	21	Coal gas (11), Carbon Monoxide (9), Carbon Dioxide (1)
Others	9	Fire(2)☐Fish Thorn (4)☐Cats(3)

Table 3. Statistics of equipment and instruments related to safety accidents and injuries in catering industry.

Equipment Type	Quantity	Related equipment
Electrical And Special Equipment	69	Pressing Machine (43), Dough Mixer (8), Meat Grinder (8), Laminating Machine (1), Oven (1), Meat Machine (1), Slicer (1), Ice Crusher (1), Pressure Cooker (2), Mixer (2), Boiler (1)
Cooking Utensils	58	Knife (48), Vegetable Pier (1), Oil Pan (6), Pot (3)
Kitchen Attached	28	Door (13), Refrigerator (5), Cabinet (2), Circuit (2), Flue (2), Stove (2), Stove Fire (1), Exhaust Fan (1)
Dining Utensils	57	Tableware (37), Hot Pot (2), Serial Sign (1), Table and Chair (6), Kettle (3), Bottle (8)
Parts	27	Glass (21), Nail (2), Iron Piece (1), Iron Plate (1), Toilet Cover (1), Diaphragm (1)
Other Subsidiary	16	Ladder (4), Mop (1), Shelf (1), Saw (1), Radiator (1), Fan (5), Tram (1), Fire Extinguisher (1), Iron Frame (1)

4 RELATIONSHIP BETWEEN THE SCENE FACTORS OF THE CATERING INDUSTRY

Through the statistics of the work safety accident scene elements of the catering industry, we found that the amount of injury, the injured person, and the injured part are more information corresponding to each record, but the amount of data of other scene elements is small. There is a large amount of information missing. Under such data conditions, it is not possible to establish a complete incident scenario description for each record. Therefore, we use the existing scenario elements and use network analysis methods to establish relationships around the types of injuries. For the same record, when two types of statistical elements appear at the same time, they are counted once on the two-node connection line. The more the two nodes count together, the stronger the correlation between the two

4.1 Type of injury and body parts

We choose the three types of scene elements of injury type, victim and injured part, as shown in Figure 3. The larger the node is, the more records are represented. The number on the line indicates the number of times the two nodes appear in the same record. The figure shows the main scenarios of the type of injury, the type of injury and the part of the body. The customer is mainly affected by the loss of control over the body balance (104), followed by the media injury (36), again the device injury (24) and other injuries (13), the injured part is more evenly distributed, including the hand (19), leg (18), head and neck (16), trunk (11), arm (6), foot (7) and property loss (16). (The number in this section indicates the number of times the node and the associated node co-occur in the same record). The main forms of injury for employees are mainly equipment injuries (314), followed by uncontrolled injuries (193) and media injuries (124), and again for other injuries (10). The injured employees are mainly concentrated in the hands (344), followed by It is a foot (79), an arm (36), a head and neck (42), a leg (36), and a torso (22)

4.2 Customer injury scenario

Because the customer and employee's form of injury is quite different, the customer and employee are analyzed separately. Focusing on the type of injury, we analyze the correlation between the location, equipment, media and damage type. The correlation results are shown in Figure 4. The space associated with the customer's uncontrolled injury is stairs (10), restaurant (8) and toilet (5). The devices related to the customer's uncontrolled injury are mainly dining utensils (4), kitchen accessories (1), other accessories (1) and parts (1). The medium associated with the customer's uncontrolled injury is primarily liquid (2). The space related to the media damage caused by the customer in the record is mainly the restaurant (3), the related equipment is mainly the dining utensils (3), and the related medium includes the liquid (11) and the gas (9). The space related to instrument damage is mainly the toilet (2), and related equipment includes dining utensils (6), parts (4), and kitchen accessories (3). There are fewer scenarios for other injuries, and the relevant media are liquid (1) and other physical states (1).

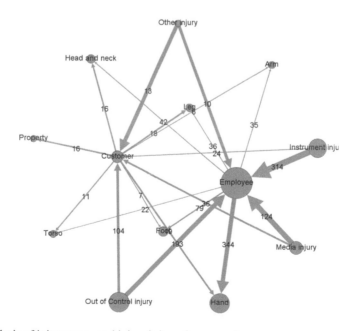

Figure 3.   Analysis of injury types and injured sites of personnel.

530

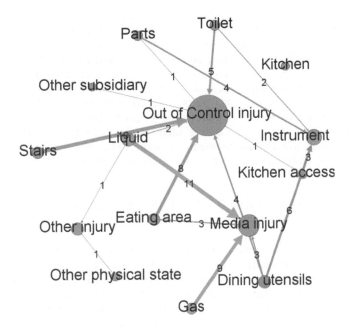

Figure 4. The relationship between customer injury and related factors.

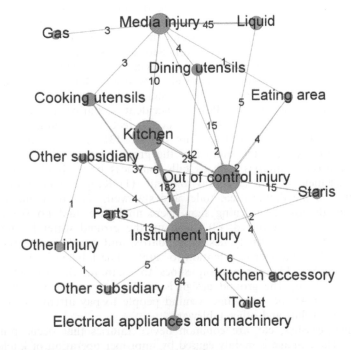

Figure 5. The relationship between injuries and related factors.

4.3 *Employee injury scenario*

The employee is the owner of the catering business unit. The scene of their injury is more complicated than the scene of customer injury. The contextual relationship of the damage form is shown in Figure 5. The main space for employees to be harmed by the equipment is the kitchen (182). Other places are less injured, specifically the restaurant

531

(2), the stairs (2) and the toilet (1). The equipment related to the injury of the equipment is mainly electrical appliances and machinery (64) and cooking utensils (37), followed by parts (13), dining utensils (12), kitchen accessories (6) and other affiliates (4). The main space for employees to be out of control injury is the kitchen (23), followed by stairs (15), restaurants (4), related equipment and facilities (15), other accessories (6), cooking utensils (5), kitchen accessories (4), parts and components (1). The space in which employees are exposed to media is mainly the kitchen (10), followed by the restaurant (1). The equipment involved is the dining utensils (4), cooking utensils (3) and kitchen accessories (2), which are related to media damage. The medium is mainly liquid (46), followed by gas (3). Other damage type association information is less.

5 ANALYSIS OF PRODUCTION DAMAGE IN CATERING INDUSTRY AND COUNTERMEASURES

Through the statistics of the scene factors of the production safety accidents in the catering industry and the correlation analysis of the accident scenes of customers and employees, we know that the production accidents in the catering industry mainly occur in the work process of the catering industry employees. And at the same time as consumers (customers) in the catering industry. There is also a certain proportion of risk during the meal. Therefore, on the one hand, the business units of the catering industry should pay attention to the safety of the employees of the unit, on the other hand, they should protect the safety of the participants of their business activities.

Physically controlled injuries such as falls are the most common type of injury in the catering industry. Physical loss of control is the result of loss of control over one's own balance, and the unfavorable environment and the unsafe behavior of the person. In the process of taking risks, a person has insufficient understanding of the dangers of the environment, failing to take precautions and making unsafe behaviors. Customer injury mainly occurred in the restaurant stairwell, restaurant dining area and toilet. The employee injuries mainly occurred in the kitchen and stairwell. Both types of injuries were related to the eating utensils. Possible accident scenarios are as follows. (1) The staff did not notice the danger of falling under the feet when holding the tableware. (2) The staff stepped on the stairs and fell into the air. (3) The space on the ground was slippery, the friction between the shoes and the ground was small, slipping and falling. To prevent uncontrolled injuries such as falls, business units should do a good job in the safety management of the space environment. (1) Keep the floor dry and level as much as possible in the dining area, toilet and stairwell. (2) Safety guidance should be carried out through obvious warning signs, personnel care, and notification of danger when safety cannot be guaranteed. (3) Clean up the ground water in time. (4) Ensure that the stair steps meet the safety design standards, and clearly notice the safety warning signs to remind people to pay attention to safety. The kitchen is the main workplace for employees. (5) Should wear non-slip shoes to clean up the ground oil stains and water stains in time. (6) The ground design is as flat as possible to reduce the embarrassing situation. (7) At the same time, remind people to pay attention to the feet while holding the dishes, dishes and other tableware.

Cuts, scratches, crushes, etc. are common device injuries that occur primarily during employee work. The damage is mainly caused by improper operation of kitchen equipment such as cooking utensils such as knives and pots, and mechanical equipment such as noodle pressing machines and meat grinders. Kitchen appliances are mainly hand-held devices and therefore have the most hand injuries. The kitchen work is done on the console, the platform has a certain height, and the instruments and items are dropped to easily hurt the feet. Possible causes of equipment damage are: employee negligence at work; employees are not skilled in equipment operation; equipment itself has safety hazards; working space is small; working space lighting conditions are poor; and many more. Therefore, the catering business unit should standardize the working space, enable employees to carry out a more comfortable working environment, train the professional skills of employees, achieve human-machine matching, cultivate employees' safety awareness, avoid overload work, use intrinsically safe equipment, and provide individual protection equipment.

Scald, fire burns, poisoning, etc. are common media injuries. The scald is mainly caused by high temperature liquid medium such as hot soup, hot oil and hot water. The liquid medium is fluid. The person may be in an unsafe position due to unsafe movements of the person, causing the medium to flow out, causing burns to the person. According to the analysis of the situational elements, appliances related to media damage include cooking utensils and dining utensils. Scalds are related to the crowdedness of the space, the slippery surface and the smoothness of the ground. The possible scenarios of scald injury are as followed. Employees operate cooking utensils such as wok and cause hot oil and hot soup to scald themselves. Employees cause burns to customers when they send hot soup to customers. Employees scald themselves when they sent hot soup. To prevent liquid media damage such as scald, the catering business unit should avoid overcrowding in the kitchen operation space, restaurant dining area, etc., reminding employees and customers to pay attention to safety when delivering hot medium tableware. Most of the poisoning is gas poisoning, mainly caused by carbon monoxide and gas. Compared with burns, the frequency of poisoning is low, but it is easy to cause group death. Gas leaks in kitchen operating rooms cause employee poisoning. When customers use charcoal hot pot, they produce carbon monoxide, causing poisoning. To prevent gas poisoning, catering business units should strengthen gas cylinder management, timely detect leaks, and reduce the use of toxic gas substances such as charcoal. Burns are mainly caused by open flames. Catering business units should strengthen the use of fire safety management.

6 CONCLUSIONS AND SUGGESTIONS

Based on the analysis of the scenarios of the catering industry business units, we analyzed the production damage of the catering industry and put forward targeted recommendations.

(1) Personnel injuries in the production process of the catering industry include employee injuries and customer injuries. In the production process, the catering business unit must do a good job of employee safety protection, on the other hand, it should ensure the safety of the dining customers.
(2) The catering industry is one of the major industries in the city, covering a wide range and high risk of shooting. Catering business units should actively purchase safety production liability insurance as a risk transfer measure for production risks.
(3) The frequency of media damage caused by uncontrolled injury, cut and other equipment injuries, scalds, etc. is very high, and the catering business unit should strengthen the corresponding damage prevention during the operation process.
(4) The safety management of the space of the catering business unit should be strengthened, including the operating space of the kitchen, the dining area of the restaurant, the stairwell, etc., to avoid crowding, slippery floors and unevenness.
(5) Intrinsically safe equipment should be used to reduce the use of unsafe substances, strengthen employee operational skills and safety awareness training, and improve the matching of human and machine. Make sure you don't hurt yourself at work, don't hurt others, and not hurt by others.

ACKNOWLEDGMENTS

This project was supported by the National Key R&D Program of China (2018YFC0809900) and Beijing Science and Technology Commission Science and Technology Plans (Z181100009018003, Z181100009018010).

REFERENCES

Beijing emergency management bureau. Guiding Opinions of the Beijing Municipal Safety Production Committee on Establishing a Pilot Work for Safety Production Liability Insurance System [EB/OL]. http://ajj.beijing.gov.cn/art/2015/3/26/art_243_6358.html

Chen Qiang & Wan Ki Chow. A Discussion of Occupational Health and Safety Management for the Catering Industry in China. International Journal of Occupational Safety and Ergonomics. 2007,13 (3): 333–339.

Gleeson. D. Health and safety in the catering industry. Occupational Medicine-Oxford, 2001, 51(6): 385–391.

Han Dan, Mu Jing, Yin Shi-jiu. Study on food safety management and consumer participation — a case study of quantitative classification of "smiling face" signs in catering services. Shandong Social Sciences, 2018(05): 160–165.

Han Yi, Zhou Hong, Mao Fei-fei. Monitoring of Disinfection Effect of Tableware in Catering Units of Wuxi City from 2013 to 2015. Chinese Journal of Disinfection, 2017, 34(09):872–873.

Liu Peng, Li Wen-tao. Food Safety Regulation of Online Meal Ordering: A Study from the Perspective of Smart Regulation. Journal of Central China Normal University (Humanities and Social Sciences), 2018, 57(01): 1–9.

Luo Gan, Fu Xiao-Man. Several common problems of emergency ventilation design of basement kitchens. Heating Ventilating & Air Conditioning, 2018, 48(04):39–41.

National Bureau of Statistics of China. GB/T4754-2017: National Economic Industry Classification [S].

National Bureau of Statistics of China. Statistical Communiqué of the People's Republic of China on 2017 National Economic and Social Development [EB/OL]. http://www.stats.gov.cn/tjsj/zxfb/201802/t20180228_1585631.html, 2018-02-28.

Song Yue, Wei Yun-fang, Zhao Yong-mei, et al. An investigation into an acute CO poisoning accident in a western restaurant kitchen. Industrial Health and Occupational Diseases, 2009, 35(02):122.

Sun-Jung Shin, Hyochung Kim and Meera Kim. Study on Status of Safety Accidents and Related Factors of the Cooks for School Foodservice in Daegu. Korean J. Food Nutr. 2017.Vol. 30. No. 6, 1299~1309.

Wei Tong-tong, Jiang Hui-ling, Wang Yu, et al. Analysis on the Fire Risk of Commercial Kitchens in China and Its Countermeasures. China Safety Science Journal, 2009, 19(04):134–139.

Xu Wen, An Dai-zhi, Sun Ru-bao, et al. Cross-sectional status of food hygiene safety in troop catering units. Military Medical Sciences, 2016, 40(08):657–660.

Yao Yao. Investigation and analysis of a small dining area fire. Fire Science and Technology, 2018, 37 (01):145–147.

Young-Woong Song, Dohyung Kee, Wook Kim. Questionnaire Survey of Accidents Occurred in Catering Kitchens for Identification of Risk Factors and Preventing Measures. J Ergon Soc Korea, 2018,37 (4):511–522.

Zhang Na, Guo Qing-qi, Han Chun-ran, et al. Application of Fuzzy Mathematics Method in Risk Assessment of Catering Microorganism Revised Draft. Journal of Chinese Institute of Food Science and Technology, 2017, 17(08):210–216.

Risk Analysis Based on Data and Crisis Response Beyond Knowledge – Huang & Nivolianitou (eds)
© 2020 Taylor & Francis Group, London, ISBN 978-0-367-25146-8

Occupational safety status and countermeasures based on casualty accidents from 2012 to 2018 in Beijing

Fucai Yu, Xuewei Ji, Aizhi Wu, Ming Wen & Yan Liu
Beijing Academy of Safety Science and Technology, Beijing, China
Institute for Public Safety Research, Department of Engineering Physics, Tsinghua University, Beijing, China

ABSTRACT: In urban production and operation activities, casualty accidents occur frequently and seriously threaten safety of urban operation and loss of life and property. In this paper, 677 casualty accidents in Beijing from 2012 to 2018 were analyzed using statistical analysis, text mining and association rules mining. The results show that: ① The number of accidents is related to the industrial structure and economic development level, with the highest proportion of accidents in Chaoyang District (19.4%) and Haidian District (14.5%), followed by the core urban area and the strong industrial area. ② The fewest casualties occurred in 2015, followed by an upward trend. There is the highest frequency of accidents in summer (July (12.4%) and August (11.7%)) and the lowest frequency in February due to Spring Festival Holiday. On the 7th and 9th days of each month, the highest frequency of accidents is only 4.4%. Such accidents mainly occur in the peak working hours (at 9-11 and 14-17) and the total frequency is up to 55.5%. ③ Because of the large-scale infrastructure construction in the stage of high-speed urbanization, the frequency in construction industry is as high as 57.6%, and related accidents include falling, electric shock and Collapse. All kinds of mechanical equipment are most dangerous factors. The casualties mainly come from Hebei and Henan around Beijing. ④ Because human error is the primary cause, strengthening staff training and safety management for enhancing staff safety awareness is essential.

1 INTRODUCTION

Urban production and operation activities are the basic social and economic activities ensuring urban operation and the lives of citizens. The unsafe factors, existing in relevant personnel, equipment, environment and management, may lead to serious loss of life and property. Based on the papers between 2001 and 2018 (Melchior et al., 2019), the analysis of cases in more than 30 countries from 1946 to 2016 shown that construction, agriculture and transportation were the economic sectors with the highest death risk. Using text mining and data processing technology to analyze historical construction accident reports and establishing a classification model of accident cause (Zhang et al., 2019), timely detection of hidden dangers and taking preventive measures can greatly reduce accident risk. According to the analysis results of building construction accidents (Shao et al., 2019), the frequency of accidents in July and August and on Monday, which are peak working hours, is highest. The accident mortality rate is higher in underdeveloped areas. The probability of high falling accidents is more than 55%. Collapse and lifting accidents are more likely to cause death. Based on PHMSA database, the development trend, causes and consequences of natural gas and dangerous liquid pipeline accidents and their casualties characteristics (Siler-Evans et al., 2014) were analyzed, and the pipeline age, grade, accident causes and fracture probability of land gas pipelines were analyzed (Lam et al., 2016). Statistical results of elevator operation accidents (Zarikas et al., 2013) in Greece from 1988 to 2009 showed that 65% of accidents were related to violations of safety protection and electrical installation rules, and the relationship between the number of deaths and the types of building and elevator was analyzed. According to the analysis results of 105 typical electrical casualties cases (Zhang et al., 2018), an index system of electrical accidents was established, and countermeasures were put forward based on the general law of accidents. Therefore, accident data are the basic information reflecting urban safety risks and determining the core risk factors. The results of relevant analysis are of great significance for determining key industries, establishing index system and improving accident prevention and control measures.

Beijing is China's political, cultural, international exchanges and science and technology innovation center, so work accidents have a wider scope, greater influence and larger risk. In order to clarify the general law of work accidents and improve the safe operation capacity, 677 casualty accidents from 2012 to 2018 in Beijing were analyzed. The temporal-spatial and industrial distribution, types, causes of accidents and other relevant information were analyzed using statistical analysis, text mining and association rules mining.

2 DATA AND METHOD

2.1 Data source

The data came from the accidents management system of Beijing Emergency Management Bureau (BEMB), which is responsible for supervising and administrating production and operation activities. Under law, the fatal and serious injury accidents should be reported promptly. The casualty accidents covered the time interval from January 1, 2012 to September 18, 2018, including 677 cases in Beijing. Meanwhile, detailed accident and casualty data contained key information, such as date, time, location, sector, type, cause and so on.

2.2 Method

With statistical analysis method and ArcGIS, the characteristic of casualty accidents is shown by various statistical charts. Text mining is an effective method obtaining the keywords describing the process of casualty accidents. A word cloud is drawn to highlight the most prominent terms. Then, the association rules among 14 important factors are identified with Apriori algorithm. Developing tool used is R 3.5.2, main packages used for algorithms design are lubridate, fmsb, tidyverse, readr, arules, arulesViz, tcltk2 and RColorBrewer for visualization. Based on the above results, corresponding countermeasures are proposed.

3 STATISTICAL ANALYSIS AND TEXT MINING

3.1 Spatial distribution of casualty accidents

According to the spatial distribution of casualty accidents in every district of Beijing (Figure 1), the most frequent accidents occurred in densely populated and economically developed Chaoyang (131, 19.4%) and Haidian (98, 14.5%). As the dominant service industry, the frequency of accidents in Xicheng (22, 3.2%) and Dongcheng (19, 2.8%) is relatively low. In addition, Changping (62, 9.1%) and Fengtai (47, 7.0%) with relatively good economy, Daxing (57, 8.4%) with developed manufacturing industry and Fangshan (51, 7.5%) with traditional petrochemical industry also have relatively more accidents. This indicates that the number of casualty accidents has a greater correlation with population density, economic level and industrial structure. Therefore, according to the characteristics of risk and industrial structure in different districts, safety supervisors with professional knowledge and corresponding safety training is necessary.

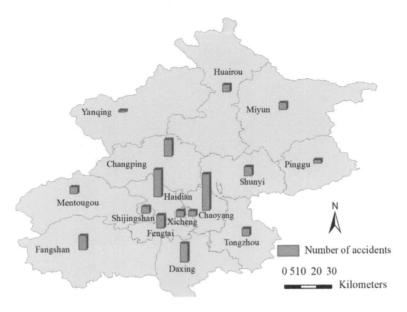

Figure 1.   Spatial distribution of casualty accidents in Beijing.

3.2   *Temporal distribution of casualty accidents*

Figure 2 shows the temporal distribution of work accidents. The number of accidents from 2012 to 2015 is relatively small, the number from 2016 to 2017 (up to 156) is rising, and the number in 2018 is declining. The frequency of casualty accidents is the highest in July (12.4%) and August (11.7%). The distribution of accidents by day is relatively average. The highest frequency is only 4.4% on 7th and 9th day. The frequency by hour is the highest in the work period of 9-11 am and 14-16 pm, and the total frequency in two periods is as high as 55.5%. Therefore, the government and enterprises should strengthen the safety management, on-site inspection and labor protection during the work period in summer. At the same time, it is necessary to optimize the supervision system, introduce advanced technical equipment and strengthen staff safety education for reducing the overall number of casualty accidents.

3.3   *Industry distribution and casualty accidents text mining*

Figure 3 depicts the distribution of casualty accidents in different industries. As Beijing is still in the stage of rapid urbanization and the scale of infrastructure construction is huge, the number of accidents in the construction industry is up to 390 (57.6%). In addition, due to the development of high-tech manufacturing industry and service industry in Beijing, the proportion of accidents in manufacturing industry (94 cases, 13.9%) and residential services, repair services (53 cases, 7.8%) is also relatively high. Therefore, improving safety management in construction site is crucial to reduce the occurrence frequency of casualty accidents through strengthening the communication among relevant authorities, for instance, Beijing Emergency Management Bureau and Beijing Municipal Commission of Housing and Urban-Rural Development.

Through text mining, the keywords depicting accident process in a word cloud are displayed (Figure 4). The larger the font size, the higher the count of word in the text. The count of black words is more than 60, and the rest words are gray.

Obviously, "Death, injured" indicating the seriousness are two most prominent words in casualty accidents. "Site, workshop" indicate that accidents mainly occur at the work site, and "worker, employees, labor" and "operation, construction,

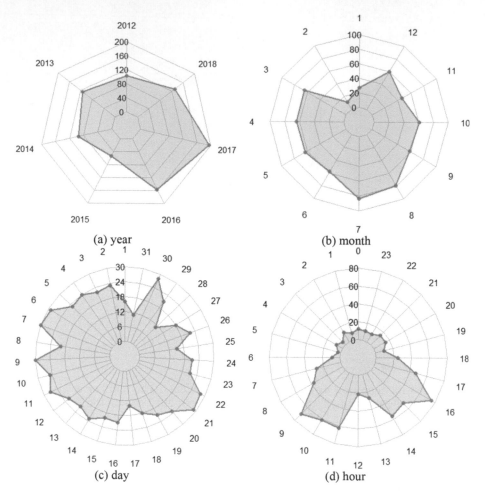

Figure 2. Number of casualty accidents by year, month, day and hour.

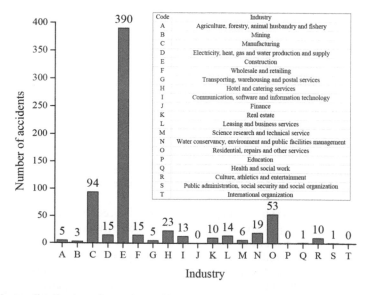

Figure 3. Industry distribution of casualty accidents.

Figure 4. Word cloud of key factors involving casualty accidents.

equipment" indicate that casualties are mainly frontline workers engaged in dangerous operations. "Rescue, hospital, first aid" indicate that people injured seriously obtain timely rescue. "Engineering, building, project" indicate that the construction industry is the highest-risk industry. "falling" and "electric shock" are two main types of accident frequently occurring in the construction industry, and falling is highly related to "elevator, roof, underground, scaffold". Operational activities include "treatment", "demolition" and so on. The casualties are mainly from "Hebei" and "Henan" which are two provinces around Beijing.

4 ASSOCIATION ANALYSIS

Apriori algorithm is a common association analysis algorithm, which can discover frequent item sets meeting special rules. With Apriori algorithm, the correlation of 14 types of factors, including the time, place, industry, type, harmful factor, cause, equipment and qualification certificate, is analyzed. The minimum length of the item sets is 4, the minimum count is 10, and the minimum confidence is 0.8. According to the above law, 547 rules, whose lift is greater than 1, are obtained. As shown in Figure 5, the maximum value of support is 0.142, the maximum value of confidence is 1, and the maximum value of lift is 29.3.

The 10 rules with the largest value of support and lift are displayed respectively. Green circle labeled different codes represents different factors, the size of red circle represents the support, and the color depth of red circle represents the lift.

As shown in Figure 6(a), the frequency of falling accidents (Typ = SG_SGLB09) occurring in building industry (Idt = E, Btyp = SG_JZGCLX12) and obtaining special operation qualification certificate (Sqc = 1) is 0.142. While, the frequency of falling accidents (Typ = SG_SGLB09), duing to violate operation regulations or labor discipline (Drc = SG_ZJY07), occurring in building industry (Idt = E) and obtaining special operation qualification certificate (Sqc = 1) is 0.105. As shown in Figure 6(b), mechanical injury accidents (Typ = SG_SGLB03) are most promoted by all kinds of mechanical equipment (Cit = SG_QYW12, Hmit = SG_ZHW14) in manufacturing industry (Idt = C).

Therefore, the construction industry is the industry with the highest frequency of casualty accidents in Beijing, and the number of falling accidents mainly caused by non-standard operations is maximum. At the same time, due to the complexity of personnel, equipment and environment, the accidents are diverse on construction site and the lift of item sets is not high.

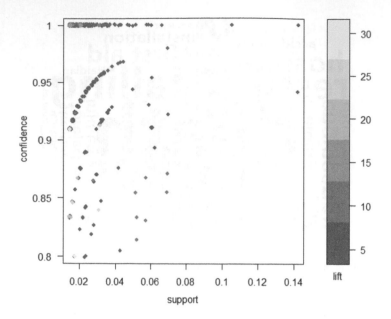

Figure 5.   Scatter plot for 547 rules.

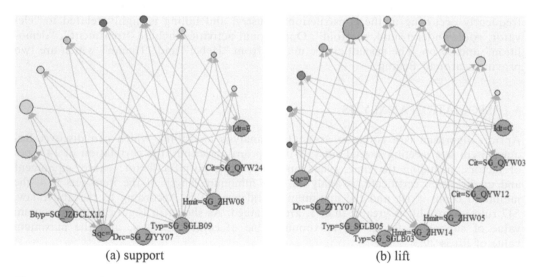

(a) support   (b) lift

Figure 6.   Ten rules with max value of support and lift respectively.

In addition, the mechanical equipment is a high-risk factor leading to mechanical injury in manufacturing industry. However, because the risk of high-tech manufacturing industry is controllable, the accident frequency is relatively low.

5   CONCLUSION

Some significant conclusions are presented through analyzing the data of casualty accidents. In Beijing, the accidents frequency is notably high in densely populated, developed and industrialized areas such as Chaoyang, Haidian, Daxing and Fangshan. Summer and heavy workload period are the time when casualty accidents occurred most frequently. Due to high-speed urbanization and industrial upgrading, more than half of accidents happened in construction activity. The most frequent accidents are falling and electric shock. The casualties, who work in construction site and workshop mainly due to faulty operation, are mainly from Hebei and Henan provinces near Beijing. So, it should be emphasized that better self-management of companies, stronger law enforcement and supervision of government are necessary to minimize casualties and property loss and even prevent such accidents.

ACKNOWLEDGMENTS

This work was supported by National Key R&D Program of China (Grant No. 2018YFC0809900) and Beijing science and technology Plan program (Z181100009018003, Z181100009018010).

REFERENCES

Lam C. and Zhou W.X. 2016. Statistical Analyses of Historical Pipeline Incident Data with Application to the Risk Assessment of Onshore Natural Gas Transmission Pipelines. International Journal of Pressure Vessels and Piping, 145: 29–40.

Melchior C. and Zanini R.R. 2019. Mortality per work accident: A literature mapping. Safety Science, 114: 72–78.

Shao B., Hu Z.G., Liu Q., Chen S. and He W.Q. 2019. Fatal accident patterns of building construction activities in China. Safety Science, 111: 253–263.

Siler-Evans K., Hanson A., Sunday C., Leonard N. and Tumminello M. 2014. Analysis of pipeline accidents in the United States from 1968 to 2009. International Journal of Critical Infrastructure Protection, 7(4): 257–269.

Zarikas V., Loupis M., Papanikolaou N. and Kyritsi C. 2013. Statistical survey of elevator accidents in Greece. Safety Science, 59: 93–103.

Zhang F., Fleyeh H., Wang X.R. and Lu M.H. 2019. Construction site accident analysis using text mining and natural language processing techniques. Automation in Construction, 99: 238–248.

Zhang Z., Wu C. and Gao K.X. 2018. Indexes design for subjective and objective scenes of electrical casualty accidents and statistical regularity. Journal of Safety Science and Technology, 14(1): 185–192. (in Chinese).

Risk Analysis Based on Data and Crisis Response Beyond Knowledge – Huang & Nivolianitou (eds)
© 2020 Taylor & Francis Group, London, ISBN 978-0-367-25146-8

Response of *Paramecium caudatum* to the stress of tetracycline and tetracycline hydrochloride

Minglei Du, Ye Zhao, Xiang Wang, Weibin Zheng, Chang Ge & Ying Chen*
Laboratory of Protozoology, Heilongjiang Key Laboratory of Biodiversity of Aquatic Organisms, College of Life Science and Technology, Harbin Normal University, Harbin, China

Nanqi Ren**
State Key Laboratory of Urban Water Resource and Environment, Harbin Institute of Technology, Harbin, China

ABSTRACT: This paper explored the acute response of *Paramecium caudatum* to the stress of tetracycline (TC) and tetracycline hydrochloride (HTC). The half-lethal concentration (LC50), the half-inhibitory concentration (EC50), and the lowest-observed-effect concentration (LOEC) were determined, and the changes of the monoclonal growth curve stressed by the half-inhibitory concentration (EC50) antibiotics were observed. The results showed that the values of 24h-LOEC and 24h-LC50 of TC were a little higher than HTC, but the 24h-EC50 was a little lower than HTC. The response of the enzyme activity of superoxide dismutase(SOD), peroxidase (POD), catalase (CAT) to 24h-EC50 indicated that TC and HTC have a strong inhibitory effect on SOD enzyme activity, but a weak inhibitory effect on CAT and POD enzyme activity. The changes in the monoclonal growth curve of *P. caudatum* suggested that TC and HTC significantly inhibited the division and reproduction of *P. caudatum*. This research provided a basis for assessing the environmental risk of tetracycline antibiotics as well as a reference for the scientific and rational use of antibiotics.

Keywords: tetracycline antibiotics, *P. caudatum*, the toxic effect, environmental risk

1 INTRODUCTION

Since penicillin was discovered by Fleming, a large number of antibiotics have been developed that can resist bacteria by continuously separating the metabolites in microorganisms. Although antibiotics have brought benefits, antibiotics abuse is becoming more and serious and a danger to human health (Liao 2016). There were more than 60 kinds of antibiotics used in animal husbandry, which played an important role in promoting the growth and development of livestock and poultry, improving the economic utilization rate of feed, and preventing and controlling animal diseases (Li 2009). At least 50% of antibiotics are used in animal husbandry and aquaculture every year around the world (Orwa 2017). A large amount of residual antibiotics was discharged into the water supply in livestock waste. Long-term exposure to antibiotics generated toxicity toward aquatic organisms and transmitted, with concentration and enrichment, through the food chain, eventually causing harm to human health. Therefore, the biological safety of antibiotics attracted the increasing attention of scientists (Zhang 2003; Lin 2007; Qiao 2015).

Studies have shown that abuse of antibiotics causes a number of drug-resistant bacteria, increasing the proportion of serious bacterial flora disorders and the metabolic toxicity of

---

*Corresponding author: lh6666@126.com, **Co-corresponding author: rnq@hit.edu.cn

Table 1. The main apparatus used in this experiment.

Instruments and equipment	Models
Constant temperature and humidity incubator	LHR-250-S
Stereomicroscope	PXS-1040VI
Ultraviolet spectrophotometer	WFZ-26A
Intelligent light incubator	HP1500GZ
A constant temperature water bath	HH-S246
High speed freezing centrifuge	GL-20G-II
Ultrasonic cell grinder	JY92-II

bacterial species (Hallingsørensen 2000). In aquatics (Robinson 2010; González-Pleiter 2013) and soil organisms (Thielebruhn 2005; Rajput 2018; Wei 2018), the toxic effects of antibiotics have been extensively reported. Wollenberger *et al.* (2000) studied the acute toxicity of nine veterinary antibiotics on fleas and discovered the 48h-EC_{50} of oxolinic acid and oxytetracycline. Thiele-Bruhn and Beck (2005) studied the toxicity of sulfapyridine and oxytetracycline on soil microbial activity and microbial biomass. Furthermore, some studies have reported the toxicity of antibiotics and antibiotic metabolites to bacteria in activated sludge and sediments (Näslund 2008; Cheng 2010).

At present, some antibiotics are widely used, including tetracycline, β-lactam, sulfonamide, macrolides, aminoglycosides and peptides, etc. (Zhang 2008; Kümmerer 2009; He 2011).Tetracycline played an important role in animal feed additives. There was 5,000–7,000 tons of tetracycline used in animal breeding per year in China. In this article, *Paramecium caudatum* was selected as the experimental subject to discuss the biological toxicity and potential environmental risk of two tetracyclines (TC and HTC). *Paramecium* has a long history as an experimental organism for toxicology research (Mayne 2017). Because protozoa is at the bottom of the aquatic food chain, residual antibiotic in protozoa will be transmitted upward along the food chain. Therefore, this investigation could supply some important information for environmental risk assessment and early warning about tetracycline antibiotics.

2 MATERIALS AND METHODS

2.1 *Experimental subject*

P. caudatum (Ciliates, Oligohymena, Membranostomata, Paramecidae, *Paramecium*) were collected in a freshwater pond near to Harbin, Heilongjiang province, China. It was cultured with 0.01 mg/Lnutrient solution of straw in a constant temperature and humidity incubator at 20–25°C for 24 hours. The metabolic waste was cleaned every 24 hours.

2.2 *Experimental drugs and instruments*

Tetracycline and tetracycline hydrochloride were of 97% purity. They were dissolved by ddH_2O to prepare an original solution with concentration 500 mg/L.

2.3 *Test of 24h acute toxicity*

The experimental concentration gradient of tetracycline and tetracycline hydrochloride on *P. caudatum* was determined by preliminary experiments (Table 2).

Six parallel experimental groups and one control group was set for every concentration gradient. Ten individual samples of *P. caudatum* were put into 0.5 ml solution as N_0, and cultured in an incubator at 20°C for 24 h. The number of ciliates survivors was counted and chose the

Table 2.　Gradient concentration of TC and HTC.

Tetracycline	Concentration gradient (mg/L)									
TC	150	100	80	56	40	15	12	10	8	6
HTC	110	70	50	42	35	25	20	15	10	5

averaged value was set as N_1, and the mortality rate was calculated according to the formula, Lethal rate (%) = $(N_0 + N_1)/N_0$ x 100%.

The same method was used for test of EC_{50} and LOEC but the indicators were the division of reproduction, and the inhibition rate was calculated with the following equation:

$$Inhibitionrate(\%) \; = \; (N_0 and N_1)/N_0 x 100\%$$

The data were processed with the Origin 2016 software, the mortality rate and concentration logarithm were analyzed by regression, and the first function was used for fitting. The 24h-$LC_{50}$was calculated according to the first function equation. The regression analysis was performed on the inhibition rate and concentration logarithm and the first function was used for fitting, then the 24h-LOEC and 24h-EC_{50} was calculated according to this equation.

2.4　Test of antioxidant enzyme activity

The 24 h-EC_{50} was used as the stress concentration for six experimental groups, the culture solution without antibiotic was used as the control group, and all groups were cultured in a constant temperature and humidity incubator at 20°C for 24 h. Then cells were picked into ddH$_2$O, centrifuged at 2000rpm to precipitate and weighed. 1500 µl phosphate buffer was added to the tube, then pulverized with an ultrasonic cell pulverizer for 3 s to release the enzyme. Finally, it was centrifuged at 4°C, 3000 rpm, for 20 min, and the supernatant was the enzyme.

The superoxide dismutase (SOD) activity was tested by the methods of Zi (1998) and Yan (2013). The catalase (CAT) activity was tested by the methods of Yan (2013). The peroxidase (POD) activity was tested by the methods of Yan (2013) and Lv (2003).

There were three parallel samples in each group. The mean value was taken as the enzyme activity data of the experimental group. Origin 2016 software was used to deal with data. SPSS19.0 software was used for one-way analysis of variance to test the significance of the difference.

2.5　Observation of monoclonal growth curve

The 24h-LOEC was used as the stress concentration, six experimental groups for each antibiotics, and one control group. Ten individuals samples of *P. caudatum* in good condition were chosen and put into the culture plate, respectively, one to each pit, and 0.5 ml solution was dropped in. They were cultured in a constant temperature and humidity incubator at 20°C. The individual samples were counted every 24h and the culture solution supplied until the senescence stage was entered.

The average was taken for every 10 parallel samples. The data was processed with Origin 2016 software, and SPSS19.0 software was used for one-way analysis of variance to test the significance of the difference.

3　RESULTS

3.1　Inhibitory tetracycline and tetracycline hydrochloride on P. caudatum in 24 h

The inhibition rate of TC and HTC on *P.caudatum* were positively correlated with the stress concentration (Table 3). Using the Origin software, we achieved the regression equation of TC and HTC.

544

Table 3.　24h inhibition toxicity of TC and HTC on *Paramecium caudatum*.

	C (mg/L)	LgC (X)	Inhibition rateY (%)
	15	1.17609	96.66
	12	1.07918	75.00
TC	10	1.00000	56.66
	8	0.90309	35.00
	6	0.77815	15.00
	25	1.39794	95.00
	20	1.30103	76.67
HTC	15	1.17609	43.33
	10	1.00000	16.67
	5	0.69897	5.00

C,stress concentration; LgC,logarithm of stress concentration.

$$TC : Y = 207.97084X - 149.66618 \ (R^2 = 0.99236).$$
$$HTC : Y = 123.08749X - 86.2187 \ (R^2 = 0.96457)$$

The regression analyses were illustrated in Figure 1. According to the equation, the theoretical values of 24h-EC$_{50}$ and 24h-LOEC of TC to *P. caudatum* were 9.12158 mg/L and 0.7227 mg/L, and the 95% confidence intervals were 8.18069~10.17068 mg/L and 0.6861~0.7588 mg/L. The 24h-EC$_{50}$ and 24h-LOEC of the HTC were 12.78439 mg/L and 0.69918 mg/L, and the 95% confidence intervals were 8.18069~10.17068 mg/L and 0.6642~0.6874 mg/L.

3.2　*Lethal toxicity tetracycline and tetracycline hydrochloride on P. caudatum in 24 h*

The 24h lethal toxicity of different concentrations of tetracycline and tetracycline hydrochloride on *P. caudatum* are shown in Table 4. Using the Origin software, we achieved the regression equation of TC and HTC.

$$TC : Y = 93.07682X - 110.00028 (R^2 = 0.98925)$$
$$HTC : Y = 161.86942X - 225.35538 \ (R^2 = 0.98741)$$

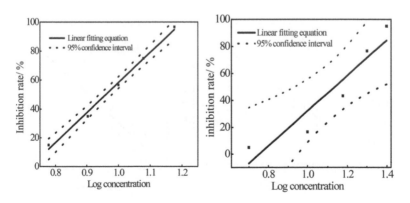

Figure 1.　Concentration logarithm and inhibition rate regression analysis of TC (left) and HTC (right) stress on *Paramecium caudatum*.

Table 4. 24h-acute lethal toxicity of TC and HTC on *Paramecium caudatum*.

	C (mg/L)	LgC (X)	MortalityY (%)
	150	2.1761	93.33
	100	2.0000	77.67
TC	80	1.9031	63.33
	56	1.7482	53.33
	40	1.6021	40.00
	100	2.00000	96.67
	70	1.84510	73.33
TCH	50	1.69897	53.33
	42	1.62325	40.00
	35	1.54407	20.00

C,stress concentration; LgC,Logarithm of stress concentration.

According to the equation, the 24h-LC$_{50}$ of TC and HTC to *P. caudatum* were 52.36125 mg/L and 50.24583 mg/L, and the 95% confidence intervals were 41.74073 ~ 65.68557 mg/L and 41.49445 ~ 60.84151 mg/L (Figure 2).

3.3 Respond of antioxidant enzyme activity

The results of three antioxidant enzyme activities under the stress of two antibiotics are shown in Figure 3. Compared with the control group, the activity of SOD enzyme decreased with the stress of antibiotics. The decreased degree was TC > HTC. The activity of the CAT enzyme was slightly decreased, and the activity of POD enzyme did not change much.

3.4 Respose of monoclonal growth curve

In the control group, the first 1–8 days were the logarithmic growth period of *P. caudatum* and the 8th ~12th days were the plateau period and the aging period started from the 12th day. The maximum average population was around 120 cells (Figure 4). There were a very similar trend of the growth curve in two stress groups. First, the population growth rate was significantly reduced and the maximum only reached 28 and 34 cells. Second, the logarithmic growth phase shortened to 7 days both in two stress experimental groups, and the plateau period was shortened to 3 days. The aging period started from the 10th day. So, the growth curve of *P. caudatum* was changed significantly by the stress of TC and HTC.

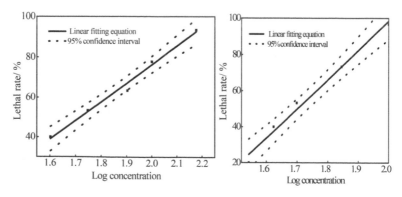

Figure 2. Regression analysis of concentration logarithm and lethal rate of TC(left) and HTC(right)on *Paramecium caudatum*.

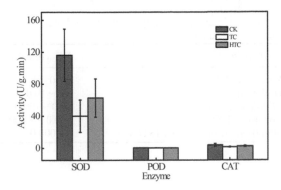

Figure 3.   SOD/POD/CAT activity of *Paramecium caudatum* in CK, TC and HTC groups.

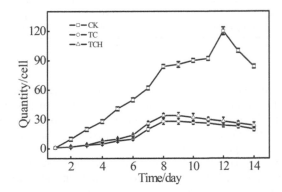

Figure 4.   Monoclonal growth curves of *Paramecium caudatum* in CK,TC and HTC groups.

4   DISCUSSION AND CONCLUSION

4.1   *Stress response of Paramecium caudatum on two antibiotics*

In recent years, researchers have carried out studies on the biological toxicity of the two tetra-cycline antibiotics (Table 5).

Compared with the *P.caudatum*, the effects concentrations of two antibiotics on other organisms were all significantly higher, except for green algae and cyan bacteria. At present, most of antibiotic toxicity tests take lethal concentration as an indicator, and less attention is paid to inhibitory concentration. However, the lethal effect of antibiotics is not significant, and low-concentration inhibition effects should be taken more attention. In this study, we found the activity of SOD enzymes of *P. caudatum* was significantly inhibited by tetracycline antibiotics of 24h-EC$_{50}$ concentration, which will inevitably lead to antioxidant capacity

Table 5.   Studies on the toxicity of tetracycline antibiotics.

Species	Lethal toxicity (mg/L)	Inhibitiontoxicity (mg/L)	Researchers
Crucian carp	322.8		Wang Huishu et al. (2008)
Zebrafish	406.0		Wang Huishu et al. (2008)
Large m.		617.2	Wang Huishu et al. (2008)
Green algae		0.09	Halling Srenson (2002)
Cyanobacteria		2.2	Halling Srenson (2002)
Elegans	167.5	82.9	Yu Zhenyang et al. (2010)

Table 6.　Criteria for toxicity evaluating and grading of chemicals to aquatic organisms.

Toxicity level	EC_{50} (mg/L)
Very toxic(VT)	<1.0
High toxic(HT)	1.0 ~ 10.0
Middle toxic(MT)	10.0 ~ 100.0
Low toxic(LT)	100.0 ~ 1 000.0
Slightly toxic or non-toxic(ST/NT)	>1 000.0

Table 7.　Risk assessment for TC and HTC to aquatic organisms.

Antibiotics	EC_{50} (mg/L)	Toxicity level
Tetracycline	9.12158	HT
Tetracycline hydrochloride	12.78439	MT

decreasing and the possibility of cell senescence and damage increasing, reducing body repair, division and proliferation ability. The experiments of the growth curve of *P. caudatum* also showed that the two antibiotics with the lowest effect concentration could cause the logarithmic phase and plateau period to shorten, the aging period to advance, and the growth rate and population to decrease significantly.

These responses of protozoa will inevitably interfere with the stability of the aquatic micro-ecological system. It is suggested that we need to consider physiological indicators and long-term effects in the environmental risk assessment of antibiotics.

4.2　Preliminary evaluation of environmental toxicity of the two antibiotics

In the 24h-acute toxicity experiments, the values of $24h-LC_{50}$ of the two antibiotics were similar, but the $24h-EC_{50}$ and 24h-LOEC were all TC < TCH. The inhibitory effect on th SOD enzyme of TC was more obvious than HTC. Therefore, the biotoxicity of TC was slightly higher than HTC.

At present, there is no clear and unified standard for the environmental safety evaluation of tetracycline antibiotics. The Hazard Classification Standards in the National Environmental Protection Agency's New Chemical Hazard Assessment Guidelines of China (Table 6) might be taken as reference. According to the Standard, the hazard levels of TC and HTC to *P. caudatum* could be determined initially (Table 7). Tetracycline is highly toxic, and tetracycline hydrochloride is a middle toxic substance.

ACKNOWLEDGMENTS

This work was supported by the Heilongjiang Province Leading Talent Echelon Reserve Leader Project and the Open Project of State Key Laboratory of Urban Water Resource and Environment, Harbin Institute of Technology (No. ES201801) and NSFC (No. 31471950).

REFERENCES

Cheng L. 2010. *Sensitive algae determination of antibiotics in acute toxicity algae infrared test*. Chongqing: Chongqing University.

González-Pleiter M., Gonzalo S., Rodea-Palomares I., et al. 2013. Toxicity of five antibiotics and their mixtures towards photosynthetic aquatic organisms: Implications for environmental risk assessment. *Water Research* 47(6):2050–2064.

Halling-Sorensen B., Sengelov G., Tjornelund J. 2002. Toxicity of tetracyclines and tetracycline degradation products to environmentally relevant bacteria, including selected tetracycline resistant bacteria *Archives of Environmental Contamination and Toxicology* 42(3):263–271.

Halling-Sørensen B., Jørgensen, S.E., et al. 2000. Algal toxicity of antibacterial agents used in intensive farming. *Chemosphere*, 40(7):731–739.

He D.C. 2011. *Study on migration accumulation and blocking technology of veterinary tetracycline antibiotics in circular agriculture.*Hunan: Hunan Agricultural University.

Kümmerer K. 2009. Antibiotics in the aquatic environment–a review–part I. *Chemosphere* 75(4):435–441.

Liao Q. 2016. Summary of the status, causes and countermeasures of antibiotic abuse in China. *World Medical Information Diges*t 16(57):41–42.

Li Z., Wang Y.J. 2009. Current status, problems and countermeasures of antibiotic use in livestock and poultry farming. *China Animal health* 11(7): 55–57.

Lin C.Z., Huang X.Z., You L.T. 2007. Preliminary investigation and countermeasures for the abuse of antibiotics in marine aquaculture. *China Food and Drug Administration* (1): 31–33.

Lu S.X. 2003. Guide to Basic Biochemistry Experiments. *China Agriculture Press*:113–142.

Mayne R., Whiting J., Adamatzky A. 2017. Toxicity and Applications of Internalised Magnetite Nanoparticles Within Live Paramecium caudatum Cells. *Bionanoscience* 8(3):1–5.

Näslund J., Hedman J.E., Agestrand C. 2008. Effects of the antibiotic ciprofloxacin on the bacterial community structure and degradation of pyrene in marine sediment. *Aquatic Toxicology* 90 (3):223–227.

Orwa J.D., Matofari J.W., Muliro P.S., et al. 2017. Assessment of sulphonamides and tetracyclines antibiotic residue contaminants in rural and periurban dairy value chains in Kenya. *International Journal of Food Contamination* 4(1):5.

Qiao Y., Wan X.H., Shen H. 2015. Research progress on antimicrobial resistance of aquatic products in China. *Chinese Journal of Antibiotics* 40(05):389–395.

Rajput V.D., Minkina T.M., Behal A., et al. 2018. Effects of zinc-oxide nanoparticles on soil, plants, animals and soil organisms: A review. *Environmental Nanotechnology Monitoring & Management*, 9:76–84.

Robinson A.A., Belden J.B., Lydy M.J. 2010. Toxicity of fluoroquinolone antibiotics to aquatic organisms. *Environmental Toxicology & Chemistry*, 24(2):423–430.

Thiele-Bruhn S., Beck I.C. 2005. Effects of sulfonamide and tetracycline antibiotics on soil microbial activity and microbial biomass. *Chemosphere* 59(4):457–465.

Wang H.Z., Luo Y., Xu W.Q., et al. 2008. Ecotoxicological effects of tetracycline and chlortetracycline on aquatic organisms. *Journal of Agro-Environment Science* 27(4): 1536–1539.

Wei Z., Wang J., Zhu L., et al. 2018. Toxicity of enrofloxacin, copper and theirinteractions on soil microbial populations and ammonia-oxidizing archaea and bacteria. *Scientific Reports* 8(1):5828.

Wollenberger L., Halling-Sørensen B., Kusk K.O. 2000. Acute and chronic toxicity of veterinary antibiotics to Daphnia magna. *Chemosphere* 40(7):723–730.

Yan Y. 2013. *Experimental study on the toxicity of important pollutants in pharmaceutical wastewater to paramecium.* Master's thesis of Harbin Normal University.

Yu Z.Y., Zhang J., Zhang H.C., et al. 2010. Acute and multi-generation toxicity of tetracycline hydrochloride to C. elegans. *Asian Journal of Ecotoxicology* 5(3):320–326.

Zhang H., Luo Y., Zhou Q.X. 2008. Advances in Research on Ecotoxicity of Tetracycline Antibiotics. *Journal of Agro-Environment Science* 27(2): 407–413.

Zhang L., Fu X.Z., Yao Y. 2003. Status and management strategies of hormones and antibiotic residues in chicken products sold in Harbin. *China Public Health Management* 19(5): 464–464.

Zi J.Q., Deng X.X. 1998. Improvement and Effect of Experimental Operation of Nitrogen Blue Tetrazole Illumination Method. *Journal of Beijing Normal University: Natural Science Edition*, 34(1): 101–104.

Risk Analysis Based on Data and Crisis Response Beyond Knowledge – Huang & Nivolianitou (eds)
© 2020 Taylor & Francis Group, London, ISBN 978-0-367-25146-8

Risk analysis of seismic hazard correlation between nuclear power plants

Qing Wu*
Institute of Geophysics, China Earthquake Administration, Beijing, China

ABSTRACT: Taking four nuclear power plants to be built in Guangdong and Fujian Coastal Area as examples, based on the seismic zoning map of China (GB18306-2015), the synthetic earthquake sequences are simulated by Monte Carlo method, and the ground motion parameters of each nuclear power plant site are obtained through attenuation relationship. By calculating the probability that multiple nuclear power plants would be affected by earthquakes that would exceed the fortification criteria at the same time, the seismic hazard correlation among nuclear power plants is quantitatively analyzed. This method can be used as a supplementary work for the site selection of nuclear power plants to improve the seismic performance of nuclear power plant cluster.

Keywords: nuclear power plant cluster, Monte Carlo method, synthetic earthquake sequences, correlation of seismic hazard

1 INTRODUCTION

The Great East Japan Earthquake occurred on 11 March, 2011. It was caused by a sudden release of energy at the interface where the Pacific tectonic plate forces its way under the North American tectonic plate. A section of the Earth's crust, estimated to be about 500 km in length and 200 km in wide, was ruptured, causing a massive earthquake with a magnitude of 9.0 and a tsunami which struck a wide area of coastal Japan. The earthquake and tsunami caused great loss of life and widespread devastation in Japan. At the Fukushima Daiichi nuclear power plant, the earthquake caused damage to the electric power supply lines to the site, and the tsunami caused substantial destruction of the operational and safety infrastructure on the site. The reactor cores in Units 1–3 overheated, the nuclear fuel melted and the three containment vessels were breached. Hydrogen was released from the reactor pressure vessels, leading to explosions inside the reactor buildings in Units 1, 3 and 4 that damaged structures and equipment and injured personnel. Radionuclides were released from the plant to the atmosphere and were deposited on land and on the ocean. The 4 other nuclear power plants along the coast were also affected to different degrees by the earthquake and tsunami. However, all operating reactor units at these 4 plants were safely shut down(IAEA, 2015).

A single earthquake event affected multiple nuclear power plants to exceed the design benchmark and caused serious accidents. This is a scenario not covered in the past nuclear power plant construction and operation practice and safety analysis, but once it happens, it may cause immeasurable consequences.

The construction of nuclear power plants in China has developed rapidly in recent years. Nuclear power plant cluster has formed in the coastal areas of Fujian and Guangdong, and the planning site selection and construction efforts are continue to be strengthened. These nuclear power plants are facing the direct threat of earthquakes above magnitude 7 in the

---

*Corresponding author: wuqing908@sina.com

Fujian-Guangdong coastal fault zone (including the Taiwan Strait seismic zone), which is highly consistent with the situation before the Fukushima Daiichi nuclear power plant accident, and the severity of the earthquake threat will further intensify with the completion of new nuclear power plants.

Therefore, in addition to Level 1 Seismic Hazard Assessment of the site itself, the site selection of new nuclear power plants also needs to consider the possibility of adjacent nuclear power plants being damaged by large earthquakes at the same time. Traditional probabilistic seismic hazard analysis method only aims at a single site, and it is difficult to reflect the possibility of multiple sites being affected by single earthquake event at the same time. Monte Carlo method can simulate the seismic sequences satisfying the spatial and temporal distribution law of earthquakes in a certain region according to the seismicity model(Gao & Pan, 1993; Musson, 1999, 2000). For each earthquake in the seismic sequences, the ground motion values of each site are calculated by attenuation relationship. Based on the statistics of specific events, the possibility of multiple sites suffering from ground motions that exceed given values at the same time is obtained.

Taking four nuclear power plants to be built along the coast of Guangdong and Fujian as examples, this paper gives the probability that at least two nuclear power plants suffering from ground motions that exceed the design benchmark at the same time through numerical statistics, and analyses the seismic hazard correlation between the sites. This work can be used as a supplement to the site selection of nuclear power plants and improve the seismic performance of nuclear power plant cluster.

2   MONTE CARLO METHOD

The Monte Carlo method uses random numbers to perform computer simulations. The basic idea is that when the number of experiments is large sufficiently, the frequency of an event appears to approximate the probability of occurrence of the event.

Based on the geophysical data of various regions in China, the seismic zoning map of China (GB18306-2015) shows the seismic zones and potential source areas, established the corresponding probability model and spatial distribution model of earthquake occurrence, and gives the basic parameters of each seismic zone. According to the basic assumptions and seismicity parameters of the zoning map, the following steps are used to synthesize the sets of earthquake sequences(Guo, 2008; Wu & Gao, 2018):

(1) Based on the assumption that the occurrence of earthquakes in seismic zones satisfies Poisson distribution, the time length T of the simulated earthquake sequence and the average annual occurrence rate ν_4 of earthquakes with magnitude 4 and above in the seismic zone should be determined firstly. Randomly generate a Poisson distribution random number L with T and ν_4 as parameters, then L is the number of earthquakes in the seismic zone for the length of time T to be simulated.

(2) Based on the assumption that the magnitude distribution of seismic zones satisfies the truncated Gutenberg-Richter relationship (magnitude-frequency relationship), and the minimum magnitude level M_0 and the maximum magnitude M_{UZ}, the magnitude of earthquakes to be simulated are determined.

The magnitude-frequency relationship is represented as:

$$\log N = a - bM \tag{1}$$

Where a and b are coefficients, N is the number of earthquakes whose magnitude is equal to or greater than M, and the initial magnitude of the zoning map is 4. The cumulative number of earthquake events is:

$$N(\mathbf{M}) = e^{a-bM} \tag{2}$$

If take $\Delta M = 0.1$, then

$$N(M) > N(M + \Delta M) \tag{3}$$

Take M = 4.1,4.2,4.3,…, M_{UZ} .Generate a random number u satisfied uniform distribution between 0 and 1. Determine whether

$$u \in \frac{N(M + \Delta M)}{N(4)} \sim \frac{N(M)}{N(4)} \tag{4}$$

If the above formula is true, the magnitude M of an earthquake event is determined.

(3) Determination of epicenter location. Firstly, the potential source area H where the earthquake located should be determined. According to the magnitude M determined in the previous step, the magnitude range d which the earthquake belongs to is determined. Because the probability $P_d(h)$ of each magnitude range locating in each potential source area is known, then generate a random number u satisfied uniform distribution between 0 and 1. Determine whether

$$u \in \sum_{h=1}^{H-1} P_d(h) \sim \sum_{h=1}^{H} P_d(h) \tag{5}$$

if so, the potential source area H where the earthquake event is located is determined. Based on the assumption that the epicenter is evenly distributed in the potential source area, a point is randomly selected in the potential source area H as the epicenter location of an earthquake.

(4) According to the probability of azimuth of the potential source area, the azimuth of the earthquake is determined.

So far, the basic elements of an earthquake have been determined. Repeat (2) ~ (4) steps until the required number L of earthquakes in the seismic zone, taking into account all possible seismic zones that may affect the site, thus determining a seismic sequence and completing one sampling.

If the time length T is set to one year, the seismic sequence obtained by one sampling is called one-year seismic sequence in this paper. The time length is set to 10 years, which is called 10-year earthquake sequence.

According to the principle of Monte Carlo method, the more samplings, the more stable the result is. But the more samplings, the more calculations. Therefore, in order to consider the accuracy of the results and the calculation quantity as a whole, it is necessary to carry out experiments with different number of samplings. When the calculation results tend to be stable, it is considered that there is no need to increase the number of samplings.

For each earthquake in seismic sequences, the peak ground acceleration(PGA) of each site is calculated by the optimal ellipse search algorithm through attenuation relationship.

3 SEISMIC HAZARD CORRELATION OF NUCLEAR POWER PLANTS TO BE BUILT

The four nuclear power plants to be built in the coastal area of Guangdong and Fujian are each regarded as a site. Figure 1 shows the spatial distribution of nuclear power plants and the potential source areas around them. Table 1 is the Seismic Fortification Criterion for the four nuclear power plants. For the design-basis earthquake, SL-2 is Ultimate Safety Earthquake and SL-1 is Operation Safety Earthquake.

According to the seismic zoning map of China (GB18306-2015), Guangdong and Fujian are located in the coastal seismic zone of South China, mainly affected by strong earthquakes in the Eastern and Western seismic zones of Taiwan. In this paper, the stochastic earthquake sequences of the coastal seismic zone of South China, the Eastern and Western seismic zones

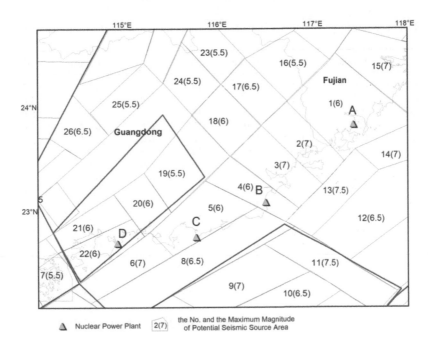

Figure 1. Spatial distribution of nuclear power plants to be built and potential seismic sources around them.

Table 1. The Seismic Fortification Criterion for the four nuclear power plants to be built.

			unit: g
NO.	Name	SL-2	SL-1
A	Zhangzhou	0.30	0.15
B	Huilai	0.29	0.145
C	Lufeng	0.20	0.10
D	Huizhou	0.25	0.125

of Taiwan are simulated. Table 2 is a list of seismicity parameters of several potential source areas around the sites.

Based on the latest potential source area division model, this paper uses Monte Carlo method to simulate one-year seismic sequences considering all seismic zones that may affect the four nuclear power plants to be built. Combining Python with C language to develop an efficient algorithm based on GPU parallel server platform. After many experiments, this paper considers that 5000000 simulations have achieved a balance in calculation amount, accuracy and stability of results. Using the attenuation model suitable for coastal areas of Guangdong and Fujian, the PGAs at the four nuclear power plant sites of each earthquake in the seismic sequences are calculated successively. The PGA attenuation relationships applicable to this area are as follows (Xiao, 2011):

When M<6.5

Long axis direction:

$$\lg Y(\mathrm{M}, \mathrm{R}) = 2.024 + 0.673M - 2.329\lg(\mathrm{R} + 2.088e^{(0.399\mathrm{M})}) \tag{6}$$

Table 2. List of seismicity parameters of potential seismic sources around the four nuclear power plants.

No.	M_{UZ}	b value	ν_4	Strike	No.	M_{UZ}	b value	ν_4	Strike
1	6.0	0.87	5.6	40°	14	7.0	0.87	5.6	40°
2	7.0	0.87	5.6	135°	15	7.0	0.87	5.6	160°
3	7.0	0.87	5.6	135°	16	5.5	0.87	5.6	135°
4	6.0	0.87	5.6	140°	17	6.5	0.87	5.6	135°
5	6.0	0.87	5.6	20°	18	6.0	0.87	5.6	135°
6	7.0	0.87	5.6	20°	19	5.5	0.87	5.6	45°
7	5.5	0.87	5.6	30°	20	6.0	0.87	5.6	30°
8	6.5	0.87	5.6	20°	21	6.0	0.87	5.6	30°
9	7.0	0.87	5.6	20°	22	6.0	0.87	5.6	30°
10	6.5	0.87	5.6	20°	23	5.5	0.87	5.6	135°
11	7.5	0.87	5.6	140°	24	5.5	0.87	5.6	135°
12	6.5	0.87	5.6	40°	25	5.5	0.87	5.6	45°
13	7.5	0.87	5.6	40°	26	6.5	0.87	5.6	45°

Short axis direction:

$$\lg Y(M, R) = 1.204 + 0.664M - 2.016\lg(R + 0.944e^{(0.447M)}) \tag{7}$$

$$\varepsilon = 0.245 \tag{8}$$

When M ≥ 6.5
Long axis direction:

$$\lg Y(M, R) = 3.565 + 0.435M - 2.329\lg(R + 2.088e^{(0.399M)}) \tag{9}$$

Short axis direction

$$\lg Y(M, R) = 2.789 + 0.420M - 2.016\lg(R + 0.944e^{(0.447M)}) \tag{10}$$

$$\varepsilon = 0.245 \tag{11}$$

Where $Y(M, R)$ is the peak ground acceleration (PGA), M is the magnitude, R is the epicentral distance, and ε is the standard deviation.

For 5000000 simulations of a 1-year earthquake sequence, if at least two nuclear power plants are simultaneously affected by ground motions exceeding the Seismic Fortification Criterion, the sequence is identified as 1. The sum of earthquake sequences identified as 1 is counted, then is divided by the total number of earthquake sequence simulations of 5000000, that is the annual exceeding probability of specific ground motions.

Through calculation, it is found that there is the possibility that the adjacent nuclear power plants will be affected by earthquakes at the same time. Table 4a~c is the annual probability

Table 4a. Annual exceedance probability of PGA at nuclear power plant A and B exceeding the design reference simultaneously.

B ≥ A ≥	SL-1(0.145g)	SL-2(0.29g)
SL-1(0.15g)	0.0000150	0.0000012
SL-2(0.30g)	0.0000012	0.0000002

Table 4b. Annual exceedance probability of PGA at nuclear power plant B and C exceeding the design reference simultaneously.

C ≥ B ≥	SL-1(0.10g)	SL-2(0.20g)
SL-1(0.145g)	0.0000194	0.0000044
SL-2(0.29g)	0.0000022	0.0000004

Table 4c. Annual exceedance probability of PGA at nuclear power plant C and D exceeding the design reference simultaneously.

D ≥ C ≥	SL-1(0.125g)	SL-2(0.25g)
SL-1(0.10g)	0.0000354	0.0000048
SL-2(0.20g)	0.0000142	0.0000016

of PGA at nuclear power plants exceeding the design reference simultaneously. It can be seen that in the case of annual exceedance probability at 10^{-4} level, the seismic hazard correlation between nuclear power plants is not high, and it is unlikely that they will be damaged at the same time by earthquake impact exceeding the fortification standard.

4 CONCLUSION AND DISCUSSION

Based on the simulation of earthquake sequences in the coastal seismic zone of South China and the Eastern and Western seismic zones of Taiwan, the possible earthquake effects on the four nuclear power plant sites to be built are calculated. Statistical analysis shows that the seismic hazard correlation among the four nuclear power plants along the coast of Guangdong and Fujian is not high. They are unlikely to be damaged by earthquakes at the same time. Basically, it is not necessary to consider the situation that they may suffer from ground motions that exceed the design benchmark simultaneously.

With the development of China's nuclear power industry, more and more nuclear power plants are planned to be built. While fully evaluating the seismic hazard of the nuclear power plant site itself, it is necessary to strengthen the consideration of external events that may exceed the design benchmark in the design and operation of the nuclear power plant, especially the case that multiple nuclear power plants encounter extreme external events at the same time, so as to find out potential safety hazards and implement improvements.

ACKNOWLEDGEMENTS

This work was jointly supported by the National key research and development program (2018YFC1504601), the key research and development program of Zhejiang Province (2018C03045) and the Special Fund of the Institute of Geophysics, China Earthquake Administration (DQJB16A02).

REFERENCES

The Director General. 2015. The Fukushima Daiichi Accident, International Atomic Energy Agency: 1.

Gao, M.T. & Pan, H. 1993. Characteristics of random field for seismic zoning results. *Acta Seismologica Sinica* 15(1): 53–60.

Musson, R. M. W. 1999. Determination of design earthquakes in seismic hazard analysis through Monte Carlo simulation. *Journal of Earthquake Engineering* 3: 463–474.

Musson, R.M.W. 2000. The use of Monte Carlo simulations for seismic hazard assessment in the U.K. *Annals of Geophysics* 43(1): 1–9.

Guo, X. 2008. Probabilistic Seismic hazard assessment based on Monte Carlo simulation. Master's Degree Thesis. Beijing: Institute of Geophysics, China Earthquake Administration:16–18.

Wu, Q. & Gao, M.T. 2018. A preliminary study on the correlativity of seismic hazard between Beijing area and Xiong'an New Area. *Seismology and Geology* 40(4): 935–943.

Xiao, L. 2011. Study on the attenuation relationship of horizontal ground motion parameters near the source of rock site. Doctor's Degree Thesis. Beijing: Institute of Geophysics, China Earthquake Administration:120.

Risk Analysis Based on Data and Crisis Response Beyond Knowledge – Huang & Nivolianitou (eds)
© 2020 Taylor & Francis Group, London, ISBN 978-0-367-25146-8

Uncertainty research for a near-field ground motion numerical simulation of the Tonghai earthquake

Zongchao Li*, Sen Qiao, Xueliang Chen, Qing Wu & Changlong Li
Institute of Geophysics, China Earthquake Administration, Beijing, China

ABSTRACT: In this paper, an empirical Green's function method is used to simulate the near-field seismic characteristics of the Tonghai Ms7.8 earthquake that occurred in 1970.The various uncertainties in the source section are considered carefully. The strong ground generation area is introduced into the source model. A mixed source model to consider the uncertainties of the asperity is also established. Through ground motion simulation with small earthquakes occurring in Tonghai, the seismic characteristics of a large earthquake could be reproduced. By considering uncertainties, a source model for predicting future destructive earthquakes is established to provide a better scientific basis for earthquake prevention, disaster reduction, earthquake fortification and urban planning.

Keywords: Tonghai earthquake, uncertainty factors, ground motion prediction, asperity, numerical simulation

1 INTRODUCTION

On January 5, 1970, an Ms7.8 earthquake occurred in Tonghai County, Yuxi City, Yunnan Province, resulting in 15,000 deaths. This earthquake caused serious landslides and geological hazards. It is the third largest earth quake in China, with more than 10,000 deaths after the Tangshan earthquake and the Wenchuan earthquake. The Tonghai earthquake occurred in the Qujiang fault zone. The surface rupture length of the main earthquake fault was about 60 km and the fault length shown by the aftershocks was 100 km. The epicenter was located at 24.06 N, 102.36 E (Figure 1). The source depth was around 13 km and the seismic moment was about 1.08 E+27 dyne.cm. The seismic intensity in the magistoseismic area was X degrees (Liu, 1999).

In August 2018, two Ms5.0 earthquakes occurred in Tonghai County. Because of special conditions during the cultural revolution, there were no seismic records during the Tonghai earthquake, and there was no accurate record of the spatial distribution of ground motion. The two Ms5.0 earthquakes were a good opportunity to use the empirical Green function method to predict the characteristics of the Tonghai earthquake in 1970. The advantage of the empirical Green function method is that the small earthquakes contained the same information, such as source, propagation path, and site condition, as for the large earthquake. So, this method could avoid the difficulty of calculating the Green function of the theory. This paper introduces the concept of a strong ground motion generating area. The asperity region is regarded as the strong earthquake generating region.

The logic tree method and statistical model will be used to study the cognitive uncertainty and random uncertainty factors of source parameters. The spatial distribution characteristics of the Tonghai earthquake will be obtained and inversion of the spatial distribution characteristics of the intensity of large earthquakes in Tonghai County could provide scientific reference for seismic design and seismic safety assessment of major projects in the area.

---

*Corresponding author: lizongchaoigo@163.com

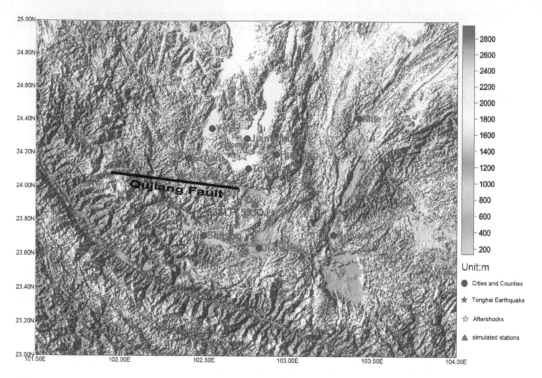

Figure 1.   The location of epicenter and cities. The black line means seismogenic fault named Qujiang.

2   EMPIRICAL GREEN FUNCTION METHOD

The empirical Green function method, which uses small earthquakes as Green functions, was first proposed by Hartzell (1978). Many seismologists, including Kanamori (1979), Dan et al. (1989), and Irikura (1983, 1986), have since revised this method. Irikura added a special time function containing a time phase delay to correct the differences in the source time functions between large earthquakes and small earthquakes. In this article, we use the version revised by Irikura to analyze the uncertainty factors. We assume that a major earthquake hypo center consists of many minor earthquake hypo centers. Accordingly, an appropriately small earthquake is chosen as the ground motion response caused by a point source, and the small earthquake is considered an empirical Green function. An empirical Green function is based on a scaling law of fault parameters for large and small events (Kanamori and Anderson 1975) in addition to the $\omega 2$ source spectra (Aki, 1967). The expressions for a synthetic main shock and aftershock (Irikura 1986; Hiore Miyake et al. 2003) are shown in Eqs. 1, 2 and 3, respectively (Miyake et al. 2003):

$$U_0(x, t) = \sum_{i=1}^{N} \sum_{j=1}^{N} \frac{r_s}{r_{ij}} F(t) * (C \cdot u_s(t)) \tag{1}$$

$$F(t) = \delta\big(t - t_{ij}\big) + \frac{1}{n'} \sum_{k=1}^{(N-1)n'} \left[\delta\left\{ t - t_{ij} - \frac{(k-1)T}{(N-1)n'} \right\} \right] \tag{2}$$

$$t_{ij} = \frac{r_{ij} - r_0}{V_s} + \frac{\xi_{ij}}{V_r} \tag{3}$$

where $U_0(x,t)$ is the synthetic record of a major earthquake, $u_s(t)$ is the Green function; r_{ij} and r_s are the hypocentral distances of the element earthquake and aftershock, respectively; r_0

denotes the distance from the site to the starting point of rupture on the fault plane of the large event; ξ_{ij} represents the coordinates (i, j) of the element source; T is the rise time for the large event, which is defined as the duration of the correction function; $F(t)$ is thecorrection function used to adjust the large and small events (Miyake et al. 2003); n' is an appropriate integer for suppressing the artificial periodicity of n and adjusting the interval of the sampling rate (Miyake *et al.*, 2003); V_s and V_r are the S-wave velocity close to the source area and the rupture velocity along the fault plane, respectively; and N and C are ratios of the fault dimensions and stress drops, respectively.

3   SOURCE PARAMETERS AND UNCERTAINTY FACTORS

The source parameters were divided into random and cognitive uncertain parameters. The random uncertain parameters were studied by statistical methods based on the empirical relationship among the source parameters, such as the seismic moment, fault rupture area, source rise time, asperity area, etc., calculated according to the magnitude of the Tonghai earthquake in 1970. The focal depth and focal mechanism also refer to the parameters in the Tonghai earthquake. Numerical simulation of the Tonghai earthquake was carried out by stations 53JQJ and 53SLP. The related information of source parameters is shown in Table 1.

The two most important parameters when using the Green's function method to predict ground motions are the stress drop ratio between a large earthquake and a small earthquake, denoted as C, and the number of sub-faults used to divide the earthquake fault plane, denoted as N (Zongchao Li, 2018).The initial rupture position was assumed to be located along the central lower part of the fault plane according to the results of a source parameter sensitivity analysis (Zongchao Li, 2018).

To calculate the parameters, the fault plane was regarded as a circle following the Brune model (Brune, 1970). The earthquake rupture process was modeled as the sudden release of shear stress along the disk-shaped fault plane. This source model is usually called Brune's model, and the values of C and N are calculated mainly according to the steps in Figure 2.

Table 1.   The source parameters of mainshock and aftershock.

Magnitude	
Mainshock	Ms7.8
Aftershock	Ms5.0
Depth	
Mainshock	13 km
Small earthquake	14.8 km
Epicenter	
Mainshock	(102.36°E,24.06°N)
Aftershock	(102.709°E,24.19°N)
Seismic moment	
Mainshock	1.076E+27dyne.cm
Small earthquake	1.181E+24dyne.cm
Focal mechanism	
Mainshock(strike/dip/slip)	297°/80°/20°
Aftershock(strike/dip/slip)	206°/77°/9°
Rupture area of mainshock	2000 km^2
Rupture length	100 km
Rupture width	20 km
Source risetime	5.1 s
S-wave velocity Vs	3.6 km/s
Rupture velocityVr	2.88 km/s

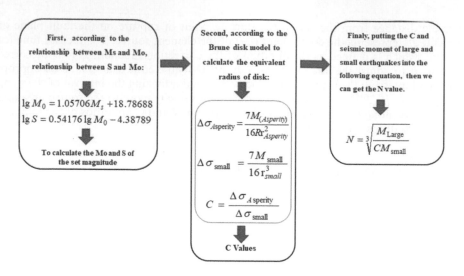

Figure 2.   Calculation process of key model parameters.

Due to the lack of historical seismic data and strong ground motion data of the Tonghai earthquake, the number and area of asperity bodies on the fault rupture surface were uncertain. This paper used a logical tree method to deal with the cognitive uncertainty of asperity. By referring to the comprehensive characteristics of the number and area of asperity in present large earthquakes and the research results of the parameters of asperity by Somerville(1999), we established the logical tree of the area, number and other parameters of asperity in the Tonghai earthquake (Figure 3, Tables 2–4).

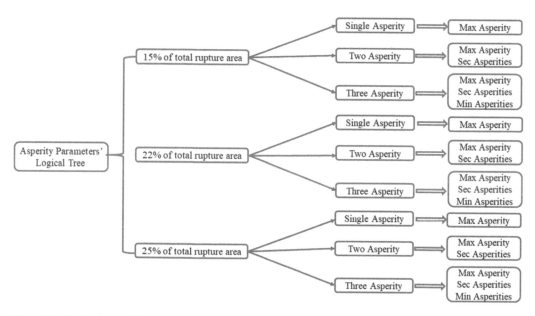

Figure 3.   Logical tree of asperity parameters.

Table 2. The parameters of single asperity.

Rate of asperity to rupture area	15%	22%	25%
Asperity area	300 km^2	440 km^2	500 km^2
Asperity seismic moment(dyne.cm)	4.73E+26	5.16E+26	5.60E+26
C value of single asperity	7.82	5.33	4.69
N value of single asperity	3.71	4.34	4.66

Table 3. The parameters of two asperities.

Rate of asperity to rupture area	15%	22%	25%
Asperity area	300 km^2	440 km^2	500 km^2
Total Mo(dyne.cm)	4.73E+26	5.16E+26	5.60E+26
Max Mo(dyne.cm)	3.16E+26	3.44E+26	3.74E+26
Sec Mo(dyne.cm)	1.57E+26	1.72E+26	1.87E+26
Max asperity area	200 km^2	294 km^2	334 km^2
Sec asperity area	100 km^2	147 km^2	167 km^2
C value of max asperity	11.74	7.98	7.03
N value of max asperity	2.84	3.31	3.56
C value of sec asperity	23.47	15.97	14.06
N value of sec asperity	1.78	2.09	2.24

Table 4. The parameters of three asperities.

Rate of asperity to rupture area	15%	22%	25%
Asperity area/km^2	300	440	500
Total Mo(dyne.cm)	4.73E+26	5.16E+26	5.60E+26
Max Mo(dyne.cm)	2.70E+26	2.92E+26	3.20E+26
Sec Mo(dyne.cm)	1.35E+26	1.48E+26	1.60E+26
Min Mo(dyne.cm)	6.80E+25	7.40E+25	8.00E+25
Max asperity area/km^2	171.6	251.4	285.7
Sec asperity area/km^2	85.8	125.7	142.9
Min asperity area/km^2	42.9	62.9	71.4
C value of max asperity	13.68	9.34	8.22
N value of max asperity	2.56	2.98	3.21
C value of sec asperity	27.35	18.67	16.4
N value of sec asperity	1.61	1.89	2.02
C value of min asperity	54.7	37.3	32.88
N value of min asperity	1.01	1.19	1.27

4 THE GROUND MOTION PREDICTED RESULTS OF THE TONGHAI EARTHQUAKE

Two stations, 53JQJ and 53SLP, were selected for the preliminary prediction of ground motion based on the uncertainty of the asperity. By combining the different numbers and areas of asperity to obtain multiple sets of asperity source models, the empirical Green function method was used to synthesize the strong ground motion of the Tonghai earthquake with two Ms5.0 earthquakes that occurred in 18 years as the Green functions. The ground motion acceleration time history obtained by the two stations under different asperity source models is shown in Figure 4, and the corresponding response spectrum of the two stations is shown in Figure 5.

By comparing the acceleration time history range and response spectrum obtained by each asperity source model, the seismic characteristics obtained by the three asperities source

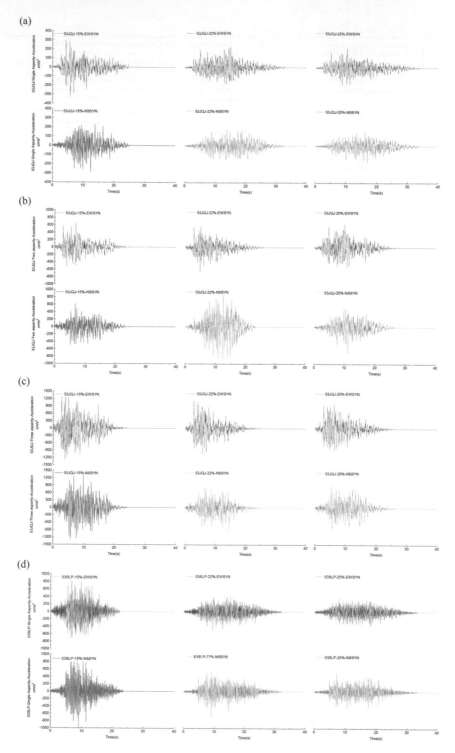

Figure 4. The acceleration time history of different asperity source models at station 53JQJ and 53SLP. (a) The acceleration time history of the single asperity model at station 53JQJ; (b) The acceleration time history of two asperity models at station 53JQJ; (c) The acceleration time history of three asperity models at station 53JQJ; (d) The accelerationtime history of the single asperity model at station 53SLP.

(e)

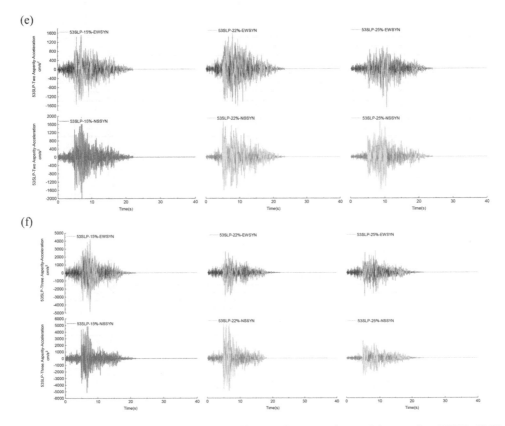

(f)

Figure 4. (Continued) (e) The accelerationtime history of two asperity models at station 53SLP; (f) The acceleration time history of three asperity models at station 53SLP.

models were too large, and the peak ground acceleration of the station 53SLP wasabout 4.0–6.0 g. The station 53JQJ, which is slightly farther away from the source, also has 1.2g peak ground acceleration. This is obviously inconsistent with the real situation. Such a large peak ground acceleration does not exist.

Through comparing and analyzing peak ground acceleration with similar magnitudes, such asthe Wenchuan earthquake, Lushan earthquake, and Nepal earthquake, combined with the distribution of seismic intensity of the Tonghai earthquake and the seismic structure of the Qujiang fault where the Tonghai earthquake occurred in a relatively independent fault,we could seethat there were two asperities on the fault plane of the Tonghai earthquake and the asperity area accounts for around15% of the total fault rupture area.The peak ground acceleration in the magistoseismic area is about 1.2–1.5g. Aprobability of 22% of the total fault rupture area with two asperities is also high. We inferred that the area of the asperities should account for about 15–22 % of the total fault rupture areas. The spatial scale of the fault rupture areas of the Qujiang fault is much smaller than that of the Longmenshan fault zone where the Wenchuan earthquake occurred, although they had the same magnitude. It releases more energy within the limited fault scale, and the destructive power of the earthquake is greater.

5 CONCLUSION AND DISCUSSION

In this paper, the uncertainty factors of seismic rupture process, including seismic moment, fault rupture area, asperity area and quantity, are considered to predict the

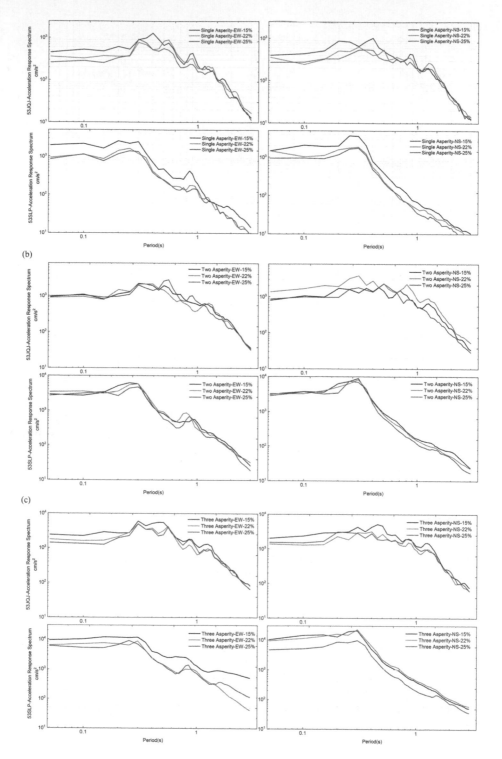

Figure 5. The acceleration response spectrum of different asperity source models at station 53J QJ and 53 SLP. (a) Single asperity model of two stations; (b) Two asperity model of two stations; (c) Three asperity model of two stations.

mainly ground motion characteristics with the empirical Green function method. By considering the uncertainty factors of the source process, especially the uncertainty factors of the asperity parameters, and comparing the characteristics of the present earthquakes' ground motion characteristic with similar magnitude, we estimated that the number of asperities on the fault plane of the Tonghai earthquake is about two. The area of the asperity should account for about 15–22% of the entire fault rupture areas. Using the logical tree method and statistical model is an effective way to deal with the uncertainty factors in the prediction process. The simulation results can reproduce the main seismic characteristics of the Tonghai earthquake.

The prediction of strong vibrations in the future must fully consider the uncertainty of the source model, especially the uncertainty factors of the asperity parameters. Studying the uncertainty of asperity can provide a more suitable source model for the prediction of destructive earthquakes in the future, so that the simulation results can better indicate the seismic characteristics of the near-field of the future earthquake. The simulation results of ground motion parameters will have better and higher reference values in seismic fortification and disaster prevention in the future. By usingthe uncertainty factors to predict the destructive earthquake in the future, the comprehensive characteristics of ground motion in the near-field are evaluated quantitatively, which provides a reference for engineering design decisions.

ACKNOWLEDGMENTS

This research was supported by the National Key R&D Program of China (Grant 2017YFC1500205) and the special fund of the Institute of Geophysics, China Earthquake Administration (Grant Number: DQJB18B20; DQJB17C08;DQJB19B06).

REFERENCES

Aki K. Scaling law of seismic spectrum. *J Geophys Res*, 1967, 72(4): 1217–1231.
Dan K., Watanabe T., Tanaka T. A semi-empirical method to synthesize earthquake ground motions based onapproximate far-field shear-wave displacement. *J Struct ConstrEng*(Transactions of AIJ), 1989(396): 27–36.
Hartzell S.H. (1978) Earthquake aftershocks as Green's functions. *Geophys Res Lett* 5(1): 1–4.
Irikura K. 1983. Semi-empirical estimation of strong ground motion during large earthquake. *Bull Disas-Prev Res*, 33:151–156.
Irikura K. 1986. Prediction of strong acceleration motion using empirical Green's function. Proceedings of the 7th Japan Earthquake Engineering Symposium. Tokyo: Architectural Institute of Japan:151−156.
Kanamori H. (1979). A semi-empirical approach to prediction of longperiod ground motions from great earthquake. *BSSA* 69(6): 1654–1670.
Kanamori H., Anderson D.L. Theoretical basis of some empirical relations in seismology. *B Seismol Soc Am*, 1975, 65(5): 1073–1095.
Li Z.C.,Gao M.T.,et al. 2018. Sensitivity analysis study of the source parameter uncertaintyfactors for predicting near-field strong ground motion. *Acta Geophys*, 66(4): 523–540.
Liu Z., *et al.*, 1999. Tonghai earthquake in 1970. Beijing: Earthquake Press.
Miyake H., Iwata T., Irikura K. 2003. Source characterization for broadband ground-motion simulation: kinematic heterogeneoussource model and strong motion generation area. *Bull SeismolSoc Am* 93(6): 2531–2545.
Somerville P., Irikura K., Graves R., et al. Characterizing crustal earthquake slip models for the prediction of strong ground motion. *Seismological Research Letters*, 1999, 70(1): 59–80.

Risk Analysis Based on Data and Crisis Response Beyond Knowledge – Huang & Nivolianitou (eds)
© 2020 Taylor & Francis Group, London, ISBN 978-0-367-25146-8

The experiences of Mexico City residents during two strong earthquakes in 2017

Jaime Santos-Reyes
Grupo de investigación: SARACS, SEPI-ESIME, Zac. Instituto Politécnico Nacional, D.F. Mexico

Tatiana Gouzeva
Grupo de investigación: SARACS, Instituto Politécnico Nacional, D.F., Mexico

ABSTRACT: This article presents some preliminary results of the experiences of Mexico City residents during the two earthquakes that hit the city in 2017. This cross-sectional study considered a sample size of N = 2,400. Most (76.3%) of the participants were at home during the earthquake on September 7, but the opposite was observed during the September 19 earthquake. For the earthquake on September 7, 45.3% of the participants of the study perceived the quake as 'very strong', and 34.1% of participants has the same perception during the earthquake on September 19. A strong association was found with sex, age, education level and the occupation of the participants and their worries about their house/building collapsing. The article gives an account of the ongoing research project.

Keywords: earthquake, seismic risk, Mexico City, resident, risk perception

1 INTRODUCTION

Disasters triggered by earthquakes usually cause a considerable amount of damage. This is particularly true when communities are not well prepared to mitigate the impacts (Turner et al., 1996; Granger et al., 1999;Yu & Tang, 2017; Maio et al., 2018; Sun et al., 2018; Timar et al., 2018;Santos-Reyes, 2019a,b). Communities are becoming extremely vulnerable given the increasing rate of earthquake occurrence in recent years. For example, the most recent earthquakes that hit Mexico City on September 7 and 19, 2017(Santos-Reyes, 2019a), and elsewhere (Isabelle et al., 2012; Tekeli-Yesil, et al., 2010). Interestingly, the pattern of occurrence of some of these earthquakes is regarded as unusual (Sarlis et al., 2018).

A vast amount of research has been conducted on seismic risk perception, the willingness to take action during an earthquake, and earthquake early warning systems (Dooley et al., 1992; Palm, 1998; Kanti & Hossain-Bhuiyan, 2010; Tekeli-Yesil et al., 2010; Isabelle et al., 2012; Cohen et al., 2013; Wei et al., 2014; Becker et al., 2017; Shapira et al., 2018; Santos-Reyes, 2016, 2019a, 2019b).All these studies have the aim of contributing to seismic risk reduction.

Regarding the unpredictable patterns of earthquake occurrence, it is worth mentioning that this was the case for the two earthquakes that hit Mexico City. First, the 8.2 magnitude of the earthquake on September 7, 2017was not expected. Second, it occurred at midnight when most of the residents of the capital city were at home, and probably sleeping. Third, the earthquake on September 19, on the other hand, occurred during the daytime, and was on the same date as the 1985 earthquake that caused death and destruction in the capital city. Fourth, the time in between these two earthquakes was twelve days. Fifth, Mexico was expecting 'the big one' with the epicenter occurring along the 'Guerrero gap' in the Pacific coast of the country, but the epicenter of the 'bigone' came from inland on September19. This is our new reality and we must learn to live with it, by building community resilience to mitigate the impact of seismic risk (Santos-Reyes, 2019b).

This article presents some preliminary findings of an on going research project on Mexico City residents' risk perceptions of these two earthquakes. In particular, we were interested in determining how residents of the capital city perceive the impact of an earthquake occurring at night and during the daytime. The paper gives an account of the preliminary findings.

2 THE EARTHQUAKES ONSEPTEMBER 7 AND 19 2017

2.1 *The 7 September earthquake*

On 7 September 2017, an earthquake of $M_{wx} = 8.2$ hit the capital city of Mexico. It is thought that the quake started in the south-east of the country. Given the magnitude of the earthquake, it has been regarded as the strongest in Mexico since 1975 ($M_w = 8.1$). The coordinates of the epicenter were 14.761° latitude N and -94.103° longitude W and the depth was 45.9 km (Santos-Reyes, 2019a).The epicenter of the quake was in the Gulf of Tehuantepec, Oaxaca, 133 km south-west of the state of Chiapas, Mexico. The earthquake occurred at 23:49:17 local time(04:49 UTM). The quake affected several states and hundreds of people were killed.

2.2 *The 19 September earthquake*

A second earthquake, of magnitude $M_w = 7.1$, hit Mexico City on September19, just twelve days later. Interestingly, the epicenter was located inland between the states of Puebla and Morelos, 120 km from Mexico City (Santos-Reyes, 2019a). As expected, several other states were affected severely in terms of human loss: according to official data, 330 people were killed, and thousands of buildings were damaged or destroyed by the strong ground movements. The earthquake occurred at 13:14:40, local time. The coordinates of the epicenter were 18.40 latitude N and -98.72 longitude W, and the depth was 57 km (Santos-Reyes, 2019a).

3 MATERIALS AND METHODS

Following the two earthquakes, a cross-sectional study was conducted to collect data for a sample size of $N = 2,400$ participants of Mexico City. To achieve this, a questionnaire was designed that covered several issues, such as seismic risk perception, earthquake early warning system performance, etc. One of the sections focused on how the respondents perceived the intensity of the quake during the occurrence of the two earthquakes. This may be regarded as a non-probability study and the survey was conducted a few weeks after the occurrences of the two earthquakes, from 4 October to 20 November 2017. SPSS, IBM, ver. 25, was employed for analyzing the collected data. The descriptive information was assessed by frequency analysis and cross-tabulations, and the statistical significance of the cross-tabulations was determined by Pearson Chi-Square test.

4 ANALYSIS AND DISCUSSION

4.1 *Locations during the earthquakes*

The following question was included in the questionnaire: Where were you during the earthquake? The possible responses were at work, street, home, shopping mall, driving, and other. The same question was given to the respondents for the two earthquakes, and the results are shown in Figure 1.For example, during the earthquake on September7 (Figure 1, left), most of the participants (76.3%) were at home, probably sleeping given that the quake occurred at 23:49. This was followed by those who were at work (9.9%), and other places (7.5%).

Figure 1 (right) shows where the respondents were during the 19 September earthquake, which occurred in the daytime: 38.9% were at work, 23.5% were in other places, 22.8% were at home, and 11.2% were on the street. In general, Figure 1 gives us an idea of where the

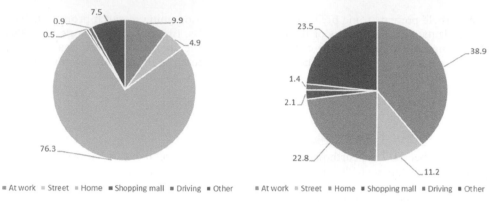

Figure 1. Where the participants were during the September 7 (left) and the September19 (right) earthquakes.

residents of the capital city were located during the occurrence of the earthquakes. As expected, most of the respondents were sleeping during the earthquake on September 7, while most were awake and out of the home during the September19 earthquake.

4.2 *Perception of the intensity of the two earthquakes*

In an attempt to assess, for example, how the respondents perceived the intensity of the quake, the following question was included in the questionnaire: How did you perceived the intensity of the earthquake? The possible responses were based on a Likert-type scale, from very weak to very strong. The results of the frequency data are shown in Figure 2 for both earthquakes.

Regarding the earthquake on September 7, 45.3% of the participants of the study perceived the quake as 'very strong' (Figure 1, left), whereas 34.1% has the same perception during the earthquake on September19 (Figure 1, right). However, when considering 'strong' and 'very strong' together, the results indicate that the earthquake on 19 September was the one felt very strongly in the city (93.6%). This may be explained because the epicenter of the earthquake was close to the capital city (about 120 km away, see section 2.2).The results presented in the subsequent sections, are related to the participants' experiences during the earthquake that occurred while they were sleeping only: the quake on September 7 2017.

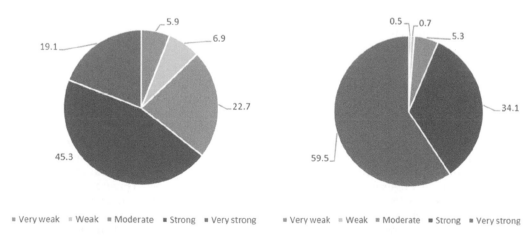

Figure 2. Responses of the participants' perception on the intensity of the September 7 (left) and the September 19 (right) earthquakes.

The destruction of buildings may be regarded as one of the main consequences of earthquakes: if the building is not earthquake-resistant, then there will be consequences in terms of human loss. In an attempt to assess how the participants perceived the vulnerability of their houses or buildings where they were during the earthquake, the following question was included in the questionnaire: How worried were you that your house/building would collapse during the earthquake? The potential response options, in a Likert-type scale were as follows: not at all, a little, neutral (or I don't know), worried, and very worried. The results are shown in Figure 3 and Table 1.

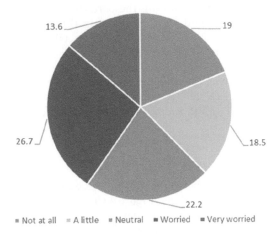

■ Not at all   ■ A little   ■ Neutral   ■ Worried   ■ Very worried

Figure 3.   Responses tothe question 'How worried were you that your house/building would collapse during the earthquake?'.

Table 1.   Relationships of the variables included in the analysis.

	Not at all*	A little*	Neutral*	Worried*	Very worried*
	N(%)	N(%)	N(%)	N(%)	N(%)
Sex					
Men	188 (41.3)	221 (49.7)	248 (46.5)	296 (46.2)	124 (38.0)
Women	267 (58.7)	224 (50.3)	285 (53.5)	345 (26.1)	202 (62.0)
Age					
≤22	206 (45.3)	183 (41.1)	241 (45.2)	212 (33.1)	79 (24.2)
23-33	125 (27.5)	120 (27.0)	171 (32.1)	190 (29.6)	85 (26.1)
34+	124 (27.3)	142 (31.9)	121 (22.7)	239 (37.3)	162 (49.7)
Education level					
Primary/Secondary	28 (6.2)	49 (11.0)	32 (6.0)	52 (8.1)	49 (15.0)
High School	295 (64.8)	282 (63.4)	352 (66.0)	394 (61.5)	162 (49.7)
Undergraduate	102 (22.4)	77 (17.3)	118 (22.1)	155 (24.2)	94 (28.8)
Postgraduate	30 (6.6)	37 (8.3)	31 (5.8)	40 (6.2)	21 (6.4)
Occupation					
Students	218 (47.9)	215 (48.3)	253 (47.5)	254 (39.6)	87 (26.7)
Public &Private Employees	142 (31.2)	144 (32.4)	181 (34.0)	261 (40.7)	160 (49.1)
Educational Sector Employees	36 (7.9)	28 (6.3)	29 (5.4)	47 (7.3)	26 (8.0)
Other	59 (13.0)	58 (13.0)	70 (13.1)	79 (12.3)	53 (16.3)

* % are given within the worries of the house/building collapsing.

Overall, 40.3% of the respondents were 'worried' to 'very worried' that their house/building was going to collapse during the earthquake (Figure 3). A relatively high percentage 'did not know'(22.2%), and 19% of the participants did not 'worry at all'. Table 1 shows the results of the relationship of the variables considered in the analysis. For example, based on Chi-square tests results, a significant relationship was found between the participants' worries about their house/building collapsing and their sex:$X2(4, N = 2,400) = 13.642, p = .009$. In particular, women were 'very worried 'about their house/building collapsing (62.0%), whereas, men were 'a little' worried (49.7%).When considering the variable age of the participants, and based on Chi-square results, there was a significant relationship between these variables:$X2 (8, N = 2,400) = 91.685, p<.001$. Participants age 34+ were 'very worried'(49.7%) about their house/building collapsing during the earthquake. Relatively young participants (≤22 years old) were not worried 'at all' (45.3%) during the earthquake, while 45.2% 'did not know'. Also, a strong association was found when considering the variable education level of the participants of the study ($X2(12, N = 2,400) = 50.229, p<.001$). Participants with an undergraduate level of education were 'very worried' during the earthquake (28.8%), and those with a basic education level (primary and secondary) were also 'very worried' (15.0%). Finally, a strong association was found when considering the variable associated with the occupations of the respondents ($X2(12, N = 2,400) = 29.187, p<.001$). Participants within the category 'other' (16.3%) and 'E&P Employees' (49.1%) were 'very worried' during the earthquake.

4.4 Actions taken during the earthquake

Knowing what actions were taken by people during the earthquake occurrence may help to ascertain whether they did the correct thing. Figure 4 shows the results of the analysis.

Figure 4 shows that 26.7% (641/2400) of the participants of the study 'left their home walking', followed by those who 'left the building walking' (11.8%; 282/2400). Interestingly, 8.8% of the respondents just stayed 'in bed' during the earthquake.

4.5 Intensity of fear during the earthquake

In general, earthquakes cause fear and may be regarded as the deadliest events on earth. We included a question related to fear in the questionnaire: How much fear did you experience during the earthquake that night? A numerical scale was used to assess the intensity of fear (0 = x no fear at all, and 10 = a lot of fear). The results showed that mean = 5.0289, SD = 3.10, and variance = 9.610.

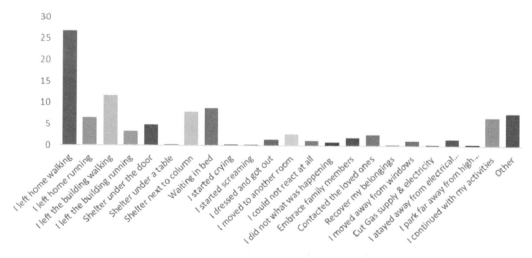

Figure 4.   Responses of the actions taken once the participants experienced the ground shaking.

5 CONCLUSION

The paper has presented some preliminary results of an ongoing research project in relation to the two strong earthquakes that hit Mexico City in 2017. Some of the results are as follows:

1) In general, most of the respondents (76.3%)were at home during the earthquake on September 7, and most were away from home during the September 19earthquake.
2) Regarding the earthquake on September 7,45.3% of the participants perceived the quake as 'very strong'. However, when considering 'strong' and 'very strong' together, the results indicate that the earthquake on September 19 was the one felt strongly in the city (93.6%).
3) Overall, 40.3% of the respondents were 'worried' to 'very worried' that their home/building was going to collapse during the earthquake. Also, a strong association was found with gender, age, education level and the occupation of the participants.
4) Around 26.7% of the participants of the study 'left home walking', followed by 11.8% who 'left the building walking'. Interestingly, 8.8% of the respondents just stayed 'in bed' during the earthquake.

ACKNOWLEDGMENTS

This research project was supported by CONACYT & SIP-IPN, under the following grants: CONACYT- No-248219 & SIP-IPN: No-20196424.

REFERENCES

Becker, J.S., Paton, D., Johnston, D.M., Ronan, K.R., McClure, J. 2017. The role of prior experience in informing and motivating earthquake preparedness, *International Journal of Disaster Risk Reduction.* 22: 179–193.

Cohen, O., Leykin, D., Lahad, M., Goldberg, A., Aharonson-Daniel, L. 2013. The conjoint community resiliency assessment measure as a baseline for profiling and predicting community resilience for emergencies, *Technol. Forecast. Social. Change* 80:1732–1741.

Dooley, D., Catalano, R., Mishra, S., Serxner, S. 1992. Earthquake preparedness: predictors in a community survey1, *J. Appl. Social. Psychol.* 22:451–470.

Granger, K., Jones, T., Leiba, M., and Scott, G. 1999. *Community Risk in Cairns: A Provisional Multi Hazard Risk Assessment.* AGSO Cities Project Report No. 1. Australian Geological Survey Organization: Canberra, Australia.

Beck, E., André-Poyaud, I., Davoine, P.A., Chardonnel, S. and Lutoff, C. 2012. Risk perception and social vulnerability to earthquakes in Grenoble (French Alps). *Journal of Risk Research*, 15(10):1245–1260.

Kanti, P.B., and Hossain-Bhuiyan, R. 2010. Urban earthquake hazard: perceived seismic risk and preparedness in Dhaka City, Bangladesh. *Disasters*, 34(2):337–359.

Maio, R., Ferreira, T.M., Vicente, R. 2018. A critical discussion on the earthquake risk mitigation of urban cultural heritage assets. *International Journal of Disaster Risk Reduction*, 27:239–247.

Palm, R. 1998. Urban earthquake hazards –The impacts of culture on perceived risk and response in the USA and Japan. *Applied Geography*, 18(I):3546.

Santos-Reyes, J. 2019a. How useful are earthquake early warnings? The case of the 2017 earthquakes in Mexico City. *International Journal of Disaster Risk Reduction.* In press. doi: 10.1016/j.ijdrr.2019.101148

Santos-Reyes, J. 2019b. *Earthquakes-Impact, Community Vulnerability, and Resilience.* IntechOpen, Croatia.

Santos-Reyes, J., Santos-Reyes, G., Gouzeva, T., & Velázquez-Martínez, D. 2016. Schoolchildren's earthquake knowledge, preparedness, and risk perception of a seismic-prone region of Mexico. *Human and Ecological Risk Assessment: An International Journal.* Vol. 23(3),494–507. DOI: 10.1080/10807039.2016.1188368

Sarlis, N.V., Skordas, E.S., Varotsos, P.A., Ramirez-Rojas, A., Flores-Márquez, E.L. 2018. Natural time analysis: On the deadly Mexico M8.2 earthquake on 7 September 2017. *Physica A* 506, 625–634.

Shapira, S., Aharonson-Daniel, L., Bar-Dayan, Y. 2018. Anticipated behavioral response patterns to an earthquake: The role of personal and household characteristics, risk perception, previous experience and preparedness. *International Journal of Disaster Risk Reduction*, 31, 1–8.

Sun, Y, Yang, Q., Chen, P. 2018. A study of spatial evolution patterns of tourist destinations disaster risks. *Journal of Risk Analysis and Crisis Response*. 8(1),35–42.

Tekeli-Yeşil, S., Dedeoğlu, N., Tanner, M., BraunFahrlaender, C., and Obrist, B. 2010. Individual preparedness and mitigation actions for a predicted earthquake in Istanbul. *Disasters*, 34(4):910−930.

Timar, L., Grimes, A., Fabling, R. 2018. That sinking feeling: The changing price of urban disaster risk following an earthquake. *International Journal of Disaster Risk Reduction*, 31, 1326–1336.

Turner, R.H., Nigg, J.M., and Paz, D.H. 1996. *Waiting for disaster: Earthquake watch in California.* Berkeley: University of California Press. USA.

Wei, H.H., Shohet, I.M., Skibniewski, M.J., Levy, R., Shapira, S., Aharonson-Daniel, L., Levi, T., Salamon, A., Levi, O. 2014. Economic feasibility analysis of pre-earthquake strengthening of buildings in a moderate seismicity/high vulnerability area, *Proc. Econ. Financ.* 18, 143–150.

Yu, X., Tang, Y. 2017. A critical review on the economics of disasters. *Journal of Risk Analysis and Crisis Response*. 7(1),27–36.

Risk Analysis Based on Data and Crisis Response Beyond Knowledge – Huang & Nivolianitou (eds)
© 2020 Taylor & Francis Group, London, ISBN 978-0-367-25146-8

Rupture direction study of two Tonghai Ms 5.0 earthquakes on August 13 and 14, 2018

Xueliang Chen*, Quanbo Luo, Mengtan Gao & Sen Qiao
Institute of Geophysics, China Earthquake Administration. Beijing, China

Shicheng Li & Jianwen Cui
Yunnan Earthquake Agency, Kunming, China

ABSTRACT: Two Ms 5.0 earthquakes occurred in Tonghai, Yunnan Province, at 01:44 am on August 13 and 03:50 am on August 14, 2018. The China Seismic Network measured the surface wave magnitudes of both earthquakes as Ms5.0, with almost the same epicenter location, and epicenter depths of 7.0 km and 6.0 km, respectively. It is generally believed that the rupture directions of these two earthquakes was NE to SW, and they are caused by NE–SW direction buried faults by relocation of earthquake sequences. Although the surface wave magnitude was the same, the moment magnitudes were different. The strong motion records obtained in the near field also clearly reflected the difference in the intensity and the rupture direction of the two earthquakes. Eliminating the site effect of the earth surface structure, the free outcrop bedrock records, which are twice the seismic incident waves of underground bedrock, were inverted by equivalent linearization method. The new distributions of ground seismic intensity on the free outcrop bedrock site were obtained. According to the distribution map, it is obvious that the August 13, 2018 Tonghai Ms5.0 earthquake has a NE–SW direction buried fault. But the August 14, 2018 Tonghai Ms 5.0 earthquake has a NW–SE direction buried fault.

Keywords: rupture direction, site response, strong ground motion records, the seismic fault

1 INTRODUCTION

An Ms 5.0 earthquake occurred in Tonghai County, Yuxi City, Yunnan Province (24.19N, 102.71°E) at 01:44 on August 13, 2018 (local time) with epicenter depth 7.0 km. Subsequently, on 03:50 on August 14, 2018, another Ms 5.0 earthquake occurred almost in the same place with an epicenter depth 6.0 km. Because the magnitude of earthquakes is relatively small and the surrounding seismic geological environment is very complex, it is very difficult to determine the seismogenic faults of these two earthquakes by seismological methods.

Historic seismic records show that 13 groups of earthquakes (15 in total) with $M \geq 5.0$ occurred in this area within a 50 km radius of the epicenter of the Tonghai Ms 5.0 earthquake. There were nine earthquakes in the M5.0–5.9 range, four earthquakes in the M6.0–6.9 range and two earthquakes in the M7.0–7.9 range. The largest earthquake was the Tonghai Ms7.8 earthquake on January 5, 1970. Two Ms 5.0 earthquakes were located in the seismic source region of the 1970 Tonghai Ms 7.8 earthquake.

*Corresponding author: xueliang_chen@aliyun.com

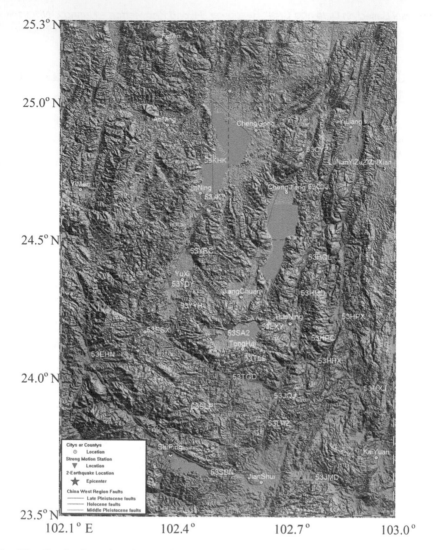

Figure 1. The distribution of stations and faults near the epicenter of the Tonghai Ms 5.0 earthquakes.

Historic seismic statistics show that Tonghai area is a relatively active strong earthquake area (Xueze Wen et al., 2011). This area is located at the south end of Sichuan-Yunnan rhombic block. There are many different direction branches: the NW–SE direction includes the YuChuan Fault, the QuJiang Fault, and the ShiPing-JianShui Fault, etc.; the NE–SW direction includes the PuDu River (middle-south section) Fault, the XiaoJiang XiZhi Fault, th XiaoJiang DongZhi Fault, and the Southern Segment of the XiaoJiang Fault, etc. See Figure 1. Complex geological structures may be the cause of the frequent occurrence of moderate and strong earthquakes in this area.

The moment magnitude reflects the difference between the August 13 and August 14 earthquakes. Using regional network data, the moment magnitudes determined by the Yunnan Earthquake Agency were Mw 5.0 and Mw 4.8 and the source depths were 10 km and 8 km, respectively.

Fortunately, there are 46 strong motion stations that include permanent stations and temporary station in this area, many of which have recorded high-quality strong motion records. This make it is possible to study rupture direction of moderate magnitude earthquakes through analyzing strong motion records. Therefore, the research results are of great significance to the development of seismological methods, the understanding of seismic mechanism, the determination of seismogenic faults and the study of regional seismic risk.

2 STRONG MOTION STATIONS, RECORDS AND PGA DISTRIBUTION

During the August 13, 2018 Ms5.0 Tonghai earthquake, 43 permanent stations in the national strong motion network obtained strong motion records. The nearest station was 2Hao Dian– 3SA2, whose epicenter distance was 3.4 km. The maximum peak ground acceleration (PGA) was 460.5 gal, which is an EW direction acceleration record at the 2Hao Dian–53SA2 station, and its instrumental intensity was $I_{instr}8.7$. The farthest station from the epicenter was JiGongXueXiao–53SJZ, and the maximum epicenter distance was 101.3 km.

Similarly, 33 stations, including three temporary stations – HeXi-53THH, DongQu-53LDQ and XingYi-53LXY – obtained strong motion records during the August 14, 2018 Ms5.0 Tonghai earthquake. The nearest station, 2HaoDian–53SA2 recorded the maximum PGA, which is 278.9 gal in the EW Direction. The instrument intensity was $I_{instr}7.9$.

Figures 2 to Figure 4 show the strong ground records for the PGA and PGA arrives time of the XiongGuan–53XJG station during the August 13 earthquake, LiShan–53TLS station during the August 13 earthquake, LiShan–53TLS station during the August 14 earthquake, respectively. From these figures, we can see that the amplitude of the 53XJG station was much bigger than at the 53TLS station during the August 13 earthquake, and that of the 53TLS station during the August 14 earthquake, while the amplitudes at the 53TLS station during the August 13 and August 14 earthquake are very similar. The quality of all strong motion records is excellent.

The amplitude, the frequency spectrum, and the duration of the strong ground motion (the records) will be significantly affected by the earth shallow-surface soil structure, which is usually called the engineering site. As a result, the PGA distributions of the records in the earth surface cannot well represent the incidence wave-field of the strong ground motion in the deeper bedrock. It is necessary to reduce the influence of site effect. As a preliminary understanding of ground motion characteristics, in

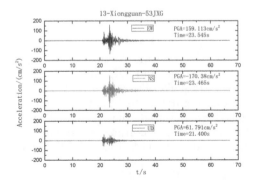

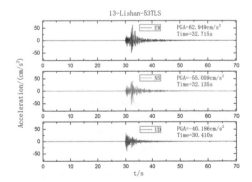

Figure 2. Strong ground records of 53XJG station, August 13, 2018 Ms 5.0 Tonghai E.

Figure 3. Strong ground record of 53TLS station, August 13, 2018 Ms 5.0 Tonghai E.

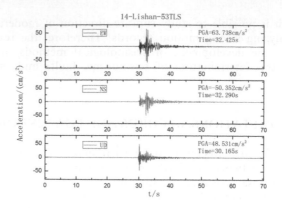

Figure 4. Strong ground record of 53TLS station, August 14, 2018 Tonghai E.

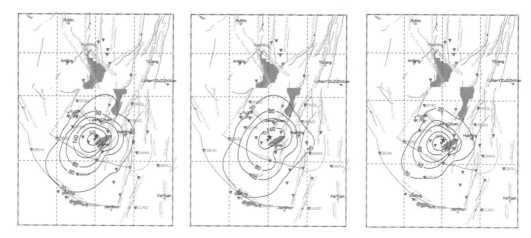

Figure 5. EW direction distribution of PGA in Aug. 13, 2018
Tonghai E.

Figure 6. NS direction distribution of PGA in Aug. 13, 2018 Tonghai E.

Figure 7. UD direction distribution of PGA in Aug. 13, 2018
Tonghai E.

Figures 5 to Figure 7 we show the EW direction distribution, NS direction distribution, and UD direction distribution of records PGA during the August 13 earthquake.
The EW direction distribution, NS direction distribution, and UD direction distribution of the PGA records during the August 14 earthquake are shown in Figures 8 to
Figure 10. Comparing the two earthquakes from Figures 5 to 10, we find that the
energy of the August 13 earthquake was indeed greater than that of the August 14
earthquake. If the shallow-surface soil structure effect is not considered, assuming that
the surface is like bedrock from the distribution shape of PGA, the NE–SW direction
hidden fault of XiaoJiangXiZhi fault zone looks like the seismogenic fault of the two
earthquakes, which is the conclusion of many scholars (Wang, Liu, et al., 2018) based
on the comprehensive analysis of the spatial distribution of the reoriented seismic
sequence and geological data. However, our results show that this understanding is not
entirely correct.

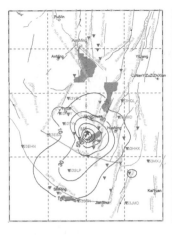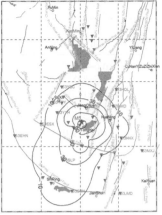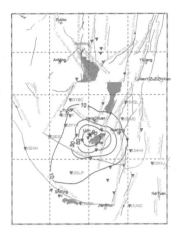

Figure 8. EW direction distribution of PGA in Aug. 14, 2018 Tonghai E.

Figure 9. NS direction distribution of PGA in Aug. 14, 2018 Tonghai E.

Figure 10. UD direction distribution of PGA in Aug. 14, 2018 Tonghai E.

3 OUTCROP BEDROCK RESPONSE BY EQUIVALENT LINEARIZATION METHOD

3.1 *Methods and site models*

We used the equivalent linearization method to analyze and invert the incidence wave-field and the seismic response of the deep bedrock (underground) for eliminating the shallow-surface site effect. The free outcrop bedrock response was two times the incidence wave-field of the deep bedrock. The basic idea of the equivalent linearization method is utilizing an equivalent shear modulus and equivalent damping instead of the shear modulus and damping ratio of all the different strain amplitude in a roughly equivalent significance of the overall dynamic effects. Because shear modulus and damping ratio has nothing to do with the strain amplitude, the whole problem changes into a linear problem (Zhenpeng, 2002). The equivalent linearization method as an expedient method is widely used in the engineering field, and a number of numerical experiments show that this method is more effective for certain problems (Liao Zhenpeng, 2002).

At present, the criteria equivalent linearization program LSSRLI is widely used in the seismic safety evaluation of the China Earthquake Administration (Xiaojun, 1989). This program is used for the seismic response (PGA and strong motion time history) of outcrop bedrock in this paper. As a calculating example, Tables 1 – Table 3 give the site model data of the Lishan borehole station–53TLS station, the site model data of Longpeng borehole station–53SLP station, and the dynamic shear modulus ratio and dynamic damping ratio of each soil type

Table 1. Site model data of the Lishan borehole station.

S.N.	Soil description	Soil type	Soil depth/m	Thickness/m	Vs/(m/s)	Density/ (kg/m^3)
1	Back fill	1	0.0~2.2	2.2	156.0	1850
2	Silty clay	2	2.2~4.5	2.3	175.0	1960
3	Sand gravel	3	4.5~5.6	1.1	192.0	2050
4	Gravelly silty clay	4	5.6~7.5	1.9	223.0	1980
5	Sand gravel	3	7.5~10.0	2.5	256.0	2050
6	Limestone	6	10.0~13.5	3.5	536.0	2200
7	Mudstone	5	13.5~14.5	1.0	453.0	2100
8	Sand	7	14.5~15.2	0.7	324.0	1770
9	Limestone	6	15.2~17.8	2.6	550.0	2300
10	Computing bedrock	10	17.8~		580.0	2400

Table 2. Site model data of the Longpeng borehole station.

S.N.	Soil description	Soil type	Soil depth/m	Thickness/m	Vs/(m/s)	Density/ (kg/m^3)
1	Back fill	1	0.0~0.5	0.5	165.0	1850
2	Clay	8	0.5~8.0	7.5	195.0	1970
3	Sandy clay	9	8.0~12.0	4.0	240.0	1940
4	Clay	8	12.0~13.6	1.6	320.0	1970
5	Sandy clay	9	13.6~16.1	2.5	380.0	1940
6	Limestone	5	16.1~19.0	2.9	536.0	2200
7	Computing bedrock	10	19.0~		580.0	2400

Table 3. The dynamic shear modulus ratio and dynamic damping ratio of each soil type corresponding to the dynamic shear strain standard value.

Soil type	γ_d	5×10^{-6}	1×10^{-5}	5×10^{-5}	1×10^{-4}	5×10^{-4}	1×10^{-3}	5×10^{-3}	1×10^{-2}
1	G_d/G_{max}	0.9600	0.9500	0.8000	0.7000	0.3000	0.2000	0.1500	0.1000
	λ_d	0.0250	0.0280	0.0300	0.0350	0.0800	0.1000	0.1100	0.1200
2	G_d/G_{max}	0.9902	0.9805	0.9097	0.8345	0.5020	0.3352	0.0916	0.0479
	λ_d	0.0182	0.0251	0.0517	0.0687	0.1153	0.1321	0.1529	0.1563
3	G_d/G_{max}	0.9900	0.9700	0.9000	0.8500	0.7000	0.5500	0.3200	0.2000
	λ_d	0.0040	0.0060	0.0190	0.0300	0.0750	0.0900	0.1100	0.1200
4	G_d/G_{max}	0.9800	0.9700	0.8400	0.7300	0.4000	0.2500	0.0700	0.0300
	λ_d	0.0120	0.0150	0.0370	0.0560	0.1120	0.1370	0.1700	0.1800
5	G_d/G_{max}	0.9981	0.9895	0.9325	0.9000	0.7832	0.7250	0.6028	0.5500
	λ_d	0.0069	0.0080	0.0120	0.0150	0.0251	0.0300	0.0412	0.0460
6	G_d/G_{max}	1.0000	1.0000	1.0000	1.0000	1.0000	1.0000	1.0000	1.0000
	λ_d	0.0040	0.0080	0.0100	0.0150	0.0210	0.0300	0.0360	0.0460
7	G_d/G_{max}	0.9650	0.9350	0.7750	0.6600	0.3000	0.2500	0.1050	0.0900
	λ_d	0.0060	0.0100	0.0300	0.0450	0.0880	0.1030	0.1240	0.1300
8	G_d/G_{max}	0.9920	0.9830	0.9200	0.8510	0.5400	0.3750	0.1150	0.0630
	λ_d	0.0227	0.0271	0.0515	0.0723	0.1469	0.1805	0.2295	0.2388
9	G_d/G_{max}	0.9962	0.9924	0.9631	0.9289	0.7231	0.5663	0.2071	0.1155
	λ_d	0.0603	0.0707	0.1021	0.1190	0.1633	0.1812	0.2085	0.2139
10	G_d/G_{max}	1.0000	1.0000	1.0000	1.0000	1.0000	1.0000	1.0000	1.0000
	λ_d	0.0000	0.0000	0.0000	0.0000	0.0000	0.0000	0.0000	0.0000

corresponding to dynamic shear strain standard value for two borehole sites. The Poisson ratio of each soil type is omitted. These data are the parameters required for calculating an equivalent linearization program.

3.2 The results

The calculated EW direction bedrock PGA distribution, NS direction bedrock PGA distribution, and UD direction bedrock PGA distribution during the August 13, 2018 Ms5.0 Tonghai earthquake are shown in Figures 11–13, respectively. Comparing Figures 5–7, the shape of bedrock PGA distribution are very different from that of earth surface PGA distribution. There is a similar understanding during the August 14 earthquake. Figures 14–16 show the calculated EW direction bedrock PGA distribution, NS direction bedrock PGA distribution, and UD direction bedrock PGA distribution during the August 14, 2018 Ms5.0 Tonghai earthquake. Compared with the shape of the earth surface PGA distribution in Figures 8–10, the shape of bedrock PGA distribution has changed significantly, especially in the regions with peak values greater than 30gal, The long axis direction of the shape changes from NE–SW direction to NW–SE direction. Therefore, these two earthquakes are especially like a group of conjugate earthquakes.

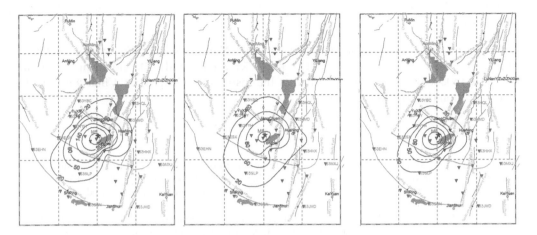

Figure 11. Bedrock PGA EW direction in Aug. 13, 2018 Tonghai E.

Figure 12. Bedrock PGA NS direction in Aug. 13, 2018 Tonghai E.

Figure 13. Bedrock PGA UD direction in Aug. 13, 2018 Tonghai E.

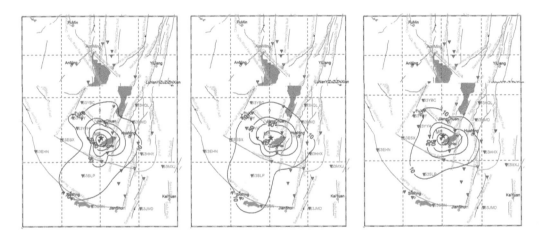

Figure 14. Bedrock PGA EW direction in Aug. 14, 2018 Tonghai E.

Figure 15. Bedrock PGA NS direction in Aug. 14, 2018 Tonghai E.

Figure 16. Bedrock PGA UD direction in Aug. 14, 2018 Tonghai E.

4 CONCLUSION AND DISCUSSION

According to the distribution maps, it is obvious that the seismogenic fault of the August 13, 2018 Ms5.0 Tonghai earthquake is NE–SW direction fault or buried fault, which may be the nearby MingXing – ErJie Fault, or a hidden fault that was parallel to it in the XiaoJiang XiZhi Fault group. This is similar to the conculsions by Guangming Wang, Zifeng Liu, et al. (2018). However, for the Ms5.0 earthquake that occurred in Tonghai on August 14, 2018, its seismogenic fault is a NW–SE direction surface-exposed fault or one buried fault, which may be the nearby YuChuan Fault, or a hidden fault that was parallel to it in the branch of the QuJiang Fault or the YuChuan Fault. From the above analysis, it can be concluded that the August 13, 2018 Ms5.0 Tonghai earthquake and the August 14, 2018 Ms5.0 Tonghai earthquake should be a pair typical conjugate earthquake.

ACKNOWLEDGMENTS

This research work was supported by the National Natural Science Foundation of China (51678537, 51278470), the National Key Research and Development Program (2017YFC1500205), and the Special Fund of the Institute of Geophysics, China Earthquake Administration (DQJB19B06).

REFERENCES

Liao Zhenpeng, Li Xiaojun. (1989). The equivalent linear method of seismic response of the surface soil layer, Seismic microzonation – theory and practice. Seismological Press: Beijing, pp. 141–153.

Liao Zhenpeng. (2002). Introduction to wave motion theories in engineering (second edition). Science Press: Beijing, pp.49–66.

Wen Xueze, Du Fang, Long Feng, et al. (2011). Tectonic dynamics and correlation of major earthquake sequences of the Xiaojiang and Qujiang-Shiping fault systems, Yunnan, China. *Science China Earth Sciences* 41(5): 713–724.

Guangming Wang, Zifeng Liu, Xiaoyan Zhao, et al. (2018). Relocation of Tonghai Ms5.0 Earthquake sequence in 2018 and discussion of it's seismogenic fault. *Journal of Seismological Research* 41(4): 503–510.

Risk Analysis Based on Data and Crisis Response Beyond Knowledge – Huang & Nivolianitou (eds)
© 2020 Taylor & Francis Group, London, ISBN 978-0-367-25146-8

Research on risk assessment methods of urban disaster chain during the Beijing Winter Olympics

Mengting Liu, Wei Zhu*, Jianchun Zheng & Qiuju You
Beijing Key Laboratory of Operation Safety of Gas, Heating and Underground Pipelines, Beijing Research Center of Urban System Engineering, Beijing Academy of Science and Technology, Beijing, China

ABSTRACT: Urban operation safety will be the fundamental requirement for the success of the 2022 Olympics and Paralympics Winter Games in Beijing. Natural disasters often affect the urban infrastructures and lead to operations risks. This paper proposes a risk assessment method of the urban disaster chain based on Bayesian networks. First, according to the risk identification of natural disasters in winter in Beijing, a risk tree for disaster chain is constructed, considering the characteristics of natural disasters' concurrency and coupling, and the relationships between natural disasters and infrastructure accidents. Then, in order to determine the hazard of the disaster chain, the Bayesian network method is employed to calculate the probability and severity of each event node in the disaster chain. Finally, taking the disaster of cold wave as an example, a quantitative risk assessment is conducted by the risk assessment method based on the index system, indicating that the method is feasible for the risk assessment of urban disaster chain.

Keywords: Na-Tech risk, disaster chain, Bayesian network, Olympics Winter Games.

1 INTRODUCTION

The 2022 Olympics and Paralympics Winter Games will be co-hosted by Beijing and Zhangjiakou. This is the largest event of ice and snow sports in the world, and an important opportunity to display China's national image, promote national development and inspire the national spirit. In terms of public safety, beside the safety of the Winter Olympics venues and stadiums, urban safety is also an important basic condition for the success of the Beijing 2022 Winter Olympics and Winter Paralympics.

Urban risks are diverse and comprehensive. A region is often affected by many disaster-causing factors, for example, a cold wave disaster is accompanied by low temperatures, high winds, heavy snow and other disasters. And natural disasters may be harmful for urban infrastructure. Therefore, the risk assessment of a single disaster is not enough to reflect the comprehensive risk for a region. It is necessary to carry out the risk assessment of multi-disasters. Xue et al. (2005) proposed a three-dimensional model of integrated urban disaster risk management: stages, types of risk, and ranks of risk, consisting of three stages and six links. Each link has a matrix pattern of different levels of management capability and different types of risk. At the same level, multiple departments can be managed collaboratively. Thus, the model reflects the evaluation value of risk and the corresponding rescue capability in each stage. Gai, Weng and Yuan (2011) projected a coupling risk assessment method for natural disasters and accidents. According to the superposed level of hazard and vulnerability, they reclassified the comprehensive risk level; and according to the coupling relation of dangerous

---

*Corresponding author: zhuweianquan@126.com

sources, the coupling times of secondary disasters are obtained and then the disaster coupling diagram in the region are drawn. Ming et al. (2013) summarized the domestic and overseas methods of the multi-disaster risk assessment for natural disasters. These methods include two types: (1) to integrate the risk factors and assess the multi-disaster risk according to the functional relationship between factors; (2) to integrate the result of single risk, first conduct the single disaster risk assessment then synthesize the risk assessment results of each single disaster. hang and Zhou (2013) studied the chain occurrence probability of geological disasters caused by earthquakes and proposed the conceptual model of disaster chain risk assessment. In term of the comprehensive risk assessment of multiple disasters, many existing methods belongs to the type of the risk result superposition, but the mutual relations between risk events are less considered.

There are secondary and coupling relationships among natural disasters and between natural disasters and accidents. Zheng and Zhang (2007) analyzed whether the annual average temperature in Beijing is correlated with extreme events, such as high temperatures, low temperatures, or strong winds. Ji, Weng and Zhao (2009) proposed a quantitative risk analysis method for the probability and consequences of the chain of emergency events. First, the method uses a probability function, hierarchical analysis and comprehensive evaluation to determine the trigger probability between events. Second, it uses methods such as compound event probability analysis and event consequence assessment model to analyze and calculate the probability and consequence of possible chain of events scenarios. Finally, quantitative risk assessment for the chain of events is conducted by the methods based on evolutionary dynamics or an index system. Liu and Wu (2015) proposed a modeling method of disaster chain risk assessment model based on complex network structures. The method uses complex network structure to characterize the evolution characteristics of the disaster chain and a Bayesian formula to calculate the joint probability distribution of disaster loss levels of each node. The existing methods of accident chain risk assessment are relatively simple in terms of risk factors and carriers. These methods do not consider the comprehensive risks of multiple disasters from the perspective of urban operation, and there are few methods to realize quantitative evaluation.

This article proposes a method of disaster chain risk assessment based on Bayesian networks. First, it identifies the risk of winter natural disaster in Beijing and put forward an index system of natural disaster risk assessment. The calculation of index "hazard "should consider the integrated consequences of multiple disaster risks. Therefore, a risk tree is constructed based on the secondary coupling characteristics of natural disasters and infrastructure accidents. Then the risk tree is transformed into Bayesian networks by structure learning and parameter learning. The trigger probability of each node in the disaster chain is calculated by the reasoning method of Bayesian networks. Combining with the severity of the consequences of each single disaster risk event, the hazard of risk can be obtained.

2 RISK IDENTIFICATION OF NATURAL DISASTERS AT THE BEIJING WINTER OLYMPICS

According to the Regulations for the Prevention of Meteorological Disasters (revised in 2017), meteorological disasters refer to the disasters caused by typhoons, rainstorms (heavy snow), cold waves, strong winds (sandstorm), low temperatures, high temperatures, drought, thunder, hail, frost and fog. In Beijing, common meteorological disasters in winter include cold waves, strong winds, low temperatures and heavy snow.

The 2022 Beijing Winter Olympics will be held on February 4–20 and the Winter Paralympics will be held on March 4–13. The venues and stadiums will be in downtown Beijing, its suburban Yanqing District, and Zhangjiakou City in Hebei Province. The average monthly temperature of Beijing and Yanqing from 1980 to 2017 is shown in Figure 1, including the annual change of average temperature in the four months of December to March, with an overall upward trend. The overall trend is closely related to the rise in winter minimum temperatures caused by global warming in recent decades, but it is still hugely volatile, and some years there have been emerged significant low temperatures.

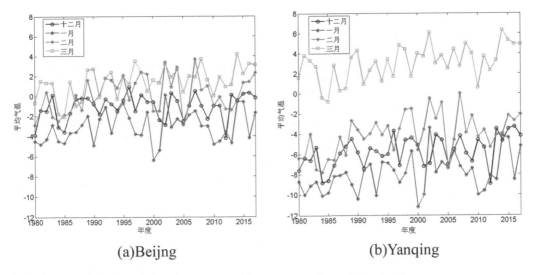

(a)Beijng   (b)Yanqing

Figure 1.　Annual change in the average monthly temperature from 1980 to 2017.

From September of one year to April of the following year, cold wave is one of the major catastrophic weather processes in Beijing, and it is also an important reason for the delay or advance of the season and abnormal weather. For example, the Beijing meteorological service issued a yellow alert for cold snap at 4:30 pm, on January 21, 2016. On January 22, affected by the strong cold air, the lowest temperature dropped around 10°C, the lowest temperature in the plains was -16~-17°C, and the lowest temperature in the mountainous area was -20~-23°C, accompanied by force4–5 north wind and gusts up to force 7.

A significant drop in temperatures is the major characteristic of a cold wave process. It is often accompanied by winds, sandstorms, blizzards and other disastrous weather. This can cause a variety of serious meteorological disasters and bring greater threats to transportation and aviation safety, which affect people's life and production. For example, on December 7, 2003, light snow fell in the downtown area of Beijing because of a strong cold air, and the roads were frozen. Traffic congestion is serious across the entire city. Freezing rain caused by cold wave led to rime on the wires. Because of the influence of wind and the weight of the rime, electrical wires and telephone wires were broken and many poles collapsed, which disrupted power transmission and communications. At the same time, winds from the cold wave caused "wind flashover" and "pollution flashover" which lead to short circuits of electrical wires. In addition, the cold wave increased the load on the power sector.

Snowstorms tend to occur in late winter and early spring. Statistics show that February is the month with the most snow days, followed by January and March. Since records began, the average annual number of heavy snow days in Beijing is only 0.4 days. Around every three years there is a heavy snow, but such heavy snow has a major impact on urban infrastructure and transportation systems. On January 4, 2010, heavy snow fell throughout the city of Beijing. The maximum snowfall in Juyongguan was 27 mm, Huairoureceived 23 mm, Changping received 16 mm, and the average snowfall downtown was 13 mm. Within 24 hours, there was the largest amount of snowfall in Beijing in January since 1951, including four records for the longest duration, the most extensive coverage, the largest precipitation and the thickest snow. On November 3, 2012, Yanqing was hit by the heaviest snowstorm since 1960. In the whole district, the average precipitation was 50.6 mm, the maximum precipitation reached 75.9 mm and the average snow depth was 48cm. After the heavy snow, the depth of snow on the road was around 50cm, and the depth of snow on the mountain was even thicker. Many trees fell and the power was cut, and thousands of people were stranded on the Badaling highway.

3 THE RISK ASSESSMENT INDEX SYSTEM OF NATURAL DISASTER FOR BEIJING WINTER OLYMPICS

According to the United Nations department of humanitarian affairs, the formula for risk assessment should be: risk (R) = hazard (H) × vulnerability (V). The framework for the public safety science and technology system can be represented by the "public safety triangle" theory put forward by academician Weicheng Fan. The three sides of the triangle are emergency, disaster carriers and emergency management. Therefore, risk assessment is not only based on risk possibility analysis and consequence analysis, but also should consider the risk tolerance of the disaster carrier to determine the level of risk.

Arisk assessment index system of natural disaster for the Beijing Winter Olympics is proposed based on an analytic hierarchy model. The first level indexes are the "hazard" of natural disaster and the "vulnerability" of disaster carriers. The second level indexes under "hazard" include the probability, strength, coverage and severity of natural disasters. The probability is determined by historical statistical data; strength is based on the features of natural disasters, for example, the strength of meteorological disaster can be determined by the levels of alert; coverage can be divided into three scales: city, district and street; and the severity of the consequences considers direct and indirect effects. Direct effect refers to the number of casualties and financial loss caused directly by natural disasters such as forest fires and geological disasters. The indirect effect can be represented by the comprehensive consequences of the risk of "natural disasters-infrastructure accidents".

This study mainly considers the scenario that natural disasters harm the urban infrastructure which will affect the progress of the Winter Olympics and the urban operation. Therefore, the disaster carriers are the urban lifeline system facilities, including the system of transportation, electric power, gas, heating power, communication, water conservancy, and so on. The second level indexes under "vulnerability" include the exposure, susceptibility and resilience of disaster carriers. The exposure of disaster carriers refers to the distribution of Olympic venues, transportation hubs and lifeline engineering and population distribution in different regions. The susceptibility of disaster carriers refers to the ability to adapt to or withstand risks that harm city infrastructure, which means the possibility that urban infrastructure can maintain business continuity under the impact of natural disasters. The susceptibility is related to the properties of urban infrastructure, such as the construction time and network structure. The resilience of disaster carriers can be measured by the emergency response force and the recovery time after the impact of natural disasters. Figure 2 shows the risk assessment index system.

Based on the index of risk assessment, the weight matrix of each index can be calculated by the analytic hierarchy process. The mathematical model of risk assessment is

$$R = \left(\sum_{i=1}^{4} m_i H_i\right) \times \left(\sum_{j=1}^{3} n_j V_j\right) \tag{1}$$

In which, m_i and n_j are the weight of hazard index and vulnerability index, respectively.

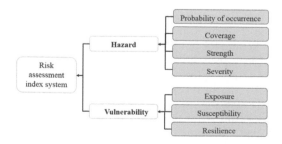

Figure 2. Risk assessment index system.

4 THE ANALYSIS OF ACCIDENT CHAINS ABOUT NATURAL DISASTERS AND URBAN INFRASTRUCTURE

Natural disasters may not only cause casualties and financial loss directly, but also cause damage or failure of urban infrastructure, which impacts the urban operation. The urban risks are typical multi-disaster risks. This means, in a specific region and specific period of time, that there are a variety of disaster factors and the consequence also contain features such as superposition and coupling (Shi, 2007). Therefore, in the natural disasters risk assessment index system of the Beijing Winter Olympics, its second level index "severity of natural disasters" is determined by the comprehensive consequences of the risk of "natural disaster-infrastructure accident". The calculation formula is as follows:

$$H_4 = \prod_{i=1}^{N} w_i X_i \qquad (2)$$

In which, N represents the number of risk events in an accident chain that may result from a natural disaster, w_i is the probability of each risk event in the chain of accidents, X_i means the severity of each risk event in the chain of events, and w_i is calculated by Bayesian networks.

Bayesian networks is the combination of graph theory and probability theory, which provides a way to visualize knowledge with graphs. It is also a technology for probabilistic reasoning. The nodes in the network represent random variables, and the directed arcs of the connecting nodes represent the mutual relations among the nodes. The strength of the relationship is expressed by conditional probability. Bayesian networks are mainly used to solve the problem of uncertainty. Taking the cold wave disaster for example, first, we studied the natural disaster – the infrastructure accident chain caused by the cold wave. The natural disaster risk tree of Beijing Winter Olympics is established, as shown in Figure 3.

Each node represents a risk event. The risk tree is transformed into Bayesian networks. All node variables have two values, $X_i = 0$ and $X_i = 1$, which represent "the event does not occur" and "the event occurs", respectively".

Next, by the parameter learning based on cases and data analysis, the conditional probability between nodes of Bayesian networks is calculated. As there are few meteorological disasters in

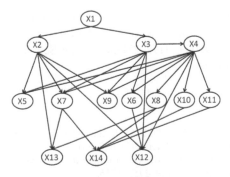

No.	Node	No.	Node
1	Cold wave	8	Gas facilities damage
2	Strong wind	9	Communication facilities damage
3	Heavy snow	10	Water facilities damage
4	Landslip	11	Heating power facilities damage
5	Sport event suspension	12	Traffic delay
6	Traffic accident	13	Fire
7	Power facilitiesdamage	14	Insufficient supply of energy and resources

Figure 3. Bayesian networks for cold wave disaster inthe Beijing Winter Olympics.

winter in Beijing the data of disasters are not sufficient, and the conditional probability table between nodes is supplemented by experts' knowledge. Finally, after determining the structure and parameters of the Bayesian networks, the reasoning is conducted. If some evidence is known, for example we know one or more risk events occur, the probability of other different risk events $p(X_i = 1)$, that is, $w_i = p(X_i = 1)$ in Equation(2), can be calculated. Moreover, the more evidence is available, the higher the certainty of the calculated results of each node.

For example, when $X_1 = 1$ the probabilities for all the rest nodes equal to 1 are as follow,

$$p(X_2 = 1, \cdots, X_{14} = 1) = (0.6, 0.3, 0.037, 0.226, 0.333, 0.133, 0.021, 0.133, 0.017, 0.021, \\ 0.342, 0.111, 0.021)$$

When inserted into Equation(2), the severity of the combined risk of a cold wave disaster is obtained.

5 CONCLUSION AND DISCUSSION

This paper considers that there are concurrent, secondary and coupling relationships between natural disasters or natural disasters and infrastructure accidents, so a method of disaster chain risk assessment based on Bayesian networks is proposed. By constructing risk trees and utilizing Bayesian networks, the trigger probability of each risk event in the disaster chain can be calculated. Combined with the severity of the consequences of each single disaster risk event, the hazard of natural disaster is obtained. It provides a way for quantitative calculation of indexes of the risk assessment index system for natural disasters in the Beijing Winter Olympics. Taking the cold wave disaster as an example, the feasibility of the method is proved. Because structure learning and parameter learning of Bayesian networks are based on data, and the sample size is larger, the uncertainty is lower. Therefore, in future studies, in addition to this knowledge and expertise, the conditional probability table could be further refined by collecting the data of disaster and questionnaires, which will make the assessment results more realistic.

ACKNOWLEDGMENTS

This project was supported by the National Key R&D Program of China (2018YFF0301003, 2018YFC080700), National Natural Science Foundation of China (7177030217),Beijing Natural Science Foundation (9192009, 4172023)and Young Scholars Program of Beijing Academy of Science and Technology.

REFERENCES

Gai, C.C., Weng W.G., and Yuan, H.Y. 2011. Multi-hazard risk assessment using GIS in urban areas. *Journal of Tsinghua University (Sci & Tech)*51(5): 627–631.

Ji, X.W., Weng, W.G., and Zhao, Q.S. 2009. Quantitative disaster chain risk analysis. *Journal of Tsinghua University (Sci & Tech)* 49(11): 1749–1752, 1756.

Liu, A.H., and Wu, C. 2015. Research on risk assessment method of disaster chain based on complex network. *Systems Engineering – Theory & Practice* 35(2): 466–472.

Ming, X.D., Xu, W., Liu, B.Y., et al.2013.An Overview of the Progress on Multi-Risk Assessment. *Journal of Catastrophology* 28(1): 126–145.

Shi, P.J. 2009. Theory and practice on disaster system research in a fifth time. *Journal of Natural Disasters* 18(5): 1–9.

Xue, Y., Huang, C.F., Zhou, J., et al. 2005. Three dimensional mode for integrated risk management of urban disaster:stage matrix pattern. *Journal of Natural Disasters* 14(6): 26–31.

Zhang, W.X., and Zhou, H.J. 2013.Conceptual model of disaster chain risk assessment: Taking Wenchuan Earthquake on 12 May 2008 as a case. *Progress in Geography* 32(1):130–138.

Zheng, Z.F., and Zhang X.L. 2007.Extreme synoptic events in Beijing and their relation with regional climate change. *Journal of Natural Disasters* 16(3): 55–59.

Risk Analysis Based on Data and Crisis Response Beyond Knowledge – Huang & Nivolianitou (eds)
© 2020 Taylor & Francis Group, London, ISBN 978-0-367-25146-8

Analysis of factors affecting construction safety management of steel structures

Anqi Lu*, Qian Wu, Na Li & Xin Hu

School of Civil Engineering, Xi'an University of Architecture and Technology, Xi'an, China

ABSTRACT: Considering the limitation and particularity of an old industrial plant renovation project, the safety management of its steel structure construction is affected by many factors. Thus, to improve safety management abilities, it is necessary to analyze the affecting factors. This study extracted 23 indicators of factors affecting construction safety management ability and established an index system. The system contains five dimensions: materials, machinery and facilities, structures, technologies, and personnel and measures. In this paper, interpretation structure modeling was used to qualitatively analyze the indicators, and an analytic network process was used to quantitatively analyze the weights of indicators. By this means, the indicators were assigned to different levels and sorted by weights. The result shows that the fundamental factors are safety operating procedures, safety technical standards and safety technology training. An old industrial plant renovation project was carried out to verify the rationality and practicability of the analytical model in the hope of providing theoretical support for similar projects.

Keywords: construction safety management, steel structure, safety management, interpretation structure modeling, analytic network process

1 INTRODUCTION

Old industrial plant renovation projects can protect urban history, continue cultural memory, save building materials and reduce energy consumption, which is an important practice of sustainable development concept (Li, 2015; Wu, 2018). In an old industrial plant renovation project, the steel structure plays an extremely important role in reinforcement and renovation projects because of its advantages of light weight, high strength, convenient construction and low environmental pollution (Li, 2006). During the construction process, steel structure construction requires more effective safety management due to changes in structural shape and stress state (Wu et al., 2017). Therefore, it is necessary to analyze the factors affecting the safety management of steel structure construction and identify the main influencing factors to provide theoretical support for improving the safety management ability of such projects.

In recent years, domestic scholars have undertaken much research on the safety management of old industrial plant renova tion. This research involves the construction safety evaluation index system (Guo, 2017), the safety supervision game (Wu, 2016), safety control elements (Li, 2016), and so on.

As far as the research object is concerned, most researchers focus on the whole construction process of old industrial plant renovation projects. There is still insufficient research on the safety management of the steel structure construction in the renovation project.

*Corresponding author: 847043782@qq.com

Thus, to explore the relationship between the influencing factors of steel structure construction safety management and sort these factors according to importance, this study established an analysis model by using interpretative structure modeling (ISM) and analytic network process (ANP). Based on an old industrial plant renovation project, the analysis model was validated, which should be helpful for designing improvement strategies for construction safety management of similar projects.

2 INDICATORS

Taking into account the characteristics of the steel structure construction in old industrial plant renovation projects and the basic principles of scientific, operability, accuracy, comprehensiveness and independence of the selection of the indicators, this study created a preliminary design by dividing the influencing factors of construction safety management into five aspects: materials, mechanical facilities, structure, technology, personnel and measures. Then, the on-site construction of steel structures in old industrial plant renovation projects was investigated for indicators. Finally, the study extracted the influencing factors of steel structure construction safety management applicable to an old industrial plant renovation project and established the index system which is as shown in Table 1, by discussing the selection of indicators with experts, scholars and field staff.

3 METHODOLOGY

3.1 Interpretative structural modeling

After fully investigating and analyzing the factors, interpretative structural modeling (ISM) was used to determine the interrelationships between the factors, and to form the adjacency

Table 1. Index system of factors affecting construction safety management of steel structures.

Primary indicators	Secondary indicators
Materials	Acceptance of materials – S1
	Machining of steel members – S2
	Coating of steel members – S3
	Storage of materials – S4
Machinery and facilities	Erection of temporary – Supports – S5
	Selection and layout of lifting machinery – S6
	Maintenance of machinery and equipment – S7
	Management of main and auxiliary equipment for installation – S8
	Protection devices and facilities – S9
Structures	Structural deformation of the original plant – S10
	Removal of the original structure – S11
	Destruction of the original steel bars – S12
	Settlement and slope of the original plant – S13
Technologies	Identification of obstacles on site – S14
	Hoisting and connection of structural – Steel members – S15
	Connection technology of new and old structures – S16
	Handling of the interface between new and old structures – S17
	Construction process monitoring – S18
Personnel & measures	Safety awareness and operational ability of personnel – S19
	Safety management organization – S20
	Safety operating procedures and technical standards – S21
	Safety technology training – S22
	Safety measures and systems – S23

matrix and reachability matrix in order to establish a hierarchy diagram that intuitively visualizes the levels and relation between the factors. The steps of ISM are as follows:

(1) Establish the reachability matrix according to the presence or absence of the relationship between any two factors.
(2) Calculate the reachable matrix from the adjacency matrix.
(3) Assign various factors to each level according to the reachability matrix.
(4) Draw the hierarchy diagram.

3.2 *Analytic network process*

The analytic network process (ANP) (Thomas 2004), which is proposed on the basis of the analytic hierarchy process (AHP), is applicable to complex systems with internal dependency and feedback. The steps of ANP are as follows:

(1) Identify goals and criteria, analyze the factors in the system and the relationship between the factors and divide the factors into different groups.
(2) Establish the network according to the goals and criteria and determine the structure of the ANP problem. Based on the ISM, establish the control layer and the network layer.
(3) Find the unweighted supermatrix.
(4) Find the weight supermatrix.
(5) Find the limit supermatrix.
(6) Determine the priority weight (Wang 2013).

Due to the huge amount of calculations involved for the ANP, this study used the Super Decisions software to complete the calculation. The Super Decisions is the only free educational software that implements AHP and ANP and was developed by the team of the creator of the method, Thomas Saaty. Its development and mantainance is sponsored by Creative Decisions Foundation which was established in 1996 by Thomas L. Saaty and his wife.

4   CASE STUDY

The old industrial plant renovation is a "five halls in one" project, that renovated old industrial plants in an original factory area and transformed them into five uses: a science and technology museum, a concert hall, a mass art gallery, a library and a youth activity center. The library was renovated from the original joint workshop. It retained most of the main structural frame and some floor slabs and renovated part of the inner courtyard into an indoor public space. The steel structure system, as the supporting structure for the new mezzanine and roof, was connected to the original structure. Especially, the library was supported by a steel frame structure and the roof was supported by a steel truss structure.

In this paper, the hierarchical structure diagram was established by ISM and the weights of the indicators determined by ANP.

Afterreading the literature and field research, the relationship between the safety management influencing factors indicators mentioned above was obtained, which constituted the following adjacency matrix A. Based on the matrix A, the reachability matrix R was obtained.

According to there achability matrix R, a hierarchy diagram was obtained, as shown in Figure 1.

After the hierarchy diagram was established, an ANP network diagram, shown in Figure 2, was established by using Super Decisions in the following way:enter the questionnaire data, conduct a consistency test, and accept the judgment matrix (the consistency ratio CR<0.1); then obtain the limit super matrix and the comprehensive weight of each factor. As shown in Table 2 the factors were sorted in the hierarchy from large to smallaccording to the comprehensive weights.

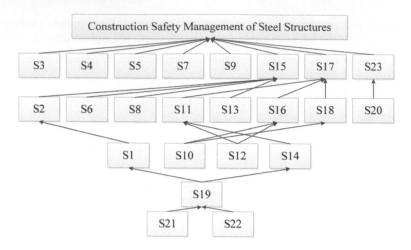

Figure 1.　The hierarchy diagram.

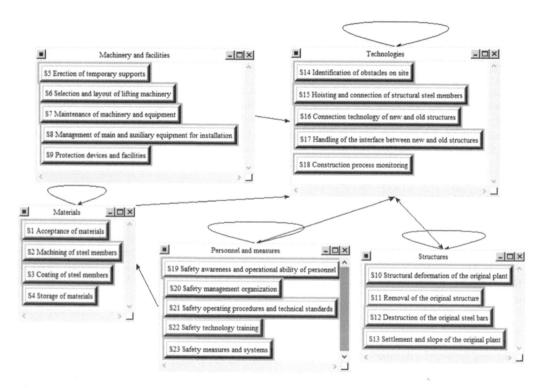

Figure 2.　ANP network diagram established by Super Decisions.

5　RESULTS AND SUGGESTIONS

According to the hierarchical structure diagram and the ranking of the factors affecting construction safety management at all levels, this studyhad the following results and related suggestions.

(1) The factors at the first level are sorted as follows: hoisting and connection of structural steel members – S15; handling of the interface between new and old structures – S17;

590

Table 2. Factors sorted according to the comprehensive weights.

Levels	Secondary indicators	Comprehensive weights
The first level	Hoisting and connection of structural steel members – S15	0.393303
	Handling of the interface between new and old structures – S17	0.169674
	Safety measures and systems – S23	0.125000
	Storage of materials – S4	0.003936
	Coating of steel members – S3	0.002346
	Erection of temporary supports – S5	0.000000
	Maintenance of machinery and equipment – S7	0.000000
	Protection devices and facilities – S9	0.000000
The second level	Connection technology of new and old structures – S16	0.078124
	Construction process monitoring – S18	0.046876
	Removal of the original structure – S11	0.044674
	Machining of steel members – S2	0.010644
	Selection and layout of lifting machinery – S6	0.000000
	Management of main and auxiliary equipment for installation – S8	0.000000
	Settlement and slope of the original plant – S13	0.000000
	Safety management organization – S20	0.000000
The third level	Identification of obstacles on site – S14	0.044674
	Acceptance of materials – S1	0.011489
	Structural deformation of the original plant – S10	0.000000
	Destruction of the original steel bars – S12	0.000000
The forth level	Safety awareness and operational ability of personnel – S19	0.069259
The fifth level	Safety operating procedures and technical standards – S21	0.000000
	Safety technology training – S22	0.000000

safety measures and systems – S23; storage of materials – S4; coating of steel members – S3; erection of temporary supports – S5;maintenance of machinery and equipment – S7; and protection devices and facilities – S9.

This level of factors directly affects the safety management of steel structure construction. There into, the factor hoisting and connection of structural steel members is the key point of construction site safety control, and the reliability of the connection is directly affected by the handling of the interface between the old and new structures, which should be taken seriously during the construction process.

(2) The factors at the second level are sorted as follows: connection technology of new and old structures – S16; construction process monitoring – S18; removal of the original structure – S11; machining of steel members – S2; selection and layout of lifting machinery – S6; management of main and auxiliary equipment for installation – S8; settlement and slope of the original plant – S13; and safety management organization – S20.

Some factors in the upper level, such as hoisting and connection of structural steel members – S15, handling of the interface between new and old structures – S17, and safety measures and systems S23, are mainly affected by the factors at this level. Among them, connection technology of new and old structures determines the key connection procedure in the steel structure construction of an old industrial plant renovation project. It is worth mentioning that construction process monitoring aims at monitoring structural deformation and settlement. Effective monitoring can help avoid safety accidents caused by collapse and components falling off.

(3) The factors at the third level are as follows:identification of obstacles on site – S14; acceptance of materials –S1; structural deformation of the original plant – S10; and destruction of the original steel bars–S12.

In an old industrial plant renovation project, due to the retention of some original structures, the identification of obstacles greatly affected the safety management of the site. In addition, the quality of the material is determined by the acceptance before construction, which affects the personal safety of the workers who use the materials on site.

(4) The following factors are at the forth level and the fifth level:safety awareness and operational ability of personnel – S19; safety operating procedures and technical standards – S21; and safety technology training – S22.

There into, the forth-level factor of safety awareness and operational ability of personnel–S19 determines the orderly or chaotic safety management of the construction site. In addition, strict compliance with safety operating procedures and technical standards and effective safety technology training can reduce the occurrence of safety accidents at its source and improve the safety management ability. The fifth-level factors of safety operating procedures and technical standards – S21 and safety technology training – S22 are the fundamental affecting factors of steel structure construction safety management. Old industrial plant renovation projects should ensure the application of operational procedures and technical standards and effectively improve the safety management capabilities and reduce potential safety hazards by safety technology training at its source.

According to the field investigation and questionnaire interview of the steel structure construction site in an old industrial plant renovation project, the analysis of the factors affecting the construction safety management based on the ISM-ANP is consistent with the actual situation, which can provide a reference for similar projects.

6 CONCLUSION AND DISCUSSION

(1) From the qualitative analysis based on ISM, the hierarchical structure diagram is established for the factors affecting construction safety management of steel structures in anold industrial plant renovation project, which visually expresses the hierarchical relationship among various factors and shows the direct and fundamental factors.

(2) From quantitative analysis based on ANP, the factors are sorted from important to unimportant according to the weights, which leads to relatively important affecting factors and provides reference for the improvement of decision-making.

(3) The combination of qualitative and quantitative analysis by ISM-AHP can not only express the hierarchical relationship between various factors, but also sort the factors according to the degree of importance. It effectively analyzes the factors affecting construction safety management of steel structures with many indicators and complicated relationships, which provides a reference for improving construction safety management abilities.

ACKNOWLEDGMENTS

This project was supported by the National Natural Science Foundation of China (51678479).

REFERENCES

Guo H.D., Chen X. and Li H.M. 2017. On the safety evaluation and the safety construction improvement of the recycling of the old industrial buildings based on SEW-UM. *Journal of Safety and Environment* 17(5): 1720–1724.
Li H.M. 2015. *The Protection and Reuse of Old Industrial Buildings*. Beijing: China Architecture & Building Press. (in Chinese).

Li Q., Guo H.D. and Fan S.J. 2016. ISM-AHP study on old Industrial buildings recycling safety control factors. *Industrial Safety and Environmental Protection* 42(02): 73–78.

Li W.H. and Fang Y.Z. 2006. A case analysis of steel structure used in reconstruction of building. *Sichuan Building Science* 32(03): 76–79.

Nripendra P.R., Daniel J.B., Abdullah M.A.B., Daniel R. and Sian R. 2019. Exploring barriers of m-commerce adoption in SMEs in the UK: Developing a framework using ISM. *International Journal of Information Management* 44: 141–153.

Raut R.D., Pragati P., Gardas B.B. and Jha M.J. 2018. Analyzing the factors influencing cloud computing adoption using three stage hybrid SEM-ANN-ISM (SEANIS) approach. *Technological Forecasting and Social Change* 134: 98–123.

Saaty T L. 2004. Decision making- the analytic hierarchy and network processes (AHP/ANP). *Journal of Systems Science & Systems Engineering* 13(1): 1–35.

Wang Y.G. and Wang C.M. 2013. Analysis of the influential factors on the airline safety performance based on ISM and ANP. *Journal of Safety and Environment* 13(4): 221–226.

Wu Q., Chen X. and Zhang Y. 2018. *Protection and Reutilization of the Old Industrial Buildings in Shaanxi.* Beijing: China Architecture & Building Press. (in Chinese).

Wu Q., Li H.M. and Chen X. 2017. *Introduction of the Crisis Management of Old Industrial Buildings.* Beijing:China Architecture & Building Press. (in Chinese).

Wu Q., Yu L., Chen X. and Du M.M. 2016. Game analysis on the safety supervision of old industrial building renovation. *Industrial Safety and Environmental Protection* 42(11): 46–49.

Xing B.J., Tang S.Q., Li N.W. and Niu L.X. 2017. Model of influencing factors for safety attention attenuation of miners with ISM coupling ANP. *Journal of Safety Science and Technology* 13(9).

Risk Analysis Based on Data and Crisis Response Beyond Knowledge – Huang & Nivolianitou (eds)
© 2020 Taylor & Francis Group, London, ISBN 978-0-367-25146-8

Unsafe behavior causation analysis of hazardous chemicals accidents based on the BN-HFACs method

Xiaowei Li & Tiezhong Liu
School of Management and Economics, Beijing Institute of Technology, Beijing, China

ABSTRACT: In this paper, BN-HFACs – the Human Factors Analysis and Classification System (HFACs) combined with Bayesian network (BN)– were proposed to analyze the human causation of hazardous chemicals accidents (HCAs). First, the traditional HFACs was revised to correspond to the characteristics of HCAs. Second, the revised HFACs was transformed into the typology of BN. Third, according to the 39 cases of HCAs in China, the original BN was obtained. Finally, failure sensitivity of every node in BN was calculated. The results show that operating errors are directly caused by human factors, while operation violation is directly caused by hazardous material environment, mechanical equipment, and human factors. At the same time, operation errors are also indirectly caused by organizational climate, hidden danger investigations, and organizational processes, while operation violations are indirectly caused by resource management, operation guidance, and planned operation. Furthermore, organizational climate has the greatest impact on unsafe behavior compared with other factors. Policy suggestions and study limitations also are discussed in this paper.

Keywords: Human factors analysis, classification system, Bayesian network, hazardous chemical accidents, human unsafe behavior

1 INTRODUCTION

Although the number of hazardous chemicals accidents (HCAs) in China has shown a downward trend in recent years (Ren and Mu, 2015), there still many casualties and economic losses caused by HCAs. Based on the analysis of 3974 HCAs in China from 2006 to 2018, it was found that 50.57% of HCAs were caused by unsafe behavior (Zhao et al., 2018). This showed that unsafe human behavior is the main factor of HCAs, therefore, it is necessary to study the unsafe behavior and the influence factors of unsafe behavior in the context of HCAs.

Human-centered is one of the two paradigms of accidents causation studies, and there are many methods of human causation analysis, such as Human Factor Analysis and Classification Systems (HFACs) (Shappell and Wiegmann, 2000), the "2–4" model, and so on. HFACs is one of the main methods adopted by scholars to study human-caused problems in various fields, including maritime (Chen et al., 2013), mining (Lenné et al., 2012), railways (Zhan et al., 2017), roads (Zhang et al., 2018), and so on. Meanwhile, come scholars use HFACs to analyze the human causation of HCAs (Theophilus et al., 2017; Zhou et al., 2017). Although many scholars have used HFACs to analyze the causation of HCAs, most have focused on single HCAs, and studies about HCAs are still relatively small compared with traffic, coal mines and other fields.

However, the HFACs method is only a qualitative analysis method, so how to conduct quantitative research on the basis of HFACs is also a common concern of scholars. Many researchers have used Bayesian network (BN) to establish BN models in the field of natural gas leakage (Yu et al., 2006) and transportation of hazardous chemicals (Zhu et al., 2016).

Other studies have conducted quantitative analysis by BN on the basis of HFACs in ship collision accidents (Wang et al., 2013), engineering (Xia et al., 2017), vapor accident (Wang et al., 2011), etc. It can be seen that the combination of HFACs and BN is suitable for the analysis of accidents caused by humans. However, the above method has rarely been applied in HCAs.

In this paper, the HFACs and BN will be combined to explore the human causation of HCAs in the context ofChina, and the key influencing factors of unsafe behaviors will be analyzed. Suggestions for improving unsafe behavior will be proposed from different perspectives.

2 METHODS OF BN-HFACS

2.1 Human factor analysis and classification systems (HFACs)

Shappell and Wiegmann (2000) proposed HFACs to analyze human errors in aviation accidents, which emphasizes that the accident causation chain could be notated as "organization influence → unsafe supervision → preconditions of unsafe behavior → unsafe behavior". Unsafe behavior includes errors and violations. Preconditions of unsafe behavior includes environmental factors, operator status, and personal factors. Unsafe supervision includes inadequate supervision, inappropriate operation plan, failure to correct known problems and illegal supervision. Organizational influence includes resource management, organizational climate, and organizational processes.

2.2 Bayesian network (BN)

Bayesian networkis a probabilistic network composed of the directed acyclic graph (DAG) and conditional probability table (CPT). There are two steps to construct BN. First, the topological relationship between random variables should be determined. Second, the CPT construction should be completed.In DAG, if there is a directed edge pointed bythe node $X_1$to the node X_2, then X_1 is the parent node of X_2, and X_2 is the child node of X_1. If there is a node thathas no parent, then it can be called a root node, and also one node can be called a leaf node if it has no child. CPT is used to represent the conditional probability distribution of nodes in DAG when nodes are discrete random variables.

2.3 Method of BN-HFACs

Bayesian networks are used to quantitatively analyze human causation in HFACs, that is, BN-HFACs. In HFACs, the factors of organization influencing are viewed as root nodes, and the unsafe behavior factors are taken as leaf nodes. The edges are connected according to the hierarchical influence of "organization influence → unsafe supervision → preconditions of unsafe behavior → unsafe behavior", then the network topology diagram of BN, also called DGA, is formed.

Suppose there are n nodes, signed as $X_0, X_1, ... X_{n-1}$, in the network, in which X_0 represents unsafe behavior, $X_1, ... X_{n-1}$ represent all factors of preconditions of unsafe behavior, unsafe supervision, and organizational influence. The actual state value of the node $X_i (i = 0, 1, \cdots n - 1)$ is represented by two state values, 0 and 1, which represent the non-failure state, such as there is no unsafe behavior, there is no unsafe supervision, and failure state, such as there is unsafe behavior, there is unsafe supervision, respectively. Therefore, the probability of occurrence of unsafe behavior can be calculated by using the joint probability distribution, inEq. (1):

$$P(X_0 = 1) = \sum_{x_1 \cdots x_n} P(X_0 = 1, X_1 = x_1, X_2 = x_2, \cdots, X_{n-1} = x_{n-1}) \tag{1}$$

In addition, BN can also be used to calculate other events' posterior probability when some events occur. Eq. (2) shows the posterior probability of other nodes when X_j is designated as a failure state.

$$P(X_i = 1 | X_j = 1) = \frac{P(X_j = 1 | X_i = 1)}{P(X_j = 1)}, 0 \leq i,j \leq n-1 \tag{2}$$

The above formula can be used to carry out causal reasoning and diagnostic reasoning. The Bayesian network can also be used to calculate the failure sensitivity of various factors in the HFACs model. In this study, failure sensitivity (FS) is defined as the change rate of the failure probability of direct consequence events caused by the failure of other causal events. The formula can be expressed as Eq. (3):

$$FS_i = \begin{cases} \frac{P(X_j=1|X_i=1)-P(X_j=1|X_i=0)}{P(X_j=1|X_i=0)}, & \Delta P \geq 0 \\ 0, & \Delta P < 0 \end{cases}, 0 \leq j < i \leq n-1 \tag{3}$$

where X_i represents cause events, which also means the lower level factors in HFACs, X_j represents direct consequence events, which also means the upper-level factors in HFACs, and ΔP represents $P(X_j = 1 | X_i = 1) - P(X_j = 1 | X_i = 0)$.

3 RESULTS

3.1 Revised HFACs for HCAs

According to the characteristics of HCAs, revised HFACs is used to describe the human causation model in the context of China, as shown in Figure 1. Among them, hazardous material environment (HME) means one of three types of environmental factors related to hazardous chemicals: chemical gas leakage, chemical reactions, and meteorological conditions. Mechanical equipment (ME) refers to the safety state of equipment with the functions of production and storage. Human factor (HF) includes risk identification

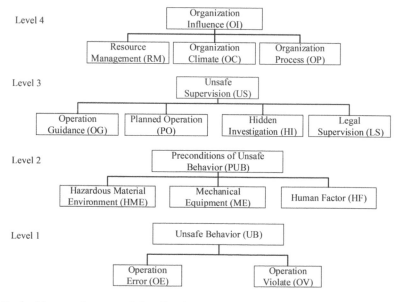

Figure 1. Revised human factors and classification systems (HFACs) for HCAs.

ability, safety awareness, safety skills, and mutual rescue consciousness. There are three aspects included in operation guidance (OG): safety instructions from managers to operators, supervision from managers regarding confined spaces, and supervision from managers to urge the operators to comply with the regulations. Planned operation (PO) mainly involves the operability of the operating procedures. Hidden investigation (HI) includes two aspects, one is the risk perception of managers and the other is that the managers do not eliminate the known hidden dangers in a timely manner. Legal supervision (LS) also divides into two facets: the management of operating tickets conforms to regulations and managers obey the rules and regulations. Resource management (RM) mainly refers to the input of safety production. Organization climate (OC) includes three aspects, enterprise safety responsibility, enterprise awareness of production safety, enterprise rules and regulations about safety. Organization process (OP) includes safety management, safety education and training, emergency management, third-party evaluation, and so on.

3.2 Transform HFACs into the topology of BN

Based on the above HFACs model, a conceptual model of BN-HFACs was constructed for the analysis of human factors HCAs. There is a four-layer BN, with unsafe behavior at the top and, in turn, preconditions, unsafe supervision and organizational influence one by one. According to the influence principle of layer-by-layer, the connection of BN was established. Unsafe behavior was regarded as the top event, that is, OE and OV are leaf nodes in BN. The organizational influence, unsafe supervision, and precondition of unsafe behavior were regarded as cause event of unsafe behavior, in which organizational influence is the parent node of BN, that is, RM, OC, and OP are parent nodes. This study assumes that all the underlying factors affect the upper-level factors, so OG, PO, HI, LS are parent nodes of HME, ME and HF, and are also child nodes of RM, OC, and OP. In addition, HME, ME, HF are parent nodes of OE and OV.

3.3 Probability matrix of BN-HFACs

3.3.1 Data collection
To determine the probability of BN nodes in different states, this study collected 89 HCAs investigation reports in China in recent years. The above 89 cases were screened according to the following three principles. First, HCAs from 2010 to the present were selected. Second, large-scale accidents with the number of deaths being equal to or greater than three were selected. Third, HCAs which directly caused by human factors were selected. Based on the above three principles, 39 cases of HCAs were collected. The main content of the data collection was the direct and indirect reasons in the investigation report of 39 HCAs. The direct reasons were operation errors and operation violations that lead directly to accidents, and the indirect reasons include organization, management, personal factors, external environmental factors, and so on.

3.3.2 Conditional probability calculation
The above factors involved in each of the 39 cases were coded and counted, and then the conditional probabilities of the upper factors under different states of the underlying factors would be calculated. So, the probability matrix of 12 nodes in the network was obtained through the following processes.

The probability matrix of Level 4 was determined by the frequency of each factor that occurred in 39 cases. For example, the number of RM occurred because the cause of 39 accidents is 9, so the failure probability of RM was defined as 0.23, and the corresponding non-failure probability was 0.77. Similarly, the failure probabilities of OC and OP were 0.56 and

0.82, respectively, and the corresponding non-failure probabilities were 0.44 and 0.18, respectively.

For nodes in Level 3, they were affected by the different states of nodes in Level 4, so the probability matrix of Level 3 was determined under consideration of layer influence. For example, OG was affected by OC and OP. Under the non-failure state of OC and OP, the failure frequency of OG is 1, and the non-failure frequency is 2, so the failure probability of OG under this condition is 0.33, and the non-failure probability is 0.67. In addition, the failure probability of OG is 0.21, 0.25 and 0.5 under the state of OC non-failure OP failure, OC failure OP non-failure and double failure of OC and OP, respectively. And if there is no corresponding state in 39 case studies, the maximum failure probability of other factors at the same level is selected as the failure probability according to the pessimistic rule. Then, the conditional probability matrix of each factor of Level 3 under the influence of Level 4 factors can be calculated. In a similar way, the conditional probability matrix of Level 2 and Level 1 was also obtained.

3.4 *Failure sensitivity analysis of BN-HFACs*

The software of GeNle 2.0 can be used to build the BN-HFACs model and calculate the sensitive value. First, the above probability matrices were input into the BN-HFACs model, and the initial BN was obtained. Second, the results of failure sensitivity analysis, as shown in Table 1, could be obtained according to Equation (3), by changing the state of each node in BN-HFACs.

Results of sensitivity analysis shown that only HF in Level 2 has the same direction failure effect on OE, and the sensitivity value is 0.09, while all three elements in Level 2 have the same direction failure effect on OV, and the order of sensitivity value is HME (0.96) > HF (0.12) > ME (0.02). In addition, for factors in Level 3 and Level 4, OC and HI also have obvious co-invalidation effects on OE, while RM, OG, and PO have obvious co-invalidation effects on OV.

For the factors of preconditions of unsafe behavior, HI and LS in Level 3 have the same direction failure effect on HME, and the sensitivity value is 0.62 and 0.13, respectively. OG and PO have the same direction failure effect on ME, and the sensitivity value is 0.85 and 0.94, respectively. Only the HI in level 3 has the same direction failure effect on HF, and the sensitivity is 0.24. In addition, for factors in Level 4, OC and OP have obvious co-effects on HME, RM has obvious effects on ME, and OC also has obvious effects on HF.

Table 1.   Failure sensitivity of nodes in BN-HFACs.

	RM	OC	OP	OG	PO	HI	LS	HME	ME	HF
OG	0.00	4.73	0.00	—	—	—	—	—	—	—
PO	0.00	1.81	0.10	—	—	—	—	—	—	—
HI	0.24	0.00	0.00	—	—	—	—	—	—	—
LS	5.20	4.67	0.00	—	—	—	—	—	—	—
HME	0.00	4.29	0.39	0.00	0.00	0.62	0.13	—	—	—
ME	1.20	0.00	0.00	0.85	0.94	0.00	0.00	—	—	—
HF	0.00	0.25	0.01	0.00	0.00	0.42	0.00	—	—	—
OE	0.00	0.34	0.08	0.00	0.00	0.24	0.00	0.00	0.00	0.09
OV	0.12	0.00	0.00	0.56	0.46	0.00	0.00	0.96	0.02	0.12

Note: columns indicate cause events, rows indicate result events.

For the factors of unsafe supervision, only OC in Level 4 has the same failure effect on OG, and the sensitivity is 4.73. OC and OP have the same failure effect on PO, and the sensitivity is 1.81 and 0.1, respectively. While RM has a significant impact on HI with a sensitivity of 0.24, RM and OC have the same impact on LS, with a sensitivity of 5.2 and 4.67, respectively.

4 DISCUSSION AND CONCLUSION

According to the above research results, we found the main factors leading to unsafe behavior in HCAs based on HFACs framework and Bayesian network. First, human factors are the most direct cause of operational errors. In other words, it is easy to lead operational errors when the operator's risk identification ability, safety awareness, safety skills and awareness of self-rescue and mutual rescue are in a poor state. Second, there are three direct causes of operation violation: the hazardous material environment, the mechanical equipment, and human factors. The hazardous material environment has the greatest impact, followed by mechanical equipment. It can be seen that employees tend to take illegal actions under the unsafe conditions of the dangerous environment. In fact, few scholars have considered the different influencing factors of operation errors and operation violations in empirical research. This study found the differences between the working environment and human factors in causing operation errors and operation violations. In addition, resource management, operation guidance, and planned operations also affect operational violations. Unlike the organizational culture that affects operational errors, the above three factors are aligned with the specific operations of enterprises or managers in security management, which also means that inappropriate and inadequate safety management will lead to employees' irregular operations.

Based on the above conclusions, there are several suggestions for hazardous chemicals enterprises to control human causation of HCAs. On one hand, when the operators of dangerous chemicals enterprises make frequent operational errors, the enterprises should pay attention to the operators' personal status, strengthen the training of operators' knowledge and skills, pay attention to hidden dangers investigations, and strengthen safety production investment. On the other hand, when the operators in the enterprises violate the rules frequently, the enterprises should examine in a timely manner the employees' working environment to judge if there are hidden dangers, and require managers to strengthen the operational guidance of operators and formulate more feasible operating procedures, and pay attention to the supervision management of managers to judge if they violate the rules to command operators. In addition, the supervision department of governments also should require hazardous chemicals enterprises to construct safety production cultures, implement the safety responsibility, enhance safety production awareness and formulate safety production related rules and regulations, and form a safe production organizational climate.

This study has great significances to enrich the theory of accident causation and change the unsafe behavior of operators in dangerous chemicals enterprises; however, it still has limitations. Only 39 HCAs cases were selected and the pessimistic principle was used to determine the probability in the non-existent state when calculating the probability matrix, which has an impact on the results. The followed study will enrich the HCAs database to obtain more objectivity and universal conclusion.

ACKNOWLEDGMENTS

This project was supported by the National Social Science Foundation of China (No. 16BGL175) and National key R & D projects of China (No. 2017YFF0209604).

REFERENCES

Chen, S.T., Wall A., Davies, P., Yang, Z.L., Wang, J. and Chou, Y.H. 2013. A Human and Organisational Factors (HOFs) analysis method for marine casualties using HFACS-Maritime Accidents (HFACS-MA). *Safety Science* 60(12): 105–114.

Lenné, M.G., Salmon, P.M., Liu, C.C. and Trotter, M. 2012. A systems approach to accident causation in mining: an application of the HFACS method. *Accident Analysis and Prevention* 48 (3): 111–117.

Ren, J.Q. and Mu Y.X. 2015. Statistical analysis and management enlightenment of hazardous chemicals accidents. *Chemical Enterprise Management* 16: 28–31.

Shappell, S.A. and Wiegmann, D.A. 2000. The Human Factors Analysis and Classification System-HFACS. *Security*. Retrieved from https://commons.erau.edu/publication/737

Theophilus, S.C., Esenowo, V.N., Arewa, A.O., Ifelebuegu, A.O., Nnadi, E.O. and Mbanaso, F.U. 2017. Human factors analysis and classification system for the oil and gas industry (HFACS-OGI). *Reliability Engineering and System Safety* 167: 168–176.

Wang, Y.F., Faghih, R.S., Hu, X.M. and Xie, M. 2011. Investigations of Human and Organizational Factors in hazardous vapor accidents. *Journal of Hazardous Materials* 191(1): 69–82.

Wang, Y.F., Xie, M., Chin, K.S. and Xiu, J.F. 2013. Accident analysis model based on Bayesian Network and Evidential Reasoning approach. *Journal of Loss Prevention in the Process Industries* 26(1): 10–21.

Xia, N., Zou, P.X.W., Liu, X. and Zhu, R. 2018. A hybrid BN-HFACS model for predicting safety performance in construction projects. *Safety Science* 101: 332–343.

Yu, P.T., Liu, Z.Y. and Chen, X.G. 2006. A Bayesian network approach to accident analysis. *Journal of Safety Science and Technology* 4(2): 45–50.

Zhan, Q., Zheng, W. and Zhao, B. 2017. A hybrid human and organizational analysis method for railway accidents based on HFACS-Railway Accidents (HFACS-RAs). *Safety Science* 91: 232–250.

Zhang, Y., Liu, T., Bai, Q., Shao W. and Wang, Q. 2018. New systems-based method to conduct analysis ofroad traffic accidents. *Transportation Research Part F: Traffic Psychology and Behaviour* 54: 96–109.

Zhao, L., Qian, Y., Hu, Q.M., Jiang, R., Li, M. and Wang, X. 2018. An Analysis of Hazardous Chemical Accidents in China between 2006 and 2017. *Sustainability* 10: 2935.

Zhou, L., Fu, G. and Xue, Y. 2017. Human and organizational factors in Chinese hazardous chemical accidents: A case study of '8.12' Tianjin Port fire and explosion using the HFACS-HC. *International Journal of Occupational Safety and Ergonomics Jose* 24(3): 329–340.

Zhu, T., Zhao, L.J. and Wang X.L. 2016. Analysis of the hazardous material transportationaccidents based on the Bayesian network method. *Journal of Safety and Environment* 16(2): 53–60.

Risk Analysis Based on Data and Crisis Response Beyond Knowledge – Huang & Nivolianitou (eds)
© 2020 Taylor & Francis Group, London, ISBN 978-0-367-25146-8

Comparative study on seismic risk of buildings in Chinese Tibet and Nepal

Changlong Li & Hongshan Lu*
Institute of Geophysics, China Earthquake Administration, Beijing, China

ABSTRACT: In this paper, we made building taxonomy for Southeastern Tibet and estimated the distribution of different types of buildings in every town based on scientific investigation and census database in this area. We built structure vulnerability model and occupancy vulnerability model for each type of building, and compared our vulnerability models with parts of vulnerability models of buildings in Nepal in the research of Nepal's seismic hazard and risk assessment. The comparison between our vulnerability models and Nepal ones shows that public and private houses in Southeastern Tibet have better seismic resistance than the concrete and UBM houses in Nepal. Old houses in Southeastern Tibet and Adobe houses in Nepal have similar vulnerability models, and vulnerability models of wood houses in both areas are also similar.

Keywords: Vulnerability, Seismic risk, Building taxonomy, Tibet, Nepal

1   INTRODUCTION

With the social modernization, the concept of Seismic Risk is being used more and more frequently. The seismic risk in a region refers to the possibility and extent of earthquake damage and disaster that the region may suffer. Seismic risk can be defined by the following equation:

$$\text{Seismic Risk} = \text{Seismic Hazard} \times \text{Vulnerability} \times \text{Exposure} \qquad (1)$$

Seismic hazard, vulnerability of man-made structures, and the exposure in the seismic hazard are the three factors of seismic risk.

Seismic risk is a hot research issue in seismology. Internationally, the Global Earthquake Model (GEM) Foundation is committed to researching global seismic risk (Silva et al., 2013). By now, seismic risk assessment has been completed and is being carried out in many places in the world such as Europe, Central Asia, the United States, South America, etc. organized by GEM (Silva et al., 2015; Wieland et al., 2015; Despotaki et al., 2017). At the same time, GEM developed *OpenQuake engine* (Silva et al., 2013), which is an open-source seismic risk calculation software and is getting more and more application in global seismic risk assessment.

There are also many other researches on seismic risk. Salgado-Gálvez et al. (2015) made a seismic risk assessment in Medellín, Colombia, and classified different levels of seismic risk in the city. Bartoli el al. (2016) made a seismic risk assessment for historic masonry towers in San Gimignano, Italy. They built vulnerability model for different types of structure, and simulated response of the structures to earthquake waves.

In China, seismic risk assessment has also been widely carried out in recent years. Wang et al. (2008) studied risk zoning of urban seismic disasters. They made an example of seismic risk zoning in Xiamen City, meshed risk grids and distinguished high seismic risk areas in Xiamen.

---

* Corresponding author: changlongli@163.com

Lu and Yu (2013) carried out a theoretical study of probability seismic risk based on the analytical function of seismic vulnerability. They used the earthquake hazard function in the form of power exponent, extended the obtained analytic function of seismic vulnerability to an analytical form of probabilistic seismic risk function, and developed the methods of seismic risk assessment. Liu et al. (2008) researched on seismic risk model in Mainland China. They collected a large number of earthquake disaster basic information, and improved seismic vulnerability model according to the actual seismic intensity distribution characteristics.

In general, the research on seismic risk assessment in various parts of the world has taken a short time and is still in its infancy and development stage. The most commonly used method of seismic risk assessment is to combine data of structures, population and their vulnerability information to give the losses result of the disaster-bearing bodies suffering earthquake disasters.

Southeastern Tibet includes Lhasa, Shannan and Nyingchi, and has a total number of 3 cities and 24 towns, as shown in Figure 1. The total population is about 2 million, accounting for two thirds of the population of Tibet. Southeastern Tibet is the most populous and economically developed part of Tibet. Meanwhile, Southeastern Tibet is located at the boundary between Eurasia Plate and India Plate, with a number of large-scale tectonic faults distributed. The occurrence of major earthquakes is frequent and seismic hazard is high. In recent years, the Chinese government has been committed to giving priority to supporting the economic construction of Tibet. Southeastern Tibet is the first region to develop. Therefore, it is of great significance to understand the seismic risk in Southeastern Tibet. In this paper, we made building taxonomy for Southeastern Tibet and investigated the distribution of every building type. We analyzed the vulnerability of every building type and made a comparison with the buildings in Nepal.

2 STUDY AREA AND DATA USED

The study area is Southeastern Tibet, as shown in Figure 1.

Building taxonomy is based on the scientific investigation by Institute of Geophysics, China Earthquake Administration. The investigation visited 13 of the 27 counties in Southeastern Tibet and more than 1,000 buildings were collected as samples. During the investigation, it was found that the distribution of buildings in the southeastern Tibet has the following characteristics:

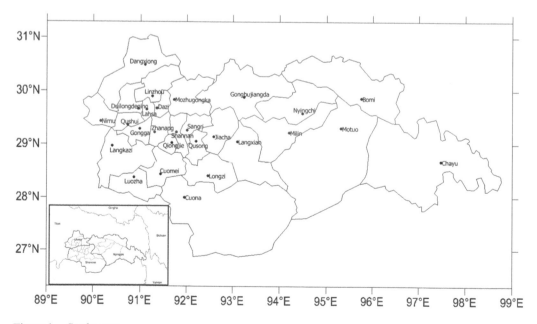

Figure 1. Study area.

(1) Population and buildings are mostly distributed in valleys, and concentrated in downtowns.
(2) Residential structures and their vulnerability characteristics are difficult to identify on satellite pictures, but can be identified by UAV aerial photography.
(3) There is a difference in housing structure and vulnerability between urban and rural residents. Urban buildings there do not have much difference between cities in Eastern China. There are many newly built private houses and old traditional Tibetan buildings in rural areas.
(4) In the forests of Nyingchi there are quite a large number of wood houses.

3 BUILDING TAXONOMY AND DISTRIBUTION INVESTIGATION

3.1 Building taxonomy

According to the scientific investigation, we divide buildings in Southeastern Tibet into four types as follows: Public; New-built Private; Old; Wood. Typical images of the four types are shown in Figure 2.

3.2 Building distribution investigation

We investigated the number of four types of building in each town using sample survey. We divided the three cities into city centers and suburbs, and divided the 24 towns into downtowns and countries. We also took samples in wood houses distributed areas. Sample areas

Figure 2. Typical images of the four building types. Top left: Public; top right: New-built Private; bottom left: Old; bottom right: Wood. Data source: Earthquake Insurance Group of Institute of Geophysics, China Earthquake Administration.

Table 1. Sampling points for building counting.

Residential Type	Sampling Point
City center	City center of Shannan
City suburbs	Niangre County, Lhasa
Downtown	Qushui Town, Lhasa
Country	Lulang County, Nyingchi
Wood distributed area	Gangdui Village, Bomi and Bangjia Village, Milin

Figure 3. UAV aerial images of each sampling point. Top left: City center of Shannan; top right: Niangre County; middle left: Qushui Town; middle right: Lulang County; bottom left: Gangdui Village, bottom right: Bangjia Village. Data source: Earthquake Insurance Group of Institute of Geophysics, China Earthquake Administration.

are shown in Table 1. The samples are taken by UAV aerial photography. Images of each sampling point are shown in Figure 3. Numbers and proportion of buildings at sampling points are shown in Table 2.

4 VULNERABILITY MODEL

We built vulnerability models for all the four types of buildings. The data we used is the building destruction reports in earthquake disasters in Tibet and surrounding areas in recent

Table 2. Numbers and proportion of buildings at sampling points.

Sampling Points	Total number	Public	Private	Old	Wood	Public proportion	Private proportion	Old proportion	Wood proportion
Niangre County	122	14	56	52	0	11.48%	45.90%	42.62%	0
Qushui Town	456	283	118	55	0	62.06%	25.88%	12.06%	0
City center of Shannan	360	219	131	10	0	60.83%	36.39%	2.78%	0.00%
Lulang County	73	2	28	41	2	2.74%	38.36%	56.16%	2.74%
Gangdui Village	21	1	7	6	7				
Bangjia Village	47	1	3	15	28				
Total of wood distributed areas	68	2	10	21	35	2.94%	14.71%	30.88%	51.47%

years, such as the 2010 Yushu M_S7.1 Earthquake (Tan et al., 2010; Qin et al., 2010; Huang et al., 2011; Bai et al., 2011); the 2015 Nepal M_S8.1 Earthquake (Gao et al., 2015; Wang et al., 2015; Qu and Yang, 2015; Zhang et al., 2016; Pan et al., 2017) and the 2017 Milin M_S6.9 Earthquake. The destruction images of these earthquakes are shown in Figure 4.

Figure 4. Images of destruction in earthquake disasters in Tibet and surrounding areas in recent years. Top left: Zhangmu after the 2015 Nepal Earthquake; top right: Jilong after the 2015 Nepal Earthquake; bottom left: Jiegu after the 2010 Yushu Earthquake; bottom right: Nyingchi after the 2017 Milin Earthquake. Data source: Gao et al. (2015); Bai et al. (2011); Qu and Yang (2015).

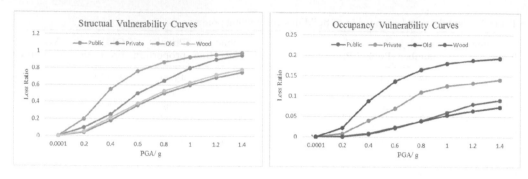

Figure 5. Vulnerability curves of the four types of buildings. Left: structural vulnerability; right: occupancy vulnerability.

To calculate casualties and economic losses caused by an earthquake respectively, we built both structural and occupancy vulnerability models. The vulnerability curves of the four types of buildings are shown in Figure 5. In which the standard deviation where PGA<0.6g is 0.3, and in other cases is 0.1.

We also compared our vulnerability models with parts of vulnerability models of buildings in Nepal built by Chaulagain et al. (2015) in the research of Nepal's seismic hazard and risk assessment. The building taxonomy of Nepal we chose to compare included Concrete, Unreinforced brick masonry (UBM), Adobe and Wood. The comparison between our vulnerability models and Nepal ones are shown in Figure 6. Figure 6 shows that public and

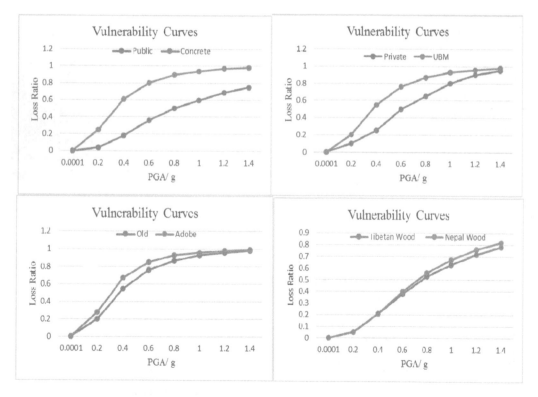

Figure 6. Vulnerability models comparison between buildings in Southeastern Tibet and Nepal. Top left: Public vs Concrete; top right: Private vs UBM; bottom left: Old vs Adobe; bottom right: Tibetan Wood vs Nepal Wood.

private houses in Southeastern Tibet have better seismic resistance than the concrete and UBM houses in Nepal. Old houses in Southeastern Tibet and Adobe houses in Nepal have similar vulnerability models, and vulnerability models of wood houses in both areas are also similar.

5 CONCLUSION AND DISCUSSION

In this paper, we built building taxonomy and vulnerability model for Southeastern Tibet using investigation data and building destruction data in history earthquake disasters, and made a comparison with the buildings in Nepal. We conclude as follows:

(1) Most private houses can resist middle-level ground motion (PGA<0.4g), but may get destroyed if the ground motion was larger. Public and wood houses have better seismic resistance, but old houses are even worse in seismic resistance.
(2) Public and private houses in Southeastern Tibet have better seismic resistance than the concrete and UBM houses in Nepal. Old houses in Southeastern Tibet and Adobe houses in Nepal have similar vulnerability models, and vulnerability models of wood houses in both areas are also similar.

The methods and results of this paper have the following points that are worth discussing:

(1) The random uncertainty is mainly from the standard deviations of the vulnerability model. The cognitive uncertainty is from the uncertainty caused by lack of cognition in the sample survey and numbering of the buildings.
(2) We only separated four types for all the buildings in Southeastern Tibet. In fact, the buildings in Southeastern Tibet are more complex than what we described. For some complex buildings, there type treated in the research increases the uncertainty of the result.
(3) The rapid development of major projects and infrastructural facilities in southeastern Tibet in recent years has attracted many people from eastern China and the population there is growing rapidly. Thus, the exposure model is expanding all the time. Our research in this paper hopes to provide a reference for understanding the present situation of seismic risk and developing disaster mitigation measures in this area.

ACKNOWLEDGEMENT

In the process of collecting data and writing this paper, the author received strong support and selfless help from the following units and individuals, and I hereby express my best gratitude: Tibet Autonomous Region Government, Tibet Earthquake Agency, Niangre County Government, Jiaerxi Village Government, Sangmu Village Government, Shannan Earthquake Agency and Nyingchi Earthquake Agency. Destruction data in the 2017 Milin Ms 6.9 Earthquake is from the Internet and Earthquake Emergency Management Team, China Earthquake Administration. This study is funded by International partnership program of Chinese Academy of Sciences (131551KYSB20160002); Earthquake Emergency Youth Key Fund (CEA_EDEM - 201816) and Fundamental Research Specific Fund (DQJB18B20; DQJB17C07; DQJB17T04).

REFERENCES

Bai GL, Xue F, Xu YZ (2011) Seismic damage analysis and reduction measures of buildings in village and town in the Yushu Earthquake. J1Xican Univ. of Arch. & Tech. (Natural Science Edition) 43(3): 309–315 (in Chinese).
Bartoli G, Betti M, Vignoli A (2016) A numerical study on seismic risk assessment of historic masonry towers: a case study in San Gimignano. Bull Earthquake Eng 14: 1475–1518.
Chaulagain H, Rodrigues H, Silva V, et al. (2015) Seismic risk assessment and hazard mapping in Nepal. Natural Hazards 78: 583–602.

Despotaki V, Burton HV, Schneider J, et al. (2017) "Back to Normal": Earthquake Recovery Modelling. California Seismic Safety Commission, CSSC Publication No. 2017-06.

Gao JR, Nima, Wen SL, et al. (2015) Characteristics of Seismic Damage by the Nepal M8.1 Earthquake in the Tibet Area of China. Technology for Earthquake Disaster Prevention 10(4): 961–968 (in Chinese).

Huang SN, Yuan YF, Meng QL, et al. (2011) Study on seismic resistance of rural houses based on earthquake damage to buildings in Yushu earthquake. World Earthquake Engineering 27(2): 77–82 (in Chinese).

Liu JF, Chen Y, Shi PJ, et al. (2008) On seismic risk assessment in Mainland China. Journal of Beijing Normal University (Natural Science) 44(5): 520–523 (in Chinese).

Lu DG, Yu XH (2013) Theoretical study of probabilistic seismic risk assessment based on analytical functions of seismic fragility. Journal of Building Structures 34(10): 41–48 (in Chinese).

Pan Y, Wang ZK, Shi SJ, et al. (2017) Investigation and analysis on seismic damage of residential buildings along the highway from Kathmandu to Zhangmu in Ms 8.1 Gorkha Earthquake. Journal of Hunan University (Natural Sciences) 44(3): 35–44 (in Chinese).

Qin ST, Li ZM, Tan M, et al. (2010) Analysis of Damage Characteristics of the M 7. 1 Yushu Earthquake of Qinghai and the Enlightenments. Journal of Catastrophology 25(3): 65–70 (in Chinese).

Qu Z, Yang YQ. (2015) Seismic damages to owner-built dwellings in the 2015 earthquake sequence in Nepal. Earthquake Engineering and Engineering Dynamics 35(4): 51–59 (in Chinese).

Salgado-Gálvez MA, Romero DZ, Velásquez CA, et al. (2015) Urban seismic risk index for Medellín, Colombia, based on probabilistic loss and casualties estimations. Natural Hazards DOI: 10.1007/s11069-015-2056-4.

Silva V, Crowley H, Pagani M, et al. (2013) Development of the OpenQuake engine, the Global Earthquake Model's open-source software for seismic risk assessment. Natural Hazards DOI 10.1007/s11069-013-0618-x.

Silva V, Crowley H, Varum H, et al. (2015) Seismic risk assessment for mainland Portugal. Bull Earthquake Eng 13:429–457.

Tan M, Li Y, Hu WH, et al. (2010) Investigation and analysis on seismic hazards on building construction of Yushu 7.1 Earthquake in Qinghai. Inland Earthquake 24(2): 173–179 (in Chinese).

Wang XQ, Huang SS, Ding X, et al. (2015) Extraction and Analysis of Building Damage Caused by Nepal Ms 8.1 Earthquake from Remote Sensing Images. Technology for Earthquake Disaster Prevention 10(3): 481–490 (in Chinese).

Wang ZT, Su JY, Ma DH, et al. (2008) Research on Risk Zoning of Urban Seismic Disasters. China Safety Science Journal 18(9): 5–10 (in Chinese).

Wieland M, Pittore M, Parolai S, et al. (2015) Towards a cross-border exposure model for the Earthquake Model Central Asia. Annals of Geophysics 58(1): S0106; doi:10.4401/ag-6663.

Zhang HY, Wang T, Lin XC, et al. (2016) Seismic damages of RC frames in Nepal Ms 8.1 Earthquake. Engineering Mechanics 33(9): 59–68 (in Chinese).

Risk Analysis Based on Data and Crisis Response Beyond Knowledge – Huang & Nivolianitou (eds)
© 2020 Taylor & Francis Group, London, ISBN 978-0-367-25146-8

Seismic safety analysis of the gas pipeline crossing through the Kezil thrust fault

Long Wang & Aiwen Liu*
Institute of Geophysics, China Earthquake Administration, Beijing, China

Yiyuan Cao
Daqing Oil Field Engineering Co., Ltd., Daqing, China

ABSTRACT: Earthquake damage investigations show that thrust faults often bring serious damage to gas pipelines, and easily cause secondary disasters such as fire and explosion. As a thin-walled shell structure, the gas supply pipeline is vulnerable to large compression deformation caused by thrust fault movement. This article studies the seismic safety of the gas pipeline crossing the Kezil thrust fault, taking the Dabei gas pipeline as an example. Through the field trench excavation of the Kezil fault, the fault plane inclination angle, the fortified fault displacement and the site soil characteristic parameters are determined. The shell finite element model has been adopted to analyze the large deformation of the gas pipeline under this thrust fault displacement, and the results are compared with the pipe's allowable tensile strain and compression strain. Finite element method results show that the main deformation of the pipe under thrust fault movement is the compression strain, and the amplitude of compression strain can be reduced with a decrease in the crossing angle. When the crossing is less than 11°, the pipe's maximum axial compression strain and tensile strain caused by the Kezil fault could be within the allowable strain range according to the requirements of seismic design code.

Keywords: gas pipeline, thrust fault, shell finite element model, seismic analysis

1 INTRODUCTION

Steel pipelines are widely used in long-distance gas supply pipeline engineering, for their high strength and adaptability to various geological and site conditions. These long-distance gas pipelines often cross seismic activity areas and there should be more focus on their seismic safety problems. The results of many earthquake disaster investigations show that the impact on the pipeline ranges from ground vibration, through sand liquefaction to active fault, that is, the dislocation of the earthquake fault caused the most serious damage to the gas pipeline.

According to the different dislocations, active faults can be divided into positive faults, reverse faults, and strike-slip faults. A steel pipeline is a thin wall shell structure; it has a certain resistance ability to tensile load, but it is susceptible to buckling failure when subjected to compression load. When passing through a normal fault, deformation of the pipeline is dominated by tensile strain, and when passing through a strike-slip fault, the deformation of the pipe could be mainly tensile strain with an appropriate crossing-fault angle. However, when the pipeline passes through the thrust fault, the deformation of the pipe is mainly compression strain. In other words, the dislocation of thrust faults poses the

*Corresponding author: liuaiwen@cea-igp.ac.cn

greatest threat to steel gas pipelines in these dislocation modes of active faults (Jin, 2010). For example, almost all of the gas pipelines crossing the thrust Chelongpu Fault were severely damaged in the 1999 Taiwan Jiji earthquake in China. A low-pressure natural gas PE pipeline buried under Jifeng Road in Wufeng town, had a diameter of 110 mm and a wall thickness of 8.3 mm, which passed vertically through the thrust fault, as shown in Figure 1. Under the upward fault movement, the gas pipeline suffered serious buckling deformation and the gas supply was interrupted. Longmenshan Fault in Sichuan province of China is also a thrust fault, with a vertical fault displacement of up to 6.2 m in the Wenchuan earthquake (Xu, 2008). Construction of the Lanchenyu oil supply pipeline began in 1998, with a total length of 1247 km, and it was recommended the pipeline should avoid crossing the Longmenshan fault based on the results of the earthquake safety evaluation. Therefore, the pipeline avoided the large fault movement and suffered only strong ground motion in the 2008 Wenchuan earthquake. The pipeline remained basically intact and continued to supply oil to the disaster area, and played an important role in the earthquake relief work (Shi, 2009).

The seismic analysis method of pipelines under fault movement includes the theoretical analysis method and the numerical analysis method. The early theoretical method ignored the bending stiffness of pipes and the lateral pressure of the surrounding soil; the fault displacement was assumed to be completely absorbed by the axial deformation of the pipeline (Newmark, 1975). Later, the improved cable-model theory and beam-model theory were suggested (Kenedy, 1977; Wang, 1985; Liu, 2002). It is worth noting that these theoretical calculation methods are all aimed at strike-slip faults. In other words, the theoretical analysis method can be used when the fault displacement is small and the pipeline is under tensile deformation. The above theoretical calculation methods are no longer available for the pipeline crossing thrust faults, so it is necessary to use the finite element model (Takada, 1998; Guo, 1999; Feng, 2001). The finite element method could also be roughly divided into two categories: beam element method and shell element method. Because the steel gas pipe is a shell structure, it is easy to produce large buckling phenomenon in its cross section under compressive load, and thus it is difficult to analyze with the beam element method. Therefore, scholars generally use the shell element method to simulate the response of buried pipelines under fault movements (Liu, 2004). Under the fault movement, soil springs in three directions are generally used to simulate the soil-pipe interaction, including the soil spring in the pipe axial direction, the horizontal lateral soil springs and the vertical soil spring. In addition to using the soil spring model to simulate the interaction between pipe and soil, some scholars directly take out buried pipelines and the surrounding soil to establish a pipe-soil 3D model, so the pipe is modeled with shell elements and the surrounding soil is modeled with body elements (Uckan, 2016).

Figure 1.   Buckling failure of gas pipeline under thrust fault movement in Jiji Earthquake.

At present, the seismic design of oil and gas pipeline engineering in China has also developed from stress design to strain design. In this article, the seismic safety problem of the gas transmission steel pipeline under the action of the reverse thrust fault is discussed by taking the case of the Dabei gas pipeline as an example, which crosses the Kezil reverse thrust fault in the western of China.

2 DETERMINATION OF KEZIL THRUST FAULT PARAMETERS

The Kezil fault is located on the northwestern edge of Tarimbasin, and generally moves to the northeast and east. The fault property is dominated by thrust movements with left-handed strike-slip. The fault plane is generally southward, and the fault inclination ranges from 10° to 80°. The length of the Kezil fault is 110km. On February 24, 1949, there was an earthquake of 7.3 magnitude at the eastern end of the fault, which caused 3,930 houses to collapse, 12 people were killed and 20 people were injured.

It is noticable that the fault parameters such as the dip angle of fault plane and the fault displacement may be different in each parts of the fault. It is necessary to determine the specific parameters of the fault at the passage of the gas pipeline through field investigation and grooving work. Based on the fault recognition of satellite images, a detailed field geological survey was carried out in the vicinity of the pipeline. According to the fault outcropping surface, a suitable position was selected for the excavation of probe and three grooves were excavated along the line of fault surface traces. The fault inclination angle and fault dislocation were clear, as shown in Figure 2. The brown soil (New Era, N_2, about 260,000 to 530,000 years ago) covers the sand and gravel layer (Late Pleistocene, Q_3, about 100,000 to 126,000 years ago). Along the fault plane, the tilt angle of the fault changes greatly, and the partial even nearly 90°, showing the basic characteristics of the thrust fault nappe structure. According to the excavation results and the existing research results on the Kezil fault zone, it can be determined that the fault movement is a pure thrust and the fortified fault displacement is 0.8 m.

3 PIPE-SOIL INTERACTION MODEL

Under the fault movement, there are interactions between the oil and gas pipeline and the surrounding soil. The interaction between the pipes and soil is generally simulated using the soil spring model, including the axial soil spring, the horizontal soil spring, and the vertical soil spring. The vertical soil spring is divided into vertical upward soil spring and vertical downward soil spring, as shown in Figure 3. The parameters of these soil springs are determined by the diameter of the pipe, the buried depth, and the type of site soil, the soil density, the soil cohesive force and the internal friction angle.

Figure 2.   Kezil thrust fault dislocation revealed by trench excavation.

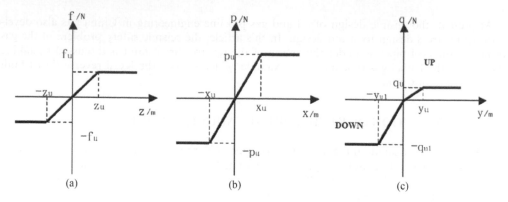

Figure 3. Soil spring models characterizing the pipe-soil interaction. (a) Axial soil spring, (b) Horizontal soil spring, (c) Vertical soil spring.

Table 1. Parameters of each direction of soil spring.

Soil spring parameters	Axial	Horizontal	Vertical (upward)	Vertical (downward)
Maximum force (N/m)	$f_s = 1.1 \times 10^4$	$P_u = 8.8 \times 10^5$	$q_u = 4.1 \times 10^4$	$q_{ul} = 2.5 \times 10^5$
Yield distance (m)	$Z_u = 0.004$	$X_u = 0.058$	$Y_u = 0.018$	$Y_{ul} = 0.051$

At the intersection of the pipeline and the Kezil fault, there is a wide pipe trench for laying and the depth of the pipeline is 1.2 M. In the buried part of the pipeline, the main surface is medium-density gravel, with a capacity of $\Omega = 21$ kN/m³, an internal friction angle $\varphi = 25°$, and a cohesive c = 11.0 kPa. With reference to the provisions of the current seismic code, the soil spring parameters of the three directions of soil interaction are calculated according to the middle dense sand as shown in Table 1.

4 ALLOWABLE STRAIN OF THE PIPE

The allowable strain of the pipe includes the allowable tensile strain and the allowable compression strain. At present, the seismic design of oil and gas pipelines has evolved from stress design to strain design (Liu, 2005). In accordance with earlier edition of oil and gas seismic design code, the allowable tensile strain of the pipeline was equal to 4%, which was the plastic yield strain of the X65 steel. Considering that surface defects may occur in the pipe welding, the allowable tensile strain is reduced to 1.29%.

The pipe wall will wrinkle due to local yield when the pipe is compressed. Thin shell folds theoretically begin at a diameter thickness ratio of 1.2. The actual cylinder will begin to wrinkle at one-half to one-quarter of the theoretical strain after the pipe test in a laboratory, but wrinkle does not mean destruction. Ignoring the severe stress concentration or weld defects, the pipeline can withstand four to six times the theoretical strain and does not rupture at the compressed folds. However, once the pipe is wrinkled, further deformation will focus on the wrinkle. For the safety of the pipeline, the allowable compression strain (ε_c) is set to the compression strain at which the pipeline begins to fold:

$$\varepsilon_c = 0.3\delta/D \tag{1}$$

where δ is the thickness of the pipe wall and D is the average diameter of the pipe. The diameter of Dabei pipeline is 508 mm, and the thickness is 12.7 mm. From the above formula, the allowable compression strain of 0.75% is obtained.

5 FINITE ELEMENT MODEL AND ANALYSIS RESULTS

In this article, the shell finite element model with an equivalent boundary is adopted to analyze the response of the Dabei pipeline under the Kezil fortified fault displacement. Assuming that the lower disk of the thrust fault remains stationary and the pipe crossing angle of 10°, the axial strain distribution along the pipeline is shown in Figure 4.

By changing the pipe crossing angle, the compression displacement in the direction of the pipe axis can be reduced, so that the compression strain reaction of the pipe is smaller than the allowable compression strain of the pipe. In this article, 13 different crossing angle cases are analyzed. The results of finite element analysis of the maximum axial tensile strain and the maximum axial compression strain are shown in Table 2. As shown in Figure 5, the results of the shell finite element model analysis show that deformation of the pipeline under thrust fault is mainly compression strain, especially when the crossing angle becomes larger. The amplitude of maximum axial compression strain in the pipeline increases rapidly with the increase of the intersection angle, and the amplitude of the maximum axial tensile strain in the pipeline tends to decrease with the increase of the crossing angle. When the Dabei pipeline passes through the Kezil fault at an angle less than or equal to 11°, both the maximum axial

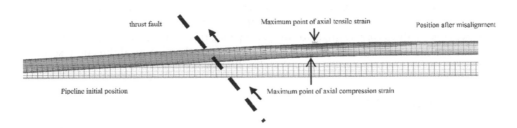

Figure 4.   Axial tensile strain distribution with the crossing angle of 10°.

Table 2.   Analysis result of pipeline crossing thrust fault with different crossing angles.

Case	Crossing angle	Maximum axial tensile strain (%)	Maximum axial compression strain (%)
1	0.02°	0.432	-0.34
2	2°	0.421	-0.38
3	6°	0.399	-0.50
4	10°	0.388	-0.66
5	11°	0.386	-0.70
6	12°	0.384	-0.75
7	13°	0.383	-0.79
8	16°	0.382	-0.89
9	20°	0.383	-1.01
10	25°	0.384	-1.18
11	30°	0.383	-1.36
12	60°	0.357	-2.09
13	90°	0.340	-2.31

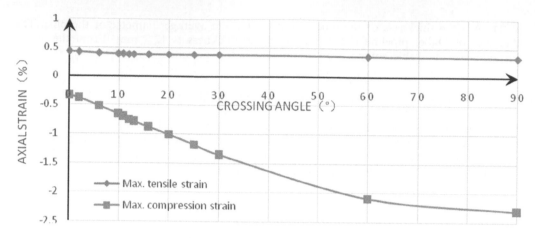

Figure 5.   Curve of axial strain with the crossing angle.

compression strain and tensile strain caused by the fault are within the corresponding allowable strain range of the pipeline.

6   CONCLUSION AND DISCUSSION

The seismic responses of the Dabei gas pipeline crossing the Kezil thrust fault with different angles were studied with shell finite element method. The results of the shell finite element analysis show that the deformation response of the pipeline under the thrust fault displacement is mainly based on compression strain. Under the condition that other parameters remain unchanged, the smaller crossing angle for the gas pipeline through the thrust fault can effectively reduce the compression displacement caused by the fault in the pipe axial direction, so that the compression strain of pipeline could be within the allowable compression strain.

ACKNOWLEDGMENTS

This project was supported by the National Natural Science Foundation of China (51778588).

REFERENCES

Feng, Q.M. and Zhao, L. 2001. Buckling analysis of buried pipes subjected to fault movements. *Earthquake Engineering and Engineering Vibration*, 21(4): 81–87. (in Chinese).
Jin, L. and Li, H.J. 2010. Nonlinear response analysis of buried pipeline crossing thrust fault. *Journal of Disaster Prevention and Mitigation Engineering*, 33(2): 130–134. (in Chinese).
Liu, A.W., Zhang, S.L., et al., 2002. A method analyzing response of buried pipeline due to earthquake movement. *Earthquake Engineering and Engineering Vibration*, 22(2): 22–27. (in Chinese).
Liu, A.W., Hu, Y.X., et al. 2004. An equivalent-boundary method for the shell analysis of buried pipelines under fault movement. *Acta Seismologica Sinica*, 26 (Supplement): 141–147. (in Chinese).
Liu, X.J. and Sun, S.P. 2005. Strain design method for underground pipeline crossing faults. *Special Structure*, 22(2): 81–85. (in Chinese).
Shi, H., Wang, L., Luan, L.B. 2009. Seismic design of Lanzhou-Chengdu-Chongqing Pipeline. *Oil and gas storage*, 28(10): 57–59. (in Chinese).
Kennedy, R.P., Chow, A.W. and William R.A. 1977. Fault movement effects on buried oil pipeline. *Transportation Engineering Journal of ASCE*, 103(5): 617–633.
Newmark, N.M. and Hall, W.J. 1975. Pipeline design to resist large fault displacement. *Proceedings of U.S. National Conference on Earthquake Engineering, Ann Arbor MI*, 416–425.

Takada, S., Liang, J.W., Li, T.Y., 1998. Shell model response of buried pipelines to large fault movements. *Journal of Structural Engineering, JSCE*, 44(3): 1637–1646.

Wang, R.L., Yeh, Y.H. 1985. A refined seismic analysis and design of buried pipeline for fault movement. *International Journal of Earthquake Engineering and Structure Dynamics*, 13(1): 75–96.

UCKAN, E., Kaya, E.W., et. al, 2016, The Performance of Thames Water Pipeline at the Kullar Fault Crossing, *International Collaboration in Lifeline Earthquake Engineering 2016 IRP 1, ASCE*, 359–365.

Guo, E.D. and Feng, Q.M. 1999. A seismic analysis method for buried steel pipe crossing fault. *Earthquake Engineering and Engineering Vibration*, 21(4): 43–47. (in Chinese).

Xu, X.W., Wen, X.Z., et al. 2008. The Ms 8.0 Wenchuan earthquake surface ruptures and its seismogenic structure. *Seimology and Geology*, 30(3): 597–629. (in Chinese).

Risk Analysis Based on Data and Crisis Response Beyond Knowledge – Huang & Nivolianitou (eds)
© 2020 Taylor & Francis Group, London, ISBN 978-0-367-25146-8

A new debris flow mitigation method based on small watershed functional map: A case study in the Shenshui watershed, China

Jun Wang & Qinghua Gong
Guangdong Open Laboratory of Geospatial Information Technology and Application, Guangzhou, China
Guangzhou Institute of Geography, Guangzhou, China

Yan Yu*
Guangdong Science and Technology Library/Guangdong Institute of Science and Technology Information and Development Strategy, Guangzhou, China

ABSTRACT: Debris flow is a common natural disaster in mountainous areas and often causes severe causalities and property loss. This paper proposed a new debris flow mitigation method based on the small watershed functional map. First, the shallow landslide was automatically identified based on remote sensing spectral characteristics, terrain slope and geometric shape. Second, debris flow run-out effects under rainfall with different return periods were analyzed by combining the debris flow numerical simulation results with different land utilization within the inundated area. Finally, the concept of "small watershed functional map" was put forward innovatively to conduct debris flow comprehensive mitigation. The effectiveness of the employed method was confirmed by performing a case study in Shenshui gully, China. The study can make a significant contribution to the literature because the proposed methods proved reasonable for assessing spatial debris flow run-out effects and can be widely employed for debris flow mitigation to save lives and property.

Keywords: debris flow, numerical simulation, run-out effect, small watershed, functional map, disaster mitigation

1 INTRODUCTION

Debris flows are rapid, gravity-induced mass movements consisting of a mixture of water, sediment, wood and anthropogenic debris that propagate along channels incised on mountain slopes and onto debris fans (Gregoretti et al. 2016). It has been reported in over 70 countries in the world and often causes severe economic losses and human casualties, seriously retarding social and economic development (Tecca and Genevois 2009; Degetto et al. 2015; Tiranti and Deangeli 2015; McCoy et al. 2012; Imaizumi et al. 2006; Hu et al. 2016; Cui et al. 2011; Dahal et al. 2009; Liu et al. 2010). Therefore, debris flow run-out effects analysis and mitigation are extremely important for preserving lives, mitigating disasters, and determining layouts for economic construction. Great efforts have been made to this issue (Stancanelli and Foti 2015; Berti and Simoni, 2007, 2014; Sakals et al. 2006; Viles et al. 2008; Armanini 2009; Tuladhar 2012; Wang et al. 2016; He and Zhai 2015; Li and Tuya 2015; Gao et al. 2016).

The numerical simulation method of debris flow not only can obtain information about the distribution of flow velocity, depth, and influence area, but can represent the debris flow inundation process. Therefore, it has been widely used in the calculation of debris flow discharge,

---

*Corresponding author: yuyan10@mails.ucas.ac.cn

prediction of inundation area, risk assessment, land use planning, and assessment of debris flow control engineering. The method was presented by researchers in the 1990s with the development of debris flow physical models and numerical computation methods. Romenshi and Toro (2004) used volume of fluid interface tracking technology to numerically simulate debris flows. Kwan et al. (2013) examined the runout characteristics of selected mobile debris flows by assuming that the amount of entrainment is proportional to the debris velocity. Chen and Zhang (2015) presented a model, EDDA (Erosion-Deposition Debris Flow Analysis), to simulate debris flow erosion, deposition and material property changes by solving the shallow water equations. Pudasaini (2016) derived a dynamical model for sub-diffusive and sub-advective fluid flow in porous media and debris material in which the solid matrix is stationary. All in all, many researches have been made to this issue, but no sound quantitative relationship has been established between land use and run-out effects so far, even though considerable losses have occurred during recent years.

In this paper, we developed a new debris flow mitigation method based on the small watershed functional map. Specifically, we first identified the shallow landslide based on remote sensing spectral characteristics, terrain slope and geometric shape. Then, we carried out debris flow run-out effect analysis via numerical simulation, compiled debris flow hazard maps under different return periods. We finally proposed the concept of "small watershed functional map" to conduct debris flow comprehensive mitigation and confirmed the effectiveness of the employed methods by performing a case study.

2 STUDY AREA DESCRIPTION

Shenshui gully, Magui town, Guangong province, China, was selected as the study area (Figure 1) which has a subtropical monsoon region and strongly influenced by the marine of climate of the South China Sea. Average annual temperature ranges from 21.3°C to 23.2°C (Wang and Xia 2012). The average annual rainfall is 2160 mm, the maximum average annual rainfall is 3150 mm, and the maximum annual rainfall of 3175 mm recorded by the Magui hydrological observation (Liu 2011). The study area is easy to form heavy rain, typhoon and

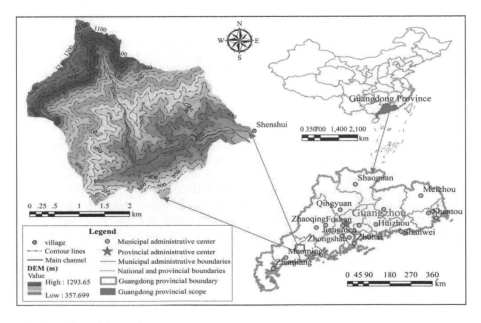

Figure 1.   Location of the study area.

other severe convective weather because of its peculiar natural and geographical conditions, leading to the collapses, shallow landslides, debris flows and other disasters occur frequently.

3 MATERIALS AND METHODOLOGY

3.1 *Shallow landslides remote sensing interpretation*

Shallow landslides directly provide the materials for debris flow initiation. Therefore, the accurate interpretation of shallow landslides is vital for debris flow simulation. Compared with other geological disasters, the shallow landslide body has the sharp distinction: (i) the rock and soil body after sliding is fully exposed, (ii) the spectral characteristic is different from the vegetation, and (iii) the spatial geometry is obviously controlled by the terrain gradient. Therefore, the shallow landslide can be automatically extracted based on its distinction. The flowchart of landslides interpretation is as follows.

(1) Separate the vegetation according to the NDVI (vegetation index) value. The vegetation on the surface of the shallow landslide was destroyed, which exposed the rock and soil sufficiently, and the NDVI value was lower than the surrounding environment. By this characteristic, the vegetation and non-vegetation area could be well separated.

(2) Remove the patches of the flat surface features by using slope gradient values. The NDVI values of the residents, farmland, mountain snow and ice, dry riverbed and lakes are lower, but the gradient of the terrain is usually gentler. Therefore, these flat surface features that have low levels of the NDVI values can be distinguished easily by slope gradient values.

(3) Distinguish the vegetation in the shadows. In the area where slope geological disasters occur frequently, the topography undulation is generally large, the negative sunlight of the mountain cannot be directly irradiated, the reflection value on the remote sensing image is low, and thus the NDVI values cannot effectively distinguish between vegetation and non-vegetation in the shadows.Therefore, the area size of broken patches can be used to eliminate the shadow of the shaded areas of the patches and to separate the vegetation in the shadows.

(4) Eliminate the pseudo-patches according to the length-width ratio of the patches. Shallow landslides generally exhibit a narrow geometric shape, and the length-width ratio can be used to further remove the pseudo-patches of which the spectral features and slope values are similar to those of shallow landslide bodies.

(5) Downward grade criterion. The most obvious feature of the shallow landslide has obvious characteristics of descending along the maximum gravity gradient. Therefore, the similarity between the long axis direction and the slope direction of the spots can be used as an important discriminating index.

3.2 *Debris flow numerical simulation and run-out effect analysis*

The numerical computer model was used to delimit the debris flow parameters in this study, with a computational fluid dynamics code based on a collocated finite-difference approach and self-programming technology. The numerical computer model requires a numerical simulation platform, debris flow attribute data, debris flow volume, and the distribution of debris flow sources. The numerical simulation platform can be obtained by surveying and mapping instruments, and by extracting the digital terrain from aerial images indirectly. The debris flow property parameters were obtained through indoor and outdoor experiments.

There are many methods for estimating the field debris flow volume, but field investigation with aerial photo or satellite image is considered the most accurate (Liu et al. 2009). Assuming a relatively gentle riverbed slope and a steady uniform debris flow, the concentration of a debris flow can be calculated from the equilibrium concentration equation (Takahashi 1980):

$$C_{d\infty} = \frac{\rho \tan\theta}{(\sigma - \rho)(\tan\phi - \tan\theta)} \qquad (1)$$

where $C_{d\infty}$ is the equilibrium concentration of a debris flow, ρ is the density of fluid in the debris flow (for water, the density is 1.0 g/cm³; for clayey water, the density is close to 1.2 g/cm³), θ is the average slope of the riverbed, σ is the density of solid particles which usually is 2.65 g/cm³, and ϕ is the angle of friction of solid particles.

Generally, the maximum loss under the given conditions is considered when conducting debris flow run-out effect assessment. The precipitation amount always controls the amount of available soil and gravel in the field. Assuming the debris flow is triggered by heavy rainfall, the debris flow total volume can be determined by the available amount of debris sources in the field and the amount of precipitation. Here, if the available debris source volume is V_s and the precipitation amount is V_w, the total volume of debris flow V_D is calculated as follows:

$$V_D = \min\left\{\frac{V_s}{C_{d\infty}}, \frac{V_w}{1 - C_{d\infty}}\right\} \qquad (2)$$

The volume of loose solid sources that can be mobilized in the field is usually determined by the results of remote sensing image data and field surveys, and the rainfall under different rainfall conditions is determined by querying hydrological manuals and hydrological calculations. After obtaining the total volume of debris flow and the corresponding debris flow source distribution from field surveys, the debris flow numerical simulation can be conducted.

3.3 *Debris flow comprehensive mitigation based on the small watershed functional map*

According to the characteristics of the debris flow watershed, the single-channel debris flow can be divided into three parts: formation area, circulation area and accumulation area. The landform, surface rolling, surface incision, surface coarseness, material condition and other characteristics of different parts are quite different. Their watershed functions of disaster prevention and mitigation are also different. Based on the above theory, the concept of "small watershed functional map" is put forward and is defined as: the sum of watershed different parts functions of disaster prevention and mitigation. According to the theoretical concept, the small watershed functional map can be divided into three divisions which are shown in Figure 2.

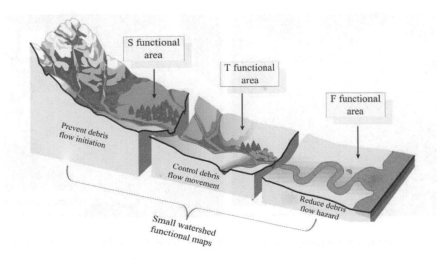

Figure 2.   The schematic diagram for debris flow small watershed functional map.

S functional area. The main commitment of this area is to prevent debris flow initiation. It is mainly the area of debris flow water source, loose material source, the gravel supply source or the debris flow initiation source. The S functional area carries out the comprehensive management of the river basin by taking measures such as slope control, channel control and slope protection, as well as administrative and ordinances measures.

T functional area. The main responsibility of this functional area is for the control of debris flow movement. Through the block, regulation and drainage and other projects, so that debris flow smoothly through or reach the designated area, to protect the residents and other affected objects. This functional area mainly emphasizes both of engineering and biological measures.

F functional area. The main commitment is of this area to reduce the debris flow hazards or damages to protected objects. It is a large area of debris flow material deposition area. By taking some early warning and precautionary measures, so that debris flow will not cause significant harm and minimum the debris flow hazards or damages.

Three functional areas are also interrelated and complement each other. In the specific application, based on the results of debris flow simulation and run-out effect analysis, the three functional are must be unified planning and layout, making it plays the greatest function to achieve the ultimate aim of debris flow comprehensive mitigation.

4 RESULTS AND DISCUSSION

4.1 Spatial distribution of the shallow landslides

The remote-sensing image of study area is shown in Figure 3(a). Using the method of shallow landslides remote sensing interpretation proposed in this paper, the spatial distribution of shallow landslides can be obtained in Figure 3(b). The interpreted result of landslide volume is 1193750 m^3 and the area covers 477500 m^2.

4.2 Debris flow numerical simulation and run-out effect analysis

With Eq. 1, the equilibrium concentration can be calculated. Using 1.2 g/cm^3 for ρ, 2.65 g/cm^3 for σ, 42° for ϕ, and 20.45° for θ, we get:

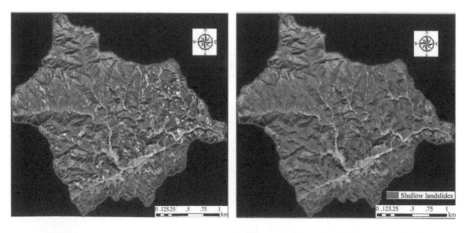

(a) The remote-sensing image   (b) The shallow landslides distribution map

Figure 3.   The remote-sensing image and spatial distribution of shallow landslide in the study area.

$$C_{d\infty} = \frac{\rho \tan \theta}{(\sigma - \rho)(\tan \phi - \tan \theta)} = \frac{1.2 \times \tan 20.45}{(2.65 - 1.2) \times (\tan 42 - \tan 20.45)} \cong 0.58 \qquad (3)$$

The amount of mobilized loose solid materials in the field is controlled by the amount of precipitation and the available volume of loose solid materials. The maximum volume of loose solid materials that can be mobilized in the field is based on the result of landslide interpretation, which is about 1193750 m³ and the area is about 477500 m². The maximum volume for debris flow occurs when all materials are mobilized, so:

$$V_D = \min\left\{\frac{V_s}{C_{d\infty}}, \frac{V_w}{1 - C_{d\infty}}\right\} = \frac{1193750}{0.58}(m^3) = 2058190 \, m^3 \qquad (4)$$

This volume is the maximum volume that cannot be increased even with more precipitation. Assume that under the extreme rainfall of 500-years return period, the loose solid materials are all mobilized and the debris flow reaches the maximum volume. Then the total volume calculated under the different return periods are given in Table 1. In this paper, the precipitation amounts under the different return periods were calculated by referring to the Guangdong Province rainstorm runoff calculation manual (1991). For the purpose of debris flow run-out effect analysis, the maximum loss under given conditions must be considered.

Based on remote sensing images and intensive field investigations, the distribution of debris flow source and the inundated outlets of debris flow were determined. The outlets were set as numerical boundary condition for these volumes of flow which were arranged at the same location, and then run the flow movement process. The fluid rheological parameter was measured by using an MCR301 rheometer. The average debris flow density was calculated to be 1.65 g/cm³, the debris flow was determined to be Bingham flow, and the yield stress and viscosity coefficient were found to be 6500 Pa and 0.5 Pa·s, respectively. The numerical simulation results for the final buried depths and inundated areas with different return periods are shown in Figure 4.

For the purpose of debris flow run-out effect analysis, the maximum loss under given conditions must be considered. In this paper, the maximum volumes of debris flows under 50-year,

Table 1. Debris flow volume under different return periods of the study area.

Return period (year)	50	100	500
Volume (m³)	400650	1005950	2058190

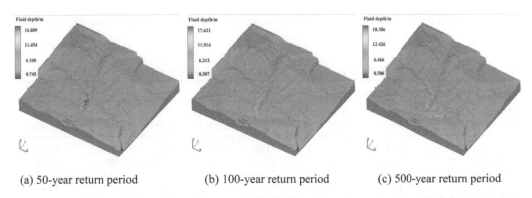

(a) 50-year return period   (b) 100-year return period   (c) 500-year return period

Figure 4. The numerical simulation results of the influence area and the final buried depth under rainfall with different return periods.

100-years and 500-year return periods were used to simulate the debris flow inundate areas and to analyze the debris flow run-out effects. The debris flow run-out effects can be analyzed by combining the simulated results and the land-use data within the influence area. Thus, the land-use data within the influence area under different return periods must be determined first, which is shown in Figure 5.

According to the characteristics of the particular local region, the standards for debris flow run-out effect analysis can be generated by combining the simulated final buried depth and the different land utilization within the influenced area, which are shown in Table 2. The level of variation in land utilization as well as the damage level and recoverability of the land are important considerations in any proposed run-out effect analysis standard, as different land utilization has different degrees of hazard. Bare land is not within the inundated area of any return periods and the water body land is not listed because of its particular characteristics. For different land utilization within the influenced area, the hazard degree is different; thus, the standard adopted for debris flow run-out effect analysis is also different. Finally, the debris flow run-out effect result is shown in Figure 6.

4.3 Spatial distribution of the shallow landslides

In the paper, the calculated maximum debris flow volume was under the extreme rainfall of 500-years return period. Thus, for the purpose of debris flow comprehensive mitigation, the maximum loss under this given condition must be considered. To carry out debris flow comprehensive mitigation based the "small watershed functional maps" proposed in this paper, the S functional area, T functional area and F functional area must be determined first. In this paper, the topographical condition and the simulated results of fluid depth and fluid velocity were used as the indexes of dividing three functional areas: (i) S functional area, v≤6 m/s,

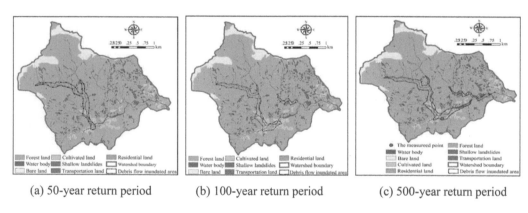

(a) 50-year return period   (b) 100-year return period   (c) 500-year return period

Figure 5. Land use within debris flow inundated area under 50-year, 100-year and 500-year return periods.

Table 2. The run-out-effect analysis standard of debris flow in the Shenshui gully watershed.

Final buried depth/m	Objects on the ground							
	Residential land	Hazard	Transporta-tion land	Hazard	Forest land	Hazard	Cultivated land	Hazard
h > 2	***	High	***	High	***	High	***	High
1 < h ≤ 2	**	Moderate	***	High	**	Moderate	***	High
0.5< h≤ 1	*	Low	**	Moderate	*	Low	**	Moderate
h ≤ 0.5	*	Low	*	Low	*	Low	*	Low

* is the damaged level: * is slightly damaged, ** is half damaged, and *** is destroyed completely.

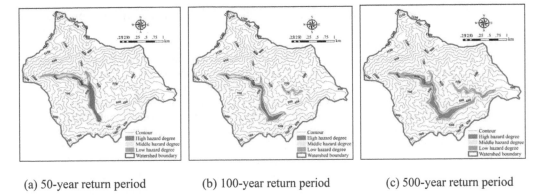

| (a) 50-year return period | (b) 100-year return period | (c) 500-year return period |

Figure 6. Debris-flow run-out-effect analysis results under 50-year, 100-year and 500-year return periods.

upstream of the watershed and slope is more than 20°; (ii) T functional area, 3<v≤6 m/s, mid-stream of the watershed and slope is between 10° and 20°; and (iii) F functional area, v<3.0 m/s, downstream of the watershed and slope less than 10°. According to the above functional partition standard of small watershed in the study area, the small watershed functional map and comprehensive mitigation measures can be obtained, seen in Figure 7.

S functional area. In the S functional area closely monitor the debris flow water source and soil source. The rainfall and runoff are monitored by self-recorded rain gauge and mud level meter. The deformation of the shallow landslides is monitored by displacement meter. The infrared camera is used to monitor the debris flow initial process. Meanwhile, pant local plants on bare land to obtain ecological protection and function of keeping water and soil. Through a series of comprehensive measures, achieve the function of preventing debris flow initiation.

T functional area. In this area, debris flow must be concentrated of the down-stream flow after the debris flow initiation. The flow velocity, mud level and movement process must be

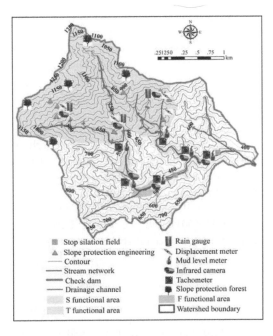

Figure 7. Debris-flow comprehensive mitigation measures in the Shenshui watershed based on the proposed method of small watershed functional map.

monitored in real time. The flow velocity is monitored by tachometer and the mud level of debris flow is monitored by ultrasonic mud level meter. The infrared camera is used to monitor the debris flow movement process. Meanwhile, the early-warning must be conducted. The debris flow channel must be set up so that the debris flow can reach the designated area, without endangering human life, property and so on. Through a series of comprehensive measures, achieve the main responsibility of controlling debris flow movement.

F functional area. The corresponding monitoring and warning, biological and engineering measures must be established to reduce the damage of debris flow on the people's lives and property. The debris flow velocity, mud level and movement process must be monitored in real time. Meanwhile, the debris flow stop siltation field is constructed to lead the debris flow into the designated area and slow down the debris flow velocity so that reduce the damage of debris flow on the people's lives and property. Besides, in the functional area, the human activities must be controlled strictly to prevent significant casualties and economic losses.

5 CONCLUSIONS

(1) This paper presents a method to identify the shallow landslides automatically based on the remote sensing spectral characteristics, terrain slope and geometric shape. Using this method, the spatial distribution of shallow landslides of the Shenshui watershed was obtained. The maximum volume of shallow landslides that can be mobilized in the field was about 1193750 m^3 and the area was about 477500 m^2.

(2) Based on the numerical simulation method, the debris flow inundated processes under different rainfall conditions in the Shenshui watershed was numerically simulated. The distribution of inundated area, fluid depth and velocity were obtained. According to the particularity of local region and the numerical simulation results, the standard for debris flow run-out effect analysis was generated by combining the simulated final buried depth and the different land utilization within the influenced area. The debris flow run-out effects under rainfall with different return periods were analyzed based on established standard.

(3)The concept of small watershed functional map was put forward innovatively to conduct debris flow comprehensive mitigation. It was divided into three parts, namely S functional area, T functional area and F functional area, and their functions of diasater mitigation were detailed analyzed. The proposed method can improve the ability of watershed comprehensive disaster mitigation and can be used as the watershed disaster mitigation paradigm systematically.

ACKNOWLEDGMENTS

This research was supported by the CRSRI Open Research Program (Program SN: CKWV2017524/KY), Natural Science Foundation of Guangdong Province (2018A030310469), the Science and Technology Project of Guangzhou City (201804010126; 201803030025), GDAS' Special Project of Science and Technology Development (2017GDASCX-0101; 2018GDASCX-0101; 2019GDASYL-0401001; 2017GDASCX-0807); the Guangdong Pro-vincial Science and Technology Program (2018B030324001) and the Water Resource Science and Technology Innovation Program of Guangdong Province (2016-15).

REFERENCES

Armanini, A., Fraccarollo, L. and Rosatti, G. 2009. Two-dimensional simulation of debris flows in erodible channels. *Computers & Geosciences*, 35(5): 993-1006.
Berti, M. and Simoni, A. 2007. Prediction of debris flow inundation areas using empirical mobility relationships. *Geomorphology*, 90: 144–161.

Berti, M. and Simoni, A. 2014. DFLOWZ: a free program to evaluate the area potentially inundated by a debris flow. *Comput. Geosci.* 67: 14–23.

Cui, P., Hu, K.H., Zhuang, J.Q., Yang, Y. and Zhang, J. 2011. Prediction of debris-flow danger area by combining hydro-logical and inundation simulation methods. *Journal of Mountain Science*, 8(1): 1-9.

Chen, H.X. and Zhang, L.M. 2015. EDDA: integrated simulation of debris flow erosion, deposition and property changes. *Geosci. Model Dev.*, 8: 829–844.

Dahal, R.K., Hasegawa, S., Nonomura, A., Yamanaka, M., Masuda, T. and Nishino, K. 2009. Failure characteristics of rainfall-induced shallow landslides in granitic terrains of Shikoku Island of Japan. *Environmental geology*, 56(7): 1295-1310.

Degetto, M., Gregoretti, C. and Bernard, M. 2015. Comparative analysis of the differences between using LiDAR cotour-based DEMs for hydrological modeling of runoff generating debris flows in the Dolomites. *Frontiers in Earth Science*, 3: 1-21.

Gao, L., Zhang, M., Chen, H.X. and Shen, P. 2016. Simulating debris flow mobility in urban settings. *Engineering Geology*, 214: 67-78.

Gregoretti, C., Degetto. M. and Boreggio, M. 2016. GIS-based cell model for simulating debris flow runout on a fan. *Journal of Hydrology*, 534: 326-340.

He, Z.Y. and Zhai G. F. 2015. Spatial Effect on public risk perception of natural disaster: a comparative study in east asia. *Journal of Risk Analysis and Crisis Response*, 3(5): 161-168.

Hu, W., Dong, X.J., Wang, G.H., van Asch T.W.J. and Hicher, P.Y. 2016. Initiation processes for run-off generated debris flows in the Wenchuan earthquake area of China. *Geomorphology*, 253: 468–477.

Imaizumi, F., Sidle, R.C., Tsuchiya, S. and Ohsaka, O. 2006. Hydrogeomorphic processes in a steep debris flow initiation zone. Geophys. Res. Lett. 33, L10404.

Kwan, J.S.H., Hui, T.H.H. and Ho, K.K.S. Modelling the motion of mobile debris flows in Hong Kong. *Landslide Science and Practice*. Springer, Berlin, pp. 29–35.

Li, S.Z. and Tuya, A. 2015. Disaster risk research literature on statistics analysis in China journal net. *Journal of Risk Analysis and Crisis Response*, 5(2): 129-140.

Liu, K.F., Li, H.C. and Hsu, Y.C. 2009. Debris flow hazard assessment with numerical simulation. *Natural Hazards*, 49(1): 137-161.

Liu, J.F., You, Y., Chen, X.Z., Pan, J.J. and Ren, H.Y. 2010. Identification of potential sites of debris flows in the upper Min River drainage, following environmental changes caused by the Wenchuan earthquake. *Journal of Mountain Science*, 3: 255-263.

Liu, J.F. 2011. Flash flood rainfall analysis of "9.21" in Caojiang, Gaozhou city. *Guangdong Water Resources and Hydropower*, S1: 33-35. (In Chinese).

Pudasaini, S. P. 2016. A novel description of fluid flow in porous and debris materials. *Eng.Geol.* 202: 62–73.

Romenski, E. and Toro, E.F. 2004. Compressible two-phase flows: two-pressure models and numerical methods. *Journal of Scientific Computing*, 13(3): 403-416.

Sakals, M.E., Innes, J.L., Wilford, D.J., Sidle, R.C. and Grant, G.E. 2006. The role of forests in reducing hydrogeomorphic hazards. *Forest Snow and Landscape Research*, 80 (1): 11–22.

Stancanelli, L.M. and Foti, E. 2015. A comparative assessment of two different debris flows propagation approaches – blind simulation on a real debris flow event. *Nat. Hazard Earth Syst. Sci.*, 15, 375–746.

Takahashi, T., Nakagawa, H., Harada, T. and Yamashiki, Y. 1992. Routing debris flows with particle segregation. *J. of Hydr. Eng.* 118 (11): 1490–1507.

Tecca, P.R. and Genevois, R. 2009. Field observations of the June 30, 2001 debris flow at Acquabona (Dolomites, Italy). *Landslides*, 6(1): 39-45.

Tuladhar, G. 2012. Disaster management system in Nepal-policy issues and solutions. *Journal of Risk Analysis and Crisis Response*, 2(3): 166-172.

Viles, H.A., Naylor, L.A., Carter, N.E.A. and Chaput, D. 2008. Biogeomorphological disturbance regimes: progress in linking ecological and geomorphological systems. *Earth Surf. Process. Landf.* 33 (9): 1419–1435.

Wang, J., Yu, Y., Wei, X.F., Gong, Q.H. and Xiong, H.X. 2016. Run-out Effects of Debris Flows Based on Numerical Simulation. *The Open Civil Engineering Journal*, 10: 859-869.

Wang, Y.X. and Xia, B. 2012. Eco-restoration strategies and measures for the soil and water conservation of Typhoon-hit areas in western Guangdong province-a case study of Magui town, Gaozhou city. *Science of Soil and Water Conservation*, 10(1): 88-93. (In Chinese).

Risk Analysis Based on Data and Crisis Response Beyond Knowledge – Huang & Nivolianitou (eds)
© 2020 Taylor & Francis Group, London, ISBN 978-0-367-25146-8

Driving factors affecting investment risk in international infrastructure construction focusing on Southeastern Asia

Ting Yuan, Pengcheng Xiang, Qianman Zhang, Zhaoying Ye & Jianbin Zhang
Faculty of Construction Management & Real Estate, Chongqing University, Chongqing, China

ABSTRACT: Chinese enterprises investing in international infrastructure markets have been experiencing greater investment risks than ever before. To relieve investment pressures, this study identified risk variables and analyzed driving factors affecting investment risk in international infrastructure construction, based on Southeastern Asian data collected from numerous official reports and websites. Fifteen representative risk variables were identified by time-weighted grey relational analysis (GRA). The hierarchical structure among risk factors was explored using the interpretative structural modeling (ISM) method and matrice d'impacts croises multiplication appliqué an classement (MICMAC) analysis. The results of the study showed that corruption of the host country, efficiency of legal frameworks in settling disputes, voice and accountability play critical driving roles in Southeastern Asia for Chinese enterprises. This study contributes a general method for the exploration of driving powers to help optimize the allocation of limited resources by Chinese investors.

Keywords: Driving factors, Infrastructure investments, Southeastern Asia, Chinese enterprises.

1 INTRODUCTION

The majority of Chinese enterprises have invested in overseas infrastructure construction as a result of the 'belt and road' strategy (Du et al., 2016). However, compared to domestic construction markets, various additional barriers should be considered when investing in international infrastructure construction markets (Eybpoosh et al., 2011). A turbulent investment environment will seriously affect the success rate of infrastructure construction for Chinese construction enterprises (Zhao et al., 2014). For example, the Colombo port project was suspended half-way, and Nepal gave up the agreement on the construction of hydropower stations signed by its former government and Chinese construction enterprises. Furthermore, different regions have different investment environments. For Chinese enterprises, it is important to carry out investment risk analysis relating to infrastructure constructions in a specific area to decrease investment challenges.

Southeastern Asia has become a popular region for investment. The proportion of investment stocks in Southeastern Asia was the largest at about 54.3%. Furthermore, the poor state of transport infrastructure in many parts of Southeastern Asia, especially railways, has limited the development of local economies. In view of the massive requirements of infrastructure construction in Southeastern Asia and investment pressures facing Chinese construction enterprises, the main objective of this study was to explore investment risks affecting infrastructure construction in Southeastern Asia for Chinese construction enterprises.

Research on the investment risks for international infrastructure construction has mainly focused on the identification of key risk indicators. Numerous key risk variables have been recognized, including the instability of governments, immaturity of legal systems (Eybpoosh et al., 2011), project desirability to the host country (Deng et al., 2014), expropriations, taxation restrictions and currency inconvertibility. However, these previous studies have had some limitations. Firstly, investment risks relating to specific regions have not received enough attention. Secondly, many previous works have ignored interrelationships between risk variables; however, Chinese construction enterprises that are eager to relieve investment pressures need to be familiar with the hierarchical structure and driving factors of investment risk (Serghei et al., 2016). Overall, despite extensive research analysis into investment risk, there has been no detailed analysis exploring the drivers of investment risk in Southeastern Asia. Therefore, research on investment risks affecting infrastructure constructions in this region are imperative to inform Chinese construction enterprises.

2 LITERATURE REVIEW

Investment risk relating to international projects has become a hot issue for research and practitioners. A number of researchers have argued that investment risk relating to international infrastructure construction relates to political risks, economic conditions of the home country, legal risks and cultural differences. Some hold the view that political risks relating to international projects include any political events that damage or impair transnational commercial profits, such as wars, revolutions, elections or internal conflicts (Ahlers et al., 2015). Another cluster of researchers argue that economic risks describe the macro-economic conditions that can impact material price, labor costs and business operational management costs. Legal risks include laws, rules and policy instability. Cultural cognitive elements emphasize cultural contexts that influence people's cognition, expectations and obligatory dimensions of social life (Effah et al., 2017).

According to definitions of investment risk, many investment risk factors relating to international projects have been identified. This study collected numerous investment risk factors by analyzing various existing risk checklists for international projects, including currency inconvertibility (Deng et al., 2014), expropriations, regional economy (Wang et al., 2016), breach of contracts, policy risks, political violence, corruptions and bribery (Chan and Owusu, 2017). Because projects are frequently carried out in different environments, they often refer to different natural environments, political and social frameworks (Chen et al., 2011). The literature review revealed that there have been few studies that have analyzed investment risk relating to international projects with a focus on Southeastern Asia.

Furthermore, investment risk indicators have close relationships with economic, legal and cultural aspects of the host country. It is critical that the investment cost-risk relationship for international infrastructure construction be determined because, whilst there has been considerable research about risk paths for international projects, such as safety risks, construction risks, etc. (Eybpoosh et al., 2011), there have been few studies focusing on investment risk structures relating to overseas projects. Therefore, attempts to identify investment risk factors have not resulted in tangible outcomes: the sources of risk and key drivers have not been identified and investment risk pressures have not been relieved.

When Chinese construction enterprises invest in international markets, resource constraints should be considered (Qazi et al., 2016). To enable the appropriate allocation of resources, it is imperative to clarify relationships between risk factors and remove obstacles in a timely fashion in order to ensure the smooth implementation of international projects.

Therefore, to identify key investment risk factors and their respective importance in Southeastern Asia, this study selected Thailand, Vietnam and the Philippines as representative countries in which to conduct the research. To some extent, key investment risk variables and driving power can make Chinese construction enterprises explore visible investment strategies and relieve investment pressures affecting infrastructure construction in Southeastern Asia.

3 RESEARCH METHODS

3.1 *Identification model of investment risk factors about international projects*

GRA was used to identify representative investment risk factors. GRA can quantify the dynamic development of a system to measure how much a factor contributes to a behavior

(Rajesh and Ravi, 2015), and recognizes key factors in small samples. GRA works well with the time-weighted method. Because the importance of different investment risk factors affecting infrastructure in Southeastern Asia will change over time, this study will apply an assessment model that considers changes in the importance of different indicators over several years.

Firstly, because different indicators have different units they were normalized to unit-free, enabling their comparability (Rajesh and Ravi, 2015). Secondly, this study argues that the best value of each indicator in different years and regions follows the guidelines below: for a positive investment risk index, this study took the maximum value of the index; for a negative investment risk index, this study took the minimum value of the index; and, the larger the GRA coefficient, the better it can measure investment risks (Li et al., 1997). This study used Equation (1) to calculate GRA coefficients.

$$r_{ij}(t) = \frac{\min\limits_{1 \le i \le m, 1 \le j \le n} \min\left(\left|v_{i0}(t) - v_{ij}(t)\right|\right) + \delta \max\limits_{1 \le i \le m, 1 \le j \le n} \max\left(\left|v_{i0}(t) - v_{ij}(t)\right|\right)}{\left(\left|v_{i0}(t) - v_{ij}(t)\right|\right) + \delta \max\limits_{1 \le i \le m, 1 \le j \le n} \max\left(\left|v_{i0}(t) - v_{ij}(t)\right|\right)} \tag{1}$$

Here, $r_{ij}(t)$ represents the GRA coefficient of the indicator i in the region j in the year t; $v_{i0}(t)$ represents the ideal value of the normalized i index in the year t; $v_{ij}(t)$ represents the value of the normalized i index in the region j in the year t; m represents the number of indicators; n represents the number of regions; δ represents the coefficient $\delta \in [0, 1]$, and is usually 0.5 according to the international practice.

Finally, investment risk factors often change more frequently due to the host country environment. This study maintained that the time weight of the indicator was greater if the year was closer to the present, adopting the time-weighted method to solve the problem according to Equations (2) and (3).

$$w_t = \lambda_t / \sum_{t=1}^{N} \lambda_t, t = 1, 2, \ldots, N \tag{2}$$

$$\lambda_t = \exp\left(-\frac{[t - N]^2}{2}\right), t = 1, 2, \ldots, N \tag{3}$$

Here, w_t represents the time weight of the year, t. Therefore, the comprehensive GRA coefficient R_i can be calculated according to Equation (4).

$$R_i = \sum_{t=t_0}^{T} w_t \left[\frac{1}{n} \sum_{j=1}^{n} r_{ij}(t)\right] \tag{4}$$

Since R_i has been corrected by the time weight, this study took indices with $R_i \ge 0.5$ as key investment risk variables affecting infrastructure construction in Southeastern Asia.

3.2 Model exploring the driving power of critical investment risk factors

The study explored the hierarchical structure among key investment risk factors to explore their respective driving power. ISM is a reliable method for exploring the hierarchy of factors. As there are limited specialists who are familiar with infrastructure construction in Southeastern Asia, it would be impossible to obtain much relevant data effectively by other methods; however, ISM can overcome these deficiencies and can analyze factors by using a small subset of a data sample.

Firstly, the adjacency matrix was developed according to the opinions of professional experts (Iyer and Sagheer, 2010). This study invited twenty professional experts to recognize relationships between factors. These experts had extensive experience and had been involved with the management of many international projects in Southeastern Asia. Secondly, the reachability matrix was obtained using Mat-lab software programming. This is a powerful tool to explore direct and indirect relationships between indicators (Iyer and Sagheer, 2010).

Then, in order to determine the level of each variable, $P(S_i)$ represents the reachable set of factors, $Q(S_i)$ represents the antecedent set of factors and $P(S_i) \cap Q(S_i)$ represents the common set of factors. Where $P(S_i) = \{S_{i(column)} | m_{ij} = 1\}$; $Q(S_i) = \{S_{j(row)} | m_{ij} = 1\}$; $L_i = \{S_i | P(S_i) \cap Q(S_i) = P(S_i), i = 0, 1, 2, \ldots, k\}$. Then, the row and column corresponding to the L_i factor were deleted from the reachable matrix M. The same step was repeated multiple times to obtain levels for all variables.

Finally, this study aimed to determine the driving power and dependence power among critical risk factors using the MICMAC method. This method has been widely used (Shen et al., 2016). Based on MICMAC theory, key factors can be classified into four categories: dependent factors, driving factors, linkage factors and autonomous factors. Dependent variables are often affected by others; driving variables have an ability to affect other factors; linkage variables are affected by some factors and influence other factors; and, autonomous variables have little driving power and little dependence power and have few links with the theme.

4 DATA COLLECTION

Data relating to Southeastern Asian countries was collected from literature, official reports and databases including Web of Science, Elsevier Science Direct, China National Knowledge Infrastructure (CNKI) database, World Bank and so on. Ten official reports and websites were used as data sources relating to the variables. A list of variables in investment risk affecting infrastructure construction in Southeastern Asia was compiled, as shown in Table 1.

Table 1. The comprehensive investment risk factors based on previous studies.

Number	Risk factors	Number	Risk factors
R1	Official exchange rate	R17	Coefficient of human inequality
R2	Tax rates	R18	Unemployment rates
R3	Culture distance	R19	Internally displaced persons
R4	Inflation	R20	Investment costs of terrorism
R5	Anti-dumping initiations	R21	Investment costs of crime and violence
R6	General government debt, % GDP	R22	Basic requirements to infrastructure
R7	Government instability	R23	Country credit rating
R8	Political relationships to China	R24	Expropriation
R9	Policy instability	R25	Strict foreign currency regulations
R10	Dealing with construction permits	R26	Restrictive labor regulations
R11	Inefficient government bureaucracy	R27	Intellectual property protection
R12	Judicial independence	R28	Investment burden to abide by local government regulations
R13	Efficiency of legal framework in settling disputes	R29	Public trust in politicians
R14	Strength of auditing and reporting standards	R30	Reasonable work ethic in national labor
R15	Corruption of host country	R31	Communication barriers
R16	Voice and accountability		

5 RESULTS

This study identified 15 key risk factors, as shown in Table 2. These risk factors require greater attention to relieve investment pressures affecting infrastructure construction in Southeastern Asia.

As shown in Figure 1, the hierarchical structure of representative investment risk factors was divided into six levels. This structure helps enable the driving powers of representative

Table 2. Representative investment risk affecting infrastructure construction in Southeastern Asia.

Identifiers	Weighted GRA coefficient	Identifiers	Weighted GRA coefficient
R24	1.000000	R25	0.692389
R4	0.825439	R9	0.667959
R13	0.804474	R6	0.647889
R19	0.802415	R7	0.626824
R16	0.742856	R28	0.616958
R31	0.722490	R11	0.601309
R15	0.721392	R21	0.565028
R18	0.714369		

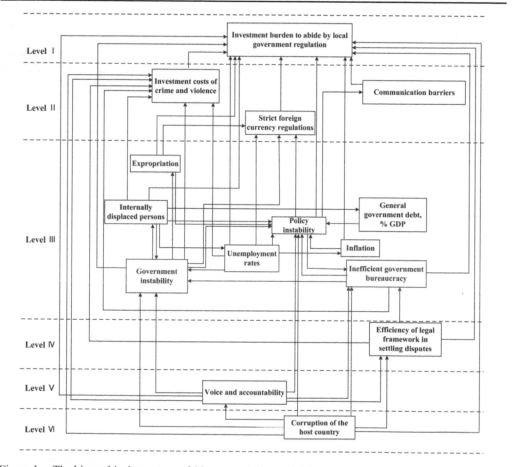

Figure 1. The hierarchical structure of 15 representative variables.

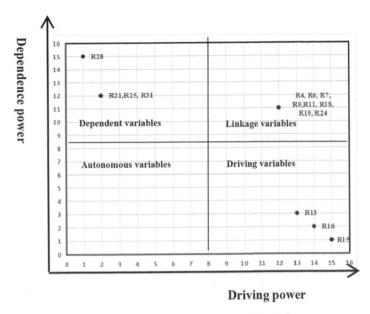

Figure 2.   Classification of representative variables by MICMAC analysis.

variables to be recognized. From Figure 1, this study discovered that most of the key risk factors were located in the third level of the structure. As shown in Figure 2, driving powers of representative risk factors were R13, R15, R16. Dependent variables identified by this study included R21, R25, R28, R31. Linkage variables included R4, R6, R7, R9, R11, R18, R19, R24. There were no autonomous variables. This showed that risk indicators identified in the study were highly relevant to the theme.

6   DISCUSSION

Based on Figure 2, driving powers included three key variables, R15, R16, R13. R15 was found to be the most important investment risk in Southeastern Asia. Officials in a host country often create administrative obstacles to serve their own interests, especially for land and construction investments (Tromme, 2016). According to Figure 1, the phenomenon of corruption is negative to construction cost. R16 was another driving power factor. In Vietnam, for example, Chinese enterprises can only negotiate construction disputes with local governments, business offices and chambers of commerce, which makes it difficult for Chinese enterprises to communicate directly with the Vietnamese Parliament or local councils. Additionally, most former officials are now leaders of local companies. Therefore, during the construction process, unclear responsibilities, policy bias and unreasonable payment rules often appear. These phenomena will increase investment costs for foreign enterprises. R13 was also a driving power factor. In the Philippines, judicial practice is complicated and time-consuming, and there is serious judicial corruption. According to results shown in Figure 1, disputes put forward by foreign enterprises cannot be solved in a timely fashion, increasing the time and capital costs.
   Based on Figure 2, linkage powers included R4, R6, R7, R9, R11, R18, R19, R24. According to Figure 1, these belong to level III. The majority of construction materials will be subject to fluctuations due to the critical role of this inflation. The unemployment rate would greatly impact government instability and policy instability relating to infrastructure projects (Figure 1). Internally displaced persons (R19) affect the economy, policies and

stability of a host country (Deng et al., 2014). If policies change frequently (R9), this can seriously affect project plans.

Dependent variables included R21, R25, R31. Communication barriers (R31) depend on the cultural distance between two countries. This study argues that this cultural distance is affected by the depth of communication in recent years: similar values and beliefs will be promoted when frequent political communications occur between two countries.

According to Figure 2, there are no autonomous variables. Autonomous factors have a weak relationship with the theme. Therefore, according to the Figure 2, all investment risks identified in this study are key investment risk factors.

7 CONCLUSION

This study identified 15 representative investment risk factors affecting the implementation of international infrastructure construction in Southeastern Asia for Chinese construction enterprises. According to the results of the study, when resources are limited, project managers should focus on three driving factors: the efficiency of the legal framework in settling disputes (R13), voice and accountability (R16), and corruption within the host country (R15). These conclusions should provide a useful reference for Chinese construction enterprises to expand the construction market in Southeastern Asia.

The primary contributions of the study are as follows. Although the study collected data from the perspective of Chinese construction enterprises and Southeastern Asian countries, this study offers a general method to explore drivers of investment risk in international infrastructure construction. The main findings of this study were that Chinese enterprises investing in Southeastern Asia need to respond promptly to driving powers. Based on Figures 1 and 2, when Chinese construction enterprises invest in Southeastern Asia, they should try to increase the transparency of communication, clarify terms of the contract and mitigate corruption by establishing goals of common interest with local enterprises.

Regardless, this study provides a reference to guide the optimal allocation of resources for overseas investment by Chinese construction enterprises, different enterprises have different abilities to withstand investment risks. Future research will be to build a solution about representative risks by analyzing characteristics of different Chinese construction enterprises.

ACKNOWLEDGEMENT

The work described in this paper was fully supported by a joint grant from Project No. 2019CDSKXYJSG0041, 2017CDJSK03XK19 supported by the Fundamental Research Funds for the Central Universities.

REFERENCES

An, T.N., Long, D.N., Long, L.H., Dang, C.N., 2015. Quantifying the complexity of transportation projects using the fuzzy analytic hierarchy process. *International Journal of Project Management.* 33(6), 1364–1376.

Ahlers, R., Budds, J., Joshi, D., Merme, V., Zwarteveen, M., 2015. Framing hydropower as green energy: assessing drivers, risks and tensions in the Eastern Himalayas. *Earth System Dynamics.* 6(1), 195–204.

Chen, Z., Li, H., Ren, H., Xu, Q., 2011. A total environmental risk assessment model for international hub airports. *International Journal of Project Management.* 29(7), 856–866.

Chan, A.P.C., Owusu, E.K., 2017. Corruption forms in the construction industry: literature review. *Journal of Construction Engineering & Management.* 143(8), 04017057.

Deng, X., Pheng, L.S., Zhao, X., 2014. Project system vulnerability to political risks in international construction projects: the case of Chinese contractors. *Project Management Journal.* 45(2), 20–33.

Du, L., Tang, W., Liu, C., Wang, S., Wang, T., Shen, W., et al., 2016. Enhancing engineer-procure-construct project performance by partnering in international markets: perspective from Chinese construction companies. *International Journal of Project Management.* 34(1), 30–43.

Eybpoosh, M., Dikmen, I., Birgonul, M.T., 2011. Identification of risk paths in international construction projects using structural equation modeling. *Journal of Construction Engineering & Management.* 137(12), 1164–1175.

Effah, E.A., Parn, E., Chan, A.P.C., Owusu-Manu, D.G., Edwards, D.J., Darko, A., 2017. Corrupt practices in the construction industry: survey of Ghanaian experience. *Journal of Management in Engineering.* 33(6), 05017006.

Iyer, K.C., Sagheer, M., 2010. Hierarchical structuring of PPP risks using interpretative structural modeling. Journal of Construction Engineering & Management. 136(2), 151–159.

Liu, J., Zhao, X., Yan, P., 2016. Risk paths in international construction projects: case study from Chinese contractors. *Journal of Construction Engineering & Management.* 142(6), 05016002.

Qazi, A., Quigley, J., Dickson, A., Kirytopoulos, K., 2016. Project complexity and risk management (Procrim): towards modelling project complexity driven risk paths in construction projects. *International Journal of Project Management.* 34(7), 1183–1198.

Rajesh, R., Ravi, V., 2015. Supplier selection in resilient supply chains: a Grey relational analysis approach. *Journal of Cleaner Production.* 86, 343–359.

Shen, L., Song, X., Wu, Y., Liao, S., Zhang, X., 2016. Interpretive structural modeling based factor analysis on the implementation of emission trading system in the Chinese building sector. *Journal of Cleaner Production.* 127, 214–227.

Serghei Floricel, John L. Michela, Sorin Piperca., 2016. Complexity, uncertainty- reduction strategies, and project performance. *International Journal of Project Management.* 34(7), 1360–1383.

Tromme, M., 2016. Corruption and corruption research in Vietnam-an overview. *Crime Law & Social Change.* 65(4), 1–20.

Wang, T., Wang, S., Zhang, L., Huang, Z., Li, Y., 2016. A major infrastructure risk-assessment framework: application to a cross-sea route project in China. *International Journal of Project Management.* 34(7), 1403–1415.

Zhao, X., Deng, X., Sui, P.L., Li, Q., 2014. Developing competitive advantages in political risk management for international construction enterprises. *Journal of Construction Engineering & Management.* 140(9), 758–782.

Analysis of compliance risks in overseas infrastructure projects under the Belt and Road Initiative

Zhaoying Ye* & Pengcheng Xiang
School of Management Science and Real Estate Management, Chongqing University, Chongqing, China

Wenwu Xiang, Feng Jin & Sanshan Wei
SINOPEC Engineering (Group) Co, Ltd., Beijing, China

ABSTRACT: Although Chinese International Construction Enterprises(ICEs) have made great efforts in overseas project risk management, an emerging new type of "compliance risk" in the rule prioritization era has been ignored. In the face of an increasingly stringent supervision environment, compliance has become a new way for Chinese enterprises to adapt to global competition. This paper analyzes the characteristics of compliance risk through the VUCA method and uses the fuzzy analytic hierarchy process (FAHP) method to quantify the risk factors on the compliance risk list.

Keywords: compliance risk management, overseas infrastructure project, the Belt and Road, FAHP, VUCA

1 INTRODUCTION

Along with the further implementation of the "go global strategy" and the promotion of the "Belt and Road Initiative", increasing numbers of Chinese enterprises aim to explore the overseas market. The countries along the B&R Initiative have become the core layout area of ICEs. Owing to the B&R Initiative, in the first half of 2018, The newly signed overseas contracts in the Belt and Road countries has reached $48.7 billion . The scale of international projects in which Chinese enterprises have participated is expanding significantly, and at the same time the complexity of projects is becoming more challenging due to the extended industrial chain, massive investment, and great differences in management environments. In regions such as Africa, South Asia, and Southeast Asia, Chinese-funded enterprises have distinct advantages. Therefore, international organizations, local authorities and their citizens are paying close attention to Chinese-funded enterprises. It is more arduous for enterprises to operate foreign projects when abiding multiple rules and doing a good job of risk management. In the operation of overseas infrastructure constructions, however, many companies are still adopting traditional management strategies. Moreover, Chinese enterprises are unfamiliar with the regulatory environment of international compliance, and the possibility of compliance risk exposure is higher than before. In case of irregularities and violations of laws, enterprises are likely to be punished by the UN, multilateral development banks or local authorities. The consequences of violation, especially in project revenue and project duration, are more than ICEs could stand, Let alone the badly corporate reputation that may be incurred.

---

*Corresponding author: Ivyy@cqu.edu.cn.

2   LITERATURE REVIEW

There are numerous ways that can be used to classify the risks for construction projects (He Zhi,1995), for instance, in accordance with the nature and attribute of risk, risk occurrences in different life cycles or different levels, or combining these (Perry, 1992; Li and Tiong, 1999; Wang, 2014). Dikmen (2000) pointed out that in the overseas project risk and market opportunity evaluation system, the risk is divided into the national level and project level. Integrated risk analysis methods are needed since the overseas projects are getting more complex. Wang (2003) divided the risks of international engineering projects in East Asia into political risk, economic risk, legal risk, contract risk, standard difference risk and cultural difference risk. There has been a transformation of overseas contracting, from traditional construction contracting such as design-bid-built (DBB) or turn-key-operate to assuming more financing responsibility of the project, such as engineering-procurement-construction+financing (EPC+F) and public-private partnership (PPP), so some researchers have studied the risk and its allocation from the perspective of contracting mode (Bing, 2005; Cheung, 2011). Early research on the analysis of compliance risk factors was confined to anti-corruption subjects (Ayres, 2009). As the risk of anti-corruption compliance exists in many industries, some scholars analyzed the causes of corruption risk in case studies, combining the characteristics of the field and industry where the corruption occurred (Ernesto, 2007; Clark, 2010). Arnold (2012) believes that organizational complexity, corporate culture, internationalization, and functional complexity would affect the level of corporate corruption (Sahin, 2010; Iwasaki I, 2012). In recent years, several scholars have considered that there are compliance risks in all stages of large-scale infrastructure project construction, from the perspective of the project life cycle (Honorati, 2007; Mohammed, 2015).

In summary, the existing research on overseas infrastructure construction project risk has achieved certain results. However, the research on compliance risk in overseas infrastructure projects needs to be enriched. To be specific, there is insufficient compliance risk identification and lack of compliance risk qualification involved in the life cycle of the project.

3   VUCA ANALYSIS OF COMPLIANCE RISK CHARACTERRISTICS

3.1   *Volatility*

The volatility of compliance risk is reflected in the sensitivity of compliance risk itself and the possibility of its risk consequences. Due to the constraints of supervision, laws and regulations from many sides, there are many triggers of compliance risk in participating in overseas infrastructure construction, and sometimes enterprises may violate regulations unconsciously. In addition, compliance risk is highly sensitive to external regulatory and policy changes, such as market-access policies. Compliance risk differs from other risks mainly because of the subject who bears the consequences. With the infrastructure projects along with the B&R countries, the political elements of the project are obvious., and with the risk factors of anti-corruption or cross-border tax compliance, the risk consequences often exceed the scope that enterprises can cope with. In some cases, the responsibility of risk response rises from the enterprise to the national level.

3.2   *Uncertainty*

The uncertainty of compliance risk is determined by the long life cycle of the project. The construction life cycle of overseas infrastructure project ranges from a few years to several decades. During this time, the possibility of violating compliance rules is greater. The Chinese-founded companies have to bear the negative consequences of the prolonged construction duration. A large number of events and activities are involved, such as the assurance of progress payments and change of exchange rate policy, and each activity is closely related to the cost, duration, quality, safety and environment of the project. Because of the unpredictability of the objective environment and the lack of historical project experience to use as a reference, the uncertainty of compliance risk has greatly increased.

3.3 *Complexity*

The complexity of compliance risk is reflected in the overseas infrastructure construction market and the treatment of compliance relationship with stakeholders. The B&R Initiative provides more opportunities for Chinese-funded enterprises to expand overseas markets. There are great differences in development level, legal system and cultural differences among countries along the B&R Initiative. As a result, it is more difficult for enterprises to complete projects, because even the same projects may vary greatly from country to country. Secondly, throughout the project process, some relationships with stakeholders can be achieved through contracts or insurance, such as consortium partners or subcontractors. However, there is no contract to regulate the relationship with non-contractual stakeholders such as local residents, trade unions and non-governmental organizations in the host country. Whether the enterprise can handle these relationships is directly related to the smooth progress of the project. Many projects are suspended due to opposition from local residents or non-governmental organizations.

3.4 *Ambiguity*

The ambiguity of compliance risk is reflected in the division of compliance risk factors and liability. Compliance risk runs through the whole project life cycle. First, individual compliance risk factors have no obvious stage characteristics for identification and analysis. Second, the relationships between compliance risk factors are highly correlated and will influence each other. This makes it more difficult to prevent and analyze compliance risks in the process. Last, there are many stakeholders in overseas infrastructure projects, and the occurrence of compliance risk events often involves multiple subjects of responsibility. The parties to the contract have to bear the consequences of compliance risk. However, due to the ambiguity of compliance risk itself, it is difficult to classify its responsibility boundary into relevant responsible parties. Therefore, the risk of compliance has unclear characteristics in the division of liability in advance, and the implementation and the assumption of liability afterward.

4 METHODOLOGY AND DATA PRESENTATION

4.1 *Risk identification*

In compliance risk identification and classification for international infrastructure projects, a global view is essential to cover the risks from all levels and every construction stage. The compliance risks identified in this paper were collected using a combination of literature review, case study, questionnaire survey, and face-to-face interviews with international project managers. This was undertaken to identify compliance risk factors as comprehensively as possible in overseas construction projects. Three categories of compliance risk factors are listed in Tables 1, 2 and 3.

4.2 *Fuzzy analytic hierarchy process*

Fuzzy analytic hierarchy process (FAHP) is a qualitative and quantitative method to evaluate multi-objective problems by combining the analytic hierarchy process (AHP) and fuzzy comprehensive evaluation. AHP is the foundation of FAHP. The key of using FAHP is to establish dual importance comparison matrix on the basis of constructing decision-making target structure, each criteria layer and the hierarchy structure of prediction indexes (Li, 2011). The ultimate objective analysis of compliance risks can be broken down into 12 risks factors (R1 to R12).

Table 1. Traditional compliance risk.

Type	Risk Factor	Sub-Indicator	Code.
Traditional Compliance Risk	Anti-Corruption Risk R1	Bribes	A1
		Accepting bribes	A2
		Joint liability	A3
		AFCA-US	A4
		UK Bribery Act	A5
		Spain II	A6
		Anti-corruption strategy in Africa (Angola)	A7
		The Beijing Declaration on Fighting Corruption (APEC)	A8
		Integrity Compliance Guidelines (The World Bank)	A9
		UN Convention Against Corruption	A10
		Convention on Combating Bribery of Foreign Public Officials' International Business Transactions (OECD)	A11
	Repression of Unfair Competition R2	Restriction of market-access policies	A12
		Fix or control prices	A13
		Divide or distribute customers, markets, territories productions	A14
		Redistrict the export or import of goods	A15
		Restrict competition or dealings with suppliers	A16
	Tax Compliance R3	Incorrect understanding of the host country's tax policy	A17
		Cross-border personal income tax planning	A18
		False report of tax returns	A19
		Tax liquidation at exit stage	A20
		Tax discrimination	A21
		Differences in tax system of host country	A22
		Change of preferential tax policy	A23
		Low information-based degree	A24
	Foreign Exchange R4	Fluctuation of international exchange rate	A25
		Profit remitted	A26
		Reduction of foreign exchange reserve in host countries	A27
		Appreciation of the RMB	A28
		Depreciation of local currency	A29
		Difficulties in converting local currency into hard currency	A30
		Government approval of foreign exchange import and export	A31
		Third-country purchases lead to an increase in hard currency expenditure (local goods are substandard)	A32

Each factor is composed of sub-indicators listed in the third volume of Table 1 (A1 to A32), Table 2 (B1 to B43) and Table 3 (C1 to C15). The FAHP steps are as follow:

Step 1. Establish hierarchical structures.

Step 2. Create fuzzy judgment matrices using pair-wise comparisons.

Step 3. Defuzz and calculate CR and fuzzy weights.

Step 4. Final ranking if the consistency meets CI < 0.1 (refer to Table 4).

Establishing a fuzzy set for R1.1Anti-corruption U={A1, A2, A3}and its correspondent evaluation level V = {V1, V2, V3, V4, V5}. The single factor evaluation matrix is shown in Table 5.

By comparing the factors of the factor set R1 in pairs, the judgment matrix is given as follows.

Calculate the maximum eigenvalue of matrix B, λ max = 3.351; weight of evaluation index, W = [0.72884 0.21617 0.05499]. Use the equation CR = CI/RI to check consistency. CR = CI/RI = 0.053/0.58 = 0.092 < 0, which means the results pass the consistency text.

Table 2. Compulsory rules compliance risk.

Type	Risk Factor	Indicator	Code.
Compulsory Rules Compliance Risk	Bidding and Contract Compliance R5	False qualification	B1
		Interim payment violation	B2
		False corporate performance information	B3
		Inconsistent staffing of key managers	B4
		Take the performance of subsidiaries as tender performance	B5
		Deficiency financial capacity	B6
		Conceal bankruptcy	B7
	Labor Employment Compliance R6	Fair employment (like language barriers)	B8
		Labor quota overrun	B9
		Salary and welfare treatment	B10
		Termination of labor contract	B11
		Respect for customary holiday	B12
		Relations with trade unions	B13
		Safety training for labor	B14
		Professional job qualification	B15
		Work permission	B16
		Professional job qualification	B17
		Workplace health and safety	B18
		Host country work visa approved	B19
		Inspection of the USPP of drugs and alcohols	B20
	Environmental Protection Compliance R7	Soil pollution	B21
		Noise pollution	B22
		Dust control in construction site	B23
		Substandard of waste water discharge	B24
		Illegal disposal of construction waste	B25
		Weak awareness of environmental protection	B26
		Opposition of local environmental NGOs	B27
		Affecting the living environment of local residents	B28
		High standard environmental regulations	B29
		Lack of environmental regulations	B30
		Destruction of eco-orientated industries (tourism)	B31
	Customs Clearance Compliance R8	Not familiar with customs rules	B32
		Insufficient preparation for customs clearance	B33
		Follow-up supervision and verification	B34
		Insufficient estimated tax amount (causes customs detention)	B35
		Transportation and logistics monopoly	B36
		Customs staff and deliberate difficulties (to solicit bribes)	B37
	Intellectual Property R9	Property ownership	B38
		Data on cross-border	B39
		Original design achievement	B40
		Original technical achievement	B41
		Trademark infringement	B42
		Unauthorized downloading of material protected	B43

Table 3. Soft rules compliance risk.

Type	Risk Factor	Indicator	Code
	Commercial Partners R10	Local customer agent	C1
		Local third legal agent	C2
		Relationships with subcontractors	C3
		Relationships with suppliers	C4
		Relationships with joint venture partner	C5
		Banquet and reception for business partners	C6
Soft Rules Compliance Risk	Social Relations R11	Personal relationships in the workplace (conflict between Chinese and local staff)	C7
		Relationships with local NGOs	C8
		Relationships with local media	C9
		Religious and cultural differences	C10
		Relationships with local communities and residents	C11
		Timely corrective measures in response to a complaint or incident	C12
		Communications with a member of a government/ legislature	C13
	Contributions R12	Undisclosed public donations	C14
		Intentional political contribution	C15

Table 4. Random index.

N	3	4	5	6	7	8	9	10	11
RI	0.58	0.96	1.12	1.24	1.32	1.41	1.45	1.49	1.51

Table 5. Evaluation matrix.

Codes	1	2	3	4	5
A1	0.00%	7.10%	14.30%	28.60%	50.00%
A2	0.00%	14.30%	21.40%	17.90%	46.40%
A3	3.60%	7.10%	21.40%	35.70%	32.10%

The results of fuzzy comprehensive evaluation are as follows:

$$A = W \cdot R = [0.72884 \quad 0.21617 \quad 0.05499] \begin{bmatrix} 1 & 7 & 9 \\ 1/7 & 1 & 7 \\ 1/9 & 1/7 & 1 \end{bmatrix} \quad (1)$$

The matrix for the other compliance risk factors and its sub-indicators were generated using the same method as that for R1-Anti-corruption. Based on the results of a series of rankings, the weights of all risks factors in each level of the hierarchy relative to the entire level directly above were obtained. After the above analytic process, the weight of each compliance risk indicator was calculated for determining the comparative importance of compliance risks in the overseas project, and the results are listed in Table 6.

Table 6. The result of FAHP.

Order	Indicator	Local Weight	Final Weight	Order	Indicator	Local Weight	Final Weight
1	A1	72.884%	12.778%	11	A6	12.243%	2.146%
2	A12	55.532%	8.931%	12	A15	12.249%	1.970%
3	A17	39.198%	6.608%	13	A13	12.239%	1.968%
4	A4	29.000%	5.084%	14	A10	10.885%	1.908%
5	A25	43.131%	4.190%	15	A14	11.586%	1.863%
6	A2	21.617%	3.790%	16	A26	18.236%	1.772%
7	A23	17.770%	2.995%	17	A18	10.331%	1.741%
8	A11	15.995%	2.804%	18	B2	28.281%	1.693%
9	A20	13.069%	2.203%	19	B32	31.301%	1.410%
10	A5	12.537%	2.198%	20	A9	7.935%	1.391%

5 DATA ANALYSIS AND CONCLUSION

Based on the FAHP method mentioned above, the comparative importance of each indicator is listed. In terms of compliance risk management, risk rank is the guidance for risk analysis and control. Table 6 shows the 20 most important compliance risk indicators. We can see that the majority are part of traditional compliance risk. Only two compulsory compliance factors, B32 (Interim payment violation) and B32 (Not familiar with customs rules), are listed in Table 5. An interim payment is a determining factor to ensure a project is on schedule. As some Chinese companies take no account of the importance of investigating pre-market customs rules, B32 not familiar with customs rules is very important.

ACKNOWLEDGMENTS

This project was supported by the Fundamental Research Funds for the Central Universities (2017CDJSK03XK19, 2019CDJSK03PY02).

REFERENCES

Adeleke FAR, Olayanju OF. The role of the judiciary in combating corruption: aiding and inhibiting factors in Nigeria. *Commonwealth Law Bulletin* 2014, 40(4): 589–607.

Ayres M, Clark J, Gourley A, et al. Anti-corruption. *International Lawyer* 2009, 70(1): 1–8.

Brown SF, Shackman J. Corruption and Related Socioeconomic Factors: A Time Series Study. *Kyklos* 2007, 60(3): 29.

Cimpoeru MV, Cimpoeru V. Budgetary transparency – an improving factor for corruption control and economic performance. *Procedia Economics and Finance*, 2015, 27: 579–586.

Clark J, Davis J, Reider-Gordon M, et al. Anti-corruption. *International Lawyer* 2010, 44(1): 451–472.

Deng X, Sui PL. Exploring critical variables that affect political risk level in international construction projects: case study from Chinese contractors. *Journal of Professional Issues in Engineering Education & Practice* 2014, 140(1): 04013002.

Ernesto Dal Bó, Martín A. Rossi. Corruption and inefficiency: theory and evidence from electric utilities. *Journal of Public Economics* 2007, 91(5–6): 939–962.

Honorati M, Mengistae T. Corruption, business environment, and small business fixed investment in India. *Social Science Electronic Publishing*, 2007: 1–35(35).

Iwasaki I, Suzuki T. The determinants of corruption in transition economies. *Economics Letters*, 2012, 114(1): 0–60.

Marjit S, Mandal B, Roy S. Trade openness, corruption and factor abundance: evidence from a dynamic panel. *Review of Development Economics*, 2014, 18(1): 45–58.

Mohammed MA, Eman Y, Hussein AH, et al. E-government factors to reduce administrative and finance corruption in Arab countries: Case study Iraqi oil sector. *Innovation & Analytics Conference & Exhibition*. AIP Publishing LLC, 2015.

Mondoro A, Dan M F, Soliman M. Optimal risk-based management of coastal bridges vulnerable to hurricanes. *Journal of Infrastructure Systems* 2017, 23(3): 40–46.

Mugarura, N. The effect of corruption factor in harnessing global anti-money laundering regimes. *Journal of Money Laundering Control* 2010, 13(3): 272–281.

Richardson GA. The Impact of Economic, Legal and Political Factors on Fiscal Corruption: A Cross-Country Study. *Social Science Electronic Publishing* 2007, 276(21): 18472–18477.

Risk Analysis Based on Data and Crisis Response Beyond Knowledge – Huang & Nivolianitou (eds)
© 2020 Taylor & Francis Group, London, ISBN 978-0-367-25146-8

Social network analysis of the relationship between risks of overseas high-speed rail projects

Jianbin Zhang, Pengcheng Xiang*, Ting Yun & Mingming Hu
School of Construction Management and Real Estate, Chongqing University, Chongqing, China

ABSTRACT: In this study, the relationship between the risks of overseas high-speed rail projects was investigated and analyzed using the social network analysis method. The relevant literature and overseas high-speed rail case data were summarized and identified, and 18 risk factors of overseas high-speed rail projects were identified. The Delphi method was used to analyze the relationship between risk factors and build a risk network. The representative indicators in the social network analysis were selected for calculation, and the calculation results were analyzed in combination with the characteristics of overseas high-speed rail projects. Finally, a risk classification based on social network analysis and the corresponding risk strategy are proposed according to the characteristics of each category of risk. This research can reveal the structural relationship and influence mechanism between risk factors of overseas high-speed rail projects and it has certain reference value.

Keywords: overseas high-speed rail project, project risk, design structure matrix, risk relationship, social network

1 INTRODUCTION

In recent years, China's high-speed rail has developed rapidly. It not only has an independent and world-class technical system, but also achieved a total operating mileage of 24,000 kilometers (Yang and Cao, 2015). With the launch of the "One Belt, One Road" strategy and comprehensive promotion, China's high-speed railway was launched to the world as a new diplomatic business card (Cui et al., 2018). However, due to the large scale of investment, complicated technology, long construction period and high safety requirements, an overseas high-speed rail project is also affected by many factors such as the political, economic, cultural, legal and natural environment of the host country. This makes the risk of overseas high-speed rail projects include a wide range of sources of risk, numerous risk factors, and more significant and complex interactions between risks (Lu, 2015). Current domestic scholars' risk research on overseas high-speed rail projects has been based on a single risk level. For example, Pan analyzed the risks of Indonesia's Jakarta to Bandung high-speed rail project from the perspective of "politicization" (Pan, 2017), but ignored the interrelationship between risks, which makes it difficult for managers to comprehensively and systematically understand the risks of overseas high-speed rail projects and reduce the efficiency of risk management. Therefore, this paper introduces social network analysis and quantitatively analyzes important nodes in the risk network from the perspectives of network density and centrality, and finds the key risk factors and key relationships of overseas high-speed rail project risks. This can help effectively manage risks and push China's high-speed rail to "go abroad"!

---

*Corresponding author: 753026984@qq.com

2 OVERSEAS HIGH-SPEED RAIL PROJECT RISK IDENTIFICATION

To study the interrelationships between risks, the existing project risks must first be identified. The article mainly identifies the risks by collecting literature or case data related to overseas high-speed rail projects. Databases, such as CNKI (China National Knowledge Infrastructure), VIP Database and Web of Science, were searched for key words such as "risk" and "high-speed", and more than 20 articles related to the topic of "overseas high-speed rail risk" were retrieved. We found that most of these documents were based on the macro level to identify the risks of overseas high-speed rail, namely political risk, economic risk, etc. (Song, 2018). Therefore, in order to further identify the mid-micro risks, the author began to consult the high-speed rail project materials that Chinese enterprises have undertaken or proposed overseas, such as "Singapore to Malaysia High-speed Railway", "Mexican High-speed Railway" and "Morocco High-speed Railway". These cases are reasoned and analyzed at the enterprise and project level for risks such as legal risk, environmental risk, capital risk, communication risk, contract risk, etc. The combination of the two methods identifies the 18 major risk factors that prevail throughout the life of the overseas high-speed rail project (Wang et al., 2016; Ge, 2016; Jia, 2017; Chang and Zhang, 2018) (Table 1).

3 OVERSEAS HIGH-SPEED RAIL PROJECT RISK NETWORK CONSTRUCTION

Building a risk network is a key step in social network analysis, and establishing a risk network is essentially about determining the interrelationship between risks. The interrelationships analyzed in the article mainly refer to the causal relationship, i.e., the occurrence of one

Table 1. Risk factors for overseas high-speed rail projects.

NO.	Risk factor	Risk description
R1	Political risk	Political turmoil and regime change, power politics interference, etc.
R2	Economic risk	Exchange rate, interest rate fluctuations, inflation or deflation, exchange restrictions
R3	Cultural risk	Different religious or cultural practices
R4	Social risk	"Anti-China" emotion, terrorism, ethnic conflict
R5	Natural risk	Natural disasters, severe weather and poor geological conditions
R6	Legal risk	Legal changes, legal loopholes, and provisions that are not conducive to foreign investment in the country
R7	Market risk	Vicious competition, human machine price fluctuations, insufficient resource supply
R8	Environmental risk	The project's pollution damage to the surrounding environment has inconvenienced surrounding residents
R9	Capital risk	Financing difficulties, the owner defaults on the project, the cost is over-run, and the project does not meet the expected return
R10	Communication risk	Unfamiliar with local language, communication barriers or ambiguity
R11	Contract risk	The contract clause is vague, the contract caused by the contract change and the counterclaim
R12	Design risk	Inaccurate geological survey, design change, slow map
R13	Information risk	Insufficient information collection in the early stage, information asymmetry, empiricism
R14	Technical risk	Technical standard differences, technical solutions are not mature
R15	Human risk	Lack of talents in international engineering, low quality of staff, and inability of employees to adapt to overseas environments
R16	Organizational risk	Unreasonable organizational structure and limited project manager capacity
R17	Coordination risk	Difficulties in communication and coordination among the various subjects of the project
R18	Operational risk	Construction efficiency is low, technology is not standardized, cutting corners

Table 2. Adjacency matrix A.

i \ j	R1	R2	R3	R4	R5	R6	R7	R8	R9	R10	R11	R12	R13	R14	R15	R16	R17	R18
R1	0	1	0	0	0	1	0	0	1	0	0	0	0	0	0	0	0	0
R2	0	0	0	0	0	0	1	0	1	0	1	0	0	0	0	0	0	0
R3	0	0	0	1	0	0	0	0	0	1	0	0	0	0	1	0	1	1
R4	0	0	0	0	0	1	0	0	0	0	0	0	0	0	0	0	1	0
R5	0	0	0	0	0	0	1	0	0	0	0	1	0	0	1	0	0	1
R6	0	0	0	0	0	0	1	0	1	0	1	0	0	0	1	0	0	0
R7	0	0	0	0	0	0	0	0	1	0	1	0	0	0	0	0	0	0
R8	0	0	0	1	0	0	0	0	1	0	0	0	0	0	0	0	0	0
R9	0	0	0	0	0	0	0	0	0	0	0	0	0	0	1	0	1	1
R10	0	0	0	0	0	0	0	0	1	0	1	1	1	0	0	0	1	1
R11	0	0	0	0	0	0	0	0	1	0	0	1	0	0	1	0	1	0
R12	0	0	0	0	0	0	0	1	1	0	0	0	0	1	0	0	0	1
R13	0	0	0	0	0	0	0	0	1	0	1	1	0	1	1	1	1	0
R14	0	0	0	0	0	0	0	1	0	0	0	0	0	0	0	0	0	1
R15	0	0	0	0	0	0	0	1	0	1	1	1	1	1	0	1	1	1
R16	0	0	0	0	0	0	0	0	1	0	1	1	0	0	0	0	1	0
R17	0	0	0	0	0	0	0	0	1	0	0	1	0	0	0	0	0	0
R18	0	0	0	0	0	0	0	1	1	0	0	0	0	0	0	0	0	0

risk factor may cause another or multiple risk factors to occur, and there are chain conduction effects between the risks. In order to analyze the causal relationship between overseas high-speed rail project risks, the author conducted a questionnaire survey with managers engaged in overseas high-speed rail projects and international engineering risk research experts. The data was collated and is summarizes in a risk relationship table in Table 2.

Table 2 show the relationship between the risks expressed in matrix ($i \times j$) form according to the design structure matrix (DSM) theory. The element of row i and the column j in the matrix is 1, indicating that thei risk factor may cause the j risk factor to occur. A value of 0 means that there is no such causal relationship (Yang, 2010). Such a matrix is called an adjacency matrix and is denoted as matrix A.

The adjacency matrix represents the direct structural relationship between different risks and does not reflect the indirect relationship between risks. Some scholars have analyzed the existence of a channel between two network nodes by introducing a reachable matrix. This idea can also help us analyze the indirect relationship of risk factors in the risk network. A complete reachable matrix can represent the direct or indirect relationship between all nodes in the network, but it also lacks practical significance. Risk can be transmitted in a chain, but maintaining this relationship requires time and energy. For example, risk A is transmitted to risk B through at least n other risk factors. The larger n is, the longer the path of A→B is. So, this conduction process needs more time and energy. Risk A will dissipate during the long conduction process, so when n is greater than a critical value, even if there is theoretically such an indirect relationship between risks A and B, it is not necessary (Wu et al., 2013). For analysis, excessive weak indirect relationship analysis will complicate the problem and will not achieve good research results, so we need to determine a suitable value of n in order to best reflect the strong indirect relationship between risks. In fact, most social network researchers believe that when n = 0, 1, the relationship between nodes is complete. Therefore, this study only performed the second-order reachability operation on the adjacency matrix, i.e., the conduction path between the two risks is not more than 2. The author used the MATLAB software to perform the operation and obtained the matrix R, as shown in Table 3.

Then the Net Draw function of UCINET software was used to visualize matrix R, and the overseas high-speed rail project risk network topology model is obtained, as shown in Figure 1.

Table 3. Reachable matrix R.

$_i\!\diagdown^{\,j}$	R1	R2	R3	R4	R5	R6	R7	R8	R9	R10	R11	R12	R13	R14	R15	R16	R17	R18
R1	0	1	0	0	0	1	0	0	1	0	0	0	0	0	0	0	0	0
R2	0	0	0	0	0	0	1	0	1	0	1	0	0	0	0	0	0	0
R3	0	0	0	1	0	0	0	0	0	1	0	0	0	1	0	1	1	1
R4	0	0	0	0	0	1	0	0	0	0	0	0	0	0	0	0	1	0
R5	0	0	0	0	0	0	1	0	0	0	0	1	0	0	1	0	0	1
R6	0	0	0	0	0	0	1	0	1	0	1	0	0	0	1	0	0	0
R7	0	0	0	0	0	0	0	0	1	0	1	0	0	0	0	0	0	0
R8	0	0	0	1	0	0	0	0	1	0	0	0	0	0	0	0	0	0
R9	0	0	0	0	0	0	0	0	0	0	0	0	0	1	0	1	1	1
R10	0	0	0	0	0	0	0	0	1	0	1	1	1	0	0	0	1	1
R11	0	0	0	0	0	0	0	0	1	1	0	0	1	0	0	0	0	1
R12	0	0	0	0	0	0	0	1	1	0	0	0	0	1	0	0	0	1
R13	0	0	0	0	0	0	0	0	1	0	1	1	0	1	1	1	1	0
R14	0	0	0	0	0	0	0	1	0	0	0	0	0	0	0	0	0	1
R15	0	0	0	0	0	0	0	1	0	1	1	1	1	1	0	1	1	1
R16	0	0	0	0	0	0	0	0	1	0	1	1	0	0	0	0	1	0
R17	0	0	0	0	0	0	0	0	1	0	0	1	0	0	0	0	0	0
R18	0	0	0	0	0	0	0	1	1	0	0	0	0	0	0	0	0	0

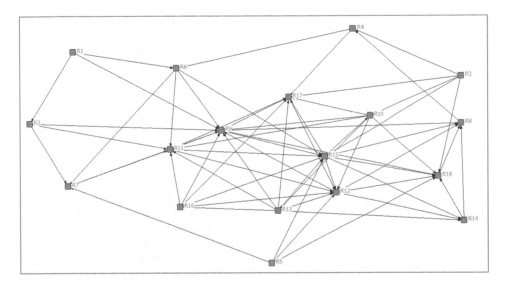

Figure 1. The overseas high-speed rail project risk network topology model.

4 SOCIAL NETWORK ANALYSIS OF RISK RELATIONSHIPS

Social network analysis covers many indicators, such as network density, centrality, faction, and clustering. Combining the characteristics of risk generation and transmission of overseas high-speed rail projects, this study selected two types of indicators: network density and centrality, because they can measure the tightness of the whole network and the importance of each risk in the network.

4.1 Network density

Density describes the degree of overall association between nodes in a network graph. It is defined as the ratio of the number of links actually owned in the graph to the maximum number of possible rows. For directed graphs, the line has directionality, so its density is calculated as follows:

$$Den = \frac{K}{N(N-1)} \tag{1}$$

Where K represents the actual number of relationships and N represents the number of nodes. According to formula (1), the density of the overseas high-speed rail project risk network can be calculated as follows:

$$Den\ (R) = \frac{\sum\limits_{i,j \in R\ \&\ i \neq j} R_{ij}}{18(18-1)} \approx 0.516 \tag{2}$$

Generally, if the network density is greater than 0.5, the network can be considered to have a large density. Combined with the analysis in Figure 1, the risk network map has no isolated points and good completeness, indicating there is a close relationship between overseas high-speed rail projects. The risk network is complex.

4.2 Node center degree

Degree centrality is a measure of how many nodes in a network are connected to the node. In a directed risk network, node centerlines include node in-degree and out-degree. The degree of node ingress represents the degree of influence of the node on other nodes, and the degree of node ingress represents the extent to which the node is affected by other nodes. In the risk network, the in-degree and out-degree of the irisk can be calculated by equations (3) and (4), respectively.

$$D_i^I = \sum_{j \in A\ \&\ i \neq j} R_{ji} \tag{3}$$

$$D_i^I = \sum_{j \in A\ \&\ i \neq j} R_{ji} \tag{4}$$

The node center degree of 18 risk factors of overseas high-speed rail project was calculated according to equations (3) and (4), and the node center degree distribution map was drawn using the in-degree and out-degree coordinate system, as shown in Figure 2. The dividing line was determined according to the half value of the sum of the maximum and minimum degrees, and the distribution map was further divided into four parts. Among them, the out-degree of R13(cultural risk) and R10(natural risk) is higher, and their in-degrees is lower, which means that such risks will have a wider impact and are rarely affected by other risk factors. In the risk network, they play the role of the risk source. In addition, environmental risk R8 and operational risk R18 have higher penetration and lower output, which means that the generation of such risk factors are mainly *caused* by other factors and rarely *affect* other factors. This is often viewed as a risk result in risk networks. In addition, in the network node centrality distribution map, special attention should be paid to the capital risk R9, the contractrisk R11 and the human risk R15. These types of risks at entry and exit are relatively high, indicating that they are at the center of the risk network and play a key role.

4.3 Overall centrality

The node center degree represents the degree of contact between a point and the local environment, reflecting the local importance of a point. To study the role of a risk factor for the entire network, the concept of the overall center degree must be introduced. Freeman judges the overall center of the node by measuring the "closeness" between points. The most intuitive

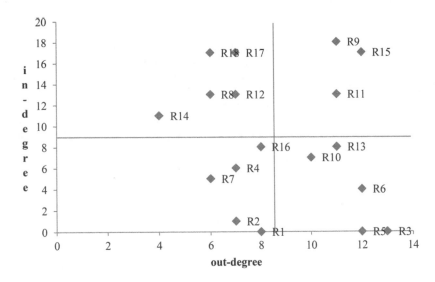

Figure 2.   Risk network node centrality distribution map.

manifestation of "proximity" is the sum of the shortcut distances between a point in the network and the other points (Sabidussi, 1996). A point where the "distance sum" is small is "close" to many other points, so the proximity and distance are reversed. The overall center of the directed graph can be calculated according to "in-closeness" and "out-closeness". This study used UCINET software to perform auxiliary calculations for "distance sum" and "proximity", as shown in Table 4.

It can be seen that the Out-Closeness risks of the entire network map are relatively small, which is mainly affected by the one-way conduction of risk. The in-closeness of the capital risk R9 and the human risk R15 are high, indicating that they are at the center of the entire risk network. In addition, the "incoming proximity" of political risk R1 and economic risk R2are low, indicating that such risks are at the source of the network, and it is difficult for other risks to affect them.

Table 4.   The closeness of risks.

	In-Farness	Out-Farness	In-Closeness	Out-Closeness
R9	17.000	89.000	100.000	19.101
R15	18.000	87.000	94.444	19.540
R17	18.000	93.000	94.444	18.280
R18	18.000	93.000	94.444	18.280
R12	21.000	91.000	80.952	18.681
R8	21.000	92.000	80.952	18.478
R11	22.000	89.000	77.273	19.101
R14	23.000	94.000	73.913	18.085
R16	26.000	91.000	65.385	18.681
R13	27.000	89.000	62.963	19.101
R10	27.000	89.000	62.963	19.101
R4	28.000	91.000	60.714	18.681
R6	32.000	86.000	53.125	19.767
R7	35.000	93.000	48.571	18.280
R2	289.000	76.000	5.882	22.368
R1	306.000	58.000	5.556	29.310
R3	306.000	69.000	5.556	24.638
R5	306.000	70.000	5.556	24.286

4.4 Intermediate center degree

There are some special risks that are often at the core of the network but have low node center degree. These types of risk are connected with the outside world only through two or three arrows, but the connection is essential for the establishment of the entire network. Because it often acts as a "man in the middle" of the two core nodes in the risk network, once it is eliminated the risk network will collapse. The middle center degree includes the middle degree of the point (Betweenness) and the middle degree of the line (Edge Betweenness). This study used UCINET software to calculate the indicators,finding four key risk factors and four groups of key relationships, as shown in Table 5.

The intermediate center degrees of capital risk R9 and human risk R15 are high, indicating that such risks are in a "hub" position in the risk network. The risk relationships related to environmental risks → legal risks, social risks → market risks also have high degrees, which acting as an intermediary and communication bridge in the risk network.

5 RISK MANAGEMENT BASED ON SOCIAL NETWORK ANALYSIS

By analyzing the risk network topology model of overseas high-speed rail projects to identify key risk factors and key risk relationships, appropriate precautionary measures to cut off the risk conduction path play an important role in curbing the spread and expansion of risks. Combined with the analysis of key risk factors, risk prevention and control can be carried out from the following aspects.

5.1 Capital risk

The indicator results of capital risk are very high through the calculation, which shows the core position of capital risk. This is because many risks will lead to an increase in the cost of the project during the construction. A shortage of funds will further affect the normal construction of the project by reducing efficiency of the workers, etc. So, the following precautions should be taken for the capital risks:

(1) In the pre-project stage, it is essential to conduct research on economic situation and stability of the host country. Fully demonstrating the economic feasibility of the project is also important.

(2) In the process of signing the contract, the floating ratio of material price and labor cost should be fully considered. The supply method of the material can be used, and the payment method of the unit price contract may be a good choice.

(3) In the process of project implementation and operation, it is necessary to fully control the unnecessary expenses caused by manpower and technology.

5.2 Human risk

The factor of human is a core factor in overseas high-speed rail projects. First Chinese enterprises should pay attention to the training of overseas engineering talents. Summarizing and sharing the existing experience of overseas projects and conducting pre-job technical training can comprehensively enhance the workers' ability with language communication,

Table 5. Between centrality of risks and relationships among risks.

Rank	Risk factor	Betweenness	Risk relationship	Edge Betweenness
1	R9	18.558	R8→R6	14.533
2	R15	18.047	R4→R7	11.143
3	R4	15.533	R18→R4	7.072
4	R8	14.433	R14→R4	6.600

organization, coordination and contract management. Second, Chinese companies should respect local customs when employing foreign workers and formulating construction plans according to the working habits of local migrant workers. Improving the construction efficiency of employees through the formulation of reward and punishment systems will reduce the risks brought about by the completion of work.

5.3 *Contract risk*

Contract risk is a key factor affecting the success or failure of overseas high-speed rail projects. The contract risk is mainly due to the lack of understanding of foreign contract specifications. Chinese enterprises always sign contracts in accordance with domestic construction experience, which will lead to contractual loopholes. So, in the early stage of signing the contract, it is necessary to investigate the laws and regulations of the host country and analyze contract issues for similar projects to avoid failures from happening again.

6 CONCLUSION

Based on the existing literature research and related cases, this paper identifies the risks existing in the investment and construction process of overseas high-speed rail projects, and classifies them into risk lists. The social network analysis method was used to find out the relationship between risks, identify key risk factors and key risk relationships, and propose corresponding management strategies. This will help Chinese companies to manage the construction of overseas high-speed rail projects, rationalize the intricate interrelationships between risks, and control key risks to successfully complete the project.

However, when constructing the overseas high-speed rail project risk network, this paper only uses 0 (no relationship between two risk factors) and 1 (a relationship between two risk factors) to qualitatively express the risk relationship, while ignoring the degree of influence between risks. It is hoped that the value of the risk relationship judgment will be more accurate in subsequent research.

REFERENCES

Chang, X. & Zhang, X.Z. 2018. Analysis of the causes of the twists and turns of the Sino-Thai high-speed railway project. *Southeast Asia, Vertical and horizontal*. 02, 46–53.

Cui, J., Zhang, J.Y. & Yu J. 2018. Research on China's "One Belt, One Road" National Railway Construction Cooperation Strategy. *International Economic Cooperation*. 05, 62–66.

Ge, M. 2016. Talking about the Difficulties of Overseas High-speed Railway Construction Organization. *Railway Construction Technology*. 01, 113–116.

Jia, D.Q. 2017. China's high-speed rail landing in Indonesia: opportunities, models and challenges. *Contemporary World*. 05, 44–47.

Lu, T.J. 2016. Problems and Path Analysis of High-speed Railway "going out". *Science and Technology Progress and Countermeasures*. 33(16), 116–118.

Pan, W.2017. Research on the political problem of China's Overseas High-speed Railway—Taking Jakarta-Bandung High-speed Railway in Indonesia as an Example. *Contemporary Asia-Pacific*. 5, 107–132, 159–160.

Song, Y.X. 2018. The Political Risks and Mechanisms of the High-speed Railway Going abroad. *Social Science Abstracts*. 01, 40–42.

Wang, S.H., Ma, R. & Guo, Q. 2016. Analysis of the structural relationship of risk factors for overseas cross-border high-speed rail investment: Taking "Singapore to Malaysia High-speed Railway" as an example. *Science and Technology Management Research*. 36(19), 224–229.

Wu, Y.N., Hu, X.L. & Zhang, S.Z. 2013. Risk Identification of PPP Project Based on ISM-HHM Method. *Journal of Civil Engineering and Management*. 30(01), 67–71.

Yang, Z. & Cao, G. 2015. The overall export status and development strategies of China's high-speed rail projects. *Business Economics Research*. 34, 133–134.

Yang, J.M. 2010. Comparison of Research Paradigms of Complex Networks and Social Networks. *Systems Engineering—Theory & Practice*. 30(11), 2046–2055.

Risk Analysis Based on Data and Crisis Response Beyond Knowledge – Huang & Nivolianitou (eds)
© 2020 Taylor & Francis Group, London, ISBN 978-0-367-25146-8

Contagion effect of associated credit risk in supply chains based on dual-channel financing

Xiaofeng Xie

School of Management and Economics, University of Electronic Science and Technology of China, Chengdu, China
General Education Department, Chengdu Neusoft University, Chengdu, China

Yang Yang

School of Economics Mathematics, Southwestern University of Finance and Economics, Chengdu, China

Jing Gu

School of Economics, Sichuan University, Chengdu, China

Qian Qian

School of Business, Sichuan Normal University, Chengdu, China

Zongfang Zhou

School of Management and Economics, University of Electronic Science and Technology of China, Chengdu, China

ABSTRACT: We consider a two-echelon supply chain with a supplier and a retailer, both with capital constraints. The capital-constrained retailer will first apply for loans from the bank, and then apply for trade credit from the supplier. Compared with the single trade credit financing mode, what are the similarities and differences in the contagion and evolution of the associated credit risk in the supply chain when the retailer adopts both trade credit and bank credit (dual-channel financing) mode. Based on Stackelberg game, this paper clarifies the contagion mechanism of the associated credit risk in the supply chain and discusses the influence of financing structure, financing channel and financing costs on the contagion effect of the associated credit risk in the supply chain under a dual-channel financing mode. Combined with the simulation analysis, it is found that compared with the single trade credit financing mode, the contagion effect is weaker under the dual-channel financing mode. The financing structure will significantly affect the contagion effect, which will increase with an increase of the proportion of trade credit financing. Both bank credit and trade credit financing costs can positively affect the contagion effect. Our research can enrich the existing credit risk management literature and provide decision support for the selection of financing methods and risk control of supply chain enterprises.

Keywords: supply chain, capital-constraint, dual-channel financing, associated credit risk, contagion effect

1 INTRODUCTION

Capital is the blood on which every enterprise lives and develops. However, most enterprises face varying degrees of capital constraints in the operation process (Burkart and Ellingen, 2004). For small and medium-sized enterprises (SMEs), debt financing is usually adopted. Previous studies have shown that because the potential financing costs of trade credit is higher than bank loans –mainly because the wholesale price provided by the core enterprises contains the possible risk premium of SMEs in the supply chain –the model of direct loans from banks to SMEs is preferred (Cunat, 2007; Zhou and Groenevelt, 2008). Based on this, SMEs in the supply chain usually give priority to bank loans, known as bank credit (Jing et al. 2012). However, due to incomplete credit records or insufficient mortgage assets of SMEs, it can be difficult to obtain adequate financial support from banks. If only relying on bank credit, SMEs in the supply chain will not be able to achieve their optimal order level, which will ultimately affect the benefits of the core enterprises and the entire supply chain. To alleviate the "ceiling dilemma" of the bank credit mentioned above and solve the operational obstacles caused by insufficient bank loans for SMEs in the supply chain, trade credit (i.e., deferred payments) provided by the core enterprises to SMEs has become an effective supplementary means (Kouvelis and Zhao, 2012, 2018). For the upstream and downstream enterprises with long-term cooperation in the supply chain, SMEs in the supply chain first apply for loans from banks, and then apply for trade credit from the core enterprises. This study examined SMEs adopting bank credit and trade credit financing at the same time, which is called supply chain dual-channel financing. (Cai et al. 2014).

The associated credit risk in the supply chain refers to the phenomenon that the credit default of the downstream SMEs causes the credit default of the upstream core enterprise or increases the probability of default. In a single trade credit financing model, the credit risk of downstream SMEs is only borne by the upstream core enterprises. Once downstream SMEs default, their credit risk will spread to the upstream core enterprises. However, in the supply chain dual-channel financing mode, the credit risk of downstream SMEs is shared by the upstream core enterprises and banks. If downstream SMEs default, their credit risk will be spread to the upstream core enterprises and banks through the trade credit and bank credit financing channels, respectively. Compared with the single trade credit financing model, the contagion mechanism and evolution trend of the associated credit risk in the supply chain under the dual-channel financing model are more complex. What are the contagion mechanisms and evolution trends of associated credit risk in the supply chain under the dual-channel financing model? What factors influence the contagion effect of the associated credit risk in the supply chain? These issues are investigated here.

In recent years, some scholars have examined the contagion effect of the associated credit risk under the supply chain scenario. Jacobson et al. (2015) and Xie et al. (2017) studied the impact of trade credit on the contagion of the associated credit risk. Yang (2011) discussed the contagion between trade credit risk and bank credit risk under the structured model. Most of the existing research was carried out under the single trade credit financing mode, and research on the associated credit risk contagion for the dual-channel financing model is still rare.

The remainder of this paper is organized as follows. In Section 2, we introduce the problem description and model assumptions. Section 3 clarifies the contagion mechanism of the associated credit risk under the dual-channel financing model. In Section 4, we construct the associated credit risk contagion effect model under the dual-channel financing mode. In Section 5, we simulate the contagion effect of the associated credit risk. Finally, we summarize our insights in Section 6.

2 PROBLEM DESCRIPTION AND MODEL ASSUMPTIONS

Considering a two-echelon supply chain consisting of a capital-constrained supplier (upstream core enterprises) and a capital-constrained retailer (downstream SMEs), the supplier (she) provides products for the retailer (he) to sell. Before the sales season (time 0), the retailer has only one ordering opportunity to meet uncertain market demands. Drawing on the setting of Jing and Seidmann (2014), neither the supplier nor the retailer has initial funding. Both the supplier and retailer need financing to meet their funding needs. Due to the lack of sufficient collateral, the retailer can only get partial financing from the bank, and the insufficient capital demand is provided by the supplier's trade credit. At time 0, after obtaining the order information of the retailer, the supplier will use the payments from the retailer and the loans from the bank to organize the production. Drawing on the general setup of Kouvelis and Zhao (2018) and Yang and Birge (2017), we describe the strategic interaction between the supplier and retailer as a Stackelberg game, with the supplier as the leader and the retailer as the follower.

At the beginning of the sales season, the retailer orders q unit products from the supplier at the wholesale price w and obtains $K = Lwq$ at an interest rates r_b. L represents the retailer loan-to-demand ratio. At the same time, the retailer applies to the supplier for deferred payments $S = (1 - L)wq$ at the interest rates r_s. As the payments K cannot meet the needs of production, the supplier obtains the financing service with the scale of $cq - K$ (c is the marginal costs of the supplier) through the bank, and the interest rate is r_a. After the supplier completes production, the retailer sells the product to the retail market at market price p. After the demand is realized, if the market is in good condition, the sales revenue of the retailer can pay off the debts of the bank and the supplier, and the supplier can also pay off the debts of the bank, and neither of them will default. Conversely, if the market situation is not ideal, the retailer has a credit risk, once the retailer defaults, the supplier may suffer losses. If the supplier receives the deferred payments from the retailer that are not sufficient to pay their banking obligations, the supplier will also default. At this point, the retailer's credit risk is transmitted to the supplier, that is, the associated credit risk in the supply chain presents a contagion effect.

Drawing on the assumptions of Kouvelis and Zhao (2012 & 2018), our other modeling assumptions are as follows:

Assumption 1(A1): No information asymmetry, i.e., all parameters are common knowledge to the supplier, retailer, and lending institutions.

Assumption 2 (A2): Due to the constraints of market demand during the loan cycle, this paper does not take into consideration the residual value of the supplier buy-backed products.

Assumption 3 (A3): The market demand x for the news vendor market is uncertain, its cumulative distribution function is $F(x)$, and the probability density is $f(x)$, among which $F(x)$ is continuous, differentiable and strictly increasing, and $\bar{F}(x) = 1 - F(x)$.

3 CONTAGION MECHANISM OF THE ASSOCIATED CREDIT RISK IN THE SUPPLY CHAIN UNDER THE DUAL-CHANNEL FINANCING MODE

The uncertainty of stochastic market demand leads to two types of credit risks of the retailer: bank credit risk caused by bank loans and trade credit risk caused by deferred payments. Drawing on Kouvelis and Zhao (2018) on the priority of bank credit and trade credit repayment, this paper assumes that the repayment order of bank credit is prior to trade credit. When the retailer's sales revenue is not enough to pay off two debts, the retailer will give priority to repaying the bank credit, and then repay the remaining funds to the trade credit. Once the retailer defaults, his credit risk will affect the supplier's credit risk through the trade credit financing channel. The retailer's default paths are shown in Figure 1.

Along the path (i): $N_R = px < K(1 + r_b)$. When $x < K(1 + r_b)/p = \delta_r^1$, the retailer will default on the bank and supplier. At this point, the supplier receives the retailer's deferred payments $N_S = 0$, and the supplier must default on the bank. i.e., if $x < \delta_r^1$, the retailer will default along the path (i), thus the supplier defaults on the bank.

Along the path (ii): $K(1 + r_b) < N_R < K(1 + r_b) + S(1 + r_s)$. When $\delta_r^1 < x < [K(1 + r_b) + S(1 + r_s)]/p = \delta_r^2$, the retailer only defaults on the supplier. At this point, the supplier has a residual claim on the retailer's sales revenue: $N_S = N_R - K(1 + r_b)$. If $N_S < (cq - K)(1 + r_a)$,

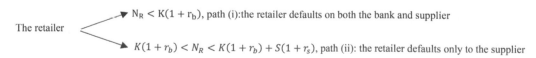

The retailer

$N_R < K(1 + r_b)$, path (i):the retailer defaults on both the bank and supplier

$K(1 + r_b) < N_R < K(1 + r_b) + S(1 + r_s)$, path (ii): the retailer defaults only to the supplier

Figure 1. All default paths for the retailer.

namely $x < [K(1 + r_b) + (cq - K)1 + r_a]/p = \delta_S$, the supplier defaults on the bank. A rational supplier obviously has $(cq - K)(1 + r_a) < S(1 + r_s)(cq - K)(1 + r_a) < S(1 + r_s)$, and thus $\delta_S < \delta_r^2$. If $\delta_r^1 < x < \delta_S$, the retailer will default along the path (ii), thus the supplier defaults on the bank.

The above analysis shows that under the dual-channel financing model of the supply chain, the credit risk of the retailer will only be transmitted to the supplier under the condition of triggering market demand and default thresholds of the retailer and supplier. When $x < \delta_r^1$, the retailer defaults along the path (i) will cause the supplier to default. When $\delta_r^1 < x < \delta_S$, the retailer defaults along the path (ii), which causes the supplier to default.

4 MODEL CONSTRUCTION

4.1 *Contagion effect of the associated credit risk in the supply chain under the dual-channel financing mode*

Definition1. The conditional probability η of the supplier's default because of the retailer's default is called the contagion intensity of the associated credit risk in the supply chain, i.e., $\eta = \text{Prob}(Z_s = 1 | Z_r = 1)$.

When $x < \delta_r^1$, the contagion intensity of the associated credit risk is simplified to

$$\eta_1 = \text{Prob}(Z_s = 1 | Z_r = 1) = 1 \tag{1}$$

When $\delta_r^1 < x < \delta_S$, the contagion intensity of the associated credit risk reaches the maximum.

$$\eta_2 = Prob(Z_s = 1 | Z_r = 1) = Prob\left(x < \delta_S | \delta_r^1 < x < \delta_r^2\right) = \frac{F(\delta_S) - F(\delta_r^1)}{F(\delta_r^2) - F(\delta_r^1)} \tag{2}$$

In the following section, the inverse solution method of the Stackelberg game is used to determine the optimal order quantity of the retailer and the optimal wholesale price of the supplier. Then, the default thresholds δ_r^1, δ_r^2 and δ_S of the supplier and retailer can be obtained. If the market demand distribution function $F(x)$ is known, the contagion intensity of the retailer's default along the path (i) and (ii) can be obtained based on the contagion intensity equations (1) and (2) of the associated credit risk.

For the retailer, positive returns can only be achieved after two debts have been liquidated. After the demand is realized, the retailer's expected profits is

$$\Pi_r = E[pmin(x, q) - K(1 + r_b) - S(1 + r_s)]^+ = p(q - \delta_r^2) - p \int_{\delta_r^2}^{q} F(x)dx \tag{3}$$

where $[a]^+ = \max(a, 0)$.

The corresponding first-order conditions is

$$\frac{\partial \pi}{\partial q} = p\bar{F}(q) - \omega\gamma\,\bar{F}(\delta_r^2),$$

Where

$$\gamma = L(1 + r_b) + (1 - L)(1 + r_s).$$

After simple calculation, the retailer's optimal order quantity q^* is satisfied:

$$\frac{\bar{F}(q^*)}{\bar{F}(\delta_r^2)} = \frac{w\gamma}{p} \tag{4}$$

653

Based on the retailer's response function Equation (4), the supplier can determine the optimal wholesale price. The expected profit of the supplier is

$$\Pi_s = E[N_S - (cq - K)(1 + r_a)]^+ = p(\delta_r^2 - \delta_S) - p\int_{\delta_S}^{\delta_r^2} F(x)dx \tag{5}$$

The corresponding first-order conditions is

$$\frac{\partial \Pi_s}{\partial \omega} = p\left(\bar{F}(\delta_r^2)\delta_r^{2\prime} - \bar{F}(\delta_s)\delta_s^{\prime}\right).$$

After simple calculation, the supplier's optimal wholesale price w^* is satisfied:

$$\frac{\bar{F}(\delta_r^2)}{\bar{F}(\delta_s)} = \frac{1}{\gamma}\left[\frac{c(1 + r_a)}{(q^*/q^{*\prime} + \omega^*)} + L(r_b - r_a)\right] \tag{6}$$

Thus, by Equations (1), (2), (4) and (6), combined with the supplier and retailer default thresholds δ_r^1, δ_r^2 and δ_S, we can determine contagion intensity η of the associated credit risk in the supply chain under the dual-channel financing mode:

$$\eta = \begin{cases} 1, \text{(i)} \\ \dfrac{F(\delta_S) - F(\delta_r^1)}{F(\delta_r^2) - F(\delta_r^1)}, \text{and} \dfrac{\bar{F}(q^*)}{\bar{F}\delta_r^2} = \dfrac{\omega^*\gamma}{p}, \text{and} \dfrac{\bar{F}(\delta_r^2)}{\bar{F}\delta_S} = \dfrac{1}{\gamma}\left[\dfrac{c(1 + r_a)}{\left(\frac{q^*}{q^{*\prime}} + \omega^*\right)} + L(r_b - r_a)\right], \text{(ii)} \end{cases} \tag{7}$$

4.2 Contagion effect of associated credit risk in supply chain under the single trade credit financing mode

When the retailer's credit quality falls below a certain level, the bank will refuse to lend him any money. At this point, the retailer loan-to-demand ratio is $L = 0$. His entire source of funding is the trade credit provided by the supplier. When the retailer uses single trade credit financing mode, his credit risk is all assumed by the supplier. Once the retailer defaults, his credit risk will be directly infected to the supplier through trade credit channel.

In the default path (ii), let the retailer loan-to-demand ratio $L = 0$, that is, the situation where the retailer only uses trade credit financing. Therefore, in a single trade credit financing model, the contagion intensity of the associated credit risk in the supply chain is

$$\eta = prob(x < \delta_S|\ \delta_r^1 < x < \delta_r^2)|_{L=0} = \frac{F(\delta_S) - F(\delta_r^1)}{F(\delta_r^2) - F(\delta_r^1)}\Big|_{L=0} \tag{8}$$

5 THE SIMULATION ANALYSIS

The following is a simulation of the above research questions using MATLAB 2017a, analyzing the intensity of contagion effect of the associated credit risk under the dual-channel financing mode. Drawing on the setting of the demand market distribution of the newsvendor by Arcelus et al. (2011) and Kouvelis and Zhao (2012), this paper assumes that market demand $x \sim U[0, D]$, then $f(x) = 1/D$, $F(x) = x/D$.

5.1 Impact of financing structure on contagion intensity

The retailer loan-to-demand ratio L measures the financing size of retailer's bank credit and trade credit, and the variable of the model is set as the retailer loan-to-demand ratio L. The

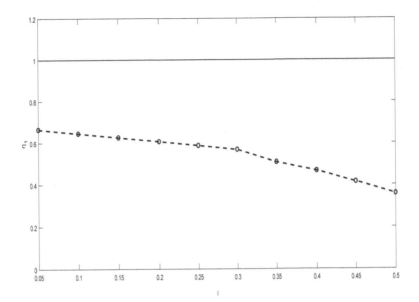

Figure 2. Variation trend of contagion intensity along with loan-to-demand ratio L.

model parameters are set at $D = 2000$, $p = 8$, $c = 3$, $r_a = 0.06, r_b = 0.08, r_s = 0.09$. The simulation result is shown in Figure 2.

In Figure 2, along with the increase of the retailer's loan-to-demand ratio L, i.e., the trade credit ratio $1 - L$ decreases, the contagion intensity of the associated credit risk in the supply chain decreases. Thus, the smaller the trade credit limits provided by the supplier to the retailer, the smaller the loss caused by the retailer's default to the supplier, and the smaller the possibility of the supplier's default. In other words, the retailer's credit risk is less contagious to the supplier. To reduce the possibility of being infected by the retailer's credit risk, the supplier can properly control the trade credit limits, reduce the potential loss caused by the retailer defaults, and then effectively control her own credit risk level.

5.2 Impact of financing channel on the contagion intensity

To investigate the impact of financing channel on the contagion intensity η, this section makes a comparative analysis of the contagion intensity η_1 and η_2 under the two default paths. In Figure 2, the contagion intensity of the retailer using only trade credit financing is $\eta_2|_{L=0}$. It is not difficult to find that, compared with the single trade credit financing, except for the default path (i), the contagion intensity is weaker under the dual-channel financing mode. This shows that when the market situation is good, the introduction of external financing channels can reduce the financing costs of the retailer, improve his sales revenue, reduce his credit risk, and reduce the possibility of the supplier being infected. Conversely, in the single trade credit financing model, only the supplier shares the credit risk of the retailer. However, in the dual-channel financing mode, the supplier and bank both share the credit risk of the retailer, so that the supplier are less likely to be infected by the credit risk of the retailer. On the whole, compared with the single trade credit financing, the contagion effect of the associated credit risk in the dual-channel financing mode is weaker.

5.3 Impact of financing costs on contagion intensity

To investigate the impact of trade credit costs r_s and bank credit costs r_b on the contagion intensity, first set the variable of the model as r_s, fixed $D = 2000$, $p = 8$, $c = 3$, $r_a = 0.06$, $r_b = 0.08$, $L = 0.5$. The simulation results are shown in Figure 3. Then, the impact

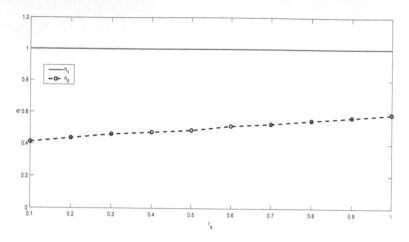

Figure 3. Variation trend of contagion intensity along with trade credit costs r_s.

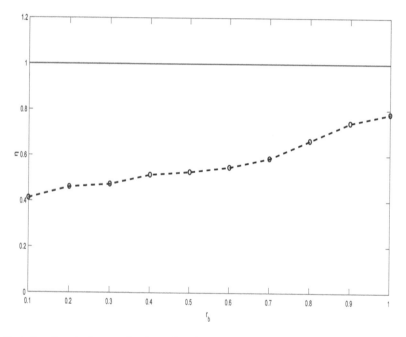

Figure 4. Variation trend of contagion intensity along with bank credit costs r_b.

of bank credit costs r_b on the contagion intensity was analyzed, and the variable of the model was set as the retailer loan rates r_b, fixed $D = 2000$, $p = 8$, $c = 3$, $r_a = 0.06$, $r_s = 0.09$, $L = 0.5$. The simulation results are shown in Figure 4.

In Figure 3, as the trade credit costs r_s increases, the contagion intensity of the associated credit risk increases gradually. The reason for this is that the increase of trade credit costs will increase the internal financing cost of the retailer, squeeze their profit margins and increase the possibility of the retailer defaults. Because the supplier's credit risk is infected by the retailer's credit risk, the probability of the retailer defaults increases, leading to the possibility of the supplier being infected.

It can be seen from Figure 4 that the higher the external financing costs of the retailer r_b, the higher the contagion intensity of the associated credit risk. This phenomenon indicates that under certain conditions, the increase of the retailer's external financing costs will also squeeze his sales profit and reduce his sales income. The retailer's reduced sales revenue pushes up his default probability, increasing the likelihood that the supplier will be infected. That is, the contagion intensity of the associated credit risk in the supply chain increases along with the external financing costs of the retailer.

6 CONCLUSIONS

Capital constraints are a common business dilemma faced by many enterprises (especially SMEs) all over the world. What are the similarities and differences in the contagion and evolution of the associated credit risk in the supply chain when the retailer is funded by both bank credit and trade credit? Based on the Stackelberg game theory, this paper clarifies the contagion mechanism of the associated credit risk in a two-echelon supply chain with a supplier and a retailer, and discusses the impact of financing structure, financing channel and financing costs on the contagion effect of the associated credit risk in the supply chain. Combined with simulation analysis, this study shows the following. (1) Compared with the single trade credit financing mode, the two-channel financing mode has a weaker contagion effect. (2) The financing structure of the retailer will significantly affect the contagion effect, which will increase with the increase of the proportion of trade credit financing. (3) Both bank credit and trade credit financing costs can positively affect the contagion effect. This study, to some extent, enriches the existing theoretical research on credit risk management, and provides a basis and reference for the selection of financing methods and risk control of supply chain enterprises.

ACKNOWLEDGMENTS

This study was supported by National Natural Science Foundation of China (Grants: 71271043 & 71701066 & 71671144 & 71871447).

REFERENCES

Arcelus, FJ, Kumar, S, Srinivasan, G. 2011. Channel coordinaton with manufacturer's return policies within a newsvendor framework. *Operation Research* 9(3):279–297.

Burkart, M, Ellingsen, T. 2004. In-kind finance: A theory of trade credit. *American Economic Review* 94(3):569–590.

Cai, GG, Chen, X, Xiao, Z. 2014. The Roles of Bank and Trade Credits: Theoretical Analysis and Empirical Evidence. *Production and Operations Management* 23(4):583–598.

Cunat, V. 2007. Trade Credit: Suppliers as Debt Collectors and Insurance Providers. *Review of Financial Studies* 20(2):491–527.

Kouvelis, P, Zhao, W. 2012. Financing the newsvendor: Supplier vs. bank, and the structure of optimal trade credit contracts. *Operations Research* 60(3):566–580.

Kouvelis, P, Zhao, W. 2018. Who should finance the supple chain? Impact of credit ratings on supply chain decisions. *Manufacturing & Service Operations Management* 20(1):19–35.

Jacobson, T, Schedvin, EV. 2015. Trade credit and the propagation of corporate failure: An empirical analysis. *Econometrica* 83(4):1315–1371.

Jing, B, Chen, X, Cai, G. 2012. Equilibrium Financing in a Distribution Channel with Capital Constraint. *Production & Operations Management* 21(6):1090–1101.

Xie, XF, Yang, Y, Zhou, ZF. 2017. The clustering analysis on data and lending strategy of supply chain with information asymmetry method. *Cluster Computing* 1(1):1–9.

Yang, Y. 2011. Credit Risk of Supply Chain Based on Credit Behavior. *Systems Engineering* 29(12):35–39.

Yang, SA, Birge, JR. 2018. Trade credit, risk sharing, and inventory financing portfolios. *Management Science* 64(8):3469–3970.

Zhou, J, Groenevelt, H. 2008. *Impacts of Financial Collaboration in a three-party Supply Chain*. Working paper, University of Rochester.

Risk Analysis Based on Data and Crisis Response Beyond Knowledge – Huang & Nivolianitou (eds)
© 2020 Taylor & Francis Group, London, ISBN 978-0-367-25146-8

Evaluation on big data industry development level of prefecture-level cities in Guizhou based on the hesitant fuzzy linguistic TOPSIS method

Pang Lujing & Zhang Mu

School of Big Data Application and Economics
Guizhou Institution for Technology Innovation & Entrepreneurship Investment, Guizhou University of Finance and Economics, Guiyang, Guizhou, China

ABSTRACT: To evaluate the development level of Guizhou's big data industry reasonably and effectively, based on the current industry developments, this study constructs an industrial-level rating index system of Guizhou's prefecture-level cities' big data development. To better reflect the influence of big data industry, the Baidu Index is added as the search engine data. First, the weight of each index is calculated by the entropy weight method, and then the development level of the big data industry in Guizhou is measured by the hesitant fuzzy language TOPSIS method. Empirical analysis shows that the TOPSIS method based on hesitant fuzzy language can better measure and analyze the development status of the big data industry and enhance the flexibility and credibility of information expression. Finally, through an evaluation of the development level of the big data industry in prefecture-level cities of Guizhou, this paper proposes relevant countermeasures and suggestions for the development of theindustry so as to continuously promote the development of the big data industry in Guizhou.

Keywords: big data, rating index, search engine, entropy weight, hesitant fuzzy language, Guizhou

1 INTRODUCTION

The development of big data and the progress of big data technology make the industry a new economic growth point. The big data industry refers to the related economic activities, mainly consisting of data production, acquisition, storage, processing, analysis and service, including data resource construction, development, sale and leasing of large data hardware and software products and related information technology services (*China Electronic Newspaper*, 20 January 2017).

There are some studies on big data industry in the literature. James (2011) believes that we can use big data creatively and effectively to improve efficiency and quality. Malik (2013) studied the principles and best practices of big data governance based on analyzing the challenges and opportunities of big data. Delia (2014) expounds the types, characteristics and development environment of the big data industry in foreign countries. Weiling (2015) studied and reflected on the strategic value of the big data industry. Lemen (2016) analyzed the characteristics of China's big data industry and the major contradictions in its development process. Qianqian et al. (2016) proposed three major contradictions in developing the big data industry through literature research and case study. Xiaoyan and Chidong (2017) took the big data industry as an example and analyzed the progress and gap of technological innovation in the intelligent industry by using panorama. Weihong (2018) analyzed the similarities and differences of policy for the big data industry at home and abroad and proposed suggestions.

Based on the existing research, most of the literature focuses on the development status and future development proposals of the big data industry. The existing research seldom involves the development level of the big data industry in China or its provinces. Based on this, Rodríguez was first to propose the concept of hesitant fuzzy linguistic set and model: hesitant fuzzy linguistic is a type of fuzzy language and fuzzy extension language. With hesitant fuzzy linguistic multiple attribute decision-making in the choice of suppliers, there is a wide range of applications in university evaluation, medical diagnosis, movie recommendation systems and other fields (Liu, 2014; Wei et al., 2015; Liao et al., 2015). The TOPSIS method is a multi-objective decision-making method and a sort method approaching the ideal solution. The Baidu Index is a data analysis platform based on netizens' behavior data. Based on hesitant fuzzy linguistic, this study combines hesitant fuzzy linguistic and TOPSIS multi-attribute decision-making methods and innovatively proposes the application of search engine data. According to the situation of industrial development, the paper propose spolicies and suggestions to better promote the development of Guizhou's big data industry, which has a certain practical and theoretical significance.

2 INTRODUCTION OF HESITAT FUZZY LINGUISTIC TOPSIS METHOD

2.1 *Preparatory knowledge of hesitant fuzzy linguistic*

Compared with precise numerical evaluation, decision-makers often use language to evaluate projects. At the same time text-free grammar provides the grammar rules that correspond the language expression term to the fuzzy linguistic set.

Definition 1: Let $S = \{s_\alpha \, \alpha = -\tau, \ldots, -1, 0, 1, \ldots, \tau\}$ be a set of linguistic terms and τ be a positive integer. If H_s is a set of finite sequential linguistic terms in s, then H_s is a hesitant fuzzy linguistic term in $x_i \in X$, $i = 1, 2, \ldots, N$ then the mathematical form of hesitant fuzzy linguistic set is

$$H_s = \{<x_i, h_s(x_i)>|x_i \in X\} \tag{1}$$

The function $H : X \rightarrow S$ denotes the possible membership degree of the element : $x_i \in X$ mapped to $A \subset X$, and $h_s(x_i)$ is the possible value in the linguistic terminology set S, and $h_s(x_i) = \{s_{\delta_l}(x_i)|s_{\delta_l}(x_i) \in S, l = 1, 2, \ldots, \# H_s\}$, $\delta_l \in \{-\tau, \ldots, -1, 0, 1, \ldots, \tau\}$ is the subscript of the linguistic term $s_{\delta_l}(x_i)$, and $\# H_s$ is the number of linguistic terms in $h_s(x_i)$. This symmetric set of linguistic terms S satisfies these conditions:

(1) Orderliness: if i>j, then $s_i > s_j$
(2) There are negative operators: $\text{Neg}(s_\alpha) = s_{-\alpha}$, especially $\text{Neg}(s_0) = s_0$.

Definition 2: To facilitate calculation, the discrete terminology hesitant fuzzy linguistic set is extended to the continuous hesitant fuzzy linguistic set $\bar{S} = \{S_\alpha|\alpha\in[-q, q]\}$ in the process of decision-making, where q is a sufficiently large positive number. As a virtual language term, $\bar{S} = \{\overline{S_\alpha} \in \bar{S}\}$ appears only in computation, satisfying top (1) and (2). If $s_\alpha, s_\beta \in \bar{S}$ and gamma$\gamma \in [0, 1]$, the following conditions are satisfied:
(1) $s_\alpha \oplus s_\beta = s_{\alpha+\beta}, s_\alpha - s_\beta = s_\alpha \oplus \text{Neg}(s_\beta)$; (2) $\gamma s_\alpha = s_{\gamma\alpha}$

Definition 3: Let E_{G_H} be the linguistic expression $ll \in S_{ll}$ generated by free grammar G_H of text into a hesitant fuzzy linguistic set H_s

- $E_{G_H}(s_g) = \{s_g|s_g \in S\}$;
- $E_{G_H}(\text{at most } s_\alpha) = \{s_g|s_g \in s, \text{and } s_g \leq s_\alpha\}$;
- $E_{G_H}(low \ than \ s_\alpha) = \{s_g|s_g \in s, \text{and } s_g < s_\alpha\}$;
- $E_{G_H}(\text{at least } s_\alpha) = \{s_g|s_g \in s, \text{and } \geq s_g s_\alpha\}$;

- $E_{G_H}(\text{great than } s_\alpha) = \{s_g | s_g \in s, \text{and } s_g > s_\alpha\};$
- $E_{G_H}(\text{between } s_\alpha \text{ and } s_\beta) = \{s_g | s_g \in s, \text{and } s_\alpha \le s_g \le s_\beta\}$

2.2 Hesitant fuzzy linguistic TOPSIS method

Let $X = \{x_1, x_2, \ldots, x_n\}$ is the scheme set $C = \{c_1, c_2, \ldots, c_m\}$, $W = \{w_1, w_2, \ldots w_n\}$ is the weight vector of attributes, which satisfies $\sum_j^n w_j = 1$ and $\sum_j^n w_j = 1$. Language Scale $S = \{s_\alpha \mid \alpha = -\tau, \ldots, -1, 0, 1, \ldots, \tau\}$. The specific steps are as follows:

Step 1: The linguistic information is transformed into hesitant fuzzy linguistic set by the text-free method and the decision matrix $R = \left(H_S^{ij}\right)_{n \times m}$ of hesitant fuzzy linguistic is obtained. Therefore, the hesitant fuzzy linguistic decision matrix can be expressed as follows:

$$R = \left(H_S^{ij}\right)_{n \times m} = \left\{ \begin{array}{cccc} H_S^{11}(x_1) & H_S^{12}(x_1) & \cdots & H_S^{1n}(x_1) \\ H_S^{21}(x_2) & H_S^{22}(x_2) & \cdots & H_S^{2n}(x_2) \\ \vdots & \vdots & \ddots & \vdots \\ H_S^{m1}(x_m) & H_S^{m1}(x_m) & \cdots & H_S^{mn}(x_m) \end{array} \right\} \tag{2}$$

For different H_S^{ij}, the number of language terms may be different. We choose to add a virtual language term to the less H_S^{ij}, so that the number of language terms contained in different H_S^{ij} is the same. Let $b = \{b_l \mid l = 1, 2, \ldots, \# b\}$ is the number of hesitant vague language, $\# b$ is the number of linguistic terms in b, b^+ and b^- are the maximum and minimum linguistic terms in b, and the optimum parameter is $\xi (0 \le \xi \le 1)$. The linguistic term $b = \xi b^+ \oplus (1 - \xi) b^-$ can be added to fewer hesitant fuzzy linguistic, we take $\xi = \frac{1}{2}$.

Step 2: Get the positive and negative ideal solution of hesitant vague linguistics. Positive ideal solution $A^+ = \left\{H_s^{1+}, H_s^{2+}, \ldots, H_s^{n+}\right\}$, negative ideal solution $A^- = \left\{H_s^{1-}, H_s^{2-}, \ldots, H_s^{n-}\right\}$.

Step 3: Calculate the weighted distances D_i^+ and D_i^-, between scheme x_i and positive and negative ideal solutions. Before calculating the weighted distance, let's look at the Euclidean distance between two sets of hesitant fuzzy linguistics. Let the extended scale $\bar{S} = \{S_i | i \in [-q, q]\}$, $H_S^{11} = \{b_1^{11}, b_2^{11}, \ldots, b_l^{11}\}$ and $H_S^{22} = \{b_1^{22}, b_2^{22}, \ldots, b_l^{22}\}$, mark $\delta(.)$ as the subscript of virtual language terminology:

$$d\left(H_S^{11}, H_S^{22}\right) = \left(\frac{1}{l} \sum_{i=1}^{l} \left(\frac{\delta(b_i^{22}) - \delta(b_i^{22})}{2q}\right)^2\right)^{\frac{1}{2}} \tag{3}$$

$$D_i^+ = \sum_{j=1}^{m} w_j d\left(H_S^{ij}, H_s^{j+}\right) \tag{4}$$

$$D_i^- = \sum_{j=1}^{m} w_j d\left(H_S^{ij}, H_s^{j-}\right) \tag{5}$$

Step 4: Relative closeness of schemes x_i to ideal solutions RC_i

$$RC_i = \frac{D_i^-}{D_i^+ + D_i^-}, \quad i = 1, 2, \ldots, n \tag{6}$$

The larger the relative closeness, the better the scheme x_i. The scheme x_i is sorted according to the relative closeness.

3 ESTABLISHMENT OF EVALUATION INDEX SYSTEM

3.1 *Index system*

Combining with the actual situation of the development of Guizhou's big data industry, this study divides the index system of the development of Guizhou's big data industry into target layer, criterion layer and index layer on the principle of scientificity, systematicness, comprehensiveness and availability.

3.2 *Data sources*

The electronic information industry provides the industrial basis for developing the big data industry and has a certain representativeness in reflecting the development level of big data industry. Therefore, we introduce the industrial scale and infrastructure of the electronic information industry as an index to evaluate the development level of the big data industry. Meanwhile, the search engine index represented by the Baidu Index reflects the influence of the big data industry. Therefore, the research data in this paper mainly comes from the statistical yearbook of the Guizhou municipalities and autonomous prefectures, the statistical yearbook of Guizhou Province, the statistical bureau of Guizhou Province, the statistical bureaus and government websites of Guizhou municipalities (2013–2017) and the "big data" search index of Guizhou Province (2013–2017) in the Baidu Index.

3.3 *Determination of index weight*

If the evaluation of the development level of the big data industry in different cities of Guizhou is carried out based on "n" evaluation indexes of "m" evaluation objects, the following original data matrix is first established:

Table 1. Evaluation system and weight of development level of the big data industry.

Target layer	Datum layer	Index level	Company	Weight
Big Data Industry	Industrial Scale	Number of Enterprises above scale	Individual	0.07
		Average Annual Number of All Employees in Enterprises	Individual	0.07
		Gross Industrial Output	Billion yuan	0.09
		Industrial Value Added	Billion yuan	0.09
		Total Assets	Billion yuan	0.10
		Main Business Income	Billion yuan	0.09
		Total profit	Billion yuan	0.06
		Accumulated investment of fixed assets completed this year	Billion yuan	0.08
	Infrastructure	Mobile phone subscribers at the end of the year	Ten thousand households	0.05
		Internet Broadband Access Users	Ten thousand households	0.06
		Number of fixed telephone subscribers at the end of the year	Ten thousand house holds	0.07
	Baidu Index	PC Search Index	Frequency	0.08
		Mobile Search Index	Frequency	0.09

Note: The number of weights worth calculating is detailed in 3.3.

$$R = \begin{bmatrix} r_{11} & r_{12} & \cdots & r_{1n} \\ r_{21} & r_{22} & \cdots & r_{2n} \\ \vdots & \vdots & \vdots & \cdots \\ r_{m1} & r_{m2} & \cdots & r_{mn} \end{bmatrix} \tag{7}$$

(1) The raw data matrix is standardized to obtain $x_{ij} = \left(x_{ij} \right)_{mn}$
x_{ij} is the standard value of the first evaluation object on the j evaluation index, and $x_{ij} \in [0,1]$. Among them, the profitability indicators that are superior to the majority are as follows:

$$x_{ij} = \frac{r_{ij} - \min_j r_{ij}}{\max_j r_{ij} - \min_j r_{ij}} \tag{8}$$

(2) Turn x_{ij} into p_{ij}: $p_{ij} = \frac{x_{ij}}{\sum_{i=1}^{m} x_{ij}}, i = 1, 2 \ldots, m; \ j = 1, 2 \ldots, n$ \hfill (9)

(3) Define the entropy of index j as:

$$H_j = -K \sum_{i=1}^{m} p_{ij} \ln p_{ij}, \ j = 1, 2 \ldots, m, K = \frac{1}{\ln m} \tag{10}$$

(4) The entropy weight of index j is defined as:

$$w_j = \frac{1 - H_j}{\sum_{j=1}^{n} (1 - H_j)} = \frac{1 - H_j}{n - \sum_{j=1}^{n} H_j}, \ j = 1, 2 \ldots, \ n, w_j \in [0, 1], \ \text{and} \ \sum_{j}^{n} w_j = 1 \tag{11}$$

3.4 *Empirical results*

From the actual and estimated data of Guizhou municipalities at all levels, we have determined the weights of thirteen indicators. Now we have used the seven linguistic terms S = {s_{-3} = terrible, s_{-2} = very poor, s_{-1} = poor s_0 = medium s_{-1} = good, s_2 = very good, s_3 = excellent} to allow experts to evaluate the language based on the data obtained, thus obtaining a hesitant fuzzy linguistic decision matrix, positive and negative ideal values and European distance.

We can know that the overall situation of Guizhou's big data industry is constantly developing. The relatively close calculation value of Guiyang City is the highest in Guizhou Province, followed by Zunyi Cityand the lowest level of development in Southwest Guizhou. It can be seen from the calculation results of these five years that although Zunyi started slowly relative to Guiyang, its development speed is the fastest among the nine cities. Except for Guiyang and Zunyi, the level of big data development in other cities and states is generally low, which

Table 2. Ranking of big data development level of cities at all levels in Guizhou.

	2013	2014	2015	2016	2017	Average value	Ranking
Guiyang	0.345271	0.451099	0.536682	0.568805	0.627164	0.505804	1
Liupanshui	0.055265	0.054927	0.185445	0.168768	0.221462	0.137173	3
Zunyi	0.13484	0.209634	0.381904	0.552237	0.604666	0.376656	2
Anshun	0.042673	0.055271	0.110962	0.085835	0.178209	0.09459	8
Bijie	0.072515	0.067038	0.115938	0.14817	0.255398	0.131812	4
Tongren	0.050116	0.063863	0.096898	0.106885	0.186231	0.100799	6
Qianxinan	0.035201	0.037416	0.08149	0.110004	0.179537	0.08873	9
Qiandongnan	0.06494	0.068722	0.099773	0.110945	0.185669	0.10601	5
Qiannan	0.051067	0.084412	0.093	0.113879	0.157534	0.099979	7

indicates that the development of the big data industry in Guizhou Province is uneven and the differences between cities and states at different levels are obvious.

4 CONCLUSIONS

To accurately evaluate the development level of big data industry in Guizhou in the last five years, this paper firstly chose three indicators: industrial scale, infrastructure and search engine index-Baidu index, to reflect the actual development of the big data industry. Second, it objectively calculates the specific weights of thirteen attribute indicators by using the method of entropy weight. Finally, it tries to overcome the complexity of the objective world and the incompleteness of people's understanding by using the TOPSIS method based on hesitant fuzzy linguistics and obtains the evaluation results that are more consistent with the cognitive, more feasible and credible level of development of large data industry. Through the final evaluation results, we can know that the big data industry in Guizhou Province is in the golden period of continuous development, in this stage of development, Guizhou Province should grasp the opportunity of developing the big data industry, but also focus on solving the challenges in the development process.

ACKNOWLEDGMENTS

This research was financially supported by the Regional Project of National Natural Science Foundation of China (71861003) and the Second Batch Projects of Basic Research Program (Soft Science Category) in Guizhou Province in 2017 (Foundation of Guizhou-Science Cooperation [2017] 1516–1).

REFERENCES

China Electronic Newspaper. 2017. Development Plan of Big Data Industry (2016–2020). *China Electronic Newspaper*, 20 January 2017.

Delia. 2014. Research on the Development of China's Big Data Industry. *Science & Technology Progress and Policy* 31(04):56–60.

Di X, Zhang C. 2018. The Evaluation Measurement Model of Technology Innovation in Intelligent Industry. *China Soft Science* (05):39–48.

Huchang L, Xunjie G, Zeshui X. 2019. A survey of decision-making theory and methodologies of hesitant fuzzy linguistic term set. *Control and Decision*: 1–10.

Lemen C, Guanghui M, Haijuan L. 2016. Big Data Industry in China: Characteristic Analysis and Policy Suggestions. *Information Studies: Theory & Application* 39(10):5–10.

Li W, Liu G, Lu Z. 2015. A Study on the Competition Situation of Big Data Technology in China on the Basis of Patent Analysis. *Journal of Information* 34(07):65–70.

Liao HC, Xu ZS, Zeng XJ, et al. 2015. Qualitative decision making with correlation coefficients of hesitant fuzzy linguistic term set. *Knowledge-Based Systems* 76:127–138.

Liu B, Rodrı́guez R M. 2014. A fuzzy envelope of hesitant fuzzy linguistic term set and its application to multicriteria decision making. *Information Sciences* 258: 220–238.

Manyika J., Chui M., Brown B., et al. 2011. *Big data: The Next Frontier for Innovation, Competition, and Productivity*. Mc Kinsey Global Institute.

Malik P. 2013. Governing big data: principles and practices. *IBM Journal of Research & Development* 57(3/4): 1–13.

Qianqian Y, Haijuan L, Lemen C. 2016. Analysis on the Major Contradictions in the Development of Big Data Industry. *Information Studies: Theory & Application* 39(10):11–15.

Rodrı́guez RM, Martı́nez L, Herrera F. 2012. Hesitant fuzzy linguistic terms sets for decision making. *IEEE Transactions on Fuzzy Systems* 20: 109–119.

Wang WL. 2015. Research and Thinking on the Strategic Value of the Big Data Industry. *Techno Economics & Management Research* (01):117–120.

Wang L, Zhou Y, Zhang Yu. 2017. Analysis on Evaluation of Low Carbon Cities and Obstacle Degree of Tianjin Based on the Entropy-weight TOPSIS Method. *Science and Technology Management Research* 37(17):239–245.

Wei CP, Zhao N, Tang XJ. 2014. Operators and comparisons of hesitant fuzzy linguistic term sets. *IEEE Transactions on Fuzzy Systems* 22(3): 575–585.

Wei C, Ge S. 2016. A Power Average Operator for Hesitant Fuzzy Linguistic Term Sets and its Application in Group Decision-making. *Journal of Systems Science and Mathematical Sciences*36 (08):1308–1317.

Weihong X, Bingdong F, Ce D. 2018. Comparative Analysis of Big Data Industry Development at Home and Abroad. *Journal of Modern Information* 38(09):113–121.

Zhang M, Zhou Z. 2008. Evaluation on the Independent Innovation Capacity of High-tech Enterprises Based on Rough Set and Entropy Weight-TOPSIS. *Mathematics in Practice and Theory* 38(24):52–58.

Zou Z, Sun J, Ren G. 2005. Study and Application on the Entropy method for Determination of Weight of Evaluating Indicators in Fuzzy Synthetic Evaluation for Water Quality Assessment. *Acta Scientiae Circumstantiae* (04):552–556.

Risk Analysis Based on Data and Crisis Response Beyond Knowledge – Huang & Nivolianitou (eds)
© 2020 Taylor & Francis Group, London, ISBN 978-0-367-25146-8

Regional differences between climate change and economic development in China in historical and future periods

Wei Gu* & Kou Bai

Key Laboratory of Environmental Change and Natural Disaster, MOE, Beijing Normal University, Beijing, China
Academy of Disaster Reduction and Emergency Management, Ministry of Civil Affairs & Ministry of Education, the Peoples' Republic of China, Beijing, China

ABSTRACT: Risk assessment of the economic impacts of climate change in the future is a key topic ininternational studies. Based on earlier research results on the nonlinear relationship between temperature and the economy in 31 provinces (historical period, 1978 to 2013), this study investigated the risk identification of the relationship between temperature and the economy in different provinces in China in the historical period and the future impacts of global warming on the economic development of different provinces. Based on the results of the risk identification, this article describes aninitial assessment of the impact of temperature rises on economic growth rate under the RCP8.5 scenario, and sheds new light onthe regional differences in the future. Lastly, the article discusses the shortcomings of analysis method used. The results of risk identification and assessment lay a certain foundation for further exploring the future impacts of climate change on the economy in China.

Keywords: regional differences, risk assessment, global warming, economic effects, temperature increases

1   INTRODUCTION

Global warming has become a key global research topic, because it will not only cause irreversible damage to the environment but have a far-reaching impact on the sustainable development of modern economic society. In addition, from the perspective of the response of modern social and economic development to global warming, increasing numbers of studies show that the relationship is nonlinear between economic growth and temperature risesin different regions in the world.

Many scholars have begun to study the impact of global warming on economic activity all over the world. It is widely believed that global warming will have a great impact on economic activity, but the effects on the economy seems inconsistent across different studies. Some studies examine the negative effects of temperature increases on economic activities. In agriculture, many new simulations around the world show that climate change has reduced wheat and corn yields by an average of 1.9% and 1.2% over the past decade (IPCC, 2014). Studies in the USA have shown a potential loss on agriculture because of climate change (Schlenker et al., 2005; Schlenker and Roberts, 2009; Hsiang et al., 2017). Scholars in China also indicate that global warming has caused reduced yield or economic losses (EC-SCNARCC, 2015; Chen et al., 2016). Other studies report on the effect of temperature on agriculture (Mendelsohn et al., 1994; Schlenker et al., 2006). Apart from agriculture, global warming also has significant impacts on

---

*Corresponding author: weigu@bnu.edu.cn

the industrial sector, especially the high-risk sectors, because most of these involve working outside and high temperatures could reduce workers' hours (Zivin and Neidell, 2014;Somanathan et al., 2015). In the United States, in high outdoors temperatures the number of hours worked decreases 0.11% (±0.004) per °C for low-risk workers and 0.53% (±0.01) per °C for outdoor laborers (Hsiang et al., 2017). Industrial output in China may decrease by 3–36% under the B1 scenario and 12–36% under the A2 scenario by 2080, respectively (Chen and Yang, 2017).

As well as the impacts on agriculture and industry, global warming also affects other sectors. For instance, temperature rises can lead to sea level rises, which will have a negative impact on coastal tourism (Gable, 1997). Moreover, it will also reduce people's outdoor leisure time (Zivin and Neidell, 2014). Every 1°C will lead to a 5.4 (±0.5) deaths per 100,000 people rise in the annual national mortality rate (Hsiang et al., 2017).

As mentioned above, it is believed that high temperatures are associated with lower economic output in many parts of the world (Hsiang, 2010; Dell et al., 2012; Deryugina and Hsiang, 2014). However, temperature increases also have positive effects on economic activity. For example, CO_2 is one of the greenhouse gases that causes global warming (IPCC, 2007), and a study by Kimball (1982) found that crop yield will increase by 33% if the concentration of CO_2 increases. Deschênes and Greenstone (2007) stated that annual profits would increase by $1.3 billion (2002$) (4%), because of climate change. According to the Third Climate Change Assessment Report in China, the yield per unit of rice for climate variations has increased by 0.56% between 1980 and 2008, excluding technological advances (EC-SCNARCC, 2015). Another study discovered that climate change will have a beneficial impact on China's macro-economy and the economy of various other sectors by 2050, considering the fertilizer effect of CO_2 (Huang et al., 2016). Previous research indicates that climate change, characterized by temperature rise, has a significant promoting effect on total crop production in Northeast China (Liu and Lin, 2007).

It can be seen that temperature rise has both harmful and beneficial effects on economic growth. At present, some scholars have begun to explore the different effects of rising temperatures on economic development. Schlenker and Roberts (2009) found a nonlinear and asymmetric relationship of temperatures and yields. The yields of corn, soybean and cotton increased with temperature: the highest threshold temperatures of the three different crops were 29°C, 30°C, and 32°C, respectively, but temperatures above these thresholds were very harmful. In China, scholars have evaluated the relationship between agriculture, industrial development and economic growth (Chen et al., 2016, 2017; Zhang et al., 2018). Burke et al (2015) found that in different countries temperature had different effects on economic production; this showed a global nonlinear function, reflecting the relationship between temperature and economic productivity. Temperature and the economy in 31 provinces in China also show an inverted U-shaped nonlinear relationship (Li et al., 2019).

Based on the above research, we explored deeply the differences in the economic development of 31 provinces, municipalities directly under the central government, and autonomous regions (hereinafter the 31 provinces) in China in the historical period 1978–2013 and the future period2020–2100. We also contrasted and analyzed the regional differences between climate change and economic development in China, which will help fully understand the positive and negative effects of global warming on different provinces and, in addition, lay a foundation for further future risk assessment of economic development in different provinces affected by climate change.

2 DATA AND METHOD

The main datasets include the annual average temperatures (hereinafter referred to as *Temp*) and the GDP per capita of 31 provinces in the historical period (1978–2013) and the future period (2020–2100). The historical data *Temp* and GDP per capita are all statistical data. The GDP per capita (CNY) were obtained from the China Statistical Yearbook and the statistical yearbooks of the individual provinces, and the *Temps*(°C) were obtained from the China Meteorological Administration (CMA). The future data is simulated data of future scenarios, which include *Temp* of the RCP8.5 scenario and the GDP per capita of the A2r scenario from 2020–2100. We selected nine global climate models to determine the *Temp* of the RCP8.5 scenario, all from multi-model simulations for the Coupled Model Inter-comparison Project Phase 5 (CMIP5), which is available from the Earth System Grid Federation (ESGF) (https://esgf-index1.ceda.ac.uk/search/cmip5-ceda/). The GDP per capita of the A2r scenario are calculated with total GDP, population, and were all taken all from the Greenhouse Gas Initiative (GGI) scenario database of the International Institute for Applied Systems Analysis (IIASA) (http://www.iiasa.ac.at/web/home/research/modelsData/models-tools-data.html). In addition, to make the GDP per capita data comparable across different provinces, we used the Gross National Expenditure Deflator (taking the price of 2010 as the base year) from the World Bank database to eliminate the national inflation. Figure 1 shows the basic information in the historical period.

Growth rate and temperature show a nonlinear relation (Burke et al., 2015). Based on this and the data in China, the relationship between economic growth rate and temperature of provinces in China was assessed (Li et al., 2019). To explore in depth the regional differences in economic growth rates (hereinafter referred to as *Growth rates*), we choose the analysis method based on previous articles (Li et al., 2019). Eq. (1) is used to evaluate the *Growth rate* of each province:

$$\ln Y_{it} = Growth\ rate_i \times Temp_i + a_i + \omega_{it} \tag{1}$$

In the equation, parameter *Y* is used to represent GDP per capita. To refrain from some of problems in the analysis process, keeping the trend and correlation of the original data unchanged, and smooth outliers of the data, the natural logarithm of the adjusted GDP per capita $\ln Y_{it}$ is used to indicate the economic condition of a particular province and it is defined as a unary linear equation, where *i* and *t* denote province and year, respectively. $Temp_i$ represents the annual average temperatures in province *i*, year *t*. All the time-invariant factors are denoted by a_i and ω_{it} is the error term.

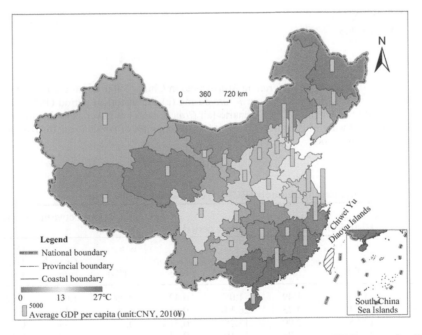

Figure 1.   The distribution of multi-year average temperatures and average GDP per capita (2010 ¥) of 31 provinces in China in the historical period.

3 RESULTS

According to Eq. (1), we calculated the *Growth rates* of 31 provinces in the historical and the future periods. In addition, the regional differences among provinces are illustrated in Figure 2 and a comparative analysis is shown in Table 1.

From Figure 2 and Table 1, it can be seen that whether in the historical or the future period, the *Growth rates* of all provinces are greater than 0, which shows that all provinces show an increasing trend and the development of each province in China is still positive, but there are large differences among them. In the historical period (Figure 2a), the average *Growth rate* of all provinces is 2.90. Among these provinces, the lowest province is Liaoning with 1.49, the highest is Fujian with 4.73, and the largest difference is 3.24, which is greater than the average *Growth rate*. In addition, Guangdong and Fujian maintain high *Growth rates*, with higher temperatures, located in the south-east coastal of China. Heilongjiang, Jilin, and Liaoning in the three north-eastern provinces all show relatively lower *Growth rates*, because of their location in the northeast of China, with lower temperatures. Several provinces in the southwest, such as Sichuan and Guangxi, also have high *Growth rates*. In the future period, it can be clearly seen from Figure 2b that although the *Growth rates* of all provinces are greater than 0, the differences among provinces are very small, and the average *Growth rate* is 0.51. The maximum is 0.65 in Hainan province and the minimum is 0.42 in Beijing. Moreover, the maximum difference is only 0.23, so it is difficult to distinguish the differences among provinces. In addition, there is also uncertainty in the simulated data of future scenarios, which is one reason for the minor differences in the future period. Therefore, we made a comparative analysis for *Growth rates* in all provinces between the historical and the future period (Figure 2c). Combined with the differences comparison in Figure 2c

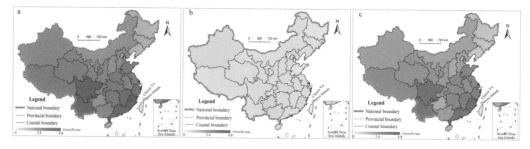

Figure 2. The distribution of *Growth rate* in 31provinces in China, 3 figures all use the same color bar. The regional difference distribution of *Growth rate* during (a) the historical period (1978–2013) and (b) the future (2010–2100) under the RCP8.5 scenario (all *Growth rates* are lower than the historical period, so little change sare shown in this figure). (c) The projected changes of *Growth rate* for the future period relative to the historical period.

Table 1. Different variations in the future (RCP8.5) compared with the past.

	Historical period		Future period		Variations between the future and the past		
	T(°C)	*Growth rate*	T(°C)	*Growth rate*	T(°C)	*Growth rate*	(Future-History)/History
Minimum	1.62	1.49	−1.05	0.42	0.29	1.03	0.69
Maximum	25.04	4.73	26.14	0.65	−6.09	4.13	0.87
Average	12.75	2.90	13.28	0.51	0.54	2.39	0.81
Maximum difference	23.42	3.24	27.19	0.23	8.75	3.10	0.18

and Table 1, it can be clearly seen that the decline of the *Growth rate* for each province between the future and the historical period is very large, with the average decline rate reaching 81%. The largest difference is 4.13, in Fujian, whose *Growth rate* decreased by more than 87% compared with the historical period. The minimum difference is Liaoning province (1.03), whose *Growth rate* also decreased by more than 69%. In addition, Figure 2c shows roughly the same distribution as the historical period (Figure 2a). In the regions with high *Temps*, this has a great impact on *Growth rate* in the future period. Moreover, global warming has an obvious inhibitory effect on regional economic development. It can be seen that the economy will still show a growth trend with temperature increases under a RCP8.5 scenario in the future, but the *Growth rates* of all provinces will be affected and will slow down. It is generally suspected that provinces with high *Temps* are more obviously inhibited by the economic development.

4 CONCLUSION AND DISCUSSION

The major conclusions of this research are as follows. (1) Whether in the historical or the future period, the *Growth rates* of all provinces are greater than 0. In the historical period, the economic development of each province shows a good development trend. The average *Growth rate* is 2.90. In the future, the *Growth rates* of different provinces will show little difference. Compared with the historical period, the *Growth rates* in the future reduce greatly, with an average reduction of more than 80%. (2) There are large regional differences in all provinces in the historical period. There are also large regional differences between the future and the historical period. Overall, it shows that the *Growth rates* of the eastern provinces are slightly higher than that of western provinces, and southern provinces are slightly higher than northern provinces.

Throughout this article, we analyze the regional differences of economic development under climate change in China in the historical and the future period. This provides references for further economic impact analysis under the future scenarios and lays a certain foundation for further risk assessment of economic impacts under climate change on a global scale.

In this paper, we evaluate the impact of temperature change on the macroeconomics of various provinces in China, but do not take into account other factors such as precipitation and wind. After all, in the context of global warming, the frequency of global extreme weather occurs will increase, which will result in all climatic factors changing. Changes in these factors will affect the economic development of countries around the world, as well as in various provinces in China. In addition, the economic data involved are multi-factor comprehensive, while in this article only the GDP per capita of each province in China is considered. However, for each province and even the country as a whole, it is very difficult to distinguish the factors that affect the economy, for the following reasons. (1) The impact of climate change on the economy is influenced by long-term and implicit perspectives. In order to highlight and evaluate the importance of global warming to the impact of economic climate, we separate it. Moreover, in future research, we will also consider the impact of multiple factors on economic development, as well as the feedback effects between economic development and climate change. (2) Today, the statistics on economic development are mostly comprehensive economic indicators such as GDP, and GDP per capita, but there is no the economic data generated by the economic impact of a certain factor.

When calculating the future *Growth rates* of all provinces, the prediction data of future scenarios are used. For these simulated data, there are uncertainties due to considering different factors. This is one of the reasons why the differences in Figure 2b are not obvious. Therefore, in the future study and risk assessment, it is necessary to select more refined simulation data of future scenarios for in-depth evaluation and analysis.

ACKNOWLEDGMENTS

This work was supported by the National Key Research and Development Program China (No. 2016YFA0602403); Beijing Municipal Natural Science Foundation (No. 9172010); National Natural Science Foundation of China (No. 41775103).

REFERENCES

Burke, M., Hsiang, S.M. and Miguel, E. 2015. Global non-linear effect of temperature on economic production. *Nature.* 527(7577): 235.

Chen, S., Chen, X. and Xu, J. 2016. Impacts of climate change on agriculture: evidence from China. *J. Environ. Econ. Manag.* 76(8):105–124.

Chen, X. and Yang, L.2017.Temperature and Industrial output: firm-level evidence from China. *J. Environ. Econ. Manag.* https://doi.org/10.1016/j.jeem.2017.07.009

Deschênes, O. and Greenstone, M. 2007. The economic impacts of climate change: evidence from agricultural output and random fluctuations in weather. *Am. Econ. Rev.* 97(1):354–385.

Dell, M., Jones, B.F. and Olken, B.A. 2012. Temperature shocks and economic growth: evidence from the last half century. *Am. Econ. J. Macroecon.* 4(3):66–95.

Deryugina, T. and Hsiang, S.M. 2014. Does the environment still matter? Daily temperature and income in the United States. *NBER Working Paper 20750.* National Bureau of Economic Research: Cambridge, MA.

EC-SCNARCC. 2015. *The third national assessment report on climate change.* Science Press: Beijing, 1–108.

Gable, F.J. 1997. Climate change impacts on Caribbean coastal areas and tourism. *J. Coastal Res.* 27 (SPECIAL), 49–69.

Hsiang, S.M. 2010. Temperatures and cyclones strongly associated with economic Production in the Caribbean and Central America. *Proc. Natl. Acad. Sci.* 107(35): 15367–15372.

Hsiang, S., Kopp, R., Jina, A., Rising, J., Delgado, M., Mohan, S., Rasmussen, D.J., Muir-Wood, R., Wilson, P., Oppenheimer, M., Larsen, K. and Houser, T. 2017. Estimating economic damage from climate change in the united states. *Science.* 356(6345): 1362.

Huang, D.L., Li, X.M. and Ju, S.P. 2016. Climate Change Affecting Grain Production, Consumption and Economic Growth in China: Based on the Agricultural CGE Model. *Chinese Agricultural Science Bulletin.* 32(20):165–176. (in Chinese).

IPCC. 2007. *Climate change 2007: the physical science basis.* Contribution of Working Group I to the Fourth Assessment Report of the Intergovernmental Panel on Climate Change. Cambridge University Press: Cambridge.

IPCC. 2014. *Climate Change 2014: Impacts, Adaptation, and Vulnerability.* Contribution of Working Group II to the Fifth Assessment Report of the Intergovernmental Panel on Climate Change. Cambridge University Press: Cambridge.

Kimball, B.A. 1982. Carbon dioxide and agricultural yield: an assemblage and analysis of 430 prior observations. *Agron. J.* 75(5): 779–788.

Li, N., Bai, K., Zhang, Z.T., Feng, J.L., Chen, X. and Liu, L. 2019. The nonlinear relationship between temperature changes and economic development for individual provinces in China. *Theor. Appl.Climatol.* https://doi.org/10.1007/s00704-018-2744-6

Liu, Y. and Lin, E. 2007. Effects of climate change on agriculture in different regions of china. *Adv. Climate Change Res.* 3(4):229–233. (in Chinese).

Mendelsohn, R., Nordhaus, W.D. and Shaw, D. 1994. The impact of global warming on agriculture: a ricardian analysis. *Am. Econ. Rev.* 84(4):753–771.

Schlenker, W., Hanemann, W.M. and Fisher, A.C. 2005. Will US agriculture really benefit from global warming? Accounting for irrigation in the hedonic approach. *Am. Econ. Rev.* 95(1):395–406.

Schlenker, W., Hanemann, W.M. and Fisher, A.C. 2006. The impact of global warming on US agriculture: an econometric analysis of optimal growing conditions. *Rev. Econ. Stat.* 88 (1): 113–125.

Somanathan, E., Somanathan, R., Sudarshan, A. and Tewari, M. 2015. The impact of temperature on productivity and labor supply: Evidence from Indian manufacturing. Working paper.

Schlenker, W. and Roberts, M.J. 2009. Nonlinear temperature effects indicate severe damages to U.S. crop yields under climate change. *P. Nat. Acad. Sci. USA.* 106(37):15594–15598.

Zhang, P., Deschenes, O., Meng, K. and Zhang, J.J. 2018. Temperature effects on productivity and factor reallocation: Evidence from a half million Chinese manufacturing plants. *J. Environ. Econ. Manag.* 88: 1–17.

Zivin, J.G. and Neidell, M. 2014. Temperature and the allocation of time: Implications for climate change. *J.Labor Econ.* 32(1): 1–26.

Research on the associated features between different countries and terrorist attacks

Guohui Li, Lizeng Zhao, Weiping Han, Ge Guo & Ying Wang
Tianjin Fire Research Institute of Ministry of Emergency Management, Tianjin, China

ABSTRACT: Terrorist attacks have become serious threats across the world and have their own characteristics. It is helpful to reveal their basic patterns and characteristics by thorough research on terrorist attacks based on historical accident cases. The research will be conducted using data from the Global Terrorism Database. The multivariate statistical analysis is introduced to explore the main features of terrorist attacks in different countries. The influencing factors of terrorist attacks include weapon type, target, attack types and attack time. This work will highlight the associations between the countries and these factors, and present the distinguishing features of terrorism in each country. The spatial-temporal distribution characteristics of different types of terrorist attack will also be analyzed. This research can be useful to prevent terrorist attacks and reduce the casualties and property damage caused.

Keywords: terrorist attack, correspondence analysis, influencing factor, spatial-temporal distribution

1 INTRODUCTION

Terrorism is the common enemy of all mankind (Jackson et al. 2013; Gill et al. 2013). Terrorist attacks threaten people's lives and affect social development. From a statistical perspective, there are over 190 countries in the world, and 137 counties have suffered terrorist attacks in the four years between 2014 and 2017. In this period, there were 56,359 terrorist attacks around the word, which killed 144,659 people. With the development of the international economy, terrorist attacks will continue to grow. Therefore, it is necessary to carry out research on terrorist attacks.

Terrorism is characterized by intelligence. As the global counter-terrorism situation changes, the attack type, target, weapon and time by terrorists are also changing (Santifort et al. 2012, Ackerman et al. 2012). In addition, terrorist attacks show different characteristics in different countries. The relationship between the terrorist attacks and the influencing factors need to be further studied, and the information hidden behind the data of terrorist attacks can be revealed based on the method of statistical analysis and data mining. This study examined the distribution of terrorist attacks in different countries, based on the data of terrorist attacks, using the method of multivariate statistical analysis, to provide technical support for counter-terrorism.

2 DATA AND METHODS

2.1 Data

The research has been conducted using data from the Global Terrorism Database (GTD), which is an open-source database including information on terrorist events around the world (Lafree et al. 2007). Information is available on the date and location of the incident, the weapons used and nature of the target, and (when identifiable) the group or individual responsible. During the four years between 2014 and 2017, there were 56,359 terrorist attacks around the word, which killed 144,659 people.

The influencing factors include attack month, attack weapon, attack target and attack type. Statistical analysis shows that 20 countries suffered 89.4% of the total terrorist attacks in 2014–2017. Therefore, the study mainly focused on these top 20 countries, as shown in Table 1.

In the database, weapons are recorded in 13 categories. In the subsequent analysis, some categories are dismissed for a lack of data or meaninglessness. Only five categories of weapons were retained: chemical weapons, firearms, explosives, incendiary weapons and melee.

The database offers 22 categories for attack targets. Because of sparse data, ten attack targets are reserved: business, government (general), police, military, educational institution, private citizens and property, religious figures/institutions, terrorists/non-state militia, transportation, and utilities. Nine attack types are as follows: assassination, armed assault, bombing/explosion, hijacking, hostage taking (barricade incident), hostage taking (kidnapping), facility/infrastructure attack, unarmed assault, and unknown.

All 12 months are considered.

2.2 Methods

Correspondence analysis (CA) is a multivariate statistics method based on the analysis of a contingency table through the row and column profiles, and it can explore the relationships among categorical variables by transforming multi-dimensional variables to two-dimensional scatter plots (Hoffman et al. 1986). The method has been introduced to study the relationship between fatalities and the influencing factors of terrorist attacks (Li et al. 2014). However, association between the countries and the factors has not been studied systematically. The association between variables can be explored via the scatter diagram. The CA is based on

Table 1. Contingency table by country and weapon.

| No. | Country | Total number of deaths | Weapon | | | | |
			Chemical	Firearms	Explosives	Incendiary	Melee
1	Iraq	12510	37	1373	9896	102	42
2	Afghanistan	6783	24	2119	3243	150	60
3	Pakistan	4977	4	1908	2861	47	14
4	India	3735	3	1184	1649	381	198
5	Philippines	2642	0	1484	872	128	13
6	Somalia	2506	1	853	1203	34	43
7	Nigeria	2369	1	1287	793	121	41
8	Yemen	2178	1	586	1072	11	7
9	Libya	1883	0	598	856	35	31
10	Ukraine	1654	0	297	1131	50	17
11	Egypt	1601	0	419	1041	38	23
12	Syria	1538	7	104	1177	9	24
13	Turkey	1239	0	319	745	58	11
14	Thailand	1209	0	338	796	68	4
15	Bangladesh	728	0	60	435	157	56
16	West Bank and Gaza Strip	619	2	101	150	80	229
17	Sudan	594	0	381	30	28	49
18	Colombia	593	1	144	321	40	2
19	Democratic Republic of the Congo	565	0	256	19	47	42
20	United Kingdom	444	1	58	176	179	20

contingency tables. First, the normalized transformation is performed on the data. Then the singular value of the matrix is calculated, and the factor load matrix can be established. The information at each dimension can be determined by singular value decomposition. Finally, the scatter diagram is obtained based on the standardization of the scores of the row and column.

The row profiles correspond to the relative frequencies of the different factors within each country, and the column profiles are the relative frequencies of different countries within each factor, as shown in Table 1. The number of dimensions is equal to the minimum of (m-1, n-1) when decomposing the contingency table, where m and n is the variable number of row and column. All the dimensions can explain the total information of the variables. If the first two dimensions can explain 80% of the total information of the variable, the corresponding relationship of variables can be reflected by the two-dimensional diagram. In the following analysis, the first two dimension can explain most of the information. Therefore, a two-dimensional scatter plot is given for each factor. Dim 1 represents the scores of each variable in dimension 1. Dim 2 represents the scores of each variable in dimension 2. The scores in dimension can be calculated based on the matrix of contingency table.

The biggest advantage of the CA is dimensionality reduction. Interpreting the scatter plot follows these principles: (1) if the two variables are close to each other, it indicates that they have similar distribution; (2) if the row and column variables are close to each other, it indicates that the row variable has a large weight on the column variable; (3) the further the variable is from the origin, the more accurate the variables are explained; (4) the origin represents the center of the average distribution (Hrdle et al. 2007).

3   RESULTS AND DISCUSSION

3.1   *The relationship between the attack type and country*

Figure 1 shows the results by attack type, that the eight categories are far from the origin, and there is a significant correspondence between these factors. The United Kingdom is strongly associated with unarmed assault. The attack type facility/infrastructures more common in Bangladesh, West Bank and Gaza Strip and India. Attacks in Thailand, Egypt, Ukraine, Iraq and Syria are related with bombing/explosion. Sudan suffers the terrorist attack of hijacking. Most of the terrorist attacks in the Philippines and Nigeria are hostage taking (barricade incident) and armed assault. Libya, Afghanistan, Yemen, and Somalia are associated with the attack type of assassination. Different types of terrorist attacks are evenly distributed in Colombia because of the original location.

3.2   *The relationship between the attack weapon and country*

The results for weapon type are shown in Figure 2. The five types of weapons are distinguished significantly. West Bank and Gaza Strip is associated with melee, and about 37% attacks are carried out this way. Terrorists tend to use incendiaries to carry out attacks in Bangladesh and the United Kingdom, accounting for 40.3% and 21.6% of the total attacks in each country, respectively. Most of the attacks with chemical weapons happened in Syria and Iraq. Out of all the countries, Iraq suffers the most terrorist attacks: globally, 97 terrorist attacks were carried out with chemical weapon in 2014–2017, and 37 of these attacks were in Iraq. Chemical weapons are lethal, and using chemical weapons in Iraq to carry out terrorist attacks should be taken very seriously. Ukraine is associated with explosives. Terrorists in Nigeria, Philippines, Sudan and Democratic Republic of the Congo tend to use firearms to carry out attacks.

3.3   *The relationship between the month and country*

If we can know the time of the high incidence of terrorist attacks, targeted measures can be taken to avoid more casualties in that critical time. The relationship between the attack month

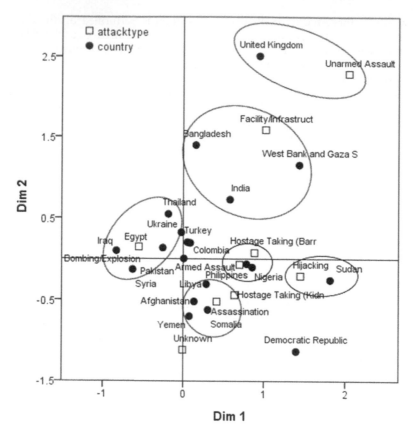

Figure 1.   Scatter graph of countries and attack types.

and country are analyzed and the results are shown in Figure 3. It can be seen that the Bangladesh and Turkey are farthest from the origin and they are best explained. The terrorist attacks in Bangladesh, Pakistan and Yemen tent to happen in January and February. Turkey and Syria have a strong association with August. Most of the attacks in Afghanistan, Thailand and India are carried out in September and October. Sudan and Democratic Republic of the Congo are far away from the origin and are related with March and June. Ukraine has a strong relationship with April and May. The Philippines, Libya, United Kingdom, Somalia and West Bank and Gaza Strip tend to suffer terrorist attacks in June and November. The time distribution of terrorist attacks may be affected by special festivals and activities, and this needs further study.

3.4   The relationship between the attack target and country

The main purpose of the terrorists is to create social panic by attacking specific targets. The terrorist attacks aim at different targets in different countries. The research on the attack targets can reveal the high-risk targets and provide effective measures to avoid casualties. The results between the country and attack target are shown in Figure 4. The countries can be divided into four categories according to the four quadrants distribution of the scatter plot. Compared to other countries, Sudan, United Kingdom, Nigeria, Syria, Iraq and Democratic Republic of the Congo tend to be associated with private citizens and property and terrorists/non-state militia:71.9% of the total attacks in Sudan and 49.3% of the total attacks in the

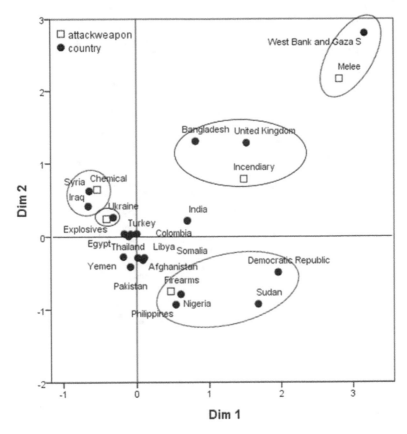

Figure 2.   Scatter graph of countries and weapon types.

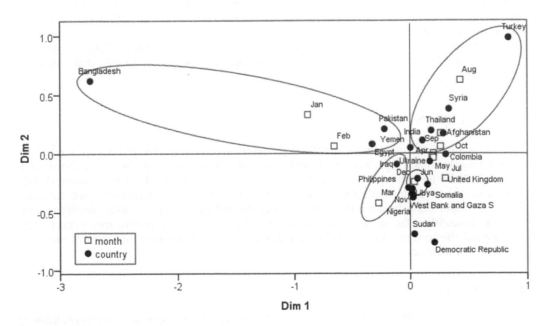

Figure 3.   Scatter graph of countries and month.

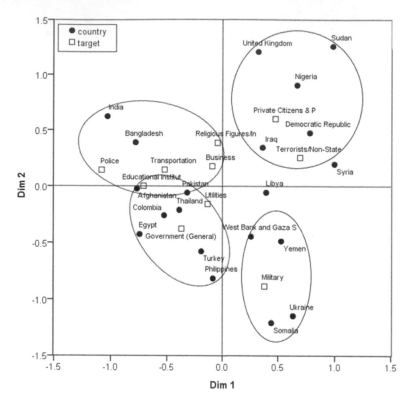

Figure 4.   Scatter graph of countries and target.

United Kingdom target private citizens and property. These countries should focus on protecting private citizens and property from attack.

Terrorists target the military in Yemen, Ukraine, Somalia and West Bank and Gaza Strip. In Ukraine and Somalia, 47.9% and 51.7% of the total attacks target the military, respectively. India and Bangladesh are associated with the targets police, transportation, business, religious figures/institutions and education institutions. In the Philippines, Turkey, Egypt, Colombia, Thailand and Pakistan, the utilities and the government (general) are targets that suffer the most terrorist attacks.

4   CONCLUSION

The associated features between the different countries and terrorist attacks are studied by considering the attack type, weapon, month and target. When we study the space distribution features of the terrorist attacks, it is clear that analysis only based on the statistical data cannot obtain accurate results. The CA can be used to explore the terrorist attack statistics, reveal the association between the countries and the influencing factors and provide the primary and secondary factors in different countries. According to the results, we can accurately understand the terrorist attacks and establish the effective prevention and control measures and decision-making.

REFERENCES

Ackerman, G., Bale, J.M. 2012. The potential for collaboration between Islamists and Western left-wing extremists: A theoretical and empirical introduction. *Dynamics of Asymmetric Conflict* 5(3): 151–171.

Gill, P., Corner, E. 2013. Disaggregating terrorist offenders: implications for research and practice. *Criminology & Public Policy* 12(1): 93–101.

Guohui, L. et al.2014. Study on correlation factors that influence terrorist attack fatalities using Global Terrorism Database. *Procedia Engineering* 84: 698–707.

Hoffman, D.L., Franke, G.R. 1986. Correspondence analysis: Graphical representation of categorical data in marketing research. *Journal of Marketing Research* 23(3): 213–217.

Hrdle, W., Simar. L. 2007. Applied multivariate statistical analysis. *Springer*.

Jackson, B.A., Bikson, T.K. 2013. Human subjects' protection and research on terrorism and conflict. *Science* 340(6131): 434–435.

Lafree G, Dugan L. 2007. Introducing the Global Terrorism Database. *Terrorism & Political Violence*, 19(2):181–204.

National Consortium for the Study of Terrorism and Responses to Terrorism (START). 2018. Global Terrorism Database [Data file]. Retrieved from https://www.start.umd.edu/gtd.

Santifort, C., Brandt, P.T. 2012. Terrorist attack and target diversity: change points and their drivers. *Journal of Peace Research* 50(1): 75–90.

Risk Analysis Based on Data and Crisis Response Beyond Knowledge – Huang & Nivolianitou (eds)
© 2020 Taylor & Francis Group, London, ISBN 978-0-367-25146-8

Using game theory for deployment of security forces in response to multiple coordinated attacks

Cheng-Kuang Wu

School of Computer Science, Zhaoqing University, Zhaoqing, China

ABSTRACT: In Paris attacks, three teams of attackers carried out a coordinated series of suicide bombings and mass shootings as diversionary attacks designed to distract the attention of the security commander, causing them to take a longer time to respond. These diversionary attacks in turn led to a greater number of casualties at the primary target, producing what is now known as the Bataclan theatre massacre. The proposed model determines the allocation of security forces for eight attacked targets in the Paris terror attacks. Two games are constructed, representing the two stages needed for the rapid deployment of security forces. One is an initial hidden-object game which is designed to determine the threat value (i.e., TV) during one attack event. Another is the security force deployment game which mobilizes the limited security forces to lay siege to the terrorist.

Keywords: Security forces deployment, Nash equilibrium, Shapley value, Threat value

1 INTRODUCTION

In Paris attacks, terrorists tend to coordinate multiple firearms assaults which, combined with diversionary attacks are designed to divert the attention of the security forces, causing them to take a longer time to respond. This in turn enables the terrorists to cause more casualties at their primary target (CNN, 2015). These diversionary attacks in turn led to a greater number of casualties at the primary target, producing what is now known as the Bataclan theatre massacre. Thus, in the modeling, the EOC (emergency operation center) needs to efficiently mobilize security forces from a non-attacked district and to improve relief efforts against a series of coordinated terror attacks.

The analytical tools provided by game theory have already found applications in a variety of research areas, where multiple agents compete and interact with each other within a specific system. In most multiagent interactions, the overall outcome depends on the choices made by all self-interested agents, with the goal being to make choices that optimize the outcome (Dixit and Skeath, 2001). Game theory also provides general mathematical techniques for analyzing goal-conflict and cooperation between two or more individuals. However, a commonly encountered problem is related to scalability. It is difficult to compute the Nash equilibrium (N.E.) from the payoff table in multi-dimensional game theory models with four players (i.e., agents). This complexity of the computations because of multiple MAS agents is a challenge for classical game theory. Thus, two game theoretic methods are applied to create a two-stage model. The proposed two-stage model uses two steps for solving the growing amounts of MAS agent work. This is a framework for deploying multiple-agents based upon the game theory which analyzes the detected attacks and feasible allocates the MAS resources.

The proposed model determines the allocation of security forces for eight attacked targets in the Paris terror attacks, especially when the available security forces are limited. Two

---

*Corresponding author: shapleyvalue@hotmail.com

games are constructed, representing the two stages needed for the rapid deployment of security forces. The first is the initial hidden-object game which is designed to determine the threat value (i.e., TV) during one attack event, the other is the security force deployment game during which the limited security forces of the response agent are mobilized to lay siege and overcome the at-tackers.

2 THE PROPOSED MODEL

There are two schematic processes included in the proposed model. For every attack, the interaction process between the attacker and the security forces commander is modeled as a non-cooperative two-person and zero-sumgame, taking into account the interaction of security measures between security forces commander and attacker. The proposed payoff functions are obtained utilizing the threat measures (such as defensive or attacking capability) for two players. The N.E. derived from these payoff functions is calculated to assign a unique threat value (TV) for each commander. Then, the second schematic process is constructed in the form of a cooperative game between all attacked events. The power index of the Shapley value is applied to calculate the marginal contribution between agents and find a mutually agreeable division of cost for security forces deployment. This study revises Owen's method (Owen, 2001) for the proposed Shapley value formula. The attacked events are grouped into coalitional groups based on the threat level so as to provide acceptable security resources deployment. The Shapley value can be used to ensure the derivation of a unique solution for each event in the security forces sharing interactions. Let $Z^h = \{z_1^h, z_2^h, z_3^h, \ldots, z_8^h\}$ be the minimum set of security forces; $\forall_i \in N$, N is the number of security forces, which is subject to the threat level h. Finally, the security forces can be deployed in all attacked events by taking the appropriate Shapley value vector.

2.1 Model assumptions

Certain assumptions must be made when considering simultaneous terror attacks and security force response scenarios within the proposed framework. The assumptions are listed below.

- During multiple terror attacks, emergencies occurring in different districts will overlap in time, so that the EOC manager will receive a series of emergency calls in response to the series of coordinated terror attacks. Player 1 is the terrorist (or terrorist group) who could be launching one of two types of attacks, a diversionary attack such as suicide bombing or shooting of innocent or unarmed people, or a primary attack to take and kill hostages. Player 2 is the security forces commander who defends the attacked district. This study assumes that each player's aim in the game is to achieve as high a payoff for him or her as possible.
- Tables 1-3 show the information concerning attack types and attack requests, security forces available in the twelve defensive districts, and crossroads that need to be passed to reach the terror targets from each of the resource sites (police agencies), as well as the population density at the eight attacked target areas.

2.2 Hidden-object game

In this model, the air combat model is applied to establish a firearms assault game. This game problem is to destroy a hidden object (Dresher, 1981). The first stage model is designed to

Table 1. Attack types and security force requests.

Types of Attacks	Requirements
Primary Attack (i.e., massacre)	8
Diversionary Attacks (i.e., suicide bombing, mass shootings)	2
Total	10

Table 2. Security forces available in the twelve attacked districts.

District	Police	Rapid Response Forces	Subtotal
A	20	20	40
B	25	20	45
C	20	20	40
D	40	20	60
E	30	20	50
F	15	20	35
G	15	20	35
H	20	20	40
I	30	20	50
J	35	20	55
K	40	20	60
L	35	20	55
Total	325	240	565

Table 3. Crossroads passed, population density in each terrorist assault area.

Event	Type of Attack	Location	District	Crossroads Passed ($r_{i,k}$)	Population Density δ_i
E1	Suicide bombing	Stade de France	**A**	14	0.8
E2	Shooting	Rue Bichat	**B**	9	0.92
E3	Suicide bombing	Stade de France	**A**	10	0.8
E4	Shooting	Rue de la Fontaine	**C**	7	0.88
E5	Shooting	Rue de Charonne	**D**	6	0.86
E6	Suicide bombing	Boulevard Voltaire	**E**	5	0.84
E7	**Shooting (Massacre)**	**Bataclan Theatre**	**F**	4	0.98
E8	Suicide bombing	Stade de France	**A**	12	0.82

determine the TV of the terror attack event. The behaviors of the terrorists and security force commanders are modeled with a two-person game theory model. We assumed that the terrorist and security force commander are rational players, forming a set of noncooperative players $A = \{a_1, a_2\}$, where a_1 is the terrorist (or group) carrying out the terror event; and a_2 is the security forces commander. The parameters for determining the measures of a primary or diversionary attack are defined in the following paragraphs.

The terrorist is player 1. The terror attack aimed at a target could be one of two types: a diversionary attack or a primary attack. S_1 denotes the set of player1's strategies: $S_1 = \{u_1, u_2\} = \{$primary attack, diversionary attack$\}$. The greater the seriousness of the attack, the more the resources (i.e., security forces) needed. W denotes the set of resource requirements for multiple attacks in one urban region. $W = \{w_{i,1}, w_{i,2}\}$. The variables $w_{i,1}$ and $w_{i,2}$ denote the number of resources required for primary and diversionary attacks, respectively (as shown in Table 1; $w_{i,1} = 2$, $w_{i,2} = 8$, $w_{i,all} = 10$) in the ith attack event. In addition, the population of the district for each gunfire event is related to the resource requirement. A terror attack on a densely populated target will generate more loss of life and thus gain greater benefit for the terrorist, while the security commander must pay a higher security cost. In contrast, a reduced number of security forces will be enough to fight in a sparsely populated target. Here, δ_i denotes the population density of the ith target in an urban district, $0 < \delta_i < 1$. The population density is the number of people in the targeted area, per 100 square meters (as shown in Table 3).

The security forces commander is player 2. In a hidden-object game, each attacked target has a security force commander, who in turn possesses resources (such as police officers or rapid response forces) with which to respond to multiple coordinated attacks. In this game, the security force commander, player 2, decides to deploy their security forces to respond to the terrorist attacks. He/she may conceal the valued object (i.e., hide the valued payoff) in primary attack and conceal the valued object in the diversionary attack. Here, S_2 denotes the sets of player 2's strategies: $S_2 = \{d_1, d_2\} = $ {hide the valued payoff in primary attack, hide the valued payoff in diversionary attack}; O denotes the set of resources of security forces available to commanders in one urban region: $O = \{o_1, o_2, \ldots, o_{12}\}$; o_i denotes the number of security forces available for the ith district; $r_{i,k}$ denotes the number of crossroads passed (as shown in Table 2-3). Crossroads are crucial for the calculation of the shortest paths for the resources to traverse from the kth depot (i.e., police agency) to the ith location of a firearms assault (i.e., target) (Fiedrich, et al.,2000). The more crossroads the security forces must pass, the more effort they have to spend to respond to the terrorist event. In contrast, if they have to pass fewer crossroads, they could respond more quickly, thereby reducing the loss of life, and gaining more benefits for player 2.

This study is based on the Paris attack scenario for which there was a primary attack or diversionary attack in one terror attack event. In this game, the security force commander, player 2, decides to deploy their security forces to respond to the terrorist attacks. He/she may hide the valued payoffs in primary attack and hide the valued payoffs in the diversionary attack. The two players make their strategy decisions simultaneously. A 2 × 2 payoff matrix for this game is created based on the two players' strategies and interactions; see Table 4.

According to Dresher's air combat model, the proposed 2 × 2 payoff matrix game has no pure strategic N.E (Dresher, 1981). A mixed Nash equilibrium pair (r^*, q^*) exists in the normal form game if this game has no pure strategy N.E., which is an optimal solution. Player 2's expected payoff is computed when player 1 and player 2 use mixed strategies r and q, respectively. The mixed N.E. for the probability vector is $r^* = \{r^*(u_1), r^*(u_2)\}$ with actions $\{u_1, u_2\} = $ {primary attack, diversionary attack} by the terrorist and the vector $q^* = \{q^*(d_1), q^*(d_2)\}$ with actions $\{d_1, d_2\} = $ {{hide the valued payoffs in primary attack, hide the valued payoffs in diversionary attack} by the security forces commander. The optimal strategies are always mixed, and are the same for the two players: makes a primary attack with a probability

$$ r_i^*(u_1) = q_i^*(d_1) = \frac{w_{i,1} - o_i}{[(r_{i,k} - 1) \times (o_i - w_{i,2})] - (o_i - w_{i,1})}, \quad i \in 1, \ldots, 8; \ k \in 1, \ldots, 6. \quad (1) $$

Player 1 (i.e., terrorist) gains a positive value (+), because player 1 benefits from player 2's resource responses. Player 2 (i.e., security force commander) pays the negative value (−), which means that player 2 loses profit as a consequence of player 1's attack. When payoff for the response agent (i.e., player 2) is expected to be small, the terrorist threat is large. Now, in the equation (2), let v_i be the ith threat value of a terror attack, which is the expected payoff for player 1.

$$ v_i = \sum_{j=1}^{2} \sum_{k=1}^{2} r_i^*(u_j) \times q_i^*(d_k) \times \pi_1(u_j, d_k), i \in 1, \ldots, 8. \quad (2) $$

Table 4. Payoff matrix for the hidden-object game for one attack event.

Terrorist	Security forces commander	
	Hide the valued payoffs in the primary attack d_1	Hide the valued payoffs in the diversionary attack d_2
Primary attack u_1	$\left(\frac{w_{i,all}}{o_i - w_{i,1}}\right) \times \left(\frac{r_{i,k}-1}{r_{i,k}}\right) \times \delta_i$	0
Diversionary attack u_2	0	$-\left(\frac{w_{i,all}}{o_i - w_{i,2}}\right) \times \left(\frac{1}{r_{i,k}}\right) \times \delta_i$

Therefore, v_i is derived from the expected payoff of the two players' optimal strategies and represents the TV of the ith terror attack in the first game model. In the next stage, the values v_i are applied to compute the Shapley value for each attacked event within a cooperative game.

2.3 Security force deployment game

This section likens the interactions of all terror firearms assaults in an urban region to the playing of a cooperative game. An efficient method is needed to decide the number and priority of the deployment of security forces to various response events when multiple attacks occur simultaneously. The Shapley value is an index of power for cost allocation (Roth and Robert, 1979). We utilize the concept of the majority coalition of party voting game to deal with security force reallocation. A majority of voters can pass any bill in that majority game. The power of a voter will depend on how crucial that voter is to the formation of a winning coalition (Dixit and Skeath, 2001). The Shapley value can provide a measure of the power of voter which represents the power index. In our second game, we compute the Shapley value of each firearms assault using the all threat values (TVs) in an urban region. When the sum of the TVs of firearms assaults passes the threshold of majority level, they can form a winning coalition and enable to compute the power index of each firearms assault.

We define y: $V \rightarrow R+$ as a one-to-one function by assigning a positive real number to each element of v and $y(0) = 0$, $V = \{v_1, v_2,...,v_i\}$, $i \in n$. The terrorist attack for response security forces deployment is based on the concept of the majority of threat level h_H, where m_H represents the corresponding threshold values. Given the output vector of all threat values, the level h, of the response region is equal to h_H, if the sum of the threat values is greater than or equal to m_H:

$$h_H \text{ if } \sum_{i=1}^{n} v_i m_H, \quad m_H = v_{Mini} + \left(\frac{v_{Max} - v_{Mini}}{2}\right) \tag{3}$$

The threat values of all firearms assaults can be grouped into the majority of level h. This is divided by 2 from the maximum value v_{Max} to minimum value v_{Mini}. All firearms assaults can be modeled as an N-person game with $X = \{1, 2,..., N\}$, which includes the set of players (i.e., terror firearms assaults) and each subset $V \subset N$, where $v_i \neq 0$, $\forall v_i \in C$ is called a coalition. The Shapley value of the ith element of the terror attack event vector is defined by

$$\omega(i) = \sum_{\substack{C \subset X \\ i \in c}}^{n} \frac{(c-1)!(n-c)!}{n!} [y(C) - y(C - \{i\})] \tag{4}$$

$$\Rightarrow \omega(i) = \sum_{\substack{C' \subset X \\ i \in c'}}^{n} \frac{(c-1)!(n-c)!}{n!} \tag{5}$$

Equation (4) can be simplified to Equation (5) because the term $y(C)-y(C-\{i\})$ will always have a value of 0 or 1, taking the value 1 whenever C' is a winning coalition. If C' is not a winning coalition, the terms $C-\{i\}$ and $y(C)$ are 0 (Owen, 2001). Hence, the Shapley value is $\omega(i)$, where C' denotes the winning coalitions with $\sum^{v_i} m_f, i \in C'$. The Shapley value of the ith attacked event output indicates the relative threat value for the thresholds m_H (i.e., threat levels). Therefore, a vector of Shapley values shows the strength of all attacked events. The numbers of security forces reallocated to the ith attacked event (i.e., firearms assault) are defined by

$$e(i) = \omega(i) \times O_{Total} \tag{6}$$

which O_{Total} is a total number of security forces in an urban region. $e(i)$ is derived from the Shapley value of the i th firearms assault $\omega(i)$ multiply by O_{Total}. Finally, the commander can reallocate security forces from all districts to evenly lay siege to terrorists attacked targets.

3 SIMULATION EXPERIMENTS

In this study, it is hypothesized that the EOC divides the city of Paris into twelve districts, each with a police agency which provides security forces and monitoring during multiple terror attacks. The information used in the simulation regarding security resource availability, emergency requests, average times taken to handle two types of attack in one terror attack, and the population density of the targets areas is shown in Tables 1–3. We assumed that the terrorists simultaneously attack eight targets in the city of Paris. The payoff matrix of the zero-sum hidden-object game for each terror attack is modeled using Table 4. Each terror attack is modeled as a non-cooperative game and sets of simulated threat measures for the terrorist and the security force commander are generated. The hypothetical parameter numbers in eight attacks are given for modeling of the payoff matrix, and sets of security measures for commander and terrorist are randomly generated. Then, eight TVs (threat values) for terror attacks are calculated using Equations (1) – (2) (see Figure 1 (a)).

The experimental results for the modeling of eight terror assault events indicate that the threat values of the terror attacks decrease if the amount of security forces available is increased even if the number of crossroads passed and population density at the target area is fixed (see Figure 1, b). It appears that using fewer diversionary attacks to draw the attention of security forces, will lower the expected payoff for the terrorist. As can be seen in Figure 2(a), the threat value of the terror attack will increase if there is an increase in the population density of the attacked target, and security force availability for each firearms assault event and the number of crossroads passed are fixed. This confirms that a terrorist firearms assault on a densely populated target would generate a greater loss of life, which increases the threat value of the terror attack. In addition, the threat value of the terror attacks decreases when the numbers of crossroads passedare increased and the security force availability for each firearms assault event and the population density of target is fixed, as seen in Figure 2(b). This result confirms that increasing the time taken to handle terror attacks decreases the threat value. This also helps to explain why terrorists might choose to stage a firearm assault a long way away from the primary (main) attack. Diverting the attention of the security forces, makes them take a longer time to respond and causes more casualties at their primary target.

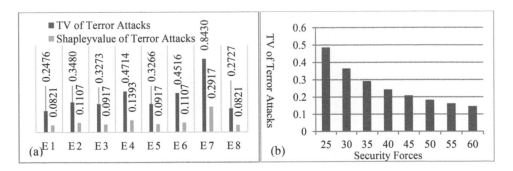

Figure 1.   (a) Shows 8 threat values and Shapley values for terror attacks. (b) Shows how security force availability affects the TV of the terror attacks.

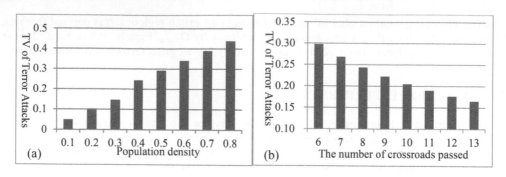

Figure 2.   (a) The effect of population density on the TV. (b) The effect of the number of crossroads passed on the TV.

4   CONCLUSIONS

The proposed model is applied to develop a plan for the reallocation of security forces for simultaneous diversionary and primary attacks. The first hidden-object game creates the threat value occurring in each firearms assault in an urban region. The second security force deployment game utilizes these threat values to compute the Shapley value for each firearms assault given the majority of the threat levels. The Shapley values are then utilized to build up a rating system for allocating all of the response security forces. In this way, we can provide a feasible and equitable cost allocation. The future works will demonstrate the suitability of the proposed game theory models for the allocation of security forces.

REFERENCES

2015 Paris Terror Attacks Fast Facts. (2016, November 30). CNN. Retrieved from http://edition.cnn. com/2015/12/08/europe/2015-paris-terror-attacks-fast-facts/
Dixit, A. & Skeath, S. 2001. Games of Strategy. W.W. Norton & Company.
Dresher M. 1981. *The Mathematics of Game of Strategy Theory and Application.* Dover Publication, Inc., New York.
Fiedrich, F., Gehbauer, F., & Rickers, U. 2000. Optimized resource allocation for emergency response after earthquake disasters. *Safety Science* 35(1): 41–57.
Owen, G. 2001. *Game Theory*, pp.265. 3rd Ed. New York, NY: Academic Press.
Roth, Alvin E. and Robert E. Verrecchia, 1979. The Shapley Value as Applied to Cost Allocation: A Reinterpretation, *Journal of Accounting Research* 17: 295–303.

Risk Analysis Based on Data and Crisis Response Beyond Knowledge – Huang & Nivolianitou (eds)
© 2020 Taylor & Francis Group, London, ISBN 978-0-367-25146-8

Modeling of scientific collaboration network of megaprojects based on risk analysis

Jinyang Dong, Tiezhong Liu & Zhixiang Li
School of Management and Economics, Beijing Institute of Technology, Beijing, China

ABSTRACT: This paper studies how to construct an efficient, stable and sustainable scientific collaboration network model in order to complete the tasks of megaprojects. The theories of complex network and risk analysis were used to summarize and construct this model. According to the analysis, megaprojects are now facing both internal and external risks. As a result, modals of the projects should follow three principles described as: the oriented technology demand, the applied project's characteristics and satisfied collaboration needs. Also, the hierarchical structure, teams' suitable participation and platform for collaboration have been held as particular useful in fulfilling the three principles and making the scientific collaboration network of megaprojects become more efficient, stable and sustainable.

Keywords: megaproject, collaboration network, internal and external risks, principle

1 INTRODUCTION

As a result of the continuous uprising of China's overall national strength, a series of megaprojects such as Shenzhou Spacecraft and Beidou Satellite Navigation System have been accomplished successfully. It has become obvious for example, that as all teams of each megaproject put more and more focuses on their own interests, the organizational management mode has definitely changed from national mobilization under the planned economy into a more flexible mode of project under the market economy. In the process of organizing and managing megaprojects, new types of risks and challenges have arisen with the conflict of interests, which have prompted some strong teams either drop out of the megaprojects or shift to other projects, thus significantly crippling the megaproject's innovation capability. It is now commonly agreed that not only are megaprojects of great significance in politics, economy, science, technology and military, they are also symbols of a country's comprehensive national strength and are often endowed with such a mission to prevail. In short, all mentioned above have made it suddenly imperative for us to find the most effective ways to build an efficient, stable and sustainable scientific collaboration network for megaprojects.

At present, the research on the management of megaprojects mainly centers on organizational management mode that was built on the theory of Systems Engineering and Complex System (Jiang et al 2013; Nie et al. 2014), Organization Interface (Li and Du, 2012), and Collaboration (He and Wang, 2018). While most scholars proposed that network models should be established from the perspective of management mode (He et al. 2013; Kong et al. 2013), few have taken the risks concern into their model. The risk analysis of projects is a relatively mature field. Some scholars have done researches on the risks on R&D projects (Wang, 2018; Liu, 2019) and megaprojects (Bruzelius, Flyvbjerg, and Rothengatter, 2002; Diéguez, Sanchez-Cazorla, and Alfalla-Luque, 2014). While researches on megaprojects with the view of complex network are still needed to be introduced.

In sum, this study proposes a scientific collaboration network model of megaprojects from the perspective of risk analysis. First, risks were analyzed from its network characteristics; Then, the construction principles were summarized to establish the network model. All studies aim to find new ideas for efficient, stable and sustainable operation of megaprojects.

2 THEORETICAL BASIS ON SCIENTIFIC COLLABORATION NETWORK CONSTRUCTION OF MEGAPROJECTS

2.1 Characteristic analysis based on complex system theory

The scientific collaboration network of megaprojects is a complex system with the following characteristics: 1) *Diversity*, after analysis of the research activities in megaproject's scientific collaboration network, it is acknowledged that this network consists of three types of heterogeneous subjects, i.e. the experts, research teams and coordination agencies. 2) *Integration*, megaprojects view science, technology, engineering and infrastructure construction all as its contained integrated elements. 3) *Collaboration*, because of the heavy tasks of megaprojects, individual innovation activities or small-scale cooperative innovations can no longer meet its requirements. 4) *Emergence*, because the behavior of subjects in the scientific collaboration network of megaprojects is implemented through mutual influence and technology integration, it can produce a giant innovation ability that could exceed the sum of single component's ability. 5) *Innovation*, as the mission of megaprojects is to meet the strategic needs of the country, which can be summed as leading the scientific frontiers, carrying out major scientific discovery, solving the weaknesses of national research strength and enhancing China's international competitiveness, the network, on the other hand requires innovation capability. 6) *Confrontation*, in general, as the purpose of implementing megaprojects is to maintain a competitive advantage at international level, it is necessary to compare its strategy with that of the opponents in order to achieve a better position in international system. 7) *Complexity*, as a top-notch innovation activity, megaproject's research work is a highly complex network which involves multidisciplinary scientific issues. 8) *Self-organization*, the fact that the participating teams have full right in deciding when to enter or exit the network has caused the state of the scientific collaboration network relatively unstable, thus making the network constantly adjusting and adapting to a new structure.

2.2 Applicability analysis of complex network theory

The scientific metrologists Katz and Martin (1997) define scientific collaboration as researchers of common purpose working together to produce new knowledge. Newman (2001) applied social network analysis method to natural science for the first time, and proposed the concept of scientific collaboration network. In the actual researching process, scientific collaboration can be played out in different forms, such as different authors, institutions (Cao and Pan, 2017; Zhao et al. 2018) or regions (Liu et al. 2015) appeared in the same paper, patent (Gao et al. 2019)or project (Zhang et al. 2018), which has scientific collaboration relationship. Scholars' work on scientific collaboration network at present mainly focuses on the network's topology characteristic (Brabasi et al., 2002), the scientific collaboration relationship between different level of subjects (Zhou et al. 2017; Xu et al. 2019; Liu Liang et al. 2019) and the impact of different network structures on performance (Li, 2017).

Organization of megaproject, as a complex system, can be abstracted in the form of complex network which describes the relationships and modes of collaboration. Therefore, the scientific collaboration network of megaproject is a network organization that links subjects in different relationships in order to fulfill the megaproject's tasks.

2.3 Risk characteristics of megaproject's scientific collaboration network

In the current studies, scholars generally believe that uncertainty is an important source of risks in network organizations. Usually, the uncertainty is divided into internal and external according to the source of it, which leads to internal risks and external risks (Jiang and Jiang 2015; Wang et al. 2013). While External risks are defined mostly as the impact of the external environment of the innovation network, including market, economic, technological change and policy (Per Levén et al. 2014; Zhang and Yang, 2018), the internal risks are proposed to be introduced by opportunistic behavior, dilemma of collaboration (Wei et al. 2018), over-dependence and so on. Scholars have studied and summarized the risks of R&D projects and concluded that the R&D projects mainly face risks from environment, management, finance, technology, time and quality (Liu et al. 2019). From the perspective of the entire process management, there are risks from technology, finance, and policy in the beginning, and personnel, technology, resource, and finance in the process (Wang, 2018). For megaprojects, Bruzelius, Flyvbjerg, and Rothengatter (2002) distinguish between four risks that megaprojects faced as cost risk, demand risk, financial market risk, and political risk. Diéguez, Sanchez-Cazorla, and Alfalla-Luque (2014) summarized a wider classification including design risks, political risks, contractual risks, construction risks, operation and maintenance risks, labor risks, users risks, economic risks and force majeure.

Based on previous studies, the scientific collaboration network of megaprojects is faced with both internal risks from technology, brain drain and dependence, and external risks from policy and environment. In area of technology, due to the rapid development of science and technology, the novelty and social benefits of megaproject's fruit may be lower than expected and can no longer be held as the leading technology contributor worldwide; Brain drain, which mainly refers to the loss of scientists or teams caused by head hunting or exit, may affects the entire project process and eventually lead the megaproject to failure; Dependence refers to excessive favor of some teams which would cause imbalances in resource allocation and lower the marginal efficiency of innovation output; Policy and Environment can affect the network in such degree that the fruit may no longer meet the nation's strategic needs because of the changes from policy, economy, culture, environment and other factors of the country during the long process of megaproject.

3 SCIENTIFIC COLLABORATION NETWORK CONSTRUCTION

3.1 Construction principles

According to the risks that scientific collaboration network of megaprojects is facing, the principles of oriented technology demand, applied project's characteristics and satisfied collaboration needs should be embodied in the network construction.

(1) Oriented technology demand

The establishment of megaproject's scientific collaboration network should be set up strictly in line with the megaproject's tasks and should be able to effectively absorb whole country's top strengths. Taking technology as an example, according to the modeling from demand to structure, from top to bottom, our decomposing should start from identifying technical requirements of megaproject's fruit, to decomposing them to technical requirements in fields of technology then breaking them down to projects. Projects should pair teams that are working in the same field. Additionally, when projects are completed, the work should be integrated from the bottom up into the overall goals.

(2) Applied project's characteristics

According to the content, megaproject's projects can be divided into four different types, i.e. science, technology, infrastructure construction and engineering. China meanwhile, has five different types of teams, i.e. corporations related to military industry, firms related to

687

military organization, Chinese Academy of Sciences, universities and general firms. To be more specific, corporations related to military industry is the backbone of innovation by systematic integration and expert in technology research; Firms related to military organization have the advantage on experimental engineering work; Chinese Academy of Sciences is good at science and applied basic research and is an active supporter in enhancing China's innovation capability; Universities, with its complete disciplines, abundant talent resources and convenience of foreign exchanges, are essential sources of innovation; General firms enjoy more flexible mechanisms, can respond faster, display more proactive service awareness and make lower product costs, each having its own advantages. For the sake of avoiding risks from brain drain and dependence, every type of teams must be fully utilized.

(3) Satisfied collaboration needs

Innovation is the spark of thoughts' collision. Megaproject's scientific collaboration network must rely on collective intelligence to achieve its tasks, which results in requirements of a platform for subjects to find suitable partners to collaborate. This kind of suitability is reflected in three aspects: First, *Suitable field of knowledge*, i.e. the source must be doing the same field of research with teams who are in strong demand in guidance, must understand the team's work and make effective communication; Second, *Suitable amount of knowledge*, meaning the source needs higher level of knowledge to provide reliable guidance; Third, *Suitable background*, meaning that in consideration of academic norms, intellectual property and business secrets, the source must share ideas within suitable boundary .

3.2 *Network construction*

According to the three principles of network construction, the scientific collaboration network of megaprojects is shown in Figure 1. As it shows: 1) While the upper layer is the subnet of integrating composed of coordinators and consultants in different fields of technology, the lower layer represents the subnet of researching which consists of executors; 2) All coordinators, consultants and executors can be collectively referred to as subjects, all of which are distributed with individual technical attributes; 3) All coordinators maintain a coordinative relationship with the consultants in the same layer of subnet and a cross-layer coordinative

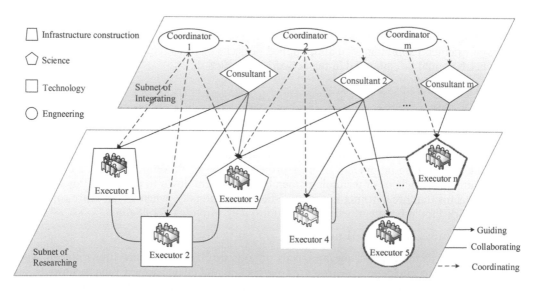

Figure 1.   Scientific collaboration network of megaprojects.

relationship with executors; 4) While all consultants form cross-layer guiding relationship with the executors, all executors are connected through project-based scientific collaboration.

Abide by the technical main line, the "field-project" hierarchical network structure is constructed and the subjects are distributed with suitable attributes, which is consistent with the successful typical megaprojects like manned space engineering. This hierarchical structure supports the scientific collaboration network of megaprojects by meeting the orientation of technical demand, and reducing brain drain by letting people do the right job and thus creating a more stable network.

Consultants and coordinators select research executors in the subnet of researching. For instance, the yellow team represents team from corporations related to military industry, the red team represents team from the firms related to military organization, the green team represent teams from the Chinese Academy of Sciences, the blue team represents team from universities, and the gray team represents team from general firms. In general, there are four different types of projects, including science, technology, infrastructure construction and engineering. As described in Figure 1, all five types of teams should be used more wisely. For example, science projects should pay more attention to the original innovation ability and consider choosing universities and the Chinese Academy of Sciences; the infrastructure construction projects should absorb more general firms into the system for introducing competition and reducing costs; the technology projects should provide more opportunities to corporations related to military industry for their technical advantages on systematic integration; the engineering projects should fully consider firms related to military organization' complete facilities and work content. Only upon above considerations, the applied project's characteristics principles can be followed and an efficient scientific collaboration network would be constructed.

In the subnet of integrating, the consultants and coordinators should know where are the strong teams and keep the list of all the megaproject's teams, which can make it possible for the consultants to point out the potential partners for the executors who have the collaboration needs, and the coordinators to coordinate executors with potential partners to communicate through private contact or public meetings. Consultants and coordinators should work together at an integrated platform in the subnet of integrating to solve the executors' collaboration needs. This platform in the scientific collaboration network of megaprojects would fulfil principle on the satisfied collaboration needs and foster innovation.

4 CONCLUSION AND DISCUSSION

This study constructs a scientific collaboration network model of megaprojects based on risk analysis. First, the characteristics of megaprojects are analyzed from the perspective of complex system. Second, based on the research of predecessor's network organization and project risk, the risks that scientific collaboration network of megaprojects is facing were analyzed. Third, the network construction principles were summarized to clarify problems that our model needs to solve. Finally, the scientific collaboration network model of megaproject was proposed. This model was constructed to provide a reference for managing megaprojects, help keep the organization stable and improve the efficiency of scientific research.

REFERENCES

Barabâsi, A.L., Jeong, H., Néda, Z., Ravasz, E., Schubert, A., & Vicsek, T. 2002. Evolution of the social network of scientific collaborations. *Physica A: Statistical mechanics and its applications, 311*(3-4): 590–614.
Bruzelius, N., Flyvbjerg, B., & Rothengatter, W. 2002. Big decisions, big risks. improving accountability in mega projects. *Transport Policy*, 9(2): 143–154.

Cao, Z.P. and Pan, Q.L. 2017. Study of Scientific Research Innovation and Cooperation between Universities in China Based on the Social Network Analysis of Co-authors in 2014. *Science and Technology Management Research* 1: 93–98.

Dai, R.W. 2002. Research on System Science and System Complexity. *Journal of System Simulation* 14(11): 1411–1416.

Flyvbjerg, B., Bruzelius, N., and Rothengatter, W. 2003. *Megaprojects and risk: An anatomy of ambition.* Cambridge University Press.

Flyvbjerg, B. (2014). What you should know about megaprojects and why: An overview. *Project management journal*, 45(2): 6–19.

Gao, L.Z., Tang, H. and Liu, G.F. 2019. Evolutionary Characteristics of the Industry-University-Institute Patent Cooperation Network Structure of Colleges and Universities in Beijing, Tianjin and Hebei. *Library & Information Studies* 12(1): 96–105.

He, X.W., Hou, G.M. and Wang, Y. 2013. The Research on Construction of Organization Coordination Network of Major Scientific and Technological Projects. *Forum on Science and Technology in China* 2: 38–44.

He, X.W. and Wang, R.R. 2018. The game relationship's analysis across the cooperative-innovation enterprises oriented towards major projects of science and technology. *Scientific Management Research* 36(4): 56–59,72.

Irimia-Diéguez, A.I., Sanchez-Cazorla, A., and Alfalla-Luque, R. 2014. Risk management in megaprojects. *Procedia-Social and Behavioral Sciences, 119*: 407–416.

Jiang, X. and Jiang, F.F. 2015. Uncertainty, Alliance Risk Management, and Alliance Performance. *Journal of Industrial Engineering/Engineering Management* 29(3): 180–190.

Jiang, Y., Kong, D.C. and Zou, R. 2013. Research on Organization and Management Mode of Complex Major Science and Technology Project. *Science Technology and Industry* 13(11): 167–172.

Nie, N., Zhou, J. and Zhu, Z.T. 2014. Simulation Study of Large-scale Engineering Organization Management Strategy Based on NK Model. *Science & Technology and Economy* 27(3): 101–105.

Li, C.H. and Du, Y.W. 2012. The identification of key cooperative interface for major S&T programs. *Science Research Management* 33(7):121–128.

Li, C.J. 2011. Research on a Complex System Management Methodology of Large-scale Science Technique Engineering. *Chinese Journal of Management Science* 19(Special Issue): 148–151.

Liu, G.W., Li, L. and Chen, W. 2019. Research on Risk Control in the Implementation of Scientific Research Projects. *Technology Wind* 6: 241–242.

Liu, L., Luo, T. and Cao, J.M. 2019. A study of the multi-scale scientific collaboration patterns based on complex networks. *Science Research Management* 40(1): 191–198.

Li, W.C, He J. and Dong, J.C. 2017. Comparison of the Influences of International Collaboration and Domestic Collaboration on Research Output from the Perspective of Network Embeddedness: A Case of Chinese Stem Cell Research Institutions. *science of science and management of S.&.T* 2017, 38(1): 98–107.

Liu, Y., Jiang, H.J., Fan, W., Yan, Z. and Ye, X.T. 2015. Study on the structure and evolution characteristics of international collaborative innovation network of nano science and technology. *Science Research Management* 36(2): 41–49.

Newman M E J. 2001. Scientific collaboration networks. I. Network construction and fundamental results. *Physical review E* 64(1): 016131.

Newman M E J. 2001. Scientific collaboration networks. II. Shortest paths, weighted networks, and centrality. *Physical review E* 64(1): 016132.

Song, X.F. 2003. Complexity, Complex System, and the Science of Complexity. *Bulletin of National Natural Science Foundation of China* 5: 8–15.

Wang, A.J. 2018. Management and Risk Control of Major Scientific Research Projects. *Power Station Auxiliary Equipment* 39(2): 48–51.

Wang, C.C., Cai, N. and Huang, C. 2013. The Effects of the Multi-Layer Network Structures on the Proliferation of Cluster Risks. *Journal of Chongqing University(Social Science Edition)*27(5): 54–59.

Wei, L. Dang, X.H. and Cheng, L. 2018. The Ambidexterity of Uncertainty and Effect on Vulnerability of Technological Innovation Network: Network Routine as a Mediator. *Business Review* 30(7): 66–78.

Xu, X. Yang, J.Y. and Li, D. 2019. International Scientific Collaboration Network Evolution under Science Funding Support: Data Analysis Based on WOS Physics. *Journal of Intelligence* 38(2): 48, 49–55.

Zhao, R.Y., Wang, X.L. and Qi, Y.K. 2019. Study on Scientific Collaboration Network and Evolution of Construction of World—class Universities in China. *Journal of Modern Information* 39(3): 132–143.

Zhang, S.Q., Gao, X., Guo, J.J., Shi, Q.Q. and Gu, J.H. 2018. A research on the collaboration network of scientific project. *Science Research Management* 39(5): 86–93.

Zhang, Y.L. and Yang, N.D. 2018. Modeling and Simulation of the Control Method against Risk Propagation in Research and Development Network. *Journal of System & Management* 27(3): 500–511.

Zhang, Y.L. and Yang, N.D.2015. Analysis of R&D Project Schedule Risk Based on the Organization-task Network: With the Organization Failure as a Risk Factor. *Chinese Journal of Management Science* 23(2): 99–107.

Zhou, J.M., Huang, Y., Kong, X.M., Zhu, D.H. and Wang, X.F. 2017. Cooperation Network Analysis of China's Library and Information Science Field. *Technology Intelligence Engineering* 3(6): 103–115.

Author Index